GREENHOUSE GAS CARBON DIOXIDE MITIGATION

Science and Technology

GREENHOUSE GAS CARBON DIOXIDE MITIGATION

Science and Technology

Martin M. Halmann
Meyer Steinberg

CRC Press
Taylor & Francis Group
Boca Raton London New York

CRC Press is an imprint of the
Taylor & Francis Group, an **informa** business

CRC Press
Taylor & Francis Group
6000 Broken Sound Parkway NW, Suite 300
Boca Raton, FL 33487-2742

First issued in paperback 2019

© 1999 by Taylor & Francis Group, LLC
CRC Press is an imprint of Taylor & Francis Group, an Informa business

No claim to original U.S. Government works

ISBN-13: 978-1-56670-284-3 (hbk)
ISBN-13: 978-0-367-40023-1 (pbk)

Library of Congress Cataloging-in-Publication Data

Catalog information may be obtained from the Library of Congress

Visit the Taylor & Francis Web site at
http://www.taylorandfrancis.com

and the CRC Press Web site at
http://www.crcpress.com

Table of Contents

Acknowledgments

The author (Steinberg) of Chapters 1 through 5 and 7 and part of Chapter 6 gratefully acknowledges the assistance of a number of his colleagues over the years in the Department of Applied Science at Brookhaven National Laboratory, in assembling the information presented. These colleagues include Dr. Hsing Cheng, Mr. Anthony Albanese, Mr. Fred Horne, and Mr. Yuanji Dong. The author wishes to express his appreciation to Mr. Fred Koomanoff, who at one time headed the Carbon Dioxide Research Division of the Basic Energy Science Department of the U.S. Department of Energy, for allowing him to perform CO_2 mitigation technology studies at a time when there was little interest in the subject until the world was alerted to the global warming problem in 1988. These studies formed the basis for much work that was subsequently conducted and which is included in this volume. The author is grateful to Mr. Perry Bergman, Project Engineer at the Federal Energy Technology Center in Pittsburgh of the U.S. Department of Energy, who was instrumental in supporting the author's subsequent continued studies. The author is also thankful to Mr. Frank Princiotta and Mr. Robert Borgwardt for supporting the development of the CO_2 mitigation processes described in Chapter 7. The author acknowledges the assistance of Ms. Barbara Roland for typing and struggling with the manuscript. Finally, the author wishes to thank his wife, Ruth Steinberg, for so ably handling his correspondence with his coauthor and for her patience, encouragement, and support.

The author (Halmann) of Chapters 8 through 14 and part of Chapter 6 wishes to thank his wife, Miryam, for her patience and encouragement.

The Authors

Professor Martin Mordehai Halmann has been Professor Emeritus at the Weizmann Institute of Science, Rehovot, Israel, since 1990. He has been a member of the scientific staff there since 1949. He studied chemistry at Hebrew University, Jerusalem and received his M.Sc. degree in 1949 and Ph.D. in 1952.

Professor Halmann's main research interests (in chronological order) have been chemical reaction mechanisms, isotope effects on chemical reactions and on Franck–Condon factors, chemical effects of nuclear transformations, photochemistry of organic phosphorus compounds, photoelectrochemical and photocatalytic reduction of carbon dioxide using semiconductors, and photocatalytic oxidation of inorganic and organic compounds in water. His previous books on *Chemical Fixation of Carbon Dioxide* and *Photodegradation of Water Pollutants* were published by CRC Press in 1993 and 1995, respectively.

Meyer Steinberg is a Senior Chemical Engineer. Until recently, he was Head of the Process Sciences Division of the Department of Applied Science at Brookhaven National Laboratory and presently is a Senior Chemical Engineer in the Department of Advanced Technology. He is a director of engineering of the HYNOL Corporation and since 1990 has been engaged in consulting practice, especially on greenhouse gas CO_2 mitigation technologies.

Mr. Steinberg worked on the Manhattan District's Atom Bomb Project at Oak Ridge and Los Alamos from 1944 to 1946 and then as a Chemical Engineer in the heavy chemical and metallurgical industry and in the development of rocket fuel from 1947 to 1957. Since 1957, he has been at the Brookhaven National Laboratory. He has contributed to the field of radiation chemical processing and chemonuclear reactors. He is an inventor and developer of concrete polymer materials. His research in nuclear energy includes nuclear waste management and safety, the use of fission and fusion reactors for synthetic carbonaceous fuel production, the linear accelerator spallator for nuclear fuel production and transmutation of fission product waste, and the reduction of nuclear weapons materials. His research in fossil and solar energy involves coal conversion, desulfurization, hydropyrolysis, environmental control technologies for the global CO_2 problem, and the conversion of biomass and municipal solid waste to clean fuels.

1 The Science and the Source of the Greenhouse Effect and Global Climate Change

1.1 THE GREENHOUSE EFFECT

The earth has an atmosphere which acts like a blanket that traps heat from solar radiation coming to the earth from the sun. The average surface temperature of the earth is 15°C. If there were no atmosphere, the surface temperature of the earth can be calculated to be −19°C. This is confirmed from the surface temperature of the moon, which receives the same amount of solar radiation as the earth but has no atmosphere. This warming of the earth is the so-called greenhouse effect. The effect is a good thing because it maintains life on earth in a viable and comfortable condition.

The actual warming of the earth due to the atmospheric blanket is strongly influenced by small amounts of gases in the earth's atmosphere, particularly carbon dioxide (CO_2) and water vapor (H_2O). These gases trap heat due to their molecular structures, which absorb mainly reflected solar radiation from the earth's surface. These are the so-called greenhouse or radiative gases. The sun radiates energy equivalent to a black body at a temperature on the order of 6000 K. The wavelength of this radiation peaks in the range of 0.4 to 0.7 μ* mainly in the visible range, but has small amounts of ultraviolet radiation down to 0.1 μ and small amounts of infrared radiation up to 3 μ. The earth, on the other hand, acts as a black body, radiating energy at 15°C, in which case the wavelength of energy emanating from the earth is in the range of 4 to 100 μ. H_2O vapor strongly absorbs radiation in the range of 4 to 7 μ, while CO_2 absorbs radiation in the range of 13 to 19 μ. Thus, H_2O and CO_2 absorb a large fraction of the reflected longer wavelength solar radiation.

---

* μ = 1 micrometer = 10,000 Å = 10^{-4} cm.

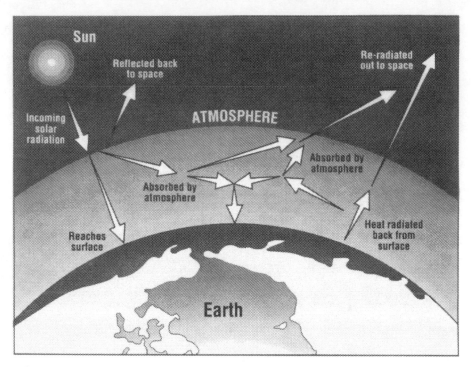

FIGURE 1.1 Some of the energy radiated at infrared wavelengths from the ground is absorbed and reradiated downward by the atmosphere — the greenhouse effect.

Some is reflected back to earth and some to outer space. An equilibrium state is reached because an energy balance is set up, where all the incoming radiation must eventually be reradiated to space. A pictorial representation of all the energy paths of incoming solar radiation and outgoing radiation is shown in Figure 1.1.

The concern is that the natural equilibrium of the CO_2 greenhouse effect is being perturbed by the increasing CO_2 concentration in the atmosphere due to human or anthropogenic activities. Furthermore, as atmospheric CO_2 increases, the global temperature increases, and this puts more water vapor into the atmosphere. Water vapor is also a very effective greenhouse gas, which tends to increase the earth's temperature even further. These radiation gas effects, which are called radiation-forcing effects, are factored into computer models to predict global climate change into the future.

Historically, it is interesting to note that Fourier in the early 1800s was the first to be concerned about a greenhouse effect. Also, Tyndall in the middle 1800s was concerned with water vapor as a greenhouse gas. In the late 1800s, Arrhenius, author of the famous chemical Fourier equation, examined the effect of water vapor and thought that CO_2 buildup due to combustion of coal in the air might be the cause.[4]

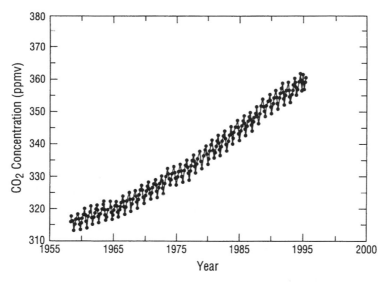

FIGURE 1.2   Monthly atmospheric CO_2 concentrations at Mauna Loa, Hawaii. (Source: ORNL-ODIAC 1995.[3])

1.2   THE CO_2 CONCENTRATION BUILDUP

Although there was some observation that the average earth surface temperature rose by 0.25°C between 1880 and 1940, it was not until the 1950s that measurements became serious and atmospheric CO_2 was suspected as the possible cause of the temperature increase. Figure 1.2 shows the monthly atmospheric CO_2 concentration measured at Mauna Loa, Hawaii. The periodic spread of CO_2 concentrations around the mean is due to seasonal changes mainly in the Northern Hemisphere. Measurements were also taken at the South Pole. Both measurements were taken far away from industrial sources of CO_2 emissions and represent well-mixed atmospheric concentrations. Then, in the 1970s, information on air trapped in ice in the Antarctic became available, tracing CO_2 concentrations back over 1000 years. This is shown in Figure 1.3, which meshes well with the recent Mauna Loa measurements. The data indicate a rather constant concentration of around 280 to 290 parts per million by volume (ppmv) CO_2 until the year 1800 and then a rapid rise in the last century, reaching 360 ppmv in 1995. These measurements indicate a current increase of 26% above pre-Industrial Revolution time.

1.3   OTHER RADIATIVE GASES

Other anthropogenic-generated radiative gases act in a manner similar to CO_2 but absorb radiation in the range of 7 to 13 μ. These include methane (CH_4), nitrous oxide (N_2O), ozone (O_3), chlorofluorocarbons (CFCs), and fluorocarbons (CFs).

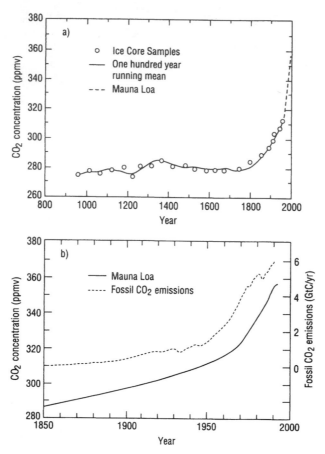

FIGURE 1.3 CO_2 concentrations over the past 1000 years from ice core records and from Mauna Loa and expanded scale since 1850. (Source: IPCC 1995,[1] p. 16.)

Table 1.1 indicates the relative importance of these radiative gases in contributing to the potential climate change. The term radiative forcing is used to quantitatively indicate the magnitude of the effect of the energy balance of the earth–atmosphere system and is usually expressed in watts per square meter (W/m^2).

These radiative gases are generated primarily by human activities, and some stay in the atmosphere for a considerable length of time. Table 1.2 indicates the concentrations, growth, and lifetime of these greenhouse gases. CO_2 gas makes up almost two-thirds of the radiative forcing and arises mainly from the combustion of fossil fuels.

1.4   EARTH–ATMOSPHERE ENERGY BALANCE

It should be noted that the total flux of energy incoming to and outgoing from the earth–atmosphere system is large compared to the direct greenhouse gas radiative

TABLE 1.1
Contribution of the Major Radiative Gases
Affecting the Earth–Atmosphere Energy Balance
(in 1992 Values)

Gas	Radiative forcing (W/m^2)	% effect
CO_2	1.56	63.6
CH_4	0.47	19.2
N_2O	0.14	5.7
CFC and others	0.28	11.5
Total	2.45	100.0

Source: IPCC 1995, p. 17.[1]

forcing, as shown in Figure 1.4. The incoming solar radiation, which amounts to 342 W/m^2, is partially reflected by clouds, the atmosphere, and the earth sea fell only — about 49% is absorbed by the earth. Some of the heat is returned to the atmosphere as sensible heating and by water evaporation, which is returned as latent heat by precipitation. The rest is radiated as thermal infrared radiation, and most of that is absorbed by the atmosphere, which in turn emits radiation both up and down, producing a greenhouse effect. Radiation lost to space also comes from cloud tops and parts of the atmosphere much colder than the surface. The radiative forcing of all the greenhouse gases only amounts to 0.72% of the incoming solar radiation; however, this relatively small value has a potentially profound effect on the earth's climate.

1.5 EARTH–ATMOSPHERE CARBON CYCLE

When dealing with CO_2 atmospheric concentration, estimates must cover the mass balances of the annual carbon budget in terms of CO_2 sources, sinks, and storage

TABLE 1.2
Greenhouse Gases, Concentrations, Growth, and Lifetime
in the Earth's Atmosphere

	CO_2	CH_4	N_2O	CFC-11
Preindustrial concentration	~280 ppmv	~700 ppbv	~275 ppbv	0
Concentration in 1994	358 ppmv	1720 ppbv	312 ppbv	268 pptv
Rate of concentration change	1.5 ppmv/yr	10 ppbv/yr	0.8 ppbv/yr	0 pptv/yr
	0.4%/yr	0.6%/yr	0.25%/yr	0%/yr
Atmospheric lifetime (years)	50–200	12	120	50

Note: ppmv = parts per million by volume, ppbv = parts per billion by volume, and pptv = parts per trillion by volume.

Source: IPCC 1995, p. 15.[1]

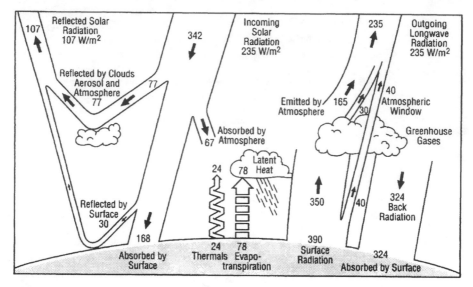

FIGURE 1.4 The earth's radiation and energy balance (values in W/m^2). (Source: IPCC 1995,[1] p. 58.)

in the atmosphere. Table 1.3 gives an estimate of the distribution of these values in gigatons or 10^9 tonnes of carbon per year [GT(C)/yr] as CO_2.

These values must be placed in the context of the total global carbon cycle and the resources in the earth–atmosphere system, as shown in Figure 1.5. The total recent annual anthropogenic emission μ of 7.1 GT(C)/yr shown in Table 1.3 is only 0.95% of the total atmospheric carbon content, and less than half, as also shown in Figure 1.3, remains stored in the atmosphere. The problem is to find out where the other half of the fossil-fuel CO_2 is stored Again, however, this relatively small

TABLE 1.3
Annual Average Anthropogenic Carbon Budget for 1980 to 1989

CO$_2$ sources

1.	Emissions from fossil-fuel combustion and cement production	5.5 ± 0.5
2.	Net emissions from changes in tropical land use	1.6 ± 1.0
3.	Total anthropogenic emissions = (1) + (2)	7.1 ± 1.1

Partitioning among reservoirs

4.	Storage in the atmosphere	3.3 ± 0.2
5.	Ocean uptake	2.0 ± 0.8
6.	Uptake by Northern Hemisphere forest regrowth	0.5 ± 0.5
7.	Inferred sink: 3 – (4 + 5 + 6)	1.3 ± 1.5

Note: CO_2 sources, sinks, and storage in the atmosphere are expressed in GT(C)/yr.

Source: IPCC 1995, p. 17.[1]

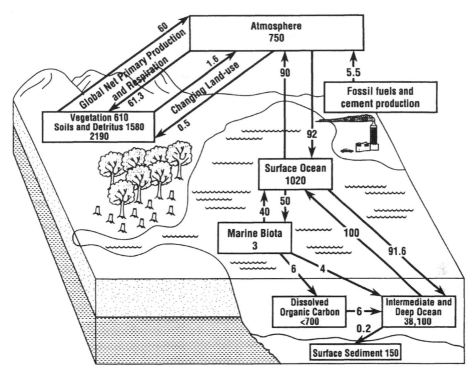

FIGURE 1.5 The global carbon cycle, showing the reservoirs [in GT(C)] and fluxes [GT(C)/yr] relevant to anthropogenic perturbation as annual averages over the period 1980 to 1989. (Source: IPCC 1995,[1] p. 77.)

increase in CO_2 in the earth's atmosphere has a potentially profound effect on global climate. Reforestation and deforestation also will increase and decrease the uptake of CO_2 from the atmosphere, and the relative effect is shown in Table 1.3.

1.6 EFFECT OF BUILDUP OF GREENHOUSE GASES

The effect of the increasing energy returned to the earth due to the buildup of greenhouse gas concentrations is an increase in the temperature of the earth's surface, in addition to heating the lower atmosphere and troposphere. The observed changes in global mean surface temperature from the 1860s to 1994 are shown in Figure 1.6 and indicate an increase of between 0.3 to 0.6°C during this period. In recent years, the earth's temperature has reached record levels. Anomalies in temperature rise occur due to volcanic action. Large amounts of matter are thrown into the stratosphere, which absorbs and reflects the incoming solar energy. This was observed in 1991 when Mount Pinatubo in the Philippines erupted. There was a cooling effect for a 2-year period, but then a return to a higher temperature occurred, as expected, due to the further buildup of greenhouse gas during the interim.

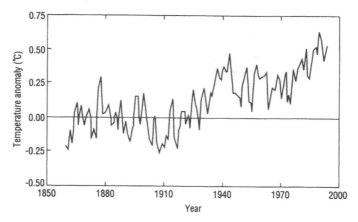

FIGURE 1.6   Observed changes in global mean temperature from 1861 to 1994. (Source: IPCC 1995,[1] p. 77.)

The main problem is to predict what will occur in the next century as the world's population continues to release CO_2 and other greenhouse gases into the atmosphere in increasing amounts. According to a consensus of the scientists on the Intergovernmental Panel on Climate Change (IPCC) Group I report of 1995, the equivalent CO_2 concentration is predicted to double within the next century and the temperature rise by the year 2100 will be in the range of 1 to 3.5°C, with a "best estimate" of 2°C.

1.7   GLOBAL EFFECTS OF TEMPERATURE INCREASE

A number of global effects are expected from a rise in temperature. These effects include (1) a rise in sea level, due to melting of the glaciers and the Antarctic ice caps, and expansion of the water in the oceans, and (2) increasing desert formation in the tropical zone and movement of improved agricultural growth into the northern temperate zones. The predicted estimates of rise in sea level by the year 2100 range between 15 cm and 90 cm, with a best estimate of 48 cm (about 1.5 ft). Coastal cities, islands, and low-lying countries such as Bangladesh and the Netherlands would be affected catastrophically as a result of being inundated by the oceans.

There is, however, great uncertainty about these effects because of other positive and negative radiative-forcing factors. The more important factors include ozone in the troposphere, which acts positively by increasing radiative forcing, and ozone in the stratosphere, which acts negatively by decreasing radiative forcing. Also in the troposphere, aerosols from sulfate emission and biomass burning act negatively, while fossil-fuel soot aerosols act positively. The cloud effect, which is driven indirectly by aerosols, can have a strong negative effect on radiative forcing. Variation in solar flux also affects radiative forcing. The level of uncertainty in these factors is presently very high, which makes it difficult to use global climate change models with a reasonable degree of confidence.

An observation by the IPCC influences the implementation of CO_2 mitigation technologies. Any agreement on a stabilized CO_2 concentration is governed more by the accumulation of anthropogenic CO_2 emissions from current emissions until the time of stabilization than by the way emissions change over this period. This means that for a given stabilized concentration value, higher emissions in earlier decades require lower emissions later on after stabilization.

The greenhouse effect can be succinctly summarized as follows. The concentration of CO_2 in the atmosphere is increasing. CO_2 is a radiative gas which reflects increasing amounts of energy returned to the earth due to its increasing concentration in the atmosphere. Changes in global climate as a result of this increasing amount of energy returned to the earth are difficult to predict because of the high level of uncertainty in understanding climatological processes.

1.8 SOURCES OF CO_2 INPUT TO THE ATMOSPHERE

Basically, the extent of anthropogenic CO_2 emissions can be traced to an increasing world population. The basic energy needs of each individual on earth include food, shelter, clothing, transportation, and industry. The energy is supplied through fossil fuel, biomass (agriculture and wood), solar energy, and nuclear energy. Most of the world's energy over the last century has come from fossil fuels — coal, oil, and gas. The per capita emission of CO_2 increases with the degree of industrialization and the efficiency with which the energy is generated and utilized. Most of the CO_2 emitted since the Industrial Revolution is attributed to the industrially developed countries and regions (e.g., North America, the former Soviet Union, Europe, Japan, and others). The potential increase in CO_2 over the next century is expected to come from the developing countries and regions (China, India, Africa, and South America). The rate at which emissions will increase will depend on how fast these industrial economies develop and how efficient the utilization of fossil fuels in these countries becomes.

Estimates of world reserves and resources of fossil fuels are shown in Table 1.4. The reserves include those conventional and unconventional deposits that have been identified as economically and technically recoverable with current technologies and prices. Resources are those conventional and unconventional deposits which are considered potentially recoverable with foreseeable advanced technological and economic developments. Gas reserves are about equal to that of oil, and coal reserves are about equal to the sum of oil and gas. The total resource of coal far exceeds that of oil and gas. However, additional possible resources of natural gas, such as methane hydrates, may be even much greater than oil and gas resources. Thus, at the current rate of consumption, hundreds of years worth of fossil fuels, especially coal, are available. Of course, the possible use of nuclear energy and renewable (solar) energy could alter global energy consumption. What is being

TABLE 1.4
Estimates of World Fossil-Fuel Energy Reserves and Resources
and Potential Equivalent CO_2 Emission

Fossil fuel	Energy reserves[a] identified (EJ)	Potential CO_2[b] emission [GT(C)]	Energy[c] resource base	Potential CO_2[b] emission [GT(C)]
Coal	25,200	637	125,500	3,175
Oil	13,100	249	24,600	467
Gas	11,700	159	36,100	491
Total	50,000	1,045[d]	185,200	4,133
Years to last at 1990 level of consumption (385 EJ/yr)	130		480	

[a] Conventional and unconventional reserves.

[b] GT(C) = 10^9 tonnes of C as CO_2 calculated with emission factors from Table 1.5.

[c] Estimates of 50% probability of resources to be discovered.

[d] Current atmospheric CO_2 content = 770 GT(C) in 1994.

Source: IPCC 1996, p. 87.[2]

considered here is the potential of altering the CO_2 content of the atmosphere by emissions due to long-term utilization of the fossil-fuel resources.

Table 1.5 lists the emissions of CO_2 from combustion of fossil fuels in terms of mass of CO_2 per unit of energy in units of kg C (as CO_2)/GJ and lb CO_2/MMBtu. Coal emits approximately 87% more CO_2 than natural gas and 34% more than oil. Oil emits 39% more than natural gas. This is related to the carbon and hydrogen

TABLE 1.5
Emissions of CO_2 from Major Sources of Fossil Fuels
per Unit of Energy Produced

Fossil fuel (elemental content)	kg(C)/GJ[a]	lb CO_2/MMBtu[b]
Coal (bituminous) ($CH_{0.8}O_{0.1}$)	24.0–25.3	203–215
Crude oil ($CH_{1.8}$)	19.0–20.3	160–172
Natural gas (CH_4)	13.6–14.0	115–119

[a] GJ = gigajoules = 10^9 joules, energy based on higher heating value.

[b] MMBtu = 10^6 Btu, energy based on higher heating value. 1.0 MMBtu = 1.05 GJ. kg CO_2/kg C = lb CO_2/lb C = 3.67. Also, GT(C)/EJ = 10^{-6} kg(C)/GJ. EJ = exajoule = 10^{18} joules.

Source: IPCC 1996, p. 80[2] and calculated from compositions and heating values of U.S. coals.

TABLE 1.6
Global Energy Consumption by Energy Source
and Equivalent CO_2 Emissions (1990)

Energy source	EJ/yr[a]	%	GT(C)/yr[b]	%
Coal	91	23.7	2.3	40.4
Oil	128	33.2	2.4	42.7
Gas	71	18.4	0.9	16.9
Nuclear	19	4.9		
Hydro	21	5.5		
Biomass	55	14.3		
Total	385	100.0	5.6	100.0

[a] $EJ = 10^{18}$ joules.

[b] $GT(C)/yr = 10^6$ tonne of C as CO_2 per year.

Source: IPCC 1996, p. 83.[2]

content of the fuel. If we apply these emission factors to the estimated fossil-fuel energy reserves and resources, we come up with the potential CO_2 emissions shown in Table 1.4. CO_2 emission based on current reserves exceeds 1000 GT(C) and is more than the current atmospheric CO_2 content of 770 GT(C), which was equivalent to 358 ppmv in 1994. Thus, the CO_2 content could double within the next century. If we take the energy resource base, CO_2 could reach almost six times its current atmospheric content.

Table 1.4 also indicates that coal makes up over 60% of the potential CO_2 emissions from the current reserves. Uncovering unconventional gas resources such as hydrates and application of nuclear and renewable (solar) energy could readily alter these estimates.

To understand current global energy consumption, Table 1.6 shows the global annual (1990) consumption by energy source, 75% of which comes from fossil fuels. Of the total equivalent CO_2 emissions of 5.6 GT(C)/yr, about equal amounts come from coal and oil, even though the energy consumed from oil is over 40% greater than that from coal.

Another important factor to understand is the distribution of CO_2 emissions from various sectors of the world economy. This is shown in Table 1.7. Power and heat generation for industry emits almost 50% of the world's CO_2 emission. The reason for this is that a large part of the world's electric power generation is based on the use of fossil-fuel coal, and coal emits the greatest amount of CO_2 per unit of energy produced. The transportation sector consumes mainly oil for gasoline production. In the residential and commercial sector, natural gas is used extensively for home heating and smaller industrial manufacturing facilities.

To indicate the major emitters of CO_2, Table 1.8 lists the top 11 countries which together emitted 75% of the total global CO_2 in 1990. The four top countries — U.S., U.S.S.R. (1988), China, and Japan — account for almost 60% of the CO_2 emitted to the atmosphere.

TABLE 1.7
Distribution of World CO_2 Emissions from the Three
Largest Sectors of Energy Consumption (1990)

Energy-consuming sector	% of world CO_2 emission
Power and heat generation from industry	47
Transportation	22
Commercial and residential	31
CO_2 emission	5.6 GT/yr

Source: IPCC 1996, p. 12.[2]

TABLE 1.8
Fossil-Fuel CO_2 Emission from
the 11 Largest Emitters (1990)

Country	GT(C)/yr
U.S.	1.39
U.S.S.R. (1985)	0.96
China (1985)	0.52
Japan	0.28
Former West Germany	0.19
U.K.	0.15
India	0.13
Poland	0.13
Canada	0.12
Italy	0.11
France	0.10
Total	4.08
World output	5.5

Source: IPCC 1996, p. 654.[2]

An overall equation which reflects the emission of carbon to the atmosphere for any country in terms of its economy, originally expressed by Dr. Yoichi Kaya[5] of Japan and subsequently modified for net carbon emissions, is as follows:

$$\text{net C} = P \times \frac{GDP}{P} \times \frac{E}{GDP} \times \frac{C}{E} - S$$

Net carbon emission = Population (P)
to atmosphere (C)

$\times$ Per capita gross domestic production (GDP/P)

$\times$ Energy generated per gross domestic production (E/GDP)

$\times$ Carbon emission per unit energy generated (C/E)

– Natural and induced removal of C as CO_2 from the atmosphere (S)

The equation reflects that greenhouse gas emission is a product of a country's population, its per capita economic output, its energy utilization efficiency, and the carbon quality of the fuel that it uses. The term S includes the natural forces for removal of C as CO_2 from the atmosphere and when mitigation technologies are applied includes the induced removal of CO_2 as well.

Thus, application of CO_2 mitigation technologies in the developed high-CO_2-emitting countries could have a significant impact on reduction of the rate of increase of CO_2 in the atmosphere. The greenhouse effect is a global problem, and emerging and developing countries could therefore benefit from the experience of CO_2 mitigation technologies in the developed countries as their own economies develop.

REFERENCES

1. IPCC 1995, *Climate Change 1995, Working Group I,* published for the Intergovernmental Panel on Climate Change, Houghton, J.T. et al., Eds., Cambridge University Press, Cambridge, 1996.
2. IPCC 1996, *Climate Change 1995, Working Group II,* published for the Intergovernmental Panel on Climate Change, Watson, R.T. et al., Eds., Cambridge University Press, Cambridge, 1996.
3. Keeling, C.D. and Whorf, T.P., ORNL/CD/AC 65 Carbon Dioxide Information Analysis Center, Oak Ridge National Laboratory, Oak Ridge, TN, 1995.
4. Weart, S.R., The discovery of the risk of global warming, *Phys. Today,* pp. 34–40, January 1997.
5. Kaya, Y. et al., A grand strategy for global warming, paper presented at the Tokyo Conference on Global Environment, September 1989.

2 Adaptive vs. Mitigative Response Strategies for Global Warming

2.1 ADAPTIVE APPROACH

The "adaptive" approach for dealing with global climate change that results in global warming basically assumes that global warming will take place and that life on earth will adapt to the effects of climate change in order to survive. The "mitigative" response assumes the application of technologies directed toward the prevention of global warming and its resulting effects. This chapter reviews the proposed adaptive responses to the impacts of global climate change resulting from the projected greenhouse effect. The severity of the impacts depends on the vulnerability or the extent to which climate change can damage the system. The adaptability depends on the degree to which adjustments can be made in practice to the effects of climate change. The impacts manifests themselves in (1) terrestrial and aquatic ecosystems, (2) hydrology and water resources, (3) food and fiber production, (4) human infrastructure, and (5) human health.[1]

2.1.1 Terrestrial and Aquatic Ecosystems

2.1.1.1 Forests

A sustained increase in global mean temperature can cause changes in regional climate that affect the growth and regeneration capacity of forests in many regions. It is expected that the greatest changes in forests will occur in high latitudes and the least in the tropics. Some forest types may disappear, while new species may be established. Because of the doubling of atmospheric CO_2, photosynthetic forest productivity could increase the standing biomass. However, it may not because of increases in pests and pathogens and increase in forest fires. The human response to forest changes would require relocation of communities that depend on forest products for both lumber and shelter.

2.1.1.2 Deserts and Desertification

Deserts are likely to become more intense, becoming hotter and drier. The soil becomes further degraded through erosion and compaction. Populations may have to relocate due to extreme desertification. Adaptation to drought may rely on the development of diversified production systems, such as ranching of animals better adapted to local conditions.

2.1.1.3 Glaciers

It is expected that between one-third and one-half of existing mountain glaciers will disappear in the next 100 years. These reduced glaciers and reduced depth of snow cover would affect the seasonal distribution of river flow and water supply to hydroelectric power and agriculture. Resulting hydrological changes and reduction in permafrost could lead to large-scale changes in navigation of rivers and lakes. Thus, the Arctic Ocean and its surrounding seas may become more navigable, which would benefit shipping. The release of methane from methane hydrates, which is expected to be abundant in the permafrost, could increase the rate of global warming.

2.1.1.4 Mountain Regions

Warmer climate in mountain regions will reduce the volume and extent of glaciers. This would affect the hydrologic system, soil stability, and the economic system. Vegetation and agriculture would shift to higher elevations. Resources of food and fuel for the mountain population could be disrupted. Mining and timber production would be affected. Another increasingly important area is recreational activities (e.g., skiing), which would also be disrupted. These effects could be difficult to adapt to, and a relocation of population may be necessary.

2.1.1.5 Lakes, Streams, and Wetlands

Increasing temperature of lakes and streams can affect the aquatic biological productivity and species survival in these regions. Runoff could be affected, and flash floods and drought could occur more frequently. The ecosystem would be changed in both the lower and higher latitudes. Declines in water level would affect drainage and stress human activities. Inland wetlands, which cover a significant fraction of the land area in many regions, could change, causing adverse ecological effects. The hydrologic effects in the form of increasing or decreasing water supplies in the wetlands would be significant, which in turn would affect waste processing and carbon storage. Although adaptation, including conservation and restoration activities, is possible, the scale of the effects with these changes would be enormous.

2.1.1.6   Coastal Systems and Oceans

Climate change is expected to raise the sea level. This could result in erosion of beaches and shores and altered tidal ranges. Loss of land habitat and major changes in the ecosystem would affect tourism and fishing economies. A rise in sea level would alter ocean circulation and vertical mixing and affect biological marine organisms. Small island states, such as in the Caribbean and the Pacific, where almost all human habitation and activity are in low-lying coastal areas, may not have the resources to relocate to higher areas. From an adaptive standpoint, low-lying countries, such as Holland and Bangladesh, have to cope with ocean intrusion. The Dutch have become famous for building dikes and seawalls to hold back the rising sea. This adaptation technology may have to be implemented by other coastal countries.

2.1.1.7   Water Resource Management

Climate change is expected to have a major impact on regional water resources. The effects will be noticed in both the ground and surface water supply, which will cause changes in domestic and industrial use, irrigation, hydropower generation, navigation, and water-based recreation. Global warming will influence precipitation, which can lead to floods and droughts. Rain and snow runoff in high altitudes will increase, while it would decrease in lower altitudes. Water supply, a problem presently encountered in many areas around the world, especially in low-lying coastal countries and islands, will become even more severe. Adaptive systems include more efficient management of existing water supplies and conservation. The wealthier countries could practice these methods at a minimum cost, whereas other countries may have to incur much greater economic and social costs to maintain their water resources.

2.1.1.8   Food and Fiber (Agriculture)

Increased CO_2 in the atmosphere could increase plant growth, increasing crop yields and agricultural productivity. This could occur mainly in the northern latitudes of the temperate zone. In the tropics and subtropics, the opposite would occur. Thus, hunger and famine could occur in tropical areas, while agricultural economies would thrive in northern latitude countries. Adaptation can include changes in management of water and irrigation, tillage methods, and crop varieties. Depending on know-how and cost, crop genetics can also be practiced. In general, adaptation strategies will be a great burden to developing countries, whereas the more affluent countries may experience a cost savings.

Livestock production will be affected because of changes in grain and pasture productivity.

2.1.1.9   Forest Products

Tropical and boreal forests are expected to decline because of global climate changes. The world's wood supplies may become increasingly inadequate due to climatic changes, as well as the practice of commercial deforestation.

2.1.1.10   Fisheries

Marine fisheries production is expected to increase in the higher latitudes, due to warmer waters, as will freshwater and agriculture production. However, losses may occur due to changes in migration routes, reproduction, and ecosystem relationships.

2.1.2   Human Infrastructure

Global climate change could have major consequences for human activities. Coastal zones could become flooded, requiring large sectors of the population to relocate. For example, it is estimated that a 50-cm (2-ft) rise in sea level could put about 46 million people at risk annually. The need for human migration in response to the effects of global climate could become a prime factor in coping with climate change. Many vulnerable human settlements are located in potentially affected areas that do not have the resources to deal with the impacts. In better economic areas, where property insurance is available, higher risk due to extremely damaging events could result in higher insurance premiums or even impossibly high insurance rates, leading to insolvency and threats to the banking system. The recent increase in the frequency of multibillion-dollar settlements for losses due to storms and fires reflects the economic worth of vulnerable areas.

2.1.2.1   Human Health

Climate change is likely to have broad adverse impacts on human health and loss of life. Exposure to thermal extremes, such as heat waves, can directly cause cardiovascular and respiratory diseases. Floods and storms directly cause deaths, injuries, and psychological disorders. Ecological disturbances (temperature, precipitation, and weather) can lead to indirect health effects. Increases in diarrheal and infectious diseases can result in increased malnutrition, hunger, injuries due to migration, and general poor mental health due to social and economic dislocations. For example, it is estimated that malaria transmission will increase by 45 to 60% in the next century due to a 3 to 5°C rise in temperature. On the other hand, temperature increase in colder regions should result in fewer cold-related deaths. Adaptive options include better protective technologies, housing, air-conditioning, water purification, vaccinations, medical attention and health care, and disaster preparedness.

2.2   SUMMARY OF ADAPTIVE RESPONSES

The following is a summary list of adaptive responses to global climate change:

1. Relocation of populations from areas threatened by climate change (e.g., coastal areas, waterways)
2. Remanage land areas
3. Construct breakwaters and dikes to protect coastal areas in populated regions

4. Change agricultural patterns
5. Provide special health care for affected areas
6. Provide for economic assistance to less developed regions
7. Practice conservation

2.3 MITIGATION RESPONSES[2]

A brief review of the mitigation responses to global warming is provided below; the responses will be discussed in more detail in subsequent chapters. Because the greatest influence of greenhouse gases is due to CO_2, these mitigative efforts are directed toward reducing CO_2 emissions from the use of fossil fuels.

1. *Improve energy efficiency.* Improving the conversion efficiencies of fossil fuels to electrical and thermal energy reduces CO_2 emission. Improving the utilization efficiencies of electrical and thermal energy also reduces CO_2 emissions.
2. *Fuel switching.* The substitution of natural gas and oil for coal reduces CO_2 emissions.
3. *Removal, recovery, and disposal of CO_2.* CO_2 can be recovered from fossil-fuel-burning power plants and engines. The fossil fuels can be disposed of in the ocean, land aquifers, depleted gas and oil wells, salt domes, and natural minerals.
4. *Utilization of CO_2.* The recovered CO_2 can be converted to materials of use to the economy, including fuels, chemicals, and construction materials.
5. *Decarbonization of fossil fuels.* Extracting carbon from fossil fuels and utilizing only the hydrogen-rich fraction of the fossil fuels can reduce CO_2 emissions significantly.
6. *Use of nonfossil energy sources.* Nuclear energy eliminates CO_2 emissions. Solar energy reduces CO_2 emissions. Hydroelectric power eliminates CO_2. Geothermal energy eliminates CO_2.
7. *Reforestation.* The planting of trees provides a means of reducing atmospheric CO_2 through photosynthesis.
8. *Utilization of biomass energy.* The use of wood and other agricultural biomass, including aquatic plants such as algae, as fuel essentially eliminates CO_2 emissions. Combustion of biomass generates CO_2, and photosynthesis to produce biomass consumes CO_2. Energy farms grow biomass as a fuel resource.
9. *Geoengineering.* The technology deals with large-scale global efforts. The use of atmospheric and space reflectants, such as volcanic dust and sulfates, can reduce solar radiation that reaches the earth. Seeding the oceans with iron could result in an increase in growth of marine organisms which reduce CO_2 in the atmosphere through photosynthesis.

REFERENCES

1. IPCC 1996, *Climate Change 1995, Working Group II,* Watson, R.T. et al., Eds., published for the Intergovernmental Panel on Climate Change, Cambridge University Press, Cambridge, 1996.
2. Steinberg, M., History of CO_2 greenhouse gas mitigation technologies, *Energy Convers. Manage.,* 33(5–8), 311–315, 1992.

3 Concepts for Controlling Atmospheric Carbon Dioxide

3.1 INTRODUCTION

As noted in Chapter 1, in order to avoid disasters, it may be necessary to mitigate or control the CO_2 concentration in the atmosphere. This can be accomplished by either removal of CO_2 from the atmosphere or avoiding emissions of CO_2 into the atmosphere. A general background and logical systematic approach can help in developing CO_2 mitigation technologies. The underlying assumption in constructing this analysis is that it does not deny the use of fossil fuel in a mitigation or control scenario. In fact, it assumes that fossil fuel will be used and that mitigation will occur either prior to, during, or subsequent to its use as a fuel. Nonfossil energy, which includes solar, renewable, and nuclear energy, can be of assistance in reducing CO_2 emissions, in conjunction with the utilization of fossil energy.

This chapter addresses mitigation of the CO_2 effect mainly by controlling the CO_2 content in the atmosphere. It does not deal with alteration of the environment to cope with the CO_2 effect. That is, the climatic and hydrological modification of the atmospheric or terrestrial environment (geoengineering) is not considered in these mitigation methods. The first step in the methodology is to devise a logic for guiding the construction of the mitigating processes and systems. It is then necessary to ascertain the physical, chemical, and biological properties required to design the mitigating systems. This is followed by an outline of the criteria for evaluating the process systems and scenarios. Descriptions of specific process system examples for obtaining the mitigating effects are then presented, and a preliminary evaluation is made for each of the process systems. In conjunction with the utilization of CO_2, a preliminary market projection is made to determine the magnitude of the effect of each of the projected uses on the control of CO_2. Finally, a priority listing of recommended research and systems analysis studies is created based on the preliminary evaluations. All options are considered, no matter how unorthodox they appear. However, this chapter should be considered as an introduction to the more relevant systems that will be discussed in detail in subsequent chapters.

3.2   LOGIC FOR CONSTRUCTING SYSTEMS AND SCENARIOS FOR MITIGATING THE ATMOSPHERIC CO_2 EFFECT

The mitigating technology for controlling the effects of atmospheric CO_2 is primarily concerned with controlling the concentration of CO_2 in the atmosphere. As such, the steps involve either restriction of emission or the removal, recovery, disposal, and reuse of CO_2. The suggested logic for guiding the construction of CO_2-mitigating or control technologies and scenarios is shown in block diagram form in Figure 3.1. The sequential steps are as follows:

1. *Control point selection.* The control points are essentially where the bulk of the CO_2 either exists or is generated and lend themselves to possible CO_2 removal sites. Four potential control points are available for regulating the concentration of CO_2 in the atmosphere: (1) the atmosphere itself; (2) the oceans; (3) the stacks and exhausts of fossil-fuel-burning plants and engines, which include (a) the power generation sector (large power plants), (b) the industrial and commercial sectors (smaller power plants and domestic heating systems, refineries, cement and chemical plants), and (c) the transportation sector (automotive engines); and (4) other terrestrial sources, such as natural wells, mines, and volcanoes. The power generation site has the advantage of providing a source of relatively high CO_2 concentration, while the atmosphere and ocean provide more flexibility for site selection and disposal or recovery and reuse options. For example, the ocean can be a source or a sink of CO_2, while the atmosphere can only be a source. Natural gas wells are readily

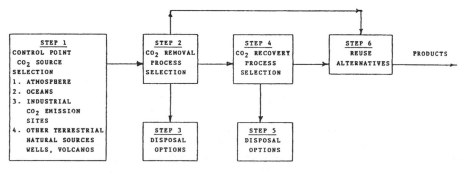

FIGURE 3.1   Logic diagram for the construction of scenarios for mitigating atmospheric carbon dioxide effects.

accessible, while volcanic sources are almost impossible to access. The choice of control point limits the available choices of CO_2 removal processes, size of the operation, number of sites required, removal efficiency, energy requirements, comparative economics, and overall feasibility of the operation.

   The types of options for the removal, recovery, disposal, and reuse of CO_2 as a function of control point are listed in Tables 3.1 through 3.5. Alternative nonfossil fuels and energy sources used in conjunction with fossil fuels are also considered potential means of controlling atmospheric CO_2.

2. *Removal process selection.* Once a control point has been selected, the next step is to survey available removal processes to determine which appear to be feasible and compatible with the selected control point.

TABLE 3.1
Options for the Removal of Carbon Dioxide as a Function of Control Point

Control point	CO_2 removal options
Atmosphere	1. Absorption by liquids 2. Adsorption by solids 3. Extraction by refrigeration 4. Increased planting of land or ocean biota — photosynthesis 5. Absorption by deep-ocean-waters process used in conjunction with ocean thermal gradient power cycle 6. Dilute-phase methanation of atmospheric CO_2 — hydrogen supplied by nonfossil energy source 7. Direct extraction from the atmosphere 8. Decomposition by low- and high-energy radiation
Oceans	1. Strip CO_2 with H_2 2. Distillation or steam stripping 3. Acidification followed by degasification 4. Precipitation as carbonates, coral formation
Major industrial CO_2 emission sources (e.g., fossil power plant stacks)	1. Absorption by liquids 2. Adsorption by solids 3. Extraction by refrigeration 4. Advanced power cycle — air separation followed by combustion of fossil fuels with oxygen and recycle CO_2 5. Capture during or subsequent to combustion by sorbents 6. Interfere with or prevent formation of CO_2 (e.g., partial combustion to CO) 7. Use of alternative energy sources
Other terrestrial sources (e.g., natural wells and volcanoes)	1. Cap wells and natural fissures

TABLE 3.2
Options for the Disposal of Carbon Dioxide as a Function of Control Point

Control point	Options for disposal of captured CO_2
Atmosphere	CO_2 captured by liquids
	1. Evaporate water, then bury residue — carbonates and caustic sorbents, mineral sorbent
	2. Deep ocean — water only
	CO_2 captured by solids, including hydrates
	1. Subterranean (wells, mines, aquifers)
	CO_2 captured by refrigeration
	1. Extraterrestrial
	2. Deep ocean
	3. Subterranean (wells, mines, aquifers)
	CO_2 captured by plants (photosynthesis)
	1. CO_2 remains integral part of plant
	CO_2 captured by deep-ocean waters
	1. Deep ocean
Oceans	CO_2 captured by distillation
	1. Extraterrestrial
	2. Deep ocean
	3. Subterranean (wells, mines, aquifers)
	CO_2 captured by acidification
	1. Extraterrestrial
	2. Deep ocean
	3. Antarctic region
	4. Subterranean (wells, mines, aquifers)
Major industrial CO_2 emission sources (e.g., fossil power plant stacks)	CO_2 captured by liquids
	CO_2 captured by solids
	CO_2 captured by refrigeration
	CO_2 removed by advanced power cycle
	1. Extraterrestrial
	2. Deep ocean
	3. Subterranean (wells, mines, aquifers)
Other terrestrial sources (e.g., natural wells and volcanoes)	1. Cap wells and fissures
	2. Mines
	3. Aquifers

Consideration is also given to altering the combustion process to limit or reduce formation of CO_2.

3. *Identification of disposal alternatives for captured CO_2.* This step involves the identification of alternatives available for disposing of the captured CO_2 (Step 2) directly, that is, without going through a CO_2 recovery step (e.g., sorbent desorption and regeneration). For example, if a naturally occurring, inexpensive, and nonpolluting CO_2 adsorbent is

TABLE 3.3
Options for the Recovery of Carbon Dioxide as a Function of Control Point

Control point	Options for recovering captured CO_2
Atmosphere	CO_2 captured by liquids
	1. Desorption by flashing
	2. Desorption by distillation
	3. Strip with gases (i.e., H_2, N_2, steam, etc.)
	4. Desorption by combined flashing and distillation or stripping
	5. Electrolytic decomposition of sodium carbonate to form sodium hydroxide, H_2, O_2, and CO_2 by nonfossil-fuel source
	CO_2 captured by solids
	1. Decomposition by heating stripping (e.g., by high-pressure steam)
	CO_2 captured by methanation
	1. Absorb methane in a solvent, then desorb by flashing, distillation, or stripping
Oceans	
Major industrial CO_2 emission sources (e.g., fossil power plant stacks)	CO_2 captured by liquids — same alternatives as for the atmosphere CO_2 captured by solids — same alternatives as for the atmosphere CO_2 captured by refrigeration
Other terrestrial sources (e.g., natural wells and volcanoes)	CO_2 removal from natural gas wells

available, the spent sorbent that contains the captured CO_2 could be buried or stored in abandoned mine shafts rather than reused. In the case of CO_2 extraction by refrigeration or distillation, no desorption step is required. A comparison between direct disposal vs. CO_2 desorption/sorbent regeneration would be made at the time of scenario evaluation, as each alternative would represent a separate scenario.

4. *CO_2 recovery process selection.* For a once-through system, the quantity of sorbent consumed in absorbing or adsorbing CO_2 would generally be very large because of the vast quantities of CO_2 that would have to be removed from the environment to effectively control the level of atmospheric CO_2. Thus, it may be advantageous, and in fact would be necessary in most cases, to desorb captured CO_2 and regenerate and reuse the sorbent. Also, direct disposal of spent sorbent may not be possible because of pollution considerations. Therefore, at this point in the formulation of a scenario, all desorption/regeneration possibilities should be identified and each selected in turn for evaluation.

5. *Identification of disposal alternatives for recovered CO_2.* This task is similar to that described under Step 3 except that CO_2 would be in the form of a relatively pure gas, liquid, or solid. Appreciable quantities of

TABLE 3.4
Options for the Disposal of Carbon Dioxide as a Function
of Control Point

Control point	Options for disposal of recovered CO_2
Atmosphere	Gaseous CO_2
	1. Extraterrestrial
	2. Deep ocean
	3. Subterranean
	Liquid CO_2
	1. Extraterrestrial
	2. Deep ocean
	3. Subterranean
	Solid CO_2
	1. Extraterrestrial
	2. Deep ocean
	3. Subterranean
Oceans	Same options as for the atmosphere
Major industrial CO_2 emission sources (e.g., fossil power plant stacks)	Same options as for the atmosphere
Other terrestrial sources (e.g., natural wells and volcanoes)	1. Store recovered CO_2 in the Antarctic

sorbent would not be present in the effluent disposal stream. Note that in the case of CO_2 extraction by refrigeration, liquid or solid CO_2 would be produced directly in the extraction step (Step 2), and therefore Steps 4 and 5 would not be required.

6. *Identification of CO_2 reuse alternatives.* The last step in the process of scenario construction involves exploring and identifying reuse possibilities for recovered CO_2 from Step 4. Several possibilities are identified in Table 3.5. One example which appears useful is to catalytically convert the recovered CO_2 to methanol by reaction with hydrogen using either a noncarbonaceous energy source or a fossil-fuel source which generates less CO_2. In some cases, it may be possible to proceed directly from CO_2 removal (Step 2) to CO_2 reuse, as indicated by the bypass line in Figure 3.1.

3.3   PHYSICAL, CHEMICAL, PHYSICAL–CHEMICAL, AND BIOLOGICAL PROPERTIES REQUIRED TO CONSTRUCT A MITIGATING PROCESS SYSTEM

In order to design CO_2-mitigating process systems, it is necessary to understand the basic physical, chemical, and biological properties related to CO_2. The mitigating

TABLE 3.5
Options for the Reuse of Carbon Dioxide as a Function of Control Point

Control point	Reuse alternatives for captured/recovered CO_2
Atmosphere	Gaseous CO_2 1. Catalytically convert CO_2 to gaseous or liquid carbonaceous fuels (methane, methanol, gasoline) by reaction with hydrogen, using nonfossil energy source (e.g., fission, fusion, solar) 2. Decomposition of CO_2 to CO using nonfossil energy source (e.g., solar and nuclear radiation and high temperature), shift CO with water to form CO_2 and H_2, subsequently convert to carbonaceous fuels 3. React CO_2 with NH_3 to form urea fertilizer 4. Use CO_2 for controlled photosynthesis in greenhouses — plant products in biosphere or burn for energy production 5. Use CO_2 as feedstock for marketable chemicals Liquid CO_2 1. Use for liquid CO_2 applications including industrial chemical uses (e.g., supercritical CO_2 solvent) 2. Use for displacing crude oil from reservoir rock 3. Same alternatives as for gaseous CO_2 after vaporization Solid CO_2 1. Dry ice applications 2. Same alternatives as for gaseous CO_2 after sublimation
Oceans	Same options as for the atmosphere
Major industrial CO_2 emission sources (e.g., fossil power plant stacks)	Same options as for the atmosphere
Other terrestrial sources (e.g., natural wells and volcanoes)	Same options for gas wells as for the atmosphere

processes can be categorized under physical, chemical, physical–chemical, and biological reaction categories. The required properties are listed under each category.

Physical properties
1. Condensation processes (compression and refrigeration)
   a. The vapor pressure of the liquid and solid, including compressibility, melting point, and critical point, over the entire range of temperature
   b. Effect of diluent gases on the condensation properties of CO_2
2. Adsorption processes
   a. Adsorption capacity properties of CO_2 on chemically nonreactive solids, including adsorption isotherms as a function of temperature and pressure over a wide range of conditions

    b. The energetics of the adsorption properties (e.g., van der Waals forces)

    c. Sorbents would include natural and synthetic sorbents (e.g., clays, zeolites, molecular sieves, etc.)

3. Absorption processes

    a. Solubility of CO_2 in various nonreactive liquid solvents as a function of temperature and CO_2 concentration

      • Inorganic solvents: fresh water, sea water, etc.

      • Organic solvents: alkane, aromatic, saturated, and unsaturated organics

Chemical properties

1. Chemical reactivity with reactive gases

    a. Hydrogen, water, ammonia, etc.

2. Chemical reactivity with reactive liquids

    a. Inorganic liquids: aqueous $NaOH$, Na_2CO_3, KOH, K_2CO_3, etc.

    b. Organic liquids: amines, monoethanolamine, diethanolamine, *n*-methylpyrrolidine, methanol, etc.

3. Chemical reactivity with reactive solids

    a. Inorganic solids: anhydrous $NaOH$, KOH, CaO, clays, silicates, zeolites, natural minerals, molecular sieves, reaction with carbon

    b. Organic solids: oxalic acid, oxylates, and carboxylates

    c. Oxidation: carbon oxidation mechanism for formation of CO and CO_2

    d. Reactivity of CO_2 on solid sorbents for use in extraction of CO_2 during and after the combustion of fossil fuels

    e. Thermal decomposition of CO_2 at various concentrations and temperatures to form reduced carbon materials: CO, C_3O_2, and C

    f. Reduction and reaction of H_2 with CO_2 at various temperatures and pressures to form CH_4, CH_3OH, CH_2O, etc.

Physical–chemical properties

1. Low-energy radiation interaction (photolysis) with CO_2 and determination of yields and distribution of products as a function of types of radiation

    a. Infrared

    b. Ultraviolet

    c. Visible light

2. High-energy radiation (radiolysis) interaction with CO_2 and determination of yields and distribution of products as a function of types of high-energy radiation

    a. Gamma radiation

    b. Electron radiation

    c. Proton and alpha particle

    d. Fission fragment radiation

3. Electrochemical reactions of CO_2

    a. Electrical discharge

TABLE 3.6
Properties of Carbon Dioxide[26]

Property	Value
Sublimation point at 101.3 kPa[a] (°C)	−78.5
Triple point at 518 kPa[b] (°C)	−56.5
Critical temperature (°C)	31.1
Critical pressure (kPa)[b]	7383
Critical density (g/L)	467
Latent heat of vaporization (J/g)[c]	
At the triple point	353.4
At 0°C	231.3
Gas density at 273 K and 101.3 kPa[a] (g/L)	1.967
Liquid density	
At −37°C (g/cc)	1.101
At 273 K (g/cc)	0.928
Gas at 298 K and 101.3 kPa[a] CO_2 (vol/vol)	0.712
Solid density at −78°C (g/cc)	1.56
Viscosity at 298 K and 101.3 kPa · s (= cP)	0.015
Heat of formation at 298 K (kJ/mol,[d] ΔH)	−393.7

[a] 101.3 kPa = 1 atm.

[b] To convert kPa to psia, multiply by 0.145.

[c] To convert J/g to Btu/lb, multiply by 0.4302.

[d] To convert kJ/mol to Btu/mol, multiply by 0.9487.

    b. Aqueous and nonaqueous electrochemistry
    c. Photoelectric effects

Biological properties

  1. The photosynthetic properties of CO_2: forestation, plant growth
  2. Marine biology of CO_2: algae growth, phytoplankton
  3. Microbial biology of CO_2
  4. Sewage sludge digestion

A listing of some of the physical properties of CO_2 is provided in Table 3.6.

3.4 CRITERIA FOR EVALUATING SYSTEMS FOR MITIGATING THE ATMOSPHERIC CARBON DIOXIDE EFFECT

The effect of atmospheric carbon dioxide on the terrestrial environment is directly due to the buildup of CO_2 concentration in the atmosphere and absorption of the infrared radiation reflected from the earth. This increase in atmospheric carbon dioxide concentration is due to both increased input to the atmosphere (e.g., increased use of fossil fuels) and decreased removal from the atmosphere (e.g., deforestation). A primary criterion for evaluating a scheme that would mitigate the effect of atmospheric CO_2 must relate to maintaining a limiting concentration of CO_2 in the atmosphere. Thus, the application of physical, chemical, or biological processes must affect the mass balance of CO_2 in the atmosphere. Following an estimate of the rate at which the proposed system removes CO_2, estimates of the energy requirement and efficiency of performing the operation must be determined. Economic estimates, which include equipment investment requirements and operating costs, must be taken into account. Finally, estimates of any other environmental side effects due to operation of the proposed CO_2-mitigating process must be specified and fully assessed.

Responses to the following questions are to be used as criteria for evaluating a CO_2-mitigating process system, guided by the logic scheme presented in the previous section.

1. Is there enough scientific and technical information available on which to base the proposed mitigating or control process system?
2. What is the magnitude of the effect of the proposed control process on the mass balance of CO_2 in the atmosphere, and what is the rate at which it can be achieved?
3. What are the mass and material requirements to conduct the proposed control system?
4. What are the energy requirements to conduct the proposed control system?
5. What is the efficiency of CO_2 removal by the proposed scheme?
6. What is the capital investment for the proposed process system?
7. What is the operating cost for the process scheme?
8. What are the effects on the environment due to the practice of the proposed CO_2-mitigating process scheme?

A number of the above questions are mutually dependent. However, the responses to these questions should provide enough information to evaluate and assess the feasibility of a specific proposed mitigating system.

3.5   SPECIFIC MITIGATING PROCESS SYSTEMS AND PRELIMINARY EVALUATION

A number of specific mitigating systems are suggested in this section to be assessed and evaluated by means of the above logic and criteria. The systems are listed under the basic type of reaction mechanism on which they depend (e.g., physical, chemical, biological, or a combination of these). All possible approaches are suggested, no matter how unorthodox a system might appear to be.

For each of the process systems, a preliminary evaluation is made to attempt to determine the relative potential value of its usefulness in addressing the CO_2 mitigation problem. The evaluation also attempts to present a statement as to the availability of scientific and technical information which forms the basis for designing the system. The brief summary evaluation is made on the basis of the following criteria:

1. Information base
2. Energy requirement
3. Capacity
4. Environmental problems
5. Potential value

In evaluating the utilization of CO_2, a brief market survey and projected future potential are presented and assessed in terms of total CO_2 emission.

It should be noted that this is by no means an exhaustive compilation, and additional schemes can be suggested as they present themselves.

3.5.1 Physical Systems

3.5.1.1 Absorption and Stripping Systems Using Liquid Solvents

Removal and recovery of CO_2 from gas streams can be accomplished by absorption in liquid solvents, both chemically reactive and nonreactive. The captured CO_2 is then removed and recovered by stripping or desorption techniques. The liquid solvents that can be used are as follows:

1. Water
2. Aqueous sodium carbonate
3. Aqueous potassium carbonate
4. Caustic (sodium hydroxide)
5. Alkanolamines (mono- and diethanolamine)
6. Other nonreactive solvents (methanol, N-methylpyrrolidine, propylene carbonate)
7. Other chemically active solvents (NH_3, alkazid, sulfinol)

Information Base. The liquid absorption/stripping process for the removal and recovery of CO_2 from gas streams is a classical chemical engineering operation. The efficiency of conducting this operation depends on the liquid solvent used. The abundant information available on the solubility and reactivity of a number of liquid absorbents for CO_2 can be utilized to design processes that can remove and recover CO_2 from power plant stack gases. This topic is discussed more fully in Chapter 5.

Energy Requirements and Capacity. An estimate of the energy requirements for scrubbing CO_2 from a power plant stack gas includes the energy for blowing the gas through the scrubbers, pumping the liquid absorbent, and providing the heat to strip out the CO_2 from the solvent and concentrate the CO_2. The energy required to separate CO_2 from stack gas using monoethanolamine solvent is about 0.27 kWh(e)/ lb CO_2[1] (equivalent to 2000 kWh/ton C). If fossil fuel is used to supply this energy, the power plant efficiency would be reduced by over 30% and the cost of power could more than double. The more recent development of newer solvents has significantly decreased this energy requirement and will be discussed more fully in Chapter 5. However, if nuclear or solar heat and power are used to provide the energy to separate the CO_2, one 500-MW(e) nuclear power reactor would be sufficient to separate all the CO_2 from a 1000-MW(e) coal-fired power plant. Thus, removal and recovery of all the CO_2 from utility and industrial coal-fired plants would require a capacity equivalent to 200 1000-MW(e) nuclear power plants. The cost of these reactors, assuming $1500/kW(e) (1990 dollars), would amount to about $300 billion. The total industrial and utility power plant CO_2 captured (0.7 × 10^9 tons C/yr) would amount to about 50% of the total CO_2 emitted in the U.S. in 1990 (1.33 × 10^9 tons C/yr).

Environmental Problems. There should be no adverse environmental effects in applying the system. However, consideration must be given to environmental impacts due to the auxiliary nonfossil-fuel power system. This would require an evaluation of the future role of nuclear power plants.

Potential Value. Absorption/stripping is a very developed operation, and other than cost, there should be no technical difficulties in operating this system. The continued search for new and improved solvents and optimization and evaluation of absorption/stripping systems applied to integrated coal-burning plants may reduce the energy requirement and cost.

3.5.1.2 Adsorption and Stripping Using Solid Adsorbents

Removal and recovery of CO_2 from gas streams can be achieved by adsorption in solid chemically reactive and nonreactive solids. The stripping or desorption effectively concentrates the CO_2 for subsequent use or disposal. The following is a partial list of solid adsorbents that may be considered for mitigating the CO_2 effect:

1. Naturally occurring sorbents (e.g., clays and zeolites)
2. Waste oil shale
3. Processed sorbents (silica, zeolites, molecular sieves)
4. Coal, char, and carbon

Recovery and reuse of the more expensive adsorbents will be necessary. However, for the naturally occurring adsorbents, it may be possible to dispose of the CO_2 and spent adsorbent in abandoned mines or even the ocean.

Information Base. On the basis of available physical/chemical information on the adsorption capacity of various adsorbents, either continuous fluid bed adsorbers/

strippers or fixed bed switching adsorption/stripping CO_2 removal/recovery systems operating on coal-fired stack gases can be designed. However, little information is available on the adsorption capacity of naturally occurring aggregates.

Energy and Capacity. A preliminary calculation based on molecular sieves indicates an energy consumption for removal of CO_2 of 0.4 kWh(e)/lb CO_2,[1] which is equivalent to an energy requirement of 2900 kWh(e)/ton C. This value is approaching the amount of power generated from a coal-burning plant (0.56 kWh(e)/lb CO_2 emitted), so that the net efficiency of a coal-burning power plant would be drastically reduced. However, a nonfossil-fuel plant could supply this energy. Assuming a 1000-MW(e) plant burning coal that contains 70% C at 40% efficiency, the rate of effluent CO_2 is 1.8×10^6 lb/hr, which is equivalent to 245 tons C/hr.

$$\text{Power required} = 2900 \text{ kWh(e)/ton C} \times 245 \text{ tons C/hr}$$

$$= 710,000 \text{ kW(e)} = 710 \text{ MW(e)}$$

Thus, a 700-MW(e) nuclear plant is needed to supply the energy to remove all the CO_2 from a 1000-MW(e) coal-fired plant. About 300 1000-MW(e) nuclear reactors would be needed to remove all the CO_2 from major utility and industrial plants (0.7 $\times 10^9$ tons C/yr), which together emit about 50% of the CO_2 in the U.S. The capital investment would amount to about $450 billion. This is a significant expenditure and may not be practical unless there is a market value to the CO_2 collected.

Environmental Problems. There should be no adverse environmental effects in applying this system. However, consideration must be given to environmental impacts due to the auxiliary nonfossil-fuel power systems. This would require an evaluation of the future role of nuclear power plants.

Potential Value. Adsorption/stripping is a viable technology. What would improve the economics for removal of CO_2 from coal-fired plants would be the discovery of a naturally occurring aggregate, either clay or a zeolite, that would have a high affinity for CO_2 and could be readily disposed of. Measurement of the adsorption capacity of various natural aggregates should be a worthwhile research effort. This work could be extended to determine the uptake of CO_2 from the atmosphere by various soils and by various man-made structures. Further evaluation of recent work on CO_2 uptake with natural minerals may be worthwhile.[23]

3.5.1.3 Diffusional Process for Separation of CO_2

Because of differences in the rates of diffusion, porous membranes can separate gaseous molecules of varying molecular weight and size. As a triatomic molecule, CO_2 will separate from diatomic O_2 and N_2.

Because of the large volume, it may be possible to diffuse air through porous natural formations such as porous rock and land strata. The CO_2 should diffuse through fine porous matter at a slower rate than O_2 and N_2, thus effecting a separation. Both dilute atmospheric and more concentrated industrial source points should be considered for diffusional separation processes. A major problem with differential processes is the pressure loss across the membrane and the pumping power required to move large volumes of gas through the membrane.

Information Base. Some information is available on membrane separation of lighter gases such as hydrogen from heavier gases such as nitrogen. Much less information is available on the diffusion separation of CO_2. Furthermore, no information is available on separation through porous naturally occurring formations such as sand or volcanic ash.

Potential Value. Because of the use of natural formations, the cost of materials would be low. On the other hand, the location of these formations may be limited. Research work on this possibility may be a worthwhile undertaking, and some work has been done in Japan on membrane separation of CO_2 from stack gases.[24]

3.5.2   Chemical Systems

3.5.2.1   Combustion of Fossil Fuel to CO at Industrial Source Point and Combination with a Nonfossil Energy Source to Produce Carbonaceous Fuels

The origin of industrial CO_2 is the combustion of carbon-containing fossil fuel with oxygen from the atmosphere. It is conceivable to burn fossil fuel to CO and then combine the CO with a nonfossil source of hydrogen to form liquid and gaseous fuels:[2]

$$C + 1/2O_2 = CO \qquad \text{(from fossil-fuel source)}$$

$$2H_2O = 2H_2 + O_2 \qquad \text{(from nonfossil energy source)}$$

$$CO + 2H_2 = CH_3OH \qquad \text{(catalytic conversion to methanol)}$$

$$nCH_3OH = (CH_2)_n + nH_2O \qquad \text{(catalytic conversion to gasoline)}$$

This system would reduce the direct emission of CO_2, for example, at a central power station. By incorporation of a nonfossil energy source to generate hydrogen, a total reduction in the direct emission of CO_2 can be achieved. However, when the methanol or gasoline is burned for automotive power, then CO_2 is emitted. Because the carbon is used twice, once in the fossil-fuel plant and then for automotive use, the overall CO_2 emission per unit of energy is significantly reduced.

Information Base. There is enough background information available to design a system to perform the above process. Gasifiers to produce CO are available. The heat produced could be used to generate electricity, which in turn can be used to electrolyze the water to produce hydrogen. Much additional energy is required to produce the necessary hydrogen to convert the CO. This energy would come from a nonfossil source, either nuclear or solar.

The electrolytic generation of hydrogen requires 18.7 kWh(e)/lb H_2 at 80% energy efficiency.[3,4] The energy required to produce 1 lb of gasoline is as follows. Note that the composition of gasoline is equivalent to CH_2.

$$\text{Electrical energy} = \frac{2 \text{ mol } H_2}{\text{mol } CH_2} \times \frac{2 \text{ lb } H_2}{\text{mol } H_2} \times \frac{18.7 \text{ kWh(e)}}{\text{lb } H_2}$$

$$\times \frac{1 \text{ mol } CH_2}{14 \text{ lb } CH_2} = 5.3 \text{ kWh(e)/lb gasoline}$$

The electrical energy produced from the partial combustion of C to CO amounts to only 0.4 kWh(e)/lb C, which is equivalent to 0.3 kWh(e)/lb gasoline. The net electrical requirement = 5.3 − 0.3 = 5.0 kWh(e)/lb gasoline.

$$\text{Electrical energy} = \frac{14 \text{ lb gasoline}}{12 \text{ lb C}} \times \frac{5.0 \times 2000 \text{ lb}}{\text{ton}}$$

$$= 11{,}600 \text{ kWh(e)/ton C fixed in gasoline}$$

Capacity. Assuming U.S. gasoline consumption of 100×10^9 gal/yr, this is equivalent to 0.3×10^9 tons of contained carbon, which can be derived from 400×10^6 tons of coal, which constitutes about 50% of the recent (1995) coal production capacity in the U.S. A conventional all-coal synthetic fuel plant would require double this amount or about 800×10^6 tons coal per year. The nuclear power capacity to produce all the gasoline in the U.S. by this means would amount to 340 1000-MW(e) power reactors. The cost of these reactors at \$1500/kW would amount to \$510 billion. The power cost would be over \$2.00/gal of gasoline plus another \$0.10/gal for the coal. Additional cost for the chemical conversion plant would be required, on the order of \$100 billion. Considering that the annual expenditure for gasoline at about \$1.00/gal amounts to \$100 billion per year, the payout time to replace petroleum with a synthetic fuel source may not be too distant.

Environmental Problems. Other than the use of nuclear power, there are no adverse environmental effects of utilizing the system. However, the future role of nuclear power must be assessed.

Potential Value. The main potential of this system is in the reduction of coal required for the production of synthetic carbonaceous fuels. Application of this process would reduce the coal requirement for synthetic fuels to 50% of the conventional all-coal process. The nonfossil energy source substitutes 50% of the coal needed in a conventional process to produce the synthetic fuel. Carbonaceous fuels from petroleum for the transportation sector presently contribute 30% of the total CO_2 emissions in the U.S. Projecting that eventually petroleum will be replaced by coal for supplying liquid carbonaceous fuel, CO_2 emissions could double to a relative value of 60% of total emissions. Utilizing the above process would tend to reduce the contribution of CO_2 emissions to 30% of current U.S. total emissions.

3.5.2.2   Isothermal Extraction of CO₂ from the Atmosphere or Stack Gases and Combination with a Nonfossil Source of Hydrogen to Form Methanol and Synthetic Fuels

An isothermal process can be devised for extraction and concentration of CO_2 from the atmosphere. A caustic solution can be used to scrub the low concentration of CO_2 from the atmosphere. The high alkalinity and chemical affinity of NaOH for CO_2 makes this an efficient CO_2 extraction process. The sodium carbonate/bicarbonate solution can then be electrolytically separated in a three-compartment cell to reform the caustic solution and produce CO_2, H_2, and oxygen. The CO_2 can then be combined with hydrogen to form methanol and subsequently catalytically dehydrated to hydrocarbon fuel (gasoline, etc.). The sequence of the reactions is as follows:

Absorption	$2NaOH + CO_2 = Na_2CO_3 + H_2O$
Electrolysis	$Na_2CO_3 + 4H_2O = 2NaOH + 3H_2 + 3/2O_2 + CO_2$
Catalytic reaction	$3H_2 + CO_2 = CH_3OH + H_2O$
Catalytic reaction	$nCH_3OH = (CH_2)_n + nH_2O$

The absorption and electrolysis take place at the same low ambient temperature, thus avoiding large heat-exchanger loads.

This system can be evaluated for both environmental control and a source of nonfossil carbonaceous fuel. The process mimics the photosynthetic process in overall effect, as oxygen is emitted from the system. This is a net CO_2-neutral system. CO_2 is absorbed from the atmosphere and is returned when the carbonaceous fuel is burned. It can also be viewed as converting nuclear energy to a liquid carbonaceous fuel.

Information Base.   There is sufficient technical information to design a process based on the above system. The step that would require additional development work is the electrolysis of sodium carbonate in a three-compartment cell to regenerate CO_2 and produce the hydrogen.

Energy Requirement.   If CO_2 is taken from the atmosphere, the total electrical energy requirement is estimated to be 2200 kWh(e)/bbl gasoline[3,4]

$$= \frac{2200 \text{ kWh(e)}}{6 \text{ lb/gal} \times 42 \text{ gal/bbl}} = \frac{8.7 \text{ kWh(e)}}{\text{lb gasoline}}$$

Energy conversion efficiency = 66%

Electrical energy per unit C converted from CO_2

$$= \frac{8.7 \text{ kWh(e)}}{\text{lb gasoline}} \times \frac{14 \text{ lb gasoline}}{12 \text{ lb C}} \times 2000 = 20,000 \frac{\text{kWh(e)}}{\text{ton C}}$$

The energy for converting CO_2 to gasoline would have to come from a nonfossil-fuel source, either nuclear or solar.

Capacity. Assuming U.S. gasoline consumption of 100×10^9 gal/yr, this is equivalent to 0.3×10^9 tons carbon, which amounts to about 30% of the total carbon emission from fossil fuel in the U.S.

The nuclear power requirement to produce all the gasoline in the U.S. by this process would amount to 600 1000-MW(e) power reactors. The cost of all the reactors at \$1500/kW would be \$900 billion. The electrical power cost for gasoline in 1990 dollars at 50 mill/kWh(e) would be $8.7 \times \$0.050 \times 6$ lb/gal = \$2.60/gal. Additional capital cost for the chemical plant could double the cost of the liquid HC fuel.

Environmental Problems. There should be no adverse environmental effects in applying this system. However, consideration should be given to environmental impacts due to the use of a nonfossil-fuel power system. This would also require an evaluation of the future role of nuclear power plants.

Potential Value. Overall, this process could reduce the total annual U.S. CO_2 emission to the atmosphere by a substantial 30%. This could also potentially be extended to total global emission, provided other countries would adopt this system. The system could be applied at power plant stacks or the CO_2 could be taken from the atmosphere. Taking CO_2 from stack gases could reduce the energy requirement by about 10% from the above figures. However, the system would require a substantial capital investment, running hundreds of billions of dollars, and would become economical with conventional fossil fuel only when gasoline costs double or a penalty is applied to the use of fossil fuel. Considering the present outlay for gasoline of over \$100 billion per year, the large capital investment to replace a fossil-fuel source of gasoline may not be prohibitive.

The chemical engineering of scrubbing the CO_2 from the atmosphere or from stack gases and the electrochemical decomposition of sodium carbonate in a three-compartment cell are indicated as research and development activities.

3.5.2.3 Reduction of CO_2 with C After Combustion and Conversion to Synthetic Fuel with a Nonfossil Source

It is possible to reduce CO_2 emission from stack gases of a fossil-fuel-burning power plant by reaction with carbon or coal. The CO formed can then be combined with a nonfossil source of hydrogen to form liquid methanol or hydrocarbon fuel.

Gasification	$1/2CO_2 + 1/2C = CO$
Electrolysis	$2H_2O = 2H_2 + O_2$
Catalytic reaction	$CO + 2H_2 = CH_3OH$
Catalytic reaction	$nCH_3OH = (CH_2)_n + nH_2O$

The system would be similar to the direct formation of CO discussed above; however, in this case, CO_2 would be allowed to form from a fossil-fuel-burning power plant, and this would subsequently be reduced to CO with the carbon in coal. The gasification of coal with CO_2 to form CO is a well-known process which is usually performed with water instead of CO_2.

Information Base. There is enough information available to operate this system. The gasification of coal is slower with CO_2 than with H_2O, but it can be operated at temperatures above 800°C.

Energy Requirements. The energy for converting the CO with electrolytic hydrogen would come from a nonfossil source, either nuclear or solar.

$$\text{Electrical energy} = \frac{2 \times 2 \times 18.7}{14 \text{ lb}} = 5.3 \text{ kWh(e)/lb gasoline}$$

$$\text{Electrical energy per unit C converted} = 5.3 \times \frac{14}{12} \times 2000 = 12,000 \frac{\text{kWh(e)}}{\text{lb C}}$$

Capacity. Assuming U.S. gasoline capacity of 100×10^9 gal/yr, this is equivalent to 0.3×10^9 tons carbon or about 400×10^6 tons of coal, which is about 50% of the recent (1990) production capacity in the U.S. If coal were to be used to produce synthetic fuel by the conventional process, the consumption would increase to 800×10^6 tons/yr, which would double the emission of CO_2. By employing this process, coal consumption would increase by 200×10^6 tons/yr or not more than 25%.

The required nuclear power plant capacity would be on the order of 350 1000-MW(e) reactors, amounting to an investment of about $525 billion. The power cost would amount to $1.50/gal plus another $0.05/gal for the coal. Additional cost for the chemical conversion plant would be required, increasing the gasoline cost by 50%. This expenditure must be balanced against the cost of replacing fossil gasoline at $100 billion per year.

Environment Problems. Although no adverse environmental problems are anticipated, consideration should be given to the impacts of the use of a nonfossil power system and the role of nuclear power in the future.

Potential Value. As stated above, the increase in emission due to production of synthetic fuel via this system could be reduced to half. Instead of CO_2 emissions increasing by 100%, the increment would be reduced to 25%, resulting in a substantial reduction in CO_2 emission.

3.5.2.4 The Extraction of CO_2 with Lime During Combustion

In the combustion of coal, lime could be added as a chemically active sorbent for CO_2.

$$CaO + C + O_2 = CaCO_3$$

If oxygen is used during the combustion at atmospheric pressure, the temperature would have to be maintained below 900°C to capture the CO_2. With air, because of the lower partial pressure of CO_2, the combustion temperature would have to be maintained below approximately 800°C. A fluidized bed combustor could best be used for this purpose. However, calcined limestone would be necessary. In so doing, the CO_2 from the calcination process requires disposal. The CaO can be regenerated by calcination at the power plant site and the concentrated CO_2 disposed of separately. Centralized calcination of limestone and disposal of CO_2, for example in deep wells, may be more efficient than calcination in smaller dispersed plants. The calcium carbonate effluent could be sent back to centralized calcination plants or could be buried in depleted mines. However, burial of the $CaCO_3$ would not produce a net reduction in CO_2 because the formation of CaO needed for the power plant CO_2 removal system must be obtained from the calcination of raw limestone which produces CO_2. This would be a zero sum system, with CO_2 being absorbed and equally emitted.

Information Base. There is sufficient information based on thermodynamic considerations to evaluate the two reaction systems.

Fluidized bed combustion	$CaO + C + O_2 = CaCO_3$
Calcination	$CaCO_3 = CaO + CO_2$
Overall	$C + O_2 = CO_2$

Fluidized bed combustion with air in the presence of lime would have to be conducted below 800°C. The power plant would operate on steam generated from the heated nitrogen originating from the air in a Rankine cycle. The calcination of the resultant $CaCO_3$ would take place at temperatures above 900°C to concentrate the CO_2.

Energy Requirement. By careful exchange of heat in the process and maintaining a strict heat balance, there should only be a loss of about 15% in thermal energy compared to a conventional coal combustion plant. This would mainly hold for on-site calcination and recovery of CO_2 for disposal.

Capacity.   Large limestone calcination plants have been built for cement manufacture. The calcination capacity for a 1000-MW(e) coal-fired power plant would amount to about 28,000 tons/day of CaO, which is equivalent to about the largest cement plant in the U.S. The mass-handling capacity of lime would be about three times the coal-handling capacity and probably would require a capital investment about equal to the power plant. In effect, the increase in capital cost would cause the cost of power generation to double. Central calcination of limestone would probably have a lower unit capital investment but would still be large.

Environmental Problems.   The mining, transporting, and handling of large amounts of limestone would result in increased problems equivalent to those associated with coal. The handling of lime and limestone dust is of importance in this respect, based on experience in cement plants.

Potential Value.   As a high-temperature removal process for CO_2, this process has some advantage. The capital investment for calcination due to the large capacity would be a large penalty. However, a further definitive systems evaluation is required.

3.5.2.5   Fuel Substitution and Decarbonization

All fossil fuels consist mainly of carbon and hydrogen in varying concentrations. Coal usually has an H/C ratio of 0.8; oil is 1.8 and natural gas 4.0. The CO_2 emissions per unit of energy decrease with increasing H/C ratio. This is the basis for fuel substitution; natural gas generates about half the CO_2 per unit of energy as coal, and oil generates about halfway between coal and gas. Table 3.7 gives the CO_2 generation from a variety of hydrocarbon fuels, including natural and derived or synthetic fuels. The ideal greenhouse fuel is hydrogen because it does not generate any CO_2. Carbon produces the highest CO_2 emission. However, there is no natural source of hydrogen. Hydrogen can be produced by electrolysis of water. The electrical energy must be supplied by either fossil fuel or a nonfossil-fuel power source. However, hydrogen can also be produced by a process called decarbonization.

Decarbonization has come to mean separating the carbon from fossil fuel prior to combustion. There are actually two processes that accomplish this. In one process, the fossil fuel (coal, oil, or gas) is gasified with steam to produce hydrogen and CO_2. The CO_2 is removed, and the remaining hydrogen is used for fuel in power plants or engines. This process is called steam reforming. To avoid emission of CO_2 to the atmosphere, the CO_2 must be recovered and sequestered.

An alternative decarbonization process removes the carbon from the fossil fuels as carbon. The carbon is not converted to CO_2, and because it is a solid, the carbon can be more easily stored or utilized as such. Decarbonization is accomplished by chemical or physical/chemical processes. The hydrogen is then utilized as a fuel with zero emission of CO_2. This system will be more fully described in Chapter 7.

TABLE 3.7
CO_2 Generation for Various Fuels[25]

Fuel type	Chemical formula	Higher heating value (Btu/lb)	CO_2 generated (lb CO_2/ lb fuel)	CO_2 generated (lb CO_2/ 1000 Btu)	Energy generated (kW(e)/lb CO_2 generated)
Coke (ashless C)	C	14,000	3.67	0.26	
Bituminous coal	$CH_{0.8}O_{0.1}$	12,700	2.59	0.20	0.55
Fuel oil and gasoline (petroleum distillate)	$(CH_2)_n$	19,600	3.14	0.16	0.70
Benzene	C_6H_6	18,000	3.38	0.19	
Acetylene	C_2H_2	21,500	3.38	0.16	
Methanol (wood alcohol)	CH_3OH	10,000	1.38	0.14	
Methanol (synthesized from coal)	CH_3OH	10,000	3.3	0.33	0.34
Biomass (wood, cellulose)	$CH_{1.44}O_{0.66}$	8,000	1.47	0.18	
Carbon monoxide	CO	4,350	1.57	0.36	0.31
Natural gas	CH_4	24,000	2.75	0.115	0.97
SNG from coal gasification	CH_4	24,000	7.9	0.33	0.34
Synthetic liquid fuels (by coal hydrogeneration)	—	—	—	0.35	
Synthetic liquid and gaseous fuels (mixed products from SRC-II process)	—	—	—	0.26	
SNG coal and nonfossil hydrogen	CH_4	24,100	2.75	0.11	1.07
Hydrogen generated by					
Coal gasification	H_2	61,000	16.5	0.27	
Natural gas reforming	H_2	61,000	7.0	0.11	
Hydrogen generated by nonfossil energy					
Hydro, solar, nuclear– electrolytic	H_2	61,000	0	0	
Geothermal–electrolytic	H_2	61,000	2.44	0.04	

3.5.3 Physical/Chemical Systems

3.5.3.1 CO_2 Recycle Power Plant

Another means of recovering CO_2 from fossil-fuel power plants is to burn coal with oxygen and recycled CO_2. The nitrogen is separated from oxygen and eliminated in an air liquefaction plant prior to combustion. Enough CO_2 is recycled to maintain a sufficiently low combustion temperature equivalent to a conventional coal/air flame. The highly concentrated CO_2, which is the only main power plant effluent, can then be directly recovered for utilization or disposal. A preliminary study of this system was performed, and it was found to be feasible. The use of CO_2 in enhanced oil recovery operations improves the economics of the system.

Potential Value. The study indicated that this method of recovery of CO_2 is one of the more efficient ones. The study indicated an energy requirement of 0.12 kWh(e)/ton of CO_2,[5] which is equivalent to 880 MW(e)/ton C removed. This requirement is one of the lowest to remove and recover CO_2. Thus, from an economic and environmental standpoint, operation of a coal-burning plant in this recycle mode, which comes closest to a completely contained system, was found worthy of further investigation.[27] Determining combustion conditions for coal–oxygen–CO_2 mixtures and improvement in separation of oxygen from air would establish the technology of this system. The CO_2 recovered from this system would still have to be disposed of or utilized. Disposal and utilization are discussed in Chapters 4 and 7.

3.5.3.2   Electrochemical Decomposition of CO_2

The high-temperature electrolysis of CO_2 to CO and oxygen is a possible method of reducing CO_2 to CO:

$$\text{Electrolysis: } CO_2 = CO + 1/2 O_2$$

The cell operates with a solid ceramic oxide electrolyte.[6] Electrodes are placed on both sides of the solid electrolyte. At elevated temperatures (>1000°C), the ceramic oxide is an ionic conductor which allows oxygen ions to migrate from the cathode to the anode to form oxygen. At the cathode, the CO_2 is reduced to oxygen ions and CO gas. The overall reaction is an electrochemical decomposition of CO_2 with the separation of CO from O_2.

As the temperature of the cell is increased, the Gibbs free energy (ΔF) is decreased due to an increase in the entropy term ($T\Delta S$). The emf of the cell is thus reduced, and the thermal energy input to make up for the total enthalpy of decomposition is increased. Thus, more direct thermal energy is utilized relative to electrolytic energy for decomposing the CO_2 and producing CO.

Information Base. There is some information on the high-temperature electrolysis of water using yttria-doped ceramic zirconia solid ionic electrolytic conductors. The mechanism for CO_2 decomposition should be similar. However, there is very little experimental information available on CO_2 decomposition using solid electrolytes.

Energy Requirements. The efficiency of conversion of thermal to chemical or CO energy should be about 50%. Thus, for a 1000-MW(e) coal-burning power plant, the thermal requirement for a solar or nuclear plant used to supply power for electrolysis would be 3500 MW(t), which is equivalent to about a 1000-MW(e) nuclear plant. The temperature requirement of about 1000°C is within the realm of possibility of the high-temperature gas-cooled nuclear reactor (HTGR). Also, the CO and O_2 would be separated and no fast quenching is necessary, as is the case for the thermal decomposition system alone. Capital investment for a high-temperature electrolysis plant should be about one-fourth that for a coal-burning power plant, on the order of $250 million.

Environmental Problems. Release of CO could be a problem due to its toxicity. Utilization of CO for synthetic fuels or organic chemical products is preferable.

Potential Value. The high-temperature electrolysis of CO_2 using solid ionic electrolyte is technically and economically more feasible than the high-temperature decomposition route and should be pursued.

3.5.3.3 Photoelectrochemical Reduction of CO_2

The photoelectrochemical reduction of CO_2 to methanol and methane is being studied on semiconductor electrodes.[7]

$$CO_2 + 2H_2O \xrightarrow{\sim\!\sim\!\sim} CH_4 + 2O_2$$

It may be possible to conduct this photoelectrochemical effect in the atmosphere with ground-based active surface electrodes, thus directly converting CO_2 to hydrocarbon species of use as liquid fuels or chemicals.

Information Base. A very limited amount of information is available on the photoelectrolysis of CO_2 to form methane. More information is available on the photoelectrolysis of water to hydrogen. It has been reported[8] that the efficiency of conversion of solar to hydrogen energy by photoelectrolysis has been improved to a value of 10%. This topic is discussed more fully in Chapter 13.

Energy Requirement and Capacity. Assuming that at least 10% of the solar energy is converted to methane as a useful fuel, this is equivalent to an energy requirement of 187,000 kWh of solar energy per ton of carbon fixed. Fixing all the carbon emitted as CO_2 from a 1000-MW(e) fossil-fuel plant would require 45,800 MWh of solar energy. These values are obtained as follows:

Heating value of CH_4 = 24,000 Btu/lb

$$= 24,000 \, \frac{Btu}{lb} \times \frac{1 \, kWh}{3413 \, Btu} \times \frac{1}{0.1 \, Eff.} \times \frac{16 \, lb \, CH_4}{12 \, lb \, C}$$

$$\times \frac{2000 \, lb}{ton} = 187,000 \, \frac{kWh}{ton \, C}$$

$$\text{Solar energy per 1000 MW(e)} = 187,000 \; \frac{kWh}{ton\ C} \times \frac{245\ tons\ C}{hr}$$

$$= 45,800 \; MWh$$

The solar energy is essentially free, but the land area required for collection and the collectors themselves are the high-cost items. Assuming a solar flux of 0.16 kW/m^2 (taking into account diurnal and cloud conditions), the land area required would amount to about 100 mi^2 or a tract of about 10 mi × 10 mi, which is a fairly large array. However, the plant would produce 34 million gal/year of gasoline, equivalent to natural gas. The cost, at an optimistic $100/kW for the solar energy plant and a 30-year depreciation, would result in a $5/gal gasoline cost, which is very high based on today's cost. The technology must await future improvements.

Environmental Problems. Other than the need to clear a large tract of land, there should be no other environmental effects of the plant.

Potential Value. It is important to improve the efficiency of the photoelectro-chemical route in order to make this system more technically and economically viable.

3.5.3.4   High-Energy Radiation Decomposition of CO_2

It is possible to irradiate CO_2 with high-energy radiation, including gamma, electron, and particle irradiation such as fission fragments, to decompose it to CO and oxygen.

$$CO_2 \xrightarrow{\ \ \ } CO + 1/2O_2$$
$$\text{rad.}$$

The continued irradiation of CO formed further decomposes to C_3O_2, known as carbon suboxide.

$$3CO \xrightarrow{\ \ \ } C_3O_2 + \tfrac{1}{2}O_2$$
$$\text{rad.}$$

The carbon suboxide is known to form a reddish polymer which may be considered for development of a useful polymer. Carbon suboxide is also known to be the anhydride of an organic acid (malonic acid) which can be a building block for the production of useful organic chemical compounds.

It is suggested that the above high-energy radiation chemistry be investigated for converting the CO_2 from stack gases to useful products and preventing emissions of CO_2 into the atmosphere.

Information Base. The irradiation of CO_2 to form CO and oxygen with high-energy ^{60}Co gamma and fission fragment radiation energy has been investigated.[9]

The formation of C_3O_2 (carbon suboxide) by irradiation of CO with ^{60}C gamma and electron accelerator energy has also been investigated.[10] The design of a process around this information would require much additional work.

Energy Requirements. At a G-value of 10, the efficiency of conversion of CO_2 to CO is about 30% of the absorbed high-energy radiation. Because fission fragment radiation is the lowest cost and largest available energy fraction of the fission process (85%), it is possible to obtain about 40% of the total fission energy as fission fragment radiation and the overall energy requirement would be reduced to 6.9 kWh/lb CO_2 or 50,600 kWh(t)/ton C fixed. For a 1000-MW(e) coal-burning electrical power station, a fission fragment reactor of 12,400 MW(e) would be required. These values have been obtained as follows:

G-value for CO formation from CO_2 = 10 molecules/100 eV

Conversion to kWh(e)/lb CO_2

$$E = \frac{121}{GEM} = \frac{1216}{10 \times 0.4 \times 44} = 6.9 \text{ kWh(t)/lb } CO_2$$

where G = G-value, E = efficiency of energy deposition, and M = molecular weight.

$$E = 6.9 \times \frac{44}{12} \times 2000 = 50,600 \text{ kWh(t)/ton C}$$

$$\text{Nuclear power/1000 MW(e) coal} = \frac{50,600 \times 245 \text{ ton C/hr}}{1000 \text{ kW/MW}} = 12,400 \text{ MW(t)}$$

A special type of thin-foil fissile fuel element would be needed. While technically possible, the reactor design and capacity would be extremely difficult to attain.

The use of high-energy (10-MeV) electrons from an electron accelerator powered by a nuclear electric power reactor would alleviate the radiation contamination problem due to fission fragment radiation. The increase in energy deposition efficiency would compensate for the electrical power needed for the accelerator, so that a similar reactor power capacity of 12,400 MW(t) would be needed to convert the CO_2 from a 1000-MW(e) coal-burning plant.

Environmental Problems. There would be a serious problem in handling fission fragment radiation to prevent contamination of the product. Electron accelerator radiation would be a more acceptable radiation source.

Potential Value. Although the radiation decomposition of CO_2 is one of the more efficient radiation processes, the need for massive amounts of radiation for the endothermic decomposition of CO_2 would require a large nuclear reactor to handle the large mass quantity of CO_2 from a coal-fired power plant stack emission. Improvement in the efficiency of production and utilization of high-energy electron radiation may be a research direction to explore in order to further improve the potential of the radiation process. Other than the production of synthetic fuels, utilization of carbon suboxide for plastics and organic chemical production may have a limited market demand, as discussed later in this chapter under utilization.

3.5.3.5 Photochemical Decomposition of CO_2

Photochemical decomposition of CO_2 can take place, forming CO, which can subsequently be photochemically converted to C_3O_2 (carbon suboxide). The photodissociation of CO_2 to CO requires UV photons. C_3O_2 can also be formed by UV photolysis. Scavengers and sensitizers can be used to enhance the effect.

$$CO_2 \xrightarrow{\hspace{1.2cm}} CO + 1/2O_2$$

UV

$$3CO \xrightarrow{\hspace{1.2cm}} C_3O_2 + 1/2O_2$$

UV

The condensed C_3O_2 polymer could then be removed for use or storage.

It is suggested that the photochemistry of CO_2 and CO be investigated for both high CO_2 concentration power plants and lower concentration decomposition of CO_2 directly from the atmosphere.

Information Base. The UV photochemical decomposition of CO_2 to CO has been studied in the region of 1295 to 1470 Å. The UV irradiation of CO to form the C_3O_2 has also been studied at 1800 Å.[11] However, much additional research would be needed to design a process. Some order-of-magnitude estimates can be made.

Energy Requirements. It is assumed that solar or nuclear power would generate electricity which would be used to excite mercury vapor lamps to generate UV. The UV radiation from these lamp sources would be absorbed in the CO_2 to cause decomposition to CO and O_2. The CO would be separated and subsequently photolyzed to form the C_3O_2. The thermal to electrical efficiency for a nuclear reactor is about 33%, and the conversion to UV is about 10%. The overall conversion of nuclear thermal to photolytic CO_2 decomposition energy is 3.3%. At a quantum yield of 1 and assuming 2250 Å absorption in CO_2 (which is equivalent to the bond energy for CO–O dissociation), the thermal energy required from the reactor is about 50 kWh(t)/lb CO_2.

Roughly another 50 kWh(t)/lb CO would be needed to generate the C_3O_2. For a 1000-MW(e) fossil-fuel plant, about 1.8×10^6 lb CO_2 per hour is generated. If a total of 100 kWh(t)/lb CO_2 is needed to convert the CO_2 to C_3O_2, a thermal power capacity of 180,000 MW(t) or 60,000 MW(e) for the nuclear reactor is required.

Clearly, this is a very large power requirement which results from the poor effi-ciency of utilization of electricity to generate UV radiation in lamps.

The energy and power requirement is derived from the following equation to dissociate molecules such as CO_2 to CO and O_2:

$$\text{Energy} = \frac{1.51 \times 10^5}{\lambda QME} = \frac{1.51 \times 10^5}{2250 \times 1 \times 44 \times 0.03} = 50 \text{ kWh(t)/lb CO}_2$$

where λ = wavelength of light (Å), Q = quantum yield , M = molecular weight, and E = overall efficiency of conversion of thermal to electrical to UV.

$$\lambda = \frac{12,400}{eV} = \frac{12,400}{5.52} = 2250 \text{ Å}$$

where

$$eV = \text{bond energy} = \frac{\Delta H}{23.07} \text{ in electron volts}$$

and ΔH = dissociation energy (kcal/mol). For CO_2 = CO + O, ΔH = 127 kcal/ mol.

Another 50 kWh(t)/lb CO_2 is assumed to be needed to photolyze CO to C_3O_2, for a total of 100 kWh(t)/lb.

For a 1000-MW(e) fossil-fuel plant, 245 T/hr of CO_2 is conserved and 1.8×10^6 lb/hr of CO_2 is generated.

The power needed to convert the CO_2 to C_3O_2 is

$$100 \frac{\text{kWh}}{\text{lb}} \times 1.8 \times 10^6 \frac{\text{lb}}{\text{hr}}$$

$$= 180,000 \text{ MW(t)} = 60,000 \text{ MW(e) at 33\% efficiency}$$

Environmental Problems. The two hazards in a plant that would employ the above process are the containment of CO and shielding operating personnel from the intense UV radiation.

Potential Value. Methods for improving the generation of UV could be sought. Investigation of the effect of the direct use of concentrated solar radiation on CO_2 could also be investigated for possible improvement in energy efficiency and eco-nomics.

3.5.3.6 Thermal Decomposition of CO_2

The thermal decomposition of recovered CO_2 can be performed to produce CO and oxygen.

$$CO_2 = CO + 1/2O_2$$

About 60% of the CO_2 is decomposed at about 2500°C, and in order to stabilize the CO to avoid a back reaction, a rapid quench to freeze the equilibrium and stabilize the CO at temperatures below 1000°C is required. The CO could then be utilized for conversion to fuels, chemicals, and polymers. Rapid separation of O_2 from CO_2 through ceramic membranes could also be considered.

Information Base. The thermodynamics and kinetics of the thermal decomposition of CO_2 are available. The quenching and stabilization of CO in the presence of O_2 are less well known, and more information is needed.

Energy Requirement. The decomposition of CO_2 to CO and $1/2O_2$ requires 68 kcal/mol of CO_2, while the combustion of C to CO_2 releases 94 kcal/mol. Thus, from an energy balance standpoint, in principle, all the CO_2 from a coal-burning plant could be reduced to CO. However, the net energy produced from the plant would amount to 28% of the original combustion of coal; 72% would be used to decompose the CO_2 to CO. In effect, the carbon in the coal would be partially burned to CO, which releases only 26 kcal/mol C or 28% of the complete combustion of carbon to CO_2. A nonfossil-fuel plant, either nuclear or solar, would be more appropriate to provide the energy to decompose the CO_2. Thus, for a 1000-MW(e) electrical generation plant with 40% efficiency, the nonfossil plant would have to have a thermal capacity of 1800 MW(t). This is very large for a solar plant. Because of the high temperatures required (2500°C), construction material problems would be very severe for a nuclear plant. Fast quenching or membrane separation is another difficult technical problem. Further details on high-temperature thermal reactions of CO_2 are reviewed in Chapter 8.

Environmental Effects. The generation of CO from CO_2 and disposal of CO to the atmosphere would tend to reduce the greenhouse effect but would not be desirable. The toxicity problem of CO in the atmosphere would be severe. Furthermore, eventually the CO in contact with atmospheric oxygen would be reoxidized to CO_2, so the effect would be temporary. The CO would have to be utilized for conversion to fuels, chemical feedstocks, or useful products (plastics, etc.).

Potential Value. The thermal decomposition of CO_2 and stabilization of CO, although possible, are technically very difficult because of the high temperatures and short time requirement for separation of the CO and oxygen.

3.5.4   Biological Systems

3.5.4.1   Photosynthetic Fixation of Atmospheric CO_2

The photosynthetic fixation of atmospheric CO_2 in plants and trees could be of great value in maintaining a CO_2 balance in the atmosphere. Deforestation has been partially responsible for the CO_2 buildup in the atmosphere. Reforestation could be a solution to the problem of the greenhouse effect as a result of continued growth in the utilization of fossil fuel by the world economy. A preliminary study suggests that the absorption and fixation of CO_2 be accomplished by allocating and planting regional forested areas to counterbalance an equivalent amount of CO_2 emitted in those regions. Further assessments of plant and forest growth as it affects the atmospheric CO_2 balance should be made.

Information Base. Although carbon balance studies are being conducted on photosynthetic processes, there has been little done on the possible use of aforestation as a CO_2 control mechanism. Further studies on tree plantations are needed to evaluate such a control measure. Data have become available on the increased yield of plant growth as a function of increased CO_2 in the atmosphere.

Capacity. A preliminary study[12] has indicated that the saving in land area for a forest plantation adjacent to a fossil-burning power plant due to the increased stack gas CO concentration is relatively small compared to planting in any area. It is estimated that a tract of 1000 km^2 (385 mi^2) planted with moderately growing trees is sufficient to fix all the CO_2 from a 1000-MW(e) coal-fired power plant over the lifetime of the plant.

Potential Value. Aforestation appears to be a feasible method of reducing the concentration of CO_2 from the atmosphere. It uses solar energy and allows an economic fixation of CO_2 from the atmosphere which does not depend on concentrated CO_2 streams. Control of dispersed sources of CO_2 is also taken into account by photosynthetic extraction of CO_2 from the atmosphere. Effort is going into improving yields of agricultural biomass to be used as fuel in power plants. The "energy farm" concept is being pursued.

3.5.4.2 Biological Reduction of CO₂ in Marine Organisms

The biological fixation of CO_2 from the atmosphere by marine organisms in the ocean, especially near the continental shelves, could extract and reduce atmospheric CO_2. The marine organisms would reduce the CO_2 to organic carbon in the marine environment, with the addition of sufficient nutrients, to keep the life cycle of the organisms active. The end product of CO_2 metabolism would be the formation of organic carbon residue on the continental shelves. Waste sewage sludge could act as a nutrient for the marine organisms.

Information Base. There is information on the biological reduction of CO_2 by marine organisms on the coastal shelf of the U.S. eastern and gulf coasts.[13] The organic C is then transported to the continental shelf, where it is further reduced and deposited as C.

Capacity. It has been suggested that the missing C in the global CO_2 cycle balance, which amounts to 0.7×10^9 tons C/yr, can be accounted for by the reduction of CO_2 and fixation by marine organisms on the continental shelves of the earth. The nutrients that drive this biological cycle come from river runoff.

Potential Value. This biological process appears to be significant because of its large capacity to reduce atmospheric CO_2. The use of sewage sludge to provide the nutrients for marine organisms can be an interesting combined waste disposal and atmospheric CO_2 fixation system. Harvesting of marine-grown biomass would be of interest. Continued investigation and development of information on marine biological fixation of CO_2 are recommended.

Environmental Impacts. There appear to be mainly beneficial environmental impacts to be derived from biological reduction in marine organisms. However, further study would be required to determine the effect on the ecological balance on the continental shelves.

3.5.5 CO_2 Disposal Schemes

After removal and recovery of CO_2, options for long-term disposal of CO_2 can be investigated in the following categories:

1. *Deep ocean.* The deep ocean is a possible reservoir for long-term storage of CO_2. Mixing and equilibration of the atmosphere take place in the ocean to a depth of about 100 m. Diffusion of CO_2 below this depth is very slow. Injection into the ocean as gas, liquid, or solid must take into account dissolution rate and buoyancy effects.[14,15,18] Further work on ocean disposal and sequestering will be described in Chapter 5.
2. *Terrestrial.* There are at least two options for burial within the earth. If a low-cost readily available solid adsorbent is found to fix CO_2 from concentrated sources, the spent sorbent may be stored in depleted or abandoned mines. An approach to using a natural sorbent based on a magnesium oxide mineral is being investigated.[23]

    Recovered CO_2 can be stored as a high-pressure gas in depleted gas and oil wells. In this connection, CO_2 is also of value for enhancing the recovery of oil from depleted oil wells. Storage capacity of much larger potential is possible in solution-excavated salt domes.[16,19] These could have the necessary capacity to store a significant fraction of the CO_2 generated by fossil-fuel utilization in the long term. CO_2 can also be sequestered in aquifers beneath land masses or beneath the ocean beds, as discussed in Chapter 5.
3. *Extraterrestrial.* A possible option is packaging CO_2 as solid, liquid, or high-pressure gas and sending the mass of CO_2 out beyond the gravitational field of the earth, either to other planets or to outer space. Although it appears that energy requirements would prohibit this disposal option, advances in space transportation missions[17] may reveal a possible feasible system.

Information Base. There are only a limited number of locations for disposal and sequestering of CO_2 on a long-term basis: (1) dissolution in the deep ocean below the thermocline, (2) storage in depleted wells and salt caverns, and (3) storage in underground aquifers. Some information has been developed on these methods;[14,15,18,19] however, more information is required on the mixing and the physical/chemical and biological effects of deep-ocean disposal. Underground structural problems due to CO_2 storage must be investigated.

Capacity, Energy, and Cost. There is sufficient capacity to dispose of all the CO_2 generated in fossil-fuel power plants in the deep ocean. The more economical method is to remove, recover, and liquefy CO_2 from power plant stacks and pump the liquid into the deep ocean. For plants located on the coast, the major energy requirement is for the removal and recovery of CO_2. Estimates indicate that for 90% removal and disposal of CO_2 from a power plant stack, the efficiency of the plant would be reduced by as much as 30%. The cost of power generation would more than double.[1]

For disposal in wells, there is sufficient capacity in depleted oil and gas wells to dispose of U.S. power plant emissions for a period of only about 10 to 15 years. Excavated salt domes and aquifers have sufficient capacity for longer term disposal of U.S. power plant CO_2 emissions. Disposal in excavated salt domes over the next 40 years might cost a minimum of $70 billion and would more than double the cost of electric power generation.[19] Injection into aquifers in certain regions may be feasible.

Environmental Effects. The environmental effects, in particular the effects on marine life, of deep-ocean disposal of CO_2 are unknown and require further study. The effects of excavation and disposal of large volumes of salt from salt domes with either fresh or ocean water require thorough investigation. Disturbance of underground aquifers by pressurized CO_2 is of concern and requires study.

Potential Value. Additional definitive evaluations of ocean, salt dome, and aquifer disposal are needed. The search for other methods of terrestrial storage, such as on natural sorbents which can be placed in abandoned mines or excavations, is worthwhile. More recent definitive studies will be discussed in subsequent chapters.

3.5.6   Utilization of CO_2

A number of products and uses can employ CO_2 recovered from the various control points (atmosphere, ocean, industrial, and terrestrial sources), some of which are as follows:[20]

1. Liquid carbon dioxide is used extensively in refrigeration systems.
2. Dry ice (solid CO_2) is used for refrigeration, mainly for foodstuffs.
3. CO_2 is used in carbonated beverages.
4. Urea, a widely used fertilizer, is produced by reacting CO_2 with ammonia:

$$2NH_3 + CO_2 = NH_2\text{--}CO\text{--}NH_2 + H_2O$$

5. Methanol for chemical and fuel use is made by reacting CO_2 with hydrogen catalytically:

$$CO_2 + 3H_2 = CH_3OH + H_2O$$

The methanol can also be dehydrated to form gasoline-like fuels:

$$nCH_3OH = nCH_2 + nH_2O$$

6. Carbamates used in inorganic chemical production are made by reacting amines and salt with CO_2:

$$CO_2 + 2RNH_2 + NaCl = RNH_2 CO_2Na + RNH_2Cl$$

7. Methane, or substitute natural gas (SNG), can be produced by reaction of CO_2 with hydrogen in the so-called catalytic methanation reaction:

$$CO_2 + 4H_2 = CH_4 + 2H_2O$$

8. CO_2 is being increasingly used in enhanced oil recovery operations.
9. CO_2 is used in fire extinguishers and inert gas-purging systems.
10. Liquid CO_2 is used in blasting systems for mining coal.
11. CO_2 is used in the production of inorganic carbonates and bicarbonates of sodium, potassium, and ammonium.
12. CO_2 can be used in secondary sewage sludge treatment. After concentrating and heating to 60°C and digestion, the treated sludge consumes CO_2, forming reduced organic materials, such as glucose, which can be used as animal feed.
13. Supercritical CO_2 has been found to be a particularly useful solvent in promoting difficult chemical reaction.[27]

Other specific uses for CO_2 are discussed in subsequent chapters.

3.5.7 Market Projection

3.5.7.1 Present Market

A U.S. market survey (1981) for gaseous, liquid, and solid CO_2 (shown in Table 3.8) indicates the major market distribution and consumption pattern. In addition, there is an over-the-fence consumption of CO_2 supplied by ammonia plants to produce urea as a major fertilizer. CO_2 consumption for urea is estimated from urea consumption, which amounted to 8×10^6 tons/yr. The total CO_2 and C equivalent consumption is then as follows:

Market consumption of CO_2	4.0×10^6 tons/yr
CO_2 for urea production	4.5×10^6
Total CO_2	8.5×10^6
Total C equivalent	2.3×10^6 tons/yr

TABLE 3.8
Major Market Consumption of CO_2: 1981[a]

Use	Distribution (% of market)	Consumption[a]	
		10^6 tons CO_2/yr	10^6 tons C/yr
Refrigeration for foodstuff	40	1.6	0.4
Carbonated beverages	20	0.8	0.3
Chemicals	20	0.8	0.2
Enhanced oil recovery	20	0.8	0.2
Total	100	4.0	1.1

[a] Estimated from *Chem Eng. News*, p. 23, July 19, 1982 and other recent references.

In 1980, the total U.S. emission of CO_2 amounted to 1.3×10^9 tons as equivalent C. This is about 25% of the world emission of 5.2×10^9 tons C/yr. Thus, the total U.S. market consumption of 2.3×10^6 tons C/yr represents only 0.18% of the U.S. total CO_2 emission. As a sink for CO_2, the market demand would have to grow by at least two factors of 10 to become a major factor in reducing man-made CO_2 emissions.

3.5.7.2 Future Projections

Estimates can then be made of the future long-term consumption of CO_2 by some major potential consumers to determine the possible impact on the CO_2 mitigation problem. These estimates are made based on the following assumptions for each use category.

Enhanced Oil Recovery (EOR). The estimated U.S. reserve of crude oil for tertiary treatment is about 300 billion barrels (*Chemical and Engineering News*, p. 1, July 26, 1982). Using CO_2 flooding, it is estimated that it takes between 1000 and 6000 SCF of CO_2 to bring up a barrel of crude. Assuming 6000 SCF, the CO_2 requirement is 0.347 ton CO_2 per barrel oil. Assuming the U.S. would meet a demand of 10×10^6 bbl/day, replacing projected imports and allowing for a decrease in domestic oil well production, the CO_2 requirement for EOR would be as follows:

$$\text{Annual } CO_2 \text{ for EOR} = 10 \times 10^6 \text{ bbl/day} \times 0.347 \text{ ton } CO_2\text{/bbl}$$
$$\times 329 \text{ operating days/yr}$$

$$= 1.14 \times 10^9 \text{ tons } CO_2\text{/yr}$$

$$= 0.31 \times 10^9 \text{ tons C equivalent/yr}$$

However, some of the CO_2 pumped into the wells dissolves in the oil and is recovered at the surface and reused in the well. It is reasonable to assume that about

50% of the CO_2 requirement for EOR would remain in the well, so that the net demand for CO_2 for EOR would be as follows:

Total C equivalent demand = $0.50 \times 0.31 \times 10^9 = 0.16 \times 10^9$ tons C/yr

A CO_2/Coal Slurry Transport Pipeline. A large potential use for CO_2 is in transporting coal through pipelines by means of a slurry of coal in liquid CO_2. The advantages of coal/CO_2 slurry transport are lower pumping power requirements and the ease of separating the CO_2 from the coal at the user end. A 70% coal/30% CO_2 slurry appears feasible (*Chemical Engineering,* p. 17, June 29, 1981). Assuming that most of the coal mined in the future at a total annual rate of a billion (10^9) tons of coal per year would be transported by liquid CO_2 slurry, the total flow of CO_2 would amount to an annual carbon equivalent of CO_2

$$= 1 \times 10^9 \text{ tons coal/yr} \times \frac{30}{70} \times \frac{12}{44}$$

$$= 0.12 \times 10^9 \text{ tons C/yr}$$

Most of the CO_2 would be recovered for recycling both for economic purposes and for prevention of loss to the atmosphere. Assuming a reasonable 10% loss of CO_2, the annual demand from an outside source would be

Total C equivalent demand = $0.1 \times 0.12 \times 10^9 = 0.012 \times 10^9$ tons C/yr

Plastics. Another possible large-scale use of CO_2 is in the polymer or plastics market. In this case, the CO_2 can be converted to CO and then used in polyesters or polyketones. Under the most optimistic assumption, it is assumed that the C in the CO_2 can be utilized in producing all the plastics in the U.S.

The 1980 total plastics consumption was 30×10^9 lb/yr. Assuming all of this quantity equals the carbon content in the plastics, the demand for carbon equivalent of the CO_2 is as follows:

$$\text{Total annual C equivalent} = \frac{3 \times 10^9 \text{ lb/yr}}{2000 \text{ lb/ton}} = 0.015 \times 10^9 \text{ tons C/yr}$$

It should be noted that this amount can permanently fix CO_2 because the solid stable plastic can be allowed to accumulate as a solid material in the terrestrial environment.

Liquid Transportation Fuels. As estimated earlier, the U.S. consumption of liquid transportation fuel (gasoline) is on the order of 100×10^9 gal/yr. The total annual carbon equivalent of producing this amount of gasoline is as follows:

Total annual C equivalent for gasoline production

$$= 100 \times 10^9 \; \frac{\text{gal/yr} \times 6 \text{ lb/gal}}{2000}$$

$$\times \frac{12 \text{ lb C}}{14 \text{ lb gasoline}} = 0.3 \times 10^9 \text{ tons C/yr}$$

Therefore, this is the amount of C that could be obtained from CO_2 recovered from the atmosphere or power plant stack gases and converted to gasoline using a nonfossil energy source.

It should be noted that SNG which can also be produced from CO_2 would be about equivalent to the amount of C fixed in the gasoline case.

CO_2 for Refrigeration. Assuming, optimistically, that the CO_2 demand for dry ice for refrigeration in the long term would increase tenfold over 1981 consumption, the annual C equivalent

$$= 10 \times 1.6 \times 10^6 \; \frac{12 \text{ tons C}}{44 \text{ tons } CO_2}$$

$$= 4.4 \times 10^6 \text{ tons C/yr}$$

$$= 0.0044 \times 10^9 \text{ tons C/yr}$$

Fertilizer Use. Assuming, optimistically, that all nitrogen fertilizer is used in the form of urea and U.S. fertilizer consumption doubles in the long-term future based on population growth, resulting in an equivalent demand for food, an estimate of the possible CO_2 consumption in urea fertilizer is determined as follows.

The 1982 consumption of NH_3, the major nitrogen fertilizer, was 17×10^6 tons/yr.[21]

$$\text{Annual } CO_2 = 17 \times 10^6 \text{ tons } NH_3 \times \frac{44 \text{ tons } CO_2}{34 \text{ tons } NH_3} \times 2 = 44 \times 10^6 \text{ tons}$$

$$\text{Annual C equivalent} = 44 \times 10^6 \text{ tons}$$

$$\times \frac{12 \text{ tons C}}{44 \text{ tons } CO_2} = 12 \times 10^6 \text{ tons}$$

$$= 0.012 \times 10^9 \text{ tons C}$$

Other Uses. Taking all other uses mainly in chemicals from the 1981 consumption and assuming a tenfold increase, another 0.002×10^9 tons C/yr consumption can be projected.

TABLE 3.9
Projected Long-Term Future Demand for CO_2

Use	Annual projected demand C equivalent of CO_2 (10^9 tons C/yr)	% of total
Enhanced oil recovery	0.160	32
CO_2 coal slurry transport	0.012	
Plastics	0.015	
Liquid transportation fuels	0.300	60
Refrigeration	0.004	
Fertilizer	0.012	
Other uses	0.002	
Total	0.505	92

Table 3.9 summarizes the projected future long-term demand for CO_2 in terms of carbon equivalent. This value of 0.505×10^9 tons C/yr is the amount of CO_2 equivalent that would be recovered from fossil-fuel power plant stacks or from the atmosphere using a nonfossil source of energy for recovery. Based on 1980 emissions, this amounts to a reduction of 39% in the U.S. emission of CO_2. However, it should be noted that 60% of this reduction comes from the application of producing liquid transportation fuels mainly consisting of methanol and gasoline. This point emphasizes again the possible impact of producing a synthetic carbonaceous fuel from CO_2 with the use of a nonfossil energy source, either solar or nuclear. All other chemical and commodity uses have less than a 5% effect on CO_2 emissions reduction.

It can also be pointed out that if SNG could also be produced from CO_2 in quantities equal to that presently consumed (on the order of 20 trillion ft^3/yr), the demand for this use would be about equivalent to that for gasoline. Thus, when oil and gas are depleted, then one possible solution to the long-term supply of carbon-containing fuel is to convert CO_2 back to synthetic carbonaceous liquid and gaseous fuels. Concerning the utilization of CO_2, unless the products of the conversion of CO_2 can be utilized on a large scale, the impact of reducing CO_2 in the atmosphere will not be significant. It appears that the utilization of CO_2 for the production of chemicals and nonfuel commodities has a small influence on the reduction of atmospheric CO_2 emissions. A number of schemes for utilization of CO_2 are examined in subsequent chapters of this book in an attempt to uncover possible new approaches to the subject.

3.6   SUMMARY

Technologies potentially useful in contributing to the CO_2 mitigation problem are summarized under each process category as follows:

Physical processes

1. Absorption and stripping systems using liquid solvents
2. Adsorption and stripping using solid adsorbents
3. Diffusional (membrane) process for separation of CO_2

Chemical processes

1. Combustion of fossil fuel to CO at utility and industrial source point and combination with a nonfossil energy source to produce carbonaceous fuel
2. Isothermal extraction of CO_2 from the atmosphere or stack gases in combination with a nonfossil source of hydrogen to form methanol and hydrocarbons
3. Reduction of CO_2 with C after combustion and conversion to synthetic fuel with a nonfossil source
4. Extraction of CO_2 with lime during combustion
5. Decarbonization of fuel by steam gasification and reforming
6. Decarbonization of fuels by removing and sequestering carbon

Physical/chemical processes

1. CO_2 recycle power plant
2. Electrochemical decomposition of CO_2
3. Photoelectrochemical decomposition of CO_2
4. High-energy radiation decomposition of CO_2
5. Photochemical decomposition of CO_2
6. Thermal decomposition of CO_2

Biological processes

1. Photosynthetic fixation of atmospheric CO_2
2. Biological reduction of CO_2 in aquatic and marine organisms

CO_2 disposal schemes

1. Depleted gas wells
2. Deep excavated salt domes
3. Deep ocean
4. Terrestrial aquifers

Utilization of CO_2

1. Synthetic carbonaceous fuels
2. Enhanced oil recovery
3. Supercritical solvent
4. Other products

A variety of additional CO_2 mitigation concepts have been presented in a series of four recent international conferences on CO_2 removal, disposal, and utilization. These are listed below. Many of these concepts fit into the general categories discussed in this chapter.

- K. Blok, W.C. Turkenburg, C.A. Hendrike, and M. Steinberg, Eds., *Proceedings of the First International Conference on Carbon Dioxide Removal* (Amsterdam, March 4 to 6, 1992), Pergamon Press, Oxford, U.K., 1992.
- J. Kondo, T. Inui, and K. Wasa, Eds., *Proceedings of the Second International Conference on Carbon Dioxide Removal* (Kyoto, October 24 to 27, 1994), Pergamon Press, Tokyo, 1995.
- H. Herzog et al., Eds., *Proceedings of the Third International Conference on Carbon Dioxide Removal* (Cambridge, MA, September 9 to 11, 1996), Pergamon Press, Boston, 1997.
- Baldur Eliasson, Chairman, Fourth International Conference on Greenhouse Gas Control Technologies (GHGT-4), August 30 to September 2, 1996, Interlaken, Switzerland, to be published.
- P.W.F. Riemer and A.Y. Smith, Eds., *Proceedings of the International Energy Agency Greenhouse Gases Mitigation Systems Conference* (London, August 22 to 25, 1995), Pergamon Press, London, 1996.
- P.W.F. Riemer, Ed., *Proceedings of the International Energy Agency Carbon Dioxide Disposal Symposium* (Oxford, March 29 to 31, 1993), Pergamon Press, Oxford, 1993.
- Y. Tamaura and T. Kamimoto, Eds., *International Symposium on CO_2 Fixation and Efficient Utilization of Energy (C and E '95)* (Tokyo, October 23 to 25, 1995), Pergamon Press, Tokyo, 1997.
- Y. Tamaura et al., Eds., *Proceedings of the International Symposium on CO_2 Fixation and Efficient Utilization of Energy* (Tokyo, November 29 to December 1, 1993), Research Center of Carbon Recycling and Utilization, Tokyo Institute of Technology, 1993.
- J. Paul and C.M. Pradier, Eds., *Carbon Dioxide Chemistry: Environmental Issues,* Royal Society of Chemistry, Cambridge, U.K., 1994.
- P.W.F. Riemer et al., Eds., *Greenhouse Gas Mitigation, Proceedings of Technologies for Activities Implemented Society* (Vancouver, Canada, May 26 to 29, 1977), Elsevier, Oxford, 1998.

A series of international conferences on CO_2 utilization have also been convened:

- Kanami Ito, Chairman, Proceedings of the International Symposium on Chemical Fixation of Carbon Dioxide (ICCDU-1), Nagoya, Japan, December 2 to 4, 1991.
- Michella Aresta, Chairman, Second International Conference on Carbon Dioxide Utilization (ICCDU-2), Bari, Italy, September 28 to 30, 1993.
- Kenneth Nicholas, Chairman, Third International Conference on Carbon Dioxide Utilization (ICCDU-3), Norman, OK, April 30 to May 4, 1995.
- Tomoyuki Inui, Chairman, Fourth International Conference on Carbon Dioxide Removal (ICCDU-4), Kyoto, Japan, September 7 to 11, 1997.

REFERENCES

1. Steinberg, M. and Albanese, A.S., Environmental control technology for atmospheric CO_2, in *Interactions of Energy and Climate 1,* Bach, Ed., D. Reidel Publishing, Boston, 1980, 521–555; BNL 27164, Brookhaven National Laboratory, Upton, NY, March 1980.
2. Steinberg, M., Synthetic Fuels from Fusion Reactors and Coal, BNL 26710, Brookhaven National Laboratory, Upton, NY, July 1979.
3. Steinberg, M., Synthetic Carbonaceous Fuels and Feedstocks from Oxides of Carbon and Nuclear Power, BNL 22785, Brookhaven National Laboratory, Upton, NY, March 1977.
4. Steinberg, M. and Baron, S., Synthetic carbonaceous fuel and feedstock using nuclear power, air and water, *Int. J. Hydrogen Energy,* 2, 189–207, 1977.
5. Horn, F.L. and Steinberg, M., A Carbon Dioxide Power Plant for Total Emission Control and Enhanced Oil Recovery, BNL 30046, Brookhaven National Laboratory, Upton, NY, August 1981.
6. Isaacs, H.S., Zirconia fuel cells and electrolyzers, in *Advances in Ceramics,* Vol. 3, American Ceramic Society, Westerville, OH, 1981.
7. Friess, P., GRI contract with SRI International on reduction of CO_2 to methanol and urethane by photo and dark reactions on semi-conductor electrodes, *GRI Dig.,* 5(2), 15, 1982; *Chem. Eng. News,* p. 28, November 22, 1982.
8. Bockris, J.O.M., *Energy: The Solar Hydrogen Alternative,* John Wiley & Sons, New York, 1975, 1915; *Science,* 218, 557, 1982.
9. Steinberg, M., Chemonuclear reactors and chemical processing, in *Advances in Nuclear Science and Technology,* Vol. 1, Konts, H., Ed., Academic Press, New York, 1962, 247–333.
10. Lind, S.C., *Radiation Chemistry of Gases,* Reinhold Publishing, New York, 1961, 112–126.
11. Luite, G., Dondes, S., and Harteck, P., Photochemical production of C_3O_2 from CO, *J. Phys. Chem.,* 44(10), 4051–4052, 1966.
12. Dang, V.D. and Steinberg, M., The Value of Forestation in Absorbing Carbon Dioxide Surrounding a Coal-Fired Power Plant, BNL 51299, Brookhaven National Laboratory, Upton, NY, August 1980.
13. Walsh, J.J., Rowe, C.T., Iverson, R.L., and McRoy, C.P., Biological export of shelf carbon is a sink of the global CO_2 cycle, *Nature,* 291(5812), 196–201, 1981.
14. Marchetti, C., Geoengineering and the CO_2 problem, *Climatic Change,* 1, 59, 1977.
15. Baes, C.F., Beall, S.E., Lee, D.W., and Marland, G., Options for the Collection and Disposal of Carbon Dioxide, ORNL 5657, Oak Ridge National Laboratory, Oak Ridge, TN, May 1980.
16. Davis, R.M., National strategic petroleum reserve, *Science,* 213, 618–622, 1981.
17. Retched, R.H., Airscooping nuclear–electric propulsion concept for advanced orbital missions, *J. Br. Interplanetary Sci.,* 31(2), 62–66, 1978.
18. Steinberg, M., Technologies for the Recovery and Disposal of Carbon Dioxide, BNL 29526, Brookhaven National Laboratory, Upton, NY, April 1981.
19. Horn, F.L. and Steinberg, M., Possible Sites for Disposal and Environmental Control of Atmospheric Carbon Dioxide, BNL 51597, Brookhaven National Laboratory, Upton, NY, September 1982.

20. Ballou, W.R., Carbon dioxide, in *Kirk-Othmer Encyclopedia of Chemical Technology*, 3rd ed., John Wiley & Sons, New York, 1978, 725–742.
21. *Chem. Eng. News*, p. 25, March 24, 1983.
22. Suda, T. et al., Development of fuel gas carbon dioxide recovery technology, in *Carbon Dioxide Chemistry Environmental Issues*, Paul, J. and Prodier, C.-M., Eds., Royal Society of Chemistry, Sweden, 1994, 222–225.
23. Lackner, K.S. et al., Carbon dioxide disposal in carbonate minerals, *Energy*, 20(22), 1153–1170, 1995.
24. Tokuda, Y. et al., Development of hollow fiber membranes for CO_2 separation, in Proceedings of the Third International Conference on Carbon Dioxide (Abstracts), September 9 to 11, 1996, 38.
25. Steinberg, M., Technologies for the Recovery and Disposal of Carbon Dioxide, BNL Report 29526, Brookhaven National Laboratory, Upton, NY, April 1981.
26. Pierratazzi, R., Carbon dioxide, in *Kirk-Othmer Encyclopedia of Chemical Technology*, 4th ed., John Wiley & Sons, New York, 1993, 35–53.
27. Abele, A.R. et al., An Experimental Program to Test the Feasibility of Obtaining Normal Performance from Combusters Using Oxygen and Recycled Gas Instead of Air, ANL Report CNSV-TM-204, Argonne National Laboratory, Argonne, IL, 1987.

4 Technologies for Improving Efficiency of Energy Conversion and Utilization and the Effects on Global CO₂ Emission

4.1 INTRODUCTION

Since the growth of atmospheric CO_2 is primarily associated with the consumption of fossil fuels, one logical response to the threat of the "greenhouse effect" would be a worldwide reduction of such consumption. A comprehensive study of global energy use, however, aimed at reducing fossil-fuel consumption to prevent or limit the rise in atmospheric CO_2, requires clarification of the enormously complex interactions among energy technology development, resource supply, global economic progress, and relevant sociopolitical activities. This is an enormously complicated task and involves large uncertainties.

A number of alternative CO_2 mitigation technologies are discussed in earlier and later chapters of this volume. Other studies concentrated on the impacts of such factors as economics, population, and energy use on projections of energy consumption and associated CO_2 releases by employing various CO_2-related global economy and energy models.[1-5] However, efficient use of energy not only would extend the period for using known finite energy resources and allow time for smooth transition to alternatives, but also would reduce the strain on the environment due to both energy production and consumption and allow more time for remedial action.

This chapter concerns the technical aspects of energy use, with emphasis on improving technology efficiency to reduce CO_2 emissions. Its objective is to explore the feasibility of future energy technologies and their potential effectiveness in lessening the buildup of atmospheric CO_2 and, thus, mitigating the projected climatic and environmental changes. The perspective, being physical rather than sociopolitical, leads to concentration on how energy is used for specific services such as heating a home, making steel, or moving a car and to avoidance of nontechnical questions such as whether energy should (or would) be saved by using mass transit rather than automobiles or multiple dwellings rather than individual homes.

4.1.1 Energy Form and Usage

Understanding of energy use must begin with differentiation of energy forms and their uses at various stages of energy flow (Figure 4.1). Energy flow starts with the extraction of primary energy resources and products; proceeds through refining and various stages of conversion, transportation, distribution, and storage; and ends with consumption of fuels and electricity by end-use technology relevant to a particular demand of energy service.

Primary energy forms such as coal and oil recovered directly from natural resources contain energy for later use as needed. For convenient and sometimes more efficient use, most primary energy must be converted into secondary energy forms such as electricity and gasoline. For instance, crude oil is refined into secondary fuels that can be used directly for end-use applications or indirectly for power generation. Secondary energy forms are readily transported and distributed, and in some cases stored in a variety of devices, and they provide the final useful energy for services such as heating a house and operating a machine.

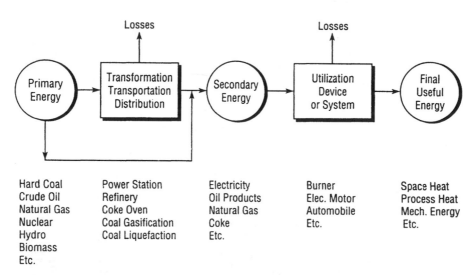

FIGURE 4.1   Schematic of energy flows.

Most energy consumption falls into four broad categories: (1) electricity generation, (2) residential and commercial use, (3) industrial use, and (4) transportation. The first category differs from the other three in that it is not an end use in itself: electricity is one of the secondary energy forms that provides final useful energy.

4.1.2 Energy Technology

Energy technology is applied over a broad, diverse spectrum of utilities, industries, devices, and activities (Table 4.1), ranging from primary resource extraction through fuel conversion and electricity generation, transmission, and distribution to the final end use that provides energy services.

Technological progress, which improves efficiencies of conversion from primary to secondary energy forms and from secondary to final energy and/or energy services, increases energy productivity. Energy efficiencies vary widely in different countries, and the differences become more apparent when examined in detail by end use. The evolution of these efficiencies, and differences among countries, indicates a considerable potential for global energy savings through improvements in end-use technologies. This means that between the two stages of energy conversion (from primary to secondary and from secondary to end use), the latter shows a greater potential for reduction in energy demand through technological improvements in efficiency. In addition, the continuing trend toward a more electrified future world and the high energy intensiveness of electricity generation demonstrate the importance of electricity-generating technologies to future fuel consumption. In this study, therefore, the term energy technology is used to refer specifically to the end use and electricity generation technologies listed in Table 4.2. End-use technology generally can be applied both to systems and to devices (home heating with a heat pump is a systems concept, whereas home heating with an improved oil burner is a device concept). Most of the technologies in Table 4.2 combine systems aspects and device aspects, with perhaps a preponderance of systems.

In this study, the energy technologies were first grouped by stage of evolution as (1) now in use, (2) emerging, and (3) advanced or far in the future The associated energy efficiency levels were evaluated in terms of (1) present, (2) achievable, and (3) theoretical maximum. The potential energy savings from introduction of efficiency improvements in the future were estimated from these values.

$$\text{Efficiency, } \eta = \frac{\text{Energy transfer of a desired kind by a device or system}}{\text{Energy input to the device or system}}$$

Energy transfer and/or input can be quantified in two ways: as work or as heat. It is called work when it exerts a force and heat when it does not. Work energy tends to change the state of motion of the object to which it is transferred. Heat energy tends to change its temperature and/or state.

The present efficiency is the value currently in practice. The achievable efficiency means the current predicted value of the present knowledge of the technology. Thus,

TABLE 4.1
Energy Technology Spectrum

Resource	Mining, fuel conditioning	Conversion	Transmission/ distribution	Storage	Tasks
Fossil fuels Coal, oil, gas, oil shale, tar, sand	Mining Surface Deep Coal cleaning	Turbines Steam Combustion Hydraulic	Fuel Track Train Tanker Pipeline	Batteries Lead acid Sodium/sulfate Lithium/metal Sulfide	Residential/commercial Space conditioning Water heating Lighting Miscellaneous electric (appliances, entertainment)
Nuclear fission ^{235}U, ^{233}Th, ^{239}Pu	Desulfurization	Diesel	Conveyor Barge	Zinc/chlorine Compressed air	
Nuclear fission D-D, D-T-Li	Liquefaction	Internal combustion	Electric Open wire	Electromagnetic	Industrial Process heat Electric-driven devices Aluminum electrolytic process
Hydropower	Gasification Low Btu, high Btu	Combined cycle Gas turbine–steam Potassium–steam Steam–ammonia	Conventional cable Cryogenic cable Superconducting cable	Flywheels	Iron and steel Nonenergy Petrochemical
Solar Terrestrial Orbiting satellite	Solvent refining Hydrogen production	Fuel cells Acid alkaline Molten carbonate	Microwave Laser	Hydrogen	Transportation Automobile Truck Air transport Bus
Wastes Agricultural Industrial Municipal	Enrichment and reprocessing	Magnetohydrodynamic Open, closed, nuclear LWR, HWR, HTGR Nuclear-gas turbine Breeder reactors		Compressed gas Liquid Metal hydride Pumped hydro Conventional Underground	Rail Ship
Geothermal Brine, hot rock, steam		LMFBR, GCFR, MSR, hybrids		Thermal	

Wood

Wind

Tidal

Thermocline

Fusion
Magnetic confinement
Laser

Solar
Concentrators
Flatplate collectors
Photovoltaic
Photosynthesis

Other
Aerogenerator
Thermonomic
Thermoelectric

TABLE 4.2

Present, Emerging, and Advanced Technologies with Associated Present, Achievable, and Theoretical Maximum Efficiencies[a,b]

	Efficiencies (%)		
	Present	Achievable	Theoretical[c]
Present technologies			
Electricity generation			
Coal steam electric (Rankine steam cycle)	33	40	67[d]
Oil steam electric	32	40	67[d]
Gas-fired electric	32	40	67[d]
Gas turbine (oil and gas)	25	35	58[e]
LWR (nuclear fission)	32	35	50[f]
Geothermal electric	9	20	48
Hydroelectric power	80	90	100
Off-peak electricity storage			
Pumped hydroelectric	75	80	100
Residential and commercial			
Space heat			
Coal	44	70	80
Oil	60[g]	72	80
Gas	60[g]	72	80
Electricity	100	100	100
Wood and waste	44	70	80
Direct thermal use of solar energy	40	50	100
District heating: integrated community energy systems	70	80	90
Water heat			
Oil	63[g]	70	80
Gas	55[g]	70	80
Electricity	100	100	100
Direct thermal use of solar energy	50	60	100
District heating	70	80	90
Industrial (process heat and miscellaneous)			
Steam boilers			
Coal	70	80	100
Oil	64	82	100
Gas	64	83	100
Wood and waste	70	80	100
Furnaces and ovens			
Coal (coke)	24	35	67
Wood and waste	24	35	67
Electricity (resistance, inductive, microwave)	100	100	100
Direct thermal use of solar energy to produce hot water and low-pressure steam	48	56	70
Transportation			
Gasoline engine (internal combustion), automobile (average mpg)	16.7	28	NA
Gas turbine, air transport (average passenger mpg)	25	40	NA
Internal combustion engine, bus, truck, and rail (average mpg)	5	6	NA
Electric rail			

TABLE 4.2
Present, Emerging, and Advanced Technologies with Associated Present, Achievable, and Theoretical Maximum Efficiencies[a,b] (continued)

	Efficiencies (%)		
	Present	Achievable	Theoretical[c]
Emerging technologies			
Electricity generation			
Atmospheric fluidized bed combustion (FBC)	34	40	70
Pressurized FBC	38	42	70
Combined cycle generation			
Pressurized FBC, combined cycle	43	50	70
Gasification, combined cycle	42	60	70
Fuel cell			
Phosphoric acid	40	50	94
Molten carbonate	45	55	94
Binary fluid cycle (e.g., Kalina cycle)	50	60	
Advanced converter reactor (LWR and GCR)	39	45	50
Breeder reactor (LMFBR and FGCR)	39	45	50
Hot-dry-rock geothermal	12	20	50
Wind	33	40	53
Solar thermal electric conversion	16	22	53
Photovoltaic cells (silicon)	16	20	90
Fusion reactor	35	40	50
Energy storage			
Batteries	70	90	100
Flywheel	70	90	100
Superconducting magnets	75	85	100
Hydrogen as energy carrier or fossil-fuel substitutes	40	60	70
Thermal technologies that accept and return energy in the form of heat or cold	80	85	100
Compressed air in underground caverns	70	95	100
Transportation			
Increasing fuel efficiency (mpg)			
Automotive gas turbine	40	46	NA
Stirling automotive engine	40	46	NA
Uncooled (adiabatic) diesel engine heavy-duty truck transport	5	6	NA
Alcohol fuels in spark ignition and diesel engine (equivalent)	30	70	NA
Advanced materials development (e.g., ceramics for high-temperature conditions) leads to a 20% fuel efficiency improvement			
Magnetically levitated trains			
Enhancing the flexibility of fuel options of engines			
Electric vehicle (miles per kWh)	2.3	3.2	NA
Hybrid vehicle (heat engine + battery system)			
Hydrogen-fueled automobile (fuel cells)			
Alternative fuels (e.g., fuels from synthetic crudes, methanol from coal)	30	37	NA

TABLE 4.2
Present, Emerging, and Advanced Technologies with Associated Present, Achievable, and Theoretical Maximum Efficiencies[a,b] (continued)

	Efficiencies (%)		
	Present	Achievable	Theoretical[c]
Conservation (savings, quads, year 2000)			
Automobile: reduction of weight		1	
Aircraft: increase load factor, shift short runs to rail service		2	
Trucks: shift to diesel, shift intercity traffic to rail		2	
Industrial (potential savings quads/yr by year 2000)			
Total achievable savings (20- to 30-year assumed range of equipment replacement lifetime)		6.0	
Total potential savings			14.5
Combustion efficiency improvement			3
Catalysis (including biocatalysis) and separation research			2
Sensors and control system development			
Carbothermic processing improvement (iron, steel, and paper industries)			1.5
Heat transfer technologies improvement			0.2
Conservation			
Energy cascading (expansion of cogeneration used to improve energy conversion systems)			1.8
Industrial waste utilization			
Waste heat recovery			4
New production technologies (e.g., robots, automation)		1.0–3.0	
Residential and commercial (conservation)			
Space heat			
Improved building design and insulation			
Residential: 50% less heating than in mid-1970s			
Commercial: 60% less heating than in mid-1970s			
Thermally activated heat pump (COP)		160	250
Air-conditioning: energy consumption can be reduced by 50% by year 2000			
Insulation			
Improved efficiency			
Water heat			
Substitution of solar energy for electricity			
Heat pump uses 50% less energy than the electric resistance water heater			
Advanced technologies			
Magnetohydrodynamic electricity generation	50	60	
Ocean thermal energy conversion	2.5	6.5	
Tidal energy			0.08
Wave energy			
Space solar energy systems			
Solar cells	7	12	
Thermal engine	12.5	22.3	

TABLE 4.2
Present, Emerging, and Advanced Technologies with Associated Present, Achievable, and Theoretical Maximum Efficiencies[a,b] (continued)

	Efficiencies (%)		
	Present	Achievable	Theoretical[c]
Advanced aqueous and nonaqueous batteries for low-cost energy storage (e.g., zinc/ferricyanide and hydrogen/nickel cells)			
Superconducting energy storage and transmission lines			

[a] Efficiency is in percent except where other units are noted.

[b] Obtained from References 8, 10, 13, 14, and 17 to 20.

[c] Depending on the operating temperatures, efficiency limited by Rankine, Carnot, or other appropriate reversible thermodynamic cycle.

[d] Operating temperatures are 1200 and 95°F.

[e] Operating temperatures are 3000 and 1000°F.

[f] Operating temperatures are 600 and 70°F.

[g] For continuous burner operation, the efficiency of direct heating systems drops sharply at low load; annual average efficiency about 40 to 50%.

the achievable efficiency changes as the technology improves in the future. The theoretical efficiency is estimated from reversible thermodynamics and depends on the operating temperatures of the system. Due to the irreversibility of the energy use process, the theoretical efficiency can be viewed as the ideally, ultimately achievable efficiency that can only be approached through eventual technological advancement. When $\eta > 1$, as in a heat pump, it is usually called a coefficient of performance (COP); when $\eta < 1$, it is usually called an efficiency. For cases in which efficiency, as defined here, cannot be conveniently expressed, energy intensity is used instead. For example, in transportation, efficiency is measured in terms of mileage traveled per unit energy used. Note that all the technology efficiencies listed in Table 4.2 are subject to some uncertainty due to environmental conditions and the degree of development and of technology. The uncertainty is larger for newly developed technologies, especially for those that are expected to be developed far in the future.

4.2 METHODOLOGICAL APPROACH

4.2.1 Global Energy Demand

Carbon dioxide buildup in the atmosphere is a worldwide problem. Evaluation of the efficiency and feasibility of any energy-efficient technology in reducing fossil-fuel consumption and, thus, limiting CO_2 buildup requires a global energy model to project global future energy demand, future energy use patterns, and the mix of fuels (i.e., fossil fuels, nuclear, solar, etc.) satisfying the energy demand. The future global energy demand used in this chapter is from an energy-economic scenario developed by Edmonds and Reilly using their Institute for Energy Analysis (IEA) Energy–CO_2 Model (Case B).[6] This model is a consistent conditional representation of economic, demographic, technical, and policy factors used to project future energy production and consumption. It has four components — supply, demand, energy balance, and CO_2 emissions — and disaggregates the world into nine geopolitical regions. The first two components generate the supply of and demand for primary energy in each region. The global equilibrium of energy production and use by fuel type is obtained via the third component, energy balance. Population, labor productivity, technology efficiency, fuel supply and demand, and other relevant parameters are specified for each of the nine regions.

In the model, the primary fuels are transformed into four secondary fuels — refined liquids, refined gases, refined solids, and electricity — which are consumed by the three end-use sectors — residential and commercial, industrial, and transportation — to provide the energy services. The mix of secondary fuels in each region is determined by the relative costs of providing the services via each alternative fuel. The demands for fuels to generate electricity are determined by the relative costs of production, as are the shares of oil and gas converted from coal and biomass. In the three OECD regions — (1) U.S., (2) Canada and Western Europe, and (3) Japan, Australia, and New Zealand — the secondary energy is consumed by three end-use sectors and in the other regions — former Soviet Union, Eastern Europe, China, etc. — by a single aggregate end-use sector. The sectoral need for energy services in each region depends on the costs of energy services, income levels, and population. More details of the model are given in Reference 6.

4.2.2 Time Frame

The year 2050 target was chosen for the following reasons. The time from conception of a new/improved technology to its widespread adoption is generally 50 years or more, which might be needed to accommodate the social infrastructure changes that normally parallel technical infrastructure changes. The current technology represents that in the 1975–85 time frame. A period of about 65 years covers two to three human generations and about two power station lifetimes. With a shorter time frame, some emerging technologies, especially in the electricity generation sector, will not penetrate enough to have much effect in reducing CO_2 emissions. These electric power generation technologies are perceived to be important on a long-term basis. During a 65-year developmental transition period, most technologies will become mature (i.e., well established and available to many competing vendors), with capital costs lower than the initial costs of their first commercial versions. By the year 2050, most of the present and emerging technologies probably could be fully developed, reaching their achievable efficiencies, and become available globally. Thus, choice of the year 2050 avoids the problem of determining market shares of various technologies with similar achievable efficiencies, particularly electricity generation.

The year 2050 is also a benchmark for studying the "CO_2 doubling time" predicted by several investigators if CO_2 emissions from fossil fuels continue to grow at the 1950–73 average rate of 4.5% per year. Doubling the preindustrial atmospheric CO_2 concentration is predicted to raise mean global temperature by 1.5 to 4.5°C, as noted in an earlier chapter.

4.2.3   Screening Technologies

Since only technologies with similar energy service outputs can be compared with each other, the screening was done by sectors: (1) residential/commercial, (2) industrial, (3) transportation, and (4) electricity generation. The energy technologies were screened on the basis of their impact on reduction of fossil-fuel consumption (and thus CO_2 emissions) and their global applicability. The impact of a particular technology on the reduction of energy demand for a specific service is proportional to improvement in efficiency over that of the mid-1970s technology, and the global applicability depends on the market share of the specific energy service by fuel type, which is related to the pattern of global energy consumption. Furthermore, the market share of a particular fuel for a specific energy service is related to the demand for this fuel. Thus, the potential reduction of the demand for a particular fuel by a specific energy service resulting from use of a relevant improved technology is proportional to the product of the demand for fuel by this specific energy service and the difference between the efficiencies of the improved and the current technology. Efficiency and market shares of services for each fuel type are therefore the two determinants used in screening. Some technologies thought to have a significant impact on future CO_2 emissions are listed in order of importance for each sector in Table 4.3. The energy technologies in each sector that passed the screening can be used for the global effects of energy technologies on future energy consumption and associated CO_2 emissions.

4.2.4   Technological Effects on Energy Consumption

The method for deriving the merit of the benefits (i.e., reduction in CO_2 emission) resulting from technology improvements is a comparison of global secondary energy demand and associated CO_2 emissions with and without technology improvements for the chosen year 2050, with the year 1975 as a base. First, possible energy savings due to improved end-use technologies are estimated in terms of efficiency improvement applied to specific energy services for each end-use sector. Then the new electricity-generating technologies are introduced to determine further possibilities for saving fuel. This approach allows calculation of the possible minimum fuel requirement for each energy end-use task and therefore establishment of the potential savings of fossil fuels and associated reduction of CO_2 emissions due to technology improvements that might be adopted globally.

TABLE 4.3
**Major Energy-Saving Technologies for Energy Service
in the Energy-Consumption Sectors**

I.   Residential and commercial sector
   1. Space heating
      a. Improved building design and construction
      b. Improved space-heating equipment
         • Burner efficiencies
         • Electric heat pump
   2. Water heating
      a. Heat pump water heaters
      b. Burner efficiencies
II. Industrial sector
   1. Process heat
      a. Technologies that recuperate waste heat
         • Ceramic heat recuperators
         • New heat pump technologies for waste heat streams of 150 to 250°F
         • Improved heat exchangers for waste heat streams of >1000°F
      b. Process flow optimization
         • Energy cascading
         • Cogeneration
      c. Improved efficiencies of fossil-fuel combustion equipment
   2. Electric drive
      • Improved motors and torque converters or clutched flywheels
      • Operating motors from alternating current synthesizers or other power factor controls
III. Transportation
   1. Automobile (13.6 mpg at mid-1970s, 40 mpg for year 2050 for whole fleet of autos)
      a. Design changes in engine, transmission, rear axle, and tires
      b. Reduce energy loss from aerodynamic drag, friction, and accessory use
      c. Reduce vehicle weight by use of lightweight materials
      d. Automotive propulsion systems improvements
         • Gas turbine and Stirling engines
         • High-speed direct-injection diesel engines
      e. Electronic control of engine and transmission
   2. Truck (4.5 mpg at mid-1970s, 6.5 mpg for year 2050, 40% improvement)
      • The adiabatic diesel engine with waste heat utilization has the potential for a 40% improvement in efficiency
      • The same measures of efficiency gains apply to trucks as to cars
IV. Electricity generation from fossil fuel
   1. Combined cycle
      • Pressurized FBC — combined cycle
      • Gasification — combined cycle
   2. Fuel cell
      • Phosphoric acid
      • Molten carbonate
   3. Fluidized bed combustion
   4. Magnetohydrodynamic

TABLE 4.4
Regional Secondary Energy Demand by Fuel Type for Year 2050
(in EJ = 10^{18} J)

Region	Liquids	Gases	Solids	Electric	Total
U.S.					
Residential/commercial	7.93	14.21	0.69	18.78	41.61
Industrial	5.06	3.46	6.73	15.21	30.46
Transportation	50.55	0.0	0.0	0.11	50.66
Total	63.53	17.67	7.42	34.10	122.73
Canada/Western Europe					
Residential/commercial	12.88	13.39	6.91	19.91	53.09
Industrial	6.56	7.11	10.44	25.07	49.16
Transportation	22.04	0.0	0.35	1.59	23.97
Total	41.47	20.50	17.69	46.57	126.23
OECD Pacific					
Residential/commercial	3.64	6.74	1.45	5.58	17.41
Industrial	3.88	6.98	9.64	16.84	37.37
Transportation	10.77	0.0	0.29	0.87	11.93
Total	18.27	13.72	11.38	23.30	66.71
Former Soviet Union/ Eastern Europe	22.59	32.44	20.49	22.33	97.85
China et al.	17.41	17.12	43.79	17.83	96.17
Middle East	6.67	5.86	0.86	3.84	17.23
Africa	8.52	7.23	12.01	10.12	37.65
Latin America	22.38	8.84	6.55	23.59	61.36
South and East Asia	9.34	4.65	19.68	14.95	48.63
All regions	210.18	128.03	139.87	196.63	674.60

The data on global secondary energy demand for the year 2050 with the 1975 energy efficiency (Table 4.4) were obtained by modifying the results for future energy demand from Edmonds and Reilly's energy-economic scenario (Case B) and the use of their IEA Energy–CO_2 Model.[6] In Regions 1 to 3 (U.S., Canada and Western Europe, and OECD Pacific), energy consumption is shown for three end-use sectors, but in Regions 4 to 9, it is shown for a single aggregate sector because the breakdown of energy demand by end-use sector for each fuel type is not available for the developing countries. Investigation of the effects of improvements in end-use technologies on future global energy consumption requires detailed regional data on energy demand by function (i.e., energy service) for each fuel type and end-use sector. However, no such data are available except for the U.S. for the mid-1970s[7] (Table 4.5).

TABLE 4.5
Demand by Fuel Type for Energy Services of Three End-Use Sectors in the U.S. for Year 1977 (in EJ = 10^{18} J)

Energy service	Oil	Gases	Coal	Electricity
Residential/commercial				
Space heat	2.77 (85)[a]	4.87 (70.4)	0.79 (100)	0.5 (14.0)
Air-conditioning		0.04 (0.7)		0.64 (15.7)
Water heat	0.49 (15)	1.99 (28.9)		0.60 (14.6)
Lighting				1.23 (30.0)
Misc. electric (appliances, etc.)				1.26 (25.7)
Total	3.26 (100)	6.88 (100)	0.79 (100)	4.10 (100)
Industrial				
Process heat	5.25 (57.31)	7.41 (91.37)	2.33 (53.56)	
Electric drive				2.43 (87.1)
Aluminum electrolytic process				0.22 (7.89)
Iron and steel			1.92[b] (44.14)	0.14 (5.01)
Nonenergy petrochemical	3.91 (42.69)	0.70 (8.63)	0.10 (2.3)	
Total	9.16 (100.0)	8.11 (100.0)	4.35 (100.0)	2.79 (100.0)
Transportation				
Automobile	10.36 (51.44)			
Air transport	2.32 (11.52)			
Truck	5.43 (26.98)			
Bus and rail	1.08 (5.34)			0.063 (100)
Ship	0.95 (4.71)			
Total	20.14 (100.0)			0.063 (100)

[a] Percentage.

[b] About 90% was used in process heat for ore preparation, pig iron production, steel making, and steel finishing (slabbing, blooming, and billeting).

4.2.4.1   U.S.

In this study, the detailed energy demand data by service for each fuel type and end-use sector in the U.S. for the year 2050 was estimated by following the allocation of energy needs for services in the mid-1970s, with the patterns of energy use assumed to be constant (see nontechnical improvements in Tables 4.6, 4.7, and 4.9). Then the high-efficiency end-use technologies could be applied to the relevant energy services to obtain the associated energy needs and savings for each fuel type and sector in the U.S. for the year 2050.

4.2.4.1.1   Residential and Commercial Sector

Space conditioning accounts for more than 50% of energy consumption in the residential and commercial sector and water heating for about 20%. Lighting and miscellaneous electricity usage, for refrigerators, elevators, etc., account for about 55% of electricity consumption.

Space Heating. With regard to space heating, a building comprises two sub-systems — the shell and the heating system — where efficiency can be improved. Considerable technical potential for energy savings in space heating hinges on improvements in building design and construction, such as reduced air infiltration, better wall insulation, electronic indoor temperature control, and passive solar installations. Many studies have shown that by the year 2000, residential buildings will require 50% less energy for space heating and commercial buildings 60% less than those built in the mid-1970s.[8] A large part of these potential energy savings should be realized by the year 2050, when much of the present building stock will have been replaced.

More energy savings in space heating can be gained by replacing inefficient heating and cooling systems with improved ones. Estimates show that gas and oil combustion heating systems can be built which use about 30% less fuel than conventional ones because of improved heat transfer and waste heat recovery. An improved heat pump for space heating could provide at least three times as much heat as an electric resistance heater.[9]

These high-efficiency measures are estimated to provide a total energy savings potential in space heating of about 70% by the year 2050 compared with energy use at the mid-1970s energy efficiency levels.

Water Heating. Fuel consumption in water heating can be reduced by 30% by increasing combustion efficiency with insulating storage tanks and pipes and heat recovery. Studies[8] have shown that a heat pump water heater uses 50% less energy than an electric resistance water heater.

Lighting. Lighting consumes about 30% of the electrical energy in the residential and commercial sector. Energy savings of 50% in lighting systems can be achieved by various measures, such as increasing the conversion efficiency of electric energy to light by applying magnetic fields to fluorescent lamps, using advanced control strategies, and optimizing the integration of lighting components and systems into the total building system.[8]

Miscellaneous Electric. The efficiencies of appliances in both commercial and residential use can be improved by better insulation (e.g., for ovens and refrigerators), integration of electronic control devices, more efficient motors, etc. and can result in a saving of up to 50% in electric consumption. Electric drives such as elevators, escalators, and service automation can be made about 35% more efficient by using electronic control and better motors with appropriate sizing, etc. Thus, energy savings of 45% were estimated for energy improvements in miscellaneous electricity usage.

The energy demand and the potential fuel savings with all the above efficiency-promoting measures in the residential and commercial sector are summarized in Table 4.6. The estimated distribution of the potential energy savings, over the

TABLE 4.6

Effects of Technology Improvement on Energy Demand and Savings for Residential and Commercial Sector in the U.S. for Year 2050 (in EJ = 10^{18} J)

Service	Liquids				Gases				Solids				Electricity			
	Nontech. improv.	Tech. improv.	Fuel savings	Fuel savings (%)	Nontech. improv.	Tech. improv.	Fuel savings	Fuel savings (%)	Nontech. improv.	Tech. improv.	Fuel savings	Fuel savings (%)	Nontech. improv.	Tech. improv.	Fuel savings	Fuel savings (%)
Space heat	6.74	1.96	4.78	70.92	10.0	3.16	6.84	68.40	0.69	0.19	0.5	72.46	2.63	0.6	2.03	77.19
Air-conditioning					0.1	0.05	0.05	50.00					2.95	0.83	2.12	71.86
Water heat	1.19	0.92	0.27	22.69	4.11	2.83	1.28	31.14					2.74	1.37	1.37	50.0
Lighting													5.63	2.82	2.81	49.91
Misc. electric													4.83	2.66	2.17	44.93
Total	7.93	2.88	5.05	63.68	14.21	6.04	8.17	57.49	0.69	0.19	0.5	72.46	18.78	8.28	10.50	55.91

energy demand with mid-1970s energy efficiency levels, in the year 2050 is about 64% for liquid fuels, 57% for gaseous fuels, 72% for solid fuels, and 56% for electricity. See Appendix 4.1 for the detailed calculations.

4.2.4.1.2   Industrial Sector

Industrial energy use is complex and varied: fuels are consumed in the form of heat (for producing process steam and for raising the temperatures of materials), electricity (for operating electrical processes and machinery), and feedstocks (for making petrochemical products). Roughly 60% of U.S. industrial energy is consumed by a few energy-intensive manufacturing sectors, in particular iron and steel, chemicals, petroleum refining, cement, and nonferrous metals. Regardless of end-use purposes, the largest share of industrial energy consumption is for process heat, and smaller shares are for mechanical power and space heating.

Process Heat. Savings in energy used for process heat can be achieved by using technologies such as recuperating waste heat (with recuperators, regenerators, low-temperature heat engines), optimizing process flow (energy cascading, cogeneration), producing heat more efficiently, and improving combustion equipment.

Some efficient heat transfer technologies[8] that will significantly improve waste heat recovery are the following:

- Ceramic heat recuperators will recover per year in the U.S. an estimated 2 exajoules (EJ) of thermal energy presently wasted with corrosive and potential fouling high-temperature exhaust streams.
- Improved heat exchangers such as advanced metallic recuperators will recover per year in the U.S. an estimated 1.6 to 3.2 EJ of process heat above 1000°F.
- New heat pump technology will upgrade per year in the U.S. an estimated 1.0 to 3.0 EJ of process heat from liquid and gas waste streams at 150 to 250°F, and bottoming-cycle engines for streams at 700°F and lower, to a more useful temperature level.
- Development of organic working fluids will improve the performance of the organic Rankine cycle driven by low-temperature waste heat.

Optimization of process flow affords many opportunities for applying improved process technologies to the management of energy flow to obtain large fuel savings. One example is energy cascading (i.e., the placement of several processes that require successively lower quality energy in a sequence so that each uses the waste heat from the preceding one). Another is combining process steam production with electricity generation in systems with high electrical-to-thermal output ratios, which not only could satisfy industrial electricity needs but, more importantly, could provide by-product electric power for sale. Further opportunities for improving energy efficiency and productivity involve the application of new catalytic and separation/concentration technologies to chemical reaction and purification processes in the chemical, petroleum-refining, and pharmaceutical industries. Several physical separation techniques that avoid evaporation and freezing, such as membrane separation, have been developed which could significantly improve the efficiency. Potential energy savings from process flow optimization are estimated to be at least 2.0 EJ[8] per year in the U.S.

Combustion by various methods is a form of energy conversion of fossil fuels used by all industrial sectors. The methods differ substantially, but they generally use mainly standard industrial combustion equipment such as boilers, furnaces, and ovens to produce process combustor heat. The technical improvements will be of two major kinds: improved combustion efficiency and advanced combustor. Combustion efficiencies can be raised in many ways. For instance, blast furnace efficiency as measured by coke consumption per ton of pig iron produced can be increased by improving burden properties, increasing iron-ore pellets to sinter ratio, raising air blast temperature and top gas pressure, using more coke oven gas to replace natural gas, improving charging and blast furnace control technology, etc. In general, efficiency of combustors can be improved by using combustion air reheaters, fluidized beds, thermal insulation, electronic control, etc. Advanced combustors, including types with lean and catalytic combustion, fluidized bed and reactive porous bed combustion, and electrically or electromagnetically augmented combustion, will have a significant impact on fuel savings. At least a 30% fuel savings is expected from both kinds of combustion efficiency improvements.

Electric Drive. Industrial electric drives include pumps, compressors, conveyors, fans, presses, machine tools, etc. used by different industries or different plants within an industry. The average practical conversion efficiency of electricity into motive power at the point of use is estimated to range from 40 to 60%.[2] It can be improved by using improved motors and torque converters or clutched flywheels, operating motors from alternating current synthesizers, or other electronic power controls. The use of robots in production technologies may also reduce electricity demand. These measures could result in savings of one-third of the electricity consumed in electric drives.

Aluminum Electrolytic Process. Electricity is consumed in reduction of alumina (Hall–Heroult process) or aluminum chloride (Alcoa process) by electrolysis to primary aluminum. The Alcoa process uses about 30% less energy than the Hall–Heroult process, which was the process mostly used in the mid-1970s.[10] Development of a chemically inert anode and a corrosion-resistant cathode will reduce the large anode–cathode gap in the aluminum reduction process, resulting in higher energy efficiency and closer to steady-state operation. Plasma arc reduction of alumina or aluminum chloride is another possible approach to reducing electricity consumption. Thus, the energy saving achievable via improved technologies by the year 2050 is conservatively estimated at about one-third.

Iron and Steel. Electricity in the iron and steel industry is consumed mainly by electric furnaces in the electric arc steel-making process. Furnace productivity can be increased by preheating the scrap, by operating at higher electric poster levels, and by injecting oxygen. Energy consumption can be further reduced by direct rolling of continuously cast slabs. These measures could save at least 25% of the electricity by the year 2050.

Petrochemical Feedstock. Fossil feedstock is used to produce petrochemicals such as synthetic fibers, plastics, fertilizers, and others. Energy can be saved by more efficient use of materials because considerable energy is used both in disposing of waste and in fabricating new materials to replace those wasted. Efficiency in material usage can be raised by reducing the amount of material per unit of product and by increasing product durability, reuse, recycling, etc. The energy savings from these measures are estimated to be about 20%.

The calculated effects of increased energy efficiency due to technological improvement in industrial energy use in the U.S. for the year 2050 are summarized in Table 4.7. The energy savings by fuel type in the year 2050 compared to the mid-1970s in industry are 48% for liquids, 67% for gases, 70% for solids, and 33% for electricity. See Appendix 4.1 for detailed calculations.

4.2.4.1.3 Transportation Sector

The transportation sector obtains energy predominantly from liquid fuels and will account for about 80% of the liquid fuels consumed in the U.S. in the year 2050. Its two main activities are passenger and freight movement. Automobiles use, by far, the most transportation liquid fuel, followed by trucks, air transport, bus, and rail in approximately that order.

Automobiles. Because the automobile system loses a very large part of the energy from the fuel (more than 75%), mainly in exhaust gases, it offers large opportunities for efficiency improvement via vehicle size reduction, design changes, and improved engines.

Small automobiles use less fuel for many reasons, including lower weight, manual rather than automatic transmission, fewer power options, smaller frontal profiles (less aerodynamic drag), etc. A small car can have a fuel economy over 40 mpg compared with 20 to 30 mpg or less for larger cars. The Energy Policy and Conservation Act of 1975 requires that the average fuel economy of each manufacturer's new automobile fleet meet the minimum fuel economic standard of 27.5 mpg by 1985 and 31 mpg by 1996. However, the average fuel consumption in 1996 was less than 26 mpg because the U.S. public was purchasing more of the heavier land rover truck-like vehicles which evidently are exempt from the minimum standard.

Improvements can be made in existing engine types, transmission, rear axles, tires, etc., as well as aerodynamic design, and vehicle weight can be reduced by material substitution. The efficiency of existing engines can be increased more than 15% by reducing the vehicle power/weight ratio, varying the compression ratio, improving the carburation, reducing internal friction, eliminating idling, etc. Engine-to-transmission matching and efficiency of transmission can be improved by using higher gear ratios, continuously variable transmission, and lock-up automatic

TABLE 4.7

Effects of Technology Improvement on Energy Demand and Savings for Industrial Sector in the U.S. for Year 2050 (in EJ = 10^{18} J)

Service	Liquids				Gases				Solids				Electricity			
	Nontech. improv.	Tech. improv.	Fuel savings	Fuel savings (%)	Nontech. improv.	Tech. improv.	Fuel savings	Fuel savings (%)	Nontech. improv.	Tech. improv.	Fuel savings	Fuel savings (%)	Nontech. improv.	Tech. improv.	Fuel savings	Fuel savings (%)
Process heat	2.90	0.88	2.02	69.66	3.16	0.91	2.25	71.20	3.60	1.15	2.45	68.06				
Electric drive													13.25	8.83	4.42	33.36
Aluminum electrolytic process													1.20	0.80	0.4	33.33
Iron and steel									2.97[a]	0.75	2.22	74.75	0.76	0.57	0.19	25.00
Petroleum feedstock	2.16	1.73	0.43	19.91	0.30	0.24	0.06	20.0	0.16	0.13	0.03	18.75				
Total	5.06	2.61	2.45	48.42	3.46	1.15	2.31	66.76	6.73	2.03	4.70	69.84	15.21	10.20	5.01	32.9

[a] About 90% was used in process heat for ore preparation, pig iron production, steel making, and steel finishing (slabbing, blooming, and billeting).

gear boxes. Combining continuously variable transmission, which enables engines to run at an optimal constant speed, with regenerative braking and energy storage systems can yield a fuel savings of about 30%. Advanced radial tires, which reduce rolling resistance, will give about a 5% fuel savings. Of the useful energy output of an automobile, about 20% is lost to internal engine friction and 12% to transmission and axle friction.[7] Reducing these losses by 25% by better lubrication can lead to fuel savings of at least 10%. A 10% reduction in vehicle weight through use of lightweight composite materials could lead to about 7% improvement in fuel efficiency.[8]

Further energy savings of about 35% can result from improved automotive propulsion systems.[7] More efficient heat engines such as the gas turbine and the Stirling engine have the potential to provide up to 30% better fuel economy than the gasoline internal combustion engine. A high-speed direct-injection turbocharged diesel engine also could provide 30% more efficiency than a conventional gasoline Otto engine. Energy in exhaust gases could be recovered by, for example, turbo compound engines. Electronic control of engine and transmission and use of ceramics for high operating temperatures could lead to additional fuel savings.

All the above measures, summarized in Table 4.8, are confidently estimated to lower the fuel consumption for the whole fleet of automobiles by about 65%, to 40 mpg by the year 2050. The use of electric motor drives and fuel cells could further improve this savings.[21]

Truck. Energy consumption by trucks depends on fuel type (diesel, gasoline, and low-pressure gas), vehicle weight, and operating range. Diesel fuel will have a large share of long-range fuel use, since it is used by 95% of heavyweight long-haul trucks.

TABLE 4.8
Improvement of Fuel Economy of Automobiles
by Various New Technologies

Property	Net change in property[a]
Air drag coefficient	−25%
Average car weight	−30%
Average engine power	−35%
Better transmission	Efficiency +35%
Rolling resistance coefficient	−40%
Switch to improved diesel or Stirling engine, etc.	Efficiency +35%
Top speed	+10%
Acceleration	−10%
Average mileage	Up to 200% (×3.0)
Fuel savings	66%

[a] The effects of various measures interact and are not simply additive or multiplicative. Ways of combining various effects without double counting are given in References 11 and 12.

The adiabatic diesel engine with waste heat utilization has the potential for 40% higher efficiency than the present conventional diesel. In addition, elimination of the need for coolant and fluid lubricant components makes the adiabatic engine lighter and smaller than the present diesel of comparable power. With allowance for higher weight, trucks can benefit from the same efficiency improvement measures as automobiles. Weight can be reduced by use of composite materials, aerodynamic drag can be decreased by slipping front end design, and engine transmission can be controlled by electronic devices. In combination, the above measures are expected to increase the fuel economy of trucks by about 60%, to 6.5 mpg by the year 2050.

Air Transport. Technical measures to improve fuel efficiency in air transport consist of general use of high-bypass turbo fan engines, turboprop engines for short hauls, weight reduction by using titanium, high-strength aluminum alloys with good stress–corrosion resistance and carbon-fiber composites, microcomputer control to increase maneuverability and gust tolerance, use of wide-bodied jets, etc. These measures could improve fuel efficiency from an average of about 20 passenger miles per gallon in the mid-1970s to 40 in the year 2050, with a 50% savings in fuel.

Bus and Rail. The same measures for efficiency improvement apply to buses as to cars and trucks, and a 40% fuel savings should be achievable. In rail transport, replacement of steam by diesel locomotives, use of regenerative braking, and lowering the weight of rolling stock could lower energy use by about 40% by the year 2050. For electric rail, introduction of better electric motors, aerodynamic drag reduction, lighter stock, and regenerative braking could lower electricity demand by more than 30% by the year 2050.

Ship. Potential energy savings of 40% could be achieved for ships through technical changes including heat recovery, improved diesel engines, variable-pitch propellers, drag reduction with anti-fouling paints, etc.

The changes in fuel demand for transportation due to technology improvements are summarized in Table 4.9. The potential fuel savings are 52% of liquid fuel and 27% of electricity. Of the liquid fuel savings, 65% comes from automobiles, 16% from trucks, and 11% from air transport.

Percentages of fuel savings by fuel type for each end-use sector in the U.S. are summarized in Table 4.10. The technology improvement factor is the ratio of energy savings from technology improvements to energy demand without those improvements. The aggregate improvement factor for each fuel type is 53% for liquids, 59% for gases, 70% for solids, and 46% for electricity.

The summary in Table 4.10 reveals an important conclusion. The large differences between the energy efficiencies of mid-1970s end-use technologies and future improved technologies indicate an enormous potential for major long-term reductions in energy consumption through technological improvements. Even if the calculated savings represent an upper limit, and the actual savings achievable in practice might be somewhat lower, the improved technologies are well enough known to provide sizable and worthwhile energy savings. It is recognized, however, that the rate at which this potential fuel savings can be accomplished is subject to various

TABLE 4.9
Effects of Technology Improvement on Energy Demand and Savings for Industrial Sector in the U.S. for Year 2050 (in EJ = 10^{18} J)

Service	Liquids				Gases				Solids				Electricity			
	Nontech. improv.	Tech. improv.	Fuel savings	Fuel savings (%)	Nontech. improv.	Tech. improv.	Fuel savings	Fuel savings (%)	Nontech. improv.	Tech. improv.	Fuel savings	Fuel savings (%)	Nontech. improv.	Tech. improv.	Fuel savings	Fuel savings (%)
Automobile	26.0	8.84	17.16	66.0	0.0			0.0	0.0			0.0				
Air transport	5.82	2.91	2.91	50.0												
Truck	13.64	9.44	4.20	30.79												
Bus and rail	2.70	1.62	1.08	40.0									0.11	0.08	0.03	27.30
Ship	2.38	1.43	0.95	39.92												
Total	50.54	24.24	26.30	52.04	0.0			0.0	0.0			0.0	0.11	0.08	0.03	27.30

TABLE 4.10
Technological Improvement Factors by Fuel Type and Sector Resulting from End-Use Technology Improvements in the U.S. for Year 2050

Sector	%			
	Liquids	Gases	Solids	Electricity
Residential/commercial	63.68	57.49	72.46	55.91
Industrial	48.42	66.76	69.84	32.94
Transportation	52.04	0.0	0.0	27.30
Improvement factor, aggregate weighted	53.20	59.31	70.08	45.57

socioeconomic considerations, including the economic growth rate, social and political attitudes, institutional arrangements, and international trade and relationships, all of which are beyond the scope of this volume.

4.2.4.2　Canada and Western Europe and OECD Pacific

As mentioned earlier, detailed data are lacking on functional energy demand by fuel type in each end-use sector in regions other than the U.S. Since Canada and Western Europe and the OECD Pacific are industrially developed, with a technological basis and economic growth trends generally similar to those in the U.S., the same efficiency gains through technology improvements are thought to apply to these regions as to the U.S. Thus, the energy savings obtainable by applying higher efficiency technologies in these regions can be calculated by applying the relevant percentage fuel savings (i.e., technology improvement factor) in the U.S. (Table 4.10) for each fuel type and end-use rector. The results are shown in Tables 4.11 and 4.12. The energy savings by fuel type in Region 2 (Canada and Western Europe) over the fuel demand in the year 2050 with mid-1970s technological efficiency are 55% for liquids, 61% for gases, 70% for solids, and 43% for electricity; in Region 3 (OECD Pacific), savings are 53% for liquids, 62% for gases, 69% for solids, and 38% for electricity.

4.2.4.3　Developing Regions

As indicated earlier, only aggregate energy demand data for end-use sectors are available for the developing regions (former Soviet Union, Eastern Europe, China et al., Middle East, Africa, Latin America, South and East Asia). Analysis of the impacts of end-use technology on energy consumption is more meaningful at the sectoral than at the aggregate level. Since economic activities are by far the predominant factor in determining how energy is consumed, the pattern of energy use in industrial countries should give some indication of the long-term economic prospects and energy use for developing regions. In other words, the future patterns of energy consumption in different parts of Regions 4 to 9 at a given time will approach those of developed countries at various developmental stages. This leads

TABLE 4.11
Effects of End-Use Technology Improvements on Energy Demand and Savings in Canada and Western Europe for Year 2050 (in EJ = 10^{18} J)

Service	Liquids				Gases				Solids				Electricity			
	Nontech. improv.	Tech. improv. savings	Fuel savings	Fuel savings (%)	Nontech. improv.	Tech. improv. savings	Fuel savings	Fuel savings (%)	Nontech. improv.	Tech. improv. savings	Fuel savings	Fuel savings (%)	Nontech. improv.	Tech. improv. savings	Fuel savings	Fuel savings (%)
Residential/commercial	12.88	4.68	8.20	63.68	13.39	5.69	7.70	57.49	6.91	1.90	5.01	72.46	19.91	8.78	11.13	55.91
Industrial	6.56	3.38	3.18	48.42	7.11	2.36	4.75	66.76	10.44	3.15	7.29	69.83	25.07	16.81	8.26	32.94
Transportation	22.04	10.57	11.47	52.04	0.0				0.35	0.26	0.09	25.71	1.59	1.15	0.43	27.04
Total	41.47	18.63	22.85	55.10	20.50	8.05	12.45	60.73	17.69	5.31	12.39	70.04	46.57	26.74	19.82	42.56

TABLE 4.12

Effects of End-Use Technology Improvements on Energy Demand and Savings in OECD Pacific for Year 2050 (in EJ = 10^{18} J)

Service	Liquids				Gases				Solids				Electricity			
	Nontech. improv. savings	Tech. improv. savings	Fuel savings	Fuel savings (%)	Nontech. improv.	Tech. improv. savings	Fuel savings	Fuel savings (%)	Nontech. improv.	Tech. improv. savings	Fuel savings	Fuel savings (%)	Nontech. improv.	Tech. improv. savings	Fuel savings	Fuel savings (%)
Residential/commercial	3.64	1.32	2.32	63.68	6.74	2.87	3.87	57.42	1.45	0.3	1.05	72.46	5.58	2.46	3.12	55.91
Industrial	3.88	2.00	1.88	48.42	6.98	2.32	4.66	66.76	9.64	2.91	6.73	69.84	16.84	11.29	5.55	32.96
Transportation	10.77	5.16	5.60	52.04	0.0				0.29	0.22	0.07	24.14	0.87	0.63	0.24	27.30
Total	18.29	8.48	9.80	53.58	13.72	5.19	8.53	62.17	11.38	3.53	7.85	68.98	23.30	14.38	8.91	38.24

TABLE 4.13
The Allocation of Sectoral Energy Demand for
Developing Regions for Year 2050 Follows
the Patterns of Developed Regions in Various Years

Developing region	Developed region	Year
Former Soviet Union/Eastern Europe	U.S.	2025
China et al.	OECD Pacific	2025
Middle East	Canada/Western Europe	2025
Africa	Canada/Western Europe	2000
Latin America	U.S.	2000
South and East Asia	OECD Pacific	2000

to the assumption — with differences in available resources, stage of industrialization, etc. among developing countries taken into account — that the sectoral energy demands in developing regions can be approximated from those of the pertinent developed regions at various future times, as shown in Table 4.13. In this way, the sectoral energy demands by fuel type in Regions 4 to 9 can be obtained by partitioning the total aggregate energy demand for each fuel type in the region (Table 4.4) into sectoral energy demands according to the proportionate allocations in the corresponding developed regions. The results are shown in Table 4.14.

The smaller industrial base and the higher economic growth rate forecast for many developing countries provide substantial opportunities for introducing energy-efficient technologies at an earlier stage of development. This implies that the technological efficiency gains in the industrial regions could probably also be achieved in the developing countries in the future.

Thus, the effects of technology on energy end use in the year 2050 for Regions 4 to 9 can be calculated by applying to each region's end-use sector the relevant U.S. technology improvement factors by fuel type (Table 4.10). The resultant energy demands and the savings due to technological improvements in Regions 4 to 9 for the year 2050 are shown in Tables 4.15 to 4.20. The fuel savings by fuel type over the energy demand in the year 2050 with mid-1970s technology efficiencies for these developing regions are in the range of 53 to 55% for liquids, 60 to 63% for gases, 68 to 70% for solids, and 39 to 46% for electricity, almost the same as in OECD regions (Regions 1 to 3), that is, 53 to 55% for liquids, 59 to 62% for gases, 69 to 70% for solids, and 38 to 46% for electricity. Table 4.21 summarizes the effects of end-use technology improvements by fuel type on regional energy demands and savings and the associated global effects. Global fuel savings by fuel type for the year 2050 are 54% for liquids, 61% for gases, 70% for coal, and 43% for electricity over the energy demand with mid-1970s technology efficiencies.

The energy demand and the savings due to technological changes for the developing regions were also estimated by an alternative method in which the aggregate energy demand is not partitioned into sectoral end-use demand; the aggregate tech-

TABLE 4.14

Sectoral Secondary Energy Demands by Fuel Type in Developing Regions (Regions 4 to 9) for Year 2050 (in EJ)

Region	Liquids	Gases	Solids	Electricity	Total
Former Soviet Union					
Residential/commercial	3.07	24.62	1.71	12.70	42.10
Industrial	2.05	7.82	18.78	9.56	38.21
Transportation	17.47	0.0	0.0	0.07	17.54
Total	22.59	32.44	20.49	22.33	97.85
China et al.					
Residential/commercial	3.81	8.13	5.18	4.75	21.87
Industrial	4.09	8.99	37.97	12.44	63.51
Transportation	9.51	0.0	0.64	0.64	10.79
Total	17.41	17.12	43.79	17.83	96.17
Middle East					
Residential/commercial	2.00	3.31	0.31	1.71	7.33
Industrial	1.06	2.55	0.53	2.00	6.14
Transportation	3.61	0.0	0.02	0.13	3.76
Total	6.67	5.86	0.86	3.84	17.23
Africa					
Residential/commercial	2.77	3.60	4.46	4.70	15.30
Industrial	1.79	3.63	7.42	5.17	18.01
Transportation	3.96	0.0	0.13	0.25	4.34
Total	8.52	7.23	12.01	10.12	37.65
Latin America					
Residential/commercial	3.41	5.88	0.56	13.89	23.74
Industrial	2.68	2.96	5.99	9.64	21.27
Transportation	16.29	0.0	0.0	0.06	16.35
Total	22.38	8.84	6.55	23.59	61.36
South and East Asia					
Residential/commercial	2.17	1.83	2.20	4.33	10.53
Industrial	2.81	2.82	17.24	10.17	33.04
Transportation	4.36	0.0	0.24	0.46	5.06
Total	9.34	4.65	19.68	14.95	48.63

nology improvement factor for each fuel type from OECD regions (Regions 1 to 3) is applied to the aggregate secondary energy demands for Regions 4 to 9 (Table 4.4 or Table 4.21, under nontechnical improvement for each fuel type). The aggregate technology improvement factors for Regions 1 to 3 (Table 4.22) are obtained by dividing the total energy savings for each fuel type in Regions 1 to 3 by the total energy demands without technology improvements for that fuel type. The resultant regional energy demands and savings obtained by this approach are summarized in

TABLE 4.15
Effects of End-Use Technology Improvements on Energy Consumption in the Former Soviet Union/Eastern Europe for Year 2050 (in EJ)

Sector	Liquids			Gases			Solids			Electricity		
	Nontech. improv.	Tech. improv.	Fuel savings	Nontech. improv.	Tech. improv.	Fuel savings	Nontech. improv.	Tech. improv.	Fuel savings	Nontech. improv.	Tech. improv.	Fuel savings
Residential/commercial	3.07	1.12	1.95	24.62	10.47	14.15	1.71	0.47	1.24	12.70	5.60	7.10
Industrial	2.05	1.06	0.99	7.82	2.60	5.22	18.78	5.66	13.12	9.56	6.41	3.15
Transportation	17.47	8.38	9.09	0.0	0.0	0.0	0.0	0.0	0.0	0.07	0.05	0.02
Total	22.59	10.56	12.03	32.44	13.07	19.37	20.49	6.13	14.36	22.33	12.06	10.27
Aggregate improvements (%)			52.25			59.71			70.08			46.0

TABLE 4.16
Effects of End-Use Technology Improvements on Energy Consumption in China et al. for Year 2050 (in EJ)

Sector	Liquids			Gases			Solids			Electricity		
	Nontech. improv.	Tech. improv.	Fuel savings	Nontech. improv.	Tech. improv.	Fuel savings	Nontech. improv.	Tech. improv.	Fuel savings	Nontech. improv.	Tech. improv.	Fuel savings
Residential/commercial	3.81	1.38	2.43	8.13	3.46	4.67	5.18	1.43	3.75	4.75	2.09	2.66
Industrial	4.09	2.11	1.98	8.99	2.99	6.00	37.97	11.45	26.52	12.44	8.34	4.10
Transportation	9.51	4.56	4.95	0.0	0.0	0.0	0.64	0.48	0.16	0.64	0.47	0.17
Total	17.41	8.05	9.36	17.12	6.45	10.67	43.79	13.36	30.43	17.83	10.9	6.93
Aggregate improvements (%)			53.76			62.32			69.49			38.87

TABLE 4.17
Effects of End-Use Technology Improvements on Energy Consumption in the Middle East for Year 2050 (in EJ)

Sector	Liquids			Gases			Solids			Electricity		
	Nontech. improv.	Tech. improv.	Fuel savings	Nontech. improv.	Tech. improv.	Fuel savings	Nontech. improv.	Tech. improv.	Fuel savings	Nontech. improv.	Tech. improv.	Fuel savings
Residential/commercial	2.00	0.73	1.27	3.31	1.41	1.90	0.31	0.09	0.22	1.71	0.75	0.96
Industrial	1.06	0.55	0.51	2.55	0.85	1.70	0.53	0.16	0.37	2.00	1.34	0.66
Transportation	3.61	1.73	1.88	0.0	0.0	0.0	0.02	0.02	0.0	0.13	0.09	0.04
Total	6.67	3.01	3.66	5.86	2.26	3.60	0.86	0.27	0.59	3.54	2.18	1.66
Aggregate improvements (%)			54.87			61.47			68.60			43.23

TABLE 4.18
Effects of End-Use Technology Improvements on Energy Consumption in Africa for Year 2050 (in EJ)

Sector	Liquids			Gases			Solids			Electricity		
	Nontech. improv.	Tech. improv.	Fuel savings	Nontech. improv.	Tech. improv.	Fuel savings	Nontech. improv.	Tech. improv.	Fuel savings	Nontech. improv.	Tech. improv.	Fuel savings
Residential/commercial	2.77	1.01	1.76	3.60	1.53	2.07	4.46	1.23	3.23	4.70	2.07	2.63
Industrial	1.79	0.92	6.87	3.63	1.20	2.42	7.42	2.24	5.18	5.17	3.47	1.70
Transportation	3.96	1.90	2.06	0.0			0.13	0.10	0.03	0.25	0.18	0.07
Total	8.52	3.83	4.69	7.23	2.73	4.49	12.01	3.57	8.44	10.12	5.72	4.40
Aggregate improvements (%)			55.05			62.10			70.27			43.48

TABLE 4.19
Effects of End-Use Technology Improvements on Energy Consumption in Latin America for Year 2050 (in EJ)

Sector	Liquids			Gases			Solids			Electricity		
	Nontech. improv.	Tech. improv.	Fuel savings	Nontech. improv.	Tech. improv.	Fuel savings	Nontech. improv.	Tech. improv.	Fuel savings	Nontech. improv.	Tech. improv.	Fuel savings
Residential/commercial	3.41	1.24	2.17	5.88	3.50	3.38	0.56	0.15	0.41	12.89	6.13	7.76
Industrial	2.68	1.38	1.30	2.96	0.98	1.98	5.99	1.81	4.18	9.64	6.47	3.17
Transportation	16.29	7.81	8.48	0.0	0.0	0.0	0.0	0.0	0.0	0.06	0.04	0.02
Total	22.38	10.43	11.95	8.84	3.48	5.36	6.55	1.96	4.59	23.59	12.64	10.95
Aggregate improvements %			53.40			60.63			70.08			46.42

TABLE 4.20
Effects of End-Use Technology Improvements on Energy Consumption in South and East Asia for Year 2050 (in EJ)

Sector	Liquids			Gases			Solids			Electricity		
	Nontech. improv.	Tech. improv.	Fuel savings	Nontech. improv.	Tech. improv.	Fuel savings	Nontech. improv.	Tech. improv.	Fuel savings	Nontech. improv.	Tech. improv.	Fuel savings
Residential/commercial	2.17	0.79	1.38	1.83	0.78	1.05	2.20	0.61	1.59	4.33	1.91	2.42
Industrial	2.81	1.45	1.36	2.82	0.94	1.88	17.24	5.20	12.04	10.17	6.82	3.35
Transportation	4.36	2.09	2.27	0.0	0.0	0.0	0.24	0.18	0.06	0.46	0.33	0.13
Total	9.34	4.33	5.01	4.05	1.72	2.93	19.68	5.99	13.69	14.95	9.06	5.90
Aggregate improvements (%)			53.64			63.01			69.56			39.46

TABLE 4.21
Summary of Effects of End-Use Technology Improvements on Energy Consumption by Fuel Type and Region for Year 2050 (in EJ)

Sector	Liquids			Gases			Solids			Electricity		
	Nontech. improv.	Tech. improv.	Fuel savings	Nontech. improv.	Tech. improv.	Fuel savings	Nontech. improv.	Tech. improv.	Fuel savings	Nontech. improv.	Tech. improv.	Fuel savings
U.S.	63.53	29.73	33.80	17.67	7.19	10.48	7.42	2.22	5.20	34.10	18.56	15.54
Canada and Western Europe	41.47	18.63	22.85	20.50	8.05	12.45	17.69	5.31	12.39	46.57	26.74	19.82
OECD Pacific	18.29	8.48	9.80	13.72	5.19	8.53	11.38	3.53	7.85	23.30	14.38	8.91
Former Soviet Union/ Eastern Europe	22.60	10.56	12.03	32.44	13.07	19.37	20.49	6.13	14.36	22.33	12.06	10.27
China et al.	17.41	8.05	9.36	17.12	6.45	10.67	43.79	13.36	30.43	17.83	10.9	6.93
Middle East	6.67	3.01	3.66	5.86	2.26	3.60	0.86	0.27	0.59	3.84	2.18	1.66
Africa	8.52	3.83	4.69	7.23	2.73	4.49	12.01	3.57	8.44	10.12	5.72	4.40
Latin America	22.38	10.43	11.95	8.84	3.48	5.36	6.55	1.96	4.59	23.59	12.64	10.95
South and East Asia	9.34	4.33	5.01	4.65	1.72	2.93	19.68	5.99	13.69	14.95	9.06	5.90
All regions	210.21	97.06	113.15	128.03	50.15	77.88	139.87	42.33	97.54	196.63	112.25	84.38
Aggregate improvements (%)		53.83			60.83			69.74			42.91	

TABLE 4.22
Aggregate Technology Improvement
Factors of Various Fuel Types Due to
End-Use Technology Improvement
for Regions 1 to 3

Fuel	Technology improvement factor (%)
Liquids	53.90
Gases	60.63
Solids	69.72
Electricity	42.58

Table 4.23. Comparison of Tables 4.21 and 4.23 shows that the differences between the results for global energy needs and savings for the year 2050 with technological improvements obtained with these two approaches are negligible (less than 1%). This helps demonstrate that the trend in energy end-use patterns in developing regions tends to follow those in developed regions at various developmental stages. Thus, either result can be used for further investigation of the effects of technology on CO_2 emissions with little difference in results.

4.2.5   Electricity Generation Technologies

The measures for electricity savings discussed so far are limited to electricity utilization technologies, and the resultant electricity demand and savings are shown in the last two columns of Table 4.21. Further energy savings can be gained from new and improved electricity generation technologies such as fluidized bed combustion combined cycle, gasification combined cycle, fuel cell, magnetohydrodynamic (MHD), etc. listed in Table 4.2. In the combined cycle technology, the exhaust heat of a gas turbine is used to raise steam through a heat recovery boiler for a steam turbine; this combination can raise the overall efficiency by 50%. A fuel cell, which does not depend on prime movers driving generators, converts chemical energy directly into electrical energy, much like a battery, except that the chemicals (fuel and oxygen or air) producing the electricity are continuously fed to the cell and the waste products are continuously removed. Effort has also focused on high-temperature fuel cells that utilize low- and medium-Btu gas from gasified coal as a source of hydrogen gas for fuel. The efficiencies of fuel cells using phosphoric acid or molten carbonate as electrolyte are 50 to 55%, with a theoretical limit approaching 100%. In MHD systems, a high-temperature, high-velocity gas or plasma, seeded with small amounts of electrically charged particles (metallic ions), produces electricity directly when passed through a strong magnetic field. Using seeded electro-conducting gas at 2500 to 3000°C, open-cycle MHD generators with steam or gas turbine prime movers are expected to operate at efficiencies of 50 to 60%. The average efficiency of the improved electricity-generating technologies using fossil

TABLE 4.23
Summary of Effects of End-Use Technology Improvements on Energy Use by Fuel Type and Region for Year 2050 (in EJ)

Sector	Liquids			Gases			Solids			Electricity		
	Nontech. improv.	Tech. improv.	Fuel savings	Nontech. improv.	Tech. improv.	Fuel savings	Nontech. improv.	Tech. improv.	Fuel savings	Nontech. improv.	Tech. improv.	Fuel savings
U.S.	63.53	29.73	33.80	17.67	7.19	10.48	7.42	2.22	5.20	34.10	18.56	15.54
Canada and Western Europe	41.47	18.63	22.85	20.50	8.05	12.45	17.69	5.31	12.39	46.57	26.74	19.82
OECD Pacific	18.29	8.48	9.80	13.72	5.19	8.53	11.38	3.53	7.85	23.30	14.38	8.91
Former Soviet Union/Eastern Europe	22.60	10.42	12.18	32.44	12.77	19.67	20.49	6.20	14.29	22.33	12.82	9.51
China et al.	17.41	8.03	9.38	17.12	6.74	10.38	43.79	13.26	30.53	17.83	10.23	7.60
Middle East	6.67	3.07	3.60	5.86	2.31	3.55	0.86	0.26	0.60	3.84	2.20	1.64
Africa	8.52	3.93	4.59	7.23	2.85	4.38	12.01	3.64	8.37	10.12	5.81	4.31
Latin America	22.38	10.32	12.06	8.84	3.48	5.36	6.55	1.98	4.57	23.59	13.55	10.04
South and East Asia	9.33	4.30	5.04	4.65	1.83	2.82	19.68	5.96	13.72	14.95	8.58	6.37
All regions	210.21	96.92	113.29	128.03	50.41	77.62	139.87	42.35	97.52	196.63	112.88	83.74
Aggregate improvements (%)		53.89			60.63			69.72			42.58	

TABLE 4.24
Summary of Fuel Needs for Electricity Generation
Without Technology Improvements by Region and Fuel Type
for Year 2050 (in EJ = 10^{18} J of Primary Energy Equivalents)

Region	Oil	Gas	Coal	Nuclear	Solar	Hydro	Total
U.S.	9.85	21.90	51.70	12.91	12.91	8.19	117.46
Canada and Western Europe	8.85	20.45	66.35	23.47	23.47	16.70	159.28
OECD Pacific	8.20	9.10	38.88	9.65	9.65	4.23	79.69
Former Soviet Union/Eastern Europe	9.66	4.45	27.72	5.20	5.20	24.38	76.64
China et al.	6.31	2.85	17.71	3.33	3.33	27.66	61.18
Middle East	0.27	3.20	3.89	1.48	1.48	3.05	13.36
Africa	0.00	0.00	0.00	0.00	0.00	10.12	10.12
Latin America	8.90	4.20	26.19	4.91	4.91	31.72	80.86
South and East Asia	2.59	2.92	18.23	3.38	3.38	20.23	50.73
All regions	54.63	69.07	250.67	64.31	64.31	146.28	649.30

fuel should reach at least 50% in the year 2050. Nuclear steam technologies with high-temperature, high-efficiency reactors such as high-temperature gas-cooled reactors (HTGR), liquid-metal-cooled fast breeder reactors (LMFBR), and nuclear gas turbine combined cycles are also expected to have efficiencies of at least 40% by the year 2050. The average efficiencies of solar electric technologies such as photovoltaic solar cells and solar thermal central station energy conversion are expected to reach about 20% in the year 2050, compared with 5 to 10% today.

Since electricity can be generated through conversion of heat from fossil or nuclear fuel (as in a steam plant), chemical energy (as in a fuel cell), or solar energy (as in a hydro plant or solar cell), the shares of the various fuels in electricity generation need to be known in order to study the effects of improved technologies on energy consumption. Table 4.24 shows these shares in the year 2050 with mid-1970s electricity-generating efficiencies for various regions. In this study, the shares of fuels in electricity generation with technology improvements in the year 2050 are obtained by partitioning the electricity demand and the savings resulting from the utilization of technology improvements (Table 4.21) into shares of fossil fuels, nuclear, hydro, and solar that are the same as those in Table 4.24. The structures of electric power generation and savings for various regions obtained by this partitioning process are shown in Tables 4.25 and 4.26.

The higher efficiencies of new and improved electricity generation technologies will decrease the amounts of fuels needed to generate a given amount of electricity. The total fossil-fuel savings derived from both electricity utilization and generation improvements are shown in Table 4.27 (see Appendix 4.2 for detailed calculations). The total fuel savings are obtained by first applying the improvements in the end-use technologies and then those in generating technologies, but the results would be the same if the order were reversed.

TABLE 4.25
Summary of Electricity Savings Due to End-Use Technology Improvements by Region and Fuel Type for Year 2050 (in EJ = 10^{18} J)

Region	Oil	Gas	Coal	Nuclear	Solar	Hydro	Total
U.S.	1.30	2.90	6.84	1.71	1.71	1.08	15.54
Canada and Western Europe	1.10	2.54	8.25	2.92	2.92	2.08	19.82
OECD Pacific	0.92	1.02	4.35	1.08	1.08	0.47	8.91
Former Soviet Union/Eastern Europe	1.29	0.60	3.71	0.70	0.70	3.27	10.27
China et al.	0.71	0.32	2.01	0.38	0.38	3.13	6.93
Middle East	0.03	0.40	0.48	0.18	0.18	0.38	1.66
Africa	0.00	0.00	0.00	0.00	0.00	4.40	4.40
Latin America	1.21	0.57	3.55	0.66	0.66	4.29	10.95
South and East Asia	0.30	0.34	2.12	0.39	0.39	2.35	5.90
All regions	6.86	8.69	31.31	8.02	8.02	21.45	84.38

4.2.6 Technology Effects on Global Fossil-Fuel Demand and Savings

The consumption of fossil fuels and the savings achievable in each of the nine regions by widespread global application of the improved end-use and electricity generation technologies in the year 2050 are summarized in Table 4.28, and overall demands and savings are summarized in Table 4.29 and Figure 4.2. These data demonstrate one important conclusion of the study: It is possible, within the framework of present known improved end-use and power generation technologies, to reduce global fossil-fuel consumption by about 55% of liquids, 60% of gases, and

TABLE 4.26
Summary of Electricity Demand with End-Use Technology Improvements by Region and Fuel Type for Year 2050 (in EJ = 10^{18} J)

Region	Oil	Gas	Coal	Nuclear	Solar	Hydro	Total[a]
U.S.	1.56	3.46	8.17	2.04	2.04	1.29	18.56
Canada and Western Europe	1.49	3.43	11.14	3.94	3.94	2.80	26.74
OECD Pacific	1.48	1.64	7.01	1.74	1.74	0.76	14.38
Former Soviet Union/Eastern Europe	1.52	0.70	4.36	0.82	0.82	3.84	12.06
China et al.	1.12	0.51	3.16	0.59	0.59	4.93	10.90
Middle East	0.04	0.52	0.63	0.24	0.24	0.50	2.18
Africa	0.00	0.00	0.00	0.00	0.00	5.72	5.72
Latin America	1.39	0.66	4.09	0.77	0.77	4.96	12.64
South and East Asia	0.46	0.52	3.26	0.60	0.60	3.61	9.06
All regions	9.06	11.44	41.82	10.74	10.74	28.41	112.24

[a] Same as the second column under Electricity in Table 4.21.

TABLE 4.27
Fossil-Fuel Needs and Savings for Electricity Generation Resulting from Technology Improvement for Year 2050 (in EJ = 10^{18} J)

Fuel type (efficiency, % year, 1975/2050)[a]	Fuel needs		Fuel savings		
	Without elec. gen. improv.	With elec. gen. improv.	From elec. gen. improv.	From end-use tech. improv.	Total
Oil (32/50)	28.31	18.12	10.19	21.44	31.63
Gas (32/50)	35.75	22.88	12.87	27.16	40.03
Coal (32/50)	126.73	83.64	43.09	94.88	137.97

[a] The average fossil-fuel efficiency of conversion to electricity, excluding distribution and transmission losses.

60% of solids in the year 2050, for an overall global reduction of almost 60% of consumption with mid-1970s efficiencies. With allowance for the additional energy needed for producing and transporting the saved energy, actual fossil-fuel savings should be greater than 60%. About half of the overall savings of 500 EJ (10^{18} J) is derived from solid-fuel-related technologies. This is in line with the predicted increasing role of coal, especially in power and liquid fuels production, in the future world energy economy. The breakdown of the overall fuel savings into three end-use sectors and by fuel type is shown in Table 4.30 and Figure 4.3. The largest share of savings, about 45%, is in the industrial sector, which accounts for about 44% of secondary energy consumption (or 48% of primary). About 39% of the savings is in the residential and commercial sector, which consumes about 35% of secondary energy (or 38% of primary). The smallest share, 15%, of the savings is in the transportation sector, which uses about 21% of secondary energy (or 13% of primary). Improvements in electricity utilization and generation technologies contribute about 54% of the fuel savings in the residential and commercial sector and 45% in the industrial sector. The energy savings in the transportation sector are essentially in liquid fuel, not electricity.

Electric power is generated from six sources: three fossil fuels (liquids, gases, and solids) and three nonfossil sources (nuclear, solar, and hydro). If the regional savings (i.e., surplus) of nonfossil-fuel-related power due to efficiency improvements could be used to meet part or all of the demand for fossil-fuel-related power in a given region, then the demand for fossil fuels could be further reduced. The global saving of electricity from nonfossil fuels due to improvements in end-use and power-generating technologies is 47 EJ (Table 4.31; see Appendix 4.3 for details). Because regions do not trade nonfossil-fuel-related power, the actual replacement of power will be lower.

The regional surplus of electricity generated from nonfossil fuels due to technology improvements is shown in Table 4.32. The amount of fossil-fuel-related electricity demand that can be met by these savings is assumed to depend on the structure of fossil-fuel-related electricity demand in each region. If the surplus of

TABLE 4.28
Regional Fuel Demands by Fuel Type With and Without Technology Improvements and Energy Savings Resulting from Technology Improvements for Year 2050 (in EJ = 10^{18} J)

Region	Liquids			Gases			Solids		
	Without tech. improv.	With tech. improv.	Fuel savings	Without tech. improv.	With tech. improv.	Fuel savings	Without tech. improv.	With tech. improv.	Fuel savings
U.S.	73.38	33.77	39.61	39.57	16.14	23.43	59.12	24.77	34.35
Canada and Western Europe	50.32	22.35	27.97	40.95	16.70	24.25	84.04	35.17	48.87
OECD Pacific	26.49	12.15	14.34	22.82	9.26	13.56	50.26	22.01	28.25
Former Soviet Union/Eastern Europe	32.26	14.49	17.77	36.89	14.85	22.04	48.21	18.12	30.09
China et al.	23.72	10.88	12.84	19.97	7.73	12.24	61.50	21.72	39.78
Middle East	6.94	3.14	3.80	9.06	3.63	5.43	4.75	2.06	2.69
Africa	8.52	3.83	4.69	7.23	2.73	4.49	12.01	3.57	8.44
Latin America	31.28	14.00	17.28	13.04	5.16	7.88	32.74	13.18	19.56
South and East Asia	11.93	5.46	6.47	7.57	3.00	4.57	37.91	14.44	23.47
All regions	264.84	120.06	144.78	197.10	79.20	117.90	390.54	155.03	235.51

TABLE 4.29
Total Fossil-Fuel Demands With and Without Technology Improvements and Energy Savings Resulting from Technology Improvements for Year 2050 (in EJ = 10^{18} J)

	Energy demand				
Fuel	Without tech. improv.[a]	With tech. improv.[b]	Energy savings[c]	Percentage of energy savings	
Liquids	264.84	120.06	144.78	29.06[d]	54.67[e]
Gases	197.10	79.19	117.91	23.67	59.82
Solids	390.54	155.03	235.51	47.27	60.30
Total	852.48	354.28	498.20	100.00	58.44

[a] Sum of total for all regions given in Table 4.21 and the corresponding region in Table 4.24 by fuel type.

[b] Difference of columns 2 and 3.

[c] Sum of total for all regions given in Table 4.21 and column 5 in Table 4.27.

[d] With respect to the total savings of 498.2 EJ.

[e] With respect to the energy demand without technical improvements.

nonfossil-fuel electricity is located where the fossil-fuel-related electricity is needed (i.e., transmission and redistribution losses are negligible), then the resultant regional amounts of electricity that can be replaced, by fuel type, are those shown in Table 4.33.

For developing regions like China, Africa, and Latin America, where hydropower is abundant, the demand for electric power can be met entirely with nonfossil fuel after the energy efficiency improvements are applied. In the former Soviet Union, where hydropower is also abundant, the replacement could reach 85% of electricity generated from fossil fuels. In the developed OECD countries, where nuclear and solar energy are more widely used, the replacement amounts are up to only about 50%. The overall savings of fossil fuels due to such replacement, by fuel type, are given for each region in Table 4.34, which shows that the global amount of fossil-fuel-related electricity replaceable by nonfossil-fuel-related electricity is about 41 EJ, and the associated fossil-fuel savings is 82.8 EJ. Almost 70% of the savings is derived from coal, which is in line with the predicted increasing role of coal and the shrinking role of gases and liquids in electricity generation. The overall demand for fossil fuels and the savings due to both technology improvements and replacement of electricity from fossil fuels with that from nonfossil fuels for the year 2050 are shown in Table 4.35. The effect of such replacement, as seen by comparing the energy demands shown in Tables 4.29 and 4.35, is that global demand for fossil fuels is further reduced, from 42% of that without technology improvements to 32%. (The actual saving is larger when the additional energy needed to produce and transport the saved energy is taken into account.)

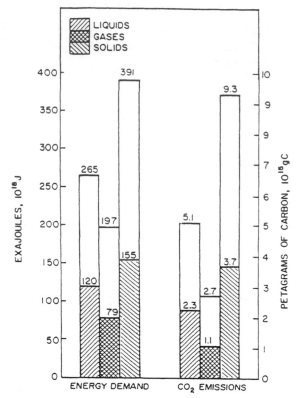

FIGURE 4.2   Global energy demand and CO_2 emissions by fuel type with and without technology improvements for year 2050. The difference between the two values in each column represents the effects of technology improvements on energy demand and CO_2 emissions.

TABLE 4.30
Global Fossil-Fuel Savings Due to Technology Improvements by End-Use Sector for Year 2050 (in EJ = 10^{18} J)

Sector	Liquids	Gases	Solids	Total	%
Residential/commercial	42.27	66.85	85.89	195.01	39.14
Industry	29.86	50.33	146.74	226.93	45.55
Transportation	72.65	0.73	2.87	76.25	15.31
Total[a]	144.78	117.91	235.51	498.20	100.00

[a] Same as energy savings column in Table 4.29.

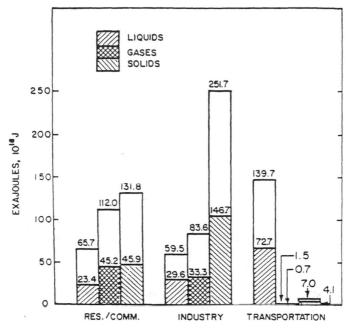

FIGURE 4.3   Global sectoral energy demand by fuel type with and without technology improvements for year 2050. The difference between the two values in each column represents the fuel savings due to technology improvements.

TABLE 4.31
Demand and Savings of Electricity Generated from Nonfossil Fuels with Technology Improvements for Year 2050 (in EJ = 10¹⁸ J)

Fuel type (efficiency, % year, 1975/2050)	Electricity demand without elec. gen. improv.	Electricity generation with elec. gen. improv.	Electricity savings		
			From elec. gen. improv.	From end-use tech. improv.	Total
Nuclear (32/40)	10.74	13.42	2.68	8.02	10.70
Solar (16/21)	10.74	14.10	3.36	8.02	11.38
Hydro (80/90)	28.41	31.96	3.55	21.45	25.00
Total	49.89	59.48	9.59	37.49	47.08

TABLE 4.32
Regional Savings of Electricity Generated from Nonfossil Fuels Due to Technology Improvements for Year 2050
(in EJ = 10^{18} J)

Region	Nuclear			Solar			Hydro		
	End-use tech. improv.	Generation tech. improv.	Total	End-use tech. improv.	Generation tech. improv.	Total	End-use tech. improv.	Generation tech. improv.	Total	Total
U.S.	0.52	1.71	2.22	0.64	1.71	2.35	0.16	1.08	1.24	5.81
Canada/Western Europe	0.98	2.92	3.90	1.23	2.92	4.15	0.35	2.08	2.43	10.48
OECD Pacific	0.43	1.08	1.52	0.54	1.08	1.62	0.09	0.47	0.57	3.71
Former Soviet Union/Eastern Europe	0.20	0.70	0.90	0.26	0.70	0.96	0.48	3.27	3.75	5.61
China et al.	0.15	0.38	0.53	0.18	0.38	0.56	0.62	3.13	3.75	4.84
Middle East	0.06	0.18	0.24	0.06	0.18	0.24	0.06	0.38	0.44	0.92
Africa	0.0	0.0	0.0	0.0	0.0	0.0	0.72	4.40	5.12	5.12
Latin America	0.19	0.66	0.85	0.24	0.66	0.90	0.62	4.29	4.91	6.66
South and East Asia	0.15	0.39	0.54	0.19	0.39	0.58	0.45	2.35	2.80	3.92
All regions	2.68	8.02	10.70	3.35	8.02	11.37	3.55	21.45	25.00	47.07

TABLE 4.33
Regional Replacement of Electricity from Fossil Fuels with Electricity from Nonfossil Fuels and Associated Fossil-Fuel Savings by Fuel Type for Year 2050 (in EJ = 10^{18} J)

Region	Liquids				Gases				Solids			
	Demand	Replace.	Final demand	Fuel savings[a]	Demand	Replace.	Final demand	Fuel savings[a]	Demand	Replace.	Final demand	Fuel savings[a]
U.S.	1.56	0.69	0.87	1.38	3.46	1.52	1.94	3.04	8.17	3.60	4.57	7.20
Canada/Western Europe	1.49	0.97	0.52	1.94	3.43	2.24	1.19	4.48	11.14	7.27	3.87	14.54
OECD Pacific	1.48	0.54	0.94	1.08	1.64	0.60	1.04	1.20	7.01	2.57	4.44	5.14
Former Soviet Union/Eastern Europe	1.52	1.30	0.22	2.60	0.70	0.60	0.1	1.20	4.36	3.72	0.64	7.44
China et al.	1.12	1.12	0.0	2.24	0.51	0.51	0.0	1.02	3.16	3.16	0.0	6.32
Middle East	0.04	0.03	0.01	0.06	0.56	0.40	0.12	0.80	0.63	0.49	0.14	0.98
Africa	0.0	0.0	0.0	0.0	0.0	0.0	0.0	0.0	0.0	0.0	0.0	0.0
Latin America	1.39	1.39	0.0	2.78	0.66	0.66	0.0	1.32	4.09	4.09	0.0	8.18
South and East Asia	0.46	0.42	0.04	0.84	0.52	0.48	0.04	0.96	3.26	3.01	0.25	6.02
All regions	9.06	6.46	2.60	12.92	11.44	7.01	4.43	14.02	41.82	27.91	13.91	55.82

[a] The improved average fossil-fuel conversion efficiency of 50% is used, without considering distribution and transmission losses.

TABLE 4.34
Replacement of Electricity from Fossil Fuels with Energy from Nonfossil Fuels and Associated Fossil-Fuel Savings for Year 2050 (in EJ = 10^{18} J)

Fuel type	Demand from fossil sources	Replacement from nonfossil sources	Final demand from fossil sources	Fuel savings from nonfossil replacement	Fuel savings (%)
Liquids	9.06	6.46	2.60	12.92	15.61
Gases	11.44	7.01	4.43	14.02	16.94
Solids	41.82	27.91	13.91	55.82	67.45
Total	62.32	41.38	20.94	82.76	100.00

4.2.7　Estimating CO_2 Emissions

The amount of CO_2 emitted is the product of the amount of a fossil fuel consumed and the amount of CO_2 emitted per unit quantity of that fuel. The carbon release per unit of energy (i.e., the CO_2 coefficient) for each fossil-fuel type is listed in Table 4.36. Regional carbon releases from the consumption of oil, gas, and coal can be calculated directly from Tables 4.29 and 4.36; the results are shown in Table 4.37. The overall CO_2 emissions, and the reductions, are summarized by region in Table 4.38. Because similar technological improvements are applied in each region, the improvements do not cause much change in the regional distribution of CO_2 emissions for the year 2050. OECD regions produce the major portions of emissions, 21% each for the U.S. and for Canada and Western Europe and 13% for Japan,

TABLE 4.35
Total Fossil-Fuel Demand and Savings Resulting from Technology Improvements and Replacement of Electricity from Fossil Fuels with Energy from Nonfossil Fuels for Year 2050 (in EJ = 10^{18} J)

Fuel type	Fuel demand Without tech. improv.	Fuel demand With tech. improv.	Fuel savings	Percentage of fuel savings	
Liquids	264.84	107.14	157.70	27.14[a]	59.55[b]
Gases	197.10	65.17	131.93	22.71	66.94
Solids	390.54	99.21	291.33	50.15	74.60
Total	852.48	271.52	580.96	100.00	68.15

[a]　With respect to the total fuel savings of 580.96 EJ.

[b]　With respect to the fuel demand with technical improvements.

TABLE 4.36
CO_2 Release Coefficients as
Carbon Release per Unit of
Energy (in Tg/EJ = 10^{12} g/10^{18} J)

Fuel	CO_2 coefficient
Liquids	19.2
Gases	13.7
Solids	23.8

Australia, and New Zealand. The former Soviet Union and Eastern Europe account for 12% and China for 13%, and the rest of the Third World countries together account for 21%.

The overall global CO_2 emissions from fossil-fuel combustion, with and without the effects of technology improvements, for the year 2050 are shown in Table 4.39 and Figure 4.2. These data indicate that in the year 2050, these emissions can be reduced, by technology improvement, to 7.1 petagrams (Pg; 10^{15} g) from 17.1 Pg of carbon (i.e., by 59%). Solids account for the largest share (56%) of the reduction in CO_2 emission, liquids for 28%, and gases for 16%. This distribution by fuel type is in line with that of CO_2 emissions with and without the effects of technology improvements because such improvements were screened on the basis of their impact on reduction of fossil-fuel consumption and thus CO_2 emissions, as discussed in Section 4.2.3.

The reductions of global CO_2 emissions by fuel type are broken down by end-use sector in Table 4.40 and Figure 4.4: they total 47% from industry, 38% from residential and commercial, and 15% from transportation. The data demonstrate one important conclusion of the study: Opportunities for reducing CO_2 emissions are immense — 57% of reduction in CO_2 from gases is in the residential and commercial sector, 62% of reduction in CO_2 from solids is in the industrial sector, and 50% of reduction in CO_2 from liquids is in the transportation sector.

Because the consumption of nuclear, solar, and hydro energy releases no carbon, CO_2 emissions can be further reduced by meeting the need for fossil-fuel-related electricity with the savings from nonfossil-fuel-related electricity. Table 4.41 shows the regional CO_2 emissions with the effects of such replacements. Industrialized countries are still the major emitters of CO_2 because some developing countries have an abundance of hydropower. The overall reduction of CO_2 emissions due to technology improvements and replacement of fossil power with nonfossil power is shown, by fuel type, in Table 4.42. Comparison of Tables 4.39 and 4.42 shows that replacement of fossil power with nonfossil power brings the reduction of CO_2 emission from 59% to 68%, from 60% to 75% for solids, from 55% to 59% for liquids, and from 60% to 62% for gases. The reductions should be larger if the CO_2 released by the fossil fuels consumed in producing the fuels saved is taken into account.

TABLE 4.37
Regional CO$_2$ Emissions from Fossil-Fuel Combustion by Fuel Type With and Without Effects of Technology Improvements for Year 2050 (in Tg C = 10^{12} g C)

Region	Liquids			Gases			Solids		
	Without tech. improv.	With tech. improv.	Reduction	Without tech. improv.	With tech. improv.	Reduction	Without tech. improv.	With tech. improv.	Reduction
U.S.	1409	648	761	542	221	320	1407	590	817
Canada/Western Europe	966	429	537	561	229	332	2000	837	1163
OECD Pacific	509	233	275	313	127	186	1196	524	672
Former Soviet Union/Eastern Europe	619	278	341	505	203	302	1147	431	716
China et al.	455	209	247	274	106	168	1464	517	947
Middle East	133	60	73	124	50	74	113	49	64
Africa	164	74	90	99	38	61	286	85	201
Latin America	601	269	332	179	71	108	779	314	465
South and East Asia	229	105	124	104	41	63	902	343	559
All regions	5085	2305	2780	2700	1085	1615	9295	3690	5605

TABLE 4.38
**Total CO_2 Emissions from Fossil-Fuel Combustion by Regions
With and Without the Effects of Technology Improvements
for Year 2050 (in Tg C = 10^{12} g C)**

Region	CO$_2$ emissions					
	Without tech. improv.	%	With tech. improv.	%	CO$_2$ emissions reduction	%
U.S.	3,358	19.66	1,460	20.62	1,898	18.98
Canada/Western Europe	3,527	20.65	1,495	21.11	2,032	20.32
OECD Pacific	2,018	11.81	885	12.50	1,133	11.33
Former Soviet Union/Eastern Europe	2,271	13.30	912	12.88	1,359	13.59
China et al.	2,193	12.84	831	11.73	1,362	13.62
Middle East	370	2.17	159	2.25	211	2.11
Africa	549	3.21	197	2.78	352	3.52
Latin America	1,559	9.13	654	9.23	905	9.05
South and East Asia	1,235	7.23	489	6.90	746	7.46
All regions	17,080	100.0	7,082	100.0	9,998	100.0

TABLE 4.39
**CO_2 Emission from Fossil-Fuel Combustion With and Without
Effects of Technology Improvement for Year 2050 (in Pg C = 10^{15} g C)**

Fuel	CO$_2$ emissions		Reduction of CO$_2$ emissions	Percentage CO$_2$ reduction	
	Without tech. improv.	With tech. improv.			
Liquids	5.08	2.30	2.78	27.8[a]	54.72[b]
Gases	2.70	1.08	1.62	16.2	60.00
Solids	9.29	3.69	5.60	56.0	60.28
Total	17.07[c]	7.07	10.00	100.0	58.58

[a] With respect to total reduction of CO_2 emission, 10.00 Pg C.

[b] With respect to CO_2 emissions without technical improvements.

[c] 4.5×10^{15} g C for Case B of Edmonds and Reilly.[6]

TABLE 4.40

Reductions of Global CO_2 Emissions Due to Technology Improvements by End-Use Sector for Year 2050 (in 10^{15} g C)

	Liquids	Gases	Solids	Total	%[a]	%[b]
Residential/commercial	0.81	0.92	2.04	3.77	37.7	22.09
Industry	0.57	0.69	3.49	4.75	47.5	27.83
Transportation	1.40	0.01	0.07	1.48	14.8	8.67
Total[c]	2.78	1.62	5.60	10.0	100.0	58.58

[a] Based on total reduction of CO_2 emissions.

[b] Based on total global CO_2 emission of 17.07×10^{15} g C.

[c] Same as reduction of CO_2 emissions column in Table 4.39.

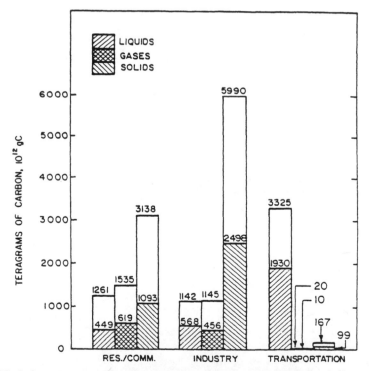

FIGURE 4.4 Global sectoral CO_2 emissions by fuel type with and without technology improvement for year 2050. The difference between the two values in each column represents the reduction in CO_2 emissions due to technology improvements.

TABLE 4.41
Total CO_2 Emissions from Fossil-Fuel Consumption by Region
With and Without Effects of Technology Improvements
and Replacement of Fossil-Related Power with Nonfossil-Related Power
for Year 2050 (in Tg C = 10^{12} g C)

Region	CO_2 emissions					
	Without tech. improv.	%	With tech. improv. + replace.	%	CO_2 emissions reduction	%
U.S.	3,358	19.66	1,219	22.95	2,139	18.18
Canada/Western Europe	3,527	20.65	1,050	19.77	2,477	21.05
OECD Pacific	2,018	11.81	725	13.65	1,293	10.99
Former Soviet Union/Eastern Europe	2,271	13.30	669	12.59	1,602	13.61
China et al.	2,193	12.84	624	11.75	1,569	13.33
Middle East	370	2.17	124	2.33	246	2.09
Africa	549	3.21	196	3.69	353	3.00
Latin America	1,559	9.13	388	7.30	1,171	9.95
South and East Asia	1,235	7.23	317	5.97	918	7.80
All regions	17,080	100.00	5,312	100.00	11,768	100.00

TABLE 4.42
CO_2 Emission from Fossil-Fuel Combustion With and Without
Effects of Technology Improvements and Replacement of Fossil-Related
Power with Nonfossil-Related Power for Year 2050 (in Pg C = 10^{15} g C)

Fuel	CO_2 emissions		Reduction of CO_2 emissions	Percentage CO_2 reduction	
	Without tech. improv.	With tech. improv. + replace.			
Liquids	5.08	2.05	3.03	26.00[a]	59.45[b]
Gases	2.70	1.03	1.81	14.37	61.85
Solids	9.29	2.36	6.93	59.63	74.60
Total	17.07	5.44	11.77	100.00	68.07

[a] With respect to total reduction of CO_2 emissions, 11.77 Pg C.

[b] With respect to CO_2 emission without technical improvements.

4.2.8   Estimates of Costs for Implementation of Improved Technologies

To assess the costs of implementing improved energy technologies, one must know (1) what types of technologies will have large impacts on future fuel consumption and associated CO_2 emissions and (2) when the technologies will become mature. Here, rather than dealing with the entire spectrum of improved technologies (Table 4.3), focus is on technologies that have a large impact on future fossil-fuel demands and related CO_2 emissions, which together contribute almost 80% of the overall reduction in CO_2 emissions from technology improvements (Table 4.43).

The largest share of energy savings in the residential and commercial sector comes from space-conditioning technologies — 70 to 80% of the fossil-fuel savings and 30 to 40% of the electricity savings; water-heating technologies account for most of the remaining fossil-fuel savings (about 20%) and for about 15% of the electricity savings. In the industrial sector, efficiency improvements in process heat provide about 80% of the fossil-fuel savings, and improvements in electric drives provide about 80% of the electricity savings. About 80% of the liquid fuel savings in the transportation sector come from automobiles and trucks, with much smaller shares from air, bus, and rail transport. Table 4.43 shows the reductions in CO_2 emissions due to technology efficiency improvements in these energy services: 18% from space conditioning and water heating, 22% from process heat and electric drives, and 7% from automobiles and trucks. The total reduction in CO_2 emissions due to improvements in these technologies is about 46% of the overall CO_2 emissions without technology improvement. This is about 80% of the reduction due to overall technology improvements (i.e., 59%).

Experience has shown that cost estimates for a new technology made early in its developmental stage are often lower, sometimes at least by a factor of two (expressed in constant dollars), than the actual costs of the first commercial version. Although development costs should generally be assessed in connection with a particular technological implementation, the great complexity of new technological

TABLE 4.43
Estimates of Major Contributors Toward CO_2 Emission Reduction in End-Use Sectors Compared with No Improvement in Technology Efficiencies for Year 2050

Energy services/end-use sector	Reduction in CO_2 emissions (%)	Reduction in CO_2 emissions as a percentage of the total 58.58% reduction
Space, water heat/residential and commercial	17.67	30.16
Process heat, electric drive/industry	21.84	37.30
Automobiles, trucks/transportation	6.65	11.35
Total	46.16	78.81

developments and the uncertainties about their market penetration make it desirable to use cost estimates that represent "mature" costs, which normally are significantly lower than the costs of the first commercial version. Also, capital cost estimates are expressed as instantaneous or overnight construction costs (i.e., capital cost estimates at a single point in time) for the year 1980.

Future implementation costs of new technologies cannot be estimated with high confidence because experience with their implementation is lacking. Costs for new technologies are generally available only for a limited number of developed regions and with a high degree of uncertainty. For the developing countries, detailed data on the costs of implementing new technologies are not available, and the lack of detailed data on energy demand and savings for each energy service in terms of fuel types and technologies makes cost estimation even more difficult. Therefore, the assessment was an order-of-magnitude estimate of the cost of implementing new energy technologies in the U.S., based on the capital cost of new technologies available in the U.S.

The capital costs for some typical major energy-saving technologies are shown in Tables 4.44 to 4.47. Table 4.44 shows the capital costs for six typical waste heat recuperators on the basis of furnace input energy savings,[13] with each system assumed to operate 5300 hr/yr. Waste heat recuperators probably represent the most typical available process heat recovery equipment. Other types, including waste heat boilers, heat pumps, bottom-cycle systems and regenerators, and kiln preheaters, can also be applied to industrial processes for metals, glass, cement, pulp and paper, and many other materials. Insofar as investments for recuperators are representative of those for other types of equipment, the average capital costs in Table 4.44 can be taken as generally applicable. Hence, the average capital cost of waste heat recovery equipment used in this study is 2.51 1980 $/GJ/yr.

Table 4.45 lists the capital cost estimates for five typical industrial cogeneration processes under various operating conditions. An average of about 70 kW of electric

TABLE 4.44
Capital Costs of Six Typical Waste Heat Recuperators,
Based on Annual Furnace Input Energy Savings

Type	Typical application	Capital cost (1980 $/GJ/yr)
Metallic radiation	Soaking pits, reheat furnaces	1.51
Ceramic tube	Forge furnaces	1.55
Flue-type convection	Reheat furnaces, aluminum remelt furnaces	1.85
Metallic plate	Aluminum reheat furnaces	2.78
Miniature metallic radiation		
Direct fired burners	Heat treating, forge furnaces	2.46
Radiant tube burners	Heat treating, forge furnaces	4.94
Average		2.51

TABLE 4.45

Incremental Capital Investment (1980 $) for Installation of Electric Power Cogeneration Plants Using By-Product Process Heat[a]

	Process pressure (psi)				
	50	200	400	200	200
	Power unit				
	Steam turbine	Steam turbine	Steam turbine	Gas turbine	Steam/ gas turbine
Heating rate to boiler (GJ/hr)	1.30	1.20	1.14	0.95[b]	1.16[c]
By-product power [kW(e)]	77	64	34	34	148
Boiler cost with power ($)	16,330	15,030	14,260	6,330	15,510
Boiler cost without power ($)	6,330	6,330	12,660	6,330	6,330
Boiler incremental cost ($)	10,000	8,700	1,600	—	9,180
Steam turbine cost (1980 $)	8,320	5,290	3,670	—	5,180
Gas turbine cost (1980 $)	—	—	—	18,140	2,160
Total incremental capital ($)	13,320	13,990	5,270	13,140	26,780
Average incremental capital (1980 $/GJ/hr)		15,660			

[a] All rates are for 1.055 GJ (10^6 Btu) per hour delivered to the process steam.

[b] Heating rate to gas turbine 1.35 GJ/hr.

[c] Heating value to gas turbine 1.59 GJ/hr.

power can be obtained for every million Btu/hr of steam required by industrial processes. Since the operating conditions in Table 4.45 are generally representative of most industrial process steam, the average incremental capital cost of 15.56 1980 $/GJ/hr is used here as a generalized cost for industrial cogeneration technologies. Table 4.46 shows generalized capital cost estimates for industrial combustion systems for various fuel types.[13] Table 4.47 gives the costs of several improved electricity generation technologies such as fluidized bed combustion, gasification com-

TABLE 4.46

Capital Costs of Process Heat Combustion Equipment (Boilers, Furnaces, etc.) and Motors, Based on Input Energy Capacity

Type	Unit capital cost (1980 $/GJ/yr)
Gas	4.72
Oil	4.92
Coal	14.22
Electricity	3.98
Blast furnace	9.86
Motor (electric drive pump, compressor, etc.)	50.0 [1980 $/kW(e)]

TABLE 4.47
Total Capital Costs for New Electricity-Generating Technologies for Mature Plants

Technology	Plant size, MW (no. of units × unit size)	Plant costs[a] (1980 $/kW)	Plant investment[b] (1980 $/kW)	Start-up, land inventory (1980 $/kW)	Capital requirement[b] (1980 $/kW)
Atmospheric fluid bed combustion	2 × 500	725 (bituminous) 700 (subbituminous)	765 735	55 50	820 785
Pressurized fluid bed combined cycle	2 × 650	780	825	50	875
Gasification combined cycle — Texaco gasifier	1 × 1000	940 (current) 880 (advanced)	975 915	55 50	1030 965
Gasification combined cycle — British Gas Council Slagging Gasifier	1 × 1000	730	760	45	805
Gasifier–fuel-cell combined cycle — Texaco Gasifier	1 × 1500	920	950	45	995
Gasifier–gas-fired boiler–combustion engineering gasifier	1 × 1000	985	1045	50	1095
MHD combined cycle	1 × 1000	1175	1295	65	1360
Dispersed fuel cell	3 × 10	555 (phosphoric acid) 540 (molten carbonate)	555 540	40 40	600 580
Advanced liquid fuel combined cycle	2 × 300	405 (distillate) 475 (residue)	410 490	45 40	455 530

[a] On the basis of overnight construction.

[b] Includes allowance for funds during construction.

TABLE 4.48

Capital Costs and Returns from Fuel Savings for Space Conditioning and Water Heading for the Residential and Commercial Sector in the U.S.

Technology	Unit capital costs[a] (1980 \$/GJ/yr)	Total capital costs (10^9 1980 \$)	Return from fuel savings (10^9 1980 \$)	ROI (%)
Space conditioning				
Improved oil burner in new building with conservation package	51.80	349	33.7	9.7
Improved gas burner in new building with conservation package	43.35	437	27.8	6.4
Improved electric heat pump in new building with conservation package	115.53	644	62.5	9.7
Subtotal		1430	124.0	8.67
Water heat				
Oil burner	23.91	28	1.9	6.71
Gas burner	23.91	98	5.2	5.25
Electric heat	23.91	65	20.6	31.48
Subtotal		192	27.7	14.41
Total		1622	151.7	9.35

[a] Based on energy input capacity, from Reference 12.

bined cycle, fuel cell, etc.[15] Power plants are generally subject to economies of scale: as plants become larger, unit capital costs decrease. The average capital requirement per unit of energy used in this study is \$910/kW.

The estimated capital costs for improved technologies in the U.S. are shown in Tables 4.48 to 4.51 for three end-use sectors (residential/commercial, industry, and

TABLE 4.49

Capital Costs and Returns from Fuel Savings for Process Heat and Electric Drive Devices for the Industrial Sector in the U.S.

Technology	Capital investment (10^9 1980 \$)	Return from fuel savings (10^9 1980 \$)	ROI (%)
Waste heat	16.0	13.51	84.44
Cogeneration	8.0	3.92	49.0
Improved combustion equipment	75.0	1.94	2.6
Improved motors (electric drive devices)	21.0	33.70[a]	160.5
Total	120.0	53.07	44.23

[a] On the basis that the industry generates half of its electricity needs and central stations supply the other half.

TABLE 4.50
Capital Costs and Returns from Fuel Savings for Automobiles and Trucks in the Transportation Sector in the U.S.

Type	Number of vehicles[a]	Capital cost[b] (10^9 1980 $)	Returns from fuel savings (10^9 1980 $)	ROI (%)
Light-duty vehicles (cars, light trucks)	3.88×10^8	2330	178	7.6
Heavy-duty trucks	3.0×10^6	180	12	6.9
Total	3.9×10^8	2510	190	7.6

[a] Assumed average miles/yr = 9600 for light-duty and 70,000 for heavy-duty.

[b] Assumed average price = $6000 for light-duty and $60,000 for heavy-duty.

transportation) and for electricity generation. The returns on investment (ROIs), based purely on fuel savings, resulting from implementing improved energy tech-nologies are also shown; these were calculated from the prices of fuels and electric-ity[16] in Tables 4.52 and 4.53. The required capital costs are about $2.2 trillion (1980 U.S. dollars) in the residential and commercial sector with an ROI of 9%, $140 billion in the industrial sector with an ROI of 41%, and $3.3 trillion in the transpor-tation sector with an ROI of 7%.

The global capital costs and ROI from fuel savings for each sector are estimated on the basis of the proportionality of the amount of energy demand and savings from the sectoral aggregate results for the U.S. The results are shown in Table 4.54.

The total capital costs for implementing major energy-efficient technologies for the year 2050 are about $6.0 and $24.0 trillion (1980 U.S. dollars), with resulting ROIs of about 8 and 10%, for the U.S. and the world, respectively. The ROI for the industrial sector is high (about 40%) because the new technologies such as waste heat recuperation, cogeneration, etc. are mainly additions to existing processes for obtaining highly efficient use of process heat. The ROIs for the transportation and

TABLE 4.51
Capital Costs and Returns from Fuel Savings for Improved Electricity-Generating Technologies in the U.S.

Fuel type	Total capital costs[a] (10^9 1980 $)	Return from fuel savings (10^9 1980 $)	ROI (%)
Oil	65.2	7.10	10.89
Gas	144.5	7.86	5.44
Coal	341.3	10.77	3.16
Total	551.0	25.73	4.7

[a] Based on $910/kW unit capital cost and operating factor of 0.65.

TABLE 4.52
U.S. Fossil-Fuel Prices for Year 1980

Fuel type	Price (1980 $)	1980 $/GJ
Oil	26.09 $/bbl	4.50[a]
Gas	2.12 $/thousand ft^3	2.16[b]
Coal	33.69 $/metric ton	1.28
Heating oil	0.978 $/gal	7.06
Heating gas	3.946 $/thousand ft^3	4.03
Gasoline	1.221 $/gal	9.64
Diesel	0.873 $/gal	5.89

[a] No. 6 residual oil price. Refiner acquisition cost of crude oil is $28.07/bbl.

[b] Includes small quantities of coke oven gas, refinery gas, and blast furnace gas.

TABLE 4.53
Costs of Fossil Fuels Delivered to Steam Electric Utility Plants, and Retail Electricity Prices, for Year 1980 in the U.S.

Fuel type	1980 ¢/MBtu	1980 ¢/kWh	1980 $/GJ
Coal	135.2		1.28
Oil	427.9		4.05
Gas	212.9		2.02
Average	189.3		1.79
Electricity			
Residential		5.36	14.83
Commercial		5.48	15.22
Industrial		3.69	10.00
Average		4.73	12.82

TABLE 4.54
Capital Costs and Returns from Fuel Savings Due to Technology Improvements for the U.S. and Worldwide in Year 2050

Sector	Capital costs (10^9 1980 $)		Returns from fuel savings (10^9 1980 $)		ROI (%)	
	U.S.	Global	U.S.	Global	U.S.	Global
Residential/commercial	2,220	11,200	200	1,130	8.8	10.0
Industrial	140	1,380	57	550	40.8	40.1
Transportation	3,330	9,120	230	640	7.0	7.0
Electricity generation	550	2,600	26	120	4.7	4.7
Total	6,240	24,300	513	2,440	8.2	10.0

electricity generation sectors are only 7 and 5%, respectively, because new electricity-generating processes and new automobile production lines have high capital costs. Table 4.54 demonstrates one important conclusion of the present study: The industrial sector is more effective than other sectors in reducing CO_2 emission (providing 50% of the overall reduction in CO_2 emissions), with low capital requirements for technological implementation (requiring 6% of the overall capital costs) and high ROI (40%).

4.3 OBSERVATIONS RELATED TO THE ESTIMATES OF REGIONAL AND GLOBAL CO$_2$ EMISSION REDUCTION DUE TO IMPROVEMENT IN EFFICIENCY OF ENERGY TECHNOLOGY

The information presented in this chapter deals with the physical rather than the sociopolitical aspects of the problem. The results indicate that more efficient end-use and electricity generation technologies, if effectively carried out, could lead to an estimated savings of about 500 EJ (10^{18} J) of fossil fuels and a reduction of 10 Pg (10^{15} g) of carbon emissions by the year 2050. Whether these energy savings and CO_2 reduction will be achieved, and to what extent, is difficult to predict because of uncertainties related to future economic growth, industrial investment activities, and infrastructure changes. The energy-efficient technologies discussed in this study require capital investment either for changes in existing stocks or for construction of new stocks. The rate and structure of economic growth will certainly affect the rate of introduction of new energy-efficient technology and equipment. The turnover in capital stock and thus in energy technology is related to investment activity, since the choice among different new technologies depends on their expected relative costs over their economic lifetimes. The technologies discussed in this study are likely to be absorbed by the market only when their benefit/cost ratios are perceived to be acceptable by investors and/or governments. Thus, the choice of future technologies and efficiencies is a function of costs and of the socioeconomic structure that determines the energy services and processes used to meet end-use sectoral needs, as well as of the supplies of various fuels.

The economic and energy position of the developing countries hinges on the capabilities of the industrialized countries to achieve a certain rate of economic growth, and the developing countries depend increasingly on growing product markets in industrialized countries. Deficit financing to stimulate economic growth in developing countries that need loans and/or official assistance from industrialized countries results in even greater interaction between developing and developed countries. The sociopolitical–economic–energy situation varies significantly among both developing and industrial countries because of large differences in environment, energy supply, labor, and other factors. The energy productivity and the costs of implementation used in this chapter are based mainly on average values available for the U.S. Obviously, the fuel demand and associated cost for a specific energy task in other countries, especially Third World countries, may deviate significantly from the average U.S. value. For a more meaningful concept of the effects of improved technology on energy consumption and CO_2 emissions, an economic–energy system study should be performed for each region, taking into account the relevant sociopolitical conditions, fuel supply and marketing, the environment, and other factors.

The economics of some of the major improved technologies are evaluated, but not the time required for their implementation. The lead time before a new technology has a reasonable market share is usually long. Investigation of market penetration rates is beyond the scope of this chapter, but the potential for higher energy productivity is driven not only by technological progress with time but also by increased cost, and sometimes depletion, of fuel, as well as increasingly important environmental constraints. For these reasons, and others discussed earlier, by the year 2050 most of the present and emerging technologies will probably be fully developed up to their achievable efficiencies and will be available globally.

It is well recognized that the calculated global fossil-fuel savings may represent an upper limit. Actually achievable savings could be lower. Nevertheless, the large margins for efficiency improvement indicate an immense potential for energy savings and reduction in CO_2 emissions. Since the energy–CO_2 issue is a global problem, and efficiencies vary widely among countries, international cooperation is needed to narrow the efficiency gap by promoting rapid and widespread absorption of energy-efficient technologies. Some developing countries already tend to employ the most advanced technologies offered by developed countries because of environmental as well as economic benefits. International cooperation includes joint research and development efforts, exchange of information and results of experience, promotion of trade in energy-efficient technologies, etc.

The potential for further reducing CO_2 emission by meeting part or all of the demand for fossil-fuel-related electricity with nonfossil-fuel electricity has also been considered. Global electricity requirements are expected to increase, primarily because of growth of the service business and increased process automation. Electricity generation is an energy-intensive process, but electricity at the end use is not only convenient but also highly energy efficient. For instance, the increasing use of electricity in computer-controlled systems has significantly increased the productivity of equipment such as cement furnaces, blast furnaces, etc. Electrification thus has important consequences for lifestyles, labor productivity, economic growth, and socioeconomic development. Most nonfossil energy such as nuclear, solar, and hydro, which releases no CO_2, is utilized in electricity generation. If the current trend toward more electrification continues, significant changes in the structure of electric power production are probable (i.e., increases in the share of nonfossil energy). Such a change in the pattern of fuel use for electricity generation will have a major impact in reducing CO_2 emissions.

Supplies of energy from new and renewable sources such as wind, geothermal, etc., although not discussed in this study, are expected to increase in the future. These are most effective when used in an ancillary capacity in an integrated energy supply system. Current experience and evaluation indicate that their future impact on overall energy consumption and CO_2 emissions will be slight compared with that of the major nonfossil energy sources, but the potential for using them together with other forms of energy to reduce costs and conserve nonrenewable energy sources in the future should be further investigated.

Uncertainties regarding achievable efficiencies, market penetration, and costs of new and improved technologies are inherent in estimating the structure of global energy demand and the resultant CO_2 emissions. It is recognized that the projected regional and the associated global energy demand by fuel type, obtained from Edmonds and Reilly's long-term economic-energy model, depends strongly on the input values of exogenous parameters such as technological efficiency, labor productivity, and the relative prices of various fuels. The model can be used to explore the consequences of alternative technological options, but it sheds no light on technological questions such as (1) which technologies have the highest probability of being leaders in increasing energy productivity and to what extent, (2) what the compatibility of technologies in various regions will be, (3) whether economic developments will permit future penetration of new or improved technologies in these regions, and (4) what degree of uncertainty is associated with both the levels of technological improvement and the anticipated market penetration. These questions are outside the scope of the present chapter, but a more detailed investigation of the effects of technological advances on CO_2 emissions will require studies to provide more meaningful input values of technological parameters in the model.

Many other investigations have resulted in suggestions for mitigating possible global CO_2 effects (e.g., stack gas scrubbing in fossil-fuel power plants). This preliminary study indicates that, of all of these, the improved energy technology approach is by far the most realistic and cost effective and, furthermore, that worldwide implementation of energy-efficient technologies not only will delay or even possibly eliminate the CO_2 warming effect but also will provide an economic ROI.

4.4   CONCLUSIONS

1. The large margins of efficiencies between present and future improved energy technologies, and the wide variations in efficiency levels among countries, indicate considerable room for major long-term reductions in energy consumption through technological improvements. This chapter makes a first-order evaluation of the potential for more effective use of energy globally and the consequent reduction of CO_2 emissions, with use of the global energy demand in the year 2050 (Table 4.4) obtained from the IEA Energy–CO_2 Model.

2. Savings achievable by more efficient end-use and electricity generation technologies, if effectively implemented, are estimated at about 145 EJ (10^{18} J) of liquids fuels, 118 EJ of gas fuels, and 236 EJ of solid fuels, with an overall fossil-fuel savings of 500 EJ, which is about 58% of the fossil-fuel demand without technology improvements for the year 2050.

3. The distribution of these fossil-fuel savings by end-use sector is 175 EJ (39%) in the residential and commercial sector, 227 EJ (46%) in the industrial sector, and 76 EJ (15%) in the transportation sector. The breakdown by fuel type for each sector is 22% liquids, 34% gases, and 44% solids in the residential and commercial sector; 13% liquids, 22% gases, and 65% solids in the industrial sector; and 95% liquids, 1% gases, and 4% solids in the transportation sector. Thus, improved technologies offer many opportunities for large savings in solid fuels in the industrial sector and in liquid fuels in the transportation sector.

4. The regional breakdown of the overall fossil-fuel savings of 500 EJ is 97 EJ (19.5%) from the U.S., 101 EJ (20%) from Canada and Western Europe, 56 EJ (11%) from the OECD Pacific, and the remaining 245 EJ (49%) from the former Soviet Union, China, and all the other developing countries. In the U.S., 35% of the savings is derived from solid fuels, 41% from liquids, and 24% from gases. In Canada and Western Europe, 48% is from solids, 28% from liquids, and 24% from gases. In the OECD Pacific, 50% is from solids, 26% from liquids, and 24% from gases. In the developing regions, 51% is from solids, 26% from liquids, and 23% from gases. Hence, except in the U.S., the general global structure of fuel savings is 50% from solid fuels, 26% from liquids, and 24% from gases. Because the U.S. has a larger demand for liquid fuels, especially in the transportation sector, than all the other regions, it has a higher potential for liquid fuel savings through technology improvement.

5. The associated overall reduction of carbon emission in the year 2050 is 10 Pg (10^{15} g), that is, carbon emissions could be reduced from 17 Pg without efficiency improvements to 7 Pg with the efficiency improvements discussed in this study — a 59% reduction. The largest share of the reduction is from solids, 5.6 Pg C, with liquids contributing 2.8 and gases 1.6 Pg C.

6. The distribution of the overall reduction of CO_2 emissions by end-use sector is 3.8 Pg C (38%) from the residential and commercial sector, 4.7 (47%) from the industrial sector, and 1.5 (15%) from the transportation sector. In the residential and commercial sector, overall CO_2 emissions from gas fuels can be reduced by 57%; in the industrial sector, overall CO_2 emissions from solid fuels can be reduced by 62%; and in the transportation sector, overall CO_2 emissions from liquid fuels can be reduced by 50% with implementation of technology efficiency improvements.

7. The regional breakdown of the overall reduction of 10 Pg C emissions is 1.9 Pg C (19%) from the U.S., 2.0 (20%) from Canada and Western Europe, 1.1 (11%) from the OECD Pacific, and the remaining 5.0 Pg C (50%) from the former Soviet Union, China, and all the other developing countries. Hence, the developing regions have the same potential for reducing CO_2 emissions as do the developed regions in the year 2050 through technology efficiency improvements.

8. The largest share of energy savings in the residential and commercial sector is from space-conditioning technologies, about 75% of the fossil-fuel savings and 35% of the electricity savings, with water-heating technologies accounting for most of the remaining fossil-fuel savings (20%) and about 15% of the electricity savings. In the industrial sector, about 80% of the fossil-fuel savings are from improvements of efficiencies in process heat, and about 80% of the savings in electricity are from improvements in electric drives. Automobiles and trucks account for about 80% of the liquid fuel savings in the transportation sector. Overall CO_2 emission in the year 2050 could be reduced by 46% compared to that without technology improvement, through improvements of these technologies; this is about 80% of the reduction due to general technology improvements.

9. The potential for further fuel savings in electricity generation was also considered: use of the savings in electricity generated from nonfossil energy (i.e., nuclear, solar, and hydro) due to technology improvement to replace demand for electricity generated from fossil fuels produces additional savings of 13 EJ of liquids, 14 EJ of gases, and 56 EJ of solids. Combining the improvements in end-use and electricity generation technologies with such replacement raises the total fossil-fuel savings in the year 2050 to 580 EJ, which is 68% of the fuel demand without the technology improvements. The associated reduction of CO_2 emission is 11.3 Pg C. This means that CO_2 emissions in the year 2050 could be reduced by 68%, compared with 59% without such replacement. If the current trend toward more electrification continues, then, since electricity generation is energy intensive, generation from nonfossil fuels will probably have a significant impact on the future structure of energy use.

10. The actual savings of fossil fuels should be larger than the estimates given if the energy needed for producing and transporting the saved energy is taken into account, and the actual reduction of CO_2 emission should be correspondingly larger.

11. An order-of-magnitude assessment of the costs of implementing improved energy technologies is made on the basis of instantaneous or overnight construction of the matured technologies. The total capital costs for technology implementation for the year 2050 are about $6.0 and $24.0 trillion (1980 U.S. dollars), resulting in total ROI based purely on fuel savings of about 8 and 10% for the U.S. and the world, respectively. The industrial sector has more potential than the other sectors for reducing CO_2 emissions (50% of the overall reduction) with low capital requirements for technology implementation (6% of overall capital requirements) and high ROI (40%). These ROI results provide an incentive for the pursuit of improvement energy technology and show that such improvement is a worthwhile goal regardless of the CO_2 emission problem.

12. The rate and extent of achieving the potential fuel savings and associated reduction of CO_2 emissions are subject to socioeconomic–political considerations which are outside the scope of this study. To provide a more meaningful picture of the effects of technologies on CO_2 emissions, subjects of uncertainty such as achievable efficiencies, market penetration, and costs of new and improved technologies should be investigated in future studies.

Undoubtedly, new and improved technologies currently unforeseen will develop in the decades ahead, and these should bring additional economic savings and further reduce CO_2 emissions.

APPENDIX 4.1: EFFECTS OF TECHNOLOGY ON ENERGY CONSUMPTION

The impact of emerging and existing technologies with efficiency improvement on energy consumption in the U.S. for the year 2050 was assessed by detailed technological analyses of the energy services used in the major energy-consuming sectors: residential and commercial, industry, and transportation. Hence, the energy technologies considered here are mainly those that provide sectoral energy end use or services. The energy demand of various services by fuel type, before introduction of efficiency improvement, for the three end-use sectors in the U.S. for the year 2050 is given in Tables 4.6, 4.7, and 4.9.

Residential and Commercial Sector

Energy Savings of Space Heat

Improved Building Design and Construction

As a result of improved building design and construction, by the year 2000 residential buildings will require 50% less and commercial buildings 60% less energy for space heating than those built in the mid-1970s. Much of the potential energy savings is expected to be realized by the year 2050, when present housing stock will have been largely replaced.

About 55% of space heat was consumed by residential buildings in the mid-1970s and the remainder by commercial buildings. With this allocation assumed to continue through the year 2050, the energy savings in space heating from improved design and construction will be 54.5% (i.e., 50% × 55% + 60% × 45% = 54.5%). Thus, the space-heating energy need in the year 2050 for new buildings will be only 45.5% of that in the mid-1970s, broken down as follows:

Liquids	6.74 EJ × 45.5%	= 3.07 EJ
Gases	10.0 EJ × 45.5%	= 4.55 EJ
Solids	0.69 EJ × 45.5%	= 0.31 EJ
Electricity	2.63 EJ × 45.5%	= 1.20 EJ

Improved Space-Heating Equipment

Improved combustion and heat transfer can result in gas and oil combustion heating systems or devices that use 30% less gas or oil than conventional designs. The final energy need for space heating, with savings due both to new buildings and higher burner efficiencies,[7] are calculated as follows:

Fuel (efficiency, %, 1975/achievable)		Fuel need (EJ)
Liquids	(46/70)	$3.07 \times 46\%/72\% = 1.96$
Cases	(50/72)	$4.55 \times 50\%/72\% = 3.16$
Coal	(44/70)	$0.31 \times 44\%/70\% = 0.19$

An electric heat pump for space heating operated at 45°F ambient provides about two to three times as much heat as an electric resistance heater with 1980 technology. A heat pump that has a COP of 3 for an ambient temperature of 45°F will have a COP of about 4 for a groundwater source temperature of 55°F. When the outdoor temperature falls below a certain value (e.g., 30°F), the heat pump becomes increasingly inefficient, and a backup heat source (e.g., electric resistance heater) is needed, but further research on the development of cascading or compound compression cycles will eliminate the need for resistance heat at all outdoor temperatures.

With backup electric resistance heat required sometimes, the electricity needed for space heating is estimated to be reducible to one-third of that for electric resistance heating in a new building. Thus, the electricity need is 1.2 EJ/3.0 = 0.4 EJ.

Because of the constraints of the built environment, the energy saving from district heating was neglected. As buildings become better insulated and combustion heating systems more efficient, heat demands for space heating become less intensive, and therefore the economics of district heat become less clear and the energy savings less attractive. Neglecting district heating avoids overestimation of the energy savings in space heating.

Air Conditioning

Typical air conditioners have a COP near 2. With improved heat exchangers to reduce the effective temperature difference across which air conditioners must operate and use of absorption air conditioners, by the year 2050 the electricity need for air conditioning is expected to be reduced by 40%. Thus, the estimated electricity need for air conditioning for the year 2050 is $2.95 \times 45.5\% \times 0.62 = 0.83$ EJ.

Water Heating

Improved water heating equipment:

Fuel (efficiency, %, 1975/achievable)[7]		Fuel need (EJ)
Liquids	(50/65)	$1.19 \times 50\%/65\% = 0.92$
Gases	(50/70)	$4.11 \times 50\%/70\% = 2.83$

Since heat pump water heaters use 50% less energy than electric resistance water heaters,[8] the estimated electricity need for water heating is 2.74 × 50% = 1.37 EJ.

With insulation of storage tanks and pipes and heat recovery from wastewater in many commercial installations, the energy needed for water heating should be even less, and the savings shown are probably feasible by the year 2050.

Lighting

About two-thirds of mid-1970s lighting energy is used by incandescent lamps and one-third by fluorescent and vapor lamps, but the latter supply about two-thirds of the useful light as they provide roughly four times as much per unit energy. Energy savings of 50% in lighting systems can be achieved by increasing their efficiency by applying magnetic fields to fluorescent lamps, using advanced control strategies, and optimally integrating lighting components into the total building system.[8,9] Thus, the estimated electricity need for lighting is 5.63 × 50% = 2.82 EJ.

Miscellaneous Electric (Appliances, Elevators, etc.)

Means for improving appliance efficiencies[8] include more efficient motors, better electronic devices, more efficient heat pumps in refrigerators and freezers, and heat insulation in refrigerators, freezers, dryers, ovens, etc., which can save about 50% of the electricity consumption. Electric drives[2] for elevators, escalators, and service automation can be made about 35% more efficient by using better motors with appropriate sizing and control electronics. A conservative estimate of the energy improvements for miscellaneous electricity usage of about 45% is assumed, even though further improvement would be possible. Thus, the electricity need for these services is 4.83 × 55% = 2.66 EJ.

The results of the above calculations for the residential and commercial sector are shown in Table 4.6.

Industry

Roughly 60% of U.S. industrial energy is consumed by a few energy-intensive processes, especially in iron and steel, chemical and petrochemical, and cement and nonferrous metal production. By end use, the largest share of industrial energy goes for process heat and smaller shares for mechanical power, space heating, and lighting. New technologies such as dielectric and microwave heating, electron beam welding, electric foil heating elements, and laser welding are expected to increase industrial energy productivity in the future.

Process Heat

The energy need for process heat can be reduced by technologies that recuperate waste heat and optimize process flow, which are conservatively assumed to save about 8 EJ of energy.[8] This saving is allocated to fuel types according to their share of process heat production:

Liquids	$8 \times 2.90/12.63 = 1.84$ EJ
Gases	$8 \times 3.16/12.63 = 2.00$ EJ
Solids	$8 \times 3.60/12.63 = 2.28$ EJ
	$8 \times 2.97/12.63 = 1.88$ EJ

With improved combustion taken into account, the fuel need for process heat is as follows:

Fuel (efficiency, %, 1975/achievable)[7]		Fuel need (EJ)
Liquids	(64/82)	$(2.90 - 1.84) \times 0.68/0.82 = 0.88$
Gases	(64/82)	$(3.16 - 2.00) \times 0.64/0.82 = 0.91$
Solids	(70/80)	$(3.60 - 2.29) \times 0.70/0.80 = 1.15$
	(24/40) (blast furnaces)	$(2.97 - 1.88) \times 0.24/0.40 = 0.75$

Electric Drive

Industrial electric drives include pumps, compressors, fans, conveyors, presses, machine tools, and so on. The average practical conversion efficiency of electricity into motive power at the point of use is estimated at 40 to 60%. It can be improved by using improved motors and torque converters or clutched flywheels, operating motors from alternating current synthesizers or other power factor controls, and so on, to save about one-third of the electricity. Thus, the final electricity need for electric drive is $13.25 \times 2/3 = 8.33$ EJ.

Aluminum Electrolytic Process

Electricity is consumed in reducing alumina (Hall–Heroult process) or aluminum chloride (Alcoa process) to primary aluminum. The Alcoa process uses about 30% less energy than the Hall–Heroult process, which was the one mostly used in the mid-1970s.[10] Plasma arc reduction of alumina or aluminum chloride offers further electricity savings. Thus, a conservative estimate of electricity savings achievable with best technologies by the year 2050 is 33%, and the electricity need is $1.20 \times 2/3 = 0.8$ EJ.

Iron and Steel

The savings in electricity usage in iron and steel making by continuous casting, electric arc furnace improvements, etc. are estimated as at least 25%[10,14] by the year 2050, and the electricity need is $0.76 \times 75\% = 0.57$ EJ.

Nonenergy Petrochemical Use (Feedstock)

Fossil feedstock comprises petrochemicals used to produce synthetic fibers, plastics, fertilizers, and other chemicals. Energy savings resulting from more efficient use of material (i.e., using less material per unit of product, increasing product durability, increasing product reuse and recycling, etc.) are estimated at about 20%. Thus, feedstock needs are as follows:

$$
\begin{array}{ll}
\text{Liquids} & 2.16 \times 80\% = 1.73 \text{ EJ} \\
\text{Gases} & 0.30 \times 80\% = 0.24 \text{ EJ} \\
\text{Solids} & 0.16 \times 80\% = 0.13 \text{ EJ}
\end{array}
$$

The results of calculations for the industrial sector are shown in Table 4.7.

Transportation Sector

Automobiles

The significant improvement in fuel consumption by automobiles through size and weight reductions, vehicle design changes, improved engines, electronic control of engine and transmission, etc. is conservatively estimated to bring the fuel economy for the whole automobile fleet to no less than 40 mpg by the year 2050. Hence, the fuel demand by automobiles is $26.0 \times 13.6/40.0 = 8.84$ EJ.

Air Transport

Technical measures such as high-bypass turbofan engines, electronic controls to maintain maneuverability, gust tolerance, and weight reduction by the use of high-strength composite materials could improve fuel efficiency from an average of 20 passenger miles per gallon in the mid-1970s to 40 in the year 2050. Hence, the fuel demand is estimated to be $5.82 \times 20/40 = 2.91$ EJ.

Trucks

With allowance for higher weights, similar means for raising efficiency can be applied to trucks as to automobiles. The average fuel efficiency of trucks could be increased by 40% from an average of 4.5 mpg in 1975 to 6.5 in 2050. The fuel demand for trucks is therefore $13.64 \times 4.5/6.5 = 9.44$ EJ.

Bus and Rail

The means of raising efficiency for buses are the same as for automobiles and trucks. For rails, technologies such as regenerative braking and lighter rolling stock could lower the energy need by about 40% by the year 2050.[2] Thus, the energy need is $2.70 \times 60\% = 1.62$ EJ.

For electric rail, improvements including better electric motors and aerodynamic drag reduction can lower the energy need by at least 30% by the year 2050. Thus, the electricity need is $0.11 \times 70\% = 0.08$ EJ.

Ship

Nearly 40% energy savings could be achieved through measures such as variable-pitch propellers, improved diesel engines, etc. Thus, the fuel need is $2.38 \times 60\% = 1.43$ EJ.

The results of the calculations for transportation are summarized in Table 4.9.

APPENDIX 4.2: CALCULATION OF SAVINGS OF FOSSIL FUELS DUE TO IMPROVEMENTS IN ELECTRICITY-GENERATING TECHNOLOGIES

The amount of fossil fuel needed to satisfy the electricity demand, neglecting transmission and distribution losses, is

$$\text{Fuel consumption} = \frac{\text{Electricity demand}}{\text{Conversion efficiency}}$$

As conversion efficiency improves, less fuel is needed to satisfy the same electricity demand, and the savings can be expressed as follows:

$$\text{Fuel savings} = \text{Fuel need before eff. improvement}$$

$$\times \left(\frac{\text{Improved eff.}}{\text{Initial eff.}} - 1 \right)$$

Example:

Global electricity demand	41.82 EJ
Initial conversion efficiency	33%
Improved conversion efficiency	50%

$$\text{Fuel need before efficiency improvements} = \frac{41.82}{0.33} = 126.73 \text{ EJ}$$

$$\text{Fuel savings} = 126.73 \left(\frac{0.5}{0.33} - 1 \right) = 43.09 \text{ EJ}$$

Fuel need with electricity-generating improvements = 126.73 − 43.09 = 83.64 EJ

Improvements in reducing transmission and distribution losses would further increase fuel savings. An advanced technology for high-load areas is superconducting cables for reducing electrical resistance losses.

APPENDIX 4.3: CALCULATION OF INCREASE IN ELECTRICITY GENERATION FROM NONFOSSIL FUEL DUE TO IMPROVEMENTS IN ELECTRICITY-GENERATING TECHNOLOGIES

The electricity generated from nonfossil fuels such as nuclear, solar, and hydro is calculated as

$$\text{Electricity generated} = \text{Fuel supply} \times \text{Conversion efficiency}$$

As conversion efficiency improves, a given amount of nonfossil fuel generates more electricity and the increase can be expressed as follows:

$$\text{Increase in elec. gen.} = \text{Elec. gen. before eff. improv.}$$

$$\times \left(\frac{\text{Improved eff.}}{\text{Initial eff.}} - 1 \right)$$

Example:

Global electricity generated from nuclear energy = 10.74 EJ
Initial conversion efficiency = 32%
Improved conversion efficiency = 40%

$$\text{Increase in elec. gen.} - 10.74 \times \left(\frac{0.4}{0.32} - 1 \right) = 2.68 \text{ EJ}$$

APPENDIX 4.4: ESTIMATES OF IMPLEMENTATION COSTS OF IMPROVED ENERGY TECHNOLOGIES

The capital costs of implementing the improved technologies for energy services were calculated by multiplying the unit capital requirements per unit energy services by the energy demand for the services. The ROIs, purely on the basis of energy savings alone, were calculated by dividing the product of unit energy price for services and the corresponding amount of energy savings by the associated capital costs. The calculations were first performed for those major improved technologies listed in Table 4.3 for each end-use sector in the U.S. The order-of-magnitude assessments of the overall capital costs of implementing improved technologies and the ROIs from fuel savings alone for the U.S. and the world were then made by extrapolating on the basis of the proportionality of the energy demands and savings from the above-calculated aggregate results. Capital cost estimates expressed here are instantaneous or overnight construction costs in constant 1980 U.S. dollars when implemented in 2050.

Residential/Commercial Sector

Space Conditioning

Assuming the unit capital costs of space conditioning in Table 4.48 for residential buildings are representative of commercial buildings, the capital cost of implementing improved space conditioning by fuel types can be estimated, given the energy demands for the service in Table 4.6, as follows:

Liquids	6.74 EJ ×	$51.8/GJ =	$349 × 10^9
Gases	10.0 EJ ×	$43.35/GJ =	$437 × 10^9
Electricity	5.58 EJ ×	$115.53/GJ =	$644 × 10^9
			$1430 × 10^9

Given the annual energy savings resulting from space-conditioning technology improvements in Table 4.6 and the associated unit energy costs in Tables 4.52 and 4.53, the ROIs are calculated as follows:

Liquids	4.78 EJ ×	$7.06/GJ/$349 × 10^9 =	33.7/349	= 9.7%
Gases	6.84 EJ ×	$4.03/GJ/$437 × 10^9 =	27.6/437	= 6.4%
Electricity	4.15 EJ ×	$15.05/GJ/$644 × 10^9 =	62.5/644	= 9.7%
		Total ROI =	123.8/1430	= 8.7%

Water Heat

With the energy demand for water heat in Table 4.6 and the associated unit capital costs for the service in Table 4.48, the capital investments needed for improved water heat technologies are as follows:

Liquids	1.19 EJ × $23.91/GJ = $28.4 × 10^9
Gases	4.11 EJ × $23.91/GJ = $98.3 × 10^9
Electricity	2.74 EJ × $23.91/GJ = $65.5 × 10^9
	$192.2 × 10^9

Following the same procedure as for space conditioning, the ROI for water heat is

Liquids	0.27 EJ ×	$7.06/GJ/$28.4 × 10^9 =	1.91/28.4	= 6.7%
Gases	1.28 EJ ×	$4.03/GJ/$98.3 × 10^9 =	5.16/98.3	= 5.3%
Electricity	1.37 EJ ×	$15.05/GJ/$65.5 × 10^9 =	20.6/65.5	= 31.5%
		Total ROI =	27.67/192.2	= 14.41%

Thus, the total capital cost and ROI in terms of constant 1980 U.S. dollars and water heat in the U.S. for the year 2050 are $1622 × 10^9 and $151 × 10^9, respectively. On the basis of the proportionality of the energy demand and savings, the overall capital costs and ROI for the residential/commercial sector in the U.S. and the world are estimated as follows:

U.S.	Capital costs	$1622 × 10^9 × 41.61 EJ/30.36 EJ = $2223 × 10^9
	ROI	$151 × 10^9 × 24.22 EJ/18.69 EJ/$2223 × 10^9 = 8.8%
Global	Capital costs	$2223 × 10^9 × 209.74 EJ/41.61 EJ = $11,205 × 10^9
	ROI	$196 × 10^9 × 138.77 EJ/24.22 EJ = 10.0%

Industry

Waste Heat Recovery

With the average capital costs per annual furnace input energy savings for waste heat recuperators given in Table 4.44 and the associated annual energy savings of about 6.2 EJ, the capital costs for implementing imported waste heat recovery technologies are

$$\$2.51/GJ \times 6.2 \; EJ = \$15.6 \times 10^9$$

The ROI is calculated as follows:

Liquids	1.43 EJ × \$4.5/GJ	=	\$6.42 × 10^9
Gases	1.55 EJ × \$2.16/GJ	=	\$3.35 × 10^9
Solids	3.22 EJ × \$1.28/GJ	=	\$4.12 × 10^9

Total ROI = \$13.89 × 10^9/\$15.6 × 10^9 = 89.0%

Cogeneration

The average incremental capital investment for cogeneration processes is \$15,660/GJ/hr, as shown in Table 4.45. For every 10^6 Btu/hr of steam required by the general industrial processes, an average of about 70 kW(e) can be generated. This by-product power displaces about 700,000 Btu/hr of fuel heating value supplied from the central station fuels. The net fuel saving for every million Btu/hr input (fuel saved by utilities less the incremental fuel consumed by the industry for power generation) is about 4.0×10^5 Btu/hr. Thus, in terms of energy savings, the incremental capital cost of cogeneration processes is \$39,150/GJ/hr. With the energy savings of about 1.8 EJ, the capital investment required for cogeneration processes is

$$\$39,150/GJ/hr \div 8760 \; hr/yr \times 1.8 \; EJ = \$8.0 \times 10^9$$

The ROI is as follows:

Liquids	(1.84 − 1.43) EJ × \$4.50/GJ	= \$1.86 × 10^9
Gases	(2.00 − 1.55) EJ × \$2.16/GJ	= \$0.97 × 10^9
Solids	(4.16 − 3.22) EJ × \$1.28/GJ	= \$1.20 × 10^9

Total ROI = \$4.03 × 10^9/\$8.0 × 10^9 = 50%

Improved Combustion Equipment

With the capital costs on the basis of input capacity of combustion equipment for process heat in Table 4.46 and the energy demands by the fuel types in Appendix 4.1, the capital investment for the improved combustion equipment is estimated to be about \$75 × 10^9.

The ROI is as follows:

Liquids	$0.18 \text{ EJ} \times \$4.5/\text{GJ}$	$= \$0.81 \times 10^9$
Gases	$0.25 \text{ EJ} \times \$2.16/\text{GJ}$	$= \$0.54 \times 10^9$
Solids	$0.51 \text{ EJ} \times \$1.28/\text{GJ}$	$= \underline{\$0.65 \times 10^9}$

$$\text{Total ROI} = \$2.00 \times 10^9/\$75 \times 10^9 = 2.7\%$$

Electric Drives

The input required to make cost estimates for the electric drive devices such as pumps include capacity (gpm), different pressure (psi), suction pressure (psi), system temperature (°F), and casing material. Since the electric drive devices are employed in a wide range of capacities, pressures, temperatures, and various materials in various industrial processes, for simplified calculations, the unit capital cost of $50/kW(e) is used here for the motor. Thus, the overall capital cost for the electric drives is

$$\$50/\text{kW(e)} \div 8760 \div 3.60 \times 10^6 \text{ J/kWh} \times 13.25 \text{ EJ} = \$21 \times 10^9$$

Assuming that the industry generates half of its electricity need and central stations supply the other half, the ROI for electric drives is

$$(2.21 \text{ EJ} \times 10.0/\text{GJ} + 2.21 \text{ EJ}/0.34 \times 1.79/\text{GJ})/\$21 \times 10^9 = 33.7/21 = 160.5\%$$

Thus, the total capital costs and the ROIs in terms of 1980 U.S. dollars for the major improved technologies in the industrial sector for the U.S. are 120×10^9 and 53.6×10^9, respectively. The overall capital cost and ROI for the industrial sector in the U.S. and the world can be estimated by extrapolating these values on the basis of the proportionality of the energy demand and savings:

U.S.	Capital costs	$\$120 \times 10^9 \times 30.46 \text{ EJ}/25.33 \text{ EJ} = \$141 \times 10^9$
	ROI	$\$53.6 \times 10^9 \times 14.47 \text{ EJ}/13.36 \text{ EJ}/\$141 \times 10^9$
		$= 57.48/141 = 40.8\%$
Global	Capital costs	$\$141 \times 10^9 \times 297.17 \text{ EJ}/30.46 \text{ EJ} = \$1375 \times 10^9$
	ROI	$(\$19.37 \times 131.57 \text{ EJ}/8.94 \text{ EJ} + \$33.70 \times$
		$34.95 \text{ EJ}/4.42 \text{ EJ}) \times 10^9/\$1375 \times 10^9$
		$= 551.54/1375 = 40.1\%$

Transportation

Light-Duty Vehicles

Since most light trucks are owned by households, and all improved automotive technologies could just as easily be applicable to light-truck markets as well, automobiles and light trucks were combined into "light-duty vehicles" in the capital cost estimates for the transportation sector. Assuming that half of the truck energy is consumed by light trucks, the number of light-duty vehicles can be estimated roughly as follows:

Energy consumed = No. of vehicles × mileage per car-year ÷ Ave. mpg
(8.84 + 4.72) EJ = No. of vehicles × 9600 miles/car-year ÷ 40 mpg
No. of vehicles = 3.88 × 108

Assuming that the average vehicle price for the light-duty vehicles is about $6000, the total capital requirement is $3.88 \times 10^8 \times \$6000 = \$2330 \times 10^9$. The ROI is

(17.16 EJ × $9.64/GJ + 2.1 EJ × $5.89/GJ)/$2.330 × 10^9
$177.8 × 10^9 /$2330 × 10^9 = 7.6%

Heavy-Duty Trucks

Following the same procedure as above, the number of heavy-duty trucks can be obtained as follows:

4.72 EJ = No. of trucks × 70,000 miles/truck-year ÷ 6.5 mpg
No. of trucks = 3.0 × 106

Assuming that the average price for a heavy-duty truck is about $60,000, the total capital requirement is $3.0 \times 10^6 \times \$60,000 = \$180 \times 10^9$. The ROI is

2.1 EJ × $5.89/GJ/$180 × 10^9 = $12.4 × 10^9/$180 × 10^9 = 6.9%

The overall capital costs and the ROI of the automobiles and trucks in the U.S. in 1980 U.S. dollars are 2510×10^9 and 190×10^9, respectively. The overall costs and ROI for the transportation sector in the U.S. and the world can be estimated by extrapolating these values on the basis of the proportionality of energy demand and savings:

U.S.	Capital costs	$2510 × 10^9 × 24.24 EJ/18.28 EJ = $3328 × 10^9
	ROI	$190 × 10^9 × 26.3 EJ/21.36 EJ/$3328 × 10^9
		= $234 × 10^9/$3328 × 10^9 = 7.0%
Global	Capital costs	$3328 × 10^9 × 66.44 EJ/24.24 EJ = $9122 × 10^9
	ROI	$234 × 10^9 × 72.1 EJ/26.3 EJ/$9122 × 10^9
		= $642 × 10^9/$3328 × 10^9 = 7.0%

Electricity Generation

Based on Table 4.47, the average capital cost for electricity generation from fossil fuel is about $910/kW. With an operating factor of 0.65 and the amount of power demand in the U.S., the capital requirement for electricity generation in the U.S. is

$$(1.56 + 3.46 + 8.17) \text{ EJ} \div \left(8760 \, \frac{\text{hr}}{\text{yr}} \times 0.65 \times 3.6 \times 10^6 \text{ J/kWh} \right)$$

$$\times \ \$910/\text{kW} = \$551 \times 10^9$$

The ROIs by fuel type in the U.S are

Liquids	1.75 EJ × $4.05/GJ =	$7.10 × 10^9
Gases	3.89 EJ × $3.46/GJ =	$7.86 × 10^9
Solids	8.42 EJ × $1.28/GJ =	$\underline{\$10.77 \times 10^9}$

Total ROI = $25.73 × 10^9/$551 × 10^9 = 4.7%

Following the same procedure as above, the global capital requirement and ROI for electricity generation are

Capital costs $\qquad$ $551 × 10^9 × (9.06 + 11.44 + 41.82) EJ/13.19 EJ
$\qquad\qquad\qquad$ = $2603 × 10^9

ROI

Liquids	10.19 EJ × $4.05/GJ = $41.27 × 10^9
Gases	12.87 EJ × $2.02/GJ = $26.00 × 10^9
Solids	43.09 EJ × $1.28/GJ = $\underline{\$55.15 \times 10^9}$

Total savings = $122 $\quad$ × 10^9

or

$$\$25.73 \times 10^9 \times 66.15 \text{ EJ}/14.06 \text{ EJ} = 121 \times 10^9$$
$$\text{ROI} = \$122 \times 10^9/\$2603 \times 10^9 = 4.7\%$$

REFERENCES

1. Anderer, J., McDonald, A., and Nakicenovic, N., *Energy in a Finite World,* International Institute of Applied Systems Analysis, Ballinger, Cambridge, MA, 1980.
2. Lovins, A.B., Lovins, L.H., Krause, F., and Bach, W., *Least Cost Energy,* Brick House Publishing, Andover, MA, 1981.
3. Siedel, S. and Keyes, D., Can We Delay a Greenhouse Warming? U.S. Environmental Protection Agency, Washington, D.C., November 1983.
4. Nordhaus, W.D. and Yoke, G.W., Future paths of energy and carbon dioxide emissions, in *Changing Climate,* National Academy Press, Washington, D.C., 1983, 87.
5. Ausubel, J.H. and Nordhaus, W.D., A review of estimates of future carbon dioxide emissions, in *Changing Climate,* National Academy Press, Washington, D.C., 1983, 153.
6. Edmonds, J.A., Reilly, J.H., Trabalka, J.R., and Reichle, D.E., An Analysis of Possible Future Atmospheric Retention of Fossil Fuel CO_2, TRO-13, DOE/OR/21400-1, U.S. Department of Energy, Washington, D.C., September 1984.

7. Reference Energy System, Year 1977, National Center for Analysis of Energy Systems, Brookhaven National Laboratory, Upton, NY, 1980.

8. Energy Conservation Multi-Year Plan FY 1986–FY 1990, U.S. Department of Energy, Washington, D.C., 1984.

9. Wolfe, H.C., Ed., *Efficient Use of Energy,* AIP Conf. Proc. No. 25, American Institute of Physics, New York, 1975.

10. Reay, D.A., *Industrial Energy Conservation,* Pergamon Press, Oxford, 1977.

11. Gray, C. and Von Hippel, F., The fuel economy of light vehicles, *Sci. Am.,* 244(5), 48–59, 1981.

12. Ross, M.H. and Williams, R.H., *Our Energy: Regaining Control,* McGraw-Hill, New York, 1981.

13. Hill, D., MARKAL Model of the U.S. Energy System, Data Base, Brookhaven National Laboratory, Upton, NY, 1984.

14. Gyftopoulos, E.P., Lazaridis, L.J., and Widmer, T.F., *Potential Fuel Effectiveness in Industry,* Ballinger, Cambridge, MA, 1974.

15. Technical Assessment Guide, EPRI P-2410-SR, Electric Power Research Institute, Palo Alto, CA, 1982.

16. Monthly Energy Review, DOE/EIA0035 (82/04), U.S. Department of Energy, Washington, D.C., April 1982.

17. Cosidine, D.Y., Ed., *Energy Technology Handbook,* McGraw-Hill, New York, 1977.

18. Bhagat, N., Ed., Technology Data Estimates, BNL 29494, Brookhaven National Laboratory, Upton, NY, 1981.

19. Shepard, M.L., Chaddock, J.B., Cocks, F.H., and Harman, C.M., *Introduction to Energy Technology,* Ann Arbor Science, Ann Arbor, MI, 1976.

20. Commoner, B., Boksenbaum, H., and Corr, M., Eds., *Energy and Human Welfare,* Vol. II, Macmillan Information, New York, 1975.

21. World Car Conference '96, Bourns College of Engineering, Center for Environmental Research and Technology, University of California, Riverside, January 21 to 24, 1996.

5 Removal, Recovery, and Disposal of Carbon Dioxide

5.1 INTRODUCTION

Three potential control points (CO_2 removal sites) are available for regulating the concentration of CO_2 in the atmosphere: (1) the atmosphere itself, (2) the surface waters of the oceans, and (3) the stacks of fossil-fuel power plants. The latter site has the advantage of providing a source of relatively high CO_2 concentration, while the first two provide more flexibility for site selection and simpler disposal or recovery and reuse logistics. A choice of control point limits the available choices of CO_2 removal processes and will reflect on the size of the operation, the number of required sites, energy requirements, the comparative economics, and the overall feasibility of operation. A most important factor is the energy required to remove, recover, and dispose of CO_2. The CO_2 concentration ranges of the three control points are listed in Table 5.1.[1] Since the concentration of CO_2 in flue gas is significantly greater (by a factor of 500) than that in the atmosphere or the ocean, the energy required to remove CO_2 from flue gas should be appreciably lower than for the other two control points. The minimum theoretical energy (the thermodynamic free energy) to separate CO_2 from each of the three sources is also listed in Table 5.1. Based on these theoretical values, it would take about three times more energy to separate CO_2 from the atmosphere than from flue gas. It should be noted that the practical energy required for separation is usually on the order of magnitude by a factor of 10 or more greater than the thermodynamic minimum.

Although removal or capture of CO_2 from a particular source is an important consideration in devising a CO_2 control system, CO_2 disposal (storage) and reuse potentials are equally or perhaps more important, so that unless a control scenario produces a net near-term decrease in the rate of buildup of atmospheric CO_2, it will be of no value. This means that the fossil-energy-generated CO_2 must not be allowed to enter the atmosphere but must be stored or recycled until large-scale acceptable nonfossil energy sources become available.

TABLE 5.1
CO_2 Concentrations at Control Points and Minimum CO_2 Separation Energies[1]

Control point	CO_2 concentration		Minimum separation energy[a] (kWh/lb CO_2)
	% by vol	% by wt	
Atmosphere	$331 \times 10^{-4\,b}$	502×10^{-4}	0.0570
Ocean	—	$100 \times 10^{-4\,c}$	0.0570
Fossil-fueled combustion equipment	7–20	11–28	0.0259–0.0179

[a] Minimum energy based on free energy of mixing: $\Delta F = RT\Sigma n_i \ln(P_1/P_1^\circ)$, where P_1° and P_1 are the initial and final partial pressures of the ith species, respectively. R is the gas constant, T is temperature in degrees absolute, and n_i is the number of moles of the ith species.

[b] 1980 atmospheric concentration.

[c] ΣCO_2 (total carbon dioxide content) $= C_{H_2CO_3} + C_{CO_2(aq)} + C_{HCO_3(T)} + C_{CO_3(T)}$.

In order to concentrate the low concentration of CO_2 in the atmosphere, large volumes of air must be processed, which requires a significant amount of energy. For example, a coal-fired power plant generating electricity to power a fan to blow air through a CO_2 scrubber at a reasonable rate and pressure drop would generate as much CO_2 as it would remove from the atmosphere. It can also be shown that the energy required to remove CO_2 from the ocean is, at a minimum, equal to or greater than that required for removal from the atmosphere. Thus, it is concluded that for removal of CO_2 from the atmosphere or the ocean, a nonfossil-fuel source is necessary. The nonfossil sources are either nuclear or solar. Solar includes fixation of CO_2 from the atmosphere to produce biomass by photosynthesis. Thus, the only current feasible method of removal of CO_2 from the atmosphere is to grow biomass — plants and trees or algae in the ocean. If some day nuclear fission is allowable or fusion becomes available, that energy source could also be used to extract CO_2 from the atmosphere.

5.2   REMOVAL AND RECOVERY OF CO_2 FROM FOSSIL-FUEL COMBUSTION

The emission of CO_2 from fossil-fuel combustion basically comes from three sectors of the economy: (1) the electrical power generation sector, (2) the industrial and domestic thermal generation sector, and (3) the transportation power sector. Roughly, about one-third of the total CO_2 generation comes from each of these three sectors. The emissions of CO_2 from the industrial, domestic, and transportation sectors are highly dispersed and generate relatively small quantities of CO_2 per unit, thus making it almost impossible to collect for disposal. The major centrally located sources of CO_2 are electric-generating fossil-fuel-fired power plants. Thus, removal and recovery concentrate on fossil-fueled power plants.

TABLE 5.2
**Energy Required to Remove and Recover CO_2 from
a Coal-Fired Power Plant**

Process	CO_2 removal efficiency (%)	kWh(e)/lb CO_2[a] recovered
Improved amine absorption/stripping integrated plant[3]	90	0.11
Oxygen/coal-fired power plant[5]	100	0.15
Amine (MEA) absorption/stripping nonintegrated plant[4]	90	0.27
Potassium carbonate absorption/stripping[4,8]	90	0.32
Molecular sieves adsorption/stripping[4]	90	0.40
Refrigeration[4]	90	0.40
Seawater absorption[4,6]	90	0.80
Pressurized fluidized bed combustor seawater pressure equalized	99	0.18
Membrane separation[7]	90	0.36

[a] Equivalent electrical energy required to recover CO_2 from flue gas of a coal-fired power plant subsequent to SO_2 removal, liquefying, and prior to pumping for terrestrial or ocean disposal of recovered CO_2.

A number of methods are available for removal and recovery of CO_2 from the flue gas from power plant stacks. These are listed in Table 5.2 together with the equivalent electrical energy required to remove CO_2 from the flue gas of a coal-fired power plant and to liquefy the CO_2 for disposal. Both thermal and electrical energy are needed to remove and recover CO_2. Because power is taken from the power plant both as steam and electrical power, the energy shown is presented as the sum of energy calculated as equivalent electrical energy.

5.2.1 Absorption/Stripping

The absorption/stripping method, using a solvent, has been practiced for a long time in the petrochemical and chemical industry. This method is a well-known unit operation in chemical engineering in refineries and for the production of ammonia and methanol. A number of solvents have been used, including alkanolamine (monoethanolamine [MEA]), alcohols (methanol), and glycols. CO_2 is absorbed from the flue gas through a packed absorption tower at a lower temperature and is heated by steam from the power plant to a higher temperature in a stripping tower to remove and concentrate the desorbed CO_2. The energy required is influenced by the concentration of the CO_2 in the flue gas. Improvements have been made recently with the MEA solvents, and energy requirements have been reduced by more than 50% with a hindered MEA and using a lower pressure drop packing for the absorption tower.[2] Furthermore, as will be shown later, the integrated plants show lower energy requirements than nonintegrated plants. The CO_2 removal operation has been integrated with the power generation plant, thus further lowering energy requirements for CO_2 recovery.

5.2.2   Adsorption/Stripping

Another well-known method of separating gases from gas streams is adsorption on solids. Typical solids include charcoal and molecular sieves. Adsorption/stripping operations have gained importance in unit operations at near isothermal conditions in the process of pressure swing adsorption in which the adsorption operation is practiced at a higher pressure to increase the partial pressure of the gas to be removed and the stripping operation is operated at a lower pressure to recover the adsorbed CO_2.[47]

5.2.3   Potassium Carbonate Absorption/Stripping

Hot potassium carbonate has been used extensively in removing acid gas from petroleum refining and coal gasification operations.[8]

5.2.4   Refrigeration (Cryogenic)

In this method, the gases are cooled down to condense the gas to be separated either to a liquid or to a solid, depending on the physical properties of the gases. In order to limit the low condensation temperature requirement, the gases are compressed to raise the partial pressure of the gas to be separated, and heat recovery is required to reduce energy requirements.

5.2.5   Membrane Separation

Diffusion of molecular gases of different sizes can be separated by porous membranes. This method avoids the need for a phase change. Various types of permeable membrane materials have been used to separate gases, including polymers, metals, and rubber composites.[8,48] Gas absorption membrane composites use an absorbing liquid on one side of a porous membrane mainly to provide a large surface-contacting area and avoid the mixing of the gas with the liquid.[9]

5.2.6   Seawater Absorption

There is a possibility of using seawater as a solvent for CO_2, and an estimate of the energy required for an absorption/stripping operation is shown in Table 5.2. Another independent estimate is shown for pumping flue gas deep into the ocean from a pressurized fluidized bed combustor (PFBC) to a depth where the partial pressure of the dissolving CO_2 is equal to the pressure of the ocean at that depth.

5.2.7   Oxygen/Coal-Fired Power Plant

A special case presents itself when considering combustion of fossil fuels to produce a pure CO_2 stream. The main reason why the CO_2 is diluted in the flue gases is due to the nitrogen in the air used for combustion. Thus, if the fossil fuel is combusted in oxygen, then the flue gas should essentially contain only CO_2. To accomplish this, oxygen is separated from the nitrogen before burning the fuel.

Figure 5.1 shows a schematic for a nominal 1000-MW(e) pulverized coal-fired power plant operating with an air liquefaction plant to produce oxygen for combustion.[5] In order to maintain a combustion temperature at the same level as with air, part of the effluent CO_2 must be recycled to the combustor. An actual test of such a system in a small industrial coal-fired boiler was conducted to show the feasibility of such a scheme.[14] The CO_2 can be sequestered, after liquefying, directly in depleted gas wells, aquifers, or in the ocean.

A review of the energy required to remove and recover CO_2 from coal-fired power plants indicates that the amine absorption/stripping system in an integrated plant is among the lowest and is used in an example below for removal and recovery of CO_2 from all fossil-fuel power plants operating in the U.S.

5.3   DISPOSAL OF CO_2

Once CO_2 is recovered from power plants, means for sequestering must be sought in order to keep it out of the atmosphere. The options are as follows:

1. Ocean disposal
2. Depleted gas wells
3. Active oil wells (enhanced oil recovery) and depleted oil wells
4. Coal beds and mines
5. Salt domes
6. Aquifers
7. Natural minerals

5.3.1   Ocean Disposal

Marchetti was one of the earliest to suggest sequestering CO_2 by dissolution in the ocean.[11] He reasoned that at the Straits of Gibraltar, the Mediterranean Sea spills over into the Atlantic Ocean and forms an underwater current into which CO_2 can be injected and would be carried down deep into the Atlantic Ocean. Investigation was further conducted by Hoffert et al.[12] taking into account that the upper layer of the ocean is in equilibrium with the atmosphere. The partial pressure of CO_2 in the atmosphere determines the solubility of gas in the ocean. Below the thermocline (about 1000 ft down, at which point the ocean temperature abruptly decreases), the concentration of dissolved CO_2 is negligible, and thus CO_2 can be pumped down and readily dissolved at the ocean depths. The capacity for dissolution of CO_2 in the ocean is adequate to absorb all the CO_2 from combustion of all the earth's resources of fossil fuels. Furthermore, if liquid CO_2 is pumped deep enough within the ocean, the density of liquid CO_2 becomes greater than the density of seawater at that depth and the liquid CO_2 can sink to the bottom floor of the ocean and form a lake. CO_2 can also form clathrates, which are solid compounds of a CO_2 molecule surrounded by about 5.75 molecules of water which aid in lowering the partial pressure of CO_2 and increase the density, thus keeping the CO_2 sequestered in the ocean.[13] There are several mechanisms for sequestering CO_2 in the deep ocean.

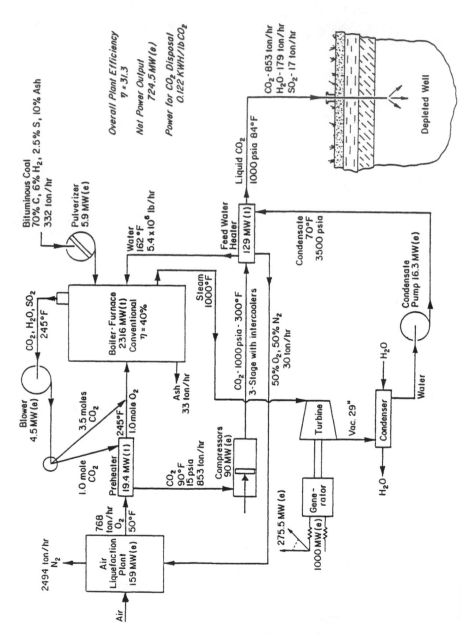

FIGURE 5.1 Oxygen coal-fired power plant with direct CO_2 disposal.

A comparison of whether to dispose of CO_2 as a gas, liquid, or solid (dry ice), below the thermocline, indicates that liquid CO_2 is the most economical form to be disposed mainly because of the lower energy expenditure for condensing and transporting. Gas requires large pumping volumes and dry ice requires large refrigeration energy loads.[6]

5.3.2 Depleted Gas Wells

For land storage, depleted natural gas wells are feasible sites where CO_2 can be sequestered.[20] Natural gas wells are normally capped off at high pressure, ranging up to several thousand pounds per square inch of pressure without leakage. Although there are hundreds of depleted wells in the world, the capacity is still limited. In terms of volume capacity, the natural gas wells can only sequester the CO_2 from natural gas combustion since one volume of natural gas combustion produces one volume of CO_2. Natural gas wells are inadequate for containing CO_2 from the combustion of the reserves of oil or coal.[5]

5.3.3 Active Oil Wells

Primary oil production removes about a third of the oil from an active oil well. To remove a substantial portion of the remaining two-thirds, enhanced oil recovery has been practiced[14] for some time. Various media, such as hot water, nitrogen, polymers, and CO_2, have been used for this purpose. Under certain conditions, CO_2 is preferred because, in addition to displacement, CO_2 dissolves in the oil and reduces its viscosity, making it easier to pump out of the geologic formation. After the oil is forced out to the surface, some of the dissolved CO_2 flashes. For economic reasons, the flashed CO_2 is recovered and recycled; thus the capacity for permanently sequestering CO_2 in oil wells is reduced. Furthermore, the compressed volume of gaseous CO_2 generated by combustion of the oil is much greater than the volume of the liquid oil, and thus only a fraction of the oil combustion CO_2 can be sequestered in oil wells.

5.3.4 Storage in Coal Mines and Deep Beds

Storage of CO_2 in mined-out and abandoned coal mine fields is not feasible because the mines cannot be readily sealed to hold the pressure; furthermore, the CO_2 volume generated by combustion of coal is much greater than the volume of the solid coal.[16]

Natural gas (methane) is released from coal beds that have been mined. In this case, the CO_2 not only is under pressure in the pores of the coal but is absorbed on the surface of the coal. Displacement of coal-bedded methane with CO_2 is being investigated, especially in deep coal deposits. Recent experiments indicate that it may be possible to pump CO_2 down into these deep coal deposits and displace the natural gas (CH_4), bringing it to the surface. Because of the higher absorptive properties of coal, it appears that twice the volume of CO_2 can be sequestered in this manner than the natural gas originally present in the coal.[15]

5.3.5 Salt Domes

Solution-mined salt domes have been used to store large quantities of oil for the national petroleum reserves. Many of these domes exist along the Gulf Coast. Thus, excavation of the salt by solution mining and placing the salt in the ocean and storing the CO_2 in the excavated domes is possible.[16] Salt domes are distributed in various places around the U.S., but would require pumping seawater to and from the ocean for solution mining and disposal of the salt in the ocean.

5.3.6 Aquifers

Deep aquifers are widespread throughout many parts of the terrestrial horizon. The deep aquifers are usually saline and are separated from the shallow aquifers which form the surface drinking water supplies. Pressurized CO_2 could displace the water as well as dissolve in the water in the geological formation. Estimates[17] indicate that a significant capacity for sequestering CO_2 is potentially possible in U.S. aquifers. A project by the Statoil Company of Norway is reported to be separating CO_2 from a gas field in the North Sea and pumping the CO_2 into an aquifer 3000 ft below the North Sea.[18,21] The experiment involves the disposal of a million tons of CO_2 per year in this manner. This is done because it is necessary to reduce the CO_2 content from 10% to 2% before sending the gas to market in pipelines. The incentive to perform this experiment is due to the fact that Norway has a carbon tax which is avoided by Statoil and which pays for the operation. Another project by Exxon at Pertamina in the Indonesian Naturna fields is being planned to sequester the CO_2 separated from the natural gas wells in deep aquifers below the South China Sea floor.[18,21] Indonesia evidently requires the sequestering of the CO_2 in order to produce the methane. The CO_2 concentration from deposits in Indonesia is as high as 70%, so that the CO_2 must be separated to produce pipeline-quality natural gas.

5.3.7 Natural Minerals

Carbonate minerals are abundantly available in the earth in the form of limestone, dolomite, or magnesite. These minerals are already carbonated and cannot be used to sequester CO_2. However, igneous rock such as that containing magnesium oxide bound to silica and alumina-forming aluminosilicates can react with CO_2 to form carbonates. Investigation of the thermodynamics, kinetics, and the process to sequester CO_2 in MgO-rich minerals indicates some feasibility in sequestering CO_2 from coal-fired plants. Handling large masses of rock and a complicated solid preparation process for reacting with CO_2 from power plants present formidable process problems.[19]

5.4 CAPACITY FOR SEQUESTERING CO_2

From the above, it is noted that the technology for the removal and recovery of CO_2 from large central power plants is in an advanced state of development. The problem lies in the disposal of the recovered CO_2 when considering the massive amounts

TABLE 5.3
Worldwide Potential Storage Capacity
for Sequestering Recovered CO_2
from Central Power Stations[14]

Storage site	Capacity in billion tonnes of CO_2
Deep ocean	5,100–>100,000
Deep aquifers	320–10,000
Depleted gas wells	500–1,100
Depleted oil wells	150–700

Note: World annual production from the use of fossil fuels in energy
production = 22 billion tonnes of CO_2.

to be kept from entering the atmosphere. It is necessary to understand the potential
capacity for storing and sequestering the recovered CO_2. There are large uncertainties in estimating the storage capacity for each of the disposal sites mentioned
above. Table 5.3, taken from the IEA greenhouse gas R&D program,[14] gives an
estimate of the capacities for the major sites. These estimates indicate that deep-
ocean disposal has at least an order of magnitude greater capacity for sequestering
CO_2 than all the other sites combined. The deep ocean in principle could sequester
the CO_2 from combustion of all the world's fossil fuel for at least the next 300 years,
which is equivalent to the recoverable coal reserves.

5.5   SYSTEM STUDY

A system study and economic analysis of the application of the absorption/stripping
system for removal and recovery of CO_2 from utility stack gases in the U.S. and
disposal of the CO_2 in appropriately applied sites as described above is given in this
section. An improved solvent process is used, and the process is integrated with
power plant operations to improve the efficiency of the combined plant.

It should be noted that this is a first-order design and cost estimation system
analysis, made primarily for the purpose of illustration and determining the order of
magnitude, feasibility, and economic costs for the removal, recovery, and disposal
of CO_2 from power plant stacks in the U.S. The base year chosen for the system
analysis is 1980, and all capacity and costs are indexed to that year.

5.5.1 CO_2 Removal and Recovery System for
Fossil-Fuel Power Plant Flue Gases

5.5.1.1   *CO_2 Emissions from Fossil Fuel*

To obtain the CO_2 emissions from fossil-fuel-burning power plants, it is necessary
to estimate the CO_2 generation associated with the consumption of various types of
fuel. The CO_2 generated in the consumption of a unit amount of coal, oil, and gas

TABLE 5.4
CO_2 Generation for Various Fossil Fuels

Fuel type	Heating value (Btu/lb)	CO_2 generated (lb CO_2/lb fuel)	CO_2 generated (lb CO_2/1000 Btu)	Energy generated (kWh(e)/lb CO_2 generated)[a]
Bituminous coal	12,700	2.59	0.204	0.55
Fuel oil and gasoline (petroleum distillate)	19,600	3.14	0.160	0.70
Natural gas	24,000	2.75	0.115	0.97

[a] Assuming a power plant efficiency of 38%.

is shown in Table 5.4. Table 5.4 also shows that the relationship between the CO_2 generated and energy production depends heavily on the fuel type. It is noted that natural gas and fuel oil produce about 50 and 80%, respectively, of the amount of CO_2 per unit energy released from burning coal.

5.5.1.2 CO_2 Removal and Recovery Using Improved Solvent Process

MEA absorption/stripping was used from Reference 1 to study CO_2 removal from stack gas because of its relatively low energy requirement. Table 5.5 shows, however, that the newer alkanolamine-based solvent, called Dow Gas/Spec FS-1, is more energy efficient than MEA for removal and recovery of CO_2.[22] Hence, Dow solvent is employed in this study to estimate the process design for the removal and recovery of CO_2 from flue gases.

A flow sheet for the recovery of CO_2 from power plant flue gases is shown in Figure 5.2. Flue gas enters the recovery system at slightly above ambient pressure and a temperature generally in excess of 250°F. First, the gas is compressed to somewhat less than 5 psig to accommodate the pressure drop through the system, and the gas is quenched to near 120°F before it enters the absorber. CO_2 is removed

TABLE 5.5
Comparison of Energy Requirements for MEA and Dow FS-1 Solvent Systems for CO_2 Removal and Recovery[a–c]

	MEA	Dow FS-1
Heat duty required (million Btu/ton of CO_2)	5.4	4.8
Electric energy (kWh(e)/ton of CO_2)	20	13

[a] Basis: 1000 tons/day of CO_2 from flue gas.

[b] Solvent concentration 20%.

[c] 90% CO_2 removal from flue gas, CO_2 concentration 8 to 15%.

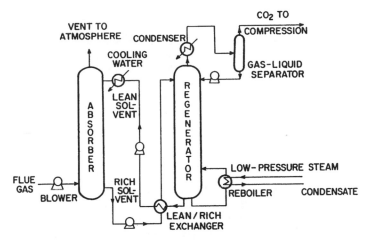

FIGURE 5.2 Flow sheet for CO_2 removal and recovery from power plant flue gas using Dow Gas/Spec FS-1.

by reacting chemically with lean Dow solvent solution, flowing countercurrent to the flue gas, in accordance with the following reaction between the amine and the CO_2:

$$R - NH_2 + H_2O + CO_2 \underset{300°F}{\overset{100°F}{\rightleftarrows}} R - NH_3HCO_3 \tag{5.1}$$

The rich solvent solution is pumped from the bottom of the absorber to a lean/rich solution heat exchanger where the cold-rich solvent is heated by the hot regenerated (lean) solvent returning from the regenerator on the way back to the absorber.

The rich solvent solution entering the top of the regenerator, where carbon dioxide is stripped from the Dow solvent solution, flows down through the column countercurrent to stripping steam generated in the solution reboiler. Elevated temperature combined with steam stripping favors the reverse of Reaction 5.1. Uncondensed steam and CO_2 exit the regenerator at the top and flow to a reflux condenser for removing water vapor. The condensate is returned to the generator as reflux while the recovered carbon dioxide is removed for compression and liquefaction.

Thus, for the removal and recovery of CO_2, thermal energy in the form of steam is needed to reboil and remove CO_2 from the solvent, and electrical energy is needed for blowers and pumps to drive the gas through the absorber and the liquid through the absorption and regeneration towers.

The liquefaction of recovered CO_2 is accomplished in a four-stage compression system equipped with coolers, water knockout drums, and a dryer. The gas is

TABLE 5.6
Energy Requirement for a 1000-Ton/Day CO_2 Recovery Plant 100 Miles From Collection Center[a]

Process	Equivalent kWh(e)/lb CO_2 recovered
Removal and recovery[b]	0.065
Liquefaction[c]	0.047
Disposal	
6-in. pipe[d]	0.0006
36-in. pipe[e]	0.001
Total	0.114

[a] See Appendix 5.1 for detailed calculation. CO_2 removal efficiency is 90%. The 1000-ton/day CO_2 recovery plant corresponds to 51-MW(e) coal-fired, 65-MW(e) oil-fired, and 90-MW(e) gas-fired power plants, respectively.

[b] Includes the equivalent thermal and electric energy required as shown in Table 5.5.

[c] Compression and liquefaction of CO_2 to 2000 psia in four stages.

[d] The pumping power required for transporting 1000 tons/day CO_2 through a 6-in. line, with a pumping efficiency of 60%.

[e] Based on a pumping rate of 15×10^6 lb/hr and pumping efficiency of 60%.

compressed to 2000 psia in the fourth stage, passed through a cooler, and liquefied in a condenser at about 80°F. The energy required to liquefy the recovered CO_2 gas amounts to 0.047 kWh(e) per pound of CO_2 recovered while 0.065 kWh(e) per lb CO_2 is needed for removing and recovering CO_2 from flue gas for a 90% removal efficiency, as shown in Table 5.6 for a 1000-ton/day CO_2 recovery plant. Appendix 5.1 gives the detailed calculation for these data. As mentioned earlier, development has progressed to reduce the energy requirements for desorbing the CO_2 absorption solvent using hindered amines as well as to reduce the pressure drop across the absorption column with improved packing material.[2]

5.5.1.3   Integration of Power Plant and CO_2 Recovery System

Figure 5.3 shows the low-pressure steam at about 100 psia and 400°F flowing to the reboiler from the low-pressure section of the power plant's turbine to supply the necessary heat for solvent regeneration. The extraction of latent heat from the low-pressure steam, which otherwise would be lost in the condenser, and the conversion of this heat to usable energy for solvent regeneration markedly reduce the energy required for the CO_2 recovery systems from power plants. As shown in Table 5.5, the heat and electrical energy required for the Dow solvent process for 90% of CO_2 recovery are 4.8×10^6 Btu/ton of CO_2 and 13 kWh(e)/ton of CO_2, respectively. Assuming a conventional power plant efficiency of 38% and boiler and reboiler

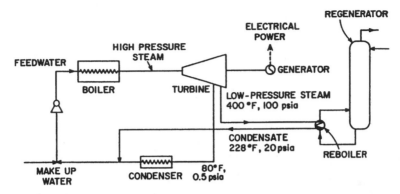

FIGURE 5.3 Integrated schematic flow diagram of power plant and reboiler for the removal and recovery of CO_2.

efficiencies of 90 and 85%, respectively, integrating the CO_2 recovery system with the power plant in this manner lowers the coal- and gas-fired power plant efficiencies to 35 and 36%, as shown in Table 5.7. In an earlier study,[1] part of the thermal energy from the fuel fed to the power plant was diverted as either equivalent steam or electrical energy to strip and recover the CO_2. As a result, the coal- and gas-fired power plant efficiencies were drastically reduced to 14 and 24%, respectively, from the original 38%. Details of the calculations are given in Appendix 5.2. The low-pressure steam from the turbine integrated with the reboiler CO_2 recovery system is the preferred system. However, this does not take into account the energy for flowing the gas through the absorption column and for liquefying the CO_2 and

TABLE 5.7
Power Plant Efficiency Comparison for Power Plants Integrated with the CO_2 Recovery System[a] Only Energy Taken into Account for Reboiler of the MEA Absorption/Stripping Operation

	Efficiency[b–d] (%)	
Fuel type	Integrated plant using low-pressure steam from power plant in reboiler	Nonintegrated plant using thermal energy from fuel directly
Coal	0.349	0.137
Oil	0.355	0.189
Gas	0.363	0.244

[a] See Appendix 5.3 for detailed calculations.

[b] Assuming a conventional power plant efficiency of 38%.

[c] Assuming the efficiencies of boiler and reboiler are 90 and 85%, respectively.

[d] Assuming the CO_2 removal efficiency is 90%.

pumping it through piping into the ocean. This is taken into account in the following section.

5.5.1.4 CO_2 Recovery Plant Costs

The CO_2 recovery plant capital cost and operating expenses depend on the concentration of CO_2 in the flue gas.[23] The conventional coal-fired power plant flue gas contains up to about 15% CO_2, while that from a natural gas-fired power plant contains about 8% CO_2. The capital cost for a plant designed to recover 1000 tons of CO_2 per day, at 8% of CO_2 input concentration, without compression and liquefaction, is estimated to be about $14 million (1980 U.S. dollars) and decreases as CO_2 concentration increases.[23] The capital cost is directly proportional to plant size in the range of 100 to 1000 tons CO_2 per day, which is equivalent to a coal-fired power plant capacity range of 5.1 to 51 MW,[2] respectively. For greater capacities, the six-tenths exponential capacity factor rule for capital cost can be applied.

The operating cost, which does not vary much with plant size in this range, is about $15 (1980 U.S. dollars) per ton of CO_2 at 8% carbon dioxide and higher and decreases as the plant size becomes larger than 1000 tons CO_2 per day.

The capital investment for the CO_2 compression and liquefaction is about $10 million (1980 U.S. dollars) for a 1000-ton/day CO_2 recovery plant.[2]

From the above cost estimates, the capital investment and electricity production costs are calculated for a typical 100-MW(e) coal-fired power plant located in the New England area, 100 miles from the coastal collection center for subsequent disposal 100 miles offshore in the ocean. Details of the estimates are given in Appendix 5.3. Table 5.8 gives the power required for the entire system as well as the cost estimate. It is noted that the power required for liquefaction and pumping liquid CO_2 amounts to as much as the equivalent power for recovery of CO_2 from the flue gas. The total power reduction of 17% shown in Table 5.8 is a lower limit. To restore the plant to its original capacity, it is necessary to make up this loss. The plant power capacity must be increased, including the power to recover the CO_2 from the makeup power. The total or "avoidance power" cost could add another 3% to the direct loss, increasing the power cost to 20% of the initial power plant capacity.

5.5.2 CO_2 Disposal Systems

The liquid carbon dioxide recovered from the power plants can be disposed of by (1) injection in the deep ocean, (2) storage in spent oil and gas wells, (3) storage in excavated salt caverns, and (4) storage in underground aquifers. In this study, liquid CO_2 is first transported by 6-in. pipeline from power plants to collection centers; from there, a 36-in. pipeline is used for final disposal in the ocean.

5.5.2.1   Ocean Disposal

In the two-box model, the ocean is represented by two reservoirs separated by a thermocline.[24] The upper reservoir consists of a surface layer about 75 m deep,

TABLE 5.8
Net Power Plant Capacity, Capital Investment, and Electricity Production Cost for a 100-MW(e) Coal-Fired Power Plant[a] with an Integrated CO_2 Control System, Located 100 Miles From a Coastal Collection Center in New England (Federal Region 1)

Process	500-m-depth ocean disposal[b]	3000-m-depth ocean disposal[b]
Power requirements for CO_2 control system	MW(e)	MW(e)
Heat required for CO_2 recovery (electricity equivalent)[c]	7.9	7.9
Electrical power required for CO_2 recovery	1.1	1.1
Electrical power required for liquefaction of CO_2	7.7	8.3
Pumping power required for transportation of CO_2 to collection center	0.3	0.9
Total power consumption	17.0	18.2
Net power plant capacity	83.0	81.8
Capital investment	$ million (1980 U.S. dollars)	
Power plant[d]	57.0	57.0
CO_2 recovery	20.6	20.6
CO_2 liquefaction[e]	15.5	15.6
Piping[f]	5.2	5.8
Total capital investment	98.3	99.0
Electricity production cost	Mills/kWh(e) (1980 U.S. dollars)	
Power plant[d]	32.0	32.0
CO_2 recovery	16.0	16.0
CO_2 liquefaction	7.3	7.6
Piping[g]	1.5	1.8
Electricity production cost	56.8	57.4

[a] See Appendix 5.3 for detailed calculation. CO_2 emission is 2182 tons/day.

[b] CO_2 pumped 100 miles off coast for 500-m-depth disposal and 200 miles off coast for 3000-m-depth disposal.

[c] Efficiency of power plant decreases to 35% from 38% for 90% CO_2 recovery.

[d] Based on capital cost of $570/kW(e) estimated from Reference 37.

[e] Estimated from Reference 6 using the six-tenths exponential capacity factor rule for capital cost.

[f] Estimated from References 36 and 40.

[g] Assuming 15% per year depreciation, 24 hr/day operation, and 350 annual operating days.

while the second or deeper layer extends from 75 m to the ocean floor. Because of wind stress, mixing occurs rapidly in the surface layer. In the deeper layer, however, mixing depends on the rates of advection and turbulence which cross vertical density gradients. Exchange between the boxes is very slow. The upper layer is close to equilibrium with the atmosphere, and thus the concentration of CO_2 in this layer depends on the concentration of CO_2 in the atmosphere. Since there is little direct exchange between the atmosphere and the second deeper ocean layer, the deep ocean is a promising vast reservoir for disposal of CO_2.

Because it is more economical to transport liquid CO_2 in pipelines, it is desirable to dispose of CO_2 in the liquefied state in the deep ocean.[1] There are two choices for selecting the distance offshore and the depth to which the liquid CO_2 must be pumped.

Although the density of liquid carbon dioxide is less than that of seawater at depths less than about 3000 m, it is much more compressible than seawater and has a much higher thermal coefficient of expansion.[25] The density of CO_2 increases quite rapidly with depth and is greater than that of seawater of similar temperature (37°F) at about 3000 m. Hence, if the liquid CO_2 is piped to a somewhat greater depth, it could be expected to form a liquid CO_2 pool and sink to the ocean floor. Ocean depths of 3000 m, which corresponds to a pressure of about 4400 psia, are accessible at about 200 miles from the shoreline of the northeast U.S., and thus a 200-mile pipeline would be required.

Another mechanism for disposal in the lower layer of the ocean is to dissolve the CO_2 using 2000-psi liquid pressure at a depth of 500 m. Since the ocean's surface layer usually extends to a depth of only about 75 m, a 500-m injection depth should be permissible for CO_2 disposal. The additional 425 m was provided to allow for dissolution of the injected CO_2 in the deep ocean and would take advantage of the natural thermocline circulation. Ocean depths of 500 m are accessible at about 100 miles from the shoreline of the northeast U.S., which is approximately the width of the continental shelf. Therefore, for disposal at 500-m depths, a 36-in. pipeline about 100 miles long is needed. On the western coast of the U.S., the continental shelf is narrower and the ocean depth at 100 miles from the shoreline is greater than 500 m.

In this study, the 100-mile main pipeline and ocean depth at 500 m were selected as adequate for disposal, as shown in Figure 5.4. Actually, if it is found necessary to pump the CO_2 200 miles offshore and down to 3000-m depths, the additional pipeline cost and compression energy requirement are not significantly greater (see Table 5.8). The increased capital investment and production cost of electricity is less than 2%. The additional energy required to compress CO_2 to 4500 psia from 2000 psia at the collection centers for 3000-m-depth ocean disposal is about 0.004 kWh(e)/lb CO_2. However, the structural problems of laying and maintaining pipe at these depths can be severe. Some experimental work is recommended to study the CO_2 concentration distribution in deep ocean for various depths of ocean disposal to determine which depth is optimum. These experiments will shed light on the dissolution behavior of CO_2 at the disposal point and how the interaction of chemical and biological processes regulates the distribution of CO_2 in the ocean. An experiment is actually being planned between the U.S. and Japan to inject CO_2 in the ocean off the coast of Hawaii and monitor the concentrations of CO_2 at various ocean depth levels.

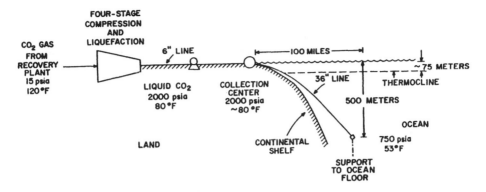

FIGURE 5.4 Schematic diagram for ocean disposal system.

Other methods of direct injection of CO_2 into the ocean have been explored, including (1) dry ice released from a ship, (2) liquid CO_2 injected from a pipe towed by a moving ship, and (3) liquid CO_2 introduced as liquid to the seafloor, producing a CO_2 lake.[18] A pictorial schematic of these methods is shown in Figure 5.4A.[18]

The hydrostatic pressure at 500 m is about 725 psia, and the temperature is less than 53°F. In this evaluation, it was assumed that the CO_2 could be cooled from about 80°F to the temperature at injection points during its journey into the deep ocean. Part of the energy acquired from the CO_2 compression is expended in overcoming the seawater static head between the CO_2 injection points and the surface of the ocean.

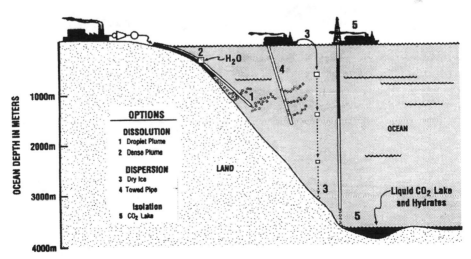

FIGURE 5.4A Ocean disposal of CO_2: five injection scenarios.[18]

5.5.2.2   Oil and Gas Wells Disposal

The availability of 12,000 spent oil and gas wells throughout the U.S.[26] suggests a possible approach to the disposal of CO_2. The pressure of depleted oil and gas wells is usually in the range of 100 to 500 psia,[27] so that CO_2 at 2000 psia can be readily injected into the abandoned wells. However, the liquid CO_2 will not remain at its initial temperature of 80°F and will slowly increase to the temperature of the surrounding well. The final supercritical temperature and pressure of the CO_2 will depend upon the well depth. An increase of about 10°F is usual for every 1000 ft of depth. For an average 10,000-ft-deep well, the CO_2 will be heated to 180°F, which will result in a final pressure of 3000 psia or more. A bottom well pressure of 4000 psia is not unusual at these depths. In fact, pressures in excess of 10,000 psia are frequently encountered at greater depths. Thus, after capping, the containment of the CO_2 under supercritical conditions in a depleted well is feasible. As an approximation, in order to determine the annual volume being made available for CO_2 disposal, it is assumed in this study that this is equal to the annual production of oil and gas. Actually, a larger volume is available during the first year of disposal because of the inventory of depleted oil and gas wells accumulated over the past 100

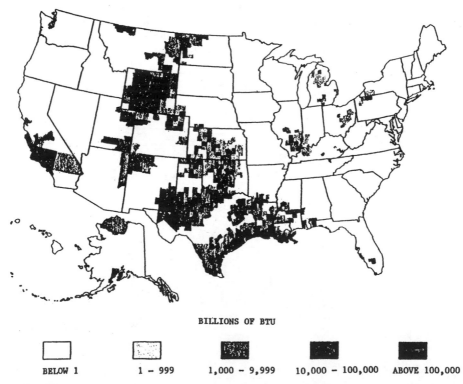

BILLIONS OF BTU

| BELOW 1 | 1 – 999 | 1,000 – 9,999 | 10,000 – 100,000 | ABOVE 100,000 |

FIGURE 5.5   Location and distribution of the annual oil production by county in the U.S.

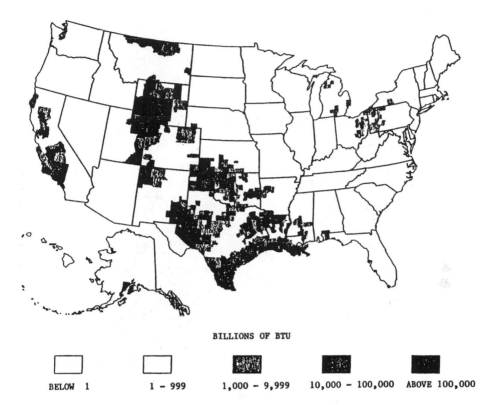

BILLIONS OF BTU

| BELOW 1 | 1 - 999 | 1,000 - 9,999 | 10,000 - 100,000 | ABOVE 100,000 |

FIGURE 5.6   Location and distribution of the annual natural gas production by county in the U.S.

years of production. Figures 5.5 and 5.6 show the location and distribution of the oil and gas annual production, respectively, in the U.S.[28]

5.5.2.3 Disposal in Salt Caverns

Salt caverns are another potential CO_2 disposal site. The mining of multimillion-barrel salt caverns for strategic petroleum reserve is routine in the U.S.[29] A solution-mining technique has been used to produce large salt caverns. Figure 5.7 shows the locations of all underground 1977 gas reservoirs and salt deposits in the U.S.[30] In addition to the Gulf area, other large salt caverns are already used to store gas in Ohio, Michigan, Pennsylvania, West Virginia, and New York. These could be used for CO_2 storage. In the West, the potential sites of salt domes include Paradox Basin in Utah, the Piercement salt masses in Colorado, salt domes in Supai Basin in Arizona, and some salt masses in North Dakota. Salt domes for gas storage can contain gas pressures in the range of 3000 psia or higher. The salt is plastic and self-seals any cracks. The cost of excavating a salt dome varies with the availability of water for solution mining and the accessibility of salt solution disposal areas, and

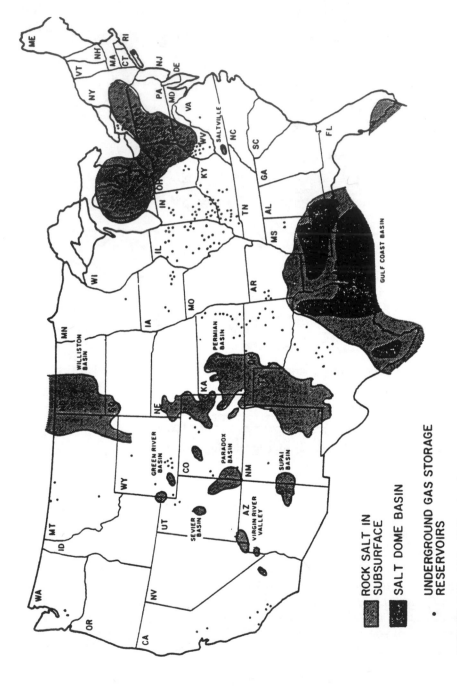

FIGURE 5.7 Major U.S. subsurface salt deposits and the location of underground gas storage reservoirs — 1977.

it has been estimated to be in the range of $1.00 to $3.50 a barrel of volume excavated.[29,30] The western sites are unsuitable for solution mining because of water shortage, and therefore only oil and gas wells are used in the West. In the Great Lakes region, however, solution mining should be feasible because of the availability of fresh water. Esthetically, putting salt in the ocean and CO_2 in the wells appears proper but may not make economic sense.

5.5.2.4 Disposal in Aquifers

Figure 5.8 shows the location of deep saline aquifers underlying the U.S.[17] The location match between power plant sources and storage of CO_2 appears better than for oil and gas reservoirs. It is estimated that as much as 65% of CO_2 removed from power plants in the U.S. can be injected into aquifers.[17] Aquifer structure for permeability porosity fracture and linkage must be investigated for acceptable sequestering. Recently, the Statoil Company of Norway has instituted a test program in the North Sea which removes CO_2 from gas production containing 10% CO_2 and reinjects the CO_2 below the seabed in aquifers to the extent of a million tons per year. One incentive is that Norway has a carbon tax of $55/ton of CO_2. Also, Exxon has been studying the removal, recovery, and disposal of CO_2 in aquifers from gas wells that contain as much as 70% CO_2 in the Naturna region of Indonesia[18] for a multibillion-dollar gas production project.

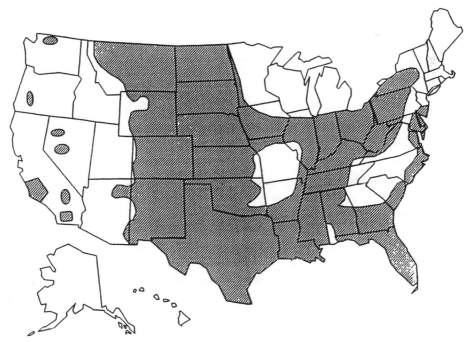

FIGURE 5.8   Saline aquifers in the U.S. based on U.S. Geological Survey.[17]

5.5.2.5 *Disposal in Coal Beds*

Consideration is being given to sequestering CO_2 in coal beds where methane exists. Methane is absorbed in coal beds and is released when mining occurs. CO_2 may be able to replace natural gas existing in deep coal deposit.[45] It is estimated that about twice the capacity of CO_2 can be sequestered than the methane content in these beds. Thus, CO_2 from power plants can be pumped down into the beds and methane can be pumped out. Presumably these deep-bedded coal seams cannot be economically mined for coal.[15] It is also estimated that the resources of natural gas in these deep coal deposits are very extensive in various parts of the world.

5.6 REGIONAL CO$_2$ DISPOSAL SYSTEMS FOR FOSSIL-FUEL POWER PLANTS

In order to study the feasibility of the disposal system for the recovered liquid CO_2 from the stacks of fossil-fuel plants in different areas of the U.S., along with their energy requirements, capital investments, and power generation cost, the design of the disposal system is laid out on a regional basis according to the U.S. Department of Energy federal regions for utilities.[31] There are ten federal regions, as shown in Figure 5.9. Data on the number of fossil fuel power plants and the power capacity in each of these regions are presented in Figure 5.10 for the 1980 base year.

Generally, the design concept for the disposal system is to pump the liquid CO_2 recovered from each of the fossil-fuel-burning power plants in the particular region via a small pipeline (6-in. diameter) to a collection center from which a larger pipeline (36-in. diameter) transports the liquid CO_2 to the ultimate disposal site (e.g., ocean depth, oil and gas wells, excavated salt domes, or aquifers). The pipeline size was chosen to optimize cost (balancing the cost of pipeline and operating cost for pumping the liquid CO_2) and for size uniformity. It is interesting to note that liquid CO_2–coal slurry pipelines have been proposed to convey finely ground coal from mine site to the coal-burning load center.[32]

5.6.1 Regional CO$_2$ Disposal Design

Design data for CO_2 disposal systems for each region are shown in Tables 5.9A through 5.9C. One notes that ocean disposal is used for coastal regions, while salt cavern, depleted oil and gas wells, and aquifer disposal are more appropriate for the inland regions, particularly Regions 5, 7, and 8. The details of the disposal systems for Regions 7 and 8 are found in Tables 5.10 and 5.11, respectively.

The recovered carbon dioxide is transported by 6-in. pipeline from each power plant to nearby collection centers where the liquid CO_2 in each region is collected for final disposal by using a 36-in. pipeline. Table 5.10 also shows the average length of 6-in. pipeline from power plants to collection centers and the locations, along with the number of 36-in. pipelines for each collection center in each region. The average 6-in. line length in each region is estimated by averaging the distance between each of the power plants and collection centers as follows:

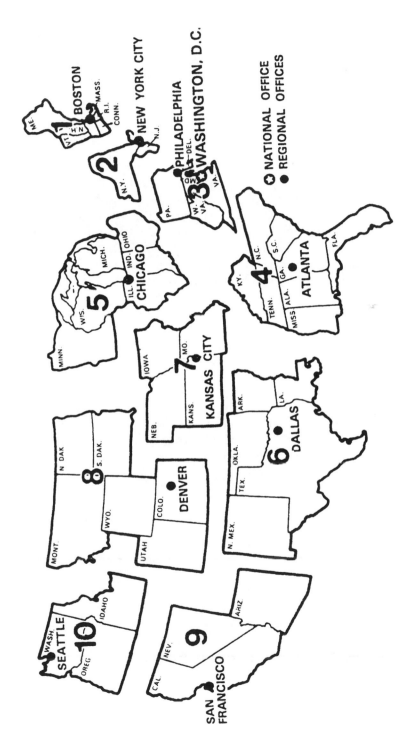

FIGURE 5.9 U.S. Department of Energy federal regions for utility operations.

Ave. 6-in. line length

$$= \frac{\displaystyle\sum_{\substack{\text{collection} \\ \text{centers}}} \left[\sum_{\substack{\text{power} \\ \text{plants}}} \left(\begin{array}{l} \text{length of pipeline from power} \\ \quad\text{plants to collection centers} \end{array} \right) \right]}{\text{total number of power plants}} \quad (5.2)$$

The power plant distributions in each region shown in Figures 5.11 through 5.20 are based on data on the location of power plants of capacities of 40 MW(e) or greater obtained from the National Emission Data System 1977 and computerized at Brookhaven National Laboratory.[33,34] Each power plant in the region was connected to a collection center with its separate 6-in. line. The number of 36-in. lines at each collection center was estimated from the rate of CO_2 collected at each center divided by the optimum capacity of a 36-in. line. An optimum capacity of 15 million lb of CO_2 per hour, which corresponds to a liquid CO_2 velocity of 11 ft/sec, was used in this study. The rate at which CO_2 is collected at the collection center is assumed to be proportional to the capacity and number of power plants piping the CO_2 generated to the center.

It should be noted in Table 5.9 that two alternative disposal systems are given for Regions 3 and 5. In Region 3, one is given for ocean disposal and the other for ocean and salt caverns. In Region 5, the designs are for ocean and salt caverns.

Once the disposal system for each region is designed, the total cost of the pipeline which includes 6- and 36-in. lines can be estimated. The pipeline cost was estimated from data given in Reference 36 and indexed to the 1980 base year used in this study and the pipeline length given in Table 5.9. For ocean disposal, pipeline capital investment per unit of net electrical production varies from a low of $55/kW(e) in Region 2 to a high of $146/kW(e) in Region 10. By disposing of the CO_2 from

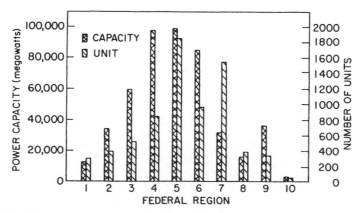

FIGURE 5.10 Distribution of number of fossil-fuel power plants and power capacity in each of the federal regions — 1980 base year.

TABLE 5.9A
CO_2 Disposal System Designs Based on the Federal Regions

Federal region	Disposal site[a]	Average length of 6-in. pipe from power plants to collection centers (miles)	Location of coastal collection centers and number of 36-in. pipelines for ocean disposal at each center
1	Ocean	86	Boston, MA (1) New London, CT (1)
2	Ocean	88	New York, NY (3) Atlantic City, NJ (1)
3	Ocean	130	New York, NY Atlantic City, NJ Baltimore, MD (5) Norfolk, VA (1)
3	Ocean and salt cavern	67	New York, NY Atlantic City, NJ Baltimore, MD (1) Norfolk, VA (1)
4	Ocean	220	Wilmington, NC (3) Savannah, GA (3) Tampa, FL (2) Mobile, AL (3)
5	Ocean	740	Baltimore, MD (11) Mobile, AL (1)

[a] Ocean disposal sites are 100 miles from the coast as shown in Figures 5.11 to 5.14, 5.16, 5.19, and 5.20.

TABLE 5.9B
CO_2 Disposal System Designs Based on the Federal Regions

Federal region	Disposal site[a]	Average length of 6-in. pipe from power plants to collection centers (miles)	Location of coastal collection centers and number of 36-in. pipelines for ocean disposal at each center
5	Salt cavern	220	0
6	Ocean and oil and gas wells	130	New Orleans, LA (1) Lake Charles, LA (1) Houston, TX (4)
7	Oil and gas wells	220	0
8	Oil and gas wells	150	0
9	Ocean	130	San Francisco, CA (1) Los Angeles, CA (2) San Diego, CA (1)
10	Ocean	150	Astoria, OR (1)

[a] Ocean disposal sites are 100 miles from the coast as shown in Figures 5.11 to 5.14, 5.16, 5.19, and 5.20.

TABLE 5.9C
CO_2 Disposal System Designs for Various Federal Regions

Federal region	Location of collection centers and number of 36-in. pipelines for oil, gas, and salt cavern disposal	Pipeline unit installed capital cost ($/kW(e) generated)
1	0	105
2	0	54.6
3	0	60.4
3	Pittsburgh, PA (2) Charleston, WV (2)	33.3
4	0	96.6
5	0	64.8
5	Columbus, OH (5) Chicago, IL (5) Lansing, MI (2)	181
6	Oklahoma City, OK (1) Carlesbad, NM (1) — Midland, TX	70.3
7	Springfield, MO (2) — Shreveport, LA Wichita, KS (2) — Dallas, TX	500
8	Gillette, WY (1) Raton, CO (2) — Lubbock, TX	194
9	0	62.7
10	0	146.1

TABLE 5.10A
CO_2 Disposal System for Federal Region 7

State	Capacity [MW(e)] (no. of units)			CO_2 recovered annually (10^6 ft³)		
	Coal	Oil	Gas	Coal	Oil	Gas
Iowa	4,284 (81)	1,456 (377)	152 (49)	1,132 (to LA)	294 (to OK)	23 (KS oil wells)
Missouri	10,914 (60)	1,628 (171)	1,669 (86)	2,885 (to TX)	329 (to LA)	250 (to OK)
Nebraska	2,345 (15)	679 (119)	523 (136)	620 (to TX)	137 (to OK)	78 (KS gas wells)
Kansas	4,025 (14)	806 (272)	3,593 (176)	1,064 (to TX)	163 (KS gas wells)	539 (to TX)

Note: Collection centers: Wichita (160 miles) → Oklahoma City, OK (200 miles) → Dallas. Springfield (360 miles) → Shreveport, LA.

TABLE 5.10B
CO$_2$ Disposal System for Federal Region 7

State	Annual capacity of wells (10^6 ft^3)	
	Oil	Gas
Iowa	0	0
Missouri	0	0
Nebraska	0	0
Kansas	37	279

TABLE 5.11A
CO$_2$ Disposal System for Federal Region 8

State	Power capacity [MW(e)] (no. of units)			CO$_2$ recovered annually (10^6 ft^3)		
	Coal	Oil	Gas	Coal	Oil	Gas
North Dakota	2263 (15)	100 (48)	2 (1)	598 (440 to WY gas wells)	20 (to ND oil and gas wells)	0.3
South Dakota	480 (7)	466 (92)	0 (0)	127 (WY oil wells)	94 (WY oil wells)	0.0
Montana	939 (4)	134 (8)	30 (3)	248 (200 to UT gas wells)	27 (MI oil and gas wells)	4.5
Wyoming	3965 (18)	82 (14)	16 (11)	1048 (to TX)	17 (WY oil and gas wells)	2.4
Colorado	4192 (30)	792 (60)	296 (39)	1108 (to TX)	160 (to CO oil and gas wells)	44.3
Utah	2224 (13)	46 (19)	31 (13)	588 (to TX)	9 (UT oil and gas wells)	4.6

Note: Collection centers: Gillette, WY (100 miles) → Wyoming oil and gas wells. Raton, CO (350 miles) → Lubbock, TX.

TABLE 5.11B
CO_2 Disposal System for Federal Region 8

State	Annual capacity of wells (10^6 ft^3)	
	Oil	Gas
North Dakota	66	112
South Dakota	0	0
Montana	26	80
Wyoming	266	453
Colorado	100	169
Utah	120	203

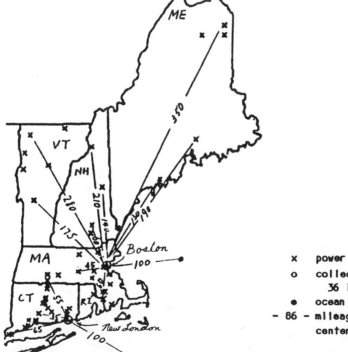

x	power plant location
o	collection center- 36 in. line
●	ocean disposal location
- 86 -	mileage to collection center- 6 in. line

FIGURE 5.11 CO_2 disposal system for Federal Region 1 — location and distribution of power plants and distance of 6-in. pipelines from power plants to collection centers for ocean disposal. (Legend in figure applies to Figures 5.12 to 5.20.)

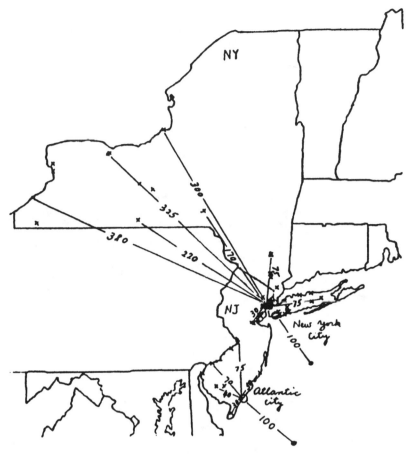

FIGURE 5.12 CO_2 disposal system for Federal Region 2 — location and distribution of power plants and distance of 6-in. pipelines from power plants to collection centers for ocean disposal.

eastern Pennsylvania and West Virginia into the salt caverns with Pittsburgh and Charleston as collection centers, respectively, the pipeline capital investment per unit of electrical production can be reduced to almost half the ocean disposal system alone. The pipeline investment for oil, gas, and salt dome disposal runs from $181/ kW(e) for Region 5 to $500/kW(e) for Region 7.

Because of the availability of fresh water from the Great Lakes and the possibility of disposing of the salt water into the Great Lakes, solution mining can be used to produce large salt caverns for CO_2 storage for the inland Federal Region No. 5. If saltwater disposal to the Great Lakes is not permitted, transport and disposal via a relatively small 6-in. line to the Atlantic Ocean is feasible. The latter was taken into account in the capital investment. The unit pipeline cost for salt caverns storage, which includes the 6-in. line for saltwater ocean disposal, is much less than one-

FIGURE 5.13 CO_2 disposal system for Federal Region 3 — location and distribution of power plants and distance of 6-in. pipelines from power plants to collection centers for ocean and salt cavern disposal.

third that of ocean disposal in this region. In general, for inland federal regions, due to the shorter pipeline distance, the pipeline costs for oil and gas wells and/or salt cavern disposal are less than those of ocean disposal. Table 5.12 shows the allocation of the CO_2 recovered from fossil-fuel power plants to the various disposal sites.

5.6.2   Energy Requirements

In Table 5.13, the number of units and the power capacity are given for coal-, oil-, and gas-fired power plants in each region. By employing the CO_2 generation data shown in Table 5.4, the rate of CO_2 generated for various fuels in each of the regions is calculated, and the power required for recovering 90% of the CO_2 generated, which includes energy required for CO_2 compression and liquefaction for each

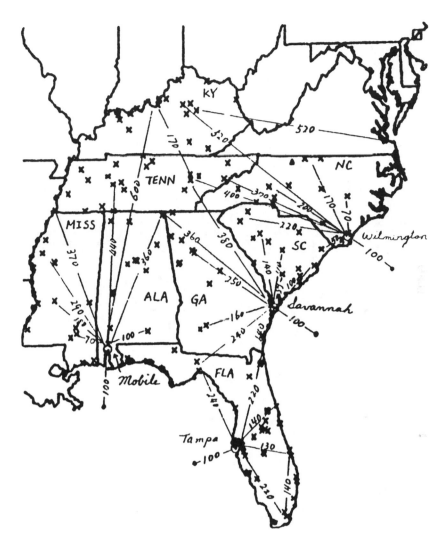

FIGURE 5.14 CO_2 disposal system for Federal Region 4 — location and distribution of power plants and distance of 6-in. pipelines from power plants to collection centers for ocean disposal.

region, can then be estimated as described in Section 5.5.1. The pumping power required to transport CO_2 from each power plant to the collection centers can also be calculated since the total length of 6-in. line has been estimated.

The main feature noted in Table 5.13 is that the total power required for the entire disposal system represents from 11 to 16% of the power-generating capacity in each of the regions. Furthermore, the power required for the removal and recovery of CO_2 is much greater than the pumping power required for the disposal of CO_2. At

FIGURE 5.15 CO_2 disposal system for Federal Region 5 — location and distribution of power plants and distance of 6-in. pipelines from power plants to collection centers for salt cavern disposal.

most (Region 4), the pumping power represents less than 5% of the removal–recovery (including liquefaction) power.

One can see from Table 5.13 that in the coal-burning regions (e.g., Regions 3 to 5), the CO_2 control system reduces the power-generating capacity by about 16% as opposed to 11% for gas-consuming regions. The average national power consumption for the total CO_2 control system designed in this study is about 15% of total national electrical-power-generating capacity. This is probably a lower limit because of the need to make up for the loss of capacity due to the recovery and disposal system. This "avoidance power cost" can add another 3% to the power consumption.

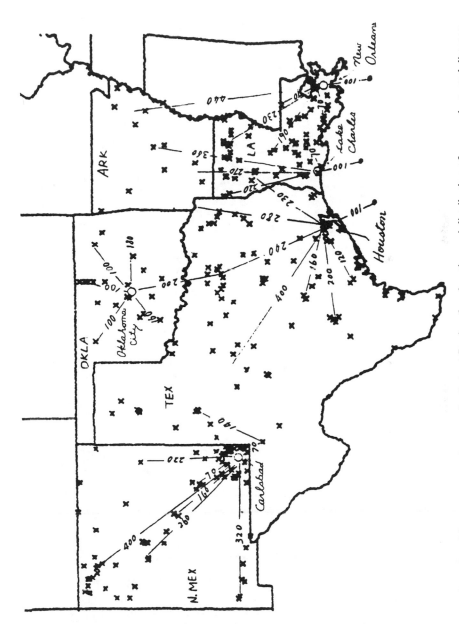

FIGURE 5.16 CO_2 disposal system for Federal Region 6 — location and distribution of power plants and distance of 6-in. pipelines from power plants to collection centers for ocean and oil and gas wells disposal.

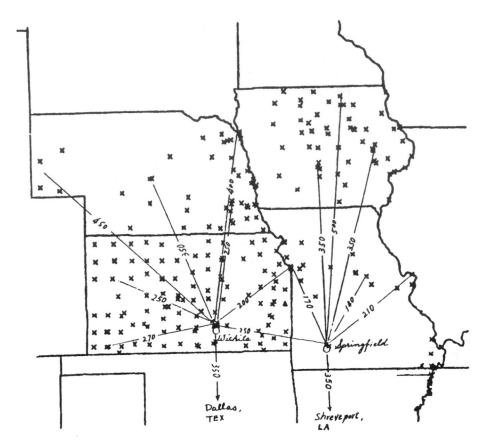

FIGURE 5.17 CO_2 disposal system for Federal Region 7 — location and distribution of power plants and distance of 6-in. pipelines from power plants to collection centers for oil and gas wells disposal.

Because the pumping power required is small, the total net power difference between ocean and salt cavern disposal for Region 5 is about 1%, despite the threshold differences in 6-in. line length.

5.6.3   Capital Investments and Power Generation Costs

Estimates of the required capital investment and the electricity production costs for power plants, for 90% of CO_2 recovery using Dow solvent, and for liquid disposal in each federal region as shown in Table 9, are presented in Tables 5.14 and 5.15, respectively. It is noted that the CO_2 control system represents an appreciable fraction of the capital investment in power plants, ranging from 70% in the oil- and gas-burning regions to as high as 150% in the coal-burning regions. A large portion

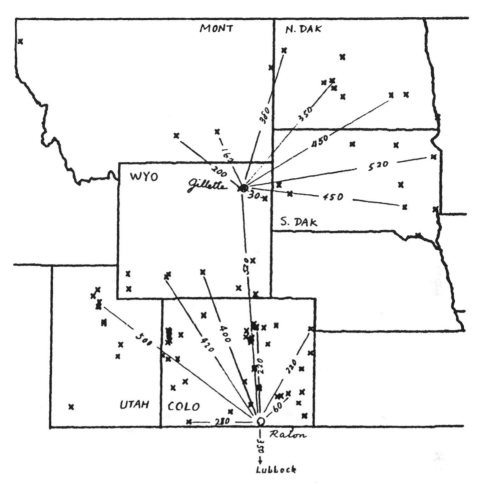

FIGURE 5.18 CO_2 disposal system for Federal Region 8 — location and distribution of power plants and distance of 6-in. pipelines from power plants to collection centers for ocean disposal.

of the control system capital investment and electricity production costs is attributed to the CO_2 removal and recovery process and a smaller portion to the disposal piping system. In such inland regions as Regions 5, 7, and 8, the capital costs for the removal, recovery, and storage of CO_2 can exceed the costs of the power plant itself because of the greater dependence on coal. On average, the CO_2 control system could double the power plant capital costs in the U.S.

The increase in power generation costs for those inland regions that normally use more coal can amount to twice the generating costs without the CO_2 control system. However, in the coastal regions, the increase in power-generating cost with CO_2 control is about 60% of the generating cost without CO_2 control. Thus, it is calcu-

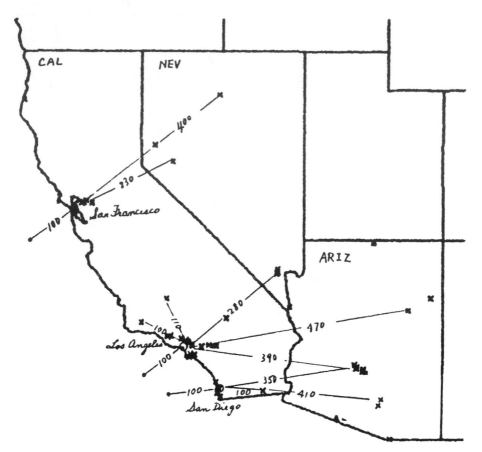

FIGURE 5.19 CO_2 disposal system for Federal Region 9 — location and distribution of power plants and distance of 6-in. pipelines from power plants to collection centers for ocean disposal.

lated that the average increase in the power generation cost could reach more than 70% the cost of conventional power plants if the CO_2 control systems were to be implemented.

5.7 SUMMARY OF FINDINGS OF THE SYSTEMS STUDY

The U.S. Department of Energy Federal Regional System, which divides the U.S. electrical-generating plant capacity into ten separate geographical areas, was utilized to design the disposal system for all power plants with a capacity above 40 MW(e). Ocean disposal was used for the U.S. coastal federal regions. Gas, oil, excavated salt wells, and deep saline aquifers were employed for the inland re-

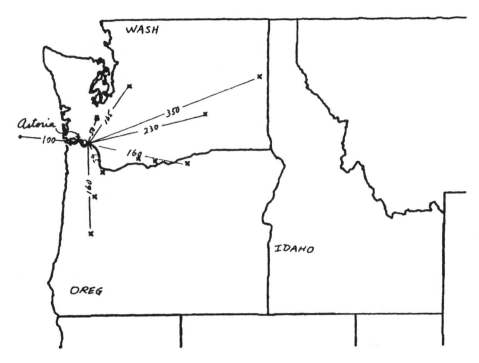

FIGURE 5.20 CO_2 disposal system for Federal Region 10 — location and distribution of power plants and distance of 6-in. pipelines from power plants to collection centers for ocean disposal.

TABLE 5.12
CO_2 Recovered from U.S. Fossil-Fuel Power Plants Allocated to the Various Disposal Sites[a]

Disposal site	Option 1: ocean and oil and gas wells		Option 2: ocean, salt dome, and oil and gas wells	
	10^9 tons/yr	%	10^9 tons/yr	%
Ocean	0.665	86	0.435	56
Oil and gas wells	0.105	14	0.105	14
Excavated salt domes	0.000	0	0.230	30
Total	0.770	100	0.770	100

[a] 90% CO_2 recovery from power plant operating 24 hr/day, 350 days/yr.

TABLE 5.13A
**Power Requirements for CO_2 Control System and Power Capacity
for Each of the Federal Regions, Base Year 1980**

Federal region fuel type	Electrical capacity [MW(e)]	Number of power plants	Rate of CO_2 generated (10^3 lb/hr)	Power required for CO_2 removal and recovery[a] [MW(e)]
Region 1				
Coal	489.2	5	890	
Oil	12,098.3	290	17,280	
Gas	8.9	9	9	
Subtotal	12,596.4	304	18,180	1,678
Region 2				
Coal	4,185.2	38	7,610	
Oil	29,731.9	342	42,470	
Gas	191.0	10	200	
Subtotal	34,108.1	390	50,280	4,638
Region 3				
Coal	42,239.6	146	76,800	
Oil	17,196.0	356	24,570	
Gas	236.6	13	240	
Subtotal	59,672.2	515	101,600	9,324
Region 4				
Coal	65,095.5	265	118,400	
Oil	26,822.5	456	38,320	
Gas	5,638.5	121	5,810	
Subtotal	97,556.5	842	162,500	14,910
Region 5				
Coal	77,640.9	557	141,200	
Oil	19,117.3	1,032	27,310	
Gas	2,006.6	263	2,070	
Subtotal	98,764.8	1,852	170,600	15,640

[a] Power requirement for CO_2 removal and recovery includes the energy (thermal and electrical) required for separating 90% of CO_2 from flue gas and that required for liquefaction of recovered CO_2. The total power required is given in MW(e) equivalents.

gions. Small (6-in.-diameter) pipelines carry the recovered liquefied CO_2 from each power plant to collection centers in each region, and from there the CO_2 is transmitted in larger lines (36-in. diameter) to the ultimate disposal site. An improved MEA solvent absorption/stripping system removes and recovers CO_2 from each power plant stack. The operating and investment cost for the entire control

TABLE 5.13B
Power Requirements for CO_2 Control System and Power Capacity for Each of the Federal Regions, Base Year 1980

Federal region fuel type	Electrical capacity [MW(e)]	Number of power plants	Rate of CO_2 generated (10^3 lb/hr)	Power required for CO_2 removal and recovery[a] [MW(e)]
Region 6				
Coal	19,673.7	42	35,770	
Oil	5,071.3	189	7,240	
Gas	60,040.4	737	61,900	
Subtotal	84,785.4	968	104,900	9,611
Region 7				
Coal	21,566.6	170	39,210	
Oil	4,553.0	934	6,500	
Gas	5,936.1	447	6,120	
Subtotal	32,055.7	1,551	51,580	4,752
Region 8				
Coal	14,063.1	87	25,570	
Oil	1,619.2	240	2,310	
Gas	347.2	57	350	
Subtotal	16,056.5	384	28,230	2,612
Region 9				
Coal	6,202.9	15	11,280	
Oil	28,903.2	305	41,290	
Gas	1,364.7	18	1,410	
Subtotal	36,470.8	338	53,980	4,974
Region 10				
Coal	1,860.0	3	3,380	
Oil	1,014.6	46	1,450	
Gas	171.0	6	180	
Subtotal	3,045.6	55	5,010	460
Total	475,112.4	7,196	747,100	68,600

[a] Power requirement for CO_2 removal and recovery includes the energy (thermal and electrical) required for separating 90% of CO_2 from flue gas and that required for liquefaction of recovered CO_2. The total power required is given in MW(e) equivalents.

system is estimated. Approximately 16% of the power plant capacity is required to drive the entire control system in those regions where coal is mainly consumed, whereas only about 11% is required where predominantly gas is used. Over 95% of the power used by the control system is needed for the removal, recovery, and

TABLE 5.13C
**Power Requirements for CO_2 Control System and Power Capacity
for Each of the Federal Regions, Base Year 1980**

Federal region fuel type	Pumping power required to dispose of CO_2 [MW(e)]	Net electrical capacity[a] [MW(e)]	% reduction in electrical capacity due to CO_2 control
Region 1			
Coal			
Oil			
Gas			
Subtotal	4	10,910	13.4
Region 2			
Coal			
Oil			
Gas			
Subtotal	41	29,430	13.7
Region 3			
Coal			
Oil			
Gas			
Subtotal	271	50,080	16.1[b]
	140	50,210	15.9[c]
Region 4			
Coal			
Oil			
Gas			
Subtotal	687	81,960	16.0
Region 5			
Coal			
Oil			
Gas			
Subtotal	633	82,490	16.5[b]
	171	82,960	16.0[d]

[a]   Net power capacity = total power capacity less the power requirement for CO_2 removal and recovery pumping power for CO_2 disposal.

[b]   Ocean disposal.

[c]   Ocean and salt cavern disposal.

[d]   Salt cavern disposal.

TABLE 5.13D
Power Requirements for CO_2 Control System and Power Capacity
for Each of the Federal Regions, Base Year 1980

Federal region fuel type	Pumping power required to dispose of CO_2 [MW(e)]	Net electrical capacity[a] [MW(e)]	% reduction in electrical capacity due to CO_2 control
Region 6			
Coal			
Oil			
Gas			
Subtotal	114	75,060	11.5
Region 7			
Coal			
Oil			
Gas			
Subtotal	9	27,300	14.8
Region 8			
Coal			
Oil			
Gas			
Subtotal	14	13,430	16.4
Region 9			
Coal			
Oil			
Gas			
Subtotal	97	31,400	13.9
Region 10			
Coal			
Oil			
Gas			
Subtotal	4	2,582	15.2
Total	1,874	404,600	14.8[b]
	1,281	405,200	14.7[c,d]

[a] Net power capacity = total power capacity less the power requirement for CO_2 removal and recovery pumping power for CO_2 disposal.

[b] Ocean disposal.

[c] Ocean and salt cavern disposal.

[d] Salt cavern disposal.

TABLE 5.14A

Capital Investment for Power Plants and CO_2 Control System for Each of the Federal Regions (in Billion [10^9] 1980 U.S. Dollars)

Region	Power plant[a]	CO_2 removal and recovery[b]	CO_2 compression and liquefaction[c]	Piping[d]	Salt cavern excavating[e]
1	7.2	2.7	2.0	1.1	0
2	19.4	7.6	5.4	1.6	0
3	34.0	15.4	11.0	3.0	0[f]
				1.7	23[g]
4	55.6	24.5	17.5	7.9	0
5	56.3	25.8	18.4	53.4	0[f]
				15.0	3.8[g]
6	48.3	15.9	11.3	5.2	0
7	18.3	7.9	5.6	13.7	0
8	9.2	4.3	3.0	2.6	0
9	20.8	8.1	5.8	2.0	0
10	1.7	0.8	0.5	0.4	0
Total	271	113	81	91	0[f]
				51	6.1[g,h]

a   At \$570/kW(e) which was estimated from Reference 37.

b   Calculated from Reference 23.

c   Calculated from Reference 22.

d   Includes 6-in. and 36-in. line costs and estimated from Reference 36.

e   At \$1.00/bbl based on Reference 38.

f   Ocean disposal.

g   Ocean and salt cavern disposal.

h   Salt cavern disposal.

liquefaction of CO_2 at the power plant sites, and the remaining 5% is used for pumping to ultimate disposal sites. The overall capital investment required for the control system ranges from 70 to 150% of the existing capital investment in the power plants, depending on the region. Thus, on average, the CO_2 control system could double the investment cost for power generation. Approximately 85 to 90% of the investment is in the absorption/stripping removal and recovery system at the power plant site. The remaining 10 to 15% is accounted for by the disposal piping and pumping investment. Depending on the region of the country, the production cost of electricity with the control system would increase by 56 to 100% over existing costs without the control system. Thus, the average cost of electricity would increase by 75%. For a 3000-m-depth ocean disposal, the increase in capacity in capital investment and production cost of electricity is only slightly higher than for a 500-m-depth disposal.

TABLE 5.14B
Capital Investment for Power Plants and CO_2 Control System for Each of the Federal Regions (in Billion [10^9] 1980 U.S. Dollars)

Region	Total CO_2 control system cost	Total cost for plant and control system	Unit cost for control system ($/kW(e) net generated)	CO_2 control system (% of power plant investment)	CO_2 control system (% of total investment)
1	5.8	13.0	532	80.6	44.6
2	14.6	34.0	496	75.3	42.9
3	29.4	63.4	587	86.5	46.4[a]
	30.4	64.4	605	89.4	47.1[b]
4	49.9	105.5	609	89.7	47.3
5	97.6	153.9	1183	173.4	63.4[a]
	63.0	112.0	759	111.9	56.2[c]
6	32.4	80.7	432	67.1	40.1
7	27.1	45.5	993	148.1	59.7
8	9.9	19.1	737	107.6	51.8
9	15.9	36.7	506	76.4	43.3
10	1.7	3.4	658	100.0	50.0
Total	284	555	673	104.8	51.2[a]
	251	514	633	92.6	48.8[b,c]

[a] Ocean disposal.

[b] Ocean and salt cavern disposal.

[c] Salt cavern disposal.

5.8 COMPARISON OF CAPTURE AND DISPOSAL COSTS

A number of studies on removal, recovery (capture), and disposal of CO_2 from coal-fired power plants have been published over the last few years.[2,9,14,42,43]

Table 5.16 lists the energy penalty and the projected costs of CO_2 capture for a number of capture technologies. The energy penalty shown includes the energy to operate the removal and recovery plant, which is also defined as the avoided CO_2 penalty. There are several plants in the U.S. that remove and recover CO_2 from coal-fired power plants[46] for use in food preservation. The cost of production of CO_2 is reported to be as low as $28/tonne. What is more difficult to assess, because of lack of experience, is the disposal cost for sequestration either in the ocean or underground in depleted gas wells, aquifers, or excavated salt domes. Disposal cost also depends on the pipeline distance from source to the disposal site. The estimates vary from $15 to $50/tonne CO_2 for 100 km distance. When added to the capture cost, the cost of electricity could increase by 50 to 100% for a solvent system to over a 200% increase for adsorption and cryogenic separation.[9,42] In general, for not too long distances, the disposal cost is usually much lower than the capture cost.

TABLE 5.15A
Electricity Production Costs of Power Plants with CO_2 Control Systems for Various Federal Regions (Mills/kWh(e) in 1980 U.S. Dollars)

Region	Power plant[a]	CO_2 removal and recovery[b]	CO_2 compression and liquefaction[c]	Piping[d]	Salt cavern excavating[e]
1	32	11.3	7.2	2.5	0.0
2	32	11.3	7.2	1.3	0.0[d]
3	32	13.8	7.9	1.3	0.0
				0.8	1.0[e]
4	32	13.5	7.8	2.3	0.0
5	32	14.1	8.0	15.5	0.0[d]
				4.3	1.0[f]
6	32	9.4	6.7	1.7	0.0
7	32	12.9	7.6	11.9	0.0
8	32	14.4	8.0	4.6	0.0
9	32	11.7	7.3	1.9	0.0
10	32	13.2	7.7	3.5	0.0
Ave. total	32	12.6	7.5	4.7	0.0[d]
				3.5	0.02[e,f]

[a] Based on the same fuel cost for all regions and estimated from Reference 37.

[b] Estimated from Reference 23.

[c] Capital charge is 15% of capital investment in CO_2 compression, liquefaction, and piping, based on 24 hr/day, 350 annual operating days.

[d] Ocean disposal.

[e] Ocean and salt cavern disposal.

[f] Salt cavern disposal.

5.9   IMPACT OF REMOVAL OF CO_2 FROM POWER PLANT STACKS ON CARBON BALANCE IN THE ATMOSPHERE

In order to note the implications of this U.S. systems study for the removal, recovery, and disposal of CO_2, the data of Woodwell et al.[35] on the carbon balance in the atmosphere can be utilized. Taking the median case, the carbon balance for 1980 in gigatons per year is shown in Table 5.17.

The U.S. is responsible for about 30% of the total world emissions of fossil-fuel carbon, and the CO_2 emitted from U.S. fossil-fuel-burning power plants amounts to about 30% of the total fossil-fuel release in the U.S. Table 5.18 gives the amount and fraction of CO_2 controlled if only the U.S. power plants practice removal, recovery, and disposal of CO_2. A value for the rest of the world is also given

TABLE 5.15B
Electricity Production Costs of Power Plants with CO_2 Control Systems for Various Federal Regions (Mills/kWh(e) in 1980 U.S. Dollars)

Region	Total CO_2 control system operating cost	Total production cost for electricity	CO_2 control system (% of power plant production cost)	CO_2 control system (% of total production cost)
1	21.0	53.0	65.6	39.6
2	19.8	51.8	61.9	38.2
3	23.0	55.0	71.9	41.8[a]
	23.5	55.5	73.4	42.3[b]
4	23.6	55.6	73.7	42.3
5	37.6	69.6	117.5	54[a]
	27.4	59.4	85.6	46.1[c]
6	17.8	49.8	55.6	35.7
7	32.4	64.4	101.2	50.3
8	27.0	59.0	84.4	45.8
9	20.9	52.9	65.3	39.5
10	24.4	56.4	76.3	43.3
Ave. total	24.7	56.7	77.3	43.0[a]
	23.8	55.8	74.3	42.3[b,c]

[a] Ocean disposal.

[b] Ocean and salt cavern disposal.

[c] Salt cavern disposal.

TABLE 5.16
Comparison of CO_2 Capture and Disposal for Coal-Fired Power Plants[a]

Capture technology	Energy penalty for capture (%)	Cost of capture ($/tonne)
Base case (no CO_2 removal)	0	0
Hindered MEA absorption (advanced solvent and integrated system	17	20
Current MEA absorption (with cogenerated steam)	35	35
Air separation flue gas recycling	30	35
Molecular sieves	80	
Membrane separation	63	45
Cryogenic fractionation	75	
Adsorption	43	84

[a] Based on various sources including References 2, 9, 14, 42, and 43.

TABLE 5.17
1980 World Carbon Balance in Gigatons C/Year for Median Case

World increment of atm CO_2	=	Fossil-fuel release	–	Ocean uptake	+	Terrestrial carbon release	Total release	Imbalance	Airborne fraction incr/release
2.5	=	5.2	–	2.0	+	3.3	8.5	4.0	30%

assuming that 30% is the fraction of the total CO_2 generated by the large fossil-fuel-burning power plants over the rest of the world.

Thus, if only the U.S. removes, recovers, and disposes of CO_2 from its fossil-fuel power plants, the reduction in the annual incremental atmospheric CO_2 content would amount to a maximum of 10%. If the entire world practices CO_2 control, the reduction would amount to a maximum of 32%. These reductions, although significant in either delaying or possibly preventing the global warming effect, will probably be difficult to achieve.

5.10 LONG-TERM STORAGE OF CO_2 IN THE OCEAN

The effect of sequestering CO_2 in the ocean is to reduce the peaking of the CO_2 concentration in the atmosphere which causes the global warming. Eventually, after a very long period of time, extending more than 1000 years, the CO_2 in the ocean will equilibrate with the atmospheric CO_2. This is illustrated in Figure 5.21, which shows the qualitative effect of the ocean disposal on CO_2 concentration, based on a constant CO_2 emission rate for 250 years (starting from preindustrial time) and then no more discharge into the atmosphere or the ocean.[18,39] Line A represents no recovery and sequestration of CO_2 from all fossil-fuel combustion. The concentration may peak up to several times the current concentration, causing potential catastrophic climate changes. The concentration will eventually decrease after emission ceases and an equilibrium is established with the ocean. Lines B1, B2, and B3 show the effect of ocean storage with increasing CO_2 sequestration in the ocean by increasing the quantity sequestered and by increasing the residence time. Line C

TABLE 5.18
U.S. and World Power Plant Contribution to Increment of
Fossil-Fuel Release (in Gigatons C/Year) — 1980 Median Case

World fossil-fuel release	Released by U.S. power plants	Fraction of release due to U.S. power plants	Released by world power plants	Fraction of release due to world power plants
5.2	0.5	10%	1.6	32%

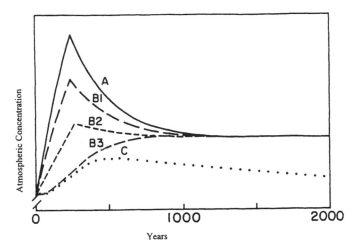

FIGURE 5.21 Qualitative effect of ocean disposal on atmospheric CO_2 concentration at an emission rate equal to that of the first 250 years of the Industrial Revolution.

(dotted) shows the potential of permanently sequestering CO_2 in the ocean due to chemical carbonate or solid CO_2 formation. Thus the benefit of sequestration in the ocean is to shave the peak of the CO_2 concentration in the atmosphere while burning fossil fuels over the next several hundred years. It may be necessary to achieve long-term permanent fixation of CO_2 in the deep ocean at depths greater than 3000 m to assure that CO_2 will remain in the ocean for thousands of years. In this respect, burying CO_2 under the ocean floor by shooting solid CO_2 torpedoes directly into the ocean bed has also been suggested, which could be an expensive proposition.[41]

5.11 PROBLEMS ASSOCIATED WITH SEQUESTERING CO$_2$ IN THE OCEAN

1. *Economic.* As seen earlier, the efficiency of power plants will be reduced and the cost of generating power from fossil fuel will increase significantly. Capital investments will be significant.
2. *Engineering problems.* The difficulty of laying deep CO_2 pipelines of large diameter to meet capacity is formidable. Towing large-diameter pipe extending to great depths in the ocean is also a challenging engineering problem. Formation of a liquid CO_2 lake on the ocean floor has never been investigated before.
3. *Environmental effects.* Injecting large quantities of CO_2 at specific points in the ocean could unbalance the ecology and the chemistry in these locations. The acidity of the ocean will increase in these locations, causing the pH to drop below 8, which is known to kill marine organisms.[40]
4. *Rapid release of sequestered CO$_2$.* It is necessary to assure that once the CO_2 is sequestered, it will not return to the atmosphere in large quantities

in relatively short times. Thermal plumes and volcanic action in the ocean could suddenly release the sequestered CO_2. This condition occurred when a lake in Africa[44] suddenly erupted, releasing huge quantities of CO_2 accumulated at the bottom of the lake, and caused the asphyxiation of thousands of animals and humans.

The social, political, and legal aspects on a national and international scale are beyond the scope of this volume but are critical to the acceptance of the disposal of CO_2 because of climate change effects. CO_2 and global climate change as it affects the use of fossil fuels might turn out to be analogous to how the problem of radioactive nuclear waste has slowed down the nuclear energy industry in the U.S.

APPENDIX 5.1: CALCULATION OF ENERGY REQUIREMENTS FOR CO$_2$ RECOVERY PLANT AS SHOWN IN TABLE 5.6

1. *Equivalent electrical energy required for removal and recovery of CO_2:*

$$4.8 \times 10^6 \ \frac{Btu}{ton} \times \frac{1 \ ton}{2000 \ lb} \times 2.934 \times 10^{-4} \ \frac{kWh(t)}{Btu}$$

$$\times \frac{(0.38 - 0.349)}{0.38} \ \frac{kWh(e)^*}{kWh(t)} + 13 \ \frac{kWh(e)}{ton \ CO_2} \times \frac{1 \ ton}{2000 \ lb}$$

$$= 0.065 \ kWh(e)/lb \ CO_2 \ recovered$$

2. *Energy required for compression and liquefaction of CO_2:*

$P_0 = 1$ atm $\qquad\qquad\qquad\qquad P_4 = 136.0$ atm (2000 psia)

Four-stage compressor

$T_0 = 120°F \qquad\qquad\qquad\qquad T_4 = 80°F$

Compression ratio $= P_1/P_0 = P_2/P_1 = P_3/P_2 = P_4/P_3 = (P_4/P_0)^{1/4} = 3.42$

Thus, the CO_2 four-stage compression can be represented as follows:

	Initial condition	First stage Comp.	First stage Cool	Second stage Comp.	Second stage Cool	Third stage Comp.	Third stage Cool	Fourth stage Comp.	Fourth stage Cool
Pressure (atm)	1.0	3.42	3.42	11.67	11.67	40	40	136	136
Temperature (°F)	120	140	30	123	30	130	30	128	27
Enthalpy (kcal/kg)	198	217.5	194	213	192.5	212	183.5	199	137.5
Entropy (kcal/kg)	1.175	1.175	1.11	1.11	1.05	1.05	0.965	0.965	0.775

---

* For integrated plant, see Appendix 5.2.

Therefore, power required by the compressors,

$$\Delta h = 217.5 - 198 + 213 - 194 + 212 - 192.5 + 199 - 183.5$$

$$= 73.5 \text{ kcal/kg} = 3.88 \times 10^{-2} \text{ kWh/lb } CO_2$$

A 2% increase in power is needed to allow for the extra power required because of the pressure drop through the intercooler. Assuming that the efficiency of the compressor is 85%, the actual energy required for compressing and dehydrating CO_2 to 2000 psia is

$$\Delta h = 3.88 \times 10^{-2} \times 1.02/0.85 = 0.047 \text{ kWh(e)/lb } CO_2$$

3. *Energy required for CO_2 disposal*: The velocity of CO_2 in 6-in. pipe is

$$\frac{1000 \text{ ton/day}}{24 \text{ hr}} \times \frac{2000 \text{ lb}}{1 \text{ ton}} \times \frac{1}{\frac{\pi}{4}\left(\frac{6}{12}\right)^2 \text{ ft}^2} \times \frac{1}{52 \text{ lb/ft}^3}$$

$$\times \frac{1 \text{ hr}}{3600 \text{ sec}} = 2.27 \text{ ft/sec}$$

and

Reynolds number, $N_{Re} = \dfrac{0.5 \times 2.37 \times 52}{0.528 \times 10^{-3} \times 6.72 \times 10^{-2}} = 1.66 \times 10^6$

Fanning friction factor, $f = 0.0027$ (from *Plant Design Handbook*, 1980)

Pressure drop $\Delta p = 2f\rho \dfrac{L}{D}\dfrac{v^2}{g_c}$

$$= 2 \times 0.0027 \times 52 \times \frac{100 \times 5280}{0.5} \times \frac{(2.27)^2}{32.2(12)^2}$$

$$= 329.5 \text{ psi}$$

Then, with a pumping efficiency of 60%,

Energy required $= 2v^2 \, fL/Dg_c\eta$

$$= 2 \times (2.27)^2 \times 0.0027 \times 100 \times 5280/(0.5 \times 32.2 \times 0.6)$$

$$= 1521 \text{ ft-lb}_f/\text{lb}_m = 0.0006 \text{ kWh(e)/lb } CO_2$$

The capacity of 36-in. line = 15×10^6 lb/hr = 21.58×10^5 gal/hr and

$$\text{Reynolds number, } N_{Re} = \frac{3 \times 11 \times 52}{0.528 \times 10^{-3} \times 6.72 \times 10^{-2}} = 4.84 \times 10^7$$

Fanning friction factor, $f = 0.0016$

$$\text{Pressure drop, } \Delta p = 2 \times 0.0016 \times 52 \times \frac{100 \times 5280}{3} \times \frac{(11)^2}{32.2 \times (12)^2}$$

$$= 760 \text{ psia}$$

Thus, the brake horsepower required for a 36-in. line is

$$\hat{W} = \frac{21.58 \times 10^5 \times 760}{1713 \times 0.6 \times 60 \times 15 \times 10^6} = 0.00177 \text{ hp-hr/lb}$$

$$= 0.001 \text{ kWh(e)/lb } CO_2$$

APPENDIX 5.2: CALCULATION OF POWER PLANT EFFICIENCY AFTER INTEGRATING POWER PLANT WITH CO_2 RECOVERY SYSTEM AS SHOWN IN TABLE 5.7

Basis: the amount of heat input to power plant boiler per ton of CO_2 generated (Q).

1. *Integrated plant using low-pressure steam from power plant in reboiler*: Assuming that all the heat required for CO_2 recovery system is supplied by the low-pressure steam, then the amount of low-pressure steam required can be found from

$$Q_a = m\,(h_A - h_C) \tag{A}$$

where Q_a = heat required for reboiler, m = amount of low-pressure steam needed, h_A = enthalpy of low-pressure steam, and h_C = enthalpy of output flow from reboiler.

Assuming the water leaving the reboiler is saturated at 20 psia, then

$$h_C = 196.27 \text{ Btu/lb}$$

The heat loss from the turbine due to extracting low-pressure steam is

$$Q_b = m\,(h_A - h_B) \tag{B}$$

where h_B = enthalpy of output steam from turbine.

Assuming the steam leaves the turbine at 0.5 psi and 80°F, then

$$h_B = 1096.4 \text{ Btu/lb}$$

Therefore, power plant efficiency,

$$E = \frac{Q - Q_b/(\eta_1 \times \eta_2)}{Q} \times 0.38$$

where η_1 = boiler efficiency, and η_2 = reboiler efficiency.
Let low-pressure steam at 100 psi and 400°F, $h_A = 1227.4$ Btu/lb,

$$= \frac{4.8 \times 10^6 \text{ Btu/ton } CO_2}{1227.4 - 196.27 \text{ Btu/lb}} = 4.655 \times 10^3 \text{ lb/ton } CO_2$$

$$Q_b = 4.655 \times 10^3 \times (1227.4 - 1096.4) = 0.61 \times 10^6 \text{ Btu/ton } CO_2$$

For coal, $Q_{coal} = 9.81 \times 10^6$ Btu/ton CO_2. Therefore,

$$E_{coal} = \frac{9.81 - 0.61/(0.91 \times 0.85)}{9.81} \times 0.38 = 0.349$$

For oil, $Q_{oil} = 12.48 \times 10^6$ Btu/ton CO_2. Therefore,

$$E_{oil} = \frac{12.48 - 0.61/(0.9 \times 0.85)}{12.48} \times 0.38 = 0.355$$

For gas, $Q_{gas} = 17.45 \times 10^6$ Btu/ton CO_2. Therefore,

$$E_{gas} = \frac{17.45 - 0.61/(0.9 \times 0.85)}{17.45} \times 0.38 = 0.363$$

2. *Nonintegrated plant using thermal energy from fuel directly*:

$$E_{coal} = \frac{9.81 - 4.8/(0.9 \times 0.85)}{9.81} \times 0.38 = 0.137$$

$$E_{oil} = \frac{12.48 - 4.8/(0.9 \times 0.85)}{12.48} \times 0.38 = 0.138$$

$$E_{gas} = \frac{17.45 - 4.8/(0.9 \times 0.85)}{17.45} \times 0.38 = 0.244$$

It should be noted that the thermal energy only to recover the CO_2 from the solvent is accounted for here. The energy for blowing the flue gas through the absorption column and liquefaction is accounted for in Appendix 5.1.

APPENDIX 5.3: CALCULATION OF NET POWER CAPACITY, CAPITAL INVESTMENT, AND ELECTRICITY PRODUCTION COST OF A 100-MW(E) COAL-FIRED POWER PLANT AS SHOWN IN TABLE 5.7 AND SUBSEQUENT COST ESTIMATES

The rate of CO_2 generation for a 100-MW(e) coal-fired power plant is

$$\frac{100 \text{ MW}}{0.55 \text{ kWh/lb } CO_2} = 181.8 \times 10^3 \text{ lb/hr}$$

1. *Energy required for CO_2 control system*
   a. Heat required at electricity equivalent for CO_2 recovery:

   $$\frac{100 \text{ MW} \times (0.38 - 0.35)}{0.38} = 7.9 \text{ MW}$$

   b. Electrical power required for CO_2 control system:

   $$181.8 \times 10^3 \text{ lb/hr} \times 0.9 \times 0.0065 \text{ kWh/lb } CO_2 = 1.1 \text{ MW}$$

   c. Electrical power required for liquefaction of CO_2:

   $$181.8 \times 10^3 \text{ lb/hr} \times 0.9 \times 0.047 \text{ kWh/lb } CO_2 = 7.7 \text{ MW}$$

   d. Pumping power required for transportation of CO_2 to collection centers:

   $$CO_2 \text{ velocity in 6-in. pipe} = \frac{181.8 \times 10^3 \times 0.9}{52 \times \dfrac{\pi}{4} \times (0.5)^2 \times 3600} = 4.45 \text{ ft/sec}$$

   $$N_{Re} = \frac{0.5 \times 4.45 \times 52}{0.528 \times 10^{-3} \times 6.72 \times 10^{-2}} = 3.25 \times 10^6$$

   Fanning friction factor, $f = 0.0024$

   $$\text{Pressure drop } \Delta p = 2\rho f \times \frac{Lv^2}{Dg_c}$$

   $$= \frac{2.52 \times 0.0024 \times 100 \times 5280 \times (4.45)^2}{0.5 \times 32.2 \times (12)^2}$$

   $$= 1125 \text{ psi}$$

Assuming the pumping efficiency is 60%, then the energy required is

$$\frac{2 \times (4.45)^2 \times 0.0024 \times 100 \times 5280}{0.5 \times 32.2 \times 0.06}$$

$$= 5195.4 \text{ ft-lb}_f/\text{lb} = 1.956 \times 10^{-3} \text{ kWh/lb}$$

Therefore, the total pumping power required for 6-in. line is

$$1.956 \times 10^{-3} \text{ kWh/lb} \times 181.8 \times 10^3 \times 0.9 = 0.32 \text{ MW}$$

The pumping power required for 36-in. line is

$$0.001 \text{ kWh/lb } CO_2 \times 181.80 \times 10^3 \text{ lb/hr} \times 0.9 = 0.16 \text{ MW}$$

In this study, part of the energy required by the CO_2 compressor is expended in providing this pumping power. Therefore, the total energy consumption is $7.9 + 1.1 + 7.7 + 0.32 = 17.0$ MW(e). The net power capacity is $100 - 17 = 83$ MW(e).

2. *Capital investment*
   a. Power plant:

$$\$570/\text{kW} \times 100 \text{ MW(e)} = \$57.0 \times 10^6$$

   b. CO_2 recovery: The capital cost for a 1000-ton/day CO_2 recovery plant is $20 million in January 1983 U.S. dollars.[23] The cost indices are 216.2 and 314 for 1980 and January 1983, respectively. Thus, the capital cost for a plant to recover 90% of 2182 tons/day CO_2 in 1980 U.S. dollars is

$$\$20 \times 10^6 \times \left(\frac{2182 \times 0.9}{1000}\right)^{0.6} \times \frac{216.2}{314} = \$20.6 \times 10^6$$

   c. CO_2 liquefaction: The capital cost for CO_2 compression and dehydration to 2000 psia of a 1000-ton/day plant was $15 million in January 1983.[22] Then the capital cost for a 1964-ton/day CO_2 liquefaction plant in 1980 U.S. dollars is

$$\$15 \times 10^6 \times \left(\frac{1964}{1000}\right)^{0.6} \times \frac{216.2}{314} = \$15.5 \times 10^6$$

   d. Piping: The costs of 6-in. and offshore 36-in. pipe in the first quarter of 1983 are $8.9/ft and $162.8/ft, respectively. Thus, the piping cost in 1980 U.S. dollars for a 100-mile line is

$$\$8.9/\,\text{ft} \times \frac{156}{191} \times 100 \times 5280 + \$162.8\,/\,\text{ft} \times \frac{156}{191}$$

$$\times 100 \times 5280 \times 2 \times \frac{181.8 \times 10^3}{181,810 \times 10^3} = \$5.2 \times 10^6$$

e. Pump:

$$\text{Bhp} = \frac{\text{gpm} \times \text{psi differential pressure}}{1713 \times \text{efficiency}}$$

$$= \frac{181.8 \times 10^3 \times 7.481 \times 1125}{60.52 \times 1713 \times 0.6} = 477\ \text{Bhp}$$

From *Process Plant Estimating Handbook* data, one can find the capital cost of a pump:

$$\$2.0 \times 10^5 \times \frac{659.6}{444.3} = \$0.3 \times 10^6$$

Compared to the piping cost, the capital investment for pumps is not significant. Thus, the total capital investment = ($57.0 + $20.6 + $15.5 + $5.2) × 10^6 = $98.3 ×$10^6$.

3. *Electricity production cost*
   a. Power plant: From Reference 37, the production cost of power plants can be estimated in 1981 dollars as follows:

Fuel expense	$4.2/GJ
Operating and maintenance labor	$2.4/GJ
Depreciation — 15%	$3.1/GJ

Thus, the production cost of power plants in 1980 dollars is

$$(\$4.2 + \$3.1 + \$2.4)\,/\,\text{GJ} \times \frac{3.6\ \text{mills/kWh}}{\$1.0/\text{GJ}} \times \frac{659.6}{721.3} = 32.0\ \text{mills/kWh}$$

b. CO_2 recovery: Based on fuel cost at $46.3/Mg (i.e., $0.021/lb coal), the energy cost for CO_2 recovery can be calculated as follows:

$$\frac{\$0.021}{\text{lb coal}} \times \frac{4.8 \times 10^6\ \text{Btu/ton CO}_2}{12,700\ \text{Btu/lb coal}} \times \frac{2182\ \text{ton} \times 0.9}{83\ \text{MW } 24\ \text{hr/day}} \times \frac{659.6}{721.3}$$

$$+ \ 32\ \text{mill/kWh} \times \frac{1.1\ \text{MW}}{83\ \text{MW}} = 7.7\ \text{mill/kWh}$$

With 15% straight-line depreciation of capital and 350 annual operating days, the operating cost for depreciation is

$$\frac{\$20.6 \times 10^6 \times 0.15}{83 \text{ MW} \times 350 \times 24} = 4.3 \text{ mill/kWh}$$

The amine circuit chemical costs \$2.39/ton CO_2,[23] and with 5% of capital for annual operating expenses, the operating cost for chemical and other operating expenses is

$$\frac{\$2.39 \times 2182 \times 659.6}{83 \text{ MW} \times 24 \times 721.3} + 1.3 = 4.0 \text{ mill/kWh}$$

Following the same approach, the operating costs for CO_2 liquefaction and piping can be calculated; the results are listed as follows:

	Mills/kWh(e)
b. CO_2 recovery	
Energy cost	7.7
Chemical and other operating expense	4.0
Depreciation — 15%	4.3
Subtotal	16.0
c. CO_2 liquefaction	
Energy cost	3.0
Labor and other operating expense	1.0
Depreciation — 15%	3.3
Subtotal	7.3
d. Piping	
Energy cost	0.1
Labor and other operating expense	0.3
Depreciation — 15%	1.1
Subtotal	1.5

Since the operating cost for pumps (<<1 mill/kWh) is negligible, the electricity production cost is

$$32.0 + 16.0 + 7.3 + 1.5 = 56.8 \text{ mills/kWh}$$

REFERENCES

1. Albanese, A.S. and Steinberg, M., Environmental Control Technology for Atmospheric Carbon Dioxide, BNL 51116, Brookhaven National Laboratory, Upton, NY, September 1979.

2. Mimura, T., Shimojo, S., Suda, T., Ijima, M., and Mitsuake, S., *Proceedings of the Second International Conference on Carbon Dioxide Removal,* ICCDR-2, Pergamon Press, Tokyo, 1995, 397–400; ICCDR-3, *Energy Convers. Manage.,* 33, Suppl. S57–S62, 1997.
3. Williams, R.H., Hydrogen from Coal with Sequestering of the Recovered CO_2, Center for Energy and Environmental Studies, Princeton, NJ, August 1990.
4. Steinberg, M., Advanced Technologies for the Recovery, Disposal and Reuse of Carbon Dioxide for Atmospheric Control, BNL 33441, Brookhaven National Laboratory, Upton, NY, July 1983.
5. Horn, F.L. and Steinberg, M., A Carbon Dioxide Power Plant for Total Emission Control and Enhanced Oil Recovery, BNL 30046, Brookhaven National Laboratory, Upton, NY, August 1981.
6. Albanese, A.S. and Steinberg, M., Environmental Control Technology for Atmospheric Carbon Dioxide, BNL 51116, Brookhaven National Laboratory, Upton, NY, September 1979.
7. Herzog, H., Golomb, D., and Zeniba, S., Feasibility, modeling and economics of sequestering power plant CO_2 emissions in the deep ocean, *Environ. Prog.,* 10(1), 64–74, 1991.
8. Probstein, R.F. and Hicks, R.E., *Synthetic Fuels,* pH Press, Cambridge, MA, 1990.
9. IEA Greenhouse Gas R&D Programme Report, Carbon Dioxide Capture from Power Stations, Cheltenham, U.K., 1994.
10. Kumar, R., Fuller, T., Kocurek, R., Teats, G., and Young, J., Tests to Produce and Recover Carbon Dioxide by Burning Coal in Oxygen and Recycling Flue Gas, ANL-CNSV-61, Argonne National Laboratory, Argonne, IL, 1987.
11. Marchetti, C., On geoengineering and the CO_2 problem, *Climate Change,* 1, 59–68, 1977.
12. Hoffert, M.I., Wey, Y.C., Callegari, A.S., and Broeker, W.S., Atmospheric response to deep sea injections of fossil fuel CO_2, *Climate Change,* 2, 53–68, 1979.
13. Handa, N. and Ohsumi, T., Eds., *Direct Ocean Disposal of Carbon Dioxide,* Terrapub, Tokyo, 1995.
14. Omerod, B., *The Disposal of Carbon Dioxide from Fossil Fuel-Fired Power Stations,* IEA Greenhouse Gas R&D Programme, Cheltenham, U.K., 1994.
15. Gunter, W.D., Gentzis, T., Rottenfusser, B.A., and Richardson, R.J.H., Deep coal bed methane in Alberta, Canada, in Proceedings of the Third International Conference on Carbon Dioxide Removal, ICCDR-3, Cambridge, MA, 1996, 217–222.
16. Horn, F.L. and Steinberg, M., A Carbon Dioxide Power Plant for Total Emissions Control and Enhanced Oil Recovery, BNL 30046, Brookhaven National Laboratory, Upton, NY, August 1981.
17. Bergman, P.D. and Winter, E.M., Disposal of Carbon Dioxide in Deep Saline Aquifers in the U.S., U.S./Japan Joint Technical Workshop, U.S. Department of Energy, State College, PA, September 30 to October 2, 1996.
18. Herzog, H., Drake, E., and Adams, E., CO_2 Capture, Reuse and Storage Technologies for Mitigating Global Climate Change — A White Paper, Final Report, Massachusetts Institute of Technology, Cambridge, MA, January 1997.
19. Lackner, K.S. and Butt, E.P., Carbon dioxide disposal in mineral carbonate, in Proceedings of the Third International Conference on Carbon Dioxide Removal, ICCDR-3, Cambridge, MA, 1996, 259–264.
20. Hendriks, C., Carbon Dioxide Removal from Coal Fired Power Plants, Doctoral thesis, Utrecht University, Netherlands, 1994.

21. Hitchen, B., Ed., *Aquifer Disposal of CO₂*, Geoscience Publishing, Alberta, Canada, 1996.
22. Kaplan, L.J., Cost-saving process recovers CO_2 from power-plant flue gas, *Chem. Eng.*, 29, 30, 1982.
23. Pauley, C.R., CO_2 recovery from flue gas, *Chem. Eng. Prog.*, 80(5), 59–62, 1984.
24. Oeschger, H., Siegenthaler, U., Schotterer, U., and Gugelmann, A., A box diffusion model to study the carbon dioxide exchange in nature, *Tellus*, 27, 168, 1975.
25. Baes, C.F., Beall, S.E., Lee, D.W., and Marland, G., The collection, disposal, and storage of carbon dioxide, in *Interactions of Energy and Climates*, Bach, W. et al., Eds., D. Reidel Publishing, 1980, 495–519.
26. Hosiba, H.H., Wilson, L.A., and Martinelli, J.W., A systematic approach to CO_2 supply by E.O.R., in The Future Supply of Nature-Made Petroleum Gas, First UNITAR Conference on Energy and Future, Meyer, R.F., Ed., 1976, 557.
27. Ikoku, C.U., *Natural Gas Engineering*, Penn Well Publishing, Tulsa, OK, 1980, 494.
28. Drysdale, F.R. and Calef, C.E., The Energetics of the United States of America: An Atlas, BNL 50501-R, Brookhaven National Laboratory, Upton, NY, October 1977.
29. Davis, R., National strategic petroleum reserve, *Science*, 213, 7, 1981.
30. Jarik, C. and Weaver, L., A Survey of Salt Deposits and Caverns, Federal Energy Agency FEA Report S-76/310, 1976.
31. Energy Data Report: Inventory of Power Plants in the United States, 1980 Annual, Assistant Administrator for Energy Data Operations, Energy Information Administration, U.S. Department of Energy, 1980.
32. Santhanam, C.J., Liquid CO_2-based slurry transport is on the move, *Chem. Eng.*, p. 50, July 11, 1983.
33. Benkovitz, C.M., Emissions Inventory Progress Report, BNL 51378, Brookhaven National Laboratory, Upton, NY, December 1980.
34. Benkovitz, C.M. and Evans, V.A., User Access to the MAP3S Source Emissions Inventory, BNL 29322, Brookhaven National Laboratory, Upton, NY, March 1977.
35. Woodwell, G.M. et al., Global deforestation: contribution to atmospheric carbon dioxide, *Science*, 222, 1081, 1983.
36. Morgan, J.M., OGJ-Morgan gas pipeline cost index, *Oil Gas J.*, p. 60, January 23, 1984.
37. Horn, F.L. and Steinberg, M., Control of carbon dioxide emissions from a power plant and use in enhanced oil recovery, *Fuel*, 61, 415, 1982.
38. Horn, F.L. and Steinberg, M., Possible Storage Sites for Disposal and Environmental Control of Atmospheric Carbon Dioxide, BNL 51597, Brookhaven National Laboratory, Upton, NY, September 1982.
39. Wilson, T.R.S., The deep ocean disposal of carbon dioxide, *Energy Convers. Manage.*, 33, 627–633, 1992, after the First Int. Conf. on Carbon Dioxide Removal (ICCDR-1), Amsterdam, Netherlands, March 4 to 6, 1992.
40. Magnesen, T. and Wahl, T., Biological Impact of Deep Sea Disposal of Carbon Dioxide, Technical Report No. 77A, The Nansen Environmental and Remote Sensing Center (NERSC), Bergen, Norway, 1993.
41. Murray, C.N., Visintini, L., Bidoglio, G., and Henry, B., Permanent storage of carbon dioxide in the marine environment: the solid CO_2 penetrator, *Energy Convers. Manage.*, 37, 1067–1072, 1995.
42. Herzog, H.J. and Drake, E.M., Carbon dioxide recovery and disposal from large energy systems, *Annu. Rev. Energy Environ.*, 21, 145–166, 1996.

43. IEA Greenhouse Gas R&D Programme Report, Carbon Dioxide Disposal from Power Plants, Cheltenham, U.K., 1994.

44. Slager, S., Silent deaths from killer lake, *Natl. Geographic,* pp. 404–420, 1987.

45. Bryer, C.W. and Guthrie, H.D., Research and Development Considerations for Sequestering CO_2 in Coal Beds, Federal Energy Technology Center, Morgantown, WV, 1997.

46. Barchas, R. and Davis, R., The Kerr-MeGee/ABB lummus crest technology for the recovery of CO_2 from stack gases, *Energy Convers. Manage.,* 33, 333–340, 1992.

47. Kikkinides, E.S., Yang, R.T., and Cho, S.H., Concentration and recovery of CO_2 from flue gas by pressure swing adsorption, *Ind. Eng. Chem. Res.,* 32, 2714–2720, 1993.

48. Morooka, S., Kuroda, T., and Kusakabe, K., Carbon Dioxide Separation from Nitrogen Using Y-Type Zeolite Membranes, in Abstracts 4th Int. Conf. on Carbon Dioxide Utilization, Kyoto, Japan, September 7 to 11, 1997, 099.

6 Bioconversion of CO_2 and Biomass Processes

Nonphotosynthetic pathways for carbon dioxide fixation are those performed by the bacterial methanogens, which are a subgroup within the kingdom of the archaea. Archaea, together with bacteria and eucarya, form the three domains of life. In anaerobic microorganisms, the Calvin cycle for CO_2 fixation does not operate, and acetyl-coenzyme A (CoA) is the central intermediate in carbon metabolism.[1-4] Considerable efforts have been made during the last decade to mimic the natural process of photosynthesis, using reaction systems which included photosensitizers, electron transfer mediators, and sometimes sacrificial electron donors.

In this chapter, the effect of concentrated CO_2 on its fixation by methanogenic bacteria and algae will be reviewed. Also, CO_2 fixation by enzymatic and biomimetic reactions will be considered.

6.1 METHANOGENIC AND ACETOGENIC BACTERIA

Nonphotosynthetic carbon dioxide fixation occurs widely in nature, by the methanogenic archaebacteria. These are obligate anaerobes that grow in freshwater and marine sediments, peats, swamps and wetlands, rice paddies, landfills, sewage sludge, manure piles, and the gut of animals. Methanogens are a major cause of the natural methane release in the environment. More than half of the methane released to the atmosphere is due to the action of methanogens, amounting to about 0.4×10^9 tons/year. These methanogenic bacteria grow optimally at temperatures between 20 and 95°C. They can use either $CO + H_2$ or $CO_2 + H_2$ as their only sources of carbon and energy, by the two reactions catalyzed by carbon monoxide dehydrogenase/ acetyl-CoA synthase (CODH/ACS), using the Wood–Ljungdahl biochemical pathway leading to acetyl-CoA:

$$CO + H_2O \Leftrightarrow CO_2 + 2H^+ + 2e^- \tag{6.1}$$

$$CH_3-CFeSP + CO + CoA \Leftrightarrow \text{acetyl-CoA} + CFeSP \qquad (6.2)$$

These are crucial reactions in the autotrophic acetyl-CoA synthesis. Both the ACS and the CODH catalytic activities are catalyzed by clusters in which nickel is bridged to (Fe_4S_4) units.[5]

The biotechnological potential of the methanogens has been reviewed by Hemming and Blotevogel[6] and Daniels.[7]

6.1.1 Thermophilic Methanogens

CO_2 fixation by methanogenic bacteria was proposed by Bugante et al.[8] as a practical approach for the upgrading of waste gases from blast furnaces, to be used as fuel for steam boilers. By methanation of the carbon oxides, their caloric value may be significantly increased. Thus, with hydrogen as reducing agent and mixtures of carbon monoxide and carbon dioxide as carbon source, methane was produced by thermophilic methanogens in a column bioreactor with 200 mL volume of media and supporters, operated at 55°C and pH 7.4. At a gas recirculation rate of 18 L/h, the rates of H_2 consumption and CH_4 production were 1380 and 300 mmol L^{-1} day^{-1}. The caloric value of the gas mixture, which originally had the composition of blast furnace gas, increased by this methanation from 755 to 6420 kcal mol^{-1}.[8]

Using a mixture of the components of low-Btu synthesis gas, CO, CO_2, and H_2, a completely biocatalytic conversion to the high-Btu methane was achieved by Klasson et al.[9] using a mixture of cultures of three bacteria.[9] The photosynthetic bacterium *Rhodospirillum rubrum* was applied to carry out the water gas-shift reaction,

$$CO + H_2O \rightarrow H_2 + CO_2 \qquad (6.3)$$

resulting in 100% conversion. Simultaneously, a mixture of two methanogens, *Methanobacterium formicicum* (which provides a high rate of hydrogen uptake but is inhibited by CO) and *Methanosarcina barkeri* (which has a smaller rate of hydrogen uptake, but is more tolerant of CO), converted CO_2 to methane:

$$4H_2 + CO_2 \rightarrow CH_4 + 2H_2O \qquad (6.4)$$

With a trickle bed reactor (a packed bed tower), the methane yield was 83% of the theoretical required by the above methanation equation. The productivity of methane achieved at flow rates above 300 mL/h was 3.4 mmol L^{-1} h^{-1}, and the mass transfer coefficient was 780 h^{-1}. While the conventional gas-phase catalytic methods for methanation of CO_2 require temperatures of 300 to 700°C and pressures of 3 to 20 atm and are sensitive to catalyst poisoning, for example, by sulfur compounds (see Chapter 10), the biological conversion with the above triculture system could be operated at 37°C, was not affected by the presence of sulfur compounds, and was less sensitive to variations in the composition of the feedstock gases.[9]

A hot spring in Kyushu, Japan, was the source for isolating the thermophilic methanogen KN-15, which could best be grown in a mineral medium containing hydrogen sulfide or sodium sulfide. The optimal temperature was found by Nishimura et al. to be 65°C and the optimal pH was 7.4 to 7.8.[10] These methanogens performed the methanation of carbon dioxide at atmospheric pressure in a gas mixture of 80% H_2, 20% CO_2, and 500 ppm of H_2S, both in batch and continuous cultures, at agitation rates of up to 1300 rpm. In batch culture, after an initial exponential growth phase, in which the cell concentration increased with $\mu_{max} = 0.62$ h^{-1}, the cell concentration increased linearly at 1.1 g dry cell/h. The cell concentration after 28 h reached the high value of 19.5 g dry cell/L. The growth limitation during the linear phase was shown to be due to the limitation of mass transfer, particularly for the gaseous hydrogen, which is only sparingly soluble in the aqueous medium. In continuous culture, washout of cells occurred at dilution rates higher than 0.5 h^{-1}.[10] In order to improve the hydrogen mass transfer from the gaseous to the liquid phase, the culture was also carried out by Nishimura et al. with the same gas mixture at pressurized conditions, with pressures up to 3.0×10^5 Pa.[11] At 3.0×10^5 Pa in batch culture, the cell concentration reached 18.5 dry cell/L after only 10 h. In continuous culture, the cell productivity was 3.0 g dry cell/L/h, and the rate of methane production reached the remarkably high value of 1.28 mol L^{-1} h^{-1}. Application of the Monod model for microbial growth, and taking the concentration of hydrogen as the limiting nutrient, cell productivity V could be related to total pressure P in the gas phase by the equation

$$V = V_{max}P/(K_s + P) \tag{6.5}$$

where V_{max} is the maximal possible cell productivity and K_s represents the Monod half-saturation constant. From the effect of gas pressure on cell productivity, values of $V_{max} = 12.8$ g dry cell/L/h and $K_s = 9.7 \times 10^5$ Pa were derived for cultures of this thermophilic methanogen.[11]

For the purification of biogas, which usually consists of a mixture of methane, carbon dioxide, hydrogen sulfide, hydrogen, and nitrogen, Strevett et al. used the chemoautotrophic methanogen *Methanobacterium thermoautotrophicum*.[12] This organism has a specific requirement for H_2S, thus completely removing this contaminant from the off-gases. The optimal growth temperature of this organism is 65 to 70°C, and the optimal pH is 7.2 to 7.6. In continuous culture growth at pH 7.0 and 62°C, the apparent maximal growth rate μ_{max} was 0.7 h^{-1}, and the apparent K_s values for CO_2 and H_2 were 5.1 and 5.3 nM, respectively. For improved mass transfer between the gaseous and liquid phases, hollow fiber membrane modules were used. The methane growth yield Y_{CH4} (grams dry weight biomass per mole of CH_4 produced) was 1.7 g mol^{-1}. In order to increase the CH_4 production while limiting the steady-state biomass concentration, growth inhibition by 1-iodopropane (PrI) was applied. This inhibitor caused an uncoupling of methanogenesis from biomass production, also resulting in a smaller requirement of H_2. In the presence of 20 μM PrI, the methane growth yield could be decreased to 0.82. This improved process doubled the CH_4 concentration, providing a purified sulfur-free biogas containing about 96% CH_4, while raising the caloric value from 21 to 35 MJ m^{-3}.[12]

6.1.2   Extremely Thermophilic Methanogens

Extremely thermophilic methanogens (hyperthermophiles) grow optimally at temperatures of 80 to 110°C, using CO_2 as their sole carbon source and molecular hydrogen or reduced sulfur compounds as electron donors. These organisms have been proposed to be the earliest and most primitive forms of life that still exist.[13] Such extremely thermophilic methanogens provide an advantage in the fermentation process due to reduced contamination risk. For the culture of the extremely thermophilic methanogen *Methanothermus fervidus,* optimal conditions were explored by Pepper and Monbouquette.[14] This organism, which had been isolated at shallow depth from a volcanic spring in Iceland, was found to have an active temperature range of 65 to 97°C and to grow optimally at 83°C. In the optimal growth medium developed, under strictly anaerobic conditions, at a pH of 6.9, with H_2 and CO_2 in the ratio 4:1, the observed doubling time of the organism was rather high, 264 min (maximal growth rate only 0.16 h^{-1}), and the maximal biomass concentration was 0.74 g dry wt/L. The CH_4 product yield coefficient was 0.52 mol CH_4/g dry wt.[14]

Tsao et al. used a continuous culture at 80°C to study the growth kinetics of the extremely thermophilic methanogen *Methanococcus jannaschii.*[15] This organism had been isolated from a deep-sea hydrothermal vent. Under a continuous gas supply of $H_2 + CO_2$ (ratio 4:1), the methane production rate increased with the rate of gas flow. The maximal specific growth rate of the organism was 0.56 h^{-1}, and washout occurred at a dilution rate of 0.66 h^{-1}. The maximal specific methane production rate was 0.32 mol g^{-1} h^{-1}.[15]

6.1.3   Bioconversion of Methane to Methanol

An attractive option for the production of useful fuel could be to follow up on the methanation of synthesis gas with the bioconversion of methane to methanol. This has been achieved by Mountfort et al.[16] and by Mehta et al.[17] both with cultures of certain bacteria and with cell-free extracts of the enzyme methane monooxygenase. Covalently immobilized cells of *Methylosinus trichosporium* were used both in batch and continuous culture to oxidize methane specifically and in high yield to methanol. DEAE–cellulose-linked cells of *M. trichosporium* were found to have highly specific methane monooxidase activity, enabling a methane oxidation rate of 66 µmol/h-mg cells. However, addition of sodium formate (20 m*M*) in the feedstream was required in order to maintain sustained methanol production.[16,17] Methanol was also produced by Patel et al. by incubating a 1:1 mixture of methane and oxygen with the soluble enzyme methane oxygenase obtained by centrifuging the cell homogenate of *Methylobacterium organophylum.* The rate of methanol formation was 93 nmol min^{-1} mg protein^{-1}.[18] With whole-cell cultures of

Methylosinus trichosporium, there was production of methanol by oxidation of methane, with a yield of 30% based on the methane utilized. The methanol yield was enhanced by the presence of chloride ions and hydrogen in the medium. The system using cell suspensions of *M. trichosporium* was suggested to be advantageous over enzyme preparations, as it avoids the need for expensive cofactors and enzyme purification.[18]

6.1.4 Thermophilic Homoacetogens

Homoacetogenic bacteria are strictly anaerobic microorganisms which catalyze the reduction of C_1 units such as CO_2 to acetate, using H_2 as the electron donor.[19] The carboxyl group of acetate was shown to be derived from CO, which is formed from CO_2 by the nickel enzyme carbon monoxide dehydrogenase. The methyl group of acetate is formed by the reduction of CO_2 in sequence first by formate dehydrogenase, followed by a series of enzymatic reactions on reduced C_1 intermediates bound to tetrahydrofolate. Acetyl-CoA is then produced from the methyl group and CO in a reaction catalyzed by carbon monoxide dehydrogenase. Acetogenesis was suggested to be involved in recycling of 10 to 20% of the carbon on earth.[20]

Continuous cultures of the thermophilic homoacetogenic bacterium *Thermoanaerobacter kivui* immobilized within a polyvinyl alcohol cryogel were found by Pusheva et al. to achieve the reaction

$$4H_2 + 2CO_2 \rightarrow CH_3COOH + 2H_2O \qquad (6.6)$$

in 100% yield.[21-23] With such cryogel beads as biocatalysts, in a thermostatically controlled flow-through reactor, the activity and mechanical strength remained constant for at least a year. At a temperature of 66°C and a constant pH of 6.4, with a continuous gas supply of O_2-free H_2–CO_2 (80:20), the production rate of acetate reached the maximal value of 0.53 g L^{-1} h^{-1}. In batch processing, the production rates were much lower. With the same bacterium (*T. kivui*) in batch culture of free-growing cells, higher acetate formation rates were observed under a gas feed of CO–H_2–CO_2 (20:64:16) than under H_2–CO_2 (80:20). This suggested the possibility of using synthesis gas, a mixture of CO, CO_2, and H_2, without separation of the components, for the production of acetic acid.[21-23]

6.2 ALGAL PHOTOSYNTHESIS

Natural photosynthesis in green plants achieves carbon dioxide fixation on a global scale. The incorporation of carbon dioxide into the biosphere by the photosynthetic action of plants and microorganisms has been estimated to amount to about 10^{11} tons CO_2 per year.[24] However, the efficiency of solar energy conversion in plant production under optimal growth conditions is only 5 to 6%. Under field conditions, even high-yielding crops (such as maize, bulrush millet, or sugarcane) convert solar energy into plant material with a maximal efficiency of 1 to 2%. Most major crops and forests achieve much lower efficiencies. The global average efficiency has been estimated as 0.15%.[25]

Photosynthesis is much more efficient in microalgae than in terrestrial C_3 and C_4 plants.[26] This high efficiency is primarily due to two factors: the action of carbonic anhydrase (CA), both extracellular and intracellular, and the CO_2-concentrating mechanism.[27] There have therefore been considerable efforts to apply intensive algal cultures, for both CO_2 fixation and the production of valuable materials. Fossil-fuel-burning power stations are often situated near seashores or estuaries. It could therefore be advantageous to search for algal species which may grow in seawater and which tolerate high CO_2 concentrations. Such algal culture may be combined with wastewater treatment, thus utilizing the high concentrations of nitrate and phosphate nutrients contained in municipal and agricultural effluents for the fixation of CO_2 emitted from power stations and steel plants.

The assimilation of CO_2 is catalyzed by ribulose-1,5-biphosphate carboxylase/oxygenase (RuBisCO), which is the most abundant protein on earth, with the double function of acting as monooxygenase when O_2 is the substrate and as carboxylase when CO_2 is the substrate.[28] As a carboxylase, this enzyme catalyzes the key reaction in the Calvin reductive pentose phosphate cycle. The oxygenase activity of RuBisCO is responsible for the loss of up to 50% of the carbon fixed in photosynthesis. The substrate specificity factor τ (which is a measure of the relative rates of carboxylation and oxygenation at any given CO_2/O_2 concentration ratio) varies widely for RuBisCO enzymes from different species. For CO_2 mitigation, it will be important to design RuBisCO enzymes in which the carboxylase activity will be considerably enhanced over the oxygenase activity.[28] The carboxylase reaction of RuBisCO is usually the rate-limiting step in photosynthesis.[29] Efforts have been made to enhance the photosynthetic activity in higher plants by gene transfer to the chloroplast, since the RuBisCO L gene is encoded in the plastid genome.[30]

In aquatic cyanobacteria and microalgae, an adaptation to a wide range of ambient CO_2 concentrations exists by the induction of an inorganic carbon concentration mechanism (CCM) at low ambient CO_2, which enables the accumulation of inorganic carbon internally in the algal cells, reaching concentrations of up to 50 mM.[31,32] These high concentrations of internal inorganic carbon are required for the operation of RuBisCO-catalyzed CO_2 assimilation. While CO_2 itself is the substrate for the carboxylation of RuBisCO, the concentration of CO_2 in the aqueous extracellular environment is usually only in the range of 10 to 12 μM. The major part of the dissolved inorganic carbon is in the form of HCO_3^-. The CCM involves the action of both some CO_2 or HCO_3^- transport system and the enzyme CA.

6.2.1   Microalgae

Batch and continuous culture of the marine green alga *Tetraselmis suecica* performed by Weiss et al. in seawater enriched with mineral nutrients and supplied with ambient air resulted in a maximal growth rate of 0.61 day^{-1}.[33] However, Laws and Berning could also maintain these algae in nutrient-enriched seawater in outdoor ponds on the island of Hawaii by the supply of pure CO_2 or electric power plant stack gases.[34] The photosynthetic efficiency was 9 to 10%, and the CO_2 utilization efficiency reached close to 100%. No adverse effects of using stack gases instead of pure oxygen were detected. Since the source of these stack gases was a power plant burning fuel oil (<0.5% sulfur), these results do not indicate if stack gases from coal-burning power stations, which may contain high concentrations of particulates or sulfur, will also be benign to algal cultures. The efficiency of CO_2 uptake depended considerably on the method of mixing the gas supply with the water. Water circulation by Archimedes screws was found to be better than paddle wheels or propellers. Air lifts considerably improved the mixing of CO_2 and air with the water. In chemostat experiments, the photosynthetic efficiencies measured with the gas supply of ambient air (0.035% CO_2), pure CO_2, and stack gases (1.5% CO_2) were 1.7, 4.6, and 5.1%, respectively.[34]

The gases discharged from power plants usually contain about 10 to 20% CO_2. Kodama et al. selected from a saline pond a species of fast-growing marine green alga, *Chlorococcum littorale*, which is tolerant of high CO_2 concentrations.[26] Under aeration with 20% CO_2, its maximal growth rate in a mineral nutrient medium occurred at an NaCl concentration of 1.5%, and even at 9% NaCl its growth rate was still 50% of the maximal. This growth rate was independent of pH in the pH range 4 to 6.5 and was adequate in the temperature range 15 to 28°C. The optimal growth rate with respect to CO_2 concentration was 5 to 10% CO_2 in air, but substantial growth rates were obtained even at 60% CO_2, after stepwise adaptation to increasing CO_2 concentrations. Under 20% CO_2 and an illumination intensity of 400 μE m^{-2} s^{-1}, the growth rate was 0.41 g dry wt L^{-1} d^{-1}.[26] The lag period of adaptation of *C. littorale* to higher CO_2 concentration was found by Pesheva et al. to be 1 to 2 days at 20% CO_2 and 3 to 6 days at 40% CO_2.[35] During this lag period, the photosynthetic activity of PSII was suppressed, while the activity of PSI was enhanced. Also, high CO_2 concentrations caused repression of CA and phosphoenol pyruvate carboxylase activities and enhanced RuBisCO activity. *C. littorale*, when adapted to anaerobic growth under mixtures of CO_2 and nitrogen, could express its hydrogenase activity. The capacity for both photohydrogen and photosynthesis O_2 evolution was found by Schnackenberg et al. to be maximal in cells grown under 5% CO_2 at pH 7.5 and 25°C.[36,37]

Using steady-state cultures of chlorella under gradually increasing CO_2 partial pressures, Pirt and Pirt could adapt these algae to photosynthetic growth in high CO_2 concentrations. The maximum specific growth rate of these adapted algae did not decrease until the CO_2 partial pressure exceeded 0.65 atm.[38]

By screening algae from hot springs and other sources, Murakami et al. selected a green alga, *Chlorella* sp. UK001, which had optimal growth temperatures in the range 35 to 40°C and grew well in air containing up to 20% CO_2. Below 30°C, its maximal growth rate was 0.32 h^{-1}.[39]

Flue gases released from a steel plant typically contain 15% CO_2 (v/v). Yun et al. stepwise adapted *C. vulgaris* to grow in batch culture supplied with 5, 10, 20, and

30% CO_2.[40,41] In experiments with flue gas from a steel plant, the growth medium was wastewater from the same steel plant polluted with a high concentration of ammonia (55 to 90 g m^{-3}) but which was deficient in phosphate. By adding phosphate (46 g m^{-3}) to this wastewater, and supplying the algae in batch culture at 27°C with the flue gas containing 15% CO_2, the simultaneous fixation of CO_2 and removal of ammonia occurred at rates of 26 g CO_2 m^{-3} h^{-1} and 0.92 g NH_3 m^{-3} h^{-1}, respectively.[40,41]

A genetically engineered marine cyanobacterium has been applied by Miyasaka et al. to the conversion of CO_2 into polyhydroxyalkanoates (PHA), useful for the production of biodegradable plastics. The unicellular marine cyanobacterium *Synechococcus* sp. was transformed to produce up to 17% of cell dry weight of the PHA, which could be extracted by chloroform. This PHA was a copolymer of β-hydroxybutyric acid, lactic acid, and other hydroxyalkanoic acids, with an average molecular weight of 1,000,000.[42]

Improved designs of photobioreactors for the optimization of high growth rate, cell density, and CO_2 utilization by the photoautotrophic cyanobacterium *Synechococcus* sp. were tested by Suh et al. The highest growth rate was obtained with an optical fiber photobioreactor under fluorescent lamp illumination.[43]

Nanba and Kawata developed a solar collector, from which light was transmitted through optical fibers into an algal culture vessel. *Chlorella* sp. was grown in this device, fixing CO_2 at a rate of 24 g CO_2 m^{-2} on a sunny day.[44]

An important utilization of the cell mass produced by algal culture could be as a component of paper. Samejima et al. developed a tubular reactor of 50-L scale for the semi-batch culture of the marine alga *Tetraselmis* sp. Tt-1, reaching a mean productivity of 20 g m^{-2} d^{-1} (12 h). The harvested algae could be mixed with kenaf pulp, producing paper of good quality.[45]

An interesting approach to ethanol production from CO_2 by microalgae was reported by Hirayama et al.[46,47] The marine microalga *Chlamydomonas* sp. YA-SH-1, which was isolated from the Red Sea, showed in photosynthetic culture a growth rate of 30 g dry mass m^{-2} d^{-1} and a starch content of 30% (dry base). After harvesting these algae, they were resuspended in either 0.4 M potassium phosphate (pH 7.7) or seawater and submitted to anaerobic self-fermentation in the dark. About 50% of the intracellular starch was converted to ethanol. The advantage of this microbial production of ethanol by comparison with the terrestrial culture via sugarcane could be the possibility of using brackish water or seawater and desert areas.[46,47]

6.2.2   Macroalgae

The growth of the marine macroalgae *Gracilaria* sp. and *G. chilensis* was found by Gao et al. to be enhanced by increasing the CO_2 concentration in the air supply from 650 ppm to 1250 ppm.[48] The enriched CO_2 in the culture caused a decrease in photorespiration. In a comprehensive review, Gao and McKinley proposed that the culture of marine macroalgae could make important contributions both to biomass production for chemicals and fuel and to CO_2 remediation.[49]

6.3 ENZYME-CATALYZED REACTIONS

Many enzymes are active in nature in catabolic processes, causing decarboxylation from organic compounds. Since these enzymes often act reversibly, they may be used in artificial systems for carboxylation (i.e., insertion of CO_2 into C–H bonds). A detailed review of enzymatic reactions for carbon dioxide utilization was presented by Aresta et al.[50,51]

6.3.1 Formate Dehydrogenase

A reversal of the natural action of the enzyme formate dehydrogenase (FDH) (which catalyzes the conversion of formate to CO_2) was achieved by Parkinson and Weaver, using the methyl viologen redox system as mediator:[52]

$$MV^{2+} + e^- = \cdot MV^+ \tag{6.7}$$

The reduction of CO_2 to HCOOH was driven by either an electrochemical reaction, on graphite foil, or photoelectrochemically, on illuminated p-InP, using an aqueous phosphate buffer. At a potential of only +0.05 V vs. the normal hydrogen electrode, and a current density of 0.6 mA/cm^2, the current efficiencies reached 80 to 93%. However, the enzyme was unstable under the conditions of the reaction.[52] This instability was shown by Kuwabata et al. to be due to the sensitivity of FDH to visible light.[53]

Enzyme-catalyzed reactions were applied by Willner et al.[54–56] to the photosynthesis of formic, malic, aspartic, and other carboxylic acids, using photosensitized regeneration of NADPH, with a bipyridinium electron relay system, in the presence of the appropriate enzymes. Quantum yields for production of formate from CO_2 and for malic acid from pyruvic acid reached 1.6 and 1.9%, respectively.

In an ingenious electrochemical system developed by Kuwabata et al., CO_2 in aqueous phosphate buffer (pH 7) containing 0.3 M NaHCO$_3$ was converted to formate using FDH as catalyst and methyl viologen or pyrroloquinoline quinone (PQQ) as electron relay.[57,58] In the presence of methanol dehydrogenase (MDH) as well, the formate was further reduced to formaldehyde and methanol. PQQ serves as a cofactor in the natural action of MDH. The electrolysis was performed in the dark with a two-compartment cell separated by a cation exchange membrane, using a glassy carbon working electrode at a constant potential –0.8 V (vs. the saturated calomel electrode [SCE]). Under appropriate conditions, with PQQ as the electron mediator, methanol was the only product, formed with up to 90% current efficiency. When only PQQ was used as the electron mediator (without methyl viologen), the production of methanol started only after an induction period. This was explained by the reduction of adsorbed PQQ on the glassy carbon electrode.[57,58]

In contrast to the sensitivity of FDH to visible light, MDH was found by Kuwabata et al. to be very stable to both visible and UV light.[58] They could thus couple the photocatalyzed reduction of CO_2 to formate (see Section 14.6.2) with the enzymatic reduction of formate to methanol catalyzed by MDH. The reaction was carried out in a CO_2-saturated aqueous dispersion (pH 7) of colloidal ZnS microcrystallites stabilized on SiO_2 (Aerosil 200CF), in the presence of 2-propanol as a hole scavenger, and with MDH and PQQ, under illumination with a 500-W high-pressure Hg arc ($\lambda > 270$ nm). The UV illumination of the dissolved PQQ in the presence of the ZnS colloid resulted in the reduction of PQQ to its reduced form, as shown by a red shift in the UV absorption peak of PQQ. The quantum efficiency for photoreduction of CO_2 to methanol at 280 nm had the remarkably high value of 5.9%.[58] This approach of combining photocatalytic reduction of CO_2 to formate with enzymatic reduction of formate to methanol could possibly become a practical method, for example, by replacing ZnS by CdS (which is photoexcited by visible light) and by replacing 2-propanol by a hole scavenger which is a waste product (e.g., sulfide ions).

6.3.2   Isocitrate Dehydrogenase

The enzyme isocitrate dehydrogenase (ICDH) was used by Sugimura et al. as electrocatalyst, with methyl viologen (MV^{2+}) as mediator, to reverse the *in vivo* metabolic oxidation of isocitric acid to CO_2 and oxoglutaric acid:

$$\text{Oxoglutaric acid} + CO_2 \leftrightarrow \text{Isocitric acid} \qquad (6.8)$$

With a glassy carbon electrode at −0.95 V vs. SCE, in aqueous tris buffer solution (pH 7), the current efficiency approached 100%. The electrochemical system did not require the presence of $NADP^+$.[59]

In another approach to bioelectrochemical CO_2 fixation, ferredoxin was used by Taniguchi to mediate the ICDH-catalyzed carboxylation of oxoglutaric acid to isocitric acid and the malic enzyme-catalyzed carboxylation of pyruvic acid to malic acid. On In_2O_3 electrodes in tris–HCl buffer solutions (pH 7.5) with 0.33 M NaCl, current efficiencies were more than 90%, with high selectivity.[60]

In an alternative process by Inoue et al., also with ICDH as the enzyme catalyst and methyl viologen as the electron mediator, but using powdered CdS as a photocatalyst, the photoassisted fixation of CO_2 into oxoglutaric acid to form citric acid was accomplished, using light of $\lambda > 390$ nm. The quantum efficiency at 410 nm was 1.2%.[61]

6.3.3   Ferredoxin-NADP+ Reductase

The photoassisted carboxylation of oxoglutaric acid was also achieved by Willner et al. with a photosensitized NADPH regeneration system, using $Ru(bpy)_3^{2+}$ as a photosensitizer, d,l-dithiothreitol as electron donor, and ferredoxin-$NADP^+$ reductase as an enzyme to recycle NADPH. Ferredoxin, which contains an iron–sulfur cluster at its redox-effective site, here serves as electron mediator.[54]

6.3.4   Pyruvate Dehydrogenase

Using pyruvate dehydrogenase as an electrocatalyst, in the presence of NADPH, CO_2 was fixed by Kuwabata et al. into acetyl-CoA, yielding pyruvic acid:[62]

$$CH_3CO-SCoA + CO_2 + NADPH \rightarrow CH_3CO-COO^- + HSCoA + NADP^+ \quad (6.9)$$

In an H-type cell separated by a Nafion membrane, with a glassy carbon electrode, using $NaHCO_3$ as the CO_2 source, the turnover number per active site of the enzyme was 500. Optimal rate and yield of pyruvate formation were at pH 5.0, at an electrode potential of –0.95 V vs. SCE.[62]

6.3.5 Phenol Carboxylase

A phenol carboxylase activity of *Pseudomonas aeruginosa* K_{172} was demonstrated by Aresta et al., converting phenol selectively to 4-OH-benzoate:

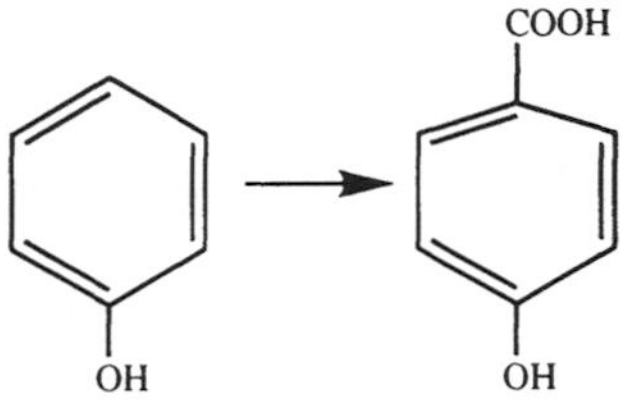

The action of this enzyme was found to be Mn(II) and Fe(II) dependent and to be stimulated by the presence of K^+ cations. By ^{14}C tracer experiments, it was shown that the active carboxylating agent is the neutral CO_2 molecule and not the bicarbonate ion.[63,64]

6.4 BIOMIMETIC REACTIONS

Various inorganic or organic compounds may mimic the catalytic action of natural enzymes.

6.4.1 Porphyrins

The natural photosynthesis by green plants was simulated by Fruge et al. with the photoreduction of CO_2 in water to formic acid and oxygen using chlorophyll-a as sensitizer.[65] An *n*-pentane extract of chlorophyll-a was electroplated on a platinized Pt foil, which was then again platinized. This foil was placed in water saturated with CO_2 and illuminated for 30 min with a 1000-W xenon arc. On the basis of mass spectrometric analysis of the gas mixture above the aqueous solution, it was concluded that the products included formic acid and that the oxygen produced was derived from the oxidation of water.[65]

A visible light-induced fixation of carbon dioxide with enolate complexes of aluminum porphyrins in benzene solutions in the presence of 1-methylimidazole was found by Inoue et al. to result in formation of β-ketocarboxylate complexes of the aluminum porphyrins, which could be converted by CH_3OH/HCl treatment into the free β-ketocarboxylic acids.[66–68] Thus, 1-phenyl-1-propanone was converted into 2-benzoyl-propanoic acid. This carboxylation of a carbonyl derivative to form a ketocarboxylic acid via an enolate as the reactive species mimics the assimilation of carbon dioxide in natural photosynthesis. Also, the photoinduced carboxylation of α,β-unsaturated esters with carbon dioxide in the presence of a methylaluminum porphyrin resulted in the formation of malonic acid derivatives. From *tert*-butyl methacrylate, ethylmethylmalonic acid mono-*tert*-butyl ester was obtained.[66–68]

6.4.2   Iron–Sulfur and Molybdenum–Sulfur Clusters

Molybdenum–iron–sulfur proteins and tungsten–iron–sulfur proteins have been shown by Thauer to be active as formylmethanofuran dehydrogenases, which catalyze the first step in the reduction of CO_2 in the methanogens *Methanobacterium thermoautotrophicum* and *M. wolfei*, producing formylmethanofuran.[69,70]

Tetranuclear iron–sulfur clusters were tested by Tezuka et al. as analogues of the Fe_4 active sites in iron–sulfur proteins, which play an important role in electron transfer in all living systems.[71,72] Iron–sulfur clusters, such as $[Fe_4S_4(SR)]^{2-}$ (where $R = C_6H_5$ or $CH_2C_6H_5$) in dimethylformamide (DMF) containing (n-Bu$_4$N)(BF$_4$) as electrolyte, mediated the electroreduction of CO_2 on mercury pool cathodes. At – 2 V (vs. SCE), HCOOH was produced at about 60% current efficiency. Other products were oxalate and CO. The presence of the iron–sulfur clusters caused a shift in the potential of carbon dioxide reduction by about 0.5 to 0.7 V in the positive direction.[71,72] Under similar conditions, but using Fe_4S_4 clusters bearing 36-member methylene backbones (which were more stable than those containing thiolates), with an Hg pool cathode and (n-Bu$_4$)(BF$_4$) as electrolyte, a current efficiency of 40% was achieved by Tomohiro et al. for the reduction of CO_2 to HCOOH.[73]

Iron–sulfur clusters also serve as attractive catalysts for the synthesis of 2-oxazolidones, which are important as pharmaceuticals and agricultural pesticides. The reactions were carried out by Kodaka et al. at 25°C in acetonitrile solutions containing ethanolamine, a thiol such as 2-mercaptopyridine, a phosphine such as triphenylphosphine or tri-n-butylphosphine, and $(Et_4N)_2[Fe_4S_4(SPh)_4]$, under an atmosphere of CO_2–O_2 (9:1). An example of such a preparation is[74]

$$\text{Ph–CHOH–CH}_2\text{–NH}_2 + \text{CO}_2 + \text{Ph}_3\text{P} \xrightarrow{\text{Thiol/F}_4\text{S}_4} \quad + \text{Ph}_3\text{PO}$$

A simulation of the action of photosynthetic bacteria in carbon dioxide fixation as well as in nitrite or nitrate reduction to molecular nitrogen was performed by Tanaka et al. in experiments with iron–sulfur clusters.[75–78] Carbon dioxide fixation was coupled with nitrite ion reduction by controlled potential electrolysis (at –1.25 V vs. SCE), using $(Bu_4N)_2[Fe_4S_4(SPh)_4]$ as electrocatalyst, in the presence of acetophenone as proton source and molecular sieve 3A as a dehydrating agent, in CO_2-saturated acetonitrile. Acetophenone was carboxylated to benzoyl-acetate, while nitrite was reduced to molecular nitrogen:[75–78]

$$8PhCOCH_3 + 2NO_2^- + 8CO_2 + 6e^- \rightarrow 8PhCOCH_2COO^- + N_2 + 4H_2O \quad (6.10)$$

In a route to α-keto acids, such as pyruvic acid, in dry acetonitrile solvent containing $(Bu_4N)(BF_4)$ and molecular sieve 3A as desiccant, iron–molybdenum and iron–sulfur clusters, $[Fe_6Mo_2S_8(SEt)_9]^{3-}$ and $[Fe_4S_4(SPh)_4]^{2-}$, were used to catalyze the reaction

$$RCO–SC_2H_5 + CO_2 + 2e^- \rightarrow RCO–COO^- + C_2H_5S^- \quad (6.11)$$

where $R = CH_3$, C_2H_5, C_6H_5. The reaction apparently involved nucleophilic attack of CO_2 bound to the iron–sulfur cluster on the $R–C(O)–SC_2H_5$ molecule. With $[Fe_6Mo_2S_8(SEt)_9]^{3-}$ as catalyst, by controlled potential electrolysis using glassy carbon electrodes at -1.55 V vs. SCE, and starting with $CH_3CO–SC_2H_5$, the current yields for production of pyruvic and formic acids were 11 and 27%, respectively. Similarly, starting from $C_2H_5CO–SC_2H_5$ and $C_6H_5CO–SC_2H_5$, the current yields of C_2H_5COOH and C_6H_5COOH were 49 and 13%, respectively. These clusters simulate the action of the enzyme pyruvate synthase.[75–78]

The same $[Fe_6Mo_2S_8(SEt)_9]^{3-}$ catalyst was used by Nagao et al. in CO_2-saturated acetonitrile to add the CO_2 moiety to methyl acrylate, $CH_2=CHC(O)OCH_3$.[79] With glassy carbon as a working electrode at -1.6 to -1.7 V vs. SCE, in the presence of Bu_4NBF_4 and molecular sieve 4A, the main products were $CH_3–CH_2–C(O)OCH_3$ (58%) and $CH_3CH[C(O)–OCH_3]C(O)OCH_3$ (13%). The mechanism proposed involved two-electron reduction of the MoFeS cluster, which interacts with CO_2 and H^+, enabling competitive nucleophilic addition of CO_2 or H^+ to methyl acrylate, followed by electrophilic attack of free CO_2 or H^+ to the olefinic carbon atoms.[79]

6.4.3 Carbodiimides and Urea Derivatives

The carboxylation of active methylene compounds with CO_2 was found by Chiba et al. to be promoted by carbodiimides and various urea derivatives, in a reaction which simulated the action of biotin as a cofactor in various enzymatic carboxylations.[80–82] Thus fluorene in dimethyl sulfoxide (DMSO) solution containing an alkali carbonate and diphenylcarbodiimide was converted into fluorene-9-carboxylic acid:

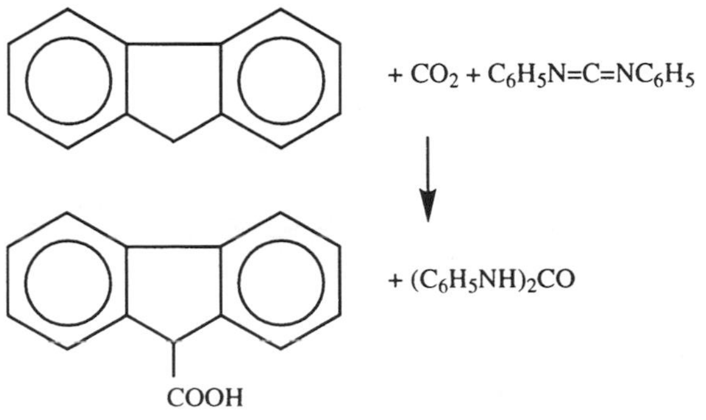

$+ CO_2 + C_6H_5N=C=NC_6H_5$

$+ (C_6H_5NH)_2CO$

COOH

The mechanism possibly involved a carbonate ion derivative of the carbodiimide

$$C_6H_5-N(CO_2^-)-C(O^-)=N-C_6H_5$$

as the carboxyl source. The yield of carboxylation of fluorene depended on the nature of the alkali carbonate and increased in the order Na < Li < Rb < K < Cs. With cesium carbonate, the yield of fluorene-9-carboxylic acid was more than 70%. Not only carbon dioxide but also the hydrogen carbonate anion were considered as the active carbon sources in the presence of carbodiimide. Other active methylene compounds which were similarly carboxylated included indene, indanone, phenylacetonitrile, acetophenone, 1-tetralone, and cyclohexanone. Instead of the carbodiimide, either 1,3-diphenylurea, acetanilide, or formanilide could be applied as activators for CO_2 in the carboxylation of active methylene compounds such as fluorene. However, the yields of carboxylation were lower than with the carbodiimide.[80–82]

6.4.4   Crown Ethers

In an even more convenient system for the carboxylation of active methylene compounds, Chiba et al. applied 18-crown-6 and potassium carbonate in DMSO solution. At room temperature, with a reaction time of only 2 hr, fluorene was carboxylated to 9-fluorene carboxylic acid (in 65% yield), cyclohexanone to 2-oxo-1-cyclohexane carboxylic acid (10%), acetophenone to benzoyl acetic acid (51%), and indene to 3-indene carboxylic acid (88%).[83]

6.4.5   Mg– and Mn–Diazadiene Complexes

As a model for the carboxylation by CO_2 in the "dark reaction" of photosynthesis and the biotin-dependent carboxylation, Walther et al. tested the activation of CO_2 by Mg– and Mn–diazadiene complexes.[84] The natural enzymatic process requires Mg for the RuBisCO reaction and Mg or Mn for the biotin-dependent carboxylation. The mechanism had been proposed to involve the intermediate formation of N-carboxylates. In the model reactions, the N-carboxylation of diazadiene–metal complexes in tetrahydrofuran (THF) solutions led to the reversible fixation of CO_2 into the diazadiene complexes, followed by CO_2 transfer to compounds with active C–H bonds, such as acetophenone and acetone. Among several diazadiene complexes studied, the most active was the bimetallic complex in which M = Mg or Mn:

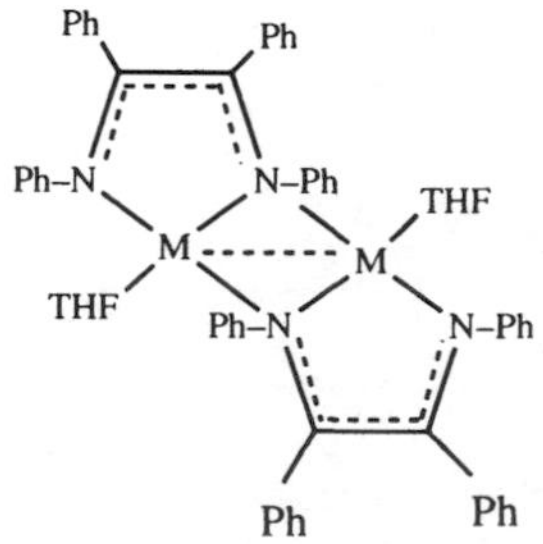

Structural investigations by IR spectroscopy and X-ray crystallography indicated that solvent THF molecules were coordinated directly to the central metal atoms and that the CO_2 group was fixed into an N-carboxylate bond. The Mg complexes were the most active, followed by those with Mn atoms. Other metal atoms were inactive. In the presence of acetophenone or acetone in THF or DMF solution, the methyl groups were carboxylated by the Mg and Mn complexes, forming (after acidification) benzoyl acetic acid or 1,3-acetone dicarboxylic acid (3-oxoglutaric acid), with yields of up to 68 or 59%, respectively:

$$PhCOCH_3 + CO_2 \rightarrow PhCOCH_2\text{–}COOH \qquad (6.12)$$

$$CH_3COCH_3 + CO_2 \rightarrow HOOC\text{–}CH_2COCH_2\text{–}COOH \qquad (6.13)$$

The similarity of the above model reactions with enzymatic carboxylation is expressed in the singular activity of magnesium metal, followed by manganese, in involving an N-carboxylate as intermediate, and in the carboxylation into active C–H bonds. In a significant difference from enzymatic reactions, in which the energy supply occurs externally by participation of ATP, these model catalysts are internally highly energetic, with their electron-rich diazadiene structure.[84]

6.4.6 Lanthanoid Complex

A complex formed by mixing lanthanum isopropoxide, La(O-*iso*-Pr)$_3$, with phenyl isocyanate, PhNCO, was found by Inoue et al. to serve as an efficient CO_2 carrier for the carboxylation of active methylene compounds, such as fluorene and phenylacetonitrile. The carboxylations of S-benzyl thiopropionate to form the thiol ester of 2-methylmalonate,

$$CH_3\text{–}CH_2\text{–}CO\text{–}S\text{–}CH_2Ph + CO_2 \rightarrow CH_3\text{–}CH(COOH)\text{–}CO\text{–}S\text{–}CH_2Ph \qquad (6.14)$$

may be considered to mimic the biological carboxylation of propionyl-CoA to 2-methylmalonyl-CoA with a biotin enzyme. This carboxylation reaction was also successful with the thioesters of phenylacetic, acetic, and isovaleric acids.[85,86]

6.4.7 Carbonic Anhydrase Models

CA plays an important role in photosynthetic CO_2 fixation.[87] CAs enhance the rate of hydration of CO_2 by a factor of up to 10^8 relative to the rate of the uncatalyzed hydration.[88] Carbon dioxide insertion into zinc–oxygen bonds is a crucial step in the CA-catalyzed hydration of CO_2 and also in the copolymerization of CO_2 and epoxides to polycarbonates catalyzed by zinc complexes.[89] In the active site of CA, a central zinc atom is coordinated in a tetrahedral configuration to three histidine imidazole groups and a water molecule, $(His_3Zn\text{–}OH)^+$. The primary step in the action of CA is deprotonation, forming an active zinc hydroxide derivative, which very rapidly reacts as a nucleophile with CO_2 to form a zinc bicarbonate intermediate, $(His_3Zn–OCO_2H)^+$. This then splits off bicarbonate, reforming the enzyme.

The Zn(II) complex of the macrocyclic tetraamine 1,4,7,10-tetraazacyclodecane ([12]ane N_4,cyclen) was shown by Zhang et al. to catalyze the hydration of CO_2 and the dehydration of the bicarbonate anion. The second-order rate constants for the hydration and dehydration reactions at 25°C were $5.8 \times 10^2\,M^{-1}\,s^{-1}$ and 4.8 $M^{-1}\,s^{-1}$, respectively. These rates were close to those of the natural carbonic anhydrase.[90,91]

Extremely active fixation of CO_2 even from air was discovered by Bazzicalupi et al. using Zn(II) and Cu(II) complexes of oxa–aza macrocycles.[92] These complexes were prepared from zinc or cupric perchlorates and a [15]aneN_3O_2 macrocycle, 1,4-dioxa-7,10,13-triazacyclopentadecane (L). Dilute aqueous solutions of these complexes (25 mM) in the pH range 8.5 to 10 readily adsorbed CO_2 from air bubbled through them. In an irreversible reaction, the complexes $\{[ZnL]_3(\mu_3\text{-}CO_3)\} \cdot (CCl_4)_4$ and $\{[CuL]_4(\mu_3\text{-}CO_3)\} \cdot (CCl_4)_4$ were formed. Potentiometric measurements indicated that in the alkaline solutions, the above complexes act as the hydroxo species, such as $(ZnLOH)^+$, which fix CO_2 by forming the hydrogen carbonate adduct $[ZnL(HCO_3)]^+$.[92]

In another model reaction mimicking the reaction of CA, Darensbourg et al. studied the reversible carboxylation of a tungsten carbonyl complex:

$$W(CO)_5OH^- + CO_2 = W(CO)_5O_2COH^- \tag{6.15}$$

The equilibrium in this reaction is very much to the right. In the carboxylated complex $W(CO)_5O_2COH^-$, the central W(0) atom was bound to five carbonyl ligands as well as to a monodentate-bound bicarbonate ligand.[93,94]

A model zinc complex, $[LZn(OH_2)]^{2+}$, where L = tris(2-benzimidazolyl-methyl)amine, was found by Nakata et al. to mimic the active site of CA. In this complex, the three imidazole ligand groups corresponded to the three histidine imidazole groups in the active site of the natural enzyme.[95,96]

6.4.8   Organophosphorus Acids

Aresta et al. succeeded in synthesizing N-arylcarbamates from arylamines and diethyl carbonate or diphenyl carbonate (DPC) in the presence of catalytic amounts of organophosphorus acids.[97] Thus, PhNHCOOPh was produced in 98% yield by reacting aniline with DPC in THF solution under N_2 for 15 h at 120°C in the presence of Ph_2POOH. The organophosphinic acid acted as a Broensted acid. In the proposed mechanism, the intermediate formation of a carbonic–phosphinic acid anhydride [e.g., $Ph_2P(O)–O–C(O)–OR$] was postulated. This mechanism bears analogy to the enzymatic reactions in which mixed anhydrides, such as carboxy phosphate,

$$^-O(HO)P(O)–O–COO^- \tag{6.16}$$

act as carboxylating agents. In living systems, the carbamate anion is produced from carboxy phosphate and ammonia with the assistance of the enzyme carbamoyl phosphate synthetase.[97]

6.4.9 Peroxocarbonates

The dioxygen complex $[RhCl(\eta^2-O_2)(P)_3]$ (where $P = PEt_2Ph$ or $PEtPh_2$) in toluene solution was found by Aresta et al. to undergo CO_2 insertion into the O–O bond, with formation of an Rh peroxocarbonate, $RhOOC(O)O(Cl)(P)_3$, followed by release of the tertiary phosphine oxide, R_3PO. This reaction mimics the action of monooxygenases as one-oxygen atom transfer agents. ^{18}O-isotopic studies indicated that the oxygen atom transferred was that which had been linked to the rhodium metal.[98]

6.5 BIOMASS PROCESSES

The basis for use of biomass as a CO_2 mitigation technology is the photosynthesis reaction of atmospheric CO_2 with water to produce plants and organisms both on land and in the ocean consisting of a carbohydrate, cellulose, algae, bacteria, etc. Thus, plants breathe in CO_2 and breathe out oxygen, and man breathes in oxygen and breathes out CO_2. The composition of plants is remarkably constant, varying within relatively narrow ranges, as shown in Table 6.1. The stoichiometric composition of wood can be represented by $CH_{1.44}O_{0.66}$. The photosynthesis reaction is then as follows:

$$CO_2 + 0.72H_2O \rightarrow CH_{1.44}O_{0.66} + 1.03O_2$$

To assist in growth, nutrients (fertilizers) that include sulfur, phosphorus, and trace metals take part in the photosynthesis reaction. The main difference in concentration occurs in the ash content of the various biomass. Agricultural waste (manure and rice hulls) generally has a higher ash content than the woods. Also, biomass has a lower sulfur content than the fossil-fuel coal.

TABLE 6.1
Composition of Biomass[98]

| | Elemental analysis (%) | | | | | |
	Softwood	Hardwood	Bagasse	Feedlot manure	Rice hulls	Grass straw
H	6.1	6.2	6.1	5.7	5.5	6.0
C	52.1	51.0	47.3	41.2	39.4	45.0
O	40.0	39.7	35.3	33.3	36.0	41.9
N	0.1	0.4	0.0	2.3	0.5	0.5
S	0.0	0.0	0.0	0.3	0.2	0.5
Ash	1.7	2.7	11.3	17.2	18.4	6.1

TABLE 6.2
Energy Value of Biomass[99]

Biomass	Higher heating value (Btu/lb)
Douglas fir	9,050
Maple	8,580
Poplar	8,920
Rice hulls	6,610
Rice straw	6,540
Manure	7,380
Green algae	12,300[106]

Biomass can be burned as fuel for various heat and power purposes or converted to hydrogen-rich fuel, all of which eventually regenerates CO_2 which is reemitted to the atmosphere by the combustion process. The energy value of biomass is shown in Table 6.2. Thus, for any process that uses biomass, there is essentially no net change in CO_2 emission, and thus biomass as an energy source is essentially CO_2 neutral. However, some fossil-fuel CO_2 emission occurs because energy is needed for planting and harvesting; when managed properly, the amount is relatively small. Table 6.3 gives some biomass growth rate yield ranges depending on region and climate, which are of importance in land use. A contributing factor to global increase in CO_2 concentration over the past decades is the cutting down of rain forests and whole forests for lumber, paper production, and industrial development, which is known as deforestation. By the same token, planting trees and forests will increase absorption and fixation of atmospheric CO_2, which is called aforestation. There have been a number of proposals, some of which have been interested in growing forests as a sink for CO_2 from fossil-fuel power plant emissions and as a means of management of CO_2 emissions. Rapid rotational crops on so-called energy farms have been proposed and modeling studies have been made on the subject.[101,102] There are even estimates that there is enough arable land in the U.S. to supply biomass as a substantial part of the fuel referred for power generation in the U.S. Land management and economics are the main factors in determining the application of growing trees and crops for energy purposes. Attempts are being made to genetically alter plants to increase the biomass yields per acre of land area.

TABLE 6.3
Biomass Growth Rate Yield Ranges[102, 114]

Biomass	Yield (dry tons/acre/yr)
Forest wood	2–5
Maize (corn)	12–25
Sugarcane	35–75

Interest has developed in cofiring coal with biomass for power generation.[103] The lower heating value and lower density of biomass compared to coal tend to lower the capacity of existing coal-fired boilers and, thus, limit the usefulness of cofiring.

Microalgae grown in large open ponds into which power plant flue gas CO_2 is dispensed can be harvested as biomass and used as an alternative fuel or converted to liquid fuel.[104,105] The higher CO_2 concentration of flue gas compared to atmospheric CO_2 increases the rate of photosynthesis; however, further increase in productivity must be achieved to overcome the economic limitations due to the requirements of large land area, favorable climate, and an ample water supply.

There are a number of excellent reviews on production of fuels from biomass using thermochemical processes, which include pyrolysis, steam gasification, hydrogasification, and partial oxidation.[99,106,107] Currently, biomass as an energy source is more expensive than the production of fossil fuels. Because of its low value and in many cases its negative disposal value (the cost to get rid of it), biomass derived from waste materials can be utilized economically as an energy source. Waste biomass includes agricultural waste (hulls, bagasse), municipal solid waste, industrial waste (sawdust, oils, tires, etc.), and animal waste (manure). Additional reviews on the latter subject can be found in References 98, 107, and 108.

Coprocessing biomass and natural gas for conversion to liquid and gaseous fuels has been proposed to improve yields and economics of conversion of biomass and reduce CO_2 emissions.[109,110] Other coprocessing systems can be found in Chapter 7.

Converting agricultural crops (corn and sugarcane) to ethanol by a fermentation process has gained usage as a motor fuel both in the U.S. and in Brazil.[111]

A proposal has been made to enhance the natural photosynthesis and CO_2 uptake in the ocean by ocean fertilization with nutrients such as iron, which is thought to limit the growth of algae in the ocean.[112] A test in the equatorial Pacific Ocean has supported the hypothesis that phytoplankton growth is limited by iron bioavailability.[113] However, the efficacy of reducing atmospheric CO_2 on a global scale in terms of capacity, economics, and environment remains controversial.

The effect of increasing concentrations of atmospheric CO_2 due to fossil-fuel combustion on the rate of photosynthesis and growth of terrestrial biomass must be considered in the overall carbon balance. There is much evidence that the photosynthetic rate, for certain plants, increases with increasing CO_2 concentrations.[114] However, the expectation of limiting global atmospheric concentration by increased CO_2 fixation is a complex issue due to the variable effect of temperature, climate, water availability, and soil conditions, and thus this topic is an important subject for ongoing research.

6.6 CONCLUSIONS ON BIOCONVERSION AND BIOMASS PROCESSES

The production of methane with methanogens is already carried out on a relatively small scale in biogas plants in various parts of the world, by anaerobic digestion of manure and municipal and agricultural waste. Large-scale methane production by methanogens using $H_2 + CO_2$ as the raw materials does not seem logical from the standpoint of CO_2 mitigation, if the H_2 will be formed by the steam reforming of natural gas (mainly methane), as is the current industrial process, and if the methane is to be burned as a fuel.

A major advantage of the bioconversion of carbon dioxide in comparison to gas–solid-phase catalytic hydrogenation (see Chapters 7 and 10) is that, in addition to methane, the methanogenic bacteria also produce valuable materials, such as enzymes, amino acids, and vitamins. Hydrogen sulfide, which is poisonous to many catalysts, is actually beneficial and consumed during bioconversion. The application of thermophilic methanogens should be particularly attractive for the treatment of blast furnace and converter gases released during the production of iron and steel. These gases contain hydrogen, carbon oxides, and the toxic and obnoxious hydrogen sulfide.

The culture of marine microalgae in seawater mixed with nutrient-rich wastewater directly supplied with power station stack gases could be an attractive option, with some of the costs offset by the production of valuable products from the biomass formed (e.g., proteins, vitamins, β-carotene).

Enzyme-catalyzed reactions are interesting, as they lead directly to larger molecules of considerable complexity, which may be more difficult to access by purely synthetic methods.

The search for models that mimic the action of the natural enzyme CA could be useful in the context of the disposal of unused CO_2. The uncatalyzed hydration of CO_2 is slow (except in highly acidic water), while the gas–liquid exchange of CO_2 between the atmosphere and water is fast. Therefore, an efficient and inexpensive catalyst for the hydration of CO_2 to bicarbonate could be a method for the fixation of CO_2 from the atmosphere.

Concerning biomass processes, the relative constancy of the elemental analysis of widely different biomass feedstock origins simplifies process design and operation. Continuing development of rapid rotational crops grown for an energy source may have an impact on reducing CO_2 emission. Cofiring fossil fuel with biomass and conversion of biomass to liquid fuel are of interest in reducing CO_2 emission.

REFERENCES

1. Woese, C.R., Magrum, L.J., and Fox, G.E., Archaebacteria, *J. Mol. Evol.*, 11, 245–252, 1978.
2. Fuchs, G. and Stupperich, E., Carbon dioxide fixation pathways in bacteria, *Physiol. Veg.*, 21, 845–854, 1983.
3. Fuchs, G., Carbon dioxide reduction by anaerobic bacteria, NATO ASI Ser. C, *Carbon Dioxide Source Carbon: Biochemical and Chemical Uses,* 206, 263–273, 1987.
4. Fuchs, G., Alternatives to the Calvin cycle and the Krebs cycle in anaerobic bacteria: pathways with carbonylation chemistry, *Colloq. Biol. Chem.*, 41, 13–20, 1990; *Chem. Abstr.*, 115, 275270b.

5. Ragsdale, S.W. and Kumar, M., Nickel-containing carbon monoxide dehydrogenase/ acetyl-CoA synthase, *Chem. Rev.*, 96, 2515–2539, 1996.

6. Hemming, A. and Blotevogel, K.H., A new pathway for carbon dioxide fixation in methanogenic bacteria, *Trends Biochem. Soc.*, 10, 198–200, 1985.

7. Daniels, L., Biotechnological potential of methanogens, *Biochem. Soc. Symp.*, 58, 181–193, 1992.

8. Bugante, E.C., Shimonura, Y., Tanaka, T., Taniguchi, M., and Oi, S., Methane production from hydrogen and carbon dioxide and monoxide in a column bioreactor of thermophilic methanogens by gas recirculation, *J. Ferment. Bioeng.*, 67, 419–421, 1989.

9. Klasson, K.T., Cowger, J.P., Ko, C.W., Vega, J.L., Clausen, E.C., and Gaddy, J.L., Methane production from synthesis gas using a mixed culture of *R. rubrum, M. barkeri* and *M. formicicum, Appl. Biochem. Biotechnol.*, 24–25, 317–328, 1990.

10. Nishimura, N., Kitaura, S., Mimura, A., and Takahara, Y., Growth of thermophilic methanogen KN-15 on H_2–CO_2 under batch and continuous conditions, *J. Ferment. Bioeng.*, 72, 280–284, 1991.

11. Nishimura, N., Kitaura, S., Mimura, A., and Takahara, Y., Cultivation of thermophilic methanogen Kn-15 on H_2–CO_2 under pressurized conditions, *J. Ferment. Bioeng.*, 73, 477–480, 1992.

12. Strevett, K.A., Vieth, R.F., and Grasso, D., Chemo-autotrophic biogas purification for methane enrichment: mechanism and kinetics, *Chem. Eng. J.*, 58, 71–79, 1995.

13. Stetter, K.O., Hyperthermophiles in the history of life, *CIBA Found. Symp.*, 202, 1–18, 1996.

14. Pepper, C.B. and Monbouquette, H.G., Issues in the culture of the extremely thermophilic methanogen, *Methanothermus fervidus, Biotechnol. Bioeng.*, 41, 970–978, 1993.

15. Tsao, J.H., Kaneshiro, S.M., Yu, S.S., and Clark, D.S., Continuous-culture of *Methanococcus jannaschii*, an extremely thermophilic methanogen, *Biotechnol. Bioeng.*, 43, 258–261, 1994.

16. Mountfort, D.O., Pybus, V., and Wilson, R., Metal ion-mediated accumulation of alcohols during alkane oxidation by whole cells of *Methylosinus trichosporium, Enzyme Microb. Technol.*, 12, 343–348, 1990.

17. Mehta, P.K., Mishra, S., and Ghose, T.K., Methanol biosynthesis by covalently immobilized cells of *Methylosinus trichosporium*: batch and continuous studies, *Biotechnol. Bioeng.*, 37, 551–556, 1991.

18. Patel, R.N., Hou, C.T., and Laskin, A.I., Microbiological Oxidation, Eur. Patent Appl. EP 88,602 (Cl. C12N9/02), September 14, 1983; *Chem. Abstr.*, 100, P33289n.

19. Diekert, G. and Wohlfarth, G., Metabolism of homoacetogens, *Leeuwenhoek Int. J. Gen. Mol. Microbiol.*, 66, 209–221, 1994.

20. Gollin, D.J., Li, X.-L., Liu, S.-M., and Ljungdahl, L.G., Primary structure of the NADP-dependent formate dehydrogenase of *Clostridium thermoaceticum*, a tungsten–selenium–iron–sulfur-containing enzyme, in Abstr. 4th Int Conf. on Carbon Dioxide Utilization, Kyoto, Japan, September 1997, O-31.

21. Rainina, E.I., Pusheva, M.A., and Ryabokon, A.M., Microbial cells immobilized in poly(vinyl alcohol) cryogels. 1. Biocatalytic reduction of CO_2 by the thermophilic homoacetogenic bacterium *Acetogenium kivui, Biotechnol. Appl. Biochem.*, 19, 321–329, 1994.

22. Ryabokon, A.M., Pusheva, M.A., Detkova, E.N., and Rainina, E.I., Reduction of CO_2 to acetate by immobilized cells of the thermophilic homoacetogenic bacterium *Thermoanaerobacter kivui, Microbiology*, 64, 657–661, 1995.

23. Kevbrina, M.V., Ryabokon, A.M., and Pusheva, M.A., Acetate formation from CO-containing gas mixtures by free and immobilized cells of the thermophilic homoacetogenic bacterium *Thermoanaerobacter kivui*, *Microbiology*, 65, 656–660, 1996.

24. Schneider, G., Carbon dioxide fixation in biology: structure and function of ribulose biphosphate carboxylase/oxygenase (Rubisco), *Carbon Dioxide Chemistry: Environ. Issues, R. Soc. Chem. Spec. Publ.*, 153, 150–159, 1994.

25. Boardman, N.K., Energy from the biological conversion of solar energy, *Philos. Trans. R. Soc. London A*, 295, 477–489, 1980.

26. Kodama, M., Ikemoto, H., and Miyachi, S., A new species of highly CO_2-tolerant fast-growing marine microalga suitable for high-density culture, *J. Mar. Biotechnol.*, 1, 21–25, 1993.

27. Miyachi, S., Kurano, N., Qiang, H., and Iwasaki, I., Carbon dioxide and microalgae, in Abstr. 4th Int. Conf. on Carbon Dioxide Utilization, Kyoto, Japan, September 1997, PL-5.

28. Watson, G.M.F. and Tabita, F.R., Microbial ribulose-biphosphate carboxylase/oxygenase. A molecule for phylogenetic and enzymological investigation, *FEMS Microbiol. Lett.*, 146, 13–22, 1997.

29. Yokota, A., Super-RuBisCo: improvement of photosynthetic performances of plants, in Abstr. 4th Int. Conf. on Carbon Dioxide Utilization, Kyoto, Japan, September 1997, KL-5.

30. Tomizawa, K.-I., Shikanai, T., Shimoide, A., Foyer, C.H., and Yokota, Y., Revertant of no-active RuBisCo tobacco mutant, SP25, obtained by chloroplast transformation method using microprojectile bombardment, in Abstr. 4th Int. Conf. on Carbon Dioxide Utilization, Kyoto, Japan, September 1997, P-083.

31. Palmqvist, K., Inorganic carbon fluxes in lichens and their photosynthesizing partners, *Carbon Dioxide Chemistry: Environ. Issues, R. Soc. Chem. Spec. Publ.*, 153, 135–141, 1994.

32. Fridlyand, L., Kaplan, A., and Reinhold, L., Quantitative evaluation of the role of a putative CO_2-scavenging entity in the cyanobacterial CO_2-concentrating mechanism, *BioSystems*, 37, 229–238, 1996.

33. Weiss, V., Gromet-Elhanan, Z., and Halmann, M., Batch and continuous culture experiments on nutrient limitations and temperature effects in the marine alga *Tetraselmis suecica*, *Water Res.*, 19, 185–190, 1985.

34. Laws, E.A. and Berning, J.L., A study of the energetics and economics of microalgal mass culture with the marine chlorophyte *Tetraselmis suecica*: implications for use of power plant stack gases, *Biotech. Bioeng.*, 37, 936–947, 1991.

35. Pesheva, I., Kodama, M., Dionisio-Sese, M.L., and Miyachi, S., Changes in photosynthetic characteristics induced by transferring air-grown cells of *Chlorococcum littorale* to high-CO_2 conditions, *Plant Cell Physiol.*, 35, 379–387, 1994.

36. Schnackenberg, J., Ikemoto, H., and Miyachi, S., Relationship between oxygen-evolution and hydrogen-evolution in a *Chlorococcum* strain with high CO_2-tolerance, *J. Photochem. Photobiol. B Biol.*, 28, 171–174, 1995.

37. Schnackenberg, J., Ikemoto, H., and Miyachi, S., Photosynthesis and hydrogen evolution under stress conditions in a CO_2-tolerant marine green alga, *Chlorococcum littorale*, *J. Photochem. Photobiol. B Biol.*, 34, 59–62, 1996.

38. Pirt, M.W. and Pirt, S.J., The influence of carbon dioxide and oxygen partial pressures on chlorella growth in photosynthetic steady-state cultures, *J. Gen. Microbiol.*, 119, 321–326, 1980.

39. Murakami, M., Yamada, F., Nishide, T., Muranaka, T., Yamaguchi, N., and Takimoto, Y., The biological CO_2 fixation using *Chlorella* sp. with high capability in fixing CO_2, in Abstr. 4th Int. Conf. on Carbon Dioxide Utilization, Kyoto, Japan, September 1997, O-33.

40. Yun, Y.S., Park, J.M., and Yang, J.W., Enhancement of CO_2 tolerance of *Chlorella vulgaris* by gradual increase of CO_2 concentration, *Biotechnol. Technol.*, 10, 713–716, 1996.

41. Yun, Y.S., Lee, S.B., Park, J.M., Lee, C.I., and Yang, J.W., Carbon dioxide fixation by algal cultivation using wastewater nutrients, *J. Chem. Technol. Biotechnol.*, 69, 451–455, 1997.

42. Miyasaka, H., Nakano, H., Akiyama, H., Kanai, S., and Hirano, M., Production of PHS (polyhydroxyalkanoate) by the genetically engineered marine cyanobacterium, in Abstr. 4th Int. Conf. on Carbon Dioxide Utilization, Kyoto, Japan, September 1997, O-19.

43. Suh, I.S., Park, C.B., Han, J.-K., and Lee, S.B., Cultivation of cyanobacterium in various types of photobioreactors for biological CO_2 fixation, in Abstr. 4th Int. Conf. on Carbon Dioxide Utilization, Kyoto, Japan, September 1997, P-041.

44. Nanba, M. and Kawata, M., CO_2-removal by a bioreactor with photosynthetic algae using solar-collecting and light-diffusing optical devices, in Abstr. 4th Int. Conf. on Carbon Dioxide Utilization, Kyoto, Japan, September 1997, P-090.

45. Samejima, Y., Hirano, A., Hon-Nami, K., Kunito, S., Masuda, K., Hasuike, M., Tsuyuki, Y., and Ogushi, Y., A marine microalga utilization for a paper: semi-batch cultivation of *Tetraselmis* sp. Tt-1 by a tubular bioreactor and its addition to whole kenaf pulp, in Abstr. 4th Int. Conf. on Carbon Dioxide Utilization, Kyoto, Japan, September 1997, P-094.

46. Hirayama, S., Ueda, R., Ogushi, Y., Hirano, A., Samejima, Y., Hon-Nami, K., and Kunito, S., Ethanol production from carbon dioxide by fermentative microalgae, in Abstr. 4th Int. Conf. on Carbon Dioxide Utilization, Kyoto, Japan, September 1997, P-096.

47. Hirano, A., Ueda, R., Hirayama, S., and Ogushi, Y., CO_2 fixation and ethanol production with microalgal photosynthesis and intracellular fermentation, *Energy*, 22, 137–142, 1997.

48. Gao, K., Aruga, Y., Asada, K., and Kiyohara, M., Influence of enhanced CO_2 on growth and photosynthesis of the red algae *Gracilaria* sp. and *G. chilensis*, *J. Appl. Phycol.*, 5, 563–571, 1993.

49. Gao, K. and McKinley, K.R., Use of macroalgae for marine biomass production and CO_2 remediation. A review, *J. Appl. Phycol.*, 6, 45–60, 1994.

50. Aresta, M. and Tommasi, I., Model systems for CO_2 reduction and carboxylation enzymes, in Int. Conf. on Carbon Dioxide Utilization, Bari, Italy, September 1993, 205–213.

51. Aresta, M., Quaranta, E., Tommasi, I., Giannoccaro, P., and Ciccarese, A., Enzymatic versus chemical carbon dioxide utilization. I. The role of metal centres in carboxylation reactions, *Gazz. Chim. Ital.*, 125, 509–538, 1995.

52. Parkinson, B.A. and Weaver, P.F., Photoelectrochemical pumping of enzymatic CO_2 reduction, *Nature*, 309, 148–149, 1984.

53. Kuwabata, S., Tsuda, R., and Yoneyama, H., Electrochemical conversion of carbon dioxide to methanol with the assistance of formate dehydrogenase and methanol dehydrogenase as biocatalysts, *J. Am. Chem. Soc.*, 116, 5437–5443, 1994.

54. Willner, I., Mandler, D., and Riklin, A., Photoinduced carbon dioxide fixation forming malic and citric acid, *J. Chem. Soc. Chem. Commun.*, pp. 1022–1024, 1986.

55. Willner, I., Mandler, D., and Maidan, R., Bio-models and artificial models for photosynthesis, *New J. Chem.*, 11, 109–121, 1987.

56. Mandler, D. and Willner, I., Photochemical fixation of carbon dioxide: enzymatic synthesis of malic, aspartic, isocitric and formic acids in artificial media, *J. Chem. Soc. Perkin Trans. II*, pp. 997–1003, 1988.

57. Kuwabata, S., Tsuda, R., Nishida, K., and Yoneyama, H., Electrochemical conversion of carbon dioxide to methanol with use of enzymes as biocatalysts, *Chem. Lett.*, pp. 1631–1634, 1993.

58. Kuwabata, S., Nishida, K., Tsuda, R., Inoue, H., and Yoneyama, H., Photochemical reduction of carbon dioxide to methanol using ZnS microcrystallite as a photocatalyst in the presence of methanol dehydrogenase, *J. Electrochem. Soc.*, 141, 1498–1503, 1994.

59. Sugimura, K., Kuwabata, S., and Yoneyama, H., Electrochemical fixation of CO_2 in oxoglutaric acid using an enzyme as an electrocatalyst, *J. Am. Chem. Soc.*, 111, 2361–2362, 1989.

60. Taniguchi, I., Electrocatalytic reduction of greenhouse gases using biofunctional metal complexes, in Proc. Int. Symp. Chemical Fixation of Carbon Dioxide, Nagoya, Japan, December 2 to 4, 1991, 81–88.

61. Inoue, H., Kubo, Y., and Yoneyama, H., Photocatalytic fixation of carbon dioxide in oxoglutaric acid using isocitrate dehydrogenase and cadmium sulfide, *J. Chem. Soc. Faraday Trans.*, 87, 553–557, 1991.

62. Kuwabata, S., Morishita, N., and Yoneyama, H., Electrochemical fixation of CO_2 in acetyl-coenzyme-A to yield pyruvic acid using pyruvate dehydrogenase complexes as an electrocatalyst, *Chem. Lett.*, pp. 1151–1154, 1990

63. Aresta, M., Fuchs, G., Quaranta, E., and Tommasi, I., Metal-ions dependence and mimetic complexes of phenol carboxylase: a new enzyme catalyzing the carboxylation of phenol using CO_2, in Proc. Int. Symp. Chemical Fixation of Carbon Dioxide, Nagoya, Japan, December 2 to 4, 1991, 353–358.

64. Aresta, M., Berloco, C., Jaruszewski, M., Quaranta, E., and Tommasi, I., Mimetic complexes of phenol carboxylase of *Pseudomonas aeruginosa*: direct ring-carboxylation of phenol, in Int. Conf. on Carbon Dioxide Utilization, Bari, Italy, September 1993, 364.

65. Fruge, D.R., Fong, G.D., and Fong, F.K., Photosynthesis of polyatomic organic molecules from carbon dioxide and water by the photocatalytic action of visible-light-illuminated platinized chlorophyll a dihydrate polycrystals, *J. Am. Chem. Soc.*, 101, 3694–3697, 1979.

66. Hirai, Y., Aida, T., and Inoue, S., Artificial photosynthesis of β-ketocarboxylic acids from carbon dioxide and ketones via enolate complexes of aluminum porphyrin, *J. Am. Chem. Soc.*, 111, 3062–3063 1989.

67. Komatsu, M., Aida, T., and Inoue, S., Novel visible-light-driven catalytic CO_2 fixation — synthesis of malonic acid derivatives from CO_2, α,β-unsaturated ester or nitrile, and diethylzinc catalyzed by aluminium porphyrins, *J. Am. Chem. Soc.*, 113, 8492–8498, 1991.

68. Inoue, S., Light-induced carbon dioxide fixation mediated by metalloporphyrin, in Proc. Int. Symp. Chemical Fixation of Carbon Dioxide, Nagoya, Japan, December 2 to 4, 1991, 201–208.

69. Thauer, R., Metal enzyme involved in CO_2 reduction by methanogenic archaea, in Int. Conf. on Carbon Dioxide Utilization, Bari, Italy, September 1993, 195–196.

70. Thauer, R., Biodiversity and unity in biochemistry, *Leeuwenhoek Int. J. Gen. Mol. Microbiol.*, 71, 21–32, 1997.

71. Tezuka, M., Yajima, T., Tsuchiya, A., Matsumoto, Y., Uchida, Y., and Hidai, M., Electroreduction of carbon dioxide catalyzed by iron–sulfur clusters $[Fe_4S_4(SR)_4]^{2-}$, *J. Am. Chem. Soc.*, 104, 6834–6836, 1982.

72. Nakazawa, M., Mizobe, Y., Matsumoto, Y., Uchida, Y., Tezuka, M., and Hidai, M., Electrochemical reduction of carbon dioxide using iron–sulfur clusters as catalyst precursors, *Bull. Chem. Soc. Jpn.*, 59, 809–814, 1986.

73. Tomohiro, T., Uoto, K., and Okuno, H., Electrochemical reduction of CO_2 catalyzed by macrocyclic Fe_4S_4 iron–sulfur clusters, *J. Chem. Soc. Chem. Commun.*, pp. 194–195, 1990.

74. Kodaka, M., Tomohiro, T., Lee, A.L., and Okuna, H., Carbon dioxide fixation forming oxazolidone coupled with a thiol/Fe_2S_4 cluster redox system, *J. Chem. Soc. Chem. Commun.*, pp. 1479–1481, 1989.

75. Tanaka, K., Wakita, R., and Tanaka, T., CO_2 fixation coupled with nitrite reduction catalyzed by $4Fe_4S$ cluster, *Chem. Lett.*, pp. 1951–1954, 1987.

76. Tanaka, K., Matsui, T., and Tanaka, T., Catalytic formation of α-keto acids by artificial CO_2 fixation, *J. Am. Chem. Soc.*, 111, 3765, 1989.

77. Tanaka, K., Carbon dioxide fixation catalyzed by FeS and MoFeS clusters, in Proc. Int. Symp. Chemical Fixation of Carbon Dioxide, Nagoya, Japan, December 2 to 4, 1991, 55–62.

78. Komeda, N., Nagao, H., Matsui, T., Adachi, G., and Tanaka, K., Electrochemical carbon dioxide fixation to thioesters catalyzed by $[Mo_2Fe_6S_8(SEt)_9]^{3-}$, *J. Am. Chem. Soc.*, 114, 3625–3630, 1992.

79. Nagao, H., Miyamoto, H., and Tanaka, K., Carbon dioxide fixation competed with proton addition to methyl acrylate, *Chem. Lett.*, pp. 323–326, 1991.

80. Chiba, K., Tagaya, H., Karasu, M., Ono, T., Hashimoto, K., and Moriwaki, Y., The carboxylation of active methylene compounds with carbon dioxide in the presence of diphenylcarbodiimide and potassium carbonate, *Bull. Chem. Soc. Jpn.*, 64, 966–970, 1991.

81. Chiba, K., Tagaya, H., Karasu, M., Ono, T., Saito, M., and Ashikagaya, A., Fixation of carbon dioxide with diphenylcarbodiimide as a model of biotin enzyme active site and a weak base: the carboxylation of fluorene under mild conditions, *Bull. Chem. Soc. Jpn.*, 64, 3738–3740, 1991.

82. Chiba, K., Tagaya, H., Karasu, M., Ishizuka, M., and Sugo, T., Carboxylation of active methylene compounds using anilide, potassium carbonate, and carbon dioxide, *Bull. Chem. Soc. Jpn.*, 67, 452–454, 1994.

83. Chiba, K., Tagaya, H., Miura, S., and Karasu, M., The carboxylation of active methylene compounds with carbon dioxide in the presence of 18-crown-6 and potassium carbonate, *Chem. Lett.*, pp. 923–926, 1992.

84. Walther, D., Ritter, U., Kempe, R., Sieler, J., and Undeutsch, B., Activation of CO_2 at transition-metal centres: simulation of enzymatic CO_2 fixation and transfer reactions by electron-rich (diazadiene) magnesium and (diazadiene) manganese complexes, *Chem. Ber.*, 125, 1529–1536, 1992.

85. Abe, H. and Inoue, S.H., Lanthanoid complex as a novel carbon dioxide carrier for the carboxylation of active methylene compounds under mild conditions, *J. Chem. Soc. Chem. Commun.*, pp. 1197–1198, 1994.

86. Inoue, S., Sugimoto, H., Ishida, N., and Shima, T., Carbon dioxide fixation with lanthanoid complex, in Abstr. 4th Int. Conf. on Carbon Dioxide Utilization, Kyoto, Japan, September 1997, P-050.

87. Badger, M.R. and Price, G.D., The role of carbonic anhydrase in photosynthesis, *Annu. Rev. Plant Physiol. Plant Mol. Biol.*, 45, 369–392, 1994.

88. Silverman, D.N., The hydration of CO_2 catalyzed by carbonic anhydrase, *Carbon Dioxide Chemistry: Environ. Issues, R. Soc. Chem. Spec Publ.*, 153, 171–178, 1994.

89. Darensbourg, D.J., Carbon dioxide insertion into metal–oxygen bonds. Relevance to copolymerization of CO_2 and epoxides, in Abstr. 3rd Int. Conf. on Carbon Dioxide Utilization, Norman, OK, May, 1995.

90. Zhang, X., van Eldik, R., Koike, T., and Kimura, E., Kinetics and mechanism of the hydration of CO_2 and dehydration of HCO_3 catalyzed by a Zn(II) complex of 1,5,9-triazacyclododecane as a model for carbonic anhydrase, in Abstr. Int. Conf. on Carbon Dioxide Utilization, Bari, Italy, September 1993, 373.

91. Zhang, X. and van Eldik, R., A functional model for carbonic anhydrase: thermodynamic and kinetic study of a tetraazacyclododecane complex of zinc, *Inorg. Chem.*, 34, 5606-5614, 1995.

92. Bazzicalupi, C., Bencini, A., Bencini, A., Bianchi, A., Corana, F., Fusi, V., Giorgi, C., Paoli, P., Paoletti, P., Valtancoli, B., and Zanchini, C., CO_2 fixation by novel copper(II) and zinc(II) complexes. A solution and solid state study, *Inorg. Chem.*, 35, 5540–5548, 1996.

93. Darensbourg, D.J., Jones, M.L.M., and Reibenspies, J.H., Synthesis and reactivity of tungsten pentacarbonyl hydroxo and bicarbonato complexes. Molecular structure of $[PPN][W(CO)_5HCO_3]$, an organometallic analog for carbonic anhydrase, *Inorg. Chem.*, 32, 4675–4676, 1993.

94. Darensbourg, D.J., Jones, M.L.M., and Reibenspies, J.H., The reversible insertion reaction of carbon dioxide with the $W(CO)_5OH^-$ anion. Isolation and characterization of the resulting bicarbonate complex $[PPN][W(CO)_5O_2COH)]$, *Inorg. Chem.*, 35, 4406–4413, 1996.

95. Nakata, K., Uddin, M.K., Ogawa, K., and Ichikawa, K., CO_2 hydration by mimic zinc complex for active site of carbonic anhydrase, *Chem. Lett.*, pp. 991–992, 1997.

96. Aresta, M., Berloco, C., and Quaranta, E., Biomimetic building-up of the carbamic moiety. The intermediacy of carboxyphosphate analogs in the synthesis of N-aryl carbamate esters from arylamines and organic carbonates promoted by phosphorus acids, *Tetrahedron*, 51, 8073–8088, 1995.

97. Aresta, M., Tommasi, I., Quaranta, E., Fragale, C., Mascetti, J., Tranquille, M., Galan, F., and Fouassier, M., Mechanism of formation of peroxocarbonates $RhOOC(O)O(Cl)(P)_3$ and their reactivity as oxygen transfer agents mimicking monooxygenases. The first evidence of CO_2 insertion into the O–O bond of $Rh(\eta^2-O_2)$ complexes, *Inorg. Chem.*, 35, 4254–4260, 1996.

98. Tillman, D.F., Fuels from waste, in *Kirk-Othmer Encyclopedia of Chemical Technology,* Vol. 11, 3rd ed., John Wiley & Sons, New York, 1980, 393–410; Vol. 12, 4th ed., 1994, 110–125.

99. Reed, T.B., *Biomass Gasification*, Noyes Data Corporation, Park Ridge, NJ, 1981.

100. Schlamadinger, B. and Marland, G., The role of forest and bioenergy strategies in the global carbon cycle, *Biomass Bioenergy,* 10(5/6), 275–300, 1996.

101. Moffat, A.S., Resurgent forests can be greenhouse gas sponges, *Science,* 277, 315–316, 1977.

102. Working Group II Intergovernmental Panel on Climate Change (IPCC), *Climate Change 1995, Impacts, Adaptations, and Mitigation of Climate Change*, Cambridge University Press, U.K., 1996, 603–609, 744–771.

103. Winslow, J., Ekman, J., and Smouse, S., Energy from waste via coal/waste co-firing, in Proc. 21st Int. Tech. Conf. on Coal Utilization and Fuel System, Clearwater, FL, March 18 to 21, 1996, 201–211.

104. Benneman, J.R., CO_2 mitigation with microalgae systems, *Energy Convers. Manage.*, 38, Suppl. S475–S479, 1997.

105. Kadam, K.L. and Sheeham, J.J., Microalgal technology for remediation of CO_2 from power plant flue gas: a technoeconomic perspective, *World Resour. Rev.*, 8(4), 493–503, 1996.

106. Klass, D.L., Fuels from biomass, in *Kirk-Othmer Encyclopedia of Chemical Technology*, Vol. 12, 4th ed., John Wiley & Sons, New York, 1994, 16–110.

107. Johansson, T.B., Kelly, H., Reddy, A.K.N., and Williams, R.H., Eds., *Renewable Energy — Sources of Fuels and Electricity*, Island Press, Washington, D.C., 1993.

108. Tchobanoglous, G., Thiesen, H., and Vigil, S., *Integrated Solid Waste Management*, McGraw-Hill, New York, 1993.

109. Dong, Y. and Steinberg, M., Hynol — an economical process of methanol production from biomass and natural gas with reduced CO_2 emission, *Int. J. Hydrogen Energy*, 22(10/11), 971–977, 1997.

110. Borgwardt, R.H., Biomass and natural gas as co-feedstock for production of fuel for fuel cell vehicles, *Biomass Bioenergy*, 12(5), 333–345, 1997.

111. Katzen, R., Madson, P.W., and Monceaux, D.A., Biomass derived ethanol and ETBE in motor fuels, paper presented at the Am. Inst. Chem. Eng. Annual Meeting, Los Angeles, November 16 to 21, 1997.

112. Kurz, K.D. and Mariev-Reimer, E., Iron fertilization of the Austral ocean. The Hamburg model assessment, *Global Biogeochem. Cycl.*, 7, 229–244, 1993.

113. Coale, K.H. et al., A massive phytoplankton bloom, induced by an ecosystem-scale iron fertilization experiment in the equatorial Pacific Ocean, *Nature*, 383, 495–501, 1996.

114. Working Group II, Intergovernmental Panel on Climate Change (IPCC), *Climate Change 1995, Impacts, Adaptation and Mitigation of Climate Change*, Cambridge University Press, U.K., 1996, 62–63, 102.

7 Decarbonization of Fossil Fuels and Conversion to Alternative Fuels

7.1 INTRODUCTION

The term "decarbonization" as applied to fossil fuels has become a byword for application to the mitigation of the global greenhouse gas problem. In the strictest sense of the word, decarbonization means the removal of carbon from fossil fuels prior to combustion. However, it has come to mean the use of fossil fuels with the avoidance of CO_2 emissions to the atmosphere. There are two approaches for achieving decarbonization of fossil fuels. One approach is to process the fossil fuel prior to combustion so as to remove the carbon as carbon and utilize only the hydrogen-rich fraction as fuel. The other approach is to convert the fossil fuel to a hydrogen-rich fuel while producing, recovering, and sequestering CO_2 prior to combustion. Decarbonization also applies to the capture, recovery, and sequestering of CO_2 after combustion, thus preventing the CO_2 from entering the atmosphere. The processes dealing with decarbonization and the production of alternative hydrogen-rich fuels are described in this chapter. The hydrogen-rich fuels can be used for central power station electricity generation or as transportation fuels.

7.2 NATURAL GAS FOR POWER PRODUCTION

Natural gas for power production is most efficiently used in combined cycle power plants. Since natural gas is a clean fuel, it can be used to produce high-temperature combustion gas for direct use in turbines. The exhaust gas from the turbine is used to raise steam in a boiler to drive a turbine. In this manner, the overall power cycle efficiency can be raised from 40% using a conventional steam plant alone to as high as 60% for the combined cycle,[1] based on lower heating value of the natural gas fuel. For reduced CO_2 emissions, the exhaust can be treated with monoethanolamine (MEA) or selexol (glycol) solvent to remove and recover the CO_2 for sequestration in nearby aquifers underground and depleted gas wells or, when near the coast, in the deep ocean. Because of the improved efficiency, CO_2 emissions for power production can be reduced by as much as 50% compared to the conventional cycle. A schematic of the natural gas combined cycle plant with CO_2 recovery and disposal is shown in Figure 7.1. By the removal and recovery of CO_2 from the exhaust gases at the back end of the power train and sequestering the CO_2, the total emission reduction can be reduced to over 90%.

7.3   NATURAL GAS FOR HYDROGEN PRODUCTION

Most of the world's hydrogen production is produced by the steam reforming of natural gas followed by water/gas shift[2] involving the following two reactions:

$$CH_4 + H_2O = CO + 3H_2$$
$$CO + H_2O = CO_2 + H_2$$

$$CH_4 + 2H_2O = CO_2 + 4H_2$$

Thus, the hydrogen comes not only from the natural gas but also from the steam. The overall process is endothermic, requiring about 60 kcal/mol methane or 15 kcal/mol H_2, not counting the inefficiency of the process. The steam reforming of natural gas takes place in a gas-fired tubular furnace at about 900°C and up to 50 atm in the presence of a nickel catalyst. The CO_2 is separated either with a solvent (MEA) or by pressure swing absorption (PSA). Schematics are shown in Figure 7.2. Conversion of the residual CO and H_2 to methane in a methanation step is also used in the presence of a nickel catalyst.[3] The CO_2 is emitted at almost a 100% concentration, and this can be readily sequestered in underground aquifers or depleted gas wells after compression and liquefaction. The hydrogen can be used in either conventional internal combustion engines or advanced highly efficient fuel cells, either for stationary electrical power production or in vehicles as an alternatives fuel. This system has been proposed by Williams as a leading CO_2 mitigation technology.[27] See Chapter 10 for further details on synthesis gas production by CO_2 reforming of methane.

7.4   COAL GASIFICATION FOR POWER PRODUCTION

Because of erosion and corrosion of turbine blades due to its ash and sulfur content, coal cannot be used directly in gas turbines for electricity production in combined cycle plants. The coal must first be gasified and the ash and sulfur removed before combusting the gas providing the energy for the gas turbines. The gasification of coal is endothermic, and because of the presence of solids, heating indirectly through tubes is very inefficient. In order to provide the endothermic energy, heating must be generated internally in the gasifier either by reacting part of the coal with oxygen

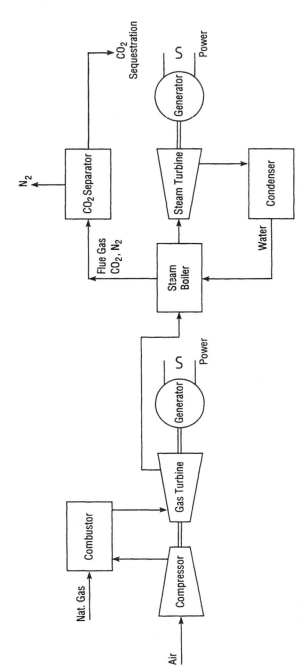

FIGURE 7.1 Natural gas combined cycle power plant. CO_2 removed and recovered from fuel gas can be sequestered underground in aquifers and depleted gas wells and in the ocean.

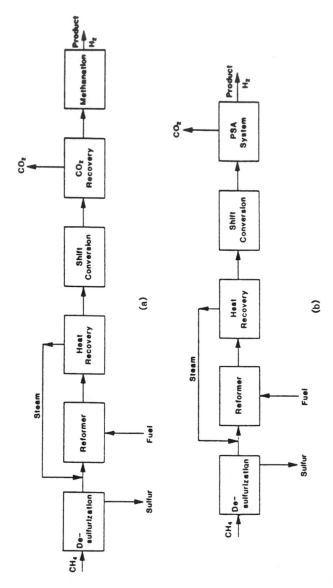

FIGURE 7.2 Flow diagram of hydrogen production from methane steam reforming: (a) conventional process and (b) with PSA modification. The CO_2 can then be sequestered in aquifers, gas wells, or the ocean.

from air or by separated oxygen to prevent dilution of the gases with nitrogen. Thus, coal gasifiers can either be air-blown or oxygen-blown. Part of the ash remains in the gasifier and is removed in some gasifiers as a molten ash and part is removed by filtering the particulates. The sulfur is either captured with limestone as a calcium sulfate in the gasifier or removed by reactive solids such as zinc titanate in a hot gas cleanup operation after the gasifiers. The coal gasification operation is represented by the following overall reaction. The gas composition distribution is limited by thermodynamic equilibrium. The gasifier always operates in a reducing atmosphere.

$$CH_{0.8}O_{0.1} + H_2O + O_2 = CO + CO_2 + CH_4 + H_2$$

$$\text{coal} \qquad\qquad\qquad \text{coal gas}$$

The CO is then shifted with steam to produce more H_2:

$$CO + H_2O = CO_2 + H_2$$

The CO_2 is removed from the coal gas by a solvent absorption/stripping operation using a solvent such as glycol (selexol). The hydrogen-rich coal gas is combusted for driving a gas turbine. The exhaust from the turbine is sent to a steam generator which raises steam for driving a lower temperature turbine. A schematic of the integrated coal gasification combined cycle (IGCC) plant is shown in Figure 7.3.[4] The combined efficiency for the IGCC plant runs about 50%. By sequestering the CO_2 in depleted wells, more than 90% reduction in CO_2 emission can be realized.

7.5   COAL GASIFICATION FOR HYDROGEN PRODUCTION

Where coal is available, but not natural gas, hydrogen can be produced from coal. Coal is gasified in a reactor in either a moving bed or a fluidized bed and is fed with steam and oxygen. Oxygen is used instead of air to prevent the dilution of the gases with nitrogen, making it easier to separate a highly concentrated CO_2 stream. The same two reactions shown above take place in the gasifier and the shifter. Figure 7.4 shows a typical plant configuration. If methanol is desired, the shift reactor is modified to adjust the H_2/CO ratio in the gases and sent to a catalytic methanol synthesis reactor.

7.6   PARTIAL OXIDATION OF OIL FOR HYDROGEN PRODUCTION

Oil or petroleum stocks are usually not considered for combined cycle power production. Perhaps this is because power plants are mainly based on coal and natural gas, whereas petroleum is mainly used for transportation fuels, diesel and gasoline. However, oil is used to produce hydrogen, especially in oil refineries. The conventional process is the partial oxidation of oil followed by water–gas shift to produce hydrogen:[2]

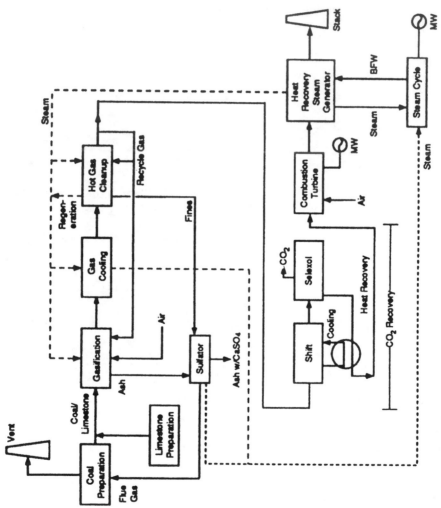

FIGURE 7.3 Integrated gasification combined cycle power system with selexol solvent CO_2 recovery.[4] The CO_2 is then ready for sequestration in underground aquifers, gas wells, or the ocean.

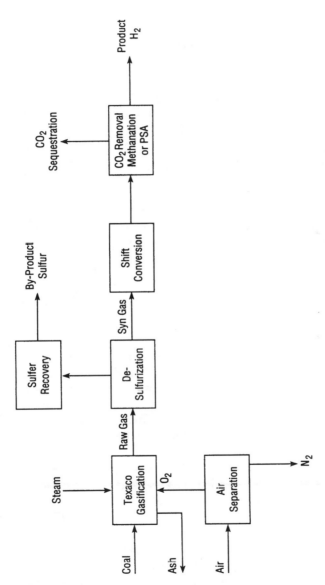

FIGURE 7.4 Coal gasification for hydrogen production. The separated CO_2 can be sequestered in gas wells, aquifers, or the ocean.

$$CH_{1.8} + H_2O + O_2 = CO + H_2 + CO$$
oil   oil gas

The energy required for the endothermic reaction is generated by the partial oxidation of the oil so the process becomes autothermal. The shift reaction is used to convert the CO with steam to hydrogen and the reaction is energy neutral.

$$CO + H_2O = CO_2 + H_2O$$

The CO_2 is separated by solvent absorption/stripping and emitted to the atmosphere or, for our purposes, sequestered in the ocean or underground. See Chapter 10 for further details on partial oxidation of hydrocarbons.

7.7 BIOMASS GASIFICATION FOR HYDROGEN PRODUCTION

In the same manner that coal can be gasified for hydrogen production, biomass, including municipal solid waste and agricultural waste, can be gasified for hydrogen production. Either oxygen-blown or air-blown gasifiers can be used. There is no need to sequester CO_2 because biomass as a feedstock is CO_2 neutral.

7.8 PRINCIPLES OF FOSSIL-FUEL DECARBONIZATION AND CO_2 MITIGATION

The above natural gas, coal, oil, and biomass conversion system to produce power gas or hydrogen for chemical or hydrogen-rich transportation fuel with capture and sequestration of the CO_2 produced is considered to be categorized under the term decarbonization of fossil fuels. Strictly speaking, the term decarbonization should refer to removal of carbon from fossil fuels producing hydrogen-rich fuel for power or transportation. Below we will discuss the decarbonization of fossil fuels for CO_2 emission reduction and with the extraction of carbon from fossil fuels and the sequestration of the elemental carbon. Coprocessing with biomass for CO_2 emission reduction fits into the scheme of CO_2 mitigation because the production and utilization of biomass as fuel are essentially CO_2 neutral.

Prior to describing the specific processes dealing with decarbonization of fossil fuel, it is important to set down several principles when considering CO_2 mitigation technologies:

1. CO_2 removal and recovery (capture) is mainly feasible from central power station sources. However, a significant energy and economic penalty is incurred.
2. Generally only one-third of the CO_2 emissions to the atmosphere comes from central power stations. Another one-third comes from industrial and domestic heat sources, and the remaining third comes from the transpor-

tation sector. The latter two sectors emit CO_2 to the atmosphere in countless small dispersed stationary and mobile sources (homes, factories, automobiles, etc.), which makes it economically almost impossible to collect for recovery or disposal.

3. The growth of biomass through solar energy photosynthesis is the only feasible method of removing the relatively small concentrations of CO_2 from the atmosphere.

4. Because carbon and carbon-rich fractions of fossil fuels are stable solids, they are much easier to dispose of and sequester than CO_2, which is a gas and can be more chemically reactive than carbon. Carbon can be sequestered or stored in landfills, mines, and the ocean bed. Carbon can also be used as a materials commodity or even burned in more efficient combined cycle plants.

5. Hydrogen-rich alternative fuels that can serve as automotive fuels include methane, hydrogen, methanol, ethanol, and dimethyl ether.

6. The key to hydrogen-rich fuel production from fossil fuels and biomass is the production of hydrogen.

7. The net fraction of hydrogen energy obtained when cracking fossil fuels to carbon and hydrogen is shown in Table 7.1. The highest hydrogen content is in natural gas and the least in coal.

8. Methanol is an ideal alternative transportation fuel because (1) it fits into the conventional infrastructure as a transportable and storable liquid, (2) it can be used in internal combustion engines with reduced polluting emissions, (3) it can be used both indirectly and directly in efficient fuel

TABLE 7.1
The Hydrogen Economy Based on Fossil Energy

Principle: • Extract hydrogen from fossil fuels
• Use hydrogen only as fuel
• Return clean carbon to ground for possible future use as CO_2 environment permits

Fossil fuel	Heat of combustion HHV (kcal/mol)	Cracking process	Atomic H fraction in fossil fuel	Heat of cracking (kcal/mol C)	Net energy in hydrogen (% of fossil fuel)
Natural gas	−212	$CH_4 \rightarrow C + 2H_2$	80	+18	64 − 5 = 59%
Petroleum					
Alkanes	−165	$CH_2 \rightarrow C + H_2$	67	+6	41 − 4 = 37%
Aromatics	−142	$CH \rightarrow C + 1/2H_2$	50	−3	24 + 1 = 25%
Coal	−116	$CH_{0.8}O_{0.08} \rightarrow C + 0.08H_2O + 0.32H_2$	43	0	19 ± 0 = 19%

Note: Heat of combustion of C = −94 kcal/mol. HHV of combustion of H_2 = −68 kcal/mol.

cells for both automotive and stationary power, and (4) it can be produced from the reaction of hydrogen and captured CO_2.

7.9   THE THERMAL CRACKING OF NATURAL GAS FOR HYDROGEN PRODUCTION

The thermal decomposition of natural gas ($CH_4 = C + 2H_2$) has been practiced for many years for the production of carbon black for rubber tire vulcanization, for pigment, and for the printing industry.[5] In the "thermal black" industrial process, the hydrogen produced is used to provide part of the thermal energy in the process. The process is practiced intermittently with the use of tandem furnaces at near atmospheric pressure. A methane–air flame is used to heat up firebrick to temperatures in the order of 1400°C (2550°F). The air is then turned off and the methane gas decomposes on the hot firebrick until the temperature drops to below 800°C (1472°F). The micron-sized carbon particulates formed are collected from the effluent gas stream in bag filters. The hydrogen-rich effluent gas is then used to heat up a second furnace in tandem with the first while the methane decomposition continues in the first furnace. Then the flow of gas is reversed so that while the first furnace is producing carbon black, the second furnace is being heated up. The process operates in tandem.

Attempts have also been made to thermally crack natural gas in a continuous fluidized bed reactor[6] for production of hydrogen. It is noted that transition metal oxide catalysts were used to increase the rate of decomposition, and even carbon itself might act as a catalyst. It appears that it should be possible to develop an efficient continuous process for the production of hydrogen with the carbon as a by-product.

The decomposition reaction is endothermic by 18 kcal/mol, so that a minimum of only 9% of the heat of combustion of methane (212 kcal/mol) is needed to drive the process. The equilibrium diagram for methane decomposition is shown in Figure 7.5. The decomposition is favored by lower pressure and higher temperatures. Above 800°C and 1 atm, the decomposition is more than 95% complete. The kinetics of the thermal decomposition of methane has been investigated at elevated pressure. The rate of decomposition is favored by higher pressure and an activation energy of 31.3 kcal/mol of CH_4 has been determined, which indicated that carbon acts as a catalyst.[8] It is noted that the methane decomposition reactor (MDR) can be heated indirectly either with methane, hydrogen, or mixtures of hydrogen and methane and even carbon. It has been suggested that a molten metal reactor might be suitable to continually decompose the methane and separate the carbon at the same time.[9] It has also been suggested that concentrated sunlight in a solar furnace can be used to thermally decompose methane.

A process for decomposing methane in a continuous manner to carbon and hydrogen using an electrical plasma torch[10] has been developed and a pilot plant has been built.

Table 7.2 compares the basic energy requirements for hydrogen production and the CO_2 emission for alternative processes. The thermal decomposition has the

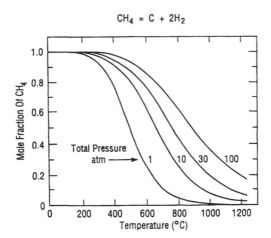

$$CH_4 = C + 2H_2$$

FIGURE 7.5 Equilibrium diagram for methane decomposition.

lowest process energy requirement per unit hydrogen produced with the least CO_2 emission. However, because the carbon produced is not used as an energy source, the thermal efficiency for hydrogen production is lower than the conventional steam reforming of methane. When using nonfossil fuel, especially for electrochemical production of hydrogen, the CO_2 emission is eliminated, but the energy requirement increases significantly. Table 7.3 gives a further comparison for hydrogen production using only fossil fuels and biomass. The highest CO_2 emission for hydrogen production comes from the use of coal by steam reforming.

An economic estimate indicates that the production of hydrogen by the thermal decomposition of methane is potentially the lowest cost process, especially if the carbon is credited as a materials commodity for such use in tires and pigments.[2] A much expanded market for carbon would be required because of the large amounts that would be generated when making a significant impact on CO_2 emissions reduction. The carbon can also be stored for later use after the rate of CO_2 emission has decreased globally. Figure 7.6 gives a process flow sheet for the Hy-C process which converts natural gas to hydrogen and coproduct carbon by the thermal decomposition of methane.

It is also of interest to point out that the carbon produced is of high purity and could advantageously be used in a gas turbine for high power plant efficiency on a combined cycle. This results in a CO_2 emission advantage over coal, which cannot be used directly in a turbine because of its sulfur and ash content.

7.10 METHANOL PRODUCTION BY THE HYDROCARB PROCESS FOR CO_2 EMISSION REDUCTION

The Hydrocarb process was originally developed to produce a coal-derived clean carbon fuel (ash- and sulfur-free) for the power industry.[11-13] The coproducts using

TABLE 7.2
Hydrogen Production: Comparison of Basic Energy Requirements and CO_2 Emission

Process	Energy efficiency Process energy	Energy efficiency Overall fuel	Process fuel	Feedstock	Ratio (mol H_2/ mol CH_4)	Energy (kcal/mol H_2) Process	Energy (kcal/mol H_2) Total	CO_2 emission (mol CO_2/ mol H_2)
1. Methane thermal decomposition $CH_4 = C + 2H_2$	(A) 80% (t)	53.5% (t)	H_2	CH_4	1.67	11.3	126.9	0.00
	(B) 80% (t)	58.1% (t)	CH_4	CH_4	1.81	11.3	117.1	0.05
	(C) 80% (t)	64.2% (t)	C	CH_4	2.00	11.3	106.0	0.12
Combined cycle power for electrical heating $CH_4 = C + 2H_2$	(D) 55% (e) / 80% (e) / 44% (t)	53.9% (t)	CH_4	CH_4	1.68	11.3	126.2	0.10
2. Methane steam reforming $CH_4 + 2H_2O = CO_2 + 4H_2$	80% (t)	94.6% (t)	CH_4	$CH_4 + H_2O$	2.95	18.8	71.9	0.34
3. Natural gas combined cycle power for water electrolysis $H_2O = H_2 + 1/2O_2$	55% (e) / 80% (e) / 44% (t)	44.0% (t)	CH_4	H_2O	1.37	154.5	154.5	0.73
4. Thermochemical cycle — coal-fired heat	60% (t)	60.0% (t)	Coal	H_2O	0.88 H_2/coal	113.3	113.3	1.14
5. Nuclear energy for electrolysis of water $H_2O + H_2 + 1/2O_2$	30% (e) / 80% (e) / 24% (t)	24.0% (t)	^{235}U	H_2O	—	283.3	283.3	0.0
6. Solar photovoltaics for electrolysis of water $H_2O = H_2 + 1/2O_2$	10% (e) / 80% (e) / 8% (t)	8.0% (t)	Solar	H_2O	—	725.0	725.0	0.0

7. Hydropower, wind power, and geothermal power for electrolysis of water $H_2O = H_2 + 1/2O_2$	80% (ϵ) 80% (ϵ) 64% (t) (thermal equiv.)	64% (t) (thermal equiv.)	Waterfall, wind, and geothermal heat	H_2O	—	106.0	106.0	0.0

TABLE 7.3
Hydrogen Production: Comparison of Basic Energy Requirements and CO_2 Emission — Fossil Fuels and Biomass

Process	Energy efficiency: Process energy	Energy efficiency: Overall fuel	Process fuel	Feedstock	Ratio (mol H_2/ (mol CH_4)	Energy (kcal/mol H_2): Process	Energy (kcal/mol H_2): Total	CO_2 emission (mol CO_2/ mol H_2)
1. Methane thermal decomposition $CH_4 = C + 2H_2$	(A) 80% (t)	53.3% (t)	H_2	CH_4	1.67	11.3	126.9	0.00
	(B) 80% (t)	58.1% (t)	CH_4	CH_4	1.81	11.3	117.1	0.05
	(C) 80% (t)	64.2% (t)	C	CH_4	2.00	11.3	106.0	0.12
Combined cycle power for electrical heating	(D) 55% (t) 80% (t) 44% (t)	53.9% (t)	CH_4	CH_4	1.68	11.3	126.2	0.10
2. Methane steam reforming $CH_4 + 2H_2O = CO_2 + 4H_2$	80% (t)	94.6% (t)	CH_4	$CH_4 + H_2O$	2.95	18.8	71.9	0.34
3. Coal steam reforming $CO_{0.8}O_{0.08} + 1.92H_2O = CO_2 + 2.32H_2$	80% (t)	92.4% (t)	Coal	Coal + H_2O	1.45 H_2/coal	21.9	73.5	0.69
4. Biomass steam reforming $CH_{1.5}O_{0.8} + 1.4H_2O = CO_2 + 2.15H_2$	80% (t)	92.0% (t)	Biomass	Biomass + H_2O	1.54 H_2/biomass	20.3	73.9	0.00

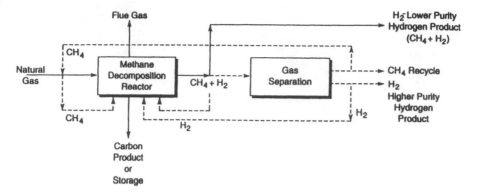

FIGURE 7.6   Hy-C process for converting natural gas to hydrogen and coproduct carbon.

coal alone as a feedstock are either hydrogen, methane, or methanol. Any carbon-aceous material can be used as feedstock in the Hydrocarb system, including the fossil fuels (coal, oil, and natural gas), biomass (wood, plants, agricultural and aquacultural products and waste, municipal solid waste), and shale oil, for conversion to carbon and hydrogen-rich by-products. For CO_2 mitigation, the reduction comes about because of two aspects: (1) by sequestering either part or all of the carbon or using it as a materials commodity but not burning it or oxidizing the carbon to CO_2 and (2) by using biomass as a feedstock which removes CO_2 from the atmosphere by photosynthesis and is thus CO_2 neutral.

The Hydrocarb process consists of three integrated chemical reaction steps and is illustrated below using biomass (wood) as feedstock.

1. *Hydrogasification of wood*:

$$a\ CH_{1.44}O_{0.06} + b\ H_2 = c\ CH_4 + d\ CO + e\ H_2O$$

   The hydrogenation reaction is usually exothermic and takes place efficiently at temperatures in the range of 800 and 900°C and at elevated pressures of 30 to 50 atm and is thermodynamically equilibrium limited. A fluidized bed reactor is usually used for hydrogasification of the solid carbonaceous material. Other reactants such as limestone can be added for removal of sulfur that may be contained in the feedstock.

2. *Methane decomposition by pyrolysis*: The process gas from the hydrogasifier containing methane is then heated in an MDR to a higher temperature, above 800°C and up to as high as 1100°C at pressures less than 50 atm. Methane is decomposed and a new equilibrium is established and carbon is deposited. The main reaction is

$$CH_4 = C + 2H_2$$

The reaction is endothermic and must be heated externally. Natural gas may be added in this reactor to provide additional hydrogen for methanol synthesis and for hydrogasification.

3. *Methanol synthesis*: The methanol synthesis reactor converts the CO and H_2 from the first two steps by the well-known catalytic synthesis process at 260°C and 50 atm pressure:

$$CO + 2H_2 = CH_3OH$$

The excess hydrogen from the catalytic methanol synthesis reactor is recycled to the hydrogasifier which completes the process.

The overall reaction can then be represented as follows:

$$CH_{1.44}O_{0.66} + 0.34CH_4 = 0.68C + 0.66CH_3OH$$

A simplified Hydrocarb process flow sheet is shown in Figure 7.7. In this process, the CO_2 extracted from the atmosphere in forming the wood feedstock by photosynthesis is partly converted back to carbon with the assistance of hydrogen from additional methane feedstock. An analysis of the Hydrocarb process for carbon and energy utilization efficiency and the generation or removal of CO_2 per unit methanol energy as a function of various feedstocks is shown in Table 7.4. Comparison is made with conventional methanol processes using the fossil-fuel feedstocks. Because of the higher hydrogen content of the natural gas, it is possible to obtain a removal of CO_2 from the atmosphere in addition to that of the gas. With coal, it is only possible to remove carbon from the coal. Furthermore, without biomass, the energy efficiency obtained with coal alone using the Hydrocarb process is only 40%, and the CO_2 emission is reduced by 60% compared to steam–oxygen reforming of coal. When using biomass and natural gas, the thermal efficiency based on fossil fuel alone can theoretically be as high as 166% and the CO_2 is reduced by 146% compared to the conventional natural gas steam-reforming process.

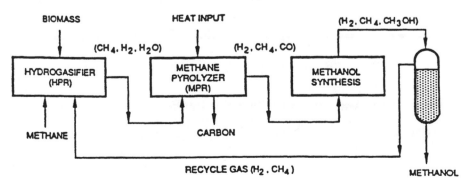

FIGURE 7.7   Simplified Hydrocarb process flow sheet.

TABLE 7.4
CO_2 Generated or Removed from the Atmosphere by Various Methanol Synthesis and Coprocessing Process Systems Using Fossil-Fuel Feedstock

Feedstock	Methanol process	Carbon utilization methanol only based on fossil-fuel feedstock only (%)	Energy utilization efficiency methanol only based on fossil-fuel feedstock only (%)	CO_2 generated (+) CO_2 removed (−) (lb CO_2/MMBtu of methanol-generated energy)
Conventional: produces CO_2				
Natural gas	Steam reforming	82	68	+170
Oil	Partial oxidation	50	64	+280
Coal — bituminous	Steam–oxygen reforming	42	64	+330
Hydrocarb: store carbon				
Bituminous coal (added H_2O)	Hydrocarb	27	40	+130
Lignite	Hydrocarb	18	30	+130
Hydrocarb coprocessing with biomass store carbon				
II Biomass + natural gas	Photosynthesis + Hydrocarb	200	166	−78
III Biomass + oil	Photosynthesis + Hydrocarb	85	115	−78
IV Biomass + bituminous coal	Photosynthesis + Hydrocarb	30	50	0

Note: Combustion of natural gas generates 110 lb CO_2/MMBtu, oil generates 160 lb/MMBtu, and coal generates 215 lb/MMBtu. Assumes 90% conversion of feedstock to methanol in Hydrocarb process.

A more detailed equilibrium-limited computer-generated simulation and economic analysis has been made[14] for the coprocessing of biomass and natural gas and optimizing the Hydrocarb process. This study indicates that a 55% reduction of CO_2 emission can be obtained when producing methanol as an alternative fuel for the transportation sector and at a competitive cost.

7.11 THE CARNOL PROCESS FOR METHANOL PRODUCTION AND CO_2 MITIGATION

The Carnol process was primarily developed to remove CO_2 from power plant flue gas, particularly coal-fired plants, and to utilize that CO_2 to produce an alternative fuel for the transportation sector as well as the domestic and industrial sectors. The basic process starts with removal and recovery of CO_2 from the power plant stack gas. Hydrogen is generated by thermal cracking of methane and sequestering the carbon. The hydrogen and CO_2 are then catalytically converted to methanol. The methanol can be used as a fuel in internal combustion engines or in fuel cells and in industrial and domestic heating systems.

The Carnol process is composed of three unit operations as follows:

1. *CO_2 removal and recovery*: Carbon dioxide is extracted from power plant stack flue gas using MEA solvent in an absorption/stripping system. As described in Chapter 5, the removal and recovery system has been improved by lowering the pressure drop across the absorption tower and using hindered amine solvents to decrease the heat required to recover the CO_2.[16] In this manner, the power required to recover 90% of the CO_2 from the flue gas of an integrated coal-fired power plant can be reduced to about 10% of the capacity of the power plant. This energy requirement can be further reduced to less than 1% when the CO_2 recovery operation is integrated with the methanol synthesis step because the exothermic heat from the methanol synthesis reaction can be used on the CO_2 stripper.

2. *Hydrogen supply*: As shown previously, the lowest energy requirement to produce hydrogen with the least CO_2 emission is by the thermal decomposition of methane:

$$CH_4 = C + 2H_2$$

   A fluidized bed reactor, a molten metal bath reactor, and an electric plasma reactor have been used to decompose methane. A solar furnace could also be used to decompose methane for hydrogen production. The carbon is sequestered or sold as a materials commodity.

   The steam reforming of natural gas for hydrogen production does not make any sense here because it produces CO_2.

   Biomass gasification is another process for hydrogen supply because it obtains hydrogen with zero CO_2 emission.

   Electrolysis of water is a conventional source of hydrogen but must use nonfossil energy supplies such as nuclear, solar, and geothermal.

3. *Methanol synthesis*: The third step in the Carnol process consists of reacting the hydrogen from step 2 with the CO_2 from step 1 in a conventional gas-phase catalytic methanol synthesis reactor:

$$CO_2 + 3H_2 = CH_3OH + H_2O$$

   This is an exothermic reaction so that the heat produced in this operation can be used to recover the stack gas CO_2 from the absorption/stripping operation described in step 1, thus reducing the energy required to recover the CO_2 from the power plant to less than 1% of the power plant capacity. This allows a distinct energy advantage over the energy penalty

requiring derating the power plant when CO_2 is disposed of by sequestering into the ocean, in which case as much as 30% of the power plant
capacity is lost. The gas-phase methanol synthesis usually takes place at
a temperature of 260°C and a pressure of 50 atm using a copper catalyst.
The synthesis can also be conducted in the liquid phase by using a slurry
of zinc catalyst at a lower temperature of 120°C and 30 atm of hydrogen
pressure so that it might be possible to combine the desorption step with
the methanol synthesis in the stripper[15] reactor. See Chapter 10 for more
details on methanol synthesis from CO_2 and H_2.

In its simplest form, the Carnol process for methanol production is a two-step
operation, as shown in Figure 7.8. When hydrogen is used to supply the energy from
the thermal decomposition of methane, then the CO_2 emission for methanol production is reduced to zero. A detailed process design and economic evaluation was
made[17] using a computer simulation model. Figure 7.8 shows the process design
combining CO_2 recovery with a liquid metal methane decomposition reactor and
liquid-phase methanol synthesis, designated as Carnol VI.

An economic study was made of the Carnol process. Table 7.5 summarizes the
production cost factors and income factors for a range of cost conditions utilizing
CO_2 from a 900-MW(e) coal-fired power plant. The capital cost for CO_2 recovery
and the methanol synthesis Carnol plant is estimated to be $961 million (1995
dollars). In terms of reducing CO_2 emission from the power plant, with $2.00 per
thousand standard cubic feet (MSCF) natural gas and a $0.55/gal methanol income,
the cost of reducing CO_2 is zero. At $3.00/MSCF natural gas and $0.45/gal income
from methanol, the CO_2 disposal cost is $47.70/ton CO_2, which is less than the
maximum estimated for ocean disposal.[9] More interesting, without any credit for
CO_2 disposal from the power plant, methanol at $0.55/gal can compete with gaso-
line at $0.76/gal (~$18/bbl oil) when natural gas is at $2.00/MSCF. Any income
from the carbon by-product makes the economics look even better. The carbon is
available as a materials commodity in such useful products as tires, as a filler in
construction materials, or as a soil conditioner.

7.11.1 CO_2 Emissions Evaluation of Entire Carnol System[17] and the Synthesis of Methanol

The entire Carnol system consists of a coal-fired power plant, a Carnol process
methanol conversion plant, and the use of methanol as a liquid automotive fuel. For
a proper evaluation, alternative methanol conversion processes must be compared.
The overall stoichiometry for various methanol production processes is given below.

1. *Carnol process for methanol synthesis*:

$$CH_4 + 0.67CO_2 = 0.67CH_3OH + 0.67H_2O + C$$

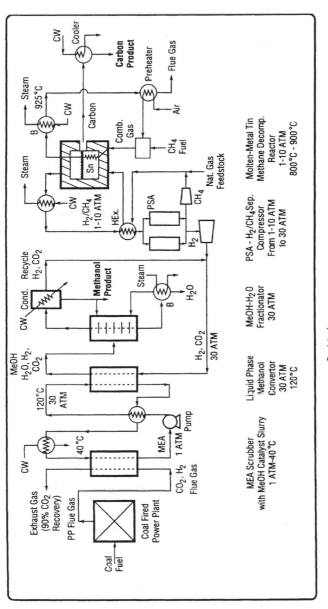

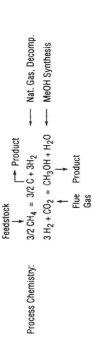

FIGURE 7.8 Carnol VI process for CO_2 mitigation technology combining CO_2 recovery from power plants with liquid metal methane decomposition and liquid-phase methanol synthesis.

TABLE 7.5
Advanced Carnol VI Preliminary Process Economics ($ 1995)

Plant size to process 90% recovery of CO_2 from 900-MW(e) nominal coal-fired power plant

90% plant factor, CO_2 rate = 611 T/hr = 4.82×10^6 tons CO_2/yr

Feedstock: natural gas rate = 2.82×10^6 T/yr = 407,000 MSCF/day

Carbon production = 2.03×10^6 T/yr

Methanol production = 3.16×10^6 T/yr = 69,300 bbl/day

Plant capital investment (IC) = 9607 T/day × 10^5 = 961×10^6

0.25 IC	Production cost factors								=	C income		Income factors			
	Natural gas		C storage		CO_2 cost							MeOH income		Cost for reducing CO_2	
10^8/yr	10^8/yr	($/MSCF)	10^8/yr	($/ton)	10^8/yr	($/ton)				10^8/yr	($/ton)	10^8/yr	($/gal)	10^8/yr	($/ton)
2.40	2.67	(2)	0.20	(10)	0	(0)				0	(0)	5.27	(0.55)	0	(0)
2.40	4.00	(3)	0.20	(10)	0	(0)				0	(0)	5.27	(0.55)	-1.34	(-27.60)
2.40	2.67	(2)	0.20	(10)	0	(0)				0	(0)	5.27	(0.55)	0	(0)
2.40	4.00	(3)	0	(0)	0	(0)				1.13	(55.60)	5.27	(0.55)	0	(0)
2.40	2.67	(2)	0.20	(10)	0	(0)				0	(0)	4.30	(0.45)	-0.97	(-20.00)
2.40	2.67	(2)	0	(0)	0	(0)				0.77	(37.90)	4.30	(0.45)	0	(0)
2.40	4.00	(3)	0.20	(10)	0	(0)				0	(0)	4.30	(0.45)	-2.30	(-47.70)
2.40	4.00	(3)	0	(0)	0	(0)				2.10	(103.00)	4.30	(0.45)	0	(0)

2. *Conventional steam-reforming methane for methanol synthesis*:

$$CH_4 + H_2O = CH_3OH + H_2$$

3. *Conventional steam reforming of methane with CO_2 addition for methanol synthesis*:

$$CH_4 + 0.67H_2O + 0.33CO_3 = 1.33CH_3OH$$

In the Carnol process, a maximum amount of CO_2 is utilized and an excess of carbon is produced. In the conventional process, no CO_2 is used and an excess of hydrogen is produced. With CO_2 addition to the conventional process, no excess of carbon or hydrogen is formed and methanol per unit natural gas is maximized.

Methanol can also be produced using biomass, and since the net CO_2 emission is zero, with CO_2 being converted to biomass by solar photosynthesis, the biomass process must also be included in the evaluation.

4. *Biomass steam gasification process for methanol synthesis*:

$$2CH_{1.4}O_{0.7} + 1.3H_2O = 1.3CH_3OH + 0.7CO$$
$$\text{wood}$$
$$\text{photosynthesis } CO_2 + 0.7H_2O = CH_{1.4}O_{0.7} + O_2$$

The alternative methanol production processes are evaluated in Table 7.6. The yield of methanol per unit of methane feedstock is shown for (1) the conventional process in two parts: (A) steam reforming of natural gas process and (B) using CO_2 addition in the conventional steam-reforming process; (2) the Carnol process in two parts: (A) using methane combustion as heat source to decompose methane for hydrogen in an MDR and (B) hydrogen combustion as heat source to decompose the methane in an MDR; and (3) steam gasification of biomass process. The Carnol process with H_2 heating and the biomass process (solar energy) reduce CO_2 to zero emission compared to the conventional process, but with losses of 35 and 47% methanol yield, respectively, compared to steam reforming. When using methane combustion in the decomposer, the Carnol process reduces CO_2 emission by 43%, while the production yield is only reduced by 26% compared to the conventional process. The conventional process with CO_2 addition (1B) is interesting because there is an increase of 32% in methanol production, although the CO_2 emission is only reduced by 23%.

Figure 7.9 links the power generation sector with the transportation sector for an entire Carnol system for reduction of CO_2 emission. The system has two feedstocks: coal to the power plant and natural gas to the methanol plant. Two products are produced: electricity from the power plant and liquid fuel for the automotive industry.

The entire Carnol system is evaluated in Table 7.7 in terms of CO_2 emissions and compared to the alternative methanol processes and to the baseline case of a con-

TABLE 7.6
Methanol Production Yields and CO_2 Emissions:
Alternative Process Comparison

Process	Production yield		CO_2 emission[a]	
	mol MeOH/ mol feedstock	% reduction from conventional	lb CO_2/ MMBtu (MeOH)	% reduction from conventional
1A Conventional process steam reforming of CH_4	0.76[b]	0	44	0
1B Conventional process with CO_2 addition	1.00	(32)[c]	34	23
2A Carnol process heating MDR with CH_4	0.56	26	25	43
2B Carnol process heating MDR with H_2	0.50	35	0	100
3 Steam gasification of biomass	0.40[d]	47	43	100

[a] CO_2 emission only from fuel production plant.

[b] Based on thermal efficiency of 64%.

[c] This represents a 32% increase in yield vs. conventional.

[d] Based on Battelle Columbus Laboratory process.

ventional coal-fired power plant and gasoline-driven automotive internal combustion engines. Methanol in a fuel cell engine is also evaluated. All the cases are normalized to emissions from a 1-MMBtu coal-fired power plant which produces CO_2 for a Carnol feedstock to supply a 1.27-MMBtu methanol plant; the methanol is then used in an automotive internal combustion engine. The last case is the use of methanol in a direct fuel cell for automotive purposes. The bases and assumptions made are listed both at the top and bottom of Table 7.7, respectively. The conclusions drawn from Table 7.7 are as follows:

1. The use of conventional methanol reduces CO_2 by 13% compared to the gasoline base case and is mainly due to the 30% improvement in efficiency of the use of methanol in internal combustion engines.
2. By addition of CO_2 recovered from the coal-fired power plant to the conventional methanol process, the CO_2 from the power plant is reduced by about 25% (161 lb/MMBtu compared to 215 lb CO_2/MMBtu) and the CO_2 emission for the entire system is reduced by 24%. It should be pointed out that the CO_2 can also be obtained from the flue gas of the reformer furnace of the methanol plant and does not need to be obtained from the coal-fired plant.

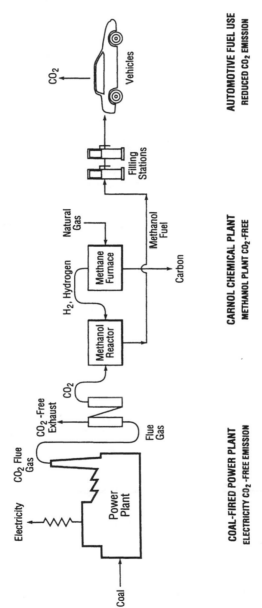

FIGURE 7.9 Integrated Carnol system configuration for CO_2 emission reduction.

TABLE 7.7
CO_2 Emission Comparison for Systems Consisting of Coal-Fired Power Plant, Fuel Process Plant, and Automotive Power Plant

Basis: 1 MMBtu for coal-fired 900-MW(e) power plant
        1.27 MMBtu of liquid fuel for internal combustion (IC) engine — other fuel efficiencies
          proportion energy up and down
        CO_2 emission units in lb CO_2/MMBtu (multiply by 0.43 for kg/GJ)

System unit	Coal-fired power plant	Fuel process plant	IC automotive power plant	Total system emission	CO_2 emission reduction (%)
Baseline case Coal-fired power plant and gasoline-driven IC engine	215	15	285	515	0
Case 1A Coal-fired power plant with conventional steam-reformed methanol plant	215	56	175[a]	448	13
Case 1B Coal-fired power plant with CO_2 addition to conventional methanol plant	161[b]	54	175	390	24
Case 2 Coal-fired power plant with Carnol process methanol plant	21[c]	32	175	228	56
Case 3 Coal-fired power plant with biomass for methanol plant	0	43	175	219	57
Case 4 Coal-fired power plant with Carnol methanol and fuel cell automotive power	11[d]	17	Fuel cell 89[e]	117	77

[a] Methanol is 30% more efficient than gasoline in IC engine.

[b] Only 25% recovery of CO_2 from coal plant is necessary for supplying CO_2 to conventional methanol plant.

[c] 90% recovery of CO_2 from coal-fired plant.

[d] Only 52% emissions of coal plant CO_2 is assigned to Carnol for fuel cells.

[e] Fuel cell is 2.5 times more efficient than conventional gasoline IC engine.

3. The Carnol process reduces the coal-fired power plant CO_2 emission by 90%, and the overall system emission is reduced by 56%. The CO_2 reduction basically comes from sequestering the carbon from the natural gas decomposition for producing the hydrogen. Also, it is noted that the carbon from coal is used twice, once for production of electricity and a second time for production of liquid fuel which is used for automotive transportation.

4. Since the use of biomass is a CO_2-neutral feedstock, there is no emission from the power plant because the production of biomass feedstock comes from an equivalent amount of CO_2 in the atmosphere which has been generated from the coal-fired power plant. Thus, the net emission comes only from burning methanol in the automotive internal combustion engine, and thus the CO_2 emission for the entire system is reduced by 57%, which is about the same as that from the Carnol system. However, it should be noted that the cost of supplying biomass feedstock is presently higher than that of natural gas feedstock and the process for conversion of biomass to methanol is less efficient than the Carnol process.

5. Another future system involves the use of fuel cells in automotive vehicles. The efficiency of fuel cells is expected to be 2.5 times greater than gasoline-driven engines.[19] Applying the Carnol process to produce methanol for fuel cell engines reduces the CO_2 emission for the entire system by as much as 77%. Furthermore, because of the huge increase in efficiency, the methanol fuel production capacity for supplying fuel cell engines by the Carnol process can be increased by 92% over that for the internal combustion engine vehicles using the same 90% of the CO_2 emissions from the coal-burning power plant.

7.12 THE HYNOL PROCESS MAXIMIZES METHANOL PRODUCTION WITH A LIMITED REDUCTION IN CO_2 EMISSION

The Hynol process was initially conceived to maximize the efficiency of production of methanol from coprocessing biomass with fossil fuels. In so doing, the CO_2 emission is decreased compared to the conventional processes for methanol production from fossil fuel. There is no need to sequester CO_2 or carbon in this process, as in the previous processes; however, the amount of CO_2 reduction is less.

7.12.1 The Hynol Process Description[20,21]

The Hynol process in general can be applied to the use of any carbonaceous feedstock, including the fossil fuels (coal, oil, and gas) and biomass (including wood, municipal and agricultural wastes, algae, and agricultural waste).

The Hynol process consists of three process reaction steps: (1) the hydrogasification of the condensed carbonaceous material (wood, coal, etc.) with hydrogen-rich re-

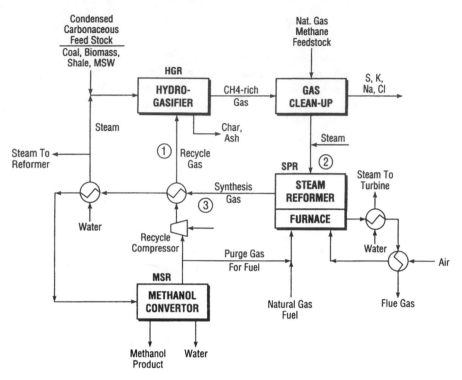

FIGURE 7.10 The Hynol process.

cycle gas to produce a methane-rich gas, (2) the steam reforming of the methane-rich gas together with the addition of the cofeedstock methane to produce carbon monoxide and hydrogen (synthesis gas), and (3) the catalytic synthesis of methanol from the carbon monoxide and hydrogen. The excess gas from the methanol synthesis reactor which is rich in hydrogen is recycled to the hydrogasifier at the head end of the process. Figure 7.10 gives a generalized flow diagram for the Hynol process. The process chemistry for each unit is as follows.

7.12.1.1 The Hydrogasifier

The main reaction taking place in the hydrogasifier (HGR) is between the condensed carbonaceous feedstock and the hydrogen in the recycle gas to produce methane:

$$C + 2H_2 = CH_4$$

This is an exothermic reaction that generates 18 kcal/mol of CH_4 produced. However, the wood (stoichiometrically $CH_{1.4}O_{0.7}$) or coal ($CH_{0.8}O_{0.1}$) contains oxygen, so that two additional reactions must be taken into account:

$$C + H_2O = CO + H_2$$

which is endothermic, absorbing about 42 kcal/mol C, and

$$CO_2 + H_2 = CO + H_2O$$

which is energetically neutral.

The conditions favoring high hydrogasification conversion of the carbonaceous material are higher pressure (30 to 50 atm) and higher temperature (800 to 900°C). A suitable design for the HGR is a fluidized bed reactor with either sand or the ash from the condensed carbonaceous material acting as the fluidizing medium. The main feature of the HGR is that it is designed so that it is self-sufficient in terms of energy and does not require any outside heat source to maintain the reaction conditions. Hydrogasification processes for biomass and coal have been developed in the past,[22,23] so there is some experience with this operation. Unconverted char from the HGR can be used as a fuel to produce steam or sequestered. In the case of biomass or coal, impurities such as sulfur will be gasified in the HGR so that a gas cleanup step is required. The sulfur can be removed as calcium sulfide with limestone:

$$CaCO_3 + H_2S = CaS + CO_2 + H_2O$$

7.12.1.2 Steam Reforming

The steam reformer, also called the steam pyrolysis reactor, combines the methane with steam to produce carbon monoxide and hydrogen. Any CO_2 present and hydrogen also produce carbon monoxide.

$$CH_4 + H_2O = CO + 3H_2$$

$$CO_2 + H_2 = CO + H_2O$$

The first reaction is highly endothermic, requiring about 60 kcal/mol of CO produced, and the second reaction is about neutral. There is much experience operating steam reformers.[24] The reactor design is usually a nickel-catalyst-packed tubular reactor which operates in the range of 30 to 50 atm and about 1000°C. The cofeedstock methane is added to the HGR together with sufficient steam to produce additional CO and H_2 synthesis gas. The heat required for the endothermic reaction is supplied by combustion of methane with air in a furnace surrounding the catalyst-packed tubular reactor. Heat exchangers are used to maintain an energy balance and produce steam for the process.

7.12.1.3 Methanol Synthesis Reactor

After cooling, the gases from the steam pyrolysis reactor essentially contain H_2 and CO in ratios exceeding three and are fed to a conventional catalytic methanol synthesis reactor (MSR) using a copper-based low-pressure catalyst.[25] The two reactions that take place in the MSR which lead to methanol are as follows:

$$CO + 2H_2 = CH_3OH$$

$$CO_2 + 3H_2 = CH_3OH + H_2O$$

Both reactions are exothermic by about 31 kcal/mol methanol. The reactor must be cooled, and the heat recovered as process steam may be used in the process to make up heat balances. A recycle ratio of 5 to 1 is used around the MSR to obtain high methanol conversions. The condensed methanol–water mixture is finally fractionated to produce fuel and chemical- or fuel-grade methanol. None of the gas is waste. All the hydrogen-rich gas from the MSR is recycled to the HGR.

7.12.2   Process Simulation and Comparison of Hynol with Conventional Methanol Processes

A process simulation computer equilibrium model was used to determine mass and energy balances and to perform a parametric analysis. When using biomass and natural gas as cofeedstock, the thermal efficiency of methanol production can reach 70%.

The conventional process for methanol production is based on the steam reforming of natural gas. The process consists of a steam reformer, shift reactor, and methanol synthesis reactor.[7] When coal or biomass is used as a feedstock, the biomass and coal are gasified with steam and oxygen in a gasifier. The resulting gas is shifted with steam to adjust the CO/H_2 ratio and is then fed to an MSR.[26]

Table 7.8[21] compares the Hynol process using cofeedstocks of biomass and natural gas with the conventional steam gasification of biomass and natural-gas-reforming plants each operating separately with the same quantities of feedstock as the Hynol plant. The Hynol plant produces over four times more methanol per unit of natural gas feedstock. The Hynol plant can also produce 15 to 38% more methanol than the two conventional plants operating independently with the same amount of feedstock as the Hynol plant. When burning methanol produced by the Hynol process, the CO_2 emission is up to 31% lower than when using the same feedstock in conventional processes.

When using the Hynol methanol in vehicles, the CO_2 emission reduction can increase to over 40% due to improved efficiency compared to gasoline.[20] Methanol fuel cells could further reduce CO_2 emission. Similar improved methanol efficiency and decreased CO_2 emission are realized with coal and natural gas as feedstock.[21] Although the CO_2 emission reduction is not as great as in the Carnol process described above, there are other advantages to the Hynol process compared to the conventional plants. The Hynol plant does not require an oxygen plant or a shift reactor, and there is no need to sequester carbon. The CO_2 emission reduction is mainly derived from the improved efficiency of methanol production and the use of biomass in the combined process.

TABLE 7.8
Comparing Biomass (BM)/Natural Gas (NG) Hynol Process with Other Conventional Biomass Gasification and Natural-Gas-Reforming Processes for Methanol Production

		Conventional process			
Factor	Hynol	BM gasification	NG reforming	BM/NG gasification[a]	BM/NG[b] gasification/ reforming
Feedstock					
Dry ash free wood (kg)	88.2	88.2		88.2	88.2
CH$_4$ (kg)	83.1	—	83.1	83.0	83.7
O$_2$ (kg)	—	44.1	—	100.0	—
Thermal efficiency (%)	71.3	52.4	64.0	51.0	61.0
Carbon conversion (%)	69.8	38.0	78.0	49.9	59.8
Methanol yield					
kg MeOH/kg wood	2.28	0.51	—	1.63	1.96
kg MeOH/kg CH$_4$	2.42	—	1.56	1.73	2.07
kg MeOH/kg total	1.17	—	—	0.79	1.01
MeOH product (kg)	201	44.8	129.6	143.8	173.0
Total (kg)	201	174.4		143.8	173.0
% increased MeOH by Hynol over conventional processes	—	15[c]		38	16
CO$_2$ emission (lb/MMBtu)	103	132		150	121
% reduced CO$_2$ by Hynol compared to conventional processes		22[d]		31	15

[a] Wood and natural gas are gasified with oxygen at 1000°C. The exit gas of the gasifier then goes through the shift reactor and the methanol synthesis reactor. The off-gas is used as fuel for oxygen plant.

[b] Wood and natural gas are gasified with steam at 800°C. The exist gas of the gasifier then goes through the steam-reforming reactor (1000°C) and the methanol synthesis reactor. The off-gas is used as one of the fuels for both gasification and steam reforming.

[c] Hynol process produces (201.1 − 174.4) × 100/174.4 = 15.3% more methanol than two separate conventional processes.

[d] Hynol process reduces (132 − 103) × 100/132 = 22% of CO$_2$ emission, compared with two separate conventional processes.

7.13 COMPARATIVE ANALYSIS OF DECARBONIZATION PROCESSES FOR METHANOL PRODUCTION

Methanol synthesis processes which do not sequester CO$_2$ and use coal as a feedstock or cofeedstock are compared in Table 7.9. The three processes discussed

TABLE 7.9
**Comparative Analysis of Methanol Production with
Reduced CO_2 Emissions Based on Coal, Natural Gas, and Biomass**

Process and reactors[a]	Feedstock	Products	MeOH thermal efficiency based on coal and natural gas (%) only[b]	CO_2 emission from MeOH combustion (lb/MMBtu)	CO_2 (% reduction from conventional processes)[c]
Hydrocarb					
HPR	Coal	MeOH + C	35	130	60
MPR	Coal + natural gas	MeOH + C	40	130	60
MSR	Coal + biomass	MeOH + C	45	~0	~100
Hynol	Coal + natural gas	MeOH	65	194	30
HPR					
SPR					
MSF					
Carnol	CO_2 from coal-fired	MeOH + C	45	~0	~100
MDR	power plant +				
MSR	natural gas				

[a] HPR = hydropyrolysis reactor, MPR = methane pyrolysis reactor, MSR = methanol synthesis reactor, SPR = steam pyrolysis reactor, and MDR = methane decomposition reactor.

[b] Conventional methanol thermal efficiency is 64%.

[c] CO_2 emission from combustion of methanol produced by natural gas reforming = 170 lb CO_2/MMBtu and produced by coal gasification = 330 lb CO_2/MMBtu.

above are compared to conventional processes in terms of thermal efficiency and CO_2 reduction. The Hydrocarb process sequesters carbon and the thermal efficiency is significantly reduced. However, by coprocessing with biomass, the thermal efficiency can be increased to 45% and the CO_2 emissions reduced to zero. The conventional thermal efficiency is 64%.

The Hynol process, which does not sequester carbon, has a high thermal efficiency of 65%, but the CO_2 emissions reduction is only 30% compared to conventional processes. The Hynol process is higher in thermal efficiency than conventional processes.

The Carnol process, which takes CO_2 from a coal-fired power plant, can obtain almost complete CO_2 emission reduction while thermal efficiency is maintained at about 45%. However, it should be pointed out that the Carnol process also produces electrical power in addition to methanol, so the comparison is not strictly comparable.

A conclusion that can be drawn from this analysis is that if deep reduction in CO_2 is to be obtained, some sacrifice in thermal efficiency results.

REFERENCES

1. Rini, M.J. and Eliasson, B., Reducing CO_2 emissions through advanced technology, in Proc. of the 22nd Int. Technical Conf. on Coal Utilization and Fuel Systems, Clearwater, FL, March 16 to 19, 1997, 415–416.

2. Cheng, H.C. and Steinberg, M., Modern and Prospective Technologies for Hydrogen Production from Fossil Fuels, BNL-41833, Brookhaven National Laboratory, Upton, NY, May 1988; *Int. J. Hydrogen Energy,* 14(1), 797–820, 1989.

3. Twigg, M.V., Ed., *Catalyst Handbook,* 2nd ed., Wolfe Publishing, Frome, England, 1989.

4. Doctor, R.D., Molburg, J.C., Thimmapuram, P., Berry, G.F., Livengood, C.D., and Johnson, R.A., Gasification combined cycle: carbon dioxide recovery, transport and disposal, in Proc. of the Int. Energy Agency Carbon Dioxide Disposal Symp., Riemer, W.F., Ed., Oxford, U.K., March 29 to 31, 1993, 1113–1120.

5. Donnet, J.B., *Carbon Black,* Marcel Dekker, New York, 1976, 16–18.

6. Pohleny, J.B. and Scott, N.H., Method of Hydrogen Production by Catalytic Decomposition of a Gaseous Hydrogen Stream, U.S. Patent 3,284,161 assigned to Universal Oil Products Co., November 8, 1966; *Chem. Eng.,* 69, 90–91, 1962.

7. Steinberg, M., The Hy-C Process (Thermal Decomposition of Natural Gas) Potentially the Lowest Cost Source of Hydrogen with the Least CO_2 Emission, BNL-61364, Brookhaven National Laboratory, Upton, NY, December 1994.

8. Kobayashi, A. and Steinberg, M., The Thermal Decomposition of Methane in a Tubular Reactor, BNL-47159, Brookhaven National Laboratory, Upton, NY, January 1992.

9. Steinberg, M., The Carnol Process for CO_2 Mitigation from Power Plants and the Transportation Sector, BNL-62835, Brookhaven National Laboratory, Upton, NY, December 1995.

10. Gaudernack, B. and Lynum, S., Hydrogen from natural gas without release of CO_2 to the atmosphere, in Proc. of the 11th World Hydrogen Energy Conf., Cocoa Beach, FL, June 1996, 511–523.

11. Grohse, E.W. and Steinberg, M., Economical Clean Carbon and Gaseous Fuels from Coal and Other Carbonaceous Raw Materials, BNL-40485, Brookhaven National Laboratory, Upton, NY, 1987.

12. Steinberg, M. and Grohse, E.W., Production of a Clean Carbon Fuel and By-Product Gaseous and Liquid Fuels from Coal by the Hydrocarb Process System, BNL-46490, Brookhaven National Laboratory, Upton, NY, 1990.

13. Steinberg, M. and Grohse, E.W., Process for the Conversion of Carbonaceous-Fuel Options, EPA Report, Air and Engineering Research Laboratory, U.S. EPA, Research Triangle, NC, January 13, 1993.

14. Steinberg, M., Dong, Y., and Borgwardt, R.H., The Coprocessing of Fossil Fuels and Biomass for CO_2 Emission Reduction in the Transportation Sector, BNL-49732, Brookhaven National Laboratory, Upton, NY, October 1993.

15. Steinberg, M., The Carnol Process for CO_2 Mitigation from Power Plants and the Transportation Sector, BNL-62835, Brookhaven National Laboratory, Upton, NY, December 1995.

16. Mimura, T., Simayoshi, H., Suda, T., Masaki, I., and Mituoka, S., Development of energy saving-technology for flue gas carbon dioxide recovery in power plant by chemical absorption method and steam system, in Proc. of the Third Int. Conf. on

Carbon Dioxide Removal, Herzog, H.J., Ed., Cambridge, MA, September 9 to 11, 1996, 557–562; Mimiera, T., Private communication, Kansai Electric Power Co., Osaka, Japan, June 1995.

17. Steinberg, M., Methanol as an agent for CO_2 mitigation, *Energy Convers.*, 38 (Suppl.), S423–S430, 1997.

18. IEA Greenhouse Gas R&D Programs, *Greenhouse Issues*, No. 7, Cheltenham, U.K., March 1993.

19. World Car Conference '96, Bourns College of Engineering, Center for Environmental Research and Technology, University of California, Riverside, January 21 to 26, 1996.

20. Steinberg, M. and Dong, Y., Hynol — An Economic Process for Methanol Production from Biomass and Natural Gas with Reduced CO_2 Emission, BNL-49733, Brookhaven National Laboratory, Upton, NY, October 1993.

21. Steinberg, M. and Dong, Y., Process and Apparatus for the Production of Methanol from Condensed Carbonaceous Material, U.S. Patent RE 35,377, November 12, 1996.

22. Steinberg, M., The flash hydropyrolysis and methanolysis of coal with hydrogen and methane, *Int. J. Hydrogen Energy*, 12(4), 251–266, 1987.

23. Steinberg, M., Fallon, P.T., and Sundaram, S.M., Flash Pyrolysis of Biomass with Reactive and Non-Reactive Gases, BNL-34510, Brookhaven National Laboratory, Upton, NY, March 1984.

24. Redler, D.E. and Twigg, M.V., Steam reforming, in *Catalyst Handbook*, 2nd ed., Twigg M.V., Ed., Wolfe Publishing, London, 1989.

25. Lee, S., *Methanol Synthesis Technology*, CRC Press, Boca Raton, FL, 1990.

26. Wyman, C.E., Bain, R.L., Hinman, N.D., and Stevens, D.J., Ethanol and methanol from cellulosic biomass, in *Renewable Energy Sources for Fuels and Electricity*, Johanssen, T.B. et al., Eds., Island Press, Washington, D.C., 1992, 914.

27. Williams, R.H., Fuel Decarbonization for Fuel Cell Applications and Sequestration of Separated CO_2, PU/CEES Report No. 295, Center for Energy and Environmental Studies, Princeton University, Princeton, NJ, January 1996.

8 Thermochemical Reactions of CO$_2$

8.1 THERMAL DISSOCIATION

8.1.1 Dissociation and Solid Electrolyte Separation of CO$_2$ from O$_2$

Thermal splitting of carbon dioxide at high temperatures could be attractive as a means for the conversion of concentrated solar energy and its storage as a gaseous fuel. Direct dissociation of carbon dioxide is possible only at very high temperatures:[1,2]

$$CO_2 = CO + 1/2O_2 \quad \Delta G = 0 \text{ at } 3350 \text{ K} \quad (8.1)$$

Such temperatures are accessible with solar furnaces, but the separation of the dissociation products at these temperatures presents formidable problems. However, the reaction may be shifted to the right by withdrawing one of the products.

A promising approach to the separation of the oxygen produced from the carbon dioxide and carbon monoxide uses an oxygen semipermeable membrane. The extraction of the oxygen released from the CO and the unreacted CO$_2$ was achieved by Nigara and Cales using calcia-stabilized zirconia.[3] The reaction cell consisted of a closed-end tube of calcia-stabilized zirconia (2 mm thick and 10 mm inner diameter). The inner part of the tube was fed with pure CO$_2$, at a flow rate of 100 cm^3 h^{-1} and a pressure of 0.1 mPa. The whole assembly was in an electric oven at 1700 to 2100 K. A CO–CO$_2$ gas mixture was pumped through an outer compartment surrounding the semipermeable cell. Thus there was a gradient of oxygen partial pressure from inside to outside of the semipermeable cell, which depended on the CO$_2$/CO ratio in the outer compartment. At the lowest ratio tested, CO$_2$/CO = 0.01, at 1984 K, 21.5 mol% of CO$_2$ was converted to CO. With a CO$_2$ consumption of 1000 cm^3 h^{-1} at 1960 K, about 100 to 200 cm^3 h^{-1} of CO was produced. Even higher permeability, due to higher oxide ion conductivity, was achieved with a ceria-doped zirconia membrane. By using the Ca-stabilized zirconia as an oxygen-selective membrane, these authors were able to enhance the conversion of CO_2 to CO from its equilibrium value of 1.2% at 1954 K up to 21.5%. However, since they used a $CO + CO_2$ mixture to sweep out the oxygen which had permeated, the reverse reaction of $CO + 1/2O_2 \rightarrow CO_2$ reduced the actual yield.[3]

In a further study, Itoh et al. used an electrically heated yttria-stabilized zirconia (YSZ) membrane.[4] Pure atmospheric pressure CO_2 was fed into the one-end-closed YSZ tube reactor (6 mol% yttria, 2 mm thick) at various flow rates. The oxygen which permeated through the membrane to an outer compartment was swept out with argon. The progress of the reaction was monitored by gas-chromatographic analysis of the CO released from the inside of the reaction tube. The temperature at the center of the reactor was controlled at 1782 K and decreased toward both ends of the reactor. The kinetic parameters determined from the temperature effect on the reaction rates were 79.93 kcal mol^{-1} for the activation energy and 3.75×10^7 mol s^{-1} atm^{-1} m^{-3} for the preexponential factor. Relatively high final conversion of CO_2 to CO of about 1.5% was achieved only at very low CO_2 flow rates, such as 0.5 cm^3 min^{-1}. At higher flow rates, such as 110 cm^3 min^{-1}, the amount of oxygen which diffused through the membrane was not sufficient to prevent the reverse reaction of recombination with CO.[4]

8.1.2   Dissociation on Metal and Metal Oxide Surfaces

The surface chemistry of carbon dioxide interacting with metal and metal oxide surfaces has been comprehensively reviewed by Freund and Roberts.[5]

On clean rhodium surfaces such as the Rh(111) and Rh(100) faces, even at low temperature (300 K), CO_2 was found by Dubois and Somorjai to be dissociated to surface-adsorbed CO and O:[6]

$$CO_{2(g)} \rightarrow CO_{ads} + O_{ads} \qquad (8.2)$$

The dissociation probably occurred at surface defect sites and was inhibited by oxygen preadsorption. At 1 atm pressure and 500 K, CO_2 did dissociate rapidly.[7]

Carbon dioxide molecules adsorb on clean metal surfaces in the linear unreactive form of CO_2, while on oxide surfaces they react to produce stable carbonates. The adsorption and reactions of CO_2 on clean Rh, Pd, Pt, Ni, Fe, Cu, Re, Al, Mg, and Ag metal surfaces were studied by Solymosi[8,9,11,12] and Solymosi and Kliveny[10] using spectroscopic methods, particularly with respect to the formation of CO_2^-, which, depending on the metal, may undergo dissociation to CO and O or transform into adsorbed CO_3 and CO. The binding energy of adsorbed CO_2 was increased by alkali adatoms. If oxygen was preadsorbed on the metals, the formation of stable carbonate structures was favored. On Pt and Cu metal surfaces, there was mainly molecular adsorption, with hardly any dissociation of CO_2. On the other hand, CO_2 adsorption was dissociative on Fe, Ni, Re, Al, and Mg surfaces. With catalyst support systems of electron-donating character, carbon dioxide was proposed by

Freund and Messmer[13] and Messmer et al.[14] to be activated by its conversion from the linear to a bent structure, in which the C–O bond becomes more reactive. This activation of carbon dioxide, leading to the adsorbed CO_2^- species, was achieved on platinum group metals by the use of preadsorbed alkali promoters.[8-12] Such alkali adatoms lower the work function of metals by 3.4 to 4.5 eV. With low coverage of alkali adatoms, such as with potassium of up to about 0.25 monolayer, the alkali atoms donate their 4s electrons to the metal surface, resulting in an increase in both the rate of adsorption and the dissociation of CO_2. On the other hand, at about monolayer coverage, carbon dioxide adsorbs in a stable form, presumably of a K–CO_2 surface complex. At low potassium coverage on Rh(III) surfaces, carbon dioxide was already found to dissociate to CO and O at 131 to 200 K. At high potassium coverage, the K–CO_2 formed was found to be transformed above 200 K into an adsorbed carbonate-like species:

$$2CO_2(ads)^- \rightarrow CO(ads) + CO_3(ads)^{2-} \tag{8.3}$$

which decomposed to carbon dioxide only above 650 K. In the activation of adsorbed CO_2 on Pt group metals by alkali metals, the initial step was proposed to be the transfer of an electron from the surface to the π-orbital of CO_2.[8-12]

On polycrystalline silver deposited on silica, reduced under H_2 at 623 K, carbon dioxide was found by Millar et al. to be adsorbed in two forms, as shown by their partially overlapping infrared absorption bands: a weakly adsorbed species, with absorption maximum at 2346 cm^{-1}, and a more strongly adsorbed species, with absorption maximum at 2336 cm^{-1}, which was desorbed at temperatures above about 250 K. The more strongly bound CO_2 was proposed to be located at silver sites modified by the presence of subsurface oxygen, which existed near grain boundary defects. Ag/SiO_2 has important uses as a catalyst, for example, for the oxidation of methanol to formaldehyde.[15]

Using field emission microscopy and thermal desorption spectroscopy on single-crystal Rh surfaces, van Tol et al. observed that CO_2 dissociation at low surface coverages to CO_{ads} and CO_{ads} started on Rh(210) at 165 K.[16,17] However, between 235 and 295 K, CO recombined with O, reforming CO_2, which was desorbed at 340 K. At high CO_2 coverages, CO_2 dissociation started at about 190 K, simultaneous with CO_2 desorption. No dissociation of CO_2 was found on clean Pt surfaces, and only weak adsorption was found at 80 K, with desorption at around 90 K.[16,17] See also Section 10.3.4 on CO_2 adsorption and dissociation on copper catalysts in connection with methanol synthesis.

The adsorption of both carbon dioxide and ammonia at clean Cu(100) and Zn(0001) surfaces was investigated by Davies and Roberts using a combination of X-ray photoelectron and electron energy loss spectroscopy.[18] While CO_2 and NH_3, when adsorbed separately on either the Cu or Zn surfaces at low temperatures (80 K), show the electron energy loss spectra of these species (vibrational bands due to the stretching and bending modes of these molecules and also some overtone bands), when heated above about 200 K, both CO_2 and NH_3 were completely desorbed from both metal surfaces. However, when CO_2 and NH_3 were coadsorbed at the Cu(100) and Zn(0001) surfaces at 80 K and then warmed to 298 K, new bands appeared in the electron energy loss spectra. These could be assigned to the vibrational bands of a surface-bound ammonium carbamate, NH_2CO_2. This species must have been formed by breaking an N–H bond. Thus, carbon dioxide was activated by its coadsorption with ammonia.[18] See also Section 14.9 on the photoreduction of CO_2 with preadsorbed NH_3 over CeO_2–TiO_2.

Adsorption of CO_2 on polycrystalline TiO_2 (anatase form) and on Pt/TiO_2 was studied by Tanaka et al. using both Auger electron and X-ray photoelectron spectroscopy.[19] The species identified on TiO_2 were graphitic carbon, HCO_3^-, CO_3^{2-}, adsorbed CO_2, and TiC, as well as adsorbed CO on Pt. The adsorbed CO_2 molecules dissociated on Pt/TiO_2 to form CO molecules and an oxygen atom. No CO_2 dissociation occurred on unsupported Pt.

8.1.3   Insertion of Cr, Mo, and W Atoms into CO_2

The insertion of the group VI metals Cr, Mo, and W into CO_2 was performed by Souter and Andrews by depositing CO_2 from its gas mixture with argon onto a Cs window held at 6 to 7 K, while ablating the metals with 35- to 50-mJ pulses of 1064 nm from a YAG laser.[20] The dominant products in the frozen matrix from Cr, Mo, and W were the insertion products OCrCO, OMoCO, and OWCO, respectively. The structures of the products were identified by their infrared absorption bands, such as their C–O stretching frequencies, and confirmed by the isotopic shifts in these spectra when using $C^{18}O_2$ and $^{13}CO_2$.

8.2   CH$_4$ REFORMING WITH CO$_2$ PRODUCED IN METALLURGICAL AND CEMENT INDUSTRIES

Steinfeld et al. proposed combining the highly endothermic process of carbon dioxide reforming of methane,[21–24]

$$CH_4 + CO_2 \rightarrow 2CO + 2H_2 \qquad \Delta H^{\circ}_{298} = 247 \text{ kJ mol}^{-1} \qquad (8.4)$$

with two major industrial processes: blast furnace iron production, for example,

$$Fe_3O_4 + 2C \rightarrow 3Fe + 2CO_2 \qquad \Delta H^{\circ}_{298} = 333 \text{ kJ mol}^{-1} \qquad (8.5)$$

$$Fe_2O_3 + 3/2C \rightarrow 2Fe + 3/2CO_2 \qquad \Delta H^{\circ}_{298} = 235 \text{ kJ mol}^{-1} \qquad (8.6)$$

and the manufacture of lime and cement,

$$CaCO_3 \rightarrow CaO + CO_2 \qquad \Delta H^{\circ}_{298} = 178 \text{ kJ mol}^{-1} \qquad (8.7)$$

The combined processes will be even more endothermic. Thus, the combination of Equation 8.4 with Equation 8.5, 8.6, or 8.7 will result in:

$$Fe_3O_4 + 4CH_4 \rightarrow 3Fe + 4CO + 8H_2 \qquad \Delta H^\circ_{298} = 978 \text{ kJ mol}^{-1} \qquad (8.8)$$

$$Fe_2O_3 + 3CH_4 \rightarrow 2Fe + 6H_2 + 3CO \qquad \Delta H^\circ_{298} = 719 \text{ kJ mol}^{-1} \qquad (8.9)$$

$$CaCO_3 + CH_4 \rightarrow CaO + 2CO + 2H_2 \qquad \Delta H^\circ_{298} = 426 \text{ kJ mol}^{-1} \qquad (8.10)$$

In the combined processes, the CO$_2$ released in Reactions 8.5, 8.6, or 8.7 is consumed *in situ* in the synthesis gas-producing reaction (8.4). At atmospheric pressure, the reactions are essentially complete at temperatures above 1200 K, producing either solid metallic iron or CaO, as well as synthesis gas. Such temperatures are readily available with concentrated sunlight. Preliminary thermogravimetric experiments indicated the feasibility of reacting Fe$_3$O$_4$ and CaO with 1 atm methane according to Equations 8.8 and 8.10.[21-24]

In an analogous process, Steinfeld et al. achieved the combined thermal reduction of zinc oxide with the reforming of methane:[25-27]

$$ZnO + CH_4 \rightarrow Zn + 2H_2 + CO \qquad \Delta H^\circ_{298} = 313 \text{ kJ mol}^{-1} \qquad (8.11)$$

At 1200 K and 1 atm, the equilibrium in this reaction consists essentially of zinc vapor and a 2:1 mixture of hydrogen and carbon monoxide. Using a solar furnace as the energy source for the high-temperature process heat, the reaction was performed in a fluidized bed quartz reactor. Illumination was by a parabolic concentrator coupled to a compound parabolic concentrator as a secondary concentrator. To promote fluidization and to prevent sintering, the ZnO powder was applied together with Al$_2$O$_3$ grains in a 1:1 mixture. At a solar flux of 57 W/cm^2, the particles were exposed to a solar concentration of about 570 suns and reached a temperature of 1373 K. The zinc vapor formed was condensed by cooling it below its boiling point, 1180 K.[25-27]

At present, zinc is produced industrially from its ZnO ores (zincite) mainly by a carbothermic process, with coke serving both as the reductant and, by its combustion, also as the fuel for the process heat.

For reduction of MgO by CH$_4$, the corresponding overall reaction with methane is

$$MgO + CH_4 \rightarrow Mg + 2H_2 + CO \qquad \Delta H^\circ_{298} = 566 \text{ kJ mol}^{-1} \qquad (8.12)$$

The equilibrium constant for this reaction becomes unity only at the very high temperature of 1800 K.[24]

8.3 THERMOCHEMICAL CYCLES

A variety of thermochemical cycles have been proposed in which the thermal energy of solar furnaces or nuclear reactors may be used to supply the energy required for the splitting of water. Analogous thermochemical cycles may be useful for the thermochemical splitting of carbon dioxide, to achieve the overall reaction

$$CO_2 = CO + 1/2O_2 \qquad \Delta G = 0 \text{ at } 3350 \text{ K} \qquad (8.1)$$

One such cycle which has been tested by Bamberger and Robinson depended on the reaction of cerium(IV) oxide with solid or molten sodium pyrophosphate at 750 to 950°C to form sodium cerium(III) phosphate, trisodium phosphate, and oxygen:[28]

$$2CeO_2 + 3Na_4P_2O_7 \rightarrow 2Na_3Ce(PO_4)_2 + 2Na_3PO_4 + 1/2O_2 \qquad (8.13)$$

The rate of oxygen evolution in this reaction step could be significantly enhanced by adding Li_3PO_4 to the reaction mixture, thus forming a lower temperature melting eutectic composition, in the temperature range 780 to 880°C. With a molar ratio $Li_3PO_4/Na_4P_2O_7 = 0.20$, the rate of O_2 production was four times larger than in the absence of the lithium salt. In the second reaction step, also at a high temperature (750 to 900°C), the sodium cerium(III) phosphate reacted with sodium carbonate to recover cerium(IV) oxide and to produce carbon monoxide:

$$2Na_3Ce(PO_4)_2 + 3Na_2CO_3 \rightarrow 2CeO_2 + 4Na_3PO_4 + 2CO_2 + CO \qquad (8.14)$$

In a variant of this reaction, steam was added in the above reaction mixture, thus promoting the water–gas-shift reaction, with formation of hydrogen and recovery of the carbon dioxide:

$$CO + H_2O \rightarrow CO_2 + H_2 \qquad (8.15)$$

In a low-temperature recovery step (at 5 to 50°C), trisodium phosphate was treated with carbon dioxide in aqueous solution to yield disodium hydrogen phosphate and sodium bicarbonate,

$$6Na_3PO_4 + 6CO_2 + 6H_2O \rightarrow 6Na_2HPO_4 + 6NaHCO_3 \qquad (8.16)$$

which could be separated by crystallization. These two compounds could be dehydrated to $Na_4P_2O_7$ and Na_2CO_3, respectively, by moderate heating (>250 and >200°C), thus closing the cycle. In contrast to most other thermochemical cycles, the above process avoids corrosive and toxic reaction intermediates. Also, the chemicals involved in this cycle need not be of high purity and are thus relatively inexpensive.[28]

An ingenious cyclic scheme to convert CO_2 to graphitic carbon via methane was proposed by Nishiguchi et al. using two reactors in series.[29] In the first reactor, CO_2 was hydrogenated to methane over Ni (10 wt%)/SiO_2 at 573 K, reaching the equilibrium conversion of 95%:

$$CO_2 + 4H_2 \rightarrow CH_4 + 2H_2O \qquad (8.17)$$

The water was condensed off, and the methane was fed together with additional methane (natural gas) into a second reactor with the same Ni/SiO_2 catalyst at 500°C, effecting the decomposition of methane with 92% conversion:

$$CH_4 \rightarrow C + 2H_2 \qquad (18.18)$$

The high yield in the decomposition of methane was achieved by using a hydrogen-permeable Pd–Ag tube inside the second reactor, in order to draw out the hydrogen produced. The hydrogen was recycled into the first reactor. The overall process was the conversion of CO_2 into graphitic carbon and steam:[29]

$$CO_2 + CH_4 \rightarrow 2C + 2H_2O \tag{8.19}$$

Another thermochemical cycle for the conversion of CO_2 to graphite was proposed by Ehrensberger et al.[30] In the first step, at 773 K, wüstite ($Fe_{1-y}O$) reacted with CO_2, forming magnetite and CO:

$$3Fe_{1-y}O + CO_2 \rightarrow Fe_3O_4 + CO \tag{8.20}$$

In a subsequent step, CO underwent disproportionation to CO_2 and graphitic carbon, a reaction catalyzed by the Fe_3O_4 and Fe_3C formed:

$$2CO \rightarrow CO_2 + C \tag{8.21}$$

The CO_2 produced, together with new CO_2, would be recycled to the first step, while the Fe_3O_4 would be thermally decomposed to $Fe_{1-y}O$ in a solar furnace, at a temperature above 2200 K.[30]

See also Section 10.6.2 on the reduction of CO_2 by wüstite to elementary carbon.

8.4 THERMOCHEMICAL CARBOXYLATION

8.4.1 Kolbe–Schmitt Synthesis of Salicylic Acid

In the classic synthesis of salicylic acid (2-hydroxybenzoic acid) by Kolbe in 1874, dry sodium phenoxide was heated under a stream of CO_2 to 180 to 200°C. In an improved Kolbe–Schmitt reaction, the carboxylation of alkali metal aryloxides was carried out under pressurized CO_2 (about 250 psi), initially at room temperature, followed by heating to 120 to 140°C for 3 h. The first stage was suggested to involve the formation of a 1:1 adduct of the phenoxide ion with CO_2. This intermediate was isolated by Hales et al. by exposing dry sodium phenoxide to dry CO_2.[31] It was unstable to water and also was completely dissociated at 80°C under atmospheric pressure. Its structure was proposed to involve chelation of the CO_2 with the sodium ion of the phenoxide. In the subsequent high-temperature reaction, an intramolecular migration of the CO_2 group occurred, by electrophilic attack of its carbon atom onto the ortho position of the phenoxide ion, yielding sodium salicylate:[31]

$$C_6H_5ONa + CO_2 \rightarrow C_6H_5ONa \cdot CO_2 \rightarrow C_6H_4(OH)COONa \tag{8.22}$$

A solvent-free $NaOPh$–CO_2 complex was isolated by Kunert et al. by introducing CO_2 into a tetrahydrofuran solution of sodium phenoxide. Its transformation at 80°C into salicylic acid could be followed by Fourier transform infrared spectroscopy.[32]

8.4.2   Synthesis of Oxalate from Cesium Carbonate

A unique thermochemical synthesis of oxalate was achieved by Kudo et al. from CO_2 (110 atm) with CO (50 atm) in a Pyrex-lined shaking-type autoclave containing cesium carbonate.[33,34] The process occurred as a gas–solid reaction in supercritical CO_2. With 2 h continuous shaking at 380°C, followed by hydrolysis in dilute perchloric acid, the conversion and yield of cesium oxalate (based on the amount of cesium carbonate charged) were 97.5 and 90.1%, respectively. Tracer experiments with $^{13}CO_2$ showed that the added carbonyl carbon of the cesium oxalate was derived from the external CO_2 and CO (and not from the cesium carbonate). This suggested a mechanism involving as an initial step the formation of an intermediate complex, which then reacted with CO to form a precursor to oxalate:

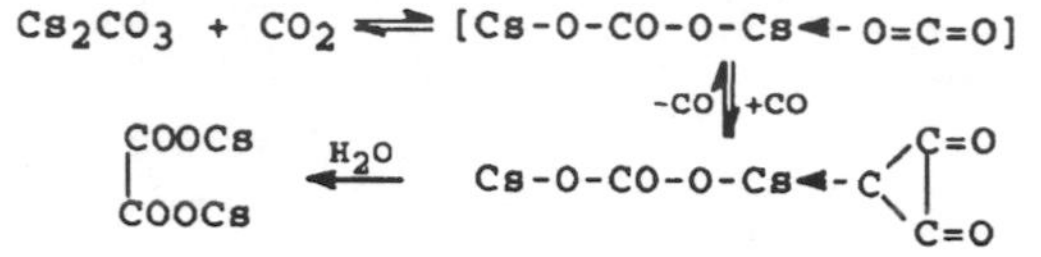

$$Cs_2CO_3 + CO_2 \rightleftharpoons [Cs-O-CO-O-Cs \blacktriangleleft -O=C=O]$$

$$\begin{array}{c} \text{COOCs} \\ | \\ \text{COOCs} \end{array} \xleftarrow{H_2O} Cs-O-CO-O-Cs \blacktriangleleft -C \begin{array}{c} C=O \\ | \\ C=O \end{array}$$

The carbonates of the other alkali metals were practically inactive in this reaction. The unique reactivity of cesium carbonate was ascribed to the large ionic radius of the cesium ion, which was thus only weakly paired with the carbonate anion. This may facilitate the coordination of the CO_2 molecule to the cesium carbonate, as shown in the above mechanism.[33,34]

8.4.3   Carboxylation of Cesium 2-Naphthoate

The carboxylation of alkali metal salts of 2-naphthoic acid (2-NC) to naphthalene di- and tricarboxylic acid salts was effected by Kudo et al. in a molten mixture of alkali metal carbonates and formate in the temperature range 350 to 420°C with CO_2 at pressures of up 500 atm.[35] Among the alkali metal salts, the order of activity for carboxylation was Cs > Rb > K, while Li and Na salts were inactive. Optimal conditions for the reaction were with a molten mixture of cesium formate (40 mmol), potassium carbonate (5 mmol), potassium-2-naphthoate (1 mmol), and CO_2 (400 atm) at 380°C for 4 h, resulting in the production of naphthalene dicarboxylate (mainly the 2,6-isomer, 2,6-NDC) and naphthalene tricarboxylate (mainly the 2,4,6-isomer, 2,4,6-NTC) in 30 and 35% yields, respectively:

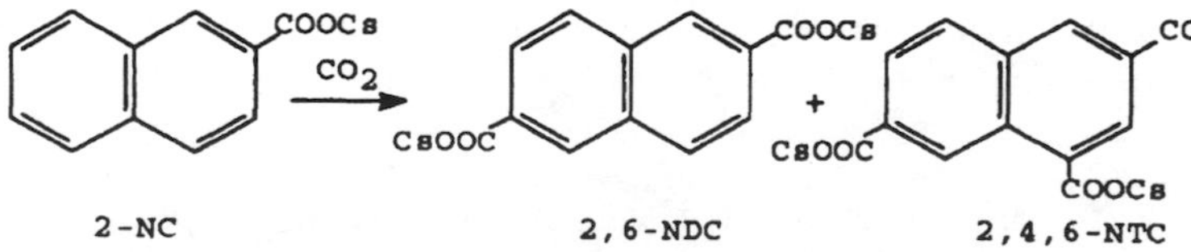

Under similar conditions, with molten cesium formate/cesium carbonate and CO_2 (400 atm) at 380°C, cesium salts of benzoic, 1-naphthoic, and 4-biphenylcarboxylic acids were carboxylated to mixtures of the corresponding di- and tricarboxylated isomers. The mechanism of these carboxylation reactions was explained by the initial formation of an addition complex between cesium carbonate and CO_2. When dry Cs_2CO_3 was contacted with CO_2 at 400 atm and 380°C for 2 h, a solid product was formed, which had a strong IR absorption band at 1660 cm^{-1}. This band, which was absent in the original Cs_2CO_3, was assigned to a carbonyl stretching mode. A tentative structure for this addition complex was proposed:[35]

8.4.4 Malonic Acid from Potassium Acetate

The reaction of anhydrous potassium acetate with CO_2 (400 atm) in the presence of anhydrous potassium carbonate in dimethylformamide solution at 250°C for 2.5 h was reported by Kudo and Takezaki to result in the formation of dipotassium malonate in 60% yield:[36]

$$CH_3COOK + CO_2 \rightarrow CH_2(COOK)_2 \qquad (8.23)$$

8.4.5 Glutaconic Acid from Cesium 3-Butenoate

Glutaconic acid is a useful intermediate in the synthesis of the important β-lactam antibiotics. Kudo et al. developed a process for the production of cesium glutaconate using the reaction of the cesium salt of 3-butenoic acid (vinylacetic acid, $H_2C=CHCH_2-COOCs$) with cesium carbonate and CO_2 (50 atm) at 10°C for 3 h:[37]

$$H_2C=CHCH_2-COOCs + CO_2 \rightarrow CsOOC-CH=CHCH_2-COOCs \quad (8.24)$$

Under these conditions, cesium glutaconate was obtained in about 55% yield. With K_2CO_3 instead of Cs_2CO_3, yields were much lower. The reaction was regioselective in that the addition of CO_2 to the 3-butenoate occurred mainly at its γ-carbon atom. The outstanding reactivity of Cs_2CO_3 for the carboxylation reaction was explained by the high electron-donating power of the cesium atom, which facilitated the dissociation of Cs_2CO_3 into Cs^+ and $CsCO_3^-$. The bulky $CsCO_3^-$ anion then abstracted a proton from the γ-carbon atom of the 3-butenoate ion.[37]

8.5 CONCLUSIONS ON THERMOCHEMICAL REACTIONS

The combined thermochemical reforming of methane to synthesis gas by the CO_2 produced in situ during the production of metals from their ores and of lime or cement from limestone may be an innovative and attractive approach to the decrease of CO_2 emission. It will be necessary to confirm whether such processes may be achieved economically on an industrial scale, possibly using concentrated sunlight as the energy source. The contribution of the world production of iron, aluminum, cement, synthesis gas, and zinc to the annual release of carbon dioxide to the atmosphere is estimated to amount to 1.2×10^{12}, 0.52×10^{12}, 0.7×10^{12}, 0.3×10^{12}, and 0.07×10^{12} kg CO_2, respectively. The coproduction of metals and synthesis gas, particularly with the application of solar energy as the source of process heat, provides the promise of substantial mitigation of CO_2 emission. The combined production of iron, lime and cement, and synthesis gas contributes about 10% to the total anthropogenic CO_2 emission.[21-27]

REFERENCES

1. Martin, L.R., Use of solar energy to reduce carbon dioxide, *Sol. Energy*, 24, 271–277, 1980.
2. Lietzke, M.H. and Mullins, C., The thermal decomposition of carbon dioxide, *J. Inorg. Nucl. Chem.*, 43, 1769–1771, 1981.
3. Nigara, Y. and Cales, B., Production of carbon monoxide by direct thermal splitting of carbon dioxide at high temperature, *Bull. Chem. Soc. Jpn.*, 59, 1997–2002, 1986.
4. Itoh, N., Sanchez, M.A.C., Xu, W.-C., Haraya, K., and Hongo, M., Application of a membrane reactor system to thermal decomposition of CO_2, *J. Membr. Sci.*, 77, 245–253, 1993.
5. Freund, H.-J. and Roberts, M.W., Surface chemistry of carbon dioxide, *Surf. Sci. Rep.*, 25, 225–273, 1996.
6. Dubois, L.H. and Somorjai, G.A., Comments on "Why CO_2 does not dissociate on Rh at low temperature" by W.H. Weinberg, *Surf. Sci. Lett.*, 128, L231, 1983.
7. Weinberg, W.H., Why CO_2 does not dissociate on Rh at low temperature, *Surf. Sci. Lett.*, 128, L224–228, 1983.
8. Solymosi, F., The bonding structure and reactions of CO_2 adsorbed on clean and promoted metal surfaces, *J. Mol. Catal.*, 65, 337–358, 1991.
9. Solymosi, F., Adsorption, bonding and reactivity of CO_2 on clean and promoted metal surfaces, in Proc. Int. Symp. Chemical Fixation of Carbon Dioxide, Nagoya, Japan, December 2 to 4, 1991, 227–236.
10. Solymosi, F. and Klivenyi, G., HREELS study on the formation of CO_2^- on a K-promoted Rh(111) surface, *Surf. Sci.*, 315, 255–268, 1994.
11. Solymosi, F., Activation and reactions of CO_2 on Rh catalysts, *Carbon Dioxide Chemistry: Environ. Issues*, R. Soc. Chem. Spec. Publ., 153, 44–54, 1994.
12. Solymosi, F., Thermal and photo-induced activation and dissociation of CO_2 on clean and promoted metal surfaces, in Abstr. 3rd Int. Conf. on Carbon Dioxide Utilization, Norman, OK, May 1995.
13. Freund, H.J. and Messmer, R.P., On the bonding and reactivity of CO_2 on metal-surfaces, *Surf. Sci.*, 172, 1–30, 1986.
14. Messmer, R.P., Freund, H.J., Schultz, P.A., and Tatar, R.C., Theoretical evidence for bent bonds in the CO_2 molecule, *Chem. Phys. Lett.*, 126, 176–180, 1986.
15. Millar, G.J., Seakins, J., Metson, J.B., Bowmaker, G.A., and Cooney, R.P., Evidence for the formation of strongly bound molecular CO_2 species on a polycrystalline silver catalyst, *J. Chem. Soc. Chem. Commun.*, pp. 525–526, 1994.
16. van Tol, M.F.H., Gielbert, A., and Nieuwenhuys, B.E., The adsorption and dissociation of CO_2 on Rh, *Appl. Surf. Sci.*, 67, 166–178, 1993.
17. van Tol, M.F.H., Gielbert, A., Wolf, R.M., Lie, A.B.K., and Nieuwenhuys, B.E., The striking difference in the behavior of Rh and Pt towards their interaction with CO_2, *Surf. Sci.*, 287, 201–207, 1993.

18. Davies, P.R. and Roberts, M.W., Activation of carbon dioxide by ammonia at Cu(100) and Zn(0001) surfaces leading to the formation of a surface carbamate, *J. Chem. Soc. Faraday Trans.*, 88, 361–368, 1992.

19. Tanaka, K., Miyahara, M., and Toyoshima, I., Adsorption of CO_2 on TiO_2 and Pt/TiO_2 studied by X-ray photoelectron spectroscopy and Auger electron spectroscopy, *J. Phys. Chem.*, 88, 3504–3508, 1984.

20. Souter, P.F. and Andrews, L., Activation of CO_2 by laser-ablated group-6 metal atoms, *Chem. Commun.*, pp. 777–778, 1997.

21. Steinfeld, A. and Thompson, G., Solar combined thermochemical processes for CO_2 mitigation in the iron, cement, and syngas industries, *Energy*, 19, 1077–1081, 1994.

22. Steinfeld, A., Frei, A., and Kuhn, P., Thermoanalysis of the combined Fe_3O_4-reduction and CH_4-reforming processes, *Metall. Mater. Trans.*, 26B, 509–515, 1995.

23. Steinfeld, A., Kuhn, P., and Tamaura, Y., CH_4-utilization and CO_2-mitigation in the metallurgical industry via solar thermo-chemistry, *Energy Convers. Manage.*, 37, 1327–1332, 1996.

24. Steinfeld, A., High-temperature solar thermochemistry for CO_2 mitigation in the extractive metallurgical industry, *Energy*, 22, 311–316, 1997.

25. Steinfeld, A., Frei, A., Kuhn, P., and Wuillemin, D., Solar thermal production of zinc and syngas via combined ZnO-reduction and CH_4-reforming processes, *Int. J. Hydrogen Energy*, 20, 793–804, 1995.

26. Steinfeld, A., Larson, C., Palumbo, R., and Foley, M., Thermodynamic analysis of the coproduction of zinc and synthesis gas using solar process heat, *Energy*, 21, 205–222, 1996.

27. Steinfeld, A., Frei, A., Kuhn, P., and Wuillemin, D., Solar thermal production of zinc and syngas via combined ZnO-reduction and CH_4-reforming processes, *Int. J. Hydrogen Energy*, 21, 243–243, 1996.

28. Bamberger, C.E. and Robinson, P.R., Thermochemical splitting of water and carbon dioxide with cerium compounds, *Inorg. Chim. Acta*, 42, 133–137, 1980.

29. Nishiguchi, H., Fukunaga, A., Miyashita, Y., Ishihara, T., and Takita, Y., Reduction of carbon dioxide to graphite carbon via methane by catalytic fixation with membrane reactor, in Abstr. 4th Int. Conf. on Carbon Dioxide Utilization, Kyoto, Japan, September 1997, O-02.

30. Ehrensberger, K., Palumbo, R., Larson, C., and Steinfeld, A., Production of carbon from carbon dioxide with iron oxides and high-temperature solar energy, *Ind. Eng. Chem. Res.*, 36, 645–648, 1997.

31. Hales, J.L., Idnes, J.I., and Lindsey, A.S., Mechanism of the Kolbe–Schmitt reaction. I. Infrared studies, *J. Chem. Soc.*, pp. 3145–3151, 1954.

32. Kunert, M., Dinjus, E., Nauck, M., and Sieler, J., Structure and reactivity of sodium phenoxide. Following the course of the Kolbe–Schmitt reaction, *Chem. Ber. Recueil*, 130, 1461–1465, 1997.

33. Kudo, K., Ikoma, F., Mori, S., Komatsu, K., and Sugita, N., Novel synthesis of oxalate from carbon dioxide and carbon monoxide in the presence of cesium carbonate, *J. Chem. Soc. Chem. Commun.*, 6, 633–634, 1995.

34. Kudo, K., Ikoma, F., Mori, S., Komatsu, K., and Sugita, N., Synthesis of oxalate from carbon monoxide and carbon dioxide in the presence of cesium carbonate, *J. Chem. Soc. Perkin Trans. 2*, pp. 679–682, 1997.

35. Kudo, K., Shima, M., Kume, Y., and Ikoma, F., Carboxylation of cesium 2-naphthoate in the alkali metal molten salts of carbonate and formate with CO_2 under high

pressure, *Sekiyu Gakkai Shi (J. Jpn. Petr. Inst.)*, 38, 40–47, 1995; *Chem. Abstr.*, 122, 160231s.

36. Kudo, K. and Takezaki, Y., Preparation of malonic acid salts, *Kogyo Kagaku Zasshi*, 70, 2147–2152, 1967; *Chem. Abstr.*, 68, 59054c.

37. Kudo, K., Ikoma, F., Mori, S., and Sugita, N., Synthesis of glutaconic acid salt from cesium 3-butenoate with carbon dioxide, *Sekiyu Gakkai Shi (J. Jpn. Petr. Inst.)*, 38, 48–51, 1995; *Chem. Abstr.*, 122, 105227v.

9 Carboxylation by CO$_2$ Insertion

9.1 INTRODUCTION

The reductive conversion of carbon dioxide to organic compounds is a relatively difficult process, in that it is both highly endothermic and requires multielectron transfer. The activation of carbon dioxide can be accomplished either by the external supply of energy (photochemical, electrical, or thermal) or by its reaction with a reactant of high free energy content. The latter category includes hydrogenation with molecular hydrogen, as described in Chapter 10. The high-temperature thermo-chemical carboxylation with CO$_2$ of alkali metal carbonates and phenolates and of various aliphatic and aromatic carboxylic acid salts was discussed in Section 8.4. The reactions with high free energy molecules, such as ammonia, amines, and organometallic compounds, are the subject of this chapter.

In a nonreductive mechanism, carbon dioxide may be readily condensed with amines, forming carbamates, and with epoxy compounds, forming cyclic carbonates. In many of these reactions, carbon dioxide may serve as a substitute for the extremely toxic phosgene.[1] The incorporation of carbon dioxide into olefinic and acetylenic hydrocarbons was achieved by their activation by transition metals, such as palladium and ruthenium, resulting in the formation of new open-chain or cyclic compounds, many of which are valuable as chemical intermediates, solvents, and pharmaceuticals. Reactions of carbon dioxide coupling include the coordination of CO$_2$ to transition metal complexes, with insertion into metal–carbon and metal–hydrogen bonds. In such catalytic reactions, carbon dioxide undergoes condensation with alkynes, alkenes, dienes, and benzene derivatives, forming pyrones, lactones, esters, and carboxylic acids, using complexes of ruthenium, rhodium, nickel, or palladium for activation of carbon dioxide. Earlier work on the insertion reactions of carbon dioxide was described in several comprehensive reviews.[2–17]

9.2 CARBAMATES AND UREA DERIVATIVES

Carbamates and urea derivatives have many important uses as pesticides, pharmaceuticals, and chemical intermediates. Processes involving the incorporation of CO_2 into the synthesis of these compounds were reviewed by Bruneau and Dixneuf.[18,19]

9.2.1 Open-Chain Carbamates

Probably the earliest example of the scientifically initiated artificial synthesis of an organic compound was Woehler's preparation of urea from ammonium cyanate. Urea, now produced on a large scale from ammonia and carbon dioxide via ammonium carbamate,

$$NH_3 + CO_2 \leftrightarrow H_2NCOOH \tag{9.1}$$

$$H_2NCOOH + NH_3 \leftrightarrow H_2N{-}COO^- \, NH_4^+ \tag{9.2}$$

is an important chemical intermediate, as well as a valuable nitrogen fertilizer.[20] The equilibrium constant for the above reactions to ammonium carbamate is highly temperature dependent. Its values are 2.08×10^5 at 25°C, 1.11×10^{-5} at 190°C, and 1.14×10^{-6} at 215°C. Thus, at high temperatures, the reaction is very strongly shifted to the left. The reaction may however be shifted to the right by dehydration, producing urea, as in its industrial production:[21]

$$H_2N{-}COO^- \, NH_4^+ = H_2NCONH_2 + H_2O \tag{9.3}$$

A primary step in the use of waste carbon dioxide, for example, from coal- or oil-fired power plants, will be to recover and concentrate this gas. Amines, such as mono-, di-, and triethanolamine, may be applied for this process, which is already now widely used for removal of carbon dioxide from natural gas. Monoethanolamine and diethanolamine react with carbon dioxide to form carbamates. The carbon dioxide is then released by heating the liquid mixture to 400 K.[22]

Monoalkylammonium N-alkylcarbamates were easily formed by saturating a solution of the amine in tetrahydrofuran (THF) with CO_2,

$$2RNH_2 + CO_2 \rightarrow (RNH_3)(O_2CNHR) \tag{9.4}$$

in which R = benzyl, allyl, *tert*-butyl, and cyclohexyl.[23,24]

In methanol solutions, carbon dioxide forms with primary and secondary amines quite stable carbamic acid amine salts; that is, the equilibrium in the equation

$$CO_2 + 2R^1R^2NH = R^1R^2NCOO^- \, (R^1R^2NH_2)^+ \tag{9.5}$$

is very much to the right.

An alternative route to carbamate salts was developed by Aresta et al. by reacting primary or secondary amines with NaBPh$_4$ and CO_2 in THF solution, yielding in addition mono- or dialkylammonium tetraphenylborates, which are useful synthetic reagents:

$$2RR'NH + NaBPh_4 + CO_2 \rightarrow (RR'NH_2)BPh_4 + RR'NCOONa \qquad (9.6)$$

Thus, $CH_2=CHCH_2NH_2$ with $NaBPh_4$ and CO_2 reacted to produce $CH_2=CHCH_2NH-$
COONa in 86% yield. The $NaBPh_4$ could be recovered by alkali treatment.[25,26]

Vaska et al. observed that the iridium complex $Ir(Cl)(CO)(Ph_3P)_2$ effectively
catalyzed the synthesis of formamide from carbon dioxide, ammonia, and hydrogen:

$$CO_2 + H_2 + NH_3 \rightarrow HC(O)NH_2 \qquad (9.7)$$

The process was carried out in homogeneous toluene or methanol solutions. The yield
of formamide increased markedly with rising pressure of the gaseous reactants.[27]

9.2.2   Urea Derivatives

Urea derivatives are important as agricultural pesticides, such as uron herbicides, as
pharmaceuticals, and as resin precursors. The common methods of synthesis involve
the very toxic phosgene and isocyanates as starting materials and intermediates. A
direct route to ureas was achieved by Fournier et al. using aliphatic and araliphatic
primary amines, which reacted with CO_2 to form ammonium carbamates. In the
presence of ruthenium complexes and a terminal alkyne, $Ru[H-C\equiv C-R']$, an *in situ*
reaction occurred, producing N,N'-disubstituted symmetrical ureas:[28]

$$2RNH_2 + CO_2 \rightarrow RNH-C(O)-NHR + H_2O \qquad (9.8)$$

In this reaction, the alkyne ruthenium intermediate acted as a dehydrating agent. Best
results were obtained by Mahé et al. with a catalyst precursor from $RuCl_3 \cdot 3H_2O$ and
n-Bu_3P, with 2-methylbut-3-yn-2-ol as the alkyne. The mixture was pressurized with
CO_2 to 5 mPa and heated to 140°C. With cyclohexylamine as the primary amine,
N,N'-dicyclohexylurea was produced during 20 h in up to 62% yield.[29]

α-Aminonitriles reacted with CO_2, neat, at room temperature, yielding disubstituted ureas. Subsequent reactions with water at room temperature yielded N-(3)-
substituted hydantoins in yields of 80 to 100%:[30]

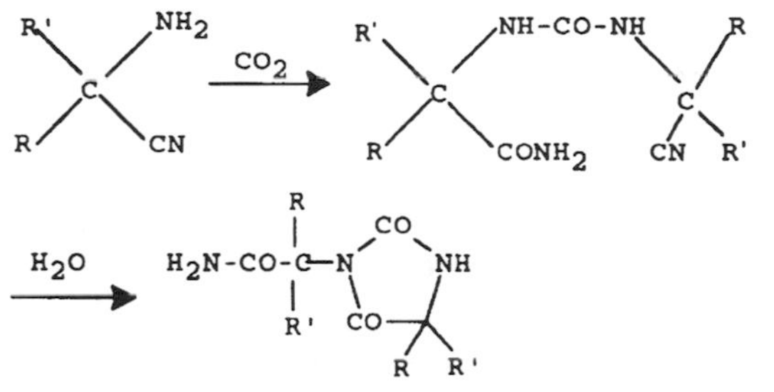

Another interesting route to substituted ureas was discovered by Nomura et al.[31]
Alkylamines in benzene solution containing small amounts of triphenylstibine oxide
(Ph_3SbO) and tetraphosphorus decasulfide (P_4S_{10}) as catalyst were pressurized with
CO_2 to 4.9 MPa and heated to 80°C for 12 h, forming 1,3-dialkylureas in high yields.

Thus, with butylamine, BuNH–CO–NHBu was produced in 88% yield. The reaction also succeeded with aromatic amines, such as aniline, but required a reaction temperature of 120°C. With diamines, the carbonylation with CO_2 led to cyclic ureas. Thus, with tetramethyl-substituted ethylene diamine, the corresponding 2-imidazolidinone was obtained:

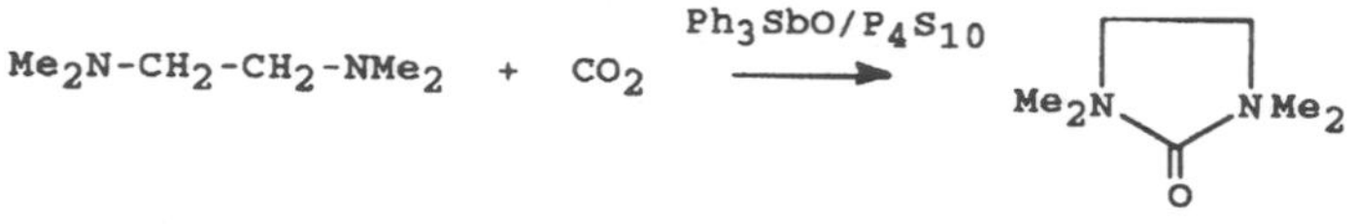

$$Me_2N-CH_2-CH_2-NMe_2 \quad + \quad CO_2 \quad \xrightarrow{Ph_3SbO/P_4S_{10}} \quad$$

The mechanism was proposed to involve the intermediate formation of carbamothioic acids.[31]

9.2.3 Urethanes and Polyurethanes

Polyurethanes are among the most important plastic materials. Annual sales in the U.S. alone amounted to more than 4.61 billion pounds in 1996.[32] Their current production depends mainly on toxic isocyanate monomers, which are formed from amines by reaction with phosgene.

In a route to urethanes (carbamate esters) which avoided the intermediate production of the very toxic phosgene and isocyanates, a palladium-complex-catalyzed synthesis was developed by McGhee and Riley using amines, carbon dioxide, and cyclic diolefins as starting components.[33] The reaction depended on (1) the activation of CO_2 by a primary or secondary amine, carried out in THF or methylene solution, in the presence of a tertiary amine base such as quinuclidine, and (2) treatment of a cyclic diolefin, such as norbornadiene, dicyclopentadiene, or 1,5-cyclooctadiene, with palladium dichloride, thus activating the diolefin by the Pd metal center toward nucleophilic attack by the carbamate anion. This reaction of the carbamate salt with the Pd diolefins was performed in a CO_2 atmosphere at low temperatures. (3) The resulting metal–carbon bond was then cleaved by protonolysis with DIPHOS [bis(diphenylphosphino)ethane] followed by $NaBH_4$, to produce the corresponding carbamate esters. Thus, norbornadiene–palladium chloride with various carbamates formed nortricyclo carbamate esters:[33]

$$RR'NH + Base + CO_2 = RR'NCO_2^- {}^+HBase$$

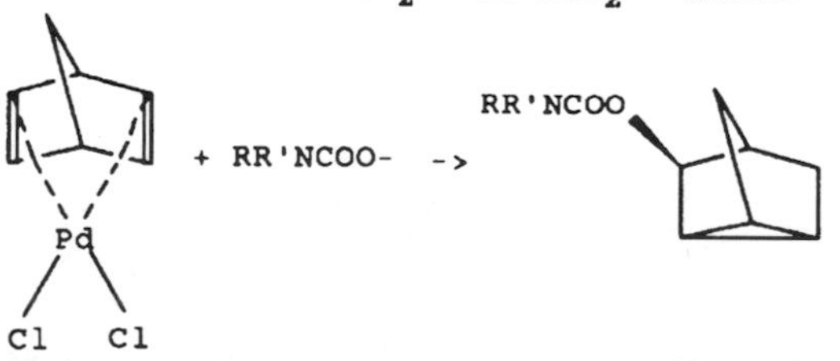

Monourethanes were also prepared by McGhee et al. from amines, alkyl chlorides, and sterically hindered strong organic bases.[34–36] A most effective base was N-cyclohexyl-N',N',N'',N''-tetramethylguanidine. The primary step in this reaction was the formation of the carbamate anion, which in the presence of the strong base acted as a powerful nucleophilic reagent in attacking the alkyl halide:

$$2RR'NH + CO_2 = RR'NCOO^- + H_2NRR' \qquad (9.9)$$

$$RR'NCOO^- + HBase + R''Cl \rightarrow RR'N{-}CO{-}OR'' \qquad (9.10)$$

Oligo- and polycarbamate esters were similarly generated by reacting diamines (such as p-phenylenediamine) with dialkyl chlorides (such as 1,4-dichlorobutane). These oligocarbamate esters were useful as prepolymers in the production of segmented polyurethanes.[34–36]

Carbon dioxide was found by Kojima et al. to be activated at room temperature and atmospheric pressure by (5,10,15,20-tetraphenyl porphinato) aluminum acetate in the presence of secondary amines and epoxides, producing dialkylcarbamic esters.[37] The Al-porphyrin acted catalytically. Thus, with 1,2-epoxypropane and diethylamine, CO_2, and the Al-porphyrin, the turnover number with respect to the formation of 2-hydroxypropyl diethylcarbamate was six. Carbamic esters were also obtained from other dialkylamines, such as piperidine and diisopropylamine. In the proposed mechanism, CO_2 was trapped in the form of an Al-carbamate on the Al-porphyrin, on the opposite side of the acetate group with respect to the porphyrin plane.[37]

Monoalkyl ammonium carbamates could be made to react with alkyl halides such as allyl bromide in the presence of crown ethers to produce carbamate esters:[23,24]

$$(RNH_3)(O_2CNHR) + R'X \rightarrow RNHC(O)OR' \qquad (9.11)$$

N-Alkylcarbamate esters $RNHC(O)OR'$ were also prepared by direct reaction of primary amines, carbon dioxide, and alkyl halides in the presence of a macrocyclic polyether, such as 18-crown-6. This method offers an attractive alternative route to the important group of carbamate esters in that it does not require the use of the very toxic phosgene or alkyl isocyanates.[23,24]

Ruthenium complexes were used by Mahé et al. to catalyze the reaction of secondary amines and terminal alkynes with CO_2, resulting in the formation of vinylcarbamates:[29]

$$R_2NH + CO_2 + H{\equiv}CR' \xrightarrow{\ [Ru]\ } R_2NCOO{-}CH{=}CHR' \qquad (9.12)$$

Bruneau et al. produced styrylcarbamates stereospecifically in 67% yield by heating phenylacetylene, diethylamine, and CO_2 (5 MPa) in toluene solution in the presence of Ru_3(norbornadiene)$(py)_2Cl_2$.[18,19,38] By reacting acetylene and pyrrolidine in acetonitrile solution with CO_2 (2 MPa) in the presence of $[RuCl_2(norbornadiene)]_n$ as catalyst, Bruneau et al. produced the corresponding vinylcarbamate in 63% yield:[38]

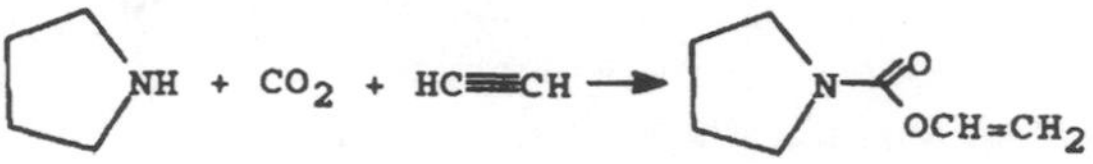

$$\text{(structure)} \quad NH + CO_2 + HC\equiv CH \longrightarrow \text{(structure)} \quad N-C\begin{smallmatrix}O\\OCH=CH_2\end{smallmatrix} \qquad (9.13)$$

Vinylcarbamates are useful monomers for the production of transparent polymers.

Secondary amines and CO_2 were found by Sasaki and Dixneuf to react with propargyl alcohols in the presence of ruthenium complexes, producing β-oxopropyl carbamates:[39]

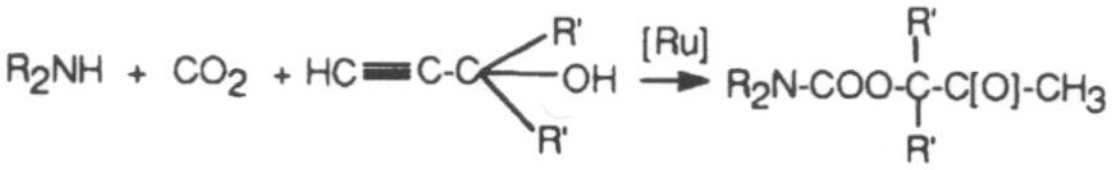

$$R_2NH + CO_2 + HC\equiv C-C\overset{R'}{\underset{R'}{\diagdown}}OH \xrightarrow{[Ru]} R_2N-COO-\overset{R'}{\underset{R'}{C}}-C[O]-CH_3 \qquad (9.14)$$

The copolymerization of CO_2 (pressurized to 300 to about 2000 psi) with 2-methylaziridine (propyleneimine) at 50°C was discovered by Super et al. to produce polyurethanes without requiring an added catalyst. The molecular weights were low, about 1000, and the glass transition temperature was about 60°C.[40]

9.2.4 Cyclic Carbamates

Cyclic carbamates (cyclic urethanes) are important intermediates in the synthesis of natural products and pharmaceuticals.

Primary aliphatic amines and carbon dioxide in methanol solution were found by Toda et al. to react with oxiranes, such as 2-aryl-2-methoxy 3-methyloxiranes, yielding the corresponding 4-aryl-4-hydroxy 5-methyloxazolidin-2-one derivatives:[41]

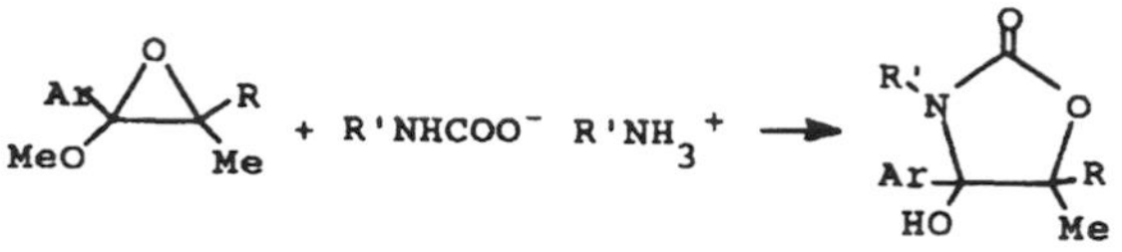

$$\text{(structure)} \quad + R'NHCOO^- \; R'NH_3^+ \longrightarrow \text{(structure)}$$

By heating disubstituted propargyl alcohols with CO_2 (5 MPa) and a primary amine in the presence of Bu_3P as catalyst (without a transition metal catalyst), Fournier et al. formed α-methylene-oxazolidinones (cyclic carbamate derivatives),[42,43]

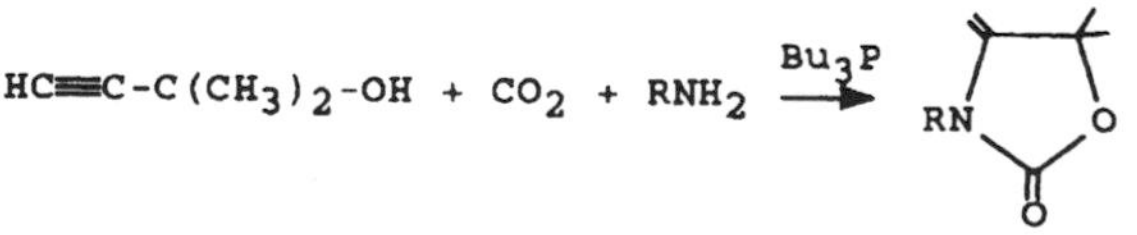

$$HC\equiv C(CH_3)_2-OH + CO_2 + RNH_2 \xrightarrow{Bu_3P} \text{(structure)}$$

A variant of the Mitsunobu reaction[44,45] was used by Kodaka et al. to achieve the condensation of ethanolamine derivatives with CO_2 to form 2-oxazolidone derivatives (cyclic urethanes).[46] 2-Oxazolidones are important as pharmaceuticals. The reaction was typically carried out by passing CO_2 through an acetonitrile solution of the substituted ethanolamine and triethylamine, producing an intermediate carbamic acid salt (a). This was then treated with a mixture of diethyl azodicarboxylate with either triphenylphosphine or tri-*n*-butylphosphine, which condensed to an N-phosphonium compound (b). In the final "Arbusov-type" reaction step, the tertiary phosphine was converted to a phosphine oxide, effecting ring closure of the carbamate to a 2-oxazolidone (c):

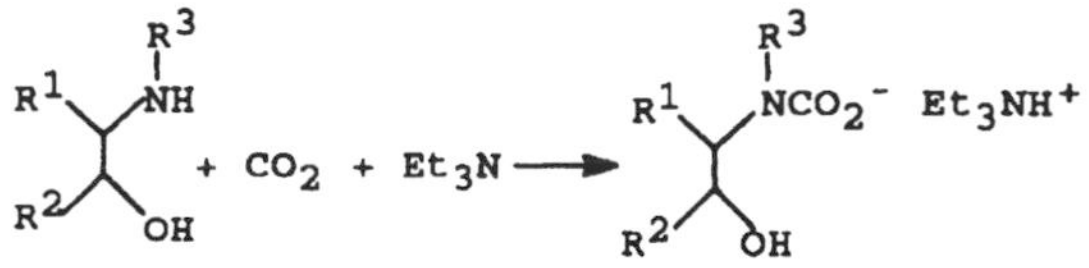

$$\text{(a)}$$

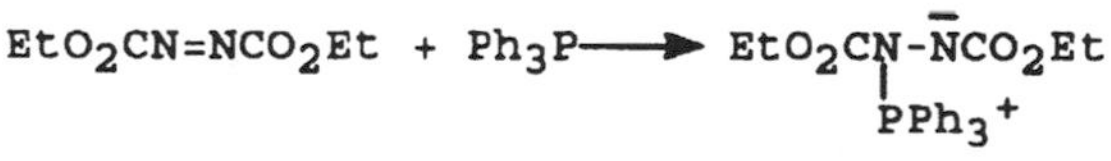

$$\text{EtO}_2\text{CN=NCO}_2\text{Et} + \text{Ph}_3\text{P} \longrightarrow \text{EtO}_2\text{CN-}\overset{-}{\text{N}}\text{CO}_2\text{Et} \quad \overset{\mid}{\text{PPh}_3^+} \qquad \text{(b)}$$

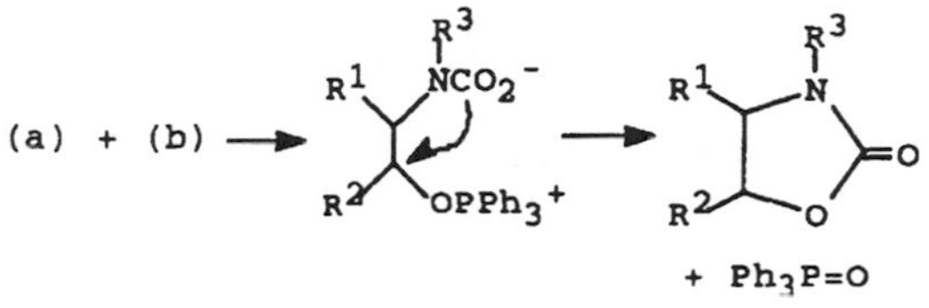

$$\text{(a)} + \text{(b)} \longrightarrow \qquad + \text{Ph}_3\text{P=O} \qquad \text{(c)}$$

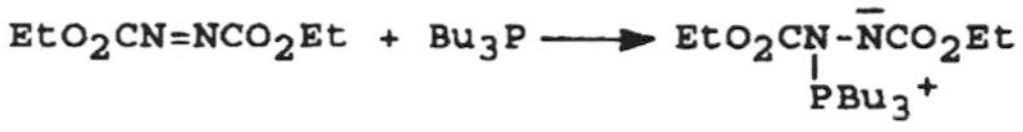

$$\text{EtO}_2\text{CN=NCO}_2\text{Et} + \text{Bu}_3\text{P} \longrightarrow \text{EtO}_2\text{CN-}\overset{-}{\text{N}}\text{CO}_2\text{Et} \quad \overset{\mid}{\text{PBu}_3^+} \qquad \text{(d)}$$

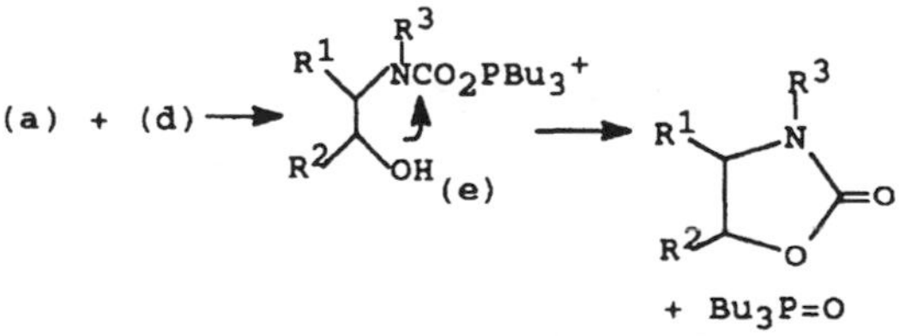

$$\text{(a)} + \text{(d)} \longrightarrow \qquad + \text{Bu}_3\text{P=O} \qquad \text{(e)}$$

Using C^{18}O$_2$ in this reaction, the mechanism of the above final step was found to depend on the phosphine applied. With Ph$_3$P as reagent, most of the phosphine oxide formed was Ph$_3$P^{16}O, indicating that the mechanism as drawn above leading to structure c was correct. On the other hand, with n-Bu$_3$P as reagent, the phosphine oxide produced was n-Bu$_3$P^{18}O. This indicated that ring closure occurred by nucleophilic attack of the hydroxyl group on the carbon atom of the carboxylate group. The proposed intermediate prior to ring closure may thus have structure e in the above scheme.[46]

Ring closure of β-aminoalcohols to 2-oxazolidones was also performed under mild conditions with more readily available starting materials. Kubota et al. reacted amino alcohol derivatives with P^{III} reagents such as Ph$_3$P or (RO)$_3$P together with CCl$_4$ and Et$_3$N in acetonitrile solution under atmospheric pressure CO$_2$ at room temperature.[47] Thus, Ph–CH(NH$_2$)–CH$_2$OH was converted into 4-phenyl-2-oxazolidone in 90% yield:

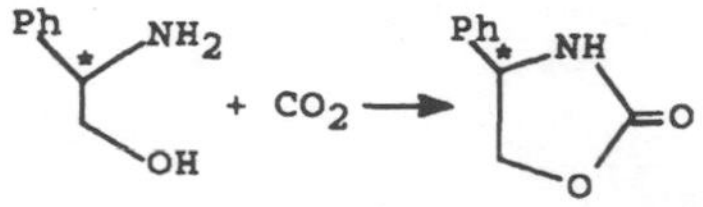

In the proposed mechanism, the active reagent was a phosphonium-type adduct of the P^{III} compound and CCl_4, which reacted with an intermediate carbamate (formed from the amino alcohol and CO_2) to produce a transient species similar to structure e in the previous scheme by Kodaka et al.[46]

9.2.5   Isocyanates from Carbamate Anions

An alternative route to isocyanates, avoiding the classic phosgene process, was developed by McGhee et al. by treating carbamate anions with strong dehydrating agents in the presence of a tertiary amine such as triethylamine.[35,48] Effective dehydrating agents were $POCl_3$, PCl_3, SO_3, P_4O_{10}, acetic anhydride, and o-sulfobenzoic acid. The reaction was successful with both primary and secondary amines, which by reaction with CO_2 (1 to 5 atm), the tertiary amine, and the dehydrating agent were converted into the corresponding isocyanates. The overall process, as shown for the reaction of n-octylamine in CH_3CN solution at 0°C, with SO_3 as dehydrating agent,

$$n\text{-}C_8H_{17}NH_2 + 2Base + CO_2 + SO_3 \rightarrow n\text{-}C_8H_{17}NCO \qquad (9.15)$$

led to the production of n-octyl isocyanate in 99% yield. The triamine triaminononane (4-aminomethyl-1,8-octanediamine) was converted to triaminononane tris-isocyanate, which could be useful as a cross-linking agent. By using halogen-free dehydrating agents, the release of acid halide waste was prevented. An important advantage of this process is that isocyanates could also be prepared from secondary amines, which was impossible by the phosgene process. The isocyanates produced may be converted immediately into urethanes. Thus, the extremely toxic methyl isocyanate, generated from methylamine, was reacted in situ with 1-naphthol to form the insecticide 1-naphthyl N-methyl carbamate.[48]

9.3   CARBONATE ESTERS

Carbonate esters may in principle be obtained by the reaction of carbon dioxide with alcohols, but the equilibrium is in favor of the dissociation of the esters:

$$ROH + CO_2 = ROCOOH \qquad (9.16)$$

9.3.1   Open-Chain Carbonates

Open-chain carbonate esters were produced by Yoshida and Ishii by reaction of CO_2 with epoxides, producing hydroxyalkyl alkyl carbonates.[49] With cyclohexene oxide and methanol, in an autoclave charged with carbon dioxide to 50 atm pressure at a temperature of 120°C for 21 h, followed by an acetylation step, the yield of 2-acetoxy cyclohexyl methyl carbonate (a) was about 3%. With increased reaction temperature, to 160°C, the yield of 2-methoxycyclohexyl acetate (b) reached 70%:

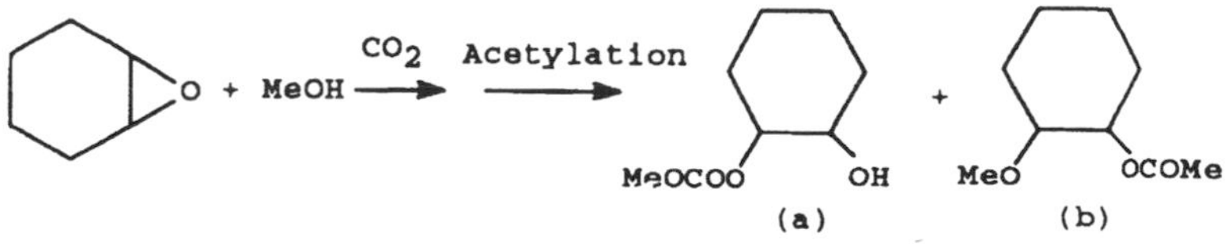

In an interesting preparation of dialkyl carbonates, McGhee and Riley reacted alcohols, CO_2, and alkyl chlorides in an aprotic polar solvent in the presence of a guanidine base:

$$ROH + CO_2 + R'Cl + Base \rightarrow RO\text{--}CO\text{--}OR' + Base \cdot HCl \qquad (9.17)$$

A useful base was N-cyclohexyl-N',N',N'',N''-tetramethylguanidine.[50]

A convenient preparation of carbonates from alcohols and CO_2 is based upon the Mitsunobu reaction.[44] The reaction was carried out by Hoffman by adding diethylazodicarboxylate to a CO_2-saturated solution of the alcohol in THF containing triphenylphosphine.[45] The overall reaction is represented by the equation

$$2ROH + CO_2 + Ph_3P + EtOOCN=NCCOOEt \rightarrow ROCOOR$$

$$+ Ph_3PO + EtOOCNH\text{--}NHCOOEt \qquad (9.18)$$

Presumably, the driving force in this reaction is the conversion of Ph_3P to Ph_3PO. By this reaction, 1-pentanol was carboxylated to dipentyl carbonate, 2-butanol to di-*sec*-butyl carbonate, 2-ethylhexanol to bis(2-ethylhexyl) carbonate, and allyl alcohol to diallyl carbonate.[45]

Dimethyl carbonate was also produced by oxidative carbonylation of methanol, using CuCl as catalyst.[51]

9.3.2 Cyclic Carbonates from Epoxides and CO_2

Cyclic carbonates are useful as aprotic polar solvents and as versatile chemical intermediates. Earlier methods for the synthesis of cyclic carbonates involved as starting materials the extremely toxic phosgene (permissible exposure <0.1 ppm).

The cycloaddition of CO_2 to epoxides to form cyclic carbonates was found by Yano et al. to occur under mild conditions, in the presence of MgO as catalyst.[52] Thus, styrene oxide in dimethylformamide (DMF) solution, with MgO which had been calcined at 400°C, was reacted with CO_2 at a pressure of 20 kg cm^{-2} and 135°C for 12 h, producing styrene carbonate in 60% yield. With the chiral (R)-styrene oxide, the product was (R)-(−)-phenyl carbonate, with complete retention of the epoxide configuration. The same reaction with propylene oxide formed propylene carbonate in 41% yield.

A highly efficient synthesis of propylene carbonate from propylene oxide (PO) and CO_2 was achieved by Zhu et al.,[53]

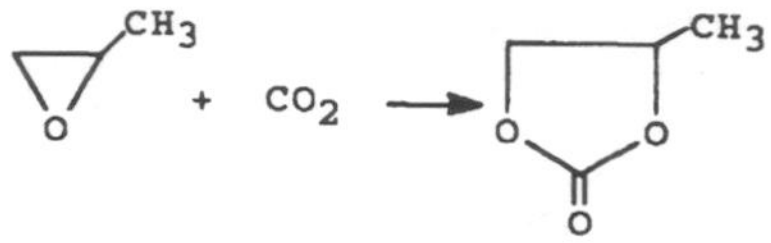

using poly(ethylene glycol) (PEG, average molecular weight $M_n = 400$) and KI as catalyst. With an initial CO_2/PO ratio of 1.2, a PO/PEG ratio of 100, a PEG/KI molar ratio of 3, at 2 MPa pressure of CO_2, and a temperature of 150°C, the yields of propylene carbonate after 2.5 and 5 h were 91 and 100%, respectively.[53] Epoxides are accessible by the "clean" catalytic oxidation of olefins with in-situ-generated hydrogen peroxide, thus avoiding the environmentally objectionable chlorohydrin process.[50]

Kihara and Endo observed that the reaction of atmospheric pressure carbon dioxide with poly(glycidyl methacrylate) films in DMF solution resulted in the quantitative conversion of the oxirane rings of the polymer into the five-member cyclic carbonates (2-oxo-1,3-dioxolan derivatives).[54] The reaction was catalyzed by alkali halides and by tetraalkylammonium salts. Best results were obtained with a mixture of NaI and Ph_3P (1.5 mol% each) as catalyst, at a temperature of 100°C:

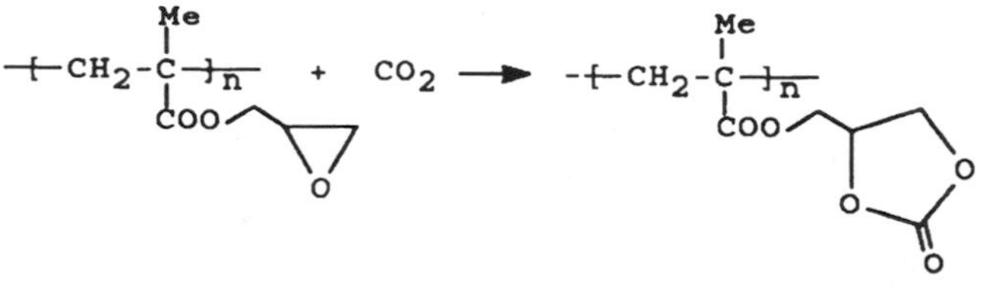

This provided an inexpensive method for introducing CO_2 into the polymer side group. The cyclic carbonate side chains could be reacted with butylamine in dimethyl sulfoxide (DMSO) solution, forming the corresponding polymethacrylate with hydroxyurethane side chains.[54]

Monomeric glycidyl methacrylate also rapidly incorporated CO_2, but the resulting monomeric 2-oxo-1,3-dioxolan-4-yl methacrylate could not be isolated in the pure state. It polymerized spontaneously, forming a high-molecular-weight polymer which was insoluble in most solvents.[55,56]

The rate of reaction of CO_2 with monomeric epoxides in N-methylpyrrolidinone solution at 100°C was found by Kihara et al. to be first order in the concentration of both the epoxide and the catalyst and to be independent of the CO_2 concentration.[57,58] Thus, the rate of incorporation of CO_2 into epoxides to form the five-member cyclic carbonates could be represented by the equation

$$-d[\text{epoxide}]/dt = k[\text{epoxide}][\text{catalyst}]^m \qquad (9.19)$$

where m depended both on the molecular weight of the epoxide and the Lewis acidity of the catalyst. The order of activity of halides was Cl > Br > I. That of the cations was Li > Na > benzyltrimethylammonium, which is their order of Lewis acidity. The proposed mechanism involved nucleophilic attack of the halide ion on the oxirane, forming a β-haloalkoxide. This reacted with CO_2 to produce an intermediate, which underwent ring closure to a cyclic carbonate.[57,58]

The same reaction was also performed by Kihara and Endo in the solid state in the presence of tertiary ammonium salts as catalysts. Optimal conditions were achieved with 6 mol% of benzyltrimethylammonium bromide dispersed into the poly(glycidyl methacrylate) films and heated at 120°C for 42 h, resulting in 86% incorporation of CO_2.[59]

The catalytic action of quarternary ammonium or phosphonium halides and alkali halides for the formation of cyclic carbonates from epoxides and CO_2 was ascribed by Darensbourg and Holtcamp to nucleophilic attack of the halide anion on the least-hindered carbon atom of the epoxy ring, together with electrophilic attack of the cation of the catalyst on the oxygen atom of the epoxy ring. Carbon dioxide then inserted itself into this cation–oxygen bond, followed by ring closure to a cyclic carbonate.[60]

The synthesis of cyclic carbonates from epoxy compounds and CO_2 was performed by Yamashita et al. using a polystyrene-bound crown ether as catalyst, in the presence of an inorganic salt.[61] Optimal conditions were at 100°C, in DMF as solvent, with 4 mol% of the polymer-supported crown ether and 2 mol% of KCl, and bubbling through the CO_2. The yields of the cyclic carbonates produced from phenyl glycidyl ether, n-butyl glycidyl ether, allyl glycidyl ether, and styrene oxide were 100, 93, 94, and 63%, respectively.

Functional polymers bearing pendant cyclic carbonate groups were prepared by Nishikubo et al. by polymerization of the cyclic carbonate (2-oxo-1,3-dioxolan-4-yl)methyl vinyl ether (OVE).[62] The monomeric OVE was synthesized by reacting glycidyl vinyl ether (GVE) in toluene solution with CO_2 using Bu_4NBr as catalyst:

$$CH_2=CH-O-CH_2-CH—CH_2 + CO_2 \longrightarrow CH_2=CH-O-CH_2-CH——CH_2$$

(GVE)   (OVE)

The polymerization of OVE by the vinyl function was performed using either free radical or cationic initiators. The pendant cyclic carbonate groups in the resulting polymer could be reacted with butylamine or benzylamine in DMSO solution, forming the corresponding pendant carbamate groups.[62] Park et al. observed that the catalytic activity of tertiary ammonium salts for the conversion of GVE with CO_2 increased with the size of the alkyl group and with the nucleophilicity of the counter anion.[63]

Baba et al. discovered that oxetanes (four-membered cyclic ethers) in the presence of organotin halide complexes incorporate CO_2, forming both trimethylene carbonate (1,3-dioxan-2-one) and polycarbonate:[64]

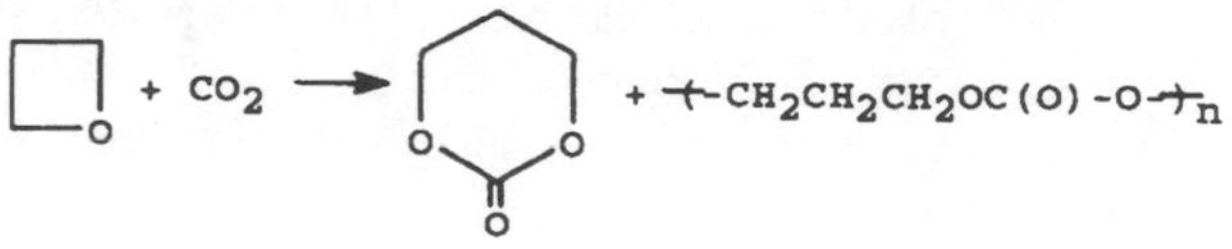

The reactions were performed at 100°C under 50 kg cm^{-2} CO_2. Optimal conditions for the selective production of the cyclic carbonate were obtained by using Bu_3SnI + hexamethylphosphoric triamide as catalyst, resulting in 100% yield. On the other hand, when Bu_2SnI_2 was used as the catalyst, polymerization occurred with 98% yield, resulting in a polymer of molecular weight $M_r = 4250$.[64]

The polymerization of the cyclic carbonate 4-methylene-1,3-dioxan-2-one

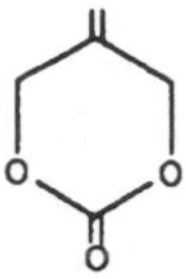

in chlorobenzene solutions was found by Takata et al. to be initiated by azobis(isobutyronitrile), resulting in a polymer ($M_m = 24600$ and $M_n = 8900$) consisting mainly of ring-opened units, as well as of vinyl-polymerized units.[65]

9.3.3   Cyclic Carbonates from Substituted Propargyl Alcohols and CO_2

The reaction of derivatives of propargylic alcohol with carbon dioxide in the presence of various transition metal complexes was observed by Inoue to lead to α-methylene cyclic carbonates. Thus, cycloaddition of CO_2 with 2-methyl-3-butyn-2-ol resulted in the cyclic carbonate[66]

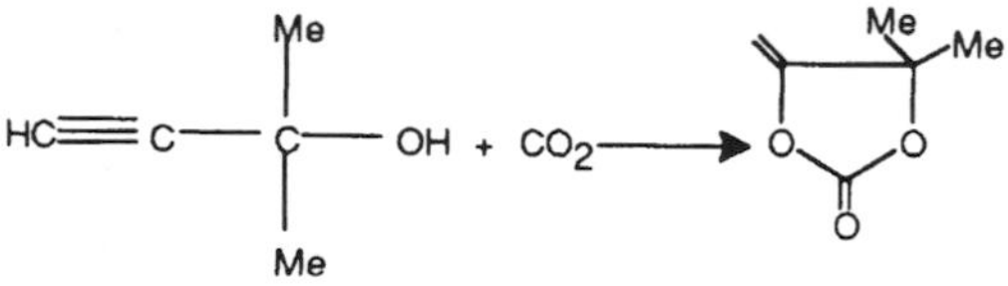

In the presence of iodobenzene and $Pd(PPh_3)_4$ as catalyst, sodium 2-methyl-3-butyn-2-olate reacted with pressurized CO_2 (10 atm) at 100°C to produce stereoselectively in 68% yield the cyclic vinylidene carbonate (E)-5-benzylidene-4,4-dimethyl-1,3-dioxolan-2-one:[67]

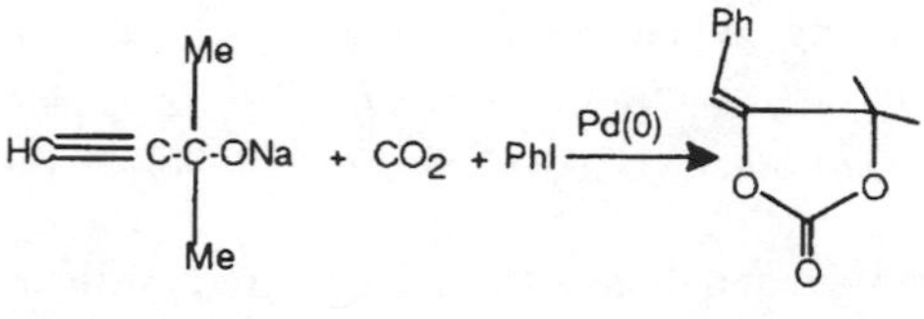

A most interesting variant of the above reaction of 2-methyl-3-butyn-2-ol was found by Inoue et al. to occur in the presence of carbon monoxide, carbon dioxide, iodobenzene, triethylamine, and transition metal–PPh_3 complexes as catalysts, such as $Pt(PPh_3)_4$.[66,68] In this reaction, carbon monoxide entered into the 2-methyl-3-butyn-2-ol molecule to produce 2,2-dimethyl-5-phenyl-3(2H)-furanone:

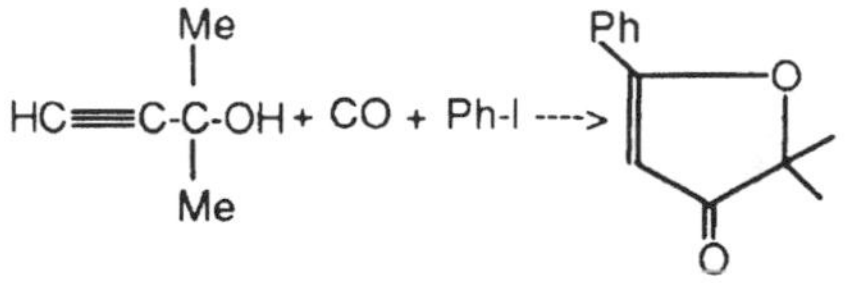

This compound, known as bullatenone, occurs in the essential oil of *Myrtus bullata,* a shrub which is endemic in New Zealand. The 3(2H)-furanone ring system is a central component of several antitumor agents and also a constituent of some flavoring compounds.[62,68]

The route to cyclic carbonates containing an unsaturated side chain was further explored by Fournier et al.[42,43,69] Disubstituted propargyl alcohols reacted with CO_2 in the presence of a phosphine catalyst to yield α-methylene cyclic carbonates. Thus, with the dimethyl-substituted propargyl alcohol, CO_2 (50 kg cm^{-2}), and a catalytic amount of tributylphosphine at 100°C for 8 h, but without solvent or metal catalyst, the 4-methylene oxazolidin-1-one was produced in 98% yield:

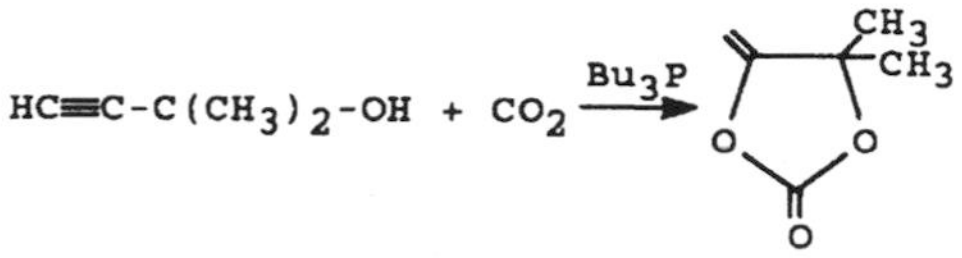

The reaction is specific for the tertiary alcohols. Thus, propargyl alcohol itself does not form cyclic carbonates under these conditions. These cyclic methylene carbonates were used by Joumier et al. to produce optically active oxazolidinones by their reaction with amino acids and oxadiazin-2-ones by their reaction with hydrazine:[70]

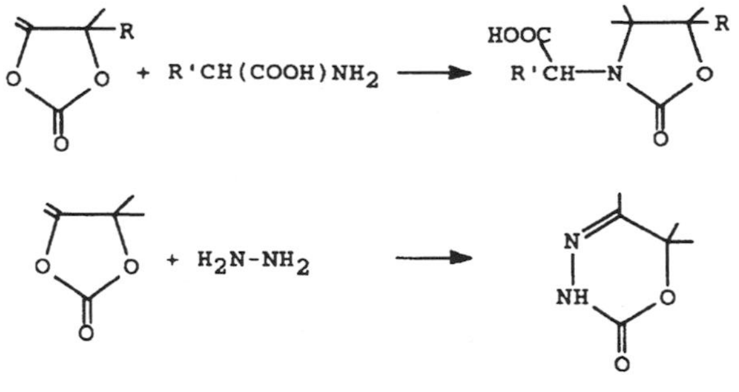

Other applications of α-methylene cyclic carbonates include their enantioselective hydrogenation with chiral Ru(II) complexes for the preparation of optically active cyclic carbonates and 1,2-diols.[71]

Oxazolidinones have also been formed directly by reacting propargyl alcohol derivatives with primary amines and CO_2 in the presence of Bu$_3$P, as described in Section 9.2.4.

9.3.4   Cyclic Carbonates from Styrene and O_2/CO_2

Styrene carbonate (1,3-benzodioxol-2-one) was produced by Aresta et al. by reacting styrene in THF solution with a 1:1 mixture of O_2 + CO_2 (total pressure 103 kPa) in the presence of rhodium complexes $RhCl(L_3)$, where L = PEt_2Ph or $PEtPh_2$:[72,73]

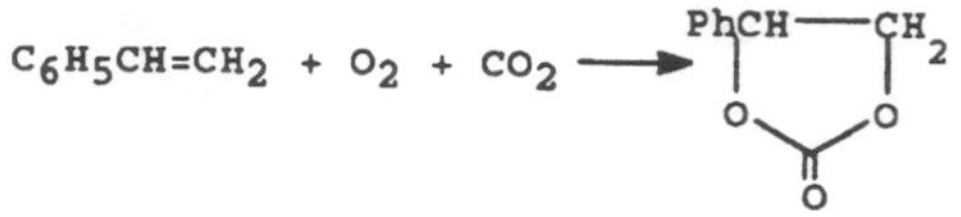

$$C_6H_5CH=CH_2 + O_2 + CO_2 \longrightarrow$$

By-products included styrene oxide, benzaldehyde, phenylacetaldehyde, and acetophenone. Up to 100 mol of styrene per mole of rhodium was converted to the cyclic carbonate, with yields of 20 to 30%. The formation of this carbonate from styrene and O_2 + CO_2 was faster than from styrene oxide and CO_2, which indicated that styrene oxide was not an intermediate in the above synthesis. The THF solvent was also oxidized under these conditions. In the mechanism proposed by Aresta et al., the role of the metal catalyst could be that of generating superoxo-radicals of the type M–O–C(O)OO·, and that peroxocarbonates acted as oxidants of olefins and ethers.[74] The use of O_2/CO_2 mixtures is of practical importance since carbon dioxide released from most combustion processes is mixed with oxygen.

9.3.5   Polycarbonates

Poly(alkylene carbonate) plastics are becoming a major application of carbonate esters.[75] They are produced currently by the Schotten–Baumann reaction of phosgene with a diol in the presence of alkali:

$$n(HOROH) + nCOCl_2 + (2n+1)NaOH \rightarrow Na[OC(O)-OR]_nOH$$

$$+ 2nNaCl + 2nH_2O \tag{9.20}$$

The polycarbonates formed with bisphenol A (4,4′-isopropylidene diphenol) as the diol have high impact strength, good transparency, and excellent resistance to outdoor exposure. Their glass transition temperature, T_g = 140°C, is much higher than most other commercial plastics.[75]

Polycarbonates have a large variety of uses, from soft drink bottles to building materials, automobile parts, and electrical components. One of the many useful applications is windows for solar heaters because of their good resistance to UV light.[76]

In the above conventional phosgene process for polycarbonate, the bisphenol A, NaOH, and phosgene are usually converted into polycarbonate by interfacial condensation between water and methylene chloride. The polycarbonate remains in the CH_2Cl_2 phase, which is washed with additional water to remove the NaCl produced. Due to the relatively high solubility of CH_2Cl_2 in water, these washings contain high concentrations of CH_2Cl_2, which is toxic and possibly carcinogenic.[77] Thus, the phosgene process is problematic not only because of the toxicity and corrosiveness of phosgene, but also because of the environmentally objectionable use of methylene chloride.

The use of the extremely toxic phosgene for the above synthesis of polycarbonates can be avoided by the copolymerization of CO_2 with epoxides. This has been achieved in reactions promoted by a variety of catalysts, including various metal complexes and salts. The mechanisms of these catalytic reactions have been comprehensively reviewed by Darensbourg and Holtcamp.[60]

Inoue et al. pioneered the alternating copolymerization of carbon dioxide and epoxides, with $Zn(CH_2CH_3)_2$ and H_2O as catalyst.[78] With this catalyst system, in benzene solution, the reaction of propylene oxide and CO_2 (atmospheric pressure) at 25°C for 5.5 h produced a small amount of methanol-insoluble product. Its elementary composition and IR and NMR data corresponded to a copolymer with a ratio of carbonate to propylene oxide units of 0.88:0.12. By increasing the CO_2 pressure to 50 to 60 atm, the yield of the copolymer was very much enhanced, and the copolymerization was then almost completely alternate, in accordance with the equation[78]

$$H_2C\!-\!\overset{R}{\overset{|}{CH}} + CO_2 \longrightarrow \left(CH_2\text{-}\overset{R}{\overset{|}{CH}}\text{-O-CO-O}\right)_n$$

For the alternating copolymerization of CO_2 and propylene oxide, zinc salts served as very active catalysts. Soga et al. prepared such catalysts from $Zn(OH)_2$ and various dicarboxylic acids. A most effective catalyst was that prepared from glutaric acid. The catalysts were used on supports, such as SiO_2, MgO, or γ-Al_2O_3. The copolymerization product was mainly poly(oxycarbonyl-oxypropylene), an alternating copolymer of CO_2 and propylene oxide.[79] Molecular weights obtained reached 100,000.[80,81]

Chlorinated hydrocarbons often have been used as solvents for the copolymerization of CO_2 and epoxides. Darensbourg et al. succeeded in performing the copolymerization of CO_2 and propylene oxide in supercritical carbon dioxide, using zinc glutarate as catalyst.[82] The substitution of supercritical CO_2 instead of the toxic methylene chloride as solvent is a major advantage, both economically and for environmental considerations. At a CO_2 pressure of 1200 psi and 60°C, the number and average molecular weights reached about 14,000 and 140,000, respectively, with high selectivity for polycarbonate vs. polyether formation. Only small amounts of the cyclic monomer propylene carbonate were produced.[82]

Remarkable catalytic activity for the copolymerization of CO_2 and epoxides was discovered by Darensbourg et al. to occur with zinc(II) phenoxides containing bulky groups in the 2 and 6 positions of the phenyl substituent.[83,84] These were (2,6-diphenylphenoxy)$_2$Zn^{II}(THF)$_2$ and (2,6-diphenylphenoxy)$_2$Zn^{II}(EtOEt)$_2$. Using these catalysts, mixtures of cyclohexene oxide and propylene oxide, under CO_2 pressure of 800 psi at 80°C for 69 h, underwent terpolymerization, reaching high turnover numbers, with only limited formation of cyclic propylene carbonate. The advantage of including cyclohexene oxide in the monomer mixture is the higher glass transitions attained in the resulting copolymer.[83,84]

The mechanism of polycarbonate formation catalyzed by Zn(II) complexes was studied by Kuran and Listos using ^{13}C NMR.[85] With diethylzinc and polyhydric phenols as catalysts, these studies indicated that the primary step was CO_2 insertion into the catalyst, followed by incorporation of propylene oxide. For enhanced copolymerization of propylene oxide with CO_2 relative to the production of cyclic carbonate, at least two zinc atoms per complex were required.[85]

In the presence of ethylzinc phenoxide or other ethylzinc compounds supported on γ-alumina, the copolymerization of propylene oxide and CO_2 was found by Listos et al. to lead to a polymer with an average molecular weight of about 30,000. Both the initiation and the propagation steps of copolymerization were proposed to involve the AlOZn–O-active bond on the surface of the supported catalyst.[86]

In order to study the mechanism of the initiation process in the zinc(II) dicarboxylate catalyzed copolymerization of CO_2 with epoxides to produce polycarbonates, Darensbourg et al. applied cadmium dicarboxylates as model catalysts.[87] This enabled using ^{113}Cd NMR to identify the bonding of the metal during the reaction. The catalyst used was $[\eta^3\text{-HB}(3\text{-Phpz})_3]\text{-Cd}^{II}$(acetate)-THF, where HB(3-Phpz) is tris-3-phenylpyrazole hydroborate. This Cd complex was less active than the zinc(II) dicarboxylates. When using this catalyst in propylene oxide or cyclohexene oxide, the THF ligand of the catalyst was replaced by these epoxide ligands. The structure of the resulting complexes was characterized by X-ray diffraction, indicating six-coordination of the cadmium atoms. By reacting the $[\eta^3\text{-HB}(3\text{-Phpz})_3]\text{-Cd}^{II}$(acetate)-THF catalyst with propylene oxide and CO_2 (at 1050 psi), cyclic carbonates were formed.[87]

Using supercritical carbon dioxide as both reactant and solvent, Super et al. performed the copolymerization of cyclohexene oxide and CO_2 with a CO_2-soluble ZnO-based catalyst.[88] This catalyst was prepared by esterification of maleic anhydride with 3,3,4,4,5,5,6,6,7,7,8,8,8-tridecafluorooctanol and reacting the resulting half-ester with ZnO. The turnover number of the catalyst was very high, reaching 400 g polymer per gram of Zn. Optimal conditions for the copolymerization were found at temperatures of 100 to 110°C, pressure = 2000 psig, with the mole fraction of cyclohexene oxide above 0.15. The polymer yield reached 69%, of which more than 90% was polycarbonate, with an average molecular weight M_w ranging from 50,000 to 180,000.[88]

Polypropylene carbonate, which is obtained by the copolymerization of the inexpensive PO with carbon dioxide, could be a very useful polymeric material. Its drawback hitherto has been insufficient thermal stability. Thus, as tested by thermogravimetry, a 5% weight loss already occurred at 215°C.[89] This thermal degradation was found by Chen et al. to be due to an "unzipping" of the polymer, releasing propylene carbonate.[90] In an effort to overcome this thermal instability, the effects of various added monomers were studied. Considerably enhanced thermal stability was attained by using maleic anhydride (MA) as an added monomer. Using PO and MA, at an MA/PO ratio of 0.72, the temperature of 5% weight loss was raised to 263°C. The catalyst used for these polymerizations was a polymer-chelated bimetallic cyanide, (polymer) $[Zn(Fe(CN)_6]_xCl_{2-3x}.$[90]

In a new process by Asahi Chemical Industry (Japan) for producing polycarbonate, Komiya et al. avoided the use of phosgene as reagent and methylene chloride as solvent by carrying out a solid-state polymerization of amorphous polymers, obtained from suitable prepolymers.[77] This enabled the production of polycarbonate of ultra-high molecular weight. The process was carried out in the following stages: (1) Bisphenol A was reacted with diphenyl carbonate (which was obtained by transesterification from dimethyl carbonate, produced by a reaction not involving phosgene), resulting in an amorphous low-molecular-weight prepolymer (M_w 2000 ~ 10,000). The phenol released was recycled to the production of diphenyl carbonate. (2) The prepolymer was crystallized either by heating it above its glass transition temperature or by treating it with a suitable solvent, such as acetone. The melting point of the crystallized prepolymer was 225°C. (3) The crystallized prepolymer underwent solid-state polymerization by heating it in an oxygen-free atmosphere at 210 to 220°C. The polycarbonates obtained had molecular weights ranging from low disk grades (M_w 15,000) to high M_w (35,000) and up to ultra-high-molecular-weight-grade polycarbonate (M_w >60,000), which has outstanding properties, such as excellent solvent and steam resistance and high transparency.[77]

9.4 CO_2–METAL CENTER COORDINATION

Carbon dioxide, while by itself is rather inert, may be activated by some reagent, usually a transition element compound, which enables coupling of carbon dioxide with suitable organic molecules. Thus, condensation products are formed containing the CO_2 moiety.[9,10] Various compounds can be produced by nonreductive fixation of carbon dioxide, such as its condensation with ethylene oxide to produce ethylene carbonate.

9.4.1   Structural Characterization

In the carbon dioxide molecule, the oxygen atoms act as weak Lewis bases, while the carbon atom is a weak Lewis acid. Thus, CO_2 can react in the coordination sphere of transition metal complexes, for example, by inserting into metal–oxygen, metal–carbon, metal–hydrogen, or metal–nitrogen bonds, forming products with one or more carbon atoms. The modes of bonding of carbon dioxide to metal centers were reviewed by Aresta et al.[14] and Gibson.[16] One case in which the carbon dioxide–metal complex was structurally characterized was the bis(tricyclophosphine) nickel complex [Ni(CO_2)(PCy$_3$)$_2$,0.75(C$_7$H$_8$), in which Cy = cyclohexyl. The complex was prepared by Aresta et al. by reacting [Ni(PCy$_3$)$_3$] with carbon dioxide in toluene. The complex was shown by IR and single-crystal study to be planar, with the CO_2 ligand in bent geometry, coordinated to the Ni atom by the carbon atom and one of the oxygen atoms, thus forming the η^2-C,O mode of bonding CO_2 to a metal center:[91,92]

M = Ni, Nb, Ta, Mo, V

A different type of bonding, presumably by η^1-C bonding of CO_2 to an axial metal coordination site,

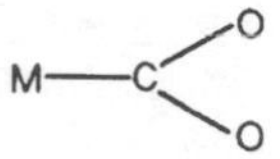

$$M-C\overset{\displaystyle O}{\underset{\displaystyle O}{<}}$$

was found by Fujita et al. in the CO_2 complexes with metal macrocycles, such as in the 1:1 complexes with cobalt macrocycles $[Co(I)L_5]^+$ (L_5 in these complexes is 5,7,7,12,14,14-hexamethyl-1,4,8,11-tetraaza cyclotetradeca-4,11-diene):[93,94]

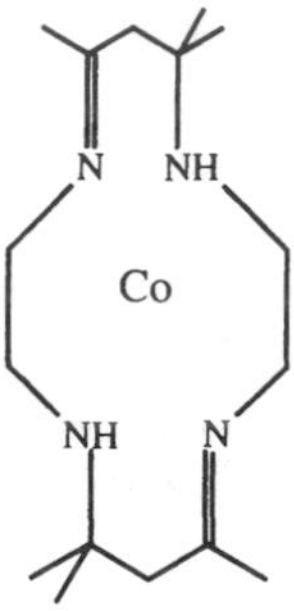

With the racemic form of this cobalt macrocycle, in acetonitrile or DMSO solutions, the CO_2 binding constant was 6×10^2 M^{-1}. The $(CoL_5CO_2)^+$ complex in acetonitrile solution underwent second-order decomposition, yielding CO and HCO_3^- as stable products.[93,94] In dry DMSO[95] and in acetonitrile, the binding of CoL^+ with CO_2 was shown to be reversible, according to the equation[96]

$$Co([14]diene)^+ + CO_2 = Co([14]diene)(CO_2)^+ \qquad (9.21)$$

See also Chapter 12 on the electrochemical reduction of carbon dioxide to carbon monoxide catalyzed by CoL^{2+} in acetonitrile–H_2O solutions.

Ito et al. isolated a crystalline carbonato complex of bis(cyclo-pentadienyl) hydridophenyl tungsten(IV), $Cp_2W(H)Ph$ (where $Cp = \eta^5$-C_5H_5), which reacted with CO_2 in wet acetone to form the mononuclear carbonato complex $Cp_2W(\eta^2$-$CO_3)$. The product was characterized by IR, ^1H, and ^{13}C NMR spectroscopy and by X-ray diffraction:[97]

$$Cp_2W(H)Ph + CO_2 + H_2O + Me_2CO \rightarrow Cp_2W(\eta^2\text{-}CO_3) + Me_2CHOH \quad (9.22)$$

Carbon dioxide was found by Bianchini and Meli to react with low-valent cobalt and rhodium complexes $(np_3)CoH$ and $(np_3)RhH$, where np_3 = tris(2-diphenyl-phosphino)ethylamine, in the presence of a solvated or complexed Lewis acid, such as the Na^+ ion.[98] On adding a THF solution of $NaBPh_4$ to a solution of $(np_3)CoH$ under a CO_2 atmosphere, the solution turned red-brown, with formation of the carbonyl complex $[(np_3)Co(CO)]BPh_4$ in 50% yield. Similarly, $(np_3)RhH$ reacted with CO_2 in the presence of Na^+ ions to produce the diamagnetic complex $[(np_3)Rh(CO)]BPh_4$ in 50% yield.

Keene et al. observed that the cobalt complex tris(2,2'-bipyridine)-cobalt(I), $Co(bpy)_3^+$ in aqueous bicarbonate solutions (pH 8.5 to 10) reacted to form carbon monoxide. The CO produced was scavenged by additional $Co(bpy)_3^+$, precipitating the insoluble $[Co(bpy)(CO)_2]_2$. A proposed intermediate in the reaction is $Co(bpy)_2(H_2O)H^{2+}$.[99]

Antiñolo et al. found that CO_2 underwent an insertion reaction into the Nb–H bond of the complex $[Nb(\eta^5\text{-}C_5H_4SiMe_3)_2H_3]$, forming a formato complex with elimination of H_2:

$$[Nb(\eta^5\text{-}C_5H_4SiMe_3)_2H_3] + CO_2 \rightarrow \{Nb(\eta^5\text{-}C_5H_4SiMe_3)_2[OC(O)H\text{-}O,O']\} \quad (9.23)$$

A similar formato complex was also prepared by the electroreduction of $[Nb(\eta_5\text{-}C_5H_4SiMe_3)_2Cl_2]$ in the presence of formic acid.[100]

Nitrosyl hydride manganese complexes were found by Nietlispach et al. to undergo a facile CO_2 insertion process, yielding formato complexes.[101] $Mn(NO)_2$-$(PMe_3)_2H$ and $Mn(NO)_2(PEt_3)_2H$ in toluene solutions reacted irreversibly with CO_2 (room temperature, 1 bar) to produce the stable complexes $Mn(NO)_2(PMe_3)_2[OC(O)H]$ and $Mn(NO)_2(PEt_3)_2[OC(O)H]$. In these formato complexes, the spectroscopic data indicated a trigonal bipyramidal structure, with the two NO groups and the formato ligand in the equatorial positions and the trialkylphosphine ligands in the axial positions.[101]

The insertion of three CO_2 molecules into the vanadium–carbon bonds of the complex $V(Mes)_3(THF)$ (where Mes = mesityl = $2,4,6\text{-}Me_3C_2$) was carried out by Vivanco.[102] The reaction was performed in THF solution of the complex under a CO_2 atmosphere at room temperature. The green crystals isolated after the reaction were shown to be tris(carboxylato) complexes, produced according to the equation

$$V(Mes)_3(THF) + 3CO_2 \rightarrow V(O_2CMes)_3 + THF \quad (9.24)$$

Various directly bonded zirconium–ruthenium heterobimetallic complexes were discovered by Casey.[103] The complexes $Cp(CO)_2M\text{-}Zr(Cl)Cp_2$, where Cp = C_5H_5 = cyclopentadienyl and M = Fe or Ru, were found by Pinkes et al. to undergo CO_2 insertion under relatively mild conditions, forming stable bimetallocarboxylates:[104]

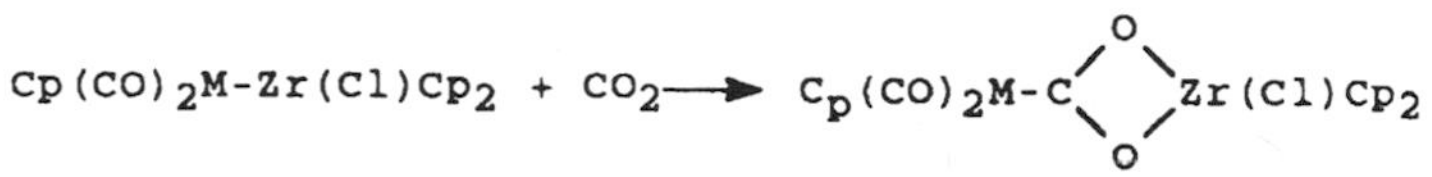

$$Cp(CO)_2M\text{-}Zr(Cl)Cp_2 + CO_2 \longrightarrow Cp(CO)_2M\text{-}C\underset{O}{\overset{O}{\diagdown}}Zr(Cl)Cp_2$$

$$M = Fe \text{ or } Ru$$

Similar bimetallocarboxylates may be important in the catalytic activation of CO_2.

The insertion of CO_2 into the Cu(I)–H, Rh(I)–H, and Rh(III)–H bonds was theoretically investigated by Sakaki and Musashi by the *ab initio* molecular orbital method.[105–107] The observed higher reactivity toward CO_2 insertion into complexes containing the Cu(I)–H bonds relative to those with Rh–H bonds was explained by the calculated activation energies E_a and reaction energies ΔE. For CO_2 insertion into $Cu(I)H(PH_3)_2$, $RH(I)H(PH_3)_3$, and $[Rh(III)H_2(PH_3)_3]^+$, values of E_a were 6.5, 21.2, and 51.3 kcal mol^{-1}, while values of ΔE were –33.5, –7.0, and –1.1 kcal mol^{-1}, respectively. The low activation energy with the Cu(I)–H complex, together with the very negative reaction energy (strong exothermicity), indicates the high reactivity of the copper hydride complexes.[105–107]

9.4.2 Reduction to Formate and Formaldehyde

Alkyl formates were synthesized by Darensbourg et al. by the hydrocondensation of alkyl halides with CO_2 and H_2 in the presence of anionic group 6 carbonyl complexes as catalysts.[108,109] The overall reaction was

$$CO_2 + H_2 + ROH \rightarrow HCO_2R + H_2O \tag{9.25}$$

Effective catalysts were $[(CO)_5CrH]^-$ and $[(CO)_5WH]^-$, which reacted reversibly with CO_2 to form metalloformate complexes, such as $[(CO)_5WCO_2H]^-$:

$$[(CO)_5)WH]^- + CO_2 \leftrightarrow [(CO)_5WCO_2H]^- \tag{9.26}$$

The reaction was performed for 24 h at 125°C, with CO_2 and H_2 each at initial pressures of 250 psi.[108,109]

When adding sodium methoxide to a tungsten pentacarbonyl complex catalyzed reaction, the reduction of CO_2 occurred even in the absence of H_2. Ovalles et al. observed that the reaction of CO_2 and $NaOCH_3$ in the presence of $[(CO)_5WCl]^-$ in THF solution resulted in the production of formaldehyde and formate.[110] The overall reaction in this case was

$$NaOCH_3 + CO_2 \rightarrow HCHO + HCOONa \tag{9.27}$$

Operating at a molar ratio of $NaOCH_3/W < 8$, at 125°C and 400 psi CO_2 for 24 h, the turnover number for HCHO production was 10 (per mole of the W complex). The proposed mechanism involved the reaction of $[(CO)_5WCl]^-$ with $NaOCH_3$ to form an alkoxide intermediate, $[(CO)_5WOCH_3]$, which underwent β-hydrogen abstraction to produce HCHO and the terminal hydride $[(CO)_5WH]^-$ in a rate-determining step. Insertion of CO_2 into the $[(CO)_5WH]^-$ complex yielded a carboxylate intermediate $[(CO)_5WO_2CH]^-$, which reacted with $NaOCH_3$ to regenerate the alkoxide complex $[(CO)_5WOCH_3]^-$, while releasing sodium formate.[110]

9.4.3 Grignard Reactions

The long-known reaction of carbon dioxide with Grignard reagents is a useful tool for regiospecific carboxylation, in which the driving force is the oxidation of one magnesium atom for each alkyl halide converted.[92] In this reaction, the electrophilic CO_2 molecules insert into the metal–carbon σ-bonds, forming carboxylates:

$$R–X + Mg \rightarrow RMgX \qquad (9.28)$$

$$RMgX + CO_2 \rightarrow R–COOMgX \qquad (9.29)$$

$$R–COOMgX + H_2O \rightarrow R–COOH + MgX(OH) \qquad (9.30)$$

In order to optimize conditions for the synthesis of aliphatic carboxylic acids, the rate constants for reaction of a series of alkyl magnesium bromides RMgBr with CO_2 were measured by Yamazaki and Hayashi.[111] These measurements were made in THF solutions at 0°C, with a large excess of the Grignard reagent (so that pseudo first-order kinetics prevailed). The resulting second-order rate constants were related to Taft's Es values, such that log (k_R/k_{Me}) was proportional to Es. The reaction constant derived from this correlation was $\rho^* = 0.37$. These data were used to estimate optimal reaction times for the carboxylation of Grignard reagents with the short-lived $^{11}CO_2$ (physical half-life 20 min), to prepare ^{11}C-labeled carboxylic acids needed as radiopharmaceuticals (for positron emission tomography).[111]

A variant of the Grignard reaction was developed by Eaton et al. using magnesium diamides, such as bis(2,2,6,6-tetramethylpiperidino)magnesium, $(TMP)_2Mg$.[112,113] With aromatic esters, such as methyl benzoate, the reaction with this "amido-Grignard" reagent, followed by CO_2 carboxylation and then esterification, resulted in the stereospecific substitution of a carboxyl group, with formation of dimethyl o-phthalate:

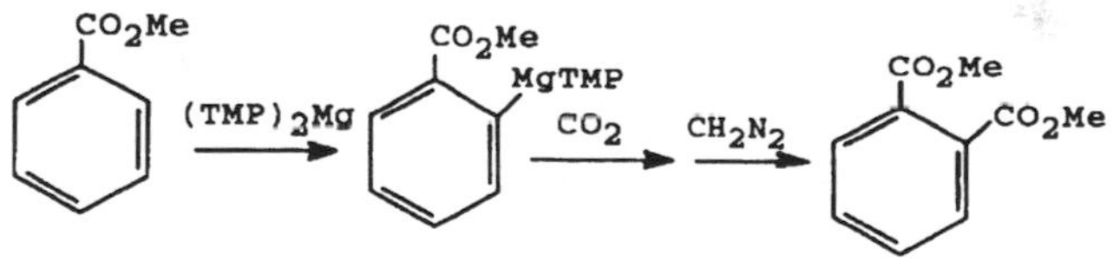

A similar reaction occurred with amide-activated strained ring systems, such as cyclopropane and cubane derivatives:[112,113]

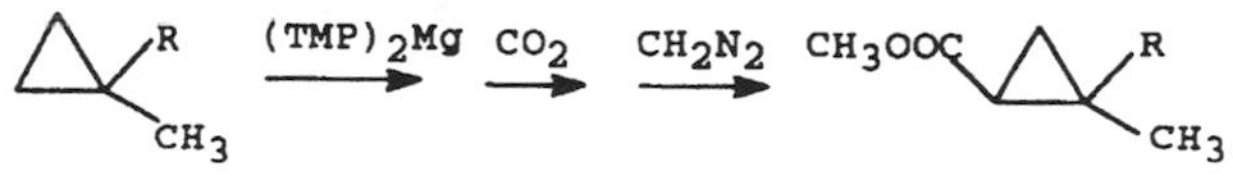

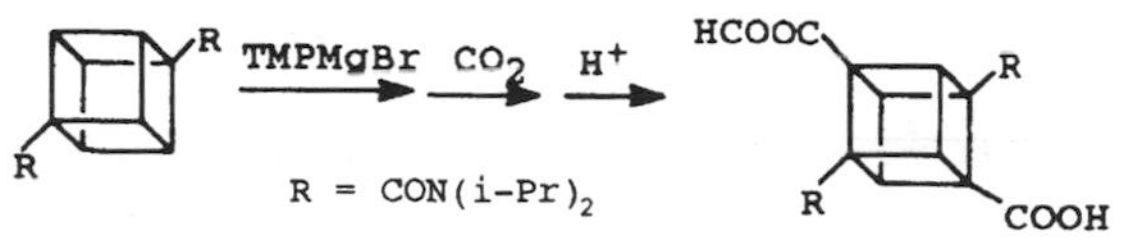

Cubane derivatives are of interest as potential pharmaceuticals and explosives.[114] See also Chapter 11 on the photochemical carboxylation of cubane.

An analogue of the Grignard reaction is the carboxylation of allylbarium halides (which were prepared by *in situ* transmetallation of organo-lithium compounds with BaI_2). This reaction was found by Yanagisawa et al. to be regioselective and stereospecific:[115]

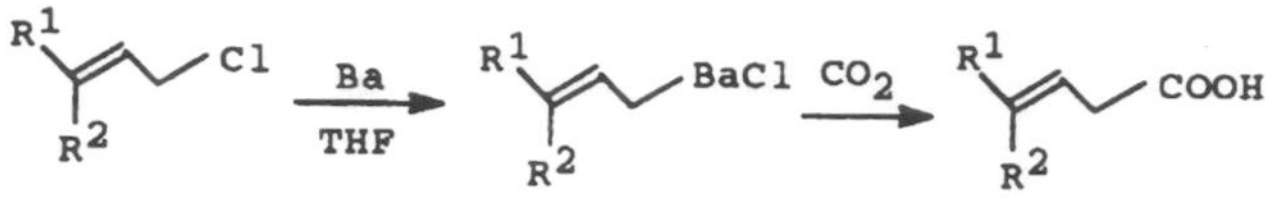

Thus, geranyl chloride was converted in 87% yield to the corresponding β,γ-unsaturated carboxylic acid.

9.4.4   Insertion into Rhenium-Alkoxide and -Arylamide Complexes

The reactivity with CO_2 of a series of rhenium complexes $(CO)_3(L_2)ReOCH_3$, $(CO)_3(L_2)ReOAr$, and $(CO)_3(L_2)ReNHC_6H_5$ (where $L = PMe_3$ or $1,2\text{-}(AsMe_2)_2C_6H_4$), was found by Simpson and Bergman to depend substantially on the nature of the complexes.[116] $(CO)_3(PMe_3)_2ReOCH_3$ in benzene solution at 25°C reacted very rapidly with CO_2, yielding the metallacarbonate insertion product in 95% yield:

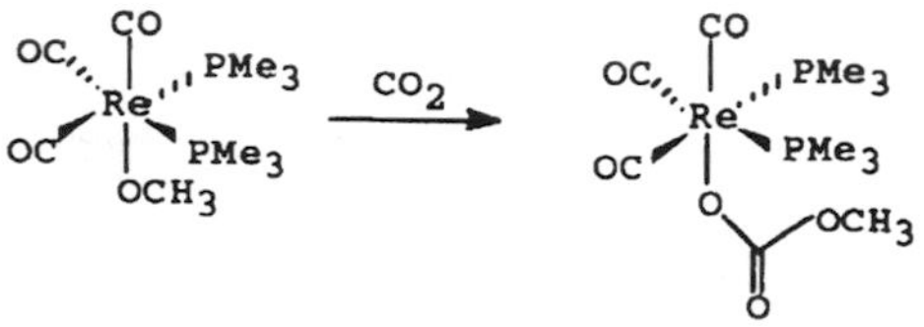

Similar arylamide complexes also reacted with CO_2, while the analogous aryloxide complexes were inert. The CO_2 insertion reaction was proposed to occur by nucleophilic attack of the coordinated methoxide oxygen atom or the arylamide nitrogen atom on the carbon atom of CO_2.[116]

9.4.5   Insertion into the Tungsten(II)-Alkoxide Bond

Buffin et al. prepared a series of new tungsten–alkoxy tricarbonyl complexes containing a tridentate chelating ligand.[117] Heating a CH_2Cl_2 solution of the W-methoxy complex at 40°C for 72 h with CO_2 at 1 atm pressure resulted in the formation of an air-stable solid tungsten(II)-η^1-carbonate complex:

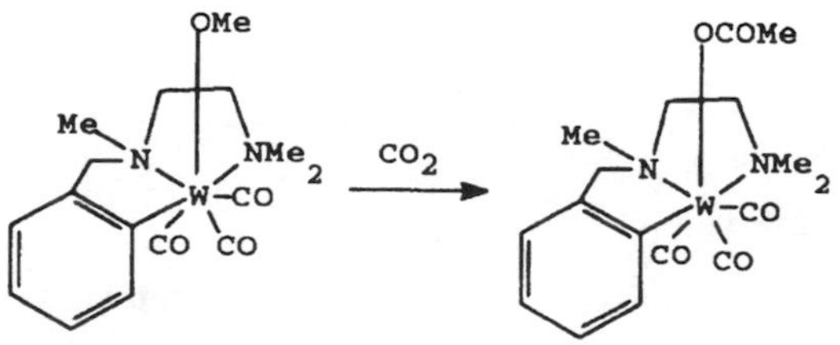

This CO_2 insertion reaction was reversible. Under vacuum, the carbonate complex released CO_2, reforming the W-methoxy complex.[117] Such complexes may be useful for CO_2 recovery.

9.4.6 Insertion into the Tungsten–Nitrogen Bond

The Cp*W(NO) complexes Cp*W(NO)(CH$_2$CMe$_3$)(NHCMe$_3$) and Cp*W(NO)-(NHCMe$_3$)Cl, where Cp* = cyclopentadienyl, were found by Legzdins et al. to undergo CO_2 insertion reactions at the W-amido bond:[118]

$$Cp*W(NO)(CH_2CMe_3)(NHCMe_3) + CO_2$$

$$\rightarrow Cp*W(NO)(\eta^2\text{-}O_2CNHCMe_3)(CH_2CMe_3) \qquad (9.31)$$

$$Cp*W(NO)(NHCMe_3)Cl + CO_2 \rightarrow Cp*W(NO)(\eta^2\text{-}O_2CNHCMe_3)Cl \qquad (9.32)$$

The reactions were performed in benzene solutions with CO_2 at 15 psig.

9.4.7 Ni-Complex-Mediated Carboxylation of Haloarenes and Styrene

Carboxylation of the phenyl group of halobenzene derivatives by CO_2 was performed by Osakada et al. using stoichiometric amounts of nickel-(COD)$_2$ (COD = 1,5-cyclobutadiene) and 2,2'-bipyridine in DMF solution.[119] Thus, bromobenzene was converted into benzoic acid and biphenyl in 55 and 21% yields, respectively. The proposed mechanism involved insertion of CO_2 into the Ni–C bond, forming a carboxylated intermediate, which upon acidification released the carboxylic acid

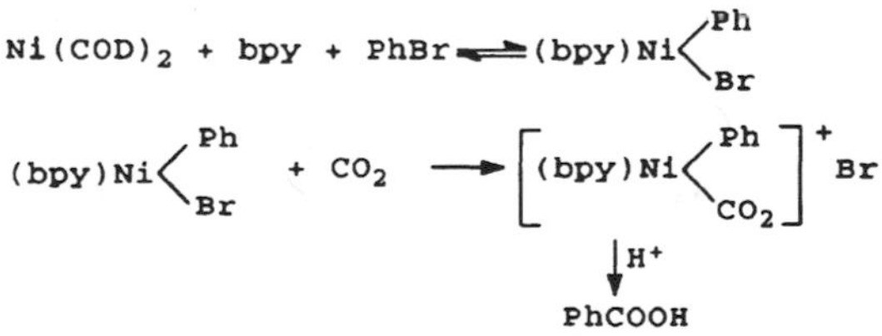

Hoberg et al. achieved the production of cinnamic acid in 70% yield by reacting styrene in THF solution with pressurized CO_2 at 85°C in the presence of the complex (DBU)Ni(0), where DBU is the bicyclic amidine 1,8-diazabicyclo[5.40]undec-7-ene. A five-member nickel(0) complex was the assumed intermediate:[120]

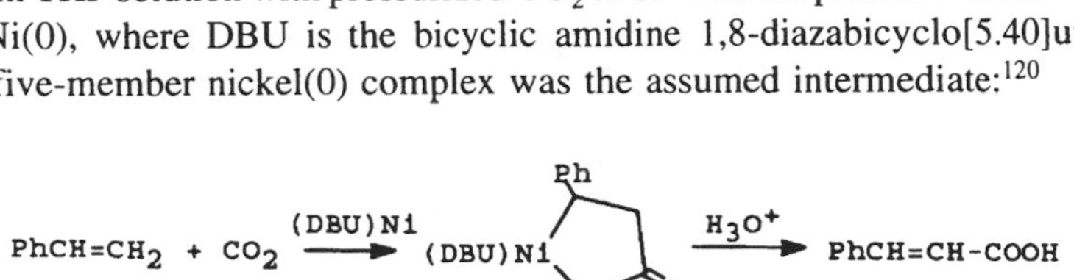

9.4.8   Ring Closure of Dienes to Lactones

Pioneering work in the field of the coupling of CO_2 into organic molecules was done by Sasaki et al. by reacting butadiene at 120°C with carbon dioxide (at 50 atm) in the presence of Pd(0)-phosphine complexes, such as $Pd(Ph_2PCH_2CH_2PPh_2)_2$, in polar aprotic solvents such as DMF.[121,122] The products were mainly butadiene oligomers (about 60%), as well as lactones (12%). In the telomerization of butadiene with CO_2, palladium complexes were found to act as specific CO_2 carriers, such as in the reaction

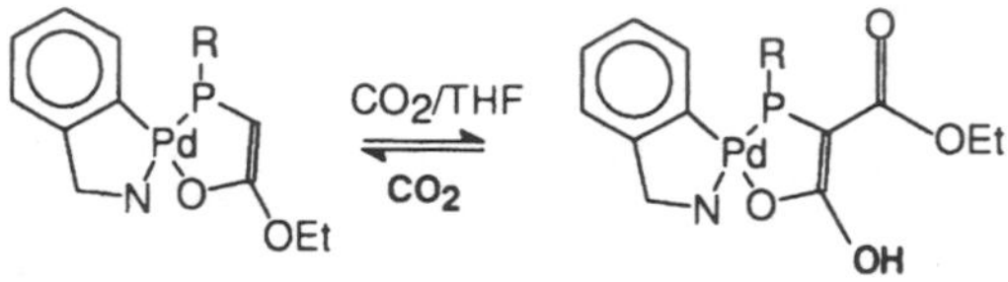

The main product of the telomerization reaction with butadiene was the six-member cyclic lactone[10]

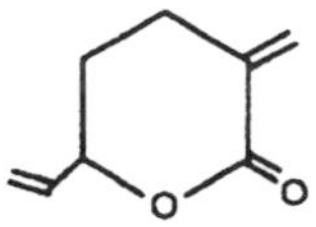

as well as carboxylic acids and esters. This telomerization reaction of butadiene produced very valuable compounds, useful as monomers or intermediates in the manufacture of polyester resins, pesticides, and plasticizers.[7,9]

9.4.9   Allenes and CO_2

In the presence of Rh(diphos)(η^6-BPh_4), where diphos = [bis(diphenylphosphino)-ethane], allene was reported by Aresta et al. to react with CO_2 to form pyrones, linear esters, and oligomers and polymers of allene.[73]

A catalyst generated from a mixture of Pd_2(dibenzylidene acetone), $CHCl_3$, and 1-(2-pyridyl)-2-(di-n-butyl-phosphino)ethane was found by Tsuda et al. to effect the cycloaddition of methoxyallene with pressurized CO_2 in acetonitrile solution at 120°C to produce (E)-5-methoxy-2-(methoxy-methylene)-4-methylene-5-pentanolide in a regio- and stereospecific reaction, in yields of up to 64%:[123]

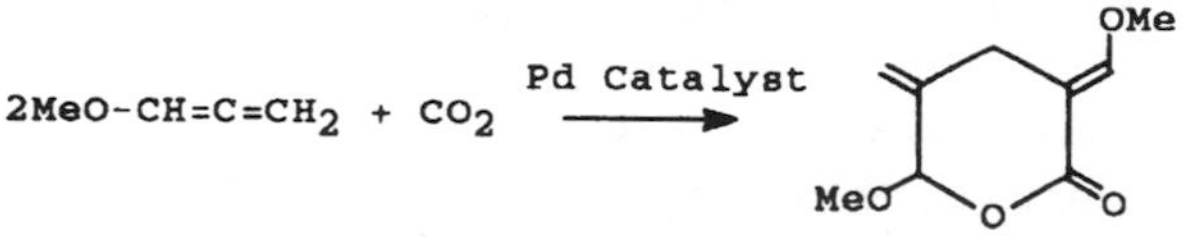

9.4.10   Carboxylation of Dienes

A unique approach for converting dienes to β,γ-unsaturated carboxylic acids was developed by Gao et al. by reacting the diene with Cp_2TiCl_2 in the presence of a Grignard reagent.[124] The Grignard reagent served to reduce the Cp_2TiCl_2 to an intermediate Cp_2TiH species, which added across the diene bond, forming an intermediate η^3-allyltitanium complex. Passing CO_2 through an ethereal solution of this complex and acidification produced the β,γ-unsaturated carboxylic acids. Thus, butadiene was carboxylated to 2-methyl-3-butenoic acid in 82% yield:

```
Cp2TiCl2 + RMgX -> Cp2TiH
Cp2TiH + CH2=CH-CH=CH2 -> Cp2(allyl)Ti-complex
                                          ↙ CO2
                     CH2=CH-CH(CH3)-COOH
```

This reaction was also applicable to cyclic dienes. Thus, cyclopentadiene was converted to 2-cyclopentene-1-carboxylic acid.[124]

An interesting method for the fixation of carbon dioxide into cyclopentadiene was discovered by Haruki et al.[125,126] In the presence of 1,8-diazabicyclo[5.4.0]undec-7-ene (DBU) and carbon dioxide (at up to 50 kg cm^{-2} pressure) in dry DMF solution, cyclopentadiene (or its mono-alkyl-substituted derivatives) was converted to dicarboxylated compounds, such as 1,3-dicarboxy cyclopentadiene:

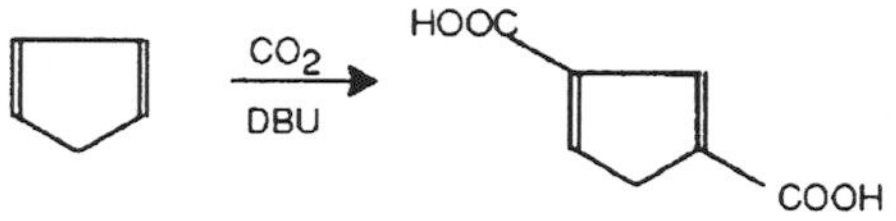

9.4.11 Carboxylation of Acetylenic Compounds

The conversion of propyne (methylacetylene) to 4,6-dimethyl-2-pyrone was performed by Albano and Aresta in acetonitrile solution with CO_2 (1 MPa) at 390 K in the presence of Rh(diphos)(η^6-BPh$_4$), where diphos = Ph$_2$PCH$_2$CH$_2$PPh$_2$.[127] A substantial fraction of the propyne underwent oligomerization with formation of trimethylbenzenes, as well as polymerization products. In more recent work, the cycloaddition of CO_2 to propyne to form 4,6-dimethyl-2-pyrone, as well as 1,3,5- and 1,2,4-trimethylbenzene was achieved by Pillai et al. using Ni, Pd, and Rh metal complexes as catalysts:[128,129]

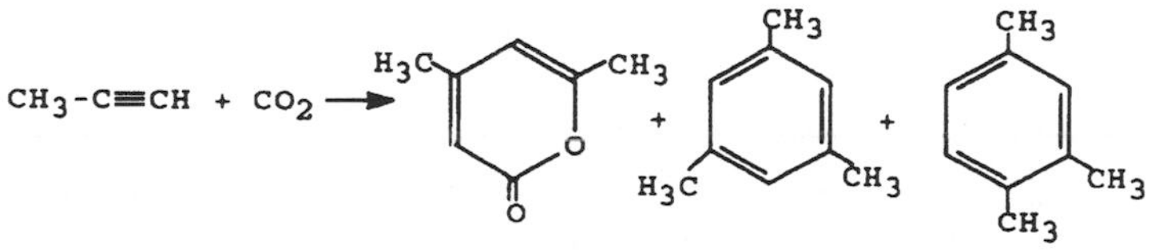

The pyrone was obtained by CO_2 incorporation into a propyne dimer, forming a cycloaddition product. Active catalysts for pyrone formation were Rh$_4$(CO)$_{12}$ and various bimetallic carbonyl clusters, such as (BzMe$_3$N)$_2$Fe$_2$Rh$_4$(CO)$_{16}$, impregnated on mildly oxidized supports. The order of activity of the supports tested was γ-Al$_2$O$_3$ > ZrO$_2$ > TiO$_2$. The process was performed at 130°C for 10 h in a stainless-steel autoclave, with a propyne/CO_2 ratio of 1:8, at a total pressure of 60 kg cm^{-2}. In the absence of CO_2, only the trimethylbenzenes were produced. FTIR measurements indicated that a key intermediate in the formation of the pyrone may be a monodentate carbonate adsorbed on the catalyst surface.[128,129]

Regioselective synthesis of O-1-(1,3-dienyl)carbamates was obtained by Höfer et al. by the addition of secondary amines and CO_2 to isopropenylacetylene in the presence of a ruthenium complex catalyst precursor containing a chelating bidentate phosphine ligand, such as $[Ph_2P(CH_2)_nPPh_2]Ru[\eta^3-CH_2=C(Me)CH_2]_2$, where n = 1 to 4:[130]

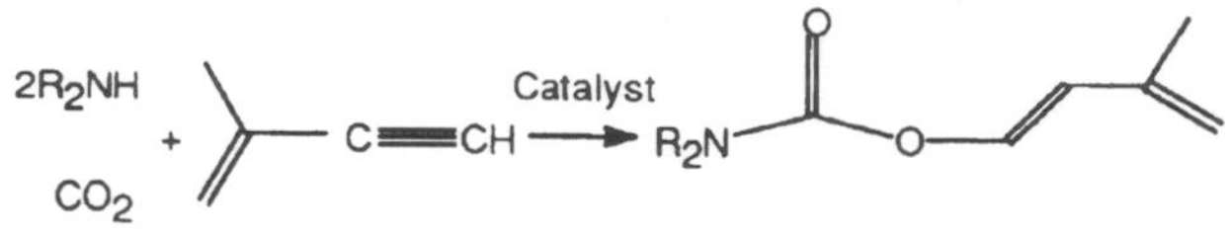

When n = 2, with the secondary amines morpholine, pyrrolidine, piperidine, and diethylamine, the yields of the O-1-(1,3-dienyl)carbamates were 62, 50, 36, and 31%, respectively. Such dienyl carbamates are very useful as substrates for the Diels–Alder reaction and as precursors for polymers and fibers.[130]

The reaction of alkynes with CO_2 to pyrone derivatives was found by Inoue et al. to be catalyzed by nickel(0) phosphine complexes.[131] Other products were cyclotrimers and -tetramers of the alkynes. Highest yields of the pyrones were obtained with $Ni(COD)_2$-$Ph_2P(CH_2)_4PPh_2$ (COD = 1,5-cyclooctadiene) as catalyst, in benzene solution, in a reaction at 120°C for 20 h, with CO_2 at 50 kg cm^{-2}. With the alkynes 3-hexyne and 4-octyne, the yields of the resulting tetraethyl- and tetrapropyl-2-pyrones were 57 and 60%, respectively. In a proposed mechanism, the reaction of two molecules of the alkyne with the Ni(0) complex led to a nickelacyclopentadiene intermediate, which underwent CO_2 insertion, forming a carboxylated complex. Intramolecular C–O coupling then released the Ni(0) complex, producing the pyrone[131]

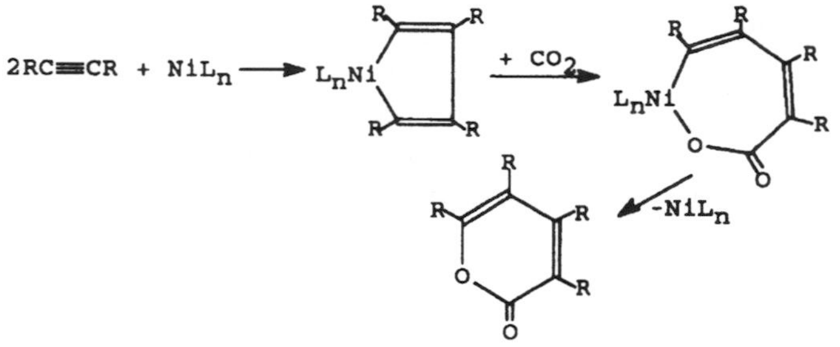

Walther et al. investigated the cyclooligomerization of CO_2 with hex-3-yne to tetraethyl-2-pyrone, which was homogeneously catalyzed by the Ni0(L) complexes Ni(TMED) and Ni(COD), where TMED = tetramethylethylene diamine and COD = cycloocta-1,5-diene.[132,133] The reaction involved the intermediate formation of a five-member metallacycle containing the CO_2 group, followed by incorporation of a second alkyne molecule:

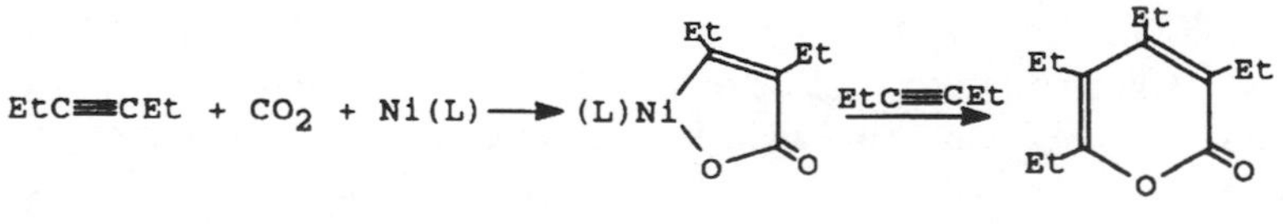

The yields of tetraethyl-2-pyrone using either Ni(TMED) or Ni(COD) were 72 and 92%, respectively. Optimal conditions for this synthesis were with the catalyst system (cyclohexyl)$_3$P-MeCN-Ni(COD)$_2$, in which MeCN acted as a bridging ligand, connected with its triple bond to one Ni center. In the reaction of CO_2 with hex-3-yne catalyzed by Ni(TMED), a side reaction also resulted in the production of CO, formed by the reduction of CO_2.[134]

9.4.12   Diynes and CO₂

Tsuda et al. achieved the nickel(0)-catalyzed fixation of carbon dioxide into diynes such as 1,6-heptadiyne or 1,7-octadiyne. Intramolecular cyclization of the diynes led to bicyclic α-pyrones fused with five- and six-member carbocycles:[135,136]

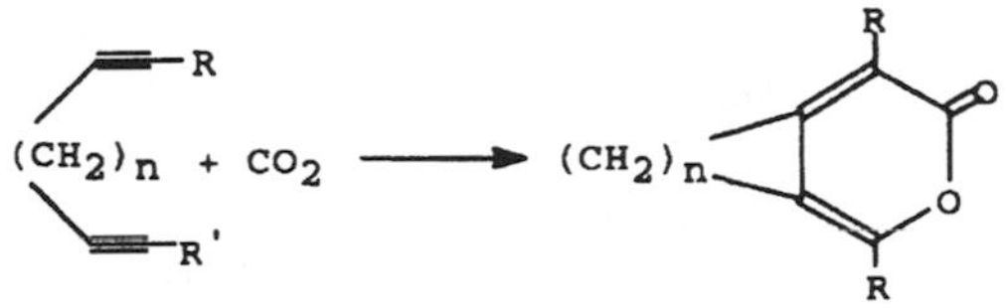

Active catalysts were Ni(0)-phosphine complexes, such as bis(cyclooctadiene)nickel-bis-(1,4-diphenylphosphino)butane. Bicyclic α-pyrones are important as synthetic intermediates.[135,136]

9.4.13   Copolymerization of Diynes and CO₂ to Polypyrones

Alternating copolymerization, in which carbon dioxide serves as a comonomer, has been achieved by Tsuda by intermolecular cyclization of acyclic or cyclic diynes, in the presence of nickel(0) catalysts, producing poly-(2-pyrones). 3,11-Tetradecadiyne was copolymerized with carbon dioxide under pressure in a solvent mixture of THF–MeCN at 110°C in the presence of the Ni(0) catalyst obtained from Ni(COD)$_2$ and a tertiary phosphine, such as PEt$_3$.[137,138]

A nickel(0) catalyst was also used by Tsuda et al. to perform the cycloaddition copolymerization of cyclic diynes with CO_2 to produce ladder poly(2-pyrone)s.[139] This reaction was performed with the diynes 1,7-cyclotridecadiyne, 1,7-cyclotetradecadiyne, and 1,8-cyclopentadecadiyne with CO_2 (50 kg cm^{-2}) in a THF–MeCN solution in the presence of Ni(COD)$_2$-tri-n-octylphosphine at 60°C for 20 h. The process with 1,7-cyclotridecadiyne may be represented by

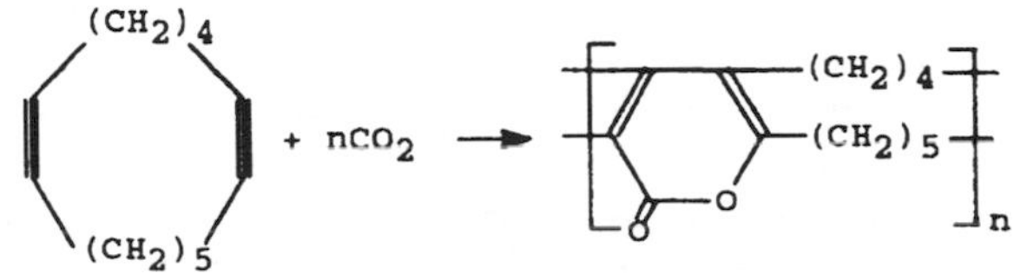

and resulted in 99% yield of a CH$_2$Cl$_2$-soluble copolymer, with molecular weight M_n = 6400 and M_w/M_n = 2.6.[139]

With ether diynes, the Ni(0)-catalyzed cycloaddition copolymerization of CO_2 resulted in poly(2-pyrone)s containing an oxyalkylene chain. The process was carried out by Tsuda et al. with the same Ni(COD)$_2$-tri-*n*-octylphosphine catalyst and 20 kg cm^{-2} CO_2 in THF–MeCN for 20 h.[140] Thus in the reaction

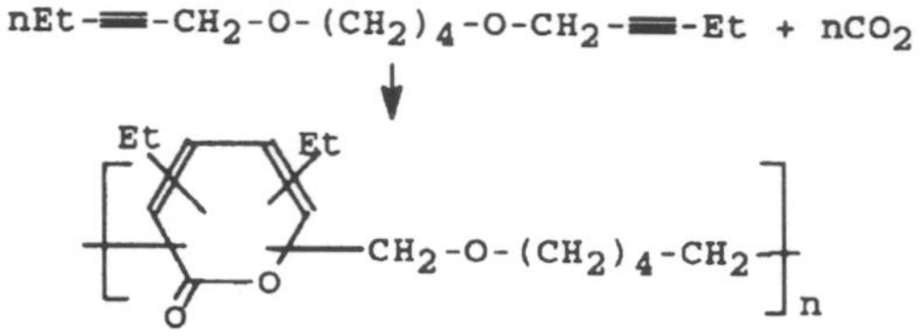
nEt-≡-CH2-O-(CH2)4-O-CH2-≡-Et + nCO2
Et
Et
CH2-O-(CH2)4-CH2
O
O
n

performed at 110°C, the product polymer was obtained in 53% yield, with a molecular weight M_n = 15,800 and M_w/M_n = 2.2.

The above CO_2 copolymerization reactions all required transition metal catalysts. Spontaneous CO_2 copolymerization without catalyst would be preferable. Tsuda et al. discovered such a reaction with a bis(ynamine), 1,4-bis(N,N-diethylaminoethynyl)-benzene.[141] This compound underwent spontaneous 1:1 copolymerization with CO_2 to produce in high-yield poly(4-pyrone)s with a molecular weight M_n > 10,000:

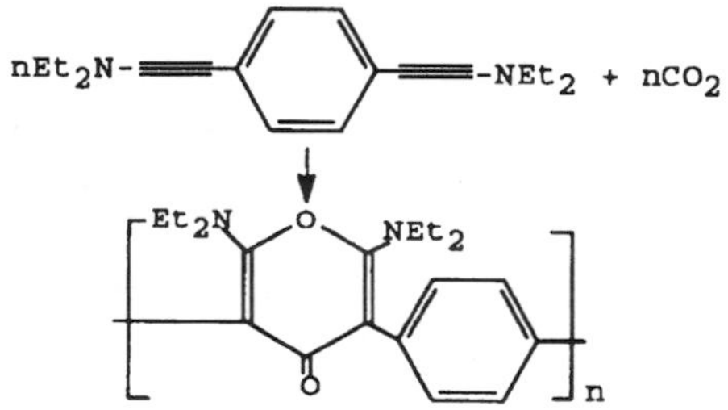
nEt2N-≡
≡-NEt2 + nCO2
Et2N
O
NEt2
O
n

The reaction was carried out at 60 to 90°C with CO_2 (50 kg cm^{-2}) in THF solution. Spectroscopic evidence indicated the presence of the 4-pyrone ring structure in the polymer.[141] The mechanism of formation of the 4-pyrone ring is not clear.

Several of the poly(2-pyrones) were reported by Tsuda to have remarkably good thermal stability.[13] Thermogravimetric analysis of the poly(2-pyrones) obtained from 3,11-tetradecadiyne, 1,7-cyclotridecadiyne, and 1,3- and 1,4-di(2-hexynyl)benzenes suffered weight loss under nitrogen only at about 420°C.

9.4.14 Friedel–Crafts Reaction

An early example of the insertion of carbon dioxide into a C–H bond was the Friedel–Crafts reaction of benzene with anhydrous $AlCl_3$, resulting in benzoic acid. The reaction was preferably carried out at pressures of 50 to 60 atm and temperatures of 80 to 150°C. The carboxylic acid was released by hydrolysis:[142]

$$C_6H_5Al_2Cl_5 + CO_2 \rightarrow C_6H_5CO_2Al_2Cl_5 \qquad (9.33)$$

This reaction could potentially be brought to a practical process:

$$C_6H_5CO_2Al_2Cl_5 + H_2O \rightarrow C_6H_5COOH + Al_2Cl_5OH \qquad (9.34)$$

9.4.15 Carboxylation of Methane to Acetic Acid

Fujiwara et al. discovered the palladium-catalyzed synthesis of acetic acid from methane and CO$_2$.[143,144] The reaction was performed by heating in an autoclave methane (40 atm) and CO$_2$ (20 atm) in trifluoroacetic acid solution containing Pd-acetate, Cu-acetate, and K$_2$S$_2$O$_8$ for 20 h at 80°C. Acetic acid was produced in 1650% yield, based on the Pd used. The mechanism of the reaction is unclear.

$$CH_4 + CO_2 \rightarrow CH_3COOH \qquad (9.35)$$

An alternative synthesis of acetic acid from CH$_4$ and CO$_2$ was described by Fujiwara et al.[145] Methane (5 atm) and CO$_2$ (20 atm) were charged into a stainless-steel autoclave containing a vanadium catalyst, VO(acac)$_2$/K$_2$S$_2$O$_8$/CF$_3$COOH, and heated at 80°C for 20 h, producing acetic acid in 97% yield (based on the CH$_4$). This reaction could potentially be brought to a practical process.

9.4.16 Dimethyl Acetylene Dicarboxylate–P^{III} Compounds with CO$_2$

The 1:1 addition compound of triphenylphosphine with dimethyl acetylene-dicarboxylate reacted with carbon dioxide to form a mixture of two carboxylate betains:[146]

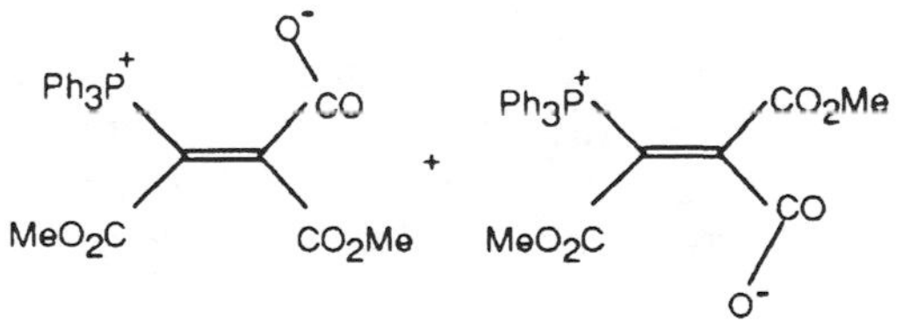

With trialkyl phosphite instead of triphenylphosphine, a different reaction took place with dimethyl acetylenedicarboxylate and carbon dioxide, producing a non-ionic cyclic phosphorane:[147]

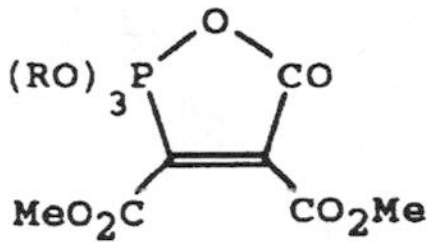

9.4.17 Insertion into Zr–H and Zr–C Bonds

Carbon dioxide reacted with hydrido-chloro-bis(cyclopentadienyl)–zirconium(IV), cp$_2$Zr(H)(Cl), by insertion into the Z–H bond, followed by release of formaldehyde:

$$cp_2Zr(H)(Cl) + CO_2 \rightarrow cp_2(Cl)Zr-O-Zr(Cl)cp_2 + HCHO \qquad (9.36)$$

In the presence of an excess of the starting complex, the formaldehyde was reduced and trapped as a methoxy ligand:[148]

$$cp_2Zr(H)(Cl) + HCHO \rightarrow cp_2Zr(OCH_3)(Cl) \qquad (9.37)$$

The insertion of CO_2 into Zr–C(Me) bonds was carried out by Kloppenburg and Petersen. The addition of two equivalents of CO_2 into $[(\eta^5\text{-}C_5Me_4)[SiMe_2(N\text{-}t\text{-}Bu)]ZrMe_2$ resulted in the formation of $\{[(\eta^5\text{-}C_5Me_4)[SiMe_2(N\text{-}t\text{-}Bu)]Zr(\eta^2\text{-}O_2CMe)(\mu\text{-}O_2CMe)\}_2$.[149]

9.4.18   Insertion into Ni–C Bond

Walther et al. observed that atmospheric pressure carbon dioxide in THF solution at –10°C reacted with 2,2′-dipyridyl-cyclooctadiene-1,5-nickel(0) [(Dipy)Ni(COD)] and bis-cyclohexyl-carbodiimide (Cy–N=C=N–Cy) to form a five-member nickel(0) chelate:[150]

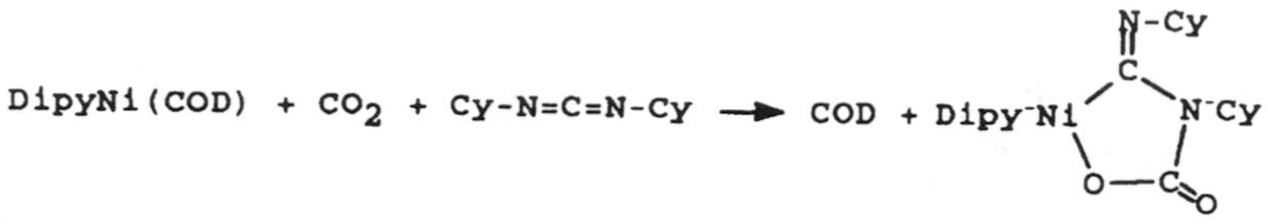

9.4.19   Insertion into Rh–Os Bond

Reduction of carbon dioxide by incorporation into a bimetallic polyhydride complex was carried out by Lundquist et al. using the rhodium–osmium complex $(COD)RhH_3$–OsP_3, where COD = 1,5-cyclooctadiene and $P_3 = (PMe_2Ph)_3$.[151] This complex in THF solution reacted completely with an excess of CO_2 (1 atm) within 8 h at 25°C, producing two complexes, as described by the stoichiometry

$$2(COD)RhH_3OsP_3 + 2CO_2 \rightarrow (COD)_2Rh_2OsP_3H_2CO_2$$

$$+ H_2Os(CO)P_3 + H_2O \qquad (9.38)$$

In the one product, $H_2Os(CO)P_3$, CO_2 had been reduced to the CO moiety. The other product, $(COD)_2Rh_2OsP_3H_2CO_2$, is a unique example of a neutral compound containing hydride as well as CO_2 ligands.[151]

9.4.20   Carboxylation of Lithiated Benzyl Alcohols

Lithium substitution at the ortho position on benzyl alcohols, followed by carboxylation with CO_2, enabled the preparation of phthalides, which are key intermediates in the synthesis of various natural products. Orito et al. caused direct ortho-lithiation of benzyl alcohol and alkoxy-substituted benzyl alcohols by reacting them at room temperature with *n*-butyl lithium in dry THF, quenching the resulting aryl lithium compound by bubbling through CO_2, and acidification with aqueous HCl, thus producing the corresponding phthalide:[152]

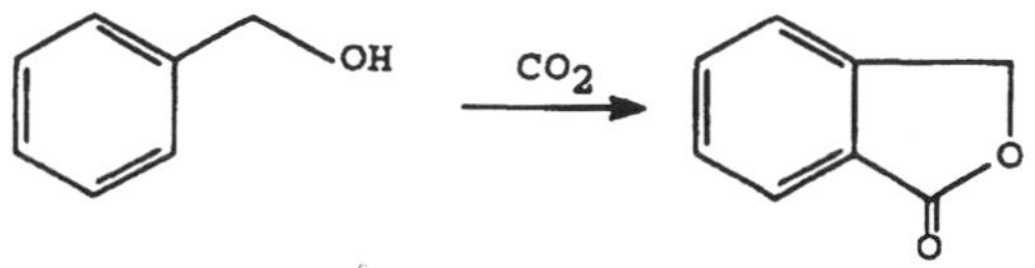

Phthalide

With some substituted benzyl alcohols, the direct lithiation occurred with low yields, and improved lithiation was achieved using *o*-bromobenzyl alcohols. Thus, for the synthesis of the phthalide-isoquinoline alcaloid bicuculline, which is active as a competitive γ-aminobutyric acid antagonist, lithiation of the *o*-bromobenzyl alcohol precursor was carried out at −78°C, followed by carboxylation and acidification:[152]

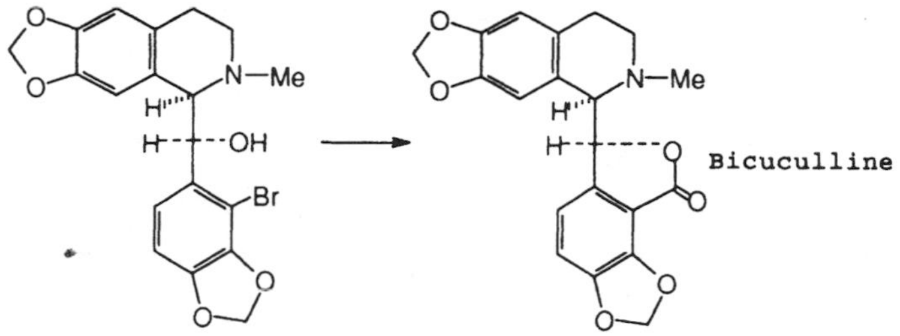

Bicuculline

Lithium derivatives of (η^6-alkylarene)tricarbonylchromium complexes were used by Kalinin et al. to achieve regioselective carboxylation with CO_2 at the benzylic position.[153] The reaction was carried out by adding a $Cr(CO)_3$ complex of the alkylarene to a solution of Et_2NLi, followed by pouring the mixture onto an excess of solid CO_2. Acidification released as products the 2-arylcarboxylic acids:

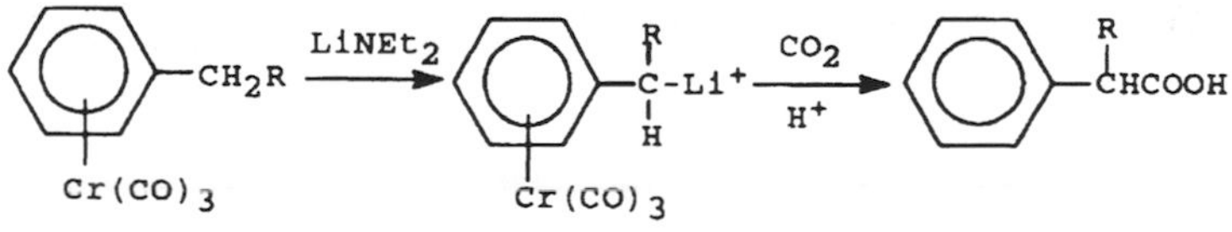

Several of these compounds are useful as nonsteroidal anti-inflammatory agents and analgesics and also serve as intermediates in the synthesis of C-norbenzomorphans.

See also Chapter 6 on the biomimetic carboxylation of active methylene compounds such as fluorene by CO_2 in the presence of K_2CO_3 and either 1,3-diphenylurea, or diphenylcarbodiimide, or an anilide (acetanilide or formanilide).

9.4.21 Ethylene Carboxylation in a Gas–Solid Reaction

An interesting coupling reaction between ethylene and CO_2 was discovered by Llorca et al. by contacting a 1:1 mixture of these gases (total pressure 42 bar) at 393

K over a platinum–tin bimetallic complex, cis-[PtCl(SnCl$_3$)(PPh$_3$)$_2$], supported on aerosol-type silica.[154] After the reaction, the product was eluted with a water–methanol mixture (1:1), yielding methyl 3-hydroxypropionate as the only detected product:

$$CH_2=CH_2 + CO_2 \rightarrow \text{Unknown intermediate} \rightarrow HOCH_2CH_2COOMe \quad (9.39)$$

The nature of the initial unknown intermediate (before the H_2O–MeOH workup) was studied by IR spectroscopy. An absorption band at 2339 cm^{-1} was assigned to CO_2 linearly coordinated to the bimetallic complex. Bands at 3098, 3076, 3008, and 2987 cm^{-1} could be assigned to an interaction between ethylene and the bimetallic complex. The simultaneous coordination of CO_2 and ethylene on the supported Pt–Sn complex was proposed to facilitate the coupling of these molecules to form the unknown intermediate.[154]

In a modification of the above work, Llorca et al. prepared silica-supported PtSn alloy catalysts, by reacting cis-[PtCl$_2$(PPh$_3$)$_2$] and SnCl$_2$ with the OH groups on the surface of partially dehydrated aerosil-type silica, followed by reduction.[155,156] With these Pt(PPh$_3$)–Sn/Si catalysts, which were shown to have a single well-defined PtSn phase, the reaction of CO_2 with ethylene in the presence of either H_2 or H_2O resulted in the production of lactic acid, as well as ethane and CO. The primary reaction in the presence of hydrogen was the reverse water–gas–shift reaction:

$$CO_2 + H_2 \rightarrow CO + H_2O \quad (9.40)$$

In a subsequent step, ethylene with CO_2 and water then produced lactic acid:

$$CO_2 + C_2H_4 + H_2O \rightarrow CH_3CHOHCOOH \quad (9.41)$$

Optimal conditions for the production of lactic acid were achieved under continuous flow over the Pt(PPh$_3$)–Sn/Si catalyst (3 wt% Pt), using a molar ratio CO_2:C_2H_4:H_2O of 1:1:1, a (flow rate)/(catalyst weight) ratio of 140 mL min^{-1} g cat^{-1}, a temperature of 423 K, and a total pressure of 35 bar, resulting in a production rate of 109 μmol min^{-1} g cat^{-1} of lactic acid.[155,156]

9.4.22　CO_2 Metathetical Exchange with Ge and Sn Bisamides

A unique activation of CO_2 by its conversion to an isocyanate and a carbodiimide was discovered by Sita et al. by reacting CO_2 at 60 psi and 25°C with pentane solutions of the divalent group 14 bisamides M[N(SiMe$_3$)$_2$]$_2$, where M = Ge or Sn.[157] In a metathetical exchange reaction, one of the oxygen atoms of the CO_2 was replaced by one of the imido groups Me$_3$SiN in the above bisamides, producing the isocyanate Me$_3$SiN=C=O. In another parallel reaction, both oxygen atoms of the CO_2 were replaced by two imido groups Me$_3$SiN, producing the carbodiimide Me$_3$SiN=C=NSiMe$_3$. The Ge and Sn compounds were converted into dimeric bisalkoxide complexes:

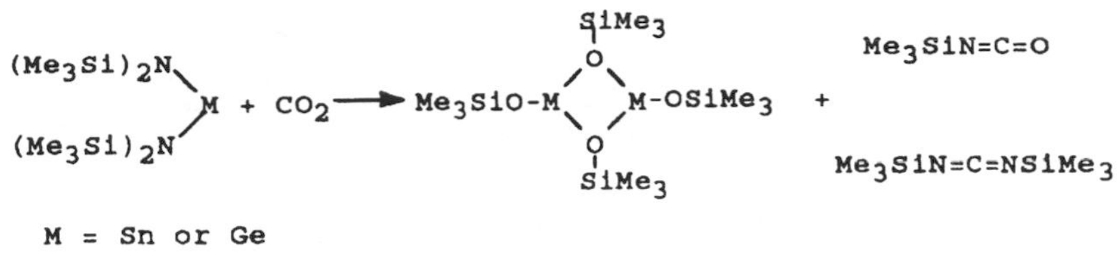

$$\begin{array}{c}
(Me_3Si)_2N \\
\diagdown \\
\hspace{2.5cm} M + CO_2 \longrightarrow Me_3SiO-M \underset{\underset{SiMe_3}{|}}{\overset{\overset{SiMe_3}{|}}{\overset{O}{\diagup\diagdown}}} M-OSiMe_3 + \\
\diagup \\
(Me_3Si)_2N
\end{array}$$

$$Me_3SiN=C=O$$

$$Me_3SiN=C=NSiMe_3$$

M = Sn or Ge

The driving force for these reactions is presumably the stronger metal–oxygen bond of the dimeric complex relative to the metal–nitrogen bond of the starting bisamide. The reaction with the Sn bisamide was essentially complete and quantitative within 10 min, yielding the isocyanate and the carbodiimide in a 4:1 ratio. With the Ge bisamide, 12 h was required for a 72% yield of the corresponding dimer. Both isocyanates and carbodiimides are very useful and highly reactive reagents. If a method could be developed for recycling the above dimeric bisalkoxide complexes to the starting bisamides, these reactions could become of practical use.[157]

9.4.23   Insertion into the Sn–C Bond

The carboxylation of allylstannanes with CO$_2$ catalyzed by Pd(0) complexes was discovered by Shi and Nicholas.[158] Effective catalysts were Pd(Ph$_3$P)$_4$ and Pd(Bu$_3$P)$_4$. Thus, allyltributyltin in dry THF solution in the presence of Pd(Ph$_3$P)$_4$ with CO$_2$ at 33 atm pressure in an autoclave was heated at 70°C for 24 h, yielding a mixture of the two isomeric organotin carboxylates:

$$Bu_3Sn-CH_2-CH=CH_2 + CO_2 \rightarrow Bu_3Sn-O-C(O)-CH_2-CH=CH_2$$

$$+ Bu_3Sn-O-C(O)-CH=CH-CH_3 \hspace{2cm} (9.42)$$

The mechanism involved insertion of CO$_2$ into the Sn–C bond of the allylstannane. With diallyldibutyltin and tetraallyltin, the same Pd(0)-catalyzed carboxylation yielded mixtures of allyltin dicarboxylates and allyltin tetracarboxylates, respectively.[158]

9.5   REVERSIBLE CO$_2$ CARRIERS

9.5.1   Cu and Zn Complexes

In a search for a reversible CO$_2$ carrier, Tsuda et al. discovered that Cu(I)phenyl-acetylide–tertiary phosphine complexes achieve reversible CO$_2$ fixation.[159] Such complexes are useful for transferring CO$_2$ to other compounds. Thus, with phenylacetylide-tri-n-butylphosphine, the reaction

$$PhC\equiv CCO_2Cu\text{-}(n\text{-}Bu_3P) \rightleftharpoons PhC\equiv CCu\text{-}(n\text{-}Bu_3P) + CO_2$$

was reversible. The equilibrium occurred readily at ambient temperature (28°C) and ordinary CO$_2$ pressure in DMF solution. The observed equilibrium values of unchanged CO_2 gas using the P^{III} ligands Bu_3P, Ph_3P, and $(MeO_3)P$ were 67, 83, and 89%, respectively. The equilibrium could be shifted by changing the temperature. Other examples of reversible carbon dioxide carriers were provided by Tsuda et al. with the cyanoacetate–phosphine complexes.[160] In the presence of n-Bu_3P as ligand, Cu(I)cyanoacetate in DMF solution underwent reversible decarboxylation,

$$NCCH_2CO_2Cu \cdot (n\text{-}Bu_3P)_x \; \rightleftharpoons \; NCCH_2Cu \cdot (n\text{-}Bu_3P)_x + CO_2$$

$$(n = 1, 2, \text{ or } 3) \qquad\qquad (9.43)$$

Ligands with higher σ-donating ability caused a shift in this equilibrium toward CO_2 fixation. The equilibrium values of the evolved CO_2 gas at 50°C in DMF solutions for $(PhO)_3P$, $(MeO)_3P$, Ph_3P, Et_3P, and n-Bu_3P ligands were 62, 55, 25, 19, and 9.8%, respectively. These carboxylation reactions involved reversible insertion of CO_2 into transition metal bonds. The $NCCH_2CO_2Cu \cdot (n\text{-}Bu_3P)_x$ complex acted as a reversible CO_2 carrier, enabling transcarboxylation to active methylene compounds. With cyclohexanone in DMF solution at 50°C, the reaction (after treatment with allyl bromide) produced allyl 1-allyl-2-oxocyclohexanone in up to 87% yield:[160]

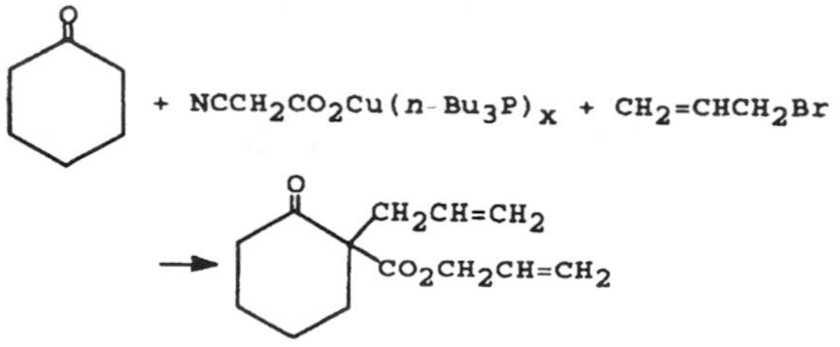

When exposing dichloromethane solutions of CuCl and various amines to an O_2/CO_2 atmosphere, Churchill et al. observed the formation of μ-carbonato dinuclear copper(II) complexes. One such complex, which was characterized in detail, was μ-carbonato-dichlorobis(N,N,N′,N′-tetramethyl-1,3-propanediamine)dicopper(II).[161]

The interest in air-stable CO_2-fixing complexes led Kitajima et al. to the search for other related complexes.[162,163] One binuclear copper complex which is both air stable and capable in toluene solution of fixing CO_2 from its low concentration in air is $\{Cu[HB(3,5\text{-}Pr_2^ipz)_3]\}_2(OH)_2$ (where $HB(3,5\text{-}Pr_2^ipz)_3$ = hydrotris-(3,5-di-*iso*propylpyrazol-1-yl)borate). The CO_2 binding was explained by nucleophilic attack of the hydroxo group of the initial complex on the electrophilic carbon of CO_2, followed by release of water:

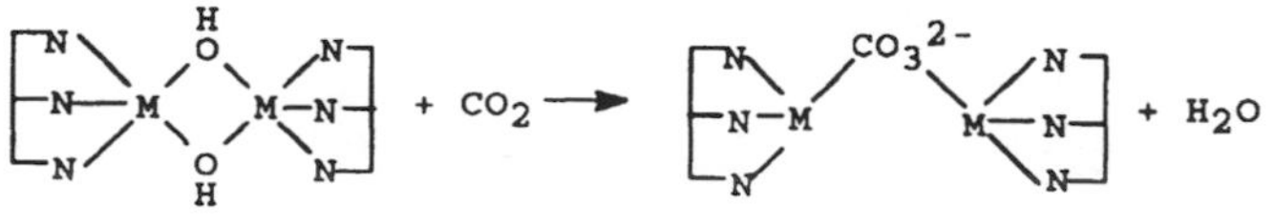

M = Mn, Fe, Co, Ni, Cu

In a comparison of the CO_2-fixing ability of hydroxo complexes of a series of first-row divalent metal ions, the order of activity was Zn > Cu > Ni ~ Co > Mn > Fe. All these complexes contained the sterically strained tris(pyrazolyl)borate ligand, $HB(3,5-Pr_2^i pz)_3$. While the Zn complex was monomeric, all the hydroxo complexes of the other metal ions had a dinuclear structure, bridged by a bis(hydroxo) unit. The Zn, Co, Ni, and Cu complexes were stable against oxygen. However, the Fe and Mn complexes were oxygen sensitive. The carbonato complexes formed by CO_2 fixation could be recycled to the hydroxo complexes by treating their toluene solutions under argon with 1 N NaOH for several hours. Thus the overall CO_2 fixation reaction may be represented by:

$$CO_2 + 2NaOH \rightarrow Na_2CO_3 + H_2O \qquad (9.44)$$

However, the authors stress that more effective catalytic systems will be required to provide a practical process. This may be achieved by the design of suitable ligands.[162,163]

9.5.2 Ni Complexes

Nickel(0)-1-azadiene complexes were found by Walther et al. to act as reversible CO_2 carriers by forming metallacyclic carbamato complexes of Ni(II).[164,165] These enabled the carboxylation of active C–H bonds, such as those of acetophenone to benzoylacetic acid. Among several complexes tested, the most active was [Ni(1-azadiene)]₂, in which the 1-azadiene had the structure

$$CH_3-N(CH_2CH_2-N=CH-CH=CH-C_6H_5)_2 \qquad (9.45)$$

This complex in oxygen-free THF solution absorbed CO_2 already at 25°C, incorporating 1 mol of CO_2 per mole of nickel. At 50°C, the CO_2 was slowly released, reforming the starting complex. The IR absorption spectrum of the CO_2 addition complex indicated the presence of the N–COOH carbamato bonding. Acetophenone was carboxylated by this complex to benzoylacetic acid in 15% yield.[164,165]

Ito and Takita developed a bis(hydroxy) dinuclear Ni(II) complex,

$$[TPANi(II)(\mu-OH)_2Ni(II)TPA](ClO_4)_2$$

where TPA = tris(pyridylmethyl)amine, which in methanol solution reacted readily with atmospheric CO_2 at room temperature, changing its color from blue to purple, while being transformed to a carbonate complex. This reaction was unaffected by the presence of oxygen. The ligated carbonate could be released by reacting with 0.1 N NaOH, reforming the bis(hydroxy) dinuclear Ni(II) complex.[166]

The nickel(II) complex of N,N-dimethyl-ethylenediamine, $Me_2N-CH_2-CH_2-NH_2$,

$$[Ni(Me_2N-CH_2-CH_2-NH_2)_3](ClO_4)_2$$

in aqueous ethanol solution was found by Tanase et al. to achieve the spontaneous fixation of carbon dioxide from air, forming the trinuclear nickel(II) complex containing three octahedral Ni(II) atoms:[167]

$$[Ni_3(Me_2N-CH_2-CH_2-NH_2)_6(CO_3)(H_2O)_4](ClO_4)_4 \qquad (9.46)$$

From crystal structure data, the carbonate ligand was shown to be joined by a hydrogen bonding network, which stabilized a bidentate carbonate. The carbonated complex had IR absorption peaks at 1626 and 1377 cm^{-1}, corresponding to the v_3 vibration of the CO_3^{2-} ion. The carbon dioxide could be released by passing nitrogen gas through a solution of the carbonated complex containing $Me_2N-CH_2-CH_2-NH_2$ and $1\ N\ ClO_4^-$ for several hours at 60°C, thus recovering the carbonate-free Ni(II) starting complex.[167]

9.5.3   Br–Mg Imido Complexes

A variety of bromomagnesium imido complexes were found by Matsumura et al. to act as effective carbon dioxide carriers, which are useful for the fixation of carbon dioxide and its transfer to active methylene compounds,[168–170] (a) the 2-morpholino imidazolino Mg(II) complex,[168] (b) the N,N-dicyclohexylamidinide Mg(II) complex,[169] and (c) the bromomagnesium thioureide complex:[170]

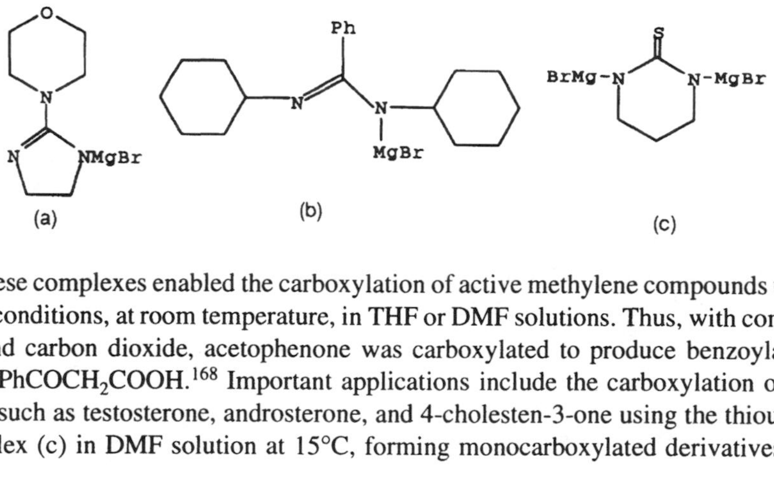

These complexes enabled the carboxylation of active methylene compounds under mild conditions, at room temperature, in THF or DMF solutions. Thus, with complex (a) and carbon dioxide, acetophenone was carboxylated to produce benzoylacetic acid, $PhCOCH_2COOH$.[168] Important applications include the carboxylation of steroids such as testosterone, androsterone, and 4-cholesten-3-one using the thioureide complex (c) in DMF solution at 15°C, forming monocarboxylated derivatives.[170]

9.5.4   Alkali Phenoxides

Organic compounds containing an active hydrogen atom were found by Bottaccio et al. to be carboxylated at room temperature by atmospheric pressure CO_2 in the presence of alkali phenoxides in hydrocarbon solvents.[171] Best results were obtained with phenoxides having bulky substituents, such as potassium 2,6-di(*tert*-butyl)-*p*-cresolate. Also, potassium phenoxides were more active than sodium phenoxides. With toluene as solvent, acetophenone was carboxylated to benzoylacetic acid, acetone to a mixture of acetoacetic and 3-oxoglutaric acids, nitromethane to nitroacetic acid, and phenylacetonitrile to phenylcyanoacetic acid.[171]

Solid sodium phenoxide dissolved in N-methyl-ε-caprolactam (NMC) formed a tetrameric complex, [(NMC)Na(OPh)]$_4$, in which the sodium ions and the phenoxide oxygen ions occupy the corners of a cube, while the NMC acts as a monodentate ligand. Walther et al. discovered that this complex served as an excellent carrier for the carboxylation of active C–H bonds with CO_2.[172] The complex fixed 0.5 mol of CO_2 per mole of its sodium phenoxide. By addition of acetone to a CO_2-saturated solution of sodium phenoxide in an excess of NMC, the sodium salt of 3-ketoglutaric acid (1,3-acetonedicarboxylic acid) was produced in 85% yield:

$$CH_3COCH_3 + 2CO_2 \rightarrow HOOC-CH_2COCH_2-COOH \qquad (9.47)$$

Similarly, acetophenone was carboxylated to benzoylacetic acid and cyclohexanone was carboxylated to cyclohexane-2,6-dicarboxylic acid.[172]

See also Chapter 6 on biomimetic carboxylations and Chapter 8 on high-temperature carboxylations.

9.5.5 Cobalt(salen) Complexes

Reversible fixation of carbon dioxide was achieved by Gambarotta et al. using bifunctional complexes containing in their structure a nucleophilic cobalt(I) and an alkali cation.[173] Complexes exhibiting such fixation were Co(R-salen)M [where salen = N,N′-ethylene-bis(salicylideneaminato), R-salen = substituted salen ligand, and M = Li, Na, K, Cs]. In solvents like tetrahydrofuran, pyridine, or toluene, the deep-green solutions of the bimetallic complexes Co(R-salen)M reversibly absorbed 1 mol of CO_2 per bimetallic unit. X-ray analysis of the adduct [Co(R-salen)MCO₂] indicated that CO_2 was anchored to the cobalt atom through a Co–C σ bond, while the oxygen atoms interacted with the alkali cation.

See Section 12.3.2.12 and 12.3.3.9 for the structure of Co-salen and for its use as mediator in the electrochemical reduction of CO_2.

9.5.6 Tertiary Amines and CO₂

The reaction of carbon dioxide with tertiary amines, such as triethanolamine (TEA), is widely used in industry, particularly for natural gas purification. In contrast to primary and secondary amines, tertiary amines do not form carbamates. Natural gas washing for removal of carbon dioxide with aqueous solutions of TEA has the added advantage of also removing hydrogen sulfide from the gaseous phase. In a kinetic study of the reaction of carbon dioxide with two tertiary amines, TEA and methyldiethanolamine in aqueous solutions, the stopped-flow method was used, with a color indicator to reveal the pH change due to the formation of the bicarbonate ion. Essentially, the amines catalyzed the hydration of carbon dioxide.[174]

In a comparative study of the absorption capacity of various amines for CO_2, Nagao et al. found that the α-amino tertiary amides were both excellent CO_2 absorbents and highly resistant toward oxidative degradation. An 1 M aqueous solution of $Me_2NCH_2CONEt_2$ had a loading capacity for CO_2 of 61%, defined as the loading difference between 40 and 100°C.[175]

See Chapter 5 for more details on industrial CO_2 recovery by solvent extraction.

9.6 PLASMA POLYMERIZATION OF PERFLUOROBENZENE AND CO_2

An interesting approach to the incorporation of carbon dioxide into organic molecules is plasma polymerization, in which thin films are formed by activation of monomer molecules in a radiofrequency-induced plasma. Mixtures of carbon dioxide and perfluorobenzene or benzene were excited by Inagaki et al. in an inductively coupled system operated at 13.56 MHz.[176] With the perfluorobenzene–CO_2 mixture, the plasma reaction produced films with carboxylic acid groups, which may have useful ion-exchange properties. With the benzene–CO_2 mixture, the resulting film contained oxygen atoms, but no carboxylic acid groups.

See also Section 9.4.5. on the reversible CO_2 insertion into the tungsten(II)–alkoxide bond.

9.7 CONCLUSIONS ON CARBOXYLATION BY CO_2 INSERTION

Among the promising products for CO_2 mitigation by chemical conversion, polycarbonate plastics have an outstanding potential, because of their excellent properties.[75] The world consumption of polycarbonates in 1979 was only 150,000 metric tons. However, in 1996, the production capacity of polycarbonates worldwide was already about 1 million tons per year, and the demand for this commodity has been growing by more than 10% per year. About one-third of the volume of sales of polycarbonates has been for building products.[76] For polyurethanes, annual sales in the U.S. amounted to more than 4.6×10^9 lb.[32] Increased production of polycarbonates and polyurethanes by processes replacing phosgene by carbon dioxide will (1) contribute directly to enhanced use of CO_2 and also (2) indirectly provide considerable mitigation of CO_2 release to the atmosphere by the substitution of the highly energy-intensive metal and cement production by plastic production.

Other materials, the enhanced use of which may contribute to CO_2 mitigation, are fuels and fuel additives such as methanol and methyl-*tert*-butyl ether and solvents such as dimethyl carbonate and ethylene carbonate.

REFERENCES

1. Aresta, M. and Quaranta, E., Carbon dioxide — a substitute for phosgene, *Chemtech*, 27, 32–40, 1997.
2. Inoue, S. and Yamazaki, N., *Organic and Bio-organic Chemistry of Carbon Dioxide*, Kodansha, Tokyo, 1982.

3. Ziessel, R., Chimie de coordination de la molecule de dioxyde de carbon: activation biologique, chimique, electrochimique et photochimique, *Nouv. J. Chim.*, 7, 613–633, 1983.

4. Behr, A., The synthesis of organic chemicals by catalytic reactions of carbon dioxide, *Bull. Soc. Chim. Belg.*, 94, 671–683, 1985.

5. Behr, A., Use of carbon dioxide in industrial organic synthesis, *Chem. Eng. Technol.*, 10, 16–27, 1987.

6. Behr, A., Carbon dioxide as an alternative C$_1$ synthetic unit: activation by transition-metal complexes, *Angew. Chem. Int. Ed. Engl.*, 27, 661–678, 1988.

7. Behr, A., Carbon dioxide as building block for fine chemicals synthesis, in *Aspects of Homogeneous Catalysis*, Ugo, R., Ed., D. Reidel, Dordrecht, 1988, 59–96.

8. Aresta, M. and Forti, G., Eds., *Carbon Dioxide as a Source of Carbon. Biochemical and Chemical Use, Proc. NATO Adv. Study Inst.*, Kluwer Academic, The Hague, 1987.

9. Braunstein, P., Matt, D., and Nobel, D., Reactions of carbon dioxide with carbon–carbon bond formation catalyzed by transition metal complexes, *Chem. Rev.*, 88, 747–764, 1988.

10. Braunstein, P., Matt, D., and Nobel, D., Carbon dioxide activation and catalytic lactone synthesis by telomerization of butadiene and CO$_2$, *J. Am. Chem. Soc.*, 110, 3207–3212, 1988.

11. Kubiak, C.P. and Ratliff, K.S., Approaches to the chemical, electrochemical and photochemical activation of carbon dioxide by transition metal complexes, *Isr. J. Chem.*, 31, 3–15, 1991.

12. Hoberg, H., One decade of transition metal activation of carbon dioxide for organic synthesis, in Int. Conf. on Carbon Dioxide Utilization, Bari, Italy, September 1993, 23–37.

13. Tsuda, T., Utilization of carbon dioxide in organic synthesis and polymer synthesis ·by the transition metal catalyzed carbon dioxide fixation into unsaturated hydrocarbons, *Gazz. Chim. Ital.*, 125, 101–110, 1995.

14. Aresta, M., Quaranta, E., Tommasi, I., Giannoccaro, P., and Ciccarese, A., Enzymatic versus chemical carbon dioxide utilization. I. The role of metal centres in carboxylation reactions, *Gazz. Chim. Ital.*, 125, 509–538, 1995.

15. Xiaoding, X. and Moulijn, J.A., Mitigation of CO$_2$ by chemical conversion: plausible chemical reactions and promising products, *Energy Fuels*, 10, 305–325, 1996.

16. Gibson, D.H., The organometallic chemistry of carbon dioxide, *Chem. Rev.*, 96, 2063–2095, 1996.

17. Leitner, W., The coordination chemistry of carbon dioxide and its relevance for catalysis: a critical survey, *Coord. Chem. Rev.*, 153, 257–284, 1996.

18. Bruneau, C. and Dixneuf, P.H., Catalytic incorporation of CO$_2$ into organic substrates. Synthesis of unsaturated carbamates, carbonates and ureas, *J. Mol. Catal.*, 74, 97–107, 1992.

19. Bruneau, C. and Dixneuf, P.H., Catalytic additions of carbon dioxide adducts to alkynes: selective synthesis of carbamates, ureas, and carbonates, in *Carbon Dioxide Fixation and Reduction in Biological and Model Systems, Proc. Royal Swedish Acad. Sci., Nobel Symp.*, Brändén, C.-I. and Schneider, G., Eds., Oxford University Press, Oxford, U.K., 1994, 131–143.

20. Mavrovic, I. and Shirley, A.R., Urea, in *Kirk-Othmer Encyclopedia of Chemical Technology*, Vol. 23, 3rd ed., Wiley-Interscience, New York, 1983, 548–575.

21. Klier, K., Catalytic conversions of CO_2: activation and hydrogenation, in Proc. Int. Symp. Chemical Fixation of Carbon Dioxide, Nagoya, Japan, December 2 to 4, 1991, 139–134.

22. Ballou, W.R., Carbon dioxide, in *Kirk-Othmer Encyclopedia of Chemical Technology*, Vol. 4, 3rd ed., Wiley-Interscience, New York, 1978, 725–742.

23. Aresta, M. and Quaranta, E., Role of the macrocyclic polyether in the synthesis of N-alkylcarbamate esters from primary amines, CO_2, and alkyl halides in the presence of crown-ethers, *Tetrahedron*, 48, 1515–1530, 1992.

24. Aresta, M. and Quaranta, E., Synthesis of carbamates from CO_2, amines and alkylating agents, in Int. Conf. on Carbon Dioxide Utilization, Bari, Italy, September 1993, 63–77.

25. Aresta, M., Dibenedetto, A., and Quaranta, E., Reaction of alkali metal tetraphenylborates with amines in the presence of CO_2: a new easy way to aliphatic and aromatic alkali metal carbamates, *J. Chem. Soc. Dalton Trans.*, pp. 3359–3363, 1995.

26. Aresta, M. and Quaranta, E., Novel, CO_2-promoted synthesis of anhydrous alkylammonium tetraphenylborates: a study on their reactivity as intramolecular and intermolecular proton-transfer agents, *J. Organomet. Chem.*, 488, 211–222, 1995.

27. Vaska, L., Schreiner, S., Felty, R.A., and Yu, J.Y., Catalytic reduction of carbon dioxide to methane and other species via formamide intermediation: synthesis and hydrogenation of $HC(O)NH_2$ in the presence of $[Ir(Cl)(CO)(Ph_3P)_2]$, *J. Mol. Catal.*, 52, L11–L16, 1989.

28. Fournier, J., Bruneau, C., Dixneuf, P.H., and Lécolier, S., Ruthenium-catalyzed synthesis of symmetrical N,N'-dialkylureas directly from carbon dioxide and amines, *J. Org. Chem.*, 56, 4456–4458, 1991.

29. Mahé, R., Sasaki, Y., Bruneau, C., and Dixneuf, P. H., Catalytic synthesis of vinyl carbamates from carbon dioxide and alkynes with ruthenium complexes, *J. Org. Chem.*, 54, 1518–1523, 1989.

30. Obrien, R.A., Worman, J.J., and Olson, E.S., Carbon dioxide in organic synthesis. Preparation and mechanism of formation of N-(3)-substituted hydantoins, *Synth. Commun.*, 22, 823–828, 1992.

31. Nomura, R., Hasegawa, Y., Ishimoto, M., Toyosaki, T., and Matsuda, H., Carbonylation of amines by carbon dioxide in the presence of an organoantimony catalyst, *J. Org. Chem.*, 57, 7339–7342, 1992.

32. Stinson, S., Polyurethane use continues to grow, *Chem. Eng. News*, 75, 22, 1997.

33. McGhee, W.D. and Riley, D.P., Palladium-mediated synthesis of urethanes from amines, carbon dioxide, and cyclic diolefins, *Organometallics*, 11, 900–907, 1992.

34. McGhee, W.D., Pan, Y., and Riley, D.P., Highly sensitive generation of urethanes from amines, carbon dioxide and alkyl chlorides, *J. Chem. Soc. Chem. Commun.*, pp. 699–700, 1994.

35. Riley, D., McGhee, W.D., and Waldman, T., Generation of urethanes and isocyanates from amines and carbon dioxide, *Am. Chem. Soc. Symp. Ser., Benign by Design. Alternative Synthetic Design for Pollution Prevention*, 577, 122–132, 1994.

36. McGhee, W.D., Riley, D.P., Christ, K., Pan, Y., and Parnas, B., Carbon dioxide as a phosgene replacement: synthesis and mechanistic studies of urethanes from amines, CO_2, and alkyl chlorides, *J. Org. Chem.*, 60, 2820–2830, 1995.

37. Kojima, F., Aida, T., and Inoue, S., Fixation and activation of carbon dioxide on aluminum porphyrin. Catalytic formation of carbamic ester from carbon dioxide, amine and epoxide, *J. Am. Chem. Soc.*, 108, 391–395, 1986.

38. Bruneau, C., Dixneuf, C.H., and Lécolier, S., Acetylene in catalysis: a one-step synthesis of vinylcarbamates with [RuCl₂(norbornadiene)], *J. Mol. Catal.*, 44, 175–178, 1988.

39. Sasaki, Y. and Dixneuf, P.H., Ruthenium-catalyzed reaction of carbon dioxide, amine and acetylenic alcohol, *J. Org. Chem.*, 52, 4389–4391, 1987.

40. Super, M.S., Parks, K.L., and Beckman, E.J., Carbon dioxide as both solvent and monomer in copolymerizations, in *Carbon Dioxide Chemistry: Environmental Issues*, Paul, J. and Pradier, C.-M., Eds., Royal Society of Chemistry, Cambridge, U.K., 1994, 396–401.

41. Toda, T., Yoshida, M., Ohshima, M., Yagi, K., and Komatsu, S., Carbon dioxide fixation reactions via carbamic acid amine salts, in Proc. Int. Symp. Chemical Fixation of Carbon Dioxide, Nagoya, Japan, December 2 to 4, 1991, 185–188.

42. Fournier, J., Bruneau, C., and Dixneuf, P.H., Phosphine catalyzed synthesis of unsaturated cyclic carbonates from carbon dioxide and propargylic alcohols, *Organometallics*, 30, 3981–3942, 1989.

43. Fournier, J., Bruneau, C., and Dixneuf, P.H., A simple synthesis of oxazolidinones in one step from carbon dioxide, *Tetrahedron Lett.*, 31, 1721–1722, 1990.

44. Mitsunobu, O. and Eguchi, M., Preparation of carboxylic esters and phosphoric esters by the activation of alcohols, *Bull. Chem. Soc. Jpn.*, 44, 3427–3430, 1971.

45. Hoffman, W.A., Convenient preparation of carbonates from alcohols and carbon dioxide, *J. Org. Chem.*, 47, 5209–5210, 1982.

46. Kodaka, M., Tomohiro, T., and Okuno, H.Y., The mechanism of the Mitsunobu reaction and its application to CO₂ fixation, *J. Chem. Soc. Chem. Commun.*, pp. 81–82, 1993.

47. Kubota, Y., Kodaka, M., Tomohiro, T., and Okuno, H.Y., Formation of cyclic urethanes from amino alcohols and carbon dioxide using phosphorus(III) reagents and halogenoalkanes, *J. Chem. Soc. Perkin Trans. 1*, pp. 5–6, 1993.

48. McGhee, W.D., Paster, M., Riley, D., Ruettimann, K., Solodar, J., and Waldman, T., Generation of organic isocyanates from amines, carbon dioxide, and electrophilic dehydrating agents. Use of *o*-sulfobenzoic acid anhydride, *Am. Chem. Soc. Symp. Ser., Green Chemistry: Designing Chemistry for the Environment*, 626, 49–58, 1996.

49. Yoshida, Y. and Ishii, S., A novel synthesis of carbonate ester from the reaction of CO₂, alcohol and epoxide, in Proc. Int. Symp. Chemical Fixation of Carbon Dioxide, Nagoya, Japan, December 2 to 4, 1991, 423–426.

50. McGhee, W.D. and Riley, D., Replacement of phosgene with carbon dioxide: synthesis of alkyl carbonates, *J. Org. Chem.*, 60, 6205–6207, 1995.

51. Rivetti, F., Romano, U., and Delledone, D., Dimethyl carbonate and its production technology, *ACS Symp. Ser., Green Chemistry*, 626, 70–80, 1996.

52. Yano, T., Matsui, H., Koike, T., Ishiguro, H., Fujihara, H., Yoshihara, M., and Maeshima, T., Magnesium oxide–catalysed reaction of carbon dioxide with an epoxide with retention of stereochemistry, *Chem. Commun.*, pp. 1129–1130, 1997.

53. Zhu, H., Chen, L.-B., and Jiang, Y.-Y., Synthesis of propylene carbonate and some dialkyl carbonates in the presence of bifunctional catalyst compositions, *Polym. Adv. Technol.*, 7, 701–703, 1996.

54. Clerici, M.G. and Ingallina, P., Clean oxidation technologies: new prospects in the epoxidation of olefins, *ACS Symp. Ser., Green Chemistry*, 626, 59–68, 1996.

55. Kihara, N. and Endo, T., Incorporation of carbon dioxide into poly(glycidyl methacrylate), *Macromolecules*, 25, 4824–4825, 1992.

56. Kihara, N. and Endo, T., Synthesis and reaction of polymethacrylate bearing cyclic carbonate moieties in the side-chain, *Macromol. Chem. Phys.*, 193, 1481–1492, 1992.

57. Kihara, N. and Endo, T., Catalytic activity of various salts in the reaction of 2,3-epoxypropyl phenyl ether and carbon dioxide under atmospheric pressure, *J. Org. Chem.*, 58, 6198–6202, 1993.

58. Sakai, T., Kihara, N., and Endo, T., Polymer reaction of epoxide and carbon dioxide. Incorporation of carbon dioxide into epoxide polymers. *Macromolecules*, 28, 4701–4706, 1995.

59. Kihara, N. and Endo, T., Solid-state catalytic incorporation of carbon dioxide into oxirane-polymer. Conversion of poly(glycidyl methacrylate) to carbonate–polymer under atmospheric pressure, *J. Chem. Soc. Chem. Commun.*, pp. 937–938, 1994.

60. Darensbourg, D.J. and Holtcamp, M.W., Catalysts for the reactions of epoxides and carbon dioxide, *Coord. Chem. Rev.*, 153, 155–174, 1996.

61. Yamashita, J., Kameyama, A., Nishikubo, T., Fukuda, W., and Tomoi, M., Addition reaction of epoxy compounds with carbon dioxide using insoluble polymer-supported crown ether complexes as catalysts, *Kobunshi Ronbunshu*, 50, 577, 1993.

62. Nishikubo, T., Kameyama, A., and Sasano, M., Synthesis of functional polymers bearing cyclic carbonate groups from (2-oxo-1,3-dioxolan-4-yl)methyl vinyl ether, *J. Polym. Sci. Polym. Chem.*, 32, 301–308, 1994.

63. Park, D.W., Moon, J.Y., Yang, J.G., Jung, S.M., Lee, J.K., and Ha, C.S., Catalytic conversion of carbon dioxide to polymer blends via cyclic carbonates, in Abstr. 4th Int. Conf. on Carbon Dioxide Utilization, Kyoto, Japan, September 1997, P-021.

64. Baba, A., Kashiwagi, H., and Matsuda, H., Reaction of carbon dioxide with oxetane catalyzed by organotin halide complexes: control of reaction by ligands, *Organometallics*, 6, 137–140, 1987.

65. Takata, T., Kanamaru, M., and Endo, T., First example of anionic polymerization with azo-containing radical initiators: anionic ring-opening polymerization of cyclic carbonate initiated by azobis(isobutyronitrile) and related azo initiators, *Macromolecules*, 29, 2315–2317, 1996.

66. Inoue, Y., Synthesis of heterocyclic compounds from CO_2 and propargylic alcohol catalyzed by transition metal complexes, in Proc. Int. Symp. Chemical Fixation of Carbon Dioxide, Nagoya, Japan, December 2 to 4, 1991, 273–280.

67. Inoue, Y., Itoh, Y., Yen, I.-F., and Imaizumi, S., Palladium(0) catalyzed carboxylative cyclized coupling of propargylic alcohol with aryl halides, *J. Mol. Catal.*, 60, L1–L3, 1990.

68. Inoue, Y., Ohuchi, K., Yen, I.-F., and Imaizumi, S., Preparation of 3(2H)-furanones from 2-propynyl alcohol, CO, and phenyl halides under CO_2 atmosphere catalyzed by transition metal complexes, *Bull. Chem. Soc. Jpn.*, 62, 3518–3522, 1989.

69. Joumier, J.-M., Fournier, J., Bruneau, C., and Dixneuf, P.H., Functional carbonates: cyclic α-methylene and β-oxopropyl carbonates from prop-2-ynyl alcohol derivatives and CO_2, *J. Chem. Soc. Perkin Trans. 1*, pp. 3271–3274, 1991.

70. Joumier, J.-M., Grainger, R., Bruneau, C., and Dixneuf, P.H., Synthesis of functional oxazolidin-2-ones and oxadiazin-2-ones in two steps from CO_2 via cyclic α-methylene carbonates, *Synlett.*, pp. 423–424, 1993.

71. Le Gendre, P., Braun, T., Bruneau, C., and Dixneuf, P.H., Preparation of optically active cyclic carbonates and 1,2-diols via enantioselective hydrogenation of α-methylenedioxolanones catalyzed by chiral ruthenium(II) complexes, *J. Org. Chem.*, 61, 8453–8455, 1996.

72. Aresta, M., Quaranta, E., and Ciccarese, A., Direct synthesis of 1,3-benzodioxol-2-one from styrene, dioxygen and carbon dioxide promoted by Rh(I), *J. Mol. Catal.*, 41, 355–359, 1987.

73. Aresta, M., Quaranta, E., and Tommasi, I., The role of metal centers in reduction and carboxylation reactions utilizing carbon dioxide, *New J. Chem.*, 18, 133–142, 1994.

74. Aresta, M., Berloco, C., Fragale, C., Quaranta, E., and Tommasi, I., Transition metal peroxocarbonates: carboxylation agents under oxidative conditions, in Abstr. Int. Conf. on Carbon Dioxide Utilization, Bari, Italy, September 1993, 363.

75. Fox, D.W., Polycarbonates, in *Kirk-Othmer Encyclopedia of Chemical Technology*, Vol. 18, 3rd ed., Wiley-Interscience, New York, 1982, 479–494.

76. Dunlop, L.H. and Desch, R., Plastic building products, in *Kirk-Othmer Encyclopedia of Chemical Technology*, Vol. 18, 3rd ed., Wiley-Interscience, New York, 1982, 87–109.

77. Komiya, K. et al., New process for producing polycarbonate without phosgene and methylene chloride, *ACS Symp. Ser., Green Chemistry*, 626, 20–32, 1996.

78. Inoue, S., Koinuma, H., and Tsuruta, T., Copolymerization of carbon dioxide and epoxide, *J. Polym. Sci. B*, 7, 287–292, 1969.

79. Soga, K., Imai, E., and Hattori, I., Alternating copolymerization of CO₂ and propylene oxide with catalysts prepared from Zn(OH)₂ and various carboxylic acids, *Polym. J.*, 13, 407–410, 1981.

80. Rokicki, A., Poly(Alkylene Carbonates) with Controlled Molecular Weights, U.S. Patent 4,943,677, 1990; *Chem. Abstr.*, 113, P192136v.

81. Motika, S.A., Rokicki, A., and Stein, B.K., Catalyst for the Copolymerization of Epoxides with CO₂, U.S. Patent 5,026,676, 1991; *Chem. Abstr.*, 115, P93209f.

82. Darensbourg, D.J., Stafford, N.W., and Katsurao, T., Supercritical carbon dioxide as solvent for the copolymerization of carbon dioxide and propylene oxide using a heterogeneous zinc carboxylate catalyst, *J. Mol. Catal. A Chem.*, 104, L1–L4, 1995.

83. Darensbourg, D.J. and Holtcamp, M.W., Catalytic activity of zinc(II) phenoxides which possess readily accessible coordination sites. Copolymerization and terpolymerization of epoxides and carbon dioxide, *Macromolecules*, 28, 7577–7579, 1995.

84. Darensbourg, D.J., Carbon dioxide insertion into metal–oxygen bonds. Relevance to copolymerization of CO₂ and epoxides, in Abstr. 3rd Int. Conf. on Carbon Dioxide Utilization, Norman, OK, May 1995.

85. Kuran, W. and Listos, T., Initiation and propagation reactions in the copolymerization of epoxide with carbon dioxide by catalysts based on diethylzinc and polyhydric phenol, *Macromol. Chem. Phys.*, 195, 977–984, 1994.

86. Listos, T., Kuran, W., and Siwicc, R., Propylene oxide homopolymerization and copolymerization with carbon dioxide by γ-alumina-supported zinc coordination catalysts, *J. Macromol. Sci. Pure Appl. Chem. A*, 32, 393–403, 1995.

87. Darensbourg, D.J., Holtcamp, M.W., Khandelwal, B., Klausmeyer, K.K., and Reibenspies, J. H., Syntheses and structures of epoxide adducts of soluble cadmium(II) carboxylates. Models for the initiation process in epoxide/CO₂ coupling reactions, *J. Am. Chem. Soc.*, 117, 538–539, 1995.

88. Super, M., Berluche, E., Costello, C., and Beckman, E.J., Copolymerization of 1,2-epoxycyclohexane and carbon dioxide using carbon dioxide as both reactant and solvent, *Macromolecules*, 30, 368–372, 1997.

89. Dixon, D.D., Ford, M.E., and Mantell, G.J., Thermal stabilization of poly(alkylene-carbonates), *J. Polym. Sci. Polym. Lett.*, 18, 131–134, 1980.

90. Chen, L.-B., Yang, S.-Y., and Peng, H., Improving thermostability of CO_2-epoxide copolymers, in Proc. Int. Symp. Chemical Fixation of Carbon Dioxide, Nagoya, Japan, December 2 to 4, 1991, 253–258.

91. Aresta, M., Nobile, C.F., Albano, V.G., Forni, E., and Manassero, M., New nickel–carbon dioxide complex: synthesis, properties and crystallographic characterization of (carbon dioxide)-bis(tricyclophosphine) nickel, *J. Chem. Soc. Chem. Commun.*, pp. 636–637, 1975.

92. Aresta, M., Quaranta, E., and Tommasi, I., The role of metal centres in reduction and carboxylation reactions utilizing carbon dioxide, in Proc. Int. Symp. Chemical Fixation of Carbon Dioxide, Nagoya, Japan, December 2 to 4, 1991, 209–226.

93. Fujita, E., Creutz, C., Sutin, N., and Szalda, D.J., Carbon dioxide activation by cobalt macrocycles — factors affecting CO_2 and CO binding, *J. Am. Chem. Soc.*, 113, 343–353, 1991.

94. Fujita, E., Creutz, C., and Sutin, N., Carbon dioxide activation by metal macrocycles, in Proc. Int. Symp. Chemical Fixation of Carbon Dioxide, Nagoya, Japan, December 2 to 4, 1991, 243–246.

95. Gangi, D.A. and Durand, R.R., Binding of carbon dioxide to cobalt and nickel tetra-aza-macrocycles, *J. Chem. Soc. Chem. Commun.*, pp. 697–699, 1986.

96. Fujita, E., Szalda, D.J., Creutz, C., and Sutin, N., Carbon dioxide activation: thermodynamics of CO_2 binding and the involvement of two cobalt centers in the reduction of CO_2 by a cobalt(I) macrocycle, *J. Am. Chem. Soc.*, 110, 4870–4871, 1988.

97. Ito, T., Sugimoto, S., Ohki, T., Nakano, T., and Osakada, K., Reaction of bis (cyclopentadienyl) hydridophenyl tungsten(IV) with carboxylic acids, carbon dioxide, and related compounds, *J. Organomet. Chem.*, 428, 69–83, 1992.

98. Bianchini, C. and Meli, A., Bifunctional activation of CO_2: a case where the acidic and basic sites are not held in the same structure, *J. Am. Chem. Soc.*, 106, 2698–2699, 1984.

99. Keene, F.R., Creutz, C., and Sutin, N., Reduction of carbon dioxide by tris(2,2'-bipyridine)cobalt(I), *Coord. Chem Rev.*, 64, 247–260, 1985.

100. Antiñolo, A., Fajardo, M., García-Yuste, S., del Hierro, I., Otero, A., Elkrami, S., Mourad, Y., and Mugnier, Y., Synthesis, electrochemistry and reactivity of formato- and acetato-niobocene complexes, *J. Chem. Soc. Dalton Trans.*, 20, 3409–3414, 1995.

101. Nietlispach, D., Bosch, H.W., and Berke, H., A comparative study of the reactivity of $Mn(NO)_2L_2H$ and $Mn(CO)_3L_2H$ complexes (L = phosphorus donor), *Chem. Ber.*, 127, 2403–2415, 1994.

102. Vivanco, M., Ruiz, J., Floriani, C., and Chiesivilla, A., Chemistry of the vanadium carbon sigma bond. 1. Insertion of carbon monoxide, isocyanides, carbon dioxide, and heterocumulenes into the V–C bond of tris(mesityl)vanadium (III), *Organometallics*, 12, 1794–1801, 1993.

103. Casey, C.P., Synthesis and reactions of directly bonded zirconium–ruthenium heterobimetallic complexes, *J. Organomet. Chem.*, 400, 205–221, 1990.

104. Pinkes, J.R., Steffey, B.D., Vites, J.C., and Cutler, A.R., Carbon dioxide insertion into the Fe–Zr and Ru–Zr bonds of the heterobimetallic complexes $Cp(CO)_2M–Zr(Cl)Cp_2$: direct production of the μ-$\eta^1(C):\eta^2(O,O')$–CO_2 compounds $Cp(CO)_2M–CO_2Zr(Cl)Cp_2$, *Organometallics*, 13, 21–23, 1994.

105. Sakaki, S. and Musashi, Y., An ab initio molecular-orbital study of insertion of CO_2 into a Rh^I–H bond, *J. Chem. Soc. Dalton Trans.*, pp. 3074–3054, 1994.

106. Sakaki, S. and Musashi, Y., Ab initio MO study of the CO_2 insertion into the Cu(I)–R bond (R = H, CH₃, or OH). Comparison between the CO_2 insertion and the C_2H_4 insertion, *Inorg. Chem.*, 34, 1914–1923, 1995.

107. Sakaki, S. and Musashi, Y., A theoretical study on CO_2 insertion into an M–H bond (M = Rh and Cu), *Int. J. Quantum Chem.*, 57, 481–491, 1996.

108. Darensbourg, D.J. and Ovalles, C., Anionic group 6B metal carbonyls as homogeneous catalysts for carbon dioxide/hydrogen activation. The production of alkyl formates, *J. Am. Chem. Soc.*, 106, 3750–3754, 1984.

109. Darensbourg, D.J., Jones, M.L.M., and Reibenspies, J.H., The reversible insertion reaction of carbon dioxide with the $W(CO)_5OH^-$ anion. Isolation and characterization of the resulting bicarbonate complex $[PPN][W(CO)_5O_2COH)]$, *Inorg. Chem.*, 35, 4406–4413, 1996.

110. Ovalles, C., Fernandez, C., and Darensbourg, D.J., Homogeneous catalytic synthesis of formaldehyde using the tungsten carbonyl complex $[(CO)_5WCl]^-$ in the presence of sodium methoxide. *J. Mol. Catal.*, 93, 125–136, 1994.

111. Yamazaki, H. and Hayashi, N., Prediction of the optimum reaction time for carbon-11 labeling reactions. The rate constants for the carboxylation of Grignard reagents, *Chem. Lett.*, pp. 525–528, 1993.

112. Eaton, P.E., Lee, C.H., and Xiong, Y., Magnesium amide bases and amido-Grignards. 1. Ortho magnesiation, *J. Am. Chem. Soc.*, 111, 8016–8018, 1989.

113. Eaton, P.E. and Lukin, K.A., Through-space amide activation of C–H bonds in triangulanes, *J. Am. Chem. Soc.*, 115, 11370–11375, 1993.

114. Bashir-Hashemi, A., Iyer, S., and Slagg, N., Cubanes and cage related molecules, *Chem. Ind.*, 14, 551–556, 1995.

115. Yanagisawa, A., Yasue, K., and Yamamoto, H., Regioselective and stereospecific synthesis of β,γ-unsaturated carboxylic acids using allylbariums, *Synlett.*, pp. 593–594, 1992.

116. Simpson, R.D. and Bergman, R.G., Comparison of the reactivity of $(CO)_3L_2ReOR$, $(CO)_3L_2ReOAr$, and $(CO)_3L_2ReNHAr$ with CO_2 and other electrophiles, *Organometallics*, 11, 4306–4315, 1992.

117. Buffin, B.P., Arif, A.M., and Richmond, T.G., Carbonyl insertion and reductive elimination chemistry of tungsten(II) alkoxides and aryloxides, *J. Chem. Soc. Chem. Commun.*, pp. 1432–1434, 1993.

118. Legzdins, P., Rettig, S.J., and Ross, K.J., Competitive reactivity of W–C, W–N, and W–O bonds at the Cp*W(NO) fragment: insertion reactions of *tert*-butyl isocyanide, *p*-tolyl isocyanate, and carbon disulfide, *Organometallics*, 13, 569–577, 1994.

119. Osakada, K., Sato, R., and Yamamoto, T., Nickel-complex-promoted carboxylation of haloarenes involving insertion of CO_2 into Ni^{II}–C bonds, *Organometallics*, 13, 4645–4647, 1994.

120. Hoberg, H., Peres, Y., and Michereit, A., C–C coupling of alkenes with CO_2 on nickel(0), *J. Organomet. Chem.*, 307, C38–C40, 1986.

121. Sasaki, Y., Inoue, Y., and Hashimoto, H., Reaction of carbon dioxide with butadiene catalysed by palladium complexes. Synthesis of 2-ethylidene-5-en-4-olide, *J. Chem. Soc.*, pp. 605–606, 1976.

122. Inoue, Y., Sasaki, Y., and Hashimoto, H., Incorporation of CO_2 in butadiene dimerization catalyzed by palladium complexes, *Bull. Chem. Soc. Jpn.*, 51, 2375–2378, 1978.

123. Tsuda, T., Yamamoto, T., and Saegusa, T., Palladium-catalyzed cycloaddition of carbon dioxide with methoxyallene, *J. Organomet. Chem.*, 429, C46–C48, 1992.

124. Gao, Y., Iijima, S., Urabe, H., and Sato, F., Carbon dioxide fixation by $Cp_2(\eta^3$-allyl)Ti complexes, *Inorg. Chim. Acta*, 222, 145–153, 1994.
125. Haruki, E., Hara, T., and Inoue, H., Synthesis of tricyclo[5.2.10^{2,6}]deca-3,6-dicarboxylic acid and its derivatives from cyclopentadiene and carbon dioxide, *Chem. Express.*, 5, 493–496, 1990.
126. Haruki, E., Hara, T., and Inoue, H., Fixation of carbon dioxide to the cyclopentadiene derivatives using DBU and CO_2, in Proc. Int. Symp. Chemical Fixation of Carbon Dioxide, Nagoya, Japan, December 2 to 4, 1991, 427–430.
127. Albano, P. and Aresta, M., Some catalytic properties of Rh(diphos)(η-BPh$_4$), *J. Organomet. Chem.*, 190, 243–246, 1980.
128. Pillai, S.M., Ohnishi, R., and Ichikawa, M., Cycloaddition of carbon dioxide to propyne over supported Rh$_4$ and Fe$_2$Rh$_4$ carbonyl cluster-derived catalysts, *J. Chem. Soc. Chem. Commun.*, pp. 246–247, 1990.
129. Pillai, S.M., Ohnishi, R., and Ichikawa, M., A mechanistic study of cycloaddition of carbon dioxide to propyne over supported Rh$_4$ and Fe$_x$ Rh$_y$ carbonyl cluster derived catalysts, *React. Kinet. Catal. Lett.*, 48, 201–208, 1992.
130. Höfer, J., Doucet, H., Bruneau, C., and Dixneuf, P.H., Ruthenium catalysed regioselective synthesis of O-1(1,3-dienyl) carbamates directly from CO_2, *Tetrahedron Lett.*, 32, 7409–7410, 1991.
131. Inoue, Y., Itoh, Y., Kazama, H., and Hashimoto, H., Reaction of dialkyl-substituted alkynes with carbon dioxide catalyzed by nickel(0) complexes. Incorporation of carbon dioxide in alkyne dimers and novel cyclotrimerization of the alkynes, *Bull. Chem. Soc. Jpn.*, 53, 3329–3333, 1980.
132. Walther, D., Schönberg, H., Dinjus, E., and Sieler, J., Activation of carbon dioxide at transition metal centers: selective cooligomerization of hexyne(-3) with the catalyst systems acetonitrile–trialkylphosphine–nickel(0) and the structure of a nickel(0) complex with side- or bound acetonitrile, *J. Organomet. Chem.*, 334, 377–388, 1987.
133. Walther, D., Bräunlich, G., Kempe, R., and Sieler, J., Activation of CO_2 on transition metal centers. Course of homogeneous catalytic formation of 2-pyrone from carbon dioxide and hex-3-yne on nickel (0) fragments, *J. Organomet. Chem.*, 436, 109–119, 1992.
134. Kempe, R., Sieler, J., Walther, D., Reinhold, J., and Rommel, K., Activation of CO_2 at transition metal centres. The route of the CO_2 reduction at nickel(0) moieties, *Z. Anorg. Allg. Chem.*, 619, 1105–1110, 1993.
135. Tsuda, T., Sumiya, R., and Saegusa, T., Nickel-mediated cycloaddition of diynes with carbon dioxide to bicyclic α-pyrones, *Synth. Commun.*, 17, 147–154, 1987.
136. Tsuda, T., Morikawa, S., Hasegawa, N., and Saegusa, T., Nickel(0)-catalyzed cycloaddition of silyl diynes with carbon dioxide to silyl bicyclic α-pyrones, *J. Org. Chem.*, 55, 2978–2981, 1990.
137. Tsuda, T., Maruta, K., and Kitaike, Y., Nickel(0)-catalyzed alternating copolymerization of carbon dioxide with diynes to poly(2-pyrones), *J. Am. Chem. Soc.*, 114, 1498–1499, 1992.
138. Tsuda, T., Nickel(0)-catalyzed 1:1 cycloaddition of diynes with carbon dioxide to poly(2-pyrone)s, in Int. Conf. on Carbon Dioxide Utilization, Bari, Italy, September 1993, 47–54.
139. Tsuda, T., Yasukawa, H., Hokazono, H., and Kitaike, Y., Nickel(0)-catalyzed cycloaddition copolymerization of cyclic diynes with carbon dioxide to ladder poly(2-pyrone)s, *Macromolecules*, 28, 1312–1315, 1995.

140. Tsuda, T., Yasukawa, H., and Komori, K., Nickel(0)-catalyzed cycloaddition copolymerization of ether diyenes with carbon dioxide to poly(2-pyrone)s, *Macromolecules*, 28, 1356–1359, 1995.

141. Tsuda, T., Hokazono, H., and Toyota, K., Efficient spontaneous 1:1 copolymerization of bis(ynamine)s with carbon dioxide to poly(4-pyrone)s, *J. Chem. Soc. Chem. Commun.*, pp. 2417–2418, 1995.

142. Thomas, C.A., *Anhydrous Aluminum Chloride in Organic Chemistry*, Reinhold Publishing, New York, 1941, 508.

143. Kurioka, M., Nakata, K., Jintoku, T., Taniguchi, Y., Takaki, K., and Fujiwara, Y., Palladium-catalyzed acetic acid synthesis from methane and carbon monoxide or dioxide, *Chem. Lett.*, p. 244, 1995.

144. Fujiwara, Y., Takaki, K., and Taniguchi, Y., Exploitation of synthetic reactions via C–H bond activation by transition metal catalysts. Carboxylation and aminomethylation of alkanes or arenes, *Synlett.*, pp. 591–599, 1996.

145. Taniguchi, Y., Kitamura, T., and Fujiwara, Y., Vanadium-catalyzed acetic acid synthesis from methane and carbon dioxide, in Abstr. 4th Int. Conf. on Carbon Dioxide Utilization, Kyoto, Japan, September 1997, P-030.

146. Johnson, A.W. and Tebby, J.C., The adducts from triphenylphosphine and dimethyl acetylene dicarboxylate, *J. Chem. Soc.*, pp. 2126–2130, 1961.

147. Griffiths, D. and Tebby, J.C., Fixation and deoxygenation of carbon dioxide to form furans using organophosphorus intermediates, *J. Chem. Soc. Chem. Commun.*, pp. 607–608, 1981.

148. Gambarotta, S., Strologo, S., Floriani, C., Chiesi-Villa, A., and Guastini, C., Stepwise reduction of CO_2 to formaldehyde and methanol: reaction of CO_2 and CO_2-like molecules with hydrido chloro bis(cyclopentadienyl)zirconium(IV), *J. Am. Chem. Soc.*, 107, 6278–6282, 1985.

149. Kloppenburg, L. and Petersen, J.L., Facile conversion of an appended silylamido to a silyloxy ligand via isocyanate elimination. Synthesis of {[(C_5Me_4)SiMe_2O]Zr(η^2-O_2CMe)(μ-O_2CMe)}$_2$ via the carboxylation of [(C_5Me_4)SiMe_2(N-*t*-Bu)]ZrMe_2, *Organometallics*, 15, 7–9, 1996.

150. Walther, D., Dinjus, E., and Herzog, V., Activation of carbon dioxide at transition metal centers — the oxidative coupling of CO_2 with carbodiimides at electron rich nickel(0), *Z. Chem.*, 23, 188, 1983.

151. Lundquist, E.G., Huffman, J.C., and Caulton, K.G., Formation of a heterometallic carbon dioxide complex with concurrent reduction of CO_2, *J. Am. Chem. Soc.*, 108, 8309–8310, 1986.

152. Orito, K., Miyazawa, M., and Suginome, H., Studies on carboxylation of alkoxy-substituted benzyl alcohols via direct lithiation and bromine–lithium exchange: synthesis of phthalides and phthalideisoquinoline alkaloids, *Tetrahedron*, 51, 2489–2496, 1995.

153. Kalinin, V.N., Cherepanov, I.A., and Moiiseev, S.K., Synthesis of 2-arylcarboxylic acids and C-norbenzomorphans via η^6-arenetricarbonylchromium complexes, *Mendeleev Commun.*, pp. 113–114, 1992.

154. Llorca, J., de la Piscina, P.R., Sales, J., and Homs, H., Activation of carbon dioxide by a silica-supported platinum tin bimetallic complex, *J. Chem. Soc. Chem. Commun.*, pp. 2555–2556, 1994.

155. Llorca, J., de la Piscina, P.R., Fierro, J.L.G., Sales, J., and Homs, N., Influence of metallic precursors on the preparation of silica-supported PtSn alloy. Characterization and reactivity in the catalytic activation of CO_2, *J. Catal.*, 156, 139–146, 1995.

156. Llorca, J., de la Piscina, P.R., Fierro, J.L.G., Sales, J., and Homs, N., Support effects on the formation of well defined PtSn alloy from a Pt–Sn bimetallic complex. Catalytic properties in the activation of CO_2, *J. Mol. Catal.*, 118, 101–111, 1997.

157. Sita, L.R., Babcock, J.R., and Xi, R., Facile metathetical exchange between carbon dioxide and the divalent group 14 bisamides $M[N(SiMe_3)_2]_2$ (M = Ge and Sn), *J. Am. Chem. Soc.*, 118, 10912–10913, 1996.

158. Shi, M. and Nicholas, K.M., Palladium-catalyzed carboxylation of allyl stannanes, *J. Am. Chem. Soc.*, 119, 5057–5058, 1997.

159. Tsuda, T., Chuja, Y., and Saegusa, T., Reversible carbon dioxide fixation by organocopper complexes, *J. Chem. Soc. Chem. Commun.*, pp. 963–964, 1975.

160. Tsuda, T., Chuja, Y., and Saegusa, T., Copper complexes acting as a reversible carbon dioxide carrier, *J. Am. Chem. Soc.*, 100, 630–632, 1978.

161. Churchill, M.R., Davies, G., El-Sayed, M.A., El-Shazly, M.F., Hutchison, J.P., Rupich, M.W., and Watkins, K.O., Synthesis, physical properties, and structural characterization of μ-carbonato-dichlorobis(N,N,N',N'-tetramethyl-1,3-propanediamine)-dicopper(II), $LCuCl(CO_3)ClCuL$, a diamagnetic initiator for the oxidative coupling of phenols by dioxygen, *Inorg. Chem.*, 18, 2296–2300, 1979.

162. Kitajima, N., Fujisawa, K., Koda, T., Hikichi, S., and Moro-oka, Y., Fixation of atmospheric CO_2 by a copper(II) complex, *J. Chem. Soc. Chem. Commun.*, pp. 1357–1358, 1990.

163. Kitajima, N., Hikichi, S., Tanaka, M., and Morooka, Y., Fixation of atmospheric CO_2 by a series of hydroxo complexes of divalent metal ions and the implication for the catalytic role of metal ion in carbonic anhydrase. Synthesis, characterization, and molecular structure of $[LM(OH)]_n$ (n = 1 or 2) and $LM(\mu = CO_3)ML$ (M(II) = Mn, Fe, Co, Ni, Zn, L = $HB(3,5-iPr_2pz)_3$), *J. Am. Chem. Soc.*, 115, 5496–5508, 1993.

164. Walther, D., Ritter, U., and Gessler, S., N-Carboxylato-metal complexes from CO_2 and synthetic potential for the formation of C–C bonds, in Int. Conf. on Carbon Dioxide Utilization, Bari, Italy, September 1993, 55–62.

165. Walther, D., Gessler, S., Ritter, U., Schmidt, A., Hamza, K., Imhof, W., Gorls, H., and Sieler, J., Organometallic CO_2 reservoirs from nickel(0)-1-azadiene-type ligands and their reactivity in the carboxylation of acetophenone, *Chem. Ber.*, 128, 281–287, 1995.

166. Ito, M. and Takita, Y., Atmospheric CO_2 fixation by dinuclear Ni(II) complex, $[TPANi(II)(\mu-OH)_2Ni(II)TPA](ClO_4)_2$ (TPA = tris(pyridylmethyl)amine), *Chem. Lett.*, pp. 929–930, 1996.

167. Tanase, T., Nitta, S., Yoshikawa, S., Kobayashi, K., Sakurai, T., and Yano, S., Spontaneous fixation of carbon dioxide in air by a nickel diamine complex: synthesis and characterization of a trinuclear nickel(II) complex with a novel hydrogen bonding system around a carbonate ligand, *Inorg. Chem.*, 31, 1058–1062, 1992.

168. Matsumura, N., Sakaguchi, Y., Ohba, T., and Inoue, H., 2-Morpholino-imidazolino magnesium(II) complex as a novel carbon dioxide carrier, *J. Chem. Soc. Chem. Commun.*, pp. 326–327, 1980.

169. Matsumura, N., Ohba, T., and Inoue, H., The function of magnesium(II) N,N'-dicyclohexylaminide complexes as a carbon dioxide carrier, *Bull. Chem. Soc. Jpn.*, 55, 3949–3950, 1982.

170. Matsumura, N., Asai, N., and Inoue, H., Fixation of carbon dioxide activated by a bromomagnesium thioureide complex, in Proc. Int. Symp. Chemical Fixation of Carbon Dioxide, Nagoya, Japan, December 2 to 4, 1991, 431–434.

171. Bottaccio, G., Marchi, M., and Chiusoli, G.P., Carboxylation of organic substrates in hydrocarbon media, *Gazz. Chim. Ital.*, 107, 499–500, 1977.
172. Walther, D., Ritter, U., Gessler, S., Sieler, J., and Kunert, M., CO$_2$ transfer by metal phenoxides: *N*-methyl-ε-caprolactam sodium phenoxide as a selective reagent for carboxylation reactions, *Z. Anorg. Allg. Chem.*, 620, 101–106, 1994.
173. Gambarotta, S., Arena, F., Floriani, C., and Zanazzi, P.F., Carbon dioxide fixation: bifunctional complexes containing acidic and basic sites working as reversible carriers, *J. Am. Chem. Soc.*, 104, 5082–5092, 1982.
174. Barth, D., Tondre, C., Lappai, G., and Delpuech, J.-J., Kinetic study of carbon dioxide reaction with tertiary amines in aqueous solutions, *J. Phys. Chem.*, 85, 3660–3667, 1981.
175. Nagao, Y., Hayakawa, A., Suzuki, H., Mitsuoka, S., Iwaki, T., Mimura, T., and Suda, T., Comparative study of various amines for the reversible absorption capacity of carbon dioxide, in Abstr. 4th Int. Conf. on Carbon Dioxide Utilization, Kyoto, Japan, September 1997, P-100.
176. Inagaki, N., Tasaka, S., and Chengfei, Z., Preparation of thin films containing sulfonic acid and carboxylic acid groups by plasma polymerization of perfluorobenzene/carbon dioxide mixtures, *Polym. Bull.*, 26, 187–191, 1991.

10 CO₂ Reforming and Hydrogenation

10.1 INTRODUCTION

The reduction of carbon dioxide to useful compounds such as fuels and chemical intermediates is inherently difficult, as it involves several steps of both electron and proton transfers. In the absence of activation by an external energy source — electrochemical, photochemical, or radiation induced — CO_2 may be activated by reduced compounds such as by hydrocarbons in the reforming reaction or by molecular hydrogen in the hydrogenation reaction. Catalysts are usually required to decrease the activation energy. The heterogeneous catalytic reactions of carbon dioxide were carefully reviewed by Fox,[1] Krylov and Mamedov,[2] and Wang et al.[3]

The direct or indirect hydrogenation of carbon dioxide may lead to several reactions, which are either kinetically or thermodynamically controlled, depending on the reaction conditions:[4]

1. The strongly endothermic reforming of methane to produce synthesis gas:

$$CH_4 + CO_2 = 2CO + 2H_2 \qquad \Delta H_{298\,K} = 247 \text{ kJ mol}^{-1} \qquad (10.1)$$

2. The exothermic methanation by the Sabatier process:

$$CO_2 + 4H_2 = CH_4 + 2H_2O \qquad \Delta H_{298\,K} = -113 \text{ kJ mol}^{-1} \qquad (10.2)$$

3. Methanol production, by the hydrogenation of both CO_2 and CO:

$$CO_2 + 3H_2 = CH_3OH(l) + H_2O(l) \qquad \Delta H_{298\,K} = -131 \text{ kJ mol}^{-1} \qquad (10.3)$$

$$CO + 2H_2 = CH_3OH(l) \qquad \Delta H_{298\,K} = -128 \text{ kJ mol}^{-1} \qquad (10.4)$$

4. The reverse water–gas-shift reaction, producing carbon monoxide and water:

$$CO_2 + H_2 = CO + H_2O(g) \qquad \Delta H_{298\,K} = 41 \text{ kJ mol}^{-1} \qquad (10.5)$$

5. The hydrogenation of CO_2 to formic acid:

$$CO_2 + H_2 = HCOOH(l) \qquad \Delta H_{298\,K} = -31.2 \text{ kJ mol}^{-1} \qquad (10.6)$$

6. The gasification of carbon by carbon dioxide (the reverse of CO disproportionation):

$$CO_2 + C = 2CO \qquad \Delta H_{298\,K} = 172.5 \text{ kJ mol}^{-1} \qquad (10.7)$$

The reforming of natural gas, followed by the hydrogenation of mixtures of carbon monoxide and carbon dioxide to methanol on various transition element catalysts, is one of the most important processes in the huge petrochemical industry. Together with carbon monoxide, carbon dioxide can also serve as the carbon source in the Fischer–Tropsch synthesis of mixtures of various aliphatic and aromatic hydrocarbons, aldehydes, ketones, and carboxylic acids. The mechanisms of these reactions have been clarified with the tools of modern surface science.

10.2   HYDROCARBON REFORMING

For the conversion of hydrocarbon feedstocks such as natural gas (mainly methane), petroleum, and coal to chemical intermediates, an important step is *reforming*. Reforming refers to the highly endothermic catalytic conversion of such hydrocarbons with either carbon dioxide or steam to produce synthesis gas (syngas), a mixture of carbon monoxide, carbon dioxide, and hydrogen. Synthesis gas may be converted by various exothermic reactions to very useful products, such as methanol, ethanol, ethylene, acetic acid, formaldehyde, phosgene, and acetone. Pure hydrogen is required for ammonia synthesis. The reforming of CH_4 with CO_2 (dry reforming) is environmentally beneficial in converting these two greenhouse gases to a useful product. The different routes available for the conversion of natural gas to synthesis gas were reviewed in detail by Fox.[1]

10.2.1   Methane CO_2 Reforming

The interaction of carbon dioxide with the widely available methane to form synthesis gas,

$$CH_4 + CO_2 = 2CO + 2H_2 \qquad \Delta H_{298\,K} = 247 \text{ kJ mol}^{-1} \qquad (10.1)$$

is most attractive for converting the two inexpensive raw materials into more valuable products. This strongly endothermic reaction requires temperatures above 700°C for high conversion. The CO$_2$ reforming of methane to H$_2$ and CO was described already by Fischer and Tropsch using various catalysts.[5] Best conditions were over Ni supported on silica, at 860°C. This reaction has recently attracted considerable interest. The advantages of CO$_2$ reforming vs. the steam reforming of methane,

$$CH_4 + H_2O = CO + 3H_2 \qquad \Delta H_{298\,K} = 206 \text{ kJ mol}^{-1} \qquad (10.8)$$

and also the problems associated with CO$_2$ reforming were reviewed by Seshan et al.[6,7] An advantage of CO$_2$ reforming of methane is the production of synthesis gas with lower remaining CH$_4$, compared with steam reforming, which leaves up to 2% of unreacted methane. High-purity synthesis gas is required for many synthetic applications. In some natural gas sources, CO$_2$ is present in high concentrations, sometimes more than 25%. It is then useful to react CO$_2$ directly with the CH$_4$. Another benefit of CO$_2$ reforming of CH$_4$ is the resulting lower H$_2$/CO ratio in the synthesis gas produced. This is necessary for the Fischer–Tropsch process, in which high H$_2$/CO ratios inhibit hydrocarbon chain growth and enhance the methanation reaction.[8,9]

10.2.2   Carbon Formation

The main problem which hindered the wide-scale introduction of CO$_2$ reforming is the formation of coke, which is thermodynamically favored except at very high temperatures, above 900°C. Carbon deposition may occur by the exothermic Boudouart reaction:[10]

$$2CO \leftrightarrow C + CO_2 \qquad \Delta H_{298\,K} = -172.5 \text{ kJ mol}^{-1} \qquad (10.9)$$

and by the endothermic cracking of methane:

$$CH_4 \leftrightarrow C + 2H_2 \qquad \Delta H_{298\,K} = 74.9 \text{ kJ mol}^{-1} \qquad (10.10)$$

Coking is less severe with steam reforming, in which the carbon may be volatilized by the reaction

$$C + H_2O \leftrightarrow CO + H_2 \qquad (10.11)$$

The stability of a series of Ni-based catalysts with different supports was tested by Turlier et al.[11] Catalyst degradation was found to be due to aging, presumably by sintering, as well as to carbon deposition. The observed sequence of stability depended on the support in the order

$$SiO_2 > Talc > SiO_2-Al_2O_3 > {\sim}ZrO_2 > Ni-Cu > Al_2O_3$$

In a comparison of the CO$_2$ reforming of methane on various nickel catalysts, Bradford and Vannice observed that Ni/TiO$_2$ and Ni/MgO were more resistant to coking than Ni/SiO_2 and Ni/C (activated carbon).[12] The Ni/MgO catalyst was stable for up to 44 h on stream. On spent Ni/SiO_2 catalysts, temperature-programmed oxidation spectra indicated a sharp peak at 806 K, which was attributed to the combustion of graphitic carbon. This carbon had been shown by transmission electron microscopy (TEM) to be deposited on the catalyst in the form of graphitic whisker fibers. The carbon deposition was facilitated by a lack of metal-support interaction in Ni/SiO_2. On the other hand, in Ni/TiO_2, a strong metal-support interaction resulted in the migration of TiO_x species onto the nickel surface, decreasing carbon deposition. With Ni/MgO, a partially reducible NiO–MgO solid solution was formed which was resistant to carbon deposition. The main contributor to carbon deposition during CO_2 reforming was proposed to be due to the exothermic carbon monoxide disproportionation reaction,

$$2CO = CO_2 + C \qquad \Delta H_{298\,K} = -172.5 \text{ kJ mol}^{-1} \qquad (10.9)$$

for which the equilibrium shifts to the left with rising temperature. Hence higher temperatures disfavor carbon deposition.[12]

10.2.3   Noble Metal Catalysts: SPARG and CALCOR Processes

With noble metal catalysts, such as Rh, Ru, and Ir, coking formation during CO_2 reforming is minimized. However, the very high cost of these metals and their low availability discouraged their large-scale industrial use. There have been considerable efforts to develop catalysts without noble metals. In the SPARG process for CO_2 reforming, part of the steam of conventional reforming has been replaced by CO_2, resulting in the production of synthesis gas with an H_2/CO_2 ratio of 1.8, which is required for many applications. Coke formation was prevented by using a partially sulfur-poisoned nickel catalyst. The SPARG process is suitable as a retrofit to existing steam-reforming plants. The product synthesis gas from this process still contains about 2.7% methane.[4]

Another commercial application of CO_2 reforming has been the development of the CALCOR process, a multistage process of reacting dry CO_2 with natural gas, liquid petroleum gas (LPG), and syngas to produce high-purity carbon monoxide that contains less than 0.1% methane.[13] In a study of this reaction (with CH_4/CO_2 ratio of 1:1, at space velocities of 3000 to 6000 h^{-1}) over alumina-supported platinum metals, in the temperature range 723 to 823 K, Solymosi et al. found that the specific activities of the catalysts decreased in the order Ru, Pd, Rh, Pt, and Ir.[14–16] This order of activity was explained to be related to the order for the dissociation of carbon dioxide to oxygen by these metals:

$$CO_2 \rightarrow CO + O \qquad (10.12)$$

Presumably the surface-adsorbed O atoms thus formed promote the dissociation of surface-adsorbed methane. The reaction products were mainly carbon monoxide and hydrogen (formed over Rh/Al_2O_3 in the ratio 2.6:1), as well as water, carbon deposits on the catalysts, and traces of ethane. In a comparison of the decomposition of CH$_4$ on Rh supported on different supports, the order of activity was Rh/Al$_2$O$_3$ > Rh/TiO$_2$ > Rh/SiO$_2$ > Rh/MgO. The activation energies for the production of hydrogen were always somewhat higher than those for the production of CO. Therefore, an increase in temperature resulted in a decrease in the CO/H$_2$ ratio. The ratio of the products indicates the participation of secondary reactions, such as the methanation of carbon dioxide and the reverse water–gas-shift reaction. An important advantage of the reforming reaction over Rh/TiO$_2$ was that only traces of carbon were deposited on the catalyst.[14–16]

In the search for optimal catalysts and supports for the CO$_2$ reforming of methane, Seshan et al. tested the reaction of CO$_2$ + CH$_4$ (2:1) at 800°C.[7,17,18] With 1 wt% Pt on several supports, the order of decreasing activity and stability was Pt/ZrO$_2$ > Pt/TiO$_2$ > Pt/γ–Al$_2$O$_3$. Using the support of choice, ZrO$_2$, the order of decreasing activity and stability with different noble metals was Pt/ZrO$_2$ ~ Rh/ZrO$_2$ > Pd/ZrO$_2$ ~ Ir/ZrO$_2$ > Ru/ZrO$_2$. In a lifetime test over 0.5 wt% Pt/ZrO$_2$ at about 700°C for 1000 h, the reforming activity was fairly stable, and there was no extensive carbon deposition. An economic evaluation of the production of syngas (for the manufacture of acetic acid) by the CO$_2$-reforming reaction with this catalyst indicated that its operating costs should be about 20% lower than steam reforming or partial oxidation of methane.[7,17,18]

Pt/ZrO$_2$ catalysts were also reported by van Keulen et al. to be very effective for the CO$_2$ reforming of methane and more stable than Pt/Al$_2$O$_3$ and Pt/TiO$_2$ catalysts.[19] The mechanism of the reaction was studied by these authors on 1 wt% Pt/ZrO$_2$, using a Temporal Analysis Products reactor system. This consisted of two high-speed pulse valves which could inject separately or together small portions of CO$_2$ and CH$_4$ (10^{14} to 10^{18} molecules per pulse) into a microreactor containing the catalyst and connected to a quadrupole mass spectrometer for monitoring the reactants and the products. Under the conditions of the experiments, carried out at temperatures of up to 700°C, no carbon deposits were formed on the catalyst. CH$_4$ and CO$_2$ were found to dissociate independently on Pt/ZrO$_2$. These experiments indicated that CH$_4$ decomposed on the Pt, producing C$_{ads}$, and H$_{ads}$. H$_{ads}$ was desorbed as H$_2$, while C$_{ads}$ reacted with an oxygen species O* produced by dissociation of CO$_2$ adsorbed on the ZrO$_2$, forming CO$_{ads}$, which was desorbed as gaseous CO. A pool of this oxygen species O* (of unknown nature, possibly OH radicals or even a carbonate species) was postulated to be in equilibrium with CO$_{2ads}$ on the surface of the catalyst

$$CO_{2ads} \Leftrightarrow CO_{ads} + O* \qquad (10.12)$$

CO$_2$ supplied oxygen to the pool, while CH$_4$ consumed this oxygen, producing either CO or CO$_2$.

Pulse experiments by Wang and Au on the reforming reaction of CH$_4$ + CO$_2$ and CD$_4$ + CO$_2$ over prereduced 0.5% Rh/SiO$_2$ at 700°C revealed a normal deuterium isotope effect. The conversion rate of CD$_4$ was about 5% lower than that of CH$_4$, indicating C–H bond breakage in the rate-determining step.[20]

In a comparison of the CO_2-reforming activity of various metals supported on an alumina-stabilized magnesia support (Mg:Al = 7:1 atomic ratio), Rostrup-Nielsen and Hansen observed the following sequence of activity: Ru, Rh > Ir > Ni, Pt, Pd. The harmful carbon formation was least severe with rhodium and ruthenium catalysts. However, sulfur-poisoned nickel catalysts were reported to be satisfactory at high operating temperatures.[21]

The mechanism of the CO_2 reforming of methane was studied by Basini and Sanfilippo using catalysts containing less than a monolayer of Rh, Ru, or Ir, supported on α-Al_2O_3, MgO, CeO_2, La_2O_3, and TiO_2, and applying very high space velocity conditions.[22] With CH_4–CO_2 (1:1) at 0.1 MPa and 750°C, and applying a contact time of 0.33 s (gas hourly space velocity [GHSV] = 11000 h^{-1}) over 0.1 wt% Rh/α-Al_2O_3, the turnover frequencies for CH_4 and CO_2 were 5.02 and 5.08 s^{-1}, respectively. Diffuse reflectance infrared Fourier transform spectra taken during the reforming reaction indicated the formation of highly reactive oxidic intermediates derived from the dissociation of CO_2, which were rapidly intraconverted into some hydridocarbonyl surface clusters. The oxidic intermediates prevent the carbon growth reactions which lead to coking.[22]

The activity of Rh catalysts for CO_2 reforming was found by Verykios et al. to depend strongly on the nature of the support, decreasing in the order YSZ > γ-Al_2O_3 $\geq$ TiO_2 > SiO_2 > La_2O_3 > MgO, where YSZ = yttria-stabilized zirconia.[23,24] Among these catalysts, Rh/YSZ and Rh/SiO_2 gave the most stable performance during 50 h of reaction. $^{13}CH_4$ and $C^{18}O_2$ tracer experiments were made on the CO_2 reforming of methane over Rh/Al_2O_3 and Rh/YSZ catalysts at 650°C to determine the origin of the intermediate carbon species accumulated on the surface of these catalysts. Over Rh/Al_2O_3, there appeared intermediate active carbon-containing species leading from CO_2 to CO, with a surface coverage of $\theta = 0.2$. However, the active carbon-containing species leading from CH_4 to CO provided only a very small surface coverage ($\theta < 0.02$). On the other hand, in the reforming reaction over Rh/YSZ, both the pathway from CO_2 to CO and from CH_4 to CO involved very small surface coverage of active carbon-containing species ($\theta < 0.02$). With respect to the active oxygen species leading to the formation of CO, the surface coverage on both Rh/Al_2O_3 and Rh/YSZ was also very small ($\theta < 0.02$). The main difference was that a large pool of lattice oxygen species was observed on Rh/YSZ at 650°C. Spillover of these oxygen species to the rhodium surface may contribute to the production of CO during the high-temperature reforming reaction. These oxygen species could explain the higher turnover frequency values observed over Rh/YSZ catalysts relative to those over Rh/Al_2O_3.[23,24]

The mechanism of CO_2 reforming on Al_2O_3-supported Rh, Ru, and Ir catalysts under normal pressure at 700°C was proposed by Mark and Maier to proceed by a stepwise process.[25] The slow rate-determining step involved dissociative adsorption of CH_4 on active metal sites, with formation of active carbon and free hydrogen:

$$CH_4 \rightarrow C_{ad} + 2H_2 \tag{10.13}$$

In a subsequent rapid step, CO$_2$ reacted with the active carbon to form two equivalents of CO:

$$C_{ad} + CO_2 \rightarrow 2CO \tag{10.14}$$

In the temperature range 750 to 850°C, with CH$_4$:CO$_2$ = 1, at atmospheric pressure and using Rh, Ru, or Ir catalysts on various supports, the activity of the catalysts could be correlated with the accessible metal surface area. The rate per unit surface area (turnover frequency) did not depend very much on the dispersion. The main function of the supports in these high-temperature reactions was to stabilize the metal surface area. Best supports were γ-Al$_2$O$_3$ and ZrO$_2$(5%)/SiO$_2$. Rhodium catalysts with these supports were stable for long reaction times.[25]

Zeolite–Y-supported rhodium complexes were found by Bhat and Sachtler to be highly active and stable catalyst precursors for the CO$_2$ reforming of CH$_4$.[26] The catalysts were prepared from aqueous solutions of [Rh(NH$_3$)$_5$Cl]Cl$_2$·6H$_2$O. The conversion of CH$_4$ reached 90% at 923 K, resulting in an H$_2$/CO ratio of about 1. The reaction was stable during 30 h on stream, and there was no coke deposition.

10.2.4 Nickel and Cobalt Catalysts

High conversions in the CO$_2$ reforming of CH$_4$ were achieved by Gadalla and Bower over Ni/Al$_2$O$_3$ in the presence of MgAl$_2$O$_4$ or Ca-aluminate.[8] MgAl$_2$O$_4$ effected the transformation of Ni/Al$_2$O$_3$ into NiAl$_2$O$_4$, which was soluble in MgAl$_2$O$_4$. Gadalla and Sommer tested the CO$_2$ reforming of methane on an industrial reforming catalyst of Ni/CaO–TiO$_2$–Al$_2$O$_3$, using a feed of CO$_2$–CH$_4$ (2.6:1) at a space velocity of 7.38 h^{-1}.[9] Almost 100% conversion was attained during 51 h on stream.

The carbon deposition during CO$_2$ reforming of CH$_4$ over Ni/Al$_2$O$_3$ catalysts was found by Horiuchi et al. to be markedly suppressed by adding basic metal oxides to the catalysts.[27] The order of decreasing effectiveness of these alkali oxides was Na$_2$O > MgO > CaO > K$_2$O. Thus, when passing CH$_4$ + CO$_2$ + Ar (15:15:60) at atmospheric pressure and 1073 K during 5 h over 10 wt% Ni/Al$_2$O$_3$ which was either without or with 10 wt% of the basic metal oxide, the amounts of carbon deposited were either 470 or <150 mg C g-cat^{-1}, respectively. When only CH$_4$ was passed over the regular Ni/Al$_2$O$_3$, a similar amount of carbon was formed as during CO$_2$ reforming, while the carbon deposition was considerably inhibited over the basic metal-treated catalyst. No carbon was formed when CO$_2$ alone was passed over the catalysts. Therefore, the carbon deposition during CO$_2$ reforming must have originated from the decomposition of CH$_4$ and not from CO$_2$. From the dependence of the rates of CO$_2$ reforming on the partial gas pressures of CH$_4$ and CO$_2$, it was possible to conclude that the surface of the Ni catalyst with basic metal oxides was abundant in adsorbed CO$_2$, thus disfavoring CH$_4$ decomposition. On the other hand, the surface of the Ni catalysts without the basic metal oxides was abundant in adsorbed CH$_4$, promoting CH$_4$ decomposition to carbon. The effect of the basic metal oxides could be due to an enhancement of the electron density of the supported Ni metal, which changed the adsorption properties of the catalysts.[27]

Osaki et al. used a "pulse surface reaction rate analysis" technique to determine the nature of the intermediate hydrocarbon species $CH_{x,ads}$ in CO_2 reforming of CH_4 over supported Co and Ni catalysts.[28–30] The experiments were performed by providing a continuous flow of CO_2 + He over the catalyst in a microreactor fitted with a quadrupole mass spectrometer. Methane, which was introduced into the reactor in brief pulses, adsorbed dissociatively on the catalyst, forming hydrogen-deficient $CH_{x,ads}$ species. The average number of hydrogen atoms x of these species produced on Ni/MgO, Ni/Al$_2$O$_3$, Ni/TiO$_2$, and Ni/SiO$_2$ had the values 2.7, 2.4, 1.9, and 1.0, respectively. The reaction of $CH_{x,ads}$ with CO_2 (or O_{ads}) was proposed to be the rate-determining step in CO_2 reforming. The decreasing number x of hydrogen atoms on the $CH_{x,ads}$ species with different supports could be correlated with increasing tendency to carbon formation on such catalysts.[28–30]

The relative turnover frequencies of the reforming reaction with Ni catalysts was found by Bradford and Vannice to decrease in the order Ni/TiO$_2$ > Ni/C > Ni/SiO$_2$ > Ni/MgO.[31] In reforming reactions with a gas feed of CO_2–CH_4–He (1:1:1.8) at space velocities of up to 200,000 h^{-1} in the temperature range 673 to 823 K, the rates of reaction as a function of the partial gas pressures of CO_2 and CH_4 could be fitted to power rate laws:

$$r = kP^a_{CH4}P^b_{CO2} \tag{10.15}$$

where a and b were the reaction orders for CH_4 and CO_2. Under the conditions of the CO_2-reforming reaction, the reverse water–gas-shift reaction was extremely close to equilibrium, and both reactions occurred simultaneously. As a result, the H_2/CO product ratio was depressed to values below 0.5, and the overall stoichiometry could be represented by:[31]

$$CH_4 + 2CO_2 \rightarrow H_2 + H_2O + 3CO \tag{10.16}$$

The activity and selectivity of Ni/SiO$_2$ and Ni/Al$_2$O$_3$ catalysts for CO_2 reforming of methane were found by Choudhary et al. to be considerably enhanced by precoating the catalysts with MgO, or CaO, or with rare earth oxides. Most effective was MgO.[32]

Improved selectivity for CO production during CO_2 reforming of CH_4 was obtained by Gronchi et al. using Ni and Rh catalysts supported on SiO$_2$ or La$_2$O$_3$. Highest activity was obtained with an Rh(0.2%)/La$_2$O$_3$ catalyst. Carbon deposition was much more severe with Ni than with Rh catalysts.[33–35]

Ruckenstein and Hu observed that NiO/MgO catalysts (prepared by impregnating MgO powder with aqueous Ni(NO$_3$)$_2$, followed by calcination in air at 800°C and reduction under H_2 at 500°C) have high activities for the CO_2 reforming of CO_2 and also excellent stability.[36–41] In these catalysts, X-ray photoelectron spectroscopy (XPS) analysis indicated that the surface was enriched in MgO and that there was electron transfer from NiO to MgO. The optimal activity was with NiO–MgO (2:10 by weight), providing a CO yield of up to 94%. With CO_2–CH_4 (1:1) at atmospheric pressure and 790°C, at a GHSV = 253,500 cm^3 g^{-1} h^{-1}, the rate of CO production was 1.9 mmol g^{-1} s^{-1} and was stable for at least 50 h. NiO was found to form a solid solution with the MgO support, which inhibited carbon deposition and was resistant to sintering. By contrast, NiO–MgO prepared by mechanical mixing of the two oxides lost its initial CO$_2$-reforming activity within a few hours. For optimal CO$_2$-reforming activity, the MgO was prepared by the decomposition of (MgCO$_3$)$_4$Mg(OH)$_2$ at 1100°C for 10 min. Ni catalysts with other supports, such as CaO, SrO, and BaO, were all found to become deactivated quite rapidly due to coking, which plugged the reactor. The order of initial CO yields with 13.6 wt% Ni on several supports decreased in the order

$$Ni/Al_2O_3 > Ni/SiO_2 > Ni/TiO_2$$

The resistance to coking of 20 wt% Ni/La$_2$O$_3$ as reforming catalyst depended on the method of preparation. When Ni(NO$_3$)$_2$ was used as the precursor, followed by calcination, the reaction of CH$_4$ + CO$_2$ (1:1) at atmospheric pressure and 790°C stopped completely within 6 h because of carbon deposition. However, the catalyst prepared from NiCl$_2$ as precursor had high stability, with a CO yield of about 50% for 26 h.[36–41]

A highly stable Ni$_{0.03}$Mg$_{0.97}$O solid solution catalyst for CO$_2$ reforming of CH$_4$, resistant to carbon deposition, was found by Tomishige et al. to consist of an Ni-carbonyl species containing bonds of C–O–Mg. At 1123 K, the CH$_4$ conversion of 80% was stable for at least 100 days. The activity and stability of this catalyst were improved by pretreatment with water, which caused the formation of Ni and Mg hydroxides.[42,43]

Kroll et al. proposed that the active phase for CO$_2$ reforming of methane on Ni/SiO$_2$ catalysts is a nickel carbide-like layer, such as Ni$_2$C or Ni$_3$C.[44,45] This layer was formed immediately upon contact with the CH$_4$ + CO$_2$ gas mixture. Thus, the active phase was suggested not to be metallic Ni. The nature of the catalytically active phase was determined by temperature-programmed hydrogenation experiments at various periods after starting the reforming reaction. Highly active forms of carbon built up within a few minutes of beginning the catalytic run. These carbon species had a C/Ni$_s$ atomic ratio of 0.33 to 0.39, where Ni$_s$ represented the number of surface Ni atoms derived by magnetic measurements. Such C/Ni$_s$ ratios corresponded to surface nickel carbides Ni$_2$C or Ni$_3$C. These surface carbon adspecies were proposed to be in equilibrium with gaseous methane. By reaction with oxygen adspecies derived from the dissociation of CO$_2$, these reactive carbon adspecies could be converted irreversibly into CO. After a prolonged period on stream (67 h), a much less reactive form of carbon with a C/Ni$_s$ ratio of 23.2 was observed. This carbon species was assigned by TEM to consist of carbonaceous filaments or whiskers. A comparison of the rates of CO$_2$ reforming of CH$_4$ and CD$_4$ showed the absence of a kinetic isotope effect, which suggested that the overall reforming reaction did not involve C–H bond breakage as the rate-determining step. The rate-determining step was therefore proposed to be the reaction C$_{ads}$ + O$_{ads}$ → CO. Optimal conditions for CO$_2$ reforming were over a catalyst of 4 wt% Ni/SiO$_2$, which had been calcined under flowing O$_2$ at 750°C for 8 h and which had not been prereduced. Over such a catalyst, the reaction of $CH_4 + CO_2 + Ar$ (15:15:60) at atmospheric pressure and 700°C, after continuous 240 h on stream, reached a methane conversion yield of about 73%. During this period, the rate of catalyst deactivation was only 0.01% h^{-1}. With a similar catalyst which had been prereduced under H_2, the rate of deactivation was much higher. Coke formation was moderate for the reaction at 700°C, but was quite severe when carrying out the CO_2 reforming at 600°C.[44,45]

In order to overcome the problem of Ni sintering during CO_2 reforming of methane over Ni catalysts, Bhattacharyya and Chang tested a nickel aluminate spinel catalyst, which was prepared from sodium aluminate and nickel nitrate.[46] The resulting $NiAl_2O_4$ had a cubic closely pack spinel structure. The reaction of $CO_2 + CH_4$ (1.23:1.00) at 300 psig, 810°C, and a space velocity of 7200 h^{-1} occurred with CH_4 and CO_2 conversions of 86 and 47%, respectively. The product ratio of H_2/CO was 1.3. No significant decrease in activity was observed during almost 100 h on stream. The $NiAl_2O_4$ catalyst, after the above reforming reaction, was found by X-ray diffraction (XRD) analysis to have been partially converted to metallic Ni and α-Al_2O_3. Under similar reforming conditions, a physically mixed NiO/Al_2O_3 catalyst was subject to very rapid deactivation.[46]

Nickel catalysts supported on pentasil-type zeolites for the CO_2 reforming of CH_4 were found by Chang et al. to have both high activity and resistance to coking, as shown by testing for up to 140 h on stream. Coke formation could be decreased considerably by pretreating the catalysts with K or Ca oxides. These KNiCa/zeolite catalysts were proposed to act by promoting the dissociative adsorption of CO_2, resulting in the oxidative removal of surface carbon species.[47,48]

The effects of Ni-crystallite size on the reactivity of carbon-containing species on Ni/SiO_2 catalysts for the methanation of $CO + H_2$ was studied by van Looij and Geus. Optimal methanation activity was observed with catalysts prepared by homogeneous deposition precipitation, with nickel crystallites of about 4-nm particle size.[49]

Ultrafine $Ni_xFe_{3-x}O_4$ particles, prepared by coprecipitation from solutions of the Ni and Fe salts, were found by Tamaura et al. to be highly effective for the methanation of CO_2. With Ni(II) ferrite (Ni/Fe molar ratio 0.15), using $H_2 + CO_2$ (4:1) at 300°C, the selectivities to CH_4 and CO were 96.4 and 2.8%, respectively.[50,51]

CH_4/CD_4 isotope effects were applied by Wang and Au to clarify the mechanism of the CO_2 reforming of methane over SiO_2-supported Ni catalysts.[52] The primary reactions leading to CO were proposed to be the dissociation of CO_2, followed by oxidation of surface-adsorbed CH_x species.

The stability of cobalt catalysts supported on active carbon or silica for the CO_2 reforming of methane was found by Guerrero-Ruiz et al. to be considerably enhanced by adding MgO as a promoter, thus decreasing the carbon deposition on the catalyst surface.[53,54] With $CH_4 + CO_2 + He$ (1:1:8) at 700°C, total pressure = 1 atm, GHSV = 1.2×10^4 h^{-1}, and using either Co–MgO(5:5)/C or Co–MgO(5:5)/SiO_2 as catalysts, the CH_4 conversions were 80 and 83%, respectively, and the selectivities to CO were 72 and 75%, respectively. Temperature-programmed desorption experiments indicated the formation of strongly adsorbed CO_2 species on the MgO surface. These species presumably reacted with surface carbon deposits, preventing the catalyst deactivation.[53,54]

The cobalt metal catalysts Co/γ-Al$_2$O$_3$ were found by Lu et al. to be highly active for CO$_2$ reforming, achieving more than 92% conversion of CH$_4$ and CO$_2$.[55] In order to identify the reaction intermediates, the CO$_2$–CH$_4$ (1:1) mixture was passed in pulses over H$_2$-prereduced 18 wt% Co/Al$_2$O$_3$. XPS and Auger electron spectroscopy (AES) measurements on the catalyst indicated that surface carbide carbon and hydrogen were formed by the dissociation of CH$_4$ on metallic cobalt species, which also abstracted oxygen atoms from CO$_2$, transferring these atoms to carbide, thus forming CO.

10.2.5 Ultrafine Single-Crystal MgO Supports

Very effective catalysts, not requiring expensive metals, were developed by Matsuura and Takayasu using ultrafine single-crystal magnesium oxide (uscMgO) as support.[56] This magnesium oxide has an extremely great surface area and also provides an excellent support effect. With 3 mol% Ni/uscMgO as catalyst system, methane conversion at 800°C was 96%. At the lower temperature of 660°C, the relative activity of different metals supported on uscMgO was tested. The order of activity was

$$Ru > Rh > Pd > Pt > Ni$$

The causes of the deactivation of Ni catalysts during CO$_2$ reforming of CH$_4$ were carefully studied by Takayasu et al. with uscMgO, SiO$_2$, and α-Al$_2$O$_3$ as supports.[57] In endurance tests carried out for 600 h at 800°C with a feed of CH$_4$ + CO$_2$ + He (15:15:70), Ni/SiO$_2$ underwent abrupt deactivation within the first 50 h, which was shown to be due to an increase in the Ni particle sizes. Ni/α-Al$_2$O$_3$ underwent severe deactivation within 100 to 200 h, accompanied by the formation of NiAl$_2$O$_4$. Ni/uscMgO underwent very slow gradual deactivation, proven to be due to the formation of carbon deposits, and without increases in the Ni particle sizes. These carbon deposits could be avoided by using as catalyst either a mechanical mixture of Ni/uscMgO + SiO$_2$ (1:1 by weight) or with only Ni/uscMgO but with an excess of CO$_2$ in the gas feed: CH$_4$ + CO$_2$ + He (15:20:65). Both methods substantially increased the catalyst lifetime. The favorable property of MgO as a support was proposed to be due to its activity as an electron donor.[57]

10.2.6 MoS$_2$ and WS$_2$ Catalysts

The sulfides of molybdenum and tungsten were found by Osaki et al. to be highly resistant to carbon formation during CO$_2$ reforming of CH$_4$.[58–60] In a comparison of MoS$_2$ and WS$_2$ with Ni/SiO$_2$ in the reaction CH$_4$ + CO$_2$ (1:1), the order of initial reactivity was Ni/SiO$_2$ >> MoS$_2$ > WS$_2$. However, in a test at 870°C, the catalytic activities of MoS$_2$ and WS$_2$ were stable with time, while Ni/SiO$_2$ lost its activity significantly within a few hours. From a study of the dependence of the reaction rates on the partial pressures of CH$_4$ and CO$_2$, it was concluded that the reaction order on Ni/SiO$_2$ catalysts was positive with respect to the CO$_2$ partial pressure and negative with respect to the CH$_4$ partial pressure. On the other hand, with the sulfide catalysts, the reaction order was negatively related to the CO_2 partial pressure and positively to the CH_4 partial pressure. Under steady-state conditions, the sulfide catalyst surfaces were almost saturated with O_{ads} due to considerable CO_2 dissociation (thus no CH_4 decomposition), while on the Ni/SiO_2 catalyst, the surface was saturated with adsorbed CH_4 (thus considerable CH_4 decomposition). While Ni/SiO_2 catalysts were sensitive to sulfur poisoning, the two sulfide catalysts were resistant to sulfur and were unaffected by introduction of H_2S during the CO_2-reforming reaction. Addition of alkaline metal salts to supported Ni catalysts inhibited the decomposition of CH_4 and therefore slowed down the carbon formation.[58-60]

10.2.7   Combined Partial Oxidation and Reforming of Methane

By including air or oxygen in the feed gas mixture of steam or CO_2 reforming of natural gas (mainly methane), three important benefits were obtained: (1) the $H_2/(CO + CO_2)$ ratio was lowered to the ratio of 2 required both for methanol and Fischer–Tropsch synthesis, (2) the exothermicity of the natural gas combustion,

$$CH_4 + 1/2O_2 \rightarrow CO + 2H_2 \qquad \Delta H_{298\,K} = -38 \text{ kJ mol}^{-1} \qquad (10.17)$$

compensated for the endothermicity of the reforming reactions, and (3) part of the carbon which would be formed was burned off. The process could thus be operated industrially in refractory-lined adiabatic fixed bed reactors.

The selective oxidation of methane with air to synthesis gas was discovered by Ashcroft et al. to be promoted by various lanthanide ruthenium oxide catalysts.[61] The reactions were performed at 777°C on CH_4–O_2–N_2 (2:1:4) at atmospheric pressure, with up to 94% conversion of CH_4 and almost 100% conversion to CO and H_2. Hickman and Schmidt performed the direct catalytic oxidation of methane with air over Pt-coated monolithic catalysts at 1100°C, using very short contact times, 10^{-2} to 10^{-4} s, and reaching high selectivities of H_2 and CO.[62]

The improvement in energy economy by the partial oxidation of natural gas was simulated by De Groote et al., operating the syngas production in a tubular fixed bed reactor with flow reversal.[63-66] The air or oxygen content of the gas feed was adjusted so that the overall partial oxidation process would be slightly exothermic. The catalyst bed (e.g., of Ni/Al_2O_3) was preheated to a uniform temperature, and the reactant gas mixture was then fed at a much lower temperature. The oxidation of the methane started at the inlet, and a temperature wave traveled in the direction of the gas flow. Just before this wave reached the outlet, the direction of the gas flow was reversed, and the wave traveled in the opposite direction. A large fraction of the carbon deposited in a semicycle could be removed during the reversed stage. The whole operation was thus adiabatic, without requiring an external heat supply. The primary step was proposed to be dissociative adsorption of CH_4 on reduced metal sites.[63-66]

A significant technical advance was achieved by Ashcroft et al. by combining the partial oxidation and the CO_2 reforming of CH_4.[67] By passing CH_4–CO_2–O_2 (49.8:48.8:1.4) at atmospheric pressure over 1% Ir/Al_2O_3 at 1050 K, the conversions of CH$_4$ and CO$_2$ were 91 and 87%, and the yields of H$_2$ and CO were 91 and 89%, respectively. Ni catalysts could also be used, but a large excess of CO$_2$ was necessary to prevent carbon deposition (e.g., CH$_4$–CO$_2$–O$_2$ [4:9:1]). By combining the exothermic partial oxidation with the endothermic CO$_2$ reforming, a thermally neutral reaction was attained.[67]

Using an Ni–CaO catalyst, Choudhary et al. performed the simultaneous catalytic partial oxidation and CO$_2$ reforming of CH$_4$, with more than 95% conversion, with more than 90% H$_2$ selectivity, and without carbon formation.[68] With the same catalyst, Choudhary and Rajput carried out simultaneous CO$_2$ and steam reforming of CH$_4$ at 800 to 850°C and space velocities of 20,000 to 30,000 cm^3 g^{-1} h^{-1}, resulting in almost quantitative conversion to syngas, with 100% selectivity for CO and H$_2$, and with very little interference by coking.[69]

The perovskite LaNiO$_3$ was found by Choudhary et al. to be an efficient catalyst for the conversion of methane to syngas, both by the partial catalytic oxidation of CH$_4$ and by the simultaneous steam, CO$_2$, and oxidative reforming reactions.[70] The catalyst precursor was prepared from aqueous solutions of the La and Ni nitrates by precipitation with Na$_2$CO$_3$. The catalytic reactions were performed at 800 to 850°C with high space velocities (contact time ~9 ms), resulting in syngas with an H$_2$/CO ratio of about 2.0, with a conversion yield of >90%, and almost 100% selectivity. By proper adjustment of the CH$_4$/CO$_2$/H$_2$O/O$_2$ ratio and temperature, an almost thermoneutral reaction could be achieved. During the reforming reaction, the reactive species was proposed to be Ni°. If the LaNiO$_3$ was modified by partial substitution of La by Ca or Sr, or by substitution of Ni by Co, the resulting material had much inferior catalytic activity and selectivity.[70]

The simultaneous combustion, reforming, water–gas shift, and carbon formation and gasification reactions were modeled by De Groote and Froment under simulated industrial conditions.[64,66,71] Such catalytic partial oxidation of methane has been applied in the Texaco and Shell processes for syngas production.

Steam–CO$_2$ mixed reforming of CH$_4$ over MgO-supported noble metals was compared by Qin et al. with the partial oxidation of CH$_4$ over the same catalyst.[72] The steam–CO$_2$ mixed reforming is advantageous in producing syngas with an H$_2$/CO ratio of 2:

$$CH_4 + 2/3H_2O + 1/3CO_2 \rightarrow 4/3CO + 8/3H_2 \qquad \Delta H_{298\,K} = 219 \text{ kJ mol}^{-1} \quad (10.18)$$

Using 0.5 wt% Rh/MgO, the partial oxidation of methane in a gas feed of CH$_4$–O$_2$–N$_2$ (3.6:1.8:7.2) was followed in a temperature-programmed reaction study. A sharp ignition of oxidation occurred at 550°C. For the mixed reforming of methane under similar conditions, with CH$_4$–H$_2$O–CO$_2$–N$_2$ (2.7:1.8:0.9:7.2), the sharp ignition onset appeared at 570°C. The simultaneous onset of ignition with both CO$_2$ and H$_2$O reforming and with partial oxidation of methane suggested a similar reaction mechanism, which was proposed to involve O$_{ad}$ as a common reaction intermediate. Over the same catalyst, the onset of the CH$_4$–D$_2$ exchange occurred at above 300°C and became very rapid at higher temperatures. Thus, on the noble metal catalyst, CH$_4$ was readily activated under reforming or partial oxidation conditions.[72]

In order to achieve the rapid synthesis of hydrogen through methane reforming, while minimizing the use of expensive rare metals, Inui et al. developed a four-component catalyst.[73-76] With Ni–Ce_2O_3–Pt–Rh (6.6:3.9:2.2:0.2 wt%), a synergistic effect of the precious metals even in such low concentrations was ascribed to their acting as portholes for hydrogen spillover. With such a catalyst, supported on alumina-wash-coated ceramic fibers, the combined combustion and reforming reaction of a mixture of CH_4–CO_2–H_2O–O_2 (60:10:10:20%) at 700°C resulted in 90% methane conversion at an extremely short contact time (5 ms) and a space velocity of 730,000 h^{-1}, with a space–time yield of hydrogen of 12,190 mol L^{-1} h^{-1}. The inclusion of a small proportion of O_2 in the gas feed provided the large amount of heat required for the endothermic reforming reaction.[73-76]

10.2.8   Partial Oxidation of Methane to Ethane and Ethene

In a variant of the carbon dioxide reforming of methane to synthesis gas over MgO- or CaO-supported catalysts, the products were found by Nishiyama and Aika to be ethane, ethene, and carbon monoxide:[77,78]

$$2CH_4 + CO_2 \rightarrow C_2H_6 + CO + H_2O \qquad (10.19)$$

$$2CH_4 + 2CO_2 \rightarrow C_2H_4 + 2CO + 2H_2O \qquad (10.20)$$

In both reactions, $\Delta G = 35$ kJ/mol CH_4 at 1073 K, which are thus energetically uphill reactions. Such oxidative coupling of methane, in which CO_2 is the oxidant, was found to be promoted by PbO/MgO and BaO/CaO. The reactions were performed with a CH_4/CO_2/O_2 (100:100:1 by volume) gas mixture. The small amount of oxygen included was designed to make the total process thermodynamically favorable. With 50% PbO/MgO and 10% BaO/CaO as catalysts, the yields of the C_2 hydrocarbons were 4.4 and 3.5%, respectively.[77,78]

10.2.9   Solar Furnace Reforming

Carbon dioxide reforming of methane was demonstrated by Levy et al. in a solar furnace, using an Engelhard catalyst (0.5% Rh on alumina). The product gas temperature was maintained at 750 to 800°C and the maximal flow rate of the gases reached 11,000 L h^{-1}. With a power input of up to 6.5 kW, the maximal methane conversion reached 85%.[79-83]

A similar carbon dioxide reforming of methane was also performed by Buck et al. with a direct absorption solar receiver/reactor, heated by a parabolic dish, applying up to 150 kW of solar power.[84,85] The catalytic system was a reticulated porous alumina foam disk coated with a rhodium catalyst. Up to 97 kW of total solar power was absorbed, bringing temperatures within the absorber to a range of 550 to 1100°C. The maximum methane conversion reached almost 70%, and the chemical efficiency reached 54%. There was no carbon deposition, but there were problems due to cracking, sintering, and deactivation of the catalyst.[84,85] In another study, by Perera et al., ruthenium and iridium supported on Eu$_2$O$_3$ were used as catalysts for carbon dioxide reforming of methane to synthesis gas, in what was proposed to be a viable solar–thermal energy system. A conversion efficiency of 90% was achieved either with 1% metal loading at 673 K or with 5% loading at 1023 K.[86]

Very effective carbon dioxide reforming was achieved by Richardson and Paripatyadar with Rh and Ru metal catalysts supported on γ-Al$_2$O$_3$. With 0.5% Rh/ Al$_2$O$_3$ at a GHSV = 103,000 h^{-1} and 800°C, the conversion of CH$_4$ reached 85.5%. Also, with this catalyst, there was no carbon deposition in the temperature range 600 to 800°C. This catalyst was successfully tested in experiments in solar receivers.[87]

Mathematical simulations were provided by Meirovitch and Segal[88] and Skocypec et al.[89] on the storage of concentrated solar energy during CO$_2$ reforming of methane.

10.2.10   Oxidation of Alkanes by CO$_2$ to Aromatics, Olefins, and Acetic Acid

10.2.10.1   Alkanes to Alkenes and Aromatics

The catalytic reduction of carbon dioxide by methane or higher hydrocarbons has the advantage of not requiring the energy- and cost-intensive production of hydrogen. The carbon dioxide reforming of propane and propene is particularly attractive. The catalytic reduction of carbon dioxide accompanying the dehydrogenation of propane on the oxide catalysts Cr$_2$O$_3$, ZnO, and Ga$_2$O$_3$ was tested by Hattori et al. using a pulse reaction technique:[90,91]

$$CO_2 + C_3H_8 \rightarrow CO + C_3H_6 + H_2O \qquad (10.21)$$

There occurred considerable decomposition of propane on the chromium oxide and zinc oxide catalysts. Also, the production of propene (propylene) was retarded by the presence of carbon dioxide. Only on gallium oxide were propane and carbon dioxide converted to propene and carbon monoxide without the formation of by-products.[90,91]

In the aromatization of lower alkanes such as propane,

$$CO_2 + C_3H_8 \rightarrow CO + \text{aromatics} + H_2O \qquad (10.22)$$

LPG fractions were selectively converted by the Cyclar process into petroleum-grade aromatics. This process became commercial in 1990 with the start-up of the first unit at the Grangemouth refinery in Scotland.[92] A typical catalyst for this aromatization of propane is Ga^{3+}-loaded ZSM-5, which enabled high selectivity to aromatics, as well as the total conversion of propane. The main products were C$_6$–C$_8$ aromatics.[93]

Carbon dioxide reduction to carbon monoxide on Zn- and Ga-loaded HZSM-5, with accompanying aromatization of ethane and propane, was studied by Hattori et al. in detail.[94–96] The Zn/HZSM-5 catalyst was obtained by refluxing an aqueous solution of $Zn(NO_3)_2$ with HZSM-5 (Si/Al ratio 1:46.6). At a reaction temperature of 823 K, the C_3H_8 conversion reached 68%, and the organic products were aromatics (51% yield), $CH_4 + C_2H_6$ (25%), $C_2H_4 + C_3H_6$ (22%), and C_4^+ (2%). The presence of carbon dioxide suppressed the formation of coke, which occurred at such high temperatures in the aromatization of propane alone on the same catalyst. The Zn/HZSM-5 catalysts were characterized by TEM and XRD. The excess of Zn on the catalyst surface was found to consist of dispersed ZnO crystallites several nanometers in diameter, which were proposed to be the catalytically active sites. The catalytic reduction of CO_2 by C_2H_6 was also effective with the Zn- or Ga-loaded HZSM-5 catalysts. Using 10% Ga/HZSM-5 at 823 K, the reaction of CO_2 + C_2H_6 resulted in 12% conversion of C_2H_6, yielding aromatics (56%), C_2H_4 (35%), and CH_4 (6%). The mechanism proposed for the aromatization of ethane and propane involved the initial dehydrogenation to the olefin, followed by oligomerization and dehydrocyclization.[94–96]

10.2.10.2 Ethylbenzene to Styrene

Styrene is an important monomer required for the production of polystyrene and copolymer blends. It is currently produced industrially by the dehydrogenation of ethylbenzene over iron oxide catalysts in the presence of a large excess of steam, a process which is very wasteful in terms of energy economy. Sugino et al. discovered a method for the dehydrogenation of ethylbenzene at 973 K by iron-oxide-based catalysts in the presence of an excess of CO_2.[97] Highest activity was achieved with a catalyst of Fe–Li (3:0.3 mmol) per gram of active carbon support. Styrene yields of up to 51% were attained. The efficient oxidative dehydrogenation of ethylbenzene with CO_2 to styrene was performed by Chang et al. using a ZSM-5-supported iron oxide catalyst.[98] The catalyst was prepared by precipitation of Fe(II) hydroxide on NaZSM-5 zeolite, resulting in an Fe_3O_4 phase which was highly dispersed in the zeolite matrix. At 600°C and W/F = 298 g h mol^{-1}, the yield of styrene was about 40%.[98] In an alternative process, Mimura et al. used an Fe/Ca/Al oxide catalyst for the dehydrogenation of ethylbenzene with excess CO_2 at 823 K. By substituting CO_2 for steam, the space–time yield of styrene was enhanced, the deactivation of the catalyst was partly suppressed, and the energy required for the production of styrene was considerably reduced.[99]

10.2.10.3 Ethylene to Divinyl and Propene

In the presence of CO_2 over a tungsten–chromium catalyst, W–Cr–O/SiO_2, ethylene was found by Mirzabekova and Mamedov to undergo highly selective conversion to divinyl (C_4H_6) and to propene, while CO_2 was reduced to CO.[100]

The lattice oxygen species generated by CO_2 on the catalyst surfaces during the CO_2 reforming of hydrocarbons was shown by Mirzabekova et al. to be readily reduced and removed by the adsorbed hydrocarbon species. This lattice oxygen was different from that generated by O_2.[101] The oxidative coupling of methane using the oxygen of CO_2 was reviewed by Aika.[102]

10.2.11 Toluene Reforming by CO_2 to Benzene

The catalytic reforming of toluene with CO_2 was found by Pillai et al. to yield benzene, CO, and H_2 by an endothermic reaction:[103]

$$C_6H_5CH_3(l) + CO_2(g) \rightarrow C_6H_6(l) + 2CO(g) + H_2(g)$$

$$\Delta H°_{298\ K} = 50.1 \text{ kcal mol}^{-1} \qquad (10.23)$$

Effective catalysts were rare-earth-promoted Pd/γ-Al_2O_3, which had been reduced under H_2 at 400°C. The reaction was performed by flowing over the catalyst toluene + CO_2 + He (1:1:3 molar ratio) at 400°C and 1 atm. The initial rates of benzene and CO formation were about 2 and 5 mmol g-cat^{-1} min^{-1}, respectively.[103]

10.2.12 Fluidized Bed Reforming of Methane

While most experiments on CO_2 reforming of CH_4 were carried out on fixed bed reactors, the reaction has also been explored in fluidized bed reactors. Olsbye et al. performed the partial oxidation of CH_4 to synthesis gas over 1.5% Ni/Al_2O_3 in a fluidized bed reactor at 700°C, using a gas feed of CH_4 + O_2 + N_2 + H_2O (2:1:2:0.5).[104] The O_2 conversion was complete after a contact time of only 8 ms. The same reactor was used by Slagtern et al. for the CO_2 reforming of CH_4, using rare-earth-promoted Ni/Al_2O_3 catalysts.[105,106] Without the rare earth promoter, the reaction of CH_4 + CO_2 + N_2 (2:2:1) over 0.15 wt% Ni/Al_2O_3 at atmospheric pressure and 800°C resulted in rapid deactivation of the catalyst, which was due both to sintering of the Ni particles and to coking. More stable operation was achieved with a lanthanum-modified catalyst. Optimal activity and stability were obtained with 2 to 5 wt% $La/0.15$ wt% Ni/Al_2O_3. Instead of the pure lanthanum, a less expensive rare earth mixture could be used (designed as Ln), with equal results as catalyst promoter. Its composition was La–Nd–Pr (66.1:25.4:7.7 atom%). The catalysts were pretreated by calcination under oxygen at 800°C and reduced under H_2 at 800°C. With 0.15% $Ni/1.7\%$ Ln/Al_2O_3 as catalyst, the initial CH_4 conversion was close to the thermodynamic equilibrium of 91.6%. Long-term tests on stream for 600 h indicated a loss in methane conversion of 30%. The rare-earth-modified catalysts had higher mechanical strength, which was particularly important for the fluidized bed operation, to minimize the attrition by loss of fines. The catalysts could by regenerated by treatment with CO_2 at 700°C. By *in situ* XRD characterization of the catalysts during the CO_2 regeneration, it was shown that this treatment probably removed coke and also caused formation of NiO. The NiO was immediately reduced to redispersed metallic Ni, which was proposed to be the active phase in the CO_2 reforming of CH_4.[105,106]

10.2.13 CO_2 Reforming in Hydrogen-Permselective Membrane Reactor

For the endothermic CO_2-reforming reaction, very high temperatures, above 800°C, are required to shift the reaction equilibrium to high conversion. An interesting approach to raise the conversion to high levels already at lower temperatures was investigated by Ioannides and Verykios by performing the reaction in membrane reactors in which the membranes were selectively permeable to hydrogen, thus shifting the reaction toward completion.[107] The reactors consisted of porous Vycor tubes, on which dense silica membranes were created by chemical vapor deposition from $SiCl_4$ and H_2O at 700°C. The hydrogen permeance of these membranes was 0.2 to 0.3 cm^3 cm^{-2} min^{-1} atm^{-1}, and the H_2/N_2 selectivity was 200 to 300. Using 0.1% Rh/SiO_2 as catalyst, with a gas feed of $CH_4 + CO_2$ (1:1), and using very low space velocities of 5 to 15 h^{-1}, the methane conversion in the membrane reactors was considerably enhanced relative to the equilibrium conversion. Thus, with a membrane reactor, methane conversions of more than 95% were already obtained at 600°C, at which the equilibrium conversion was only about 40%. Also, the hydrogen selectivity was higher than 99% at temperatures above 550°C. The improved hydrogen selectivity was explained by the effect of hydrogen removal in preventing the reverse water–gas-shift reaction,

$$CO_2 + H_2 = CO + H_2O \qquad (10.5)$$

thus improving the selectivity of the reforming reaction. The same membrane reactor was also successfully applied to the partial catalytic oxidation of methane. Regrettably, the low hydrogen permeance of the above membrane reactors prevented the application of the high space velocities required for industrial applications.[107]

Rapid hydrogen permeation with an absolute selectivity was attained by Kikuchi et al. by plating a thin Pd or Pd–Ag alloy film (about 20 m thick) on the outer surface of a porous alumina ceramic tube, which was incorporated in the catalytic reactor for CO_2 reforming of methane.[108–110] By selective removal of hydrogen from the reaction system, the equilibrium of the reforming reaction was shifted to the product side, enabling the reaction to proceed to almost 100% completion even at the low temperature of 773 K. At this temperature, Ni catalysts for the reforming reaction were useless because of extensive coking. Among noble metal catalysts, the order of activity was Rh ~ Pt > Pd > Ru > Ir. Pt catalysts supported on $Al_2O_3–La_2O_3$ and $Al_2O_3–CeO_2$ enabled effective syngas production, with very little coking, The same reactor was also successfully applied to the partial catalytic oxidation of methane to syngas at 773 K, using air as the oxidant.[108–110]

10.3   CATALYTIC HYDROGENATION

The catalytic hydrogenation of carbon dioxide to water and carbon monoxide, which is the reverse of the water–gas-shift reaction, is a mildly endothermic reaction,

$$CO_2(g) + H_2(g) = CO(g) + H_2O(g) \qquad (10.5)$$

with enthalpy and free energy changes of $\Delta H_{298 K} = 41.17$ kJ/mol and $\Delta G_{298 K} = 28.64$ kJ/mol. On the other hand, the methanation of carbon dioxide (the Sabatier reaction),

$$CO_2(g) + 4H_2(g) = CH_4(g) + 2H_2O(g) \qquad (10.2)$$

is strongly exothermic, with $\Delta H_{298\ K} = -164.91$ kJ/mol and $\Delta G_{298\ K} = -113.6$ kJ/mol.

10.3.1 Reverse Water–Gas-Shift Reaction

10.3.1.1 Heterogeneous Catalysis

Highly effective catalysts for the reverse water–gas-shift reaction were developed by Taoda et al. using finely divided molybdenum sulfide on various support materials such as TiO_2.[111] The catalyst was prepared by impregnating the support material with an ammoniacal solution of ammonium tetrathiomolybdate, followed by drying and calcining at 350°C. The activity of the supported MoS_2 at 400°C for the hydrogenation of CO_2 to CO depended on the support in the order $TiO_2 > Al_2O_3 > ZrO_2 > SiO_2 > CeO_2 > MnO_2$. The selectivity to CO production was more than 99.5%.

Iron-substituted ZSM-5 molecular sieves were studied by Kaspar et al. as catalysts for the hydrogenation of CO_2.[112] Such iron-containing zeolites were prepared by reacting sodium silicalite ($Na_2O \cdot 2SiO_2 \cdot H_2O$) with iron salts, forming iron silicalites. These materials effectively catalyzed the water–gas-shift reaction and had low selectivities for hydrocarbon formation.

In a mechanistic study of the reverse water–gas-shift reaction, a clean Cu(110) single-crystal model catalyst was tested by Ernst et al. at temperatures of 573 to 723 K.[113] The kinetic data could best be interpreted by a "surface redox" or "oxygen adatom" mechanism for both the forward and reverse water–gas-shift reactions. For the reverse water–gas-shift reaction, the proposed rate-determining step is dissociative adsorption of CO_2, with the formation of adsorbed CO (CO_a) and oxygen atoms (O_a):[113]

$$CO_2 \Leftrightarrow CO_a + O_a \qquad (10.24)$$

$$CO_a \Leftrightarrow CO \qquad (10.25)$$

$$H_2 \Leftrightarrow 2H_a \qquad (10.26)$$

$$H_a + O_a \Leftrightarrow OH_a \qquad (10.27)$$

$$H_a + OH_a \Leftrightarrow H_2O_a \qquad (10.28)$$

$$H_2O_a \Leftrightarrow H_2O \qquad (10.29)$$

CO_2 hydrogenation with high selectivity toward formation of CO was observed by Román-Martínez et al. on platinum highly dispersed on carbon.[114–116] The activity of these catalysts was very much enhanced by loading alkaline earth ions onto the catalysts. Most effective were Ca^{2+} ions, followed by Mg^{2+} ions. A highly active carbon support was obtained by carbonization of a phenolformaldehyde polymer resin at 1273 K under N_2, followed by an oxidizing treatment with HNO_3 and loading of the alkaline earths by ion exchange. Using a catalyst with 0.9 wt% Pt–2.4 wt% Ca/C and CO_2 + H_2 (1:3), space velocity = 4000 h^{-1}, pressure = 1 MPa, and temperature = 523 K, CO was produced with 98% selectivity and a turnover frequency of 345 molecules CO per surface metal site per second. The activation energy of the reaction with this catalyst was 82 kJ mol^{-1}. The function of the alkaline earth ions was proposed to enhance CO_2 chemisorption, while the Pt particles promoted H_2 dissociation. The activity of Pt supported on carbon (without alkaline earths) for the CO_2 hydrogenation was very considerably increased by heat treatment to a high temperature, up to 1200 K. This effect was suggested to be due to dissolution of carbon atoms in platinum, resulting in a change in structure.[114–116]

10.3.1.2   Homogeneous Catalysis

A homogeneous solution-phase reaction similar to the reverse water–gas-shift reaction was observed by Yoshida et al. with some Rh complexes.[117] Rhodium(I) complexes $Rh(PL_3)_3$, such as $Rh[P(i-Pr)_3]_3$, reacted with CO_2 in the presence of water to form dihydro bicarbonato complexes, such as $Rh_2(O_2COH)(i-Pr)_2$. These dihydro bicarbonato complexes reduced CO_2 to form Rh(I)carbonyl-bicarbonato complexes, in an analogue of the reverse water–gas-shift reaction:

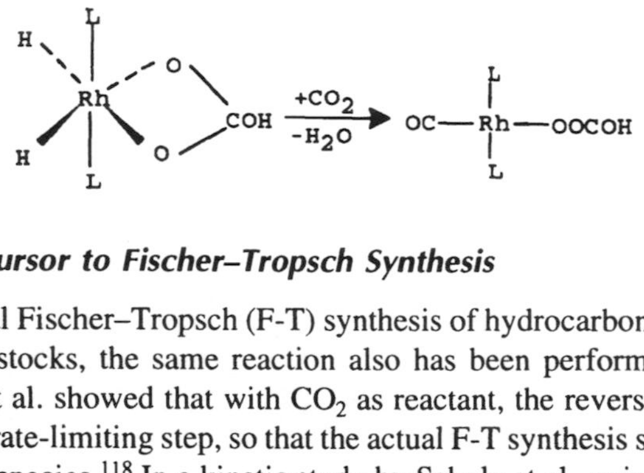

10.3.1.3   Precursor to Fischer–Tropsch Synthesis

While the original Fischer–Tropsch (F-T) synthesis of hydrocarbons applied H_2–CO mixtures as feedstocks, the same reaction also has been performed with H_2–CO_2 mixtures. Choi et al. showed that with CO_2 as reactant, the reverse water–gas-shift reaction was the rate-limiting step, so that the actual F-T synthesis seemed to involve CO as the active species.[118] In a kinetic study by Schulz et al., using Fe–Al_2O_3–Cu–K (100:13:10:25) as catalyst, with H_2–CO_2 (1:3) at 523 K and 1 MPa total pressure, during the initial 16 h on stream, the only reaction was formation of CO. Subsequently, the F-T activity started to appear, with production of hydrocarbons and oxygenates (C_{1+}).[119]

10.3.2   Methanation

Interest in the methanation reaction has been stimulated by efforts to produce synthetic methane from the synthesis gas obtained in the gasification of coal.[120] The methanation reaction is usually performed on catalysts of nickel supported on alumina, at temperatures of 300 to 400°C. It is applied as the final CO_2 and CO removal step in the purification of hydrogen to be used in ammonia synthesis, which requires methanation of the carbon oxides to below 5-ppm concentrations.[121,122]

10.3.2.1 Ni Catalysts

Ni/Al_2O_3 catalysts containing 0 to 25 wt% Ni were prepared by Aksoylu et al. by a coprecipitation method from Ni and Al nitrate solutions and were tested for their CO_2 adsorption and methanation activity in the range 433 to 533 K.[123,124] Al_2O_3 itself (without Ni) was active in CO_2 adsorption, but without methanation activity. With these Ni/Al_2O_3 catalysts, after H_2 reduction at 623 K, using a flow system of CO_2 + H_2 (1:9) at atmospheric pressure and a space velocity of 2400 h^{-1}, the selectivity of conversion to CH_4 was 99.7%. The catalytic activity increased with the Ni loading and was highest with 25 wt% Ni/Al_2O_3. However, the methane production rate per meter squared of Ni surface was highest on the 5 wt% Ni/Al_2O_3 catalyst.

The mechanism of CO_2 methanation on a silica-supported Ni catalyst was studied by Tahri et al. using the temperature-programmed surface reaction technique.[125,126] In this method, the catalyst was first exposed to CO_2 at ambient temperature, and then treated with H_2 at rising temperatures, while measuring the desorption of the products. By exposing Ni (25 wt%)/SiO_2 to He + 5% CO_2, the dissociative chemisorption of CO_2 was already detected at ambient temperature, at which the adsorbed oxygen reacted rapidly with hydrogen:

$$CO_{2(g)} \rightarrow O_{ads} + CO_{ads} \qquad (10.30)$$

The thermal reaction with hydrogen led to CO, CH_4, and H_2O. Methane and water were produced simultaneously, and with maximal rates at about 190°C. The rate-limiting step of the methanation reaction may be the dissociation of the C–O bond of CO_{ads}, assisted by either H_{ads} or another CO_{ads} species.[125,126]

Over nickel catalysts, the methanation of CO_2 proceeded more rapidly and was more selective (92 to 99%) than the methanation of CO. In a diffuse reflectance infrared Fourier transform spectroscopic (DRIFTS) study of the methanation of CO_2 over Ni/Al_2O_3 catalysts, surface intermediates were classified by Fujita et al. into weakly adsorbed species, which could be rapidly flushed off by a stream of helium, and strongly adsorbed species, which were only slowly released by He flushing.[127] The observed infrared bands included strongly adsorbed linearly bound CO (at 2050 cm^{-1}), two types of bridged structures of adsorbed CO — weakly adsorbed bridged CO (at 1910 cm^{-1}) and strongly adsorbed bridged CO (at 1830 cm^{-1}) — as well as weakly adsorbed unidentate and strongly adsorbed bidentate formate species.

Ni/MgO catalysts for the methanation of CO_2 were prepared by Nakayama et al. by treating a melt of the nitrates of Ni and Mg with citric acid, followed by several stages of heating and by calcination in air at 773 K. Highest yield of CO_2 conversion was obtained with a catalyst containing 70 wt% Ni/MgO, resulting in 100% selectivity of formation of methane.[128]

In an unusual reaction, a NiO/kieselguhr catalyst was preirradiated by Ogura et al. with a low-pressure mercury lamp. Carbon dioxide was then made to flow over it in the dark, producing preferentially methane, as well as small yields of ethane and ethene.[129]

10.3.2.2   Ru, Pd, and Rh Catalysts

The catalytic hydrogenation of CO_2 (total pressure 9.5 atm) on palladium supported on various oxides was found by Solymosi et al. to start at a measurable rate only above 520 K.[130,131] With highly dispersed Pd, the main product was methane, while on poorly dispersed Pd, methanol was formed. Pd/TiO_2 was the most effective catalyst for CO_2 hydrogenation among various Pd/oxide combinations tested. The adsorbed formate was shown not to be an intermediate in methanol production. The dissociation of adsorbed CO was proposed to be an important step in the methanation of CO_2.

On $Rh/CeO_2/SiO_2$ catalysts, with CeO_2 loading less than 6.1%, the presence of CeO_2 in a disperse phase as small crystallites on the silica surface was found by Trovarelli et al. to enhance the methanation activity, relative to Rh/SiO_2 not promoted with CeO_2.[132–135] With Rh/CeO_2 and Ru/CeO_2 as catalysts, only CH_4 was produced from $CO_2 + H_2$ (1:4) at 550 K, while with Ir, Pd, and Pt supported on CeO_2, the main product was CO, with smaller amounts of CH_4. With 1 wt% Rh/CeO_2 at 500 K, the methanation activity was 2.2 mmol CH_4 g-cat^{-1} h^{-1}. Oxygen vacancies, produced during the high-temperature reduction of CeO_2-supported noble metal catalysts, were suggested to be the driving force for CO_2 activation, with the formation of CO and the oxidation of reduced ceria.

In a comparative study of the hydrogenation of CO and CO_2 over 3.4 wt% Rh/SiO_2, Fisher and Bell used *in situ* infrared spectroscopy to identify the intermediate adsorbed species on the catalyst.[136] Under similar reaction conditions, the hydrogenation of CO_2 to form CH_4 was faster than that of CO. Also, the apparent activation energies for the $H_2 + CO_2$ and the $H_2 + CO$ reactions were 16.6 and 23.2 kcal mol^{-1}, respectively. These differences were explained by the different adsorption behavior of CO and CO_2. With both reactant gases, the catalyst surface became covered almost to saturation by a combination of linearly and bridge-bonded CO. However, the CO hydrogenation resulted in a higher total CO coverage than CO_2 hydrogenation, probably because CO_2 hydrogenation required the dissociation of CO_2 as a prior step. The rate of methane production was found to depend on the total CO coverage, passing through a maximum as the CO coverage decreased from saturation to zero. At equal CO surface coverage (which required lower gas-phase partial pressures of CO than CO_2), the rates of methane formation from CO and CO_2 were the same. The dependence of the rate of methanation of CO_2, expressed as the turnover frequency (TOF), on the temperature T and the partial gas pressure p of H_2 and CO_2 followed the power law

$$TOF = Ae^{-E/RT}p(H_2)^x p(CO_2)^y \qquad (10.31)$$

The values of the partial gas pressure dependencies for H_2 and CO_2 were x = 0.53 and y = –0.46, respectively. The unusual negative value of the pressure dependence of the rate on the partial pressure of CO_2 reflected the inverse relationship of the rate of methanation on the total CO coverage of the catalyst. The mechanism subsequent to CO$_{ad}$ formation was therefore proposed to be similar for CO and CO$_2$ hydrogenation, leading to adsorbed H$_2$CO. The dissociation of this to CH$_{2,ad}$ and O$_{ad}$ was assumed to be the rate-determining step. The latter two species were then rapidly hydrogenated to CH$_4$ and H$_2$O.[136]

A marked improvement in the catalysis of CO$_2$ methanation was achieved by Kishida et al. with Pd microparticles prepared from water-in-oil microemulsions.[137] With such Pd particles, of average size of about 3.2 nm, supported on either SiO$_2$ or ZrO$_2$, the reaction of H$_2$ + CO$_2$ + Ar (6:3:1) at a pressure of 4.9 MPa resulted in CH$_4$ as the only product. The reaction rates with this catalyst were about 80 times faster than with catalysts prepared by impregnation of the same supports with rhodium chloride, and the reactions could be performed at about 100°C lower temperatures. The same method was applied to the preparation of rhodium particles with an average size of 5 nm, supported on SiO$_2$.[138] Such catalysts were effective for the methanation of CO$_2$, but were deactivated with time on stream due to coke deposition.

The exothermic methanation of carbon dioxide with hydrogen was used by Levitan et al. to close the loop of the above-mentioned solar-furnace-driven carbon dioxide reforming (see Section 10.2.9). With the same Rh/alumina catalyst as used for the reforming, conversion of over 80% was achieved in the methanation reaction.[79]

10.3.2.3   Fe Catalysts

Suzuki et al. prepared reduced iron oxide catalysts by treating Fe$_3$O$_4$ under H$_2$ at 400 or 500°C.[139] Mixtures of CO$_2$ and H$_2$ in different proportions were passed over the catalyst at atmospheric pressure. The products included CO, CH$_4$, C$_2$H$_4$, C$_2$H$_6$, and C$_3$ compounds. Optimal production of hydrocarbons was obtained at a reaction temperature of 400°C, with a CO$_2$/H$_2$ ratio of 1:4 and a space velocity of 3000 cm^3 g^{-1} h^{-1}. The distribution of products could be interpreted by a mechanism in which the primary step was reduction to CO,

$$CO_2 + H_2 \rightarrow CO_{ads} \qquad (10.32)$$

followed by an F-T-type reaction,

$$nCO_{ads} + mH_2 \rightarrow C_nH_{2m} \qquad (10.33)$$

10.3.2.4   Co/Cu/K Catalysts

High selectivity in the hydrogenation of carbon dioxide to methane was reported by Baussart et al. with some Co/Cu/K catalysts.[140] A catalyst prepared from Co$_3$O$_4$ alone had a selectivity of 97.7% for CH$_4$ formation at the low temperature of 424 K. The presence of K in the catalyst matrix favored the production of CO. An increase in the ratio of Cu/Co in the catalyst resulted in an enhancement in the selectivity to C$_2$H$_6$.

10.3.2.5   Cobalt Foil Catalysts

Cobalt foils were tested by Fröhlich et al. as catalysts for the hydrogenation of CO_2.[141] The cobalt was activated by repeated cycles of oxidation (with 0.1% O_2 in N_2 for 0.5 h) and reduction (with H_2 for 10 h) at 873 K. The rates of hydrogenation of CO_2 to CH_4 and CO (the main products), as well as to traces of C_2H_6 and C_3H_8, increased with successive cycles of reduction/oxidation, leveling off by about 10 cycles. The repeated reduction/oxidation cycles led to deeper penetration of oxygen into the bulk of the catalyst, accompanied by increased surface roughness. Also, AES analyses indicated an increase in the sulfur content of the surface, due to segregation of small amounts of sulfur from the bulk to the surface of the catalyst. This did not cause deactivation of the catalyst. On the other hand, injection of a pulse of H_2S during CO_2 hydrogenation resulted in severe inhibition of the methanation reaction. This poisoning by sulfur occurred when a monolayer of H_2S had adsorbed on the surface. A model was proposed by Lojewska and Dziembaj to account for the variation in the rates of methanation of CO_2, based on the variation in the concentration of active centers on the cobalt foil.[142]

10.3.2.6   Ru/TiO$_2$ Catalysts

A redox catalyst of highly dispersed substoichiometric ruthenium oxide supported on titanium oxide powder was found by Thampi et al. to be effective for the methanation of carbon dioxide at ambient temperature.[143] The catalyst contained 5% RuO_2 (3.8% Ru) in 20-Å clusters on titanium dioxide (Degussa P25, about 80% anatase, Brumuer–Emmett–Teller- (BET) specific surface 55 m^2 g^{-1}). In the dark, the rate of the methanation of carbon dioxide was very low at 25°C, but increased rapidly with rising temperature. Illumination with a solar simulator (at 80 mW cm^{-1}) caused a very marked enhancement in the rate of methanation, which was initially ascribed to bandgap excitation of the titanium dioxide support. However, subsequent work by Grätzel[144] and Melsheimer et al.[145] indicated that the effect of illumination could also be explained by local heating of the catalyst material and was not an intrinsic photochemical step.

Ru/TiO$_2$ (3.8% Ru) was found by Prairie et al. to be more active as a methanation catalyst by two orders of magnitude than Ru/Al$_2$O$_3$ (3.8% Ru).[146] This was explained to be due to the higher density of active sites on Ru/TiO$_2$ relative to Ru/Al$_2$O$_3$. About 25% of the ruthenium in this catalyst was in an oxide form, and the particle size of the Ru metal was in the range of 10 to 30 Å. The reaction rate on Ru/TiO$_2$ was independent of the CO_2 concentration and was approximately one-half order in H_2 concentration. A Fourier transform infrared (FTIR) spectroscopic study revealed the formation of surface-adsorbed CO_{ads} and $HCOO^-_{ads}$ intermediates. The CO_{ads} was presumably formed by the reverse water–gas-shift reaction. These species were proposed to be intermediates toward CH_4 production. In further FTIR studies by Gupta et al. on the hydrogen-pretreated Ru/TiO$_2$ (3.8% Ru) catalyst during CO_2 methanation, the initial steps included reduction of CO_2 on Ru° sites and formation of a linear Ru–(CO)$_{ad}$ species (bands at 1950 and 1990 cm^{-1}), followed by the development of some Ru-oxycarbonyl species, such as Ru(O)CO$_{ad}$ and Ru(H)–CO$_{ad}$, which were precursors to methane. The Ru/TiO$_2$ catalyst was remarkably tolerant to oxygen. Exposure to air at room temperature did not affect its subsequent activity for methanation. Also, a CO$_2$ + H$_2$ + air (1:4.5:1.47) mixture achieved over this catalyst at 200°C the same CO$_2$ to CH$_4$ conversion (~100%) as a CO$_2$ + H$_2$ (1:4.1) mixture.[147–150]

In a further detailed study on the surface reaction kinetics of the CO$_2$ methanation reaction on 2% Ru/TiO$_2$, Marwood et al. used DRIFTS to measure the transient buildup and decay of linearly bound adsorbed CO (CO$_{ads}$) by its absorption band at 2023 cm^{-1}.[151,152] In step-up experiments at 383 K, the feed gas flowing over the catalyst (which contained 0.8 mbar H$_2$O) was switched from 20% H$_2$ in He to a mixture of 20% H$_2$ + 10% CO$_2$ in He. In sequence, CO$_{ads}$ was detected, followed by CH$_4$ and water in the gas phase. The rate of formation of CO$_{ads}$ was enhanced by increasing H$_2$ concentration and was inhibited by increasing concentration of water in the feed gas. In step-down experiments, the feed gas was switched from 20% H$_2$ + 10% CO$_2$ in He to 20% H$_2$ in He, thus measuring the rate of hydrogenation of CO$_{ads}$. By varying the concentration of H$_2$ in the feed gas, a half-order dependence on the H$_2$ concentration was observed, suggesting a first-order dependence on adsorbed hydrogen atoms (H$_{ads}$), formed by dissociation from gaseous H$_2$. Also, the rate of hydrogenation of CO$_{ads}$ was independent of the H$_2$O partial pressure. Since inhibition by water had been observed both under steady-state conditions for the overall methanation of CO$_2$ and under transient conditions for the formation of CO$_{ads}$, it was possible to conclude that the rate-limiting step of the overall methanation reaction is the formation of CO$_{ads}$. The activation energy for the overall CO$_2$ methanation on 2% Ru/TiO$_2$ was determined to be 80 kJ mol^{-1}. In a tentative proposed mechanism for CO$_2$ methanation, CO$_2$ reacts rapidly with surface hydroxyl groups, forming adsorbed hydrogen carbonate, which is rapidly hydrogenated to formate at the metal-support interface, (HCOO$^-$)$_I$. This active formate is in equilibrium with formate bound to the support, (HCOO$^-$)$_s$, which is detected by its infrared bands and which acts as a spectator in the methanation reaction. The interfacial formate then splits to form CO$_{ads}$, which is hydrogenated to CH$_4$.[151,152]

10.3.2.7 SiO$_2$-Supported Catalysts

In a comparison of the CO$_2$ hydrogenation activity of SiO$_2$-supported catalysts, the order of specific activity was found by Weatherbee and Bartholomew to be Co/SiO$_2$ > Ru/SiO$_2$ > Ni/SiO$_2$ >> Fe/SiO$_2$. However, the selectivity of formation of CH$_4$ was slightly different: Ru/SiO$_2$ > Ni/SiO$_2$ > Co/SiO$_2$ > Fe/SiO$_2$. The selectivity to C$_{2+}$ hydrocarbons was an order of magnitude lower than that to methane. Fe/SiO$_2$ was the most effective catalyst for hydrocarbon formation, with selectivities at 250°C of up to 3.9%.[153,154]

10.3.2.8 Al$_2$O$_3$-Supported Catalysts

The mechanism of methanation of carbon dioxide was clarified by Huang et al. using rate measurements over the following catalysts: Ni/Al$_2$O$_3$, Ru/Al$_2$O$_3$, Pt/Al$_2$O$_3$, and Ir/Al$_2$O$_3$. From a kinetic analysis of the rates of evolution of the products CO and

CH_4, the existence of two different pathways was indicated: A parallel reaction mechanism was observed with Ni-alumina and Ru-alumina. On the other hand, with Pt-alumina and Ir-alumina, a consecutive reaction mechanism was predominant:[155]

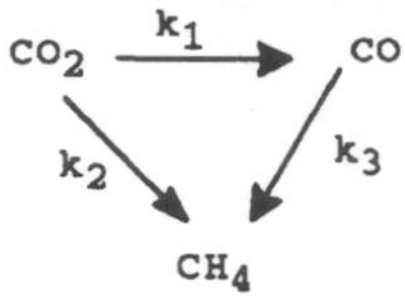

$$CO_2 \xrightarrow{k_1} CO$$

$$\begin{array}{c} k_2 \searrow \quad \swarrow k_3 \\ CH_4 \end{array} \quad \textbf{Parallel reactions}$$

$$CO_2 \xrightarrow{k_1} CO \xrightarrow{k_3} CH_4 \quad \textbf{Consecutive reactions}$$

10.3.2.9　Fe/Mn/Rh Catalysts

The hydrogenation of CO_2 on various Fe–Mn oxide catalysts was tested by Dziembaj et al. in a pulse microreactor.[156] The catalysts were prereduced in an H_2 stream at 400°C for 24 h. For the catalytically most active Fe/Mn/Rh catalyst, the bulk and surface compositions after this reduction were 54 and 30% of Fe, 45 and 69% of Mn, and 1.2 and 2% of Rh, respectively. The surface of the catalyst thus became strongly depleted in Fe (which had diffused into the bulk) and became enriched in Mn and Rh. The rates of hydrogenation of CO_2 were determined by introducing brief pulses of CO_2 into the H_2. The conversions of CO_2 at 400°C into total hydrocarbons (mainly CH_4) over Fe/Mn/Rh, Fe/Mn, and Fe/Mn/La catalysts were 58, 16, and 4.7%, respectively. With the most active Fe/Mn/Rh catalyst, the selectivities to formation of CH_4 and C_2H_6 were 99.7 and 0.3%, the only other product being water.[156]

10.3.2.10　Ni/La₂O₃/Ru Catalysts

Rapid methanation of carbon dioxide with hydrogen was achieved by Inui et al. using a composite catalyst, Ni–La$_2$O$_3$–Ru (4.3:2.5:0.7 wt%), dispersed on a spherical silica support with a meso-macro bimodal pore structure.[157–160] Total CO_2 conversion was obtained, with 100% selectivity for methane. The space–time yield of methane reached 500 mol L^{-1} h^{-1} at a temperature range around 270°C. The function of the weakly basic La$_2$O$_3$ part was to enhance the CO_2 adsorption capacity, while the presence of the Ru enhanced the adsorption of hydrogen and presumably facilitated the hydrogen spillover. By improved design of the catalyst support structure (silica-coated ceramic fibers), and using an Ni–La$_2$O$_3$–Ru (13.9:3.4:0.4 wt%) catalyst, even more rapid methanation of CO_2 was attained by Inui et al. With $CO_2 + H_2$ (12:88) at a flow rate of 10 L h^{-1}, a temperature of 450°C, and a space velocity of 226,300 h^{-1} (or contact time of 16 ms), the CO_2 conversion reached 96% and the space–time yield of methane reached 1205 mol L^{-1} h^{-1}.[157–160]

10.3.2.11　Cu/Mo₂C, Cu/Fe₃C, and Cu/WC Catalysts

The carbides Mo$_2$C, Fe$_3$C, and WC were found by Dubois et al. to be moderately effective catalysts for the hydrogenation of carbon dioxide. Thus, using copper-promoted molybdenum carbide, Cu/Mo$_2$C, with a gas mixture of Ar/H$_2$/CO$_2$ = 10.1/22.7/67.2 at 220°C, the CO$_2$ conversion was 4%, and the selectivity to methanol, dimethyl ether, methane, ethane, and CO was 31.5, 1.4, 13.5, 3.0, and 48.5%.[161]

10.3.2.12 Ni/Zr Catalysts

A glassy metal alloy, Ni$_{64}$Zr$_{36}$, was used by Schild et al. as catalyst precursor. Under CO$_2$ hydrogenation conditions (at 493 K), the amorphous alloy was transformed to a microporous solid, with metallic nickel particles contained in a ZrO$_2$ matrix. Methane was the almost exclusive CO$_2$ reduction product, along with traces of ethane.[162]

Ni/ZrO$_2$ catalysts were prepared by Hashimoto et al. by oxidation–reduction treatment of Ni–Zr alloys. The methanation activity of such catalysts was considerably enhanced by addition of rare earth elements, such as Y, Ce, and Sm.[163–165]

10.3.2.13 Bi$_x$Gd$_{1-x}$VO$_4$ Catalysts

Mixed bismuth gadolinium vanadates, Bi$_x$Gd$_{1-x}$VO$_4$, were found by Le Bras et al. to be active catalysts for CO$_2$ hydrogenation, producing mainly CO and some CH$_4$, but also traces of CH$_3$OH and C$_2^+$ hydrocarbons.[165,167] These catalysts were prepared by precipitation from ammonium vanadate, bismuth nitrate, and gadolinium nitrate. Treatment included calcination at 773 K under oxygen and prereduction under hydrogen. By reacting CO$_2$ + H$_2$ (1:3) at 2 × 10^5 Pa and 543 K over Bi$_{0.5}$Gd$_{0.5}$VO$_4$ (prereduced at 573 K), the conversion of CO$_2$ was 2 to 4%, and the selectivities to CO and CH$_4$ were 97 and 3%, respectively. In the steady state of CO$_2$ hydrogenation, the catalysts were shown by XRD, IR, and XPS to be mixtures of the oxides Bi$_2$O$_3$, V$_2$O$_3$, and Gd$_2$O$_3$, with metallic Bi° formed by reduction. Both surface Bi and V species seem to be involved in the catalytic activity.[166,167]

10.3.2.14 LaNi$_5$ Catalysts

Hydrogen storage alloys were proposed by Ando et al. to activate hydrogen and thus to be effective for the hydrogenation of CO$_2$.[168] The intermetallic compound LaNi$_5$ was prepared by arc melting the metal constituents under an argon atmosphere. After pretreatment by 1% H$_2$ in N$_2$ at 250°C, the reaction was performed by passing over the catalyst CO$_2$ + H$_2$ (1:4) under 5 mPa. The products were CH$_4$ and C$_2$H$_6$ formed with selectivities of 98 and 2%, respectively, and a turnover number (moles CO$_2$ converted to CH$_4$ per active site) of 79 × 10^3 s^{-1}. The amount of H$_2$ chemisorption on this catalyst was 44 × 10^{-3} mol g–cat^{-1}, which provided a measure of the number of Ni atoms on the catalyst surface. Among various La–Ni alloys, the methanation activity was in the order LaNi$_5$ > LaNi$_4$Al > LaNi$_4$Cr > LaNi$_4$Cu > Ni.[168]

10.3.2.15 Fe/Zr/Ru Catalysts

Highly effective CO$_2$ methanation catalysts were prepared by Hashimoto et al. from amorphous Fe valve metal alloys containing a small amount of Pt group elements.[169,170]

The alloys were formed by argon arc melting, fabricated into amorphous ribbons (20 to 30 μm thick), and activated by immersion in dilute HF solution. The activation treatment enhanced both the surface roughness and the surface concentration of Pt group elements. Very high methanation activity was attained with an amorphous alloy catalyst of Fe–Zr–Ru, which had a surface area of 0.8 m^2 g^{-1} and a surface composition of Fe, Zr, and Ru of 39, 25, and 37 at%, respectively. On this catalyst, the conversion of CO_2 + H_2 (1:4) was 100% selective to CH_4 and was measurable already at 100°C.[169,170]

10.3.2.16 Mechanochemical Activation of Catalytic Hydrogenation

Nanophase iron carbides, produced by the ball milling of elemental iron and graphite powders, were found by Trovarelli et al. to catalyze the hydrogenation of CO_2 mainly to CO, as well as to CH_4 and C_2–C_6 hydrocarbons.[171] This mechanosynthesis of the nanophase iron carbides was performed in a vibratory ball mill at room temperature on a mixture of iron powder (initially 60 m) and graphite (initially 45 m) in the atomic ratio 3:1, sealed in an N_2 atmosphere. After 15 h of milling, 94% of the iron had been converted to carbides, mainly Fe_3C, with a crystallite size of about 10 nm. The reaction was carried out in a flow microreactor over the catalyst prereduced with H_2, with CO_2 + H_2 (1:3) at 0.1 MPa and 523 K, GHSV = 15,000 to 20,000 h^{-1}. The rates of formation of CO and CH_4 were 1.4 × 10^{-3} and 0.7 × 10^{-3} mol g-cat^{-1} h^{-1}, respectively, and the selectivity to C_2–C_6 hydrocarbons was 22%. The alkene/alkane ratio in the hydrocarbons was 0.22. The same catalytic results were obtained with commercial Fe_3C powders after similar milling. The catalytic activity of these nanophase iron carbides may be due to the very high density of defects (up to 10^{19} per cubic centimeter), with a very large number of interphase boundaries.171

In contrast to the above results of milling Fe_3C prior to its application as catalyst for CO_2 hydrogenation, the rates of methanation of CO_2 by Ru, Ni, and Fe powders mixed with MgO powder were found by Mori et al. to be dramatically increased by milling these catalysts *in situ* during the hydrogenation reaction.[172] The reactions were carried out in batch mode in a stainless-steel ball mill fitted with stainless-steel balls, which was made both to rotate at 1100 rpm and to vibrate at an amplitude of 6 mm. The choice of MgO as the catalyst support was due to its basic properties, which enhanced adsorption of CO_2. Using CO_2 + H_2 (1:5) at a total initial pressure of 80 kPa, the rates of methanation were tested in the temperature range 80 to 150°C. The order of methanation activity of the various catalysts under milling conditions was Ru/MgO > Ru–Fe/MgO > Ni/MgO > Ni–Fe/MgO > Ni–Fe. The methanation yield increased with rising temperature, reaching 96% at 180°C during 1 h. The activation energy was considerably decreased by the milling. For the Ru/MgO catalyst, the activation energy for CO_2 methanation in the milling system was only 41 kJ mol^{-1}, while it was 74 kJ mol^{-1} under the same conditions but without the milling balls. Premilling of the catalyst prior to the methanation reaction had no effect on the yields of methane. The specific area of the catalyst particles was found by the BET method not to have increased significantly during the milling process.

The source for the mechanochemical activity of the milled catalyst was therefore ascribed to the continuous formation of nascent surfaces during the mechanical treatment. Without this formation of fresh surfaces, the active sites of the catalyst rapidly became deactivated by carbonaceous matter or other poisons.[172]

10.3.2.17   Methanation with Water Vapor Permselective Membrane

By continuously removing the water formed in the methanation of CO$_2$,

$$CO_2 + 4H_2 = CH_4 + 2H_2O \qquad \Delta H_{298\,K} = -113 \text{ kJ mol}^{-1} \qquad (10.2)$$

the equilibrium can be shifted toward completion of the reaction. Ohya et al. integrated a water vapor permselective membrane with a membrane catalyst (0.5% Ru/Al$_2$O$_3$).[173] The water-vapor-permeable membrane was formed by depositing 15 layers of microporous glass on porous ceramic tubing. The average diameter of this glass membrane was 3 nm. This membrane was stable at high temperatures and was tested in the temperature range 480 to 719 K, at an absolute pressure of 0.2 MPa, with a gas feed of molar ratio H$_2$/CO$_2$ = 1 to 5. At 573 K and a space velocity of 0.0308 s^{-1}, the CO$_2$ conversion in the methanation using this membrane was increased by about 18% relative to the reaction without the membrane.[173]

10.3.3   Hydrogenation of Alkaline Earth Carbonates

Since most of the carbon on earth (>99.9%) exists as carbonates such as limestone (CaCO$_3$), it may sometimes be useful to also reduce solid carbonates. In experiments by Reller et al., alkaline earth carbonates, when heated under hydrogen, underwent degradation at reaction temperatures which were lower by at least 150 K compared with the degradation under inert or oxidizing atmospheres.[174] Magnesite decomposed in a reducing atmosphere to MgO, as well as to even amounts of CO$_2$ and CO. From calcite, the main gaseous product was CO, with the CO–CO$_2$ ratio 10:1. With mixed alkaline earth/transition metal carbonates (prepared by coprecipitation from the nitrate salt solutions with Na$_2$CO$_3$) under H$_2$, the decomposition temperature was lowered even further. Thus, with Ca–Ni carbonates (10% Ni) under H$_2$, the decrease was 400 K compared with the decomposition temperature of pure CaCO$_3$ under a nonreducing atmosphere. From Ca–Co, Ca–Ni, and Mg–Ni (10% transition metal) carbonates, the predominant gaseous product was CH$_4$, with small amounts or traces of CO.[174]

By hydrogenation at 200 to 250°C, NiCO$_3$ and CoCO$_3$ were found by Tsuneto et al. to produce methane, even without any added catalysts.[175] CaCO$_3$ and BaCO$_3$ were hydrogenated to CH$_4$ in the presence of Ni or Co catalyst (0.01 g catalyst per 1 g carbonate) at 400°C. Very rapid hydrogenation occurred with 4MgCO$_3$·Mg(OH)$_2$·5H$_2$O at 300°C in the presence of a catalyst of ultrafine Ni powder, reaching a production rate of 3.2 mmol CH$_4$ h^{-1} (g substrate)$^{-1}$.

Such reactions are obviously not of interest in the context of CO$_2$ mitigation.

10.3.4   Methanol Synthesis

The catalytic hydrogenation of mixtures of carbon monoxide and carbon dioxide to produce methanol is one of the major reactions in the chemical industry. Selective hydrogenation to yield mainly methanol has been achieved with catalysts based on copper, chromium, zinc, and palladium, while methanation of carbon dioxide is the preferred reaction on catalysts of nickel supported on alumina or rhodium supported on titania. The earlier work on methanol synthesis was reviewed by Wade et al.[176] and Ziessel.[177]

10.3.4.1   ZnO Catalysts

The effect of the supports MgO, Al_2O_3, SiO_2, TiO_2, ZrO_2, and Nb_2O_5 was tested by Inoue et al. for the CO_2 + H_2 synthesis of methanol on ZnO catalysts.[178] Highest activity and selectivity for methanol production was with the system ZnO/ZrO_2 (1:9) at 360°C. With this system, methanol synthesis was more selective using CO_2 + H_2 than using CO + H_2. Acidic supports, such as Al_2O_3, TiO_2, and SiO_2, favored the dehydration of methanol to dimethyl ether. Therefore, methanol production was more selective with less acidic or even basic supports, such as ZnO, MgO, and ZrO_2.[178]

Since synthesis gas usually consists of a mixture of CO, CO_2, and H_2, there have been many studies to clarify whether CO or CO_2 is the carbon species undergoing hydrogenation over methanol synthesis catalysts. When only CO_2 was used as the carbon source, the rate of methanol synthesis on the model catalysts of Cu/ZnO was observed by Klier et al. to be slower than with CO_2–CO mixtures.[179–181] With a commercial catalyst of $Cu/Zn/Al_2O_3$ (ICI), under industrial conditions (250°C, 40 to 50 atm, a GHSV of 10,000 to 120,000 h^{-1}), the direct hydrogenation of CO_2 to methanol was achieved.

TiO_2 as a support in methanol synthesis was found by Tagawa et al. to suppress the reverse water–gas-shift reaction.[182–185] An optimal methanol synthesis catalyst had the composition $CuO–ZnO–TiO_2$ (30:30:40 by weight). Using the gas mixture H_2/CO_2 = 4:1, at 513 K, the conversion was 15.4%, the selectivity to methanol 21.9%, and the yield of methanol 3.4%. The addition of cobalt (1% of the copper component) decreased CO formation and enhanced the production of both methane and methanol. Acid supports promoted high selectivity to methanol synthesis but provided low activity. On neutral and basic supports, only CO was formed, by the reverse water–gas-shift reaction. High methanol synthesis activity was obtained on amphoteric supports, such as TiO_2, which inhibited CO formation. The addition of iron to the above $CuO–ZnO/TiO_2$ catalyst (by impregnation with iron nitrate) resulted in much increased production of hydrocarbons.[182–185]

Joo et al. tested various Cu catalysts supported on ZnO or Al_2O_3 for methanol synthesis.[186] Highest productivity was with a Cu/ZnO (1:3) catalyst with a very small Cu surface area of 0.3 m^2 g-cat^{-1}. Using CO_2–H_2 (1:3) at 400 psi, 250°C, and a feed rate of 6000 L h^{-1}, the CO_2 conversion was 7.4 mmol g-cat^{-1}, the turnover number (TON) was 1018 mol MeOH Cu-$atom^{-1}$ h^{-1}, and the selectivities to methanol and CO were 65 and 35%, respectively. The conversion of CO_2 was proportional to the surface area of copper. However, lower surface content of copper favored higher TON and selectivity to the formation of methanol.[186]

The high stability of CuO–ZnO–Al$_2$O$_3$ catalysts containing 5% Al$_2$O$_3$ for methanol synthesis from CO$_2$ was confirmed by Hirano et al. By recycling H$_2$ + CO$_2$ (3:1) at about 515 K, 9 MPa, and GHSV = 5000 h^{-1} over such a catalyst, about 90% of the supplied CO$_2$ was converted to methanol.[187]

Activated carbon as an inert support for metal catalysts has advantages in providing high catalyst dispersion and assisting in the reduction of the catalyst precursors by carbon to metal during high-temperature calcination. Instead of the usual procedure of impregnating activated carbon with the metal salts, followed by calcination, Sakata et al. impregnated sawdust with these salts and then pyrolyzed and calcined the material simultaneously.[188] For preparation of a methanol synthesis catalyst, they impregnated sawdust with an aqueous mixture of Cu and Zn nitrates, pyrolyzed under N$_2$ at 500°C and reduced with H$_2$ + N$_2$ (1:1) at 280°C. The resulting catalyst of 12.5 wt% Cu–3.4 wt% ZnO/C was used for the reaction of CO$_2$ + H$_2$ + N$_2$ (1:4:1.7) at 260°C, space velocity = 1000 h^{-1}, and total pressure = 2 MPa. The major products were CO and CH$_3$OH, with traces of CH$_4$, and the methanol production rate was 5.6 mmol g-(Cu + ZnO)$^{-1}$ h^{-1}. The high activity was ascribed to the small particle sizes of the Cu and ZnO particles, 8 and 28 nm, respectively, and to their uniform dispersion. With this catalyst, the rate of methanol production was about 2.5 times faster than with a conventionally prepared Cu–ZnO on activated carbon.[188]

The kinetics of methanol synthesis at various H$_2$, CO, and CO$_2$ compositions was measured by Chanchlani et al. over Cu/ZnO (Cu:Zn = 30:70 atomic ratio) and Cu/ZnO/Al$_2$O$_3$ (Cu:Zn:Al = 60:30:10 atomic ratio). The ternary catalyst was more active for methanol production, but also generated methane as a minor by-product (5 to 10% of the methanol yield). The kinetic data proved that methanol was formed from either CO or CO$_2$.[189]

The outstanding importance of the catalyst support was revealed by Ghazi et al. in a comparative study of the catalyst support systems Ni (3.8)–Mo (7.0)/Al$_2$O$_3$ and Ni (3.0)–Mo (7.9)/ZnO.[190] In experiments at an H$_2$/CO$_2$ ratio of 2, at 553 K and a pressure of 3 MPa, the selectivities after 7 to 8 h of reaction were with the Al$_2$O$_3$-supported catalyst: methanol, 9%; methane, 50%; and CO, 38%. With the ZnO-supported catalyst, the selectivities were methanol, 25%; methane, 1%; and CO, 73%. From the time course of the evolution of the products, the authors concluded that methanol production seemed to be mainly by the direct hydrogenation of CO$_2$, although some methanol may be derived from the hydrogenation of CO. The results showed that methanol formation was favored on the ZnO support.[154]

In a study by Sakurai and Haruta on methanol synthesis from CO$_2$ over various gold catalysts, highest methanol yield and stable operation were observed with Au (33 atom%)/ZnO–TiO$_2$. Using CO$_2$–H$_2$–Ar (23:67:10) at 250°C, a total pressure of 50 atm, and a space velocity of 3000 h^{-1} mL g-cat^{-1}, the methanol selectivity reached about 55%.[191]

High methanol selectivity requires high gas pressures and low reaction temperatures. Optimal conditions for methanol synthesis on a CuO–ZnO–Al$_2$O$_3$–Cr$_2$O$_3$ catalyst were obtained by Arakawa et al. at 7 mPa pressure, 250°C temperature, space velocity GHSV = 1800 h^{-1}, and H$_2$/CO$_2$ = 3/1, resulting in 79% selectivity for methanol production, 20% yield, and 25% CO$_2$ conversion.[192] By mixing the methanol synthesis catalyst with an acidic zeolite catalyst, Arakawa et al. succeeded in both enhancing the conversion of CO$_2$ to methanol and effecting the dehydration of some of the methanol formed to dimethyl ether. Best results were obtained with a hybrid catalyst of CuO–ZnO–Al$_2$O$_3$ (32:66:2) and mordenite (Zeolon 200H). With H$_2$/CO$_2$ = 3:1 at 240°C and 3 MPa, the conversion of CO$_2$ was 25%, and the selectivities for dimethyl ether, methanol, and CO were 55, 13, and 32%, respectively.[193]

Fujitani et al. observed that the promoting effect of metal oxides on the yield of methanol per unit surface area of Cu or Pd was Ga$_2$O$_3$ > ZnO > Cr$_2$O$_3$ > Al$_2$O$_3$ = ZrO$_2$ > SiO$_2$ for Cu catalysts and Ga$_2$O$_3$ > ZnO > Cr$_2$O$_3$ > TiO$_2$ > Al$_2$O$_3$ > ZrO$_2$ > SiO$_2$ for Pd catalysts.[194,195] Reactive frontal chromatography was used to determine the oxide species produced on the surface of the catalysts during the methanol synthesis. The specific activity of methanol synthesis increased linearly with the oxygen coverage θ up to θ = 0.16 and decreased above θ = 0.18. The support effect of metal oxides with copper catalysts was proposed to be due to the regulation of the Cu$^+$/Cu$^\circ$ ratio on the surface of the copper particles. For both Cu and Pd catalysts, the active sites are composed of Cu$^+$/Cu$^\circ$ or Pd$^\circ$/Pd$^+$ sites at the interface between the metal particles and the metal oxides.[194,195]

Saito et al. succeeded in considerably enhancing the methanol synthesis activity of Cu/ZnO catalysts for methanol synthesis by the addition of various metal oxides.[196] The highest activity was observed with a five-component catalyst, Cu/ZnO/ZrO$_2$/Al$_2$O$_3$/Ga$_2$O$_3$. On this catalyst, with H$_2$/CO$_2$ = 3, a feed gas rate of 300 mL min^{-1}, at 523 K and total pressure 5 MPa, the methanol synthesis activity was 785 g-CH$_3$OH kg-cat^{-1}, while with the plain Cu/ZnO, the activity was only 516 g-CH$_3$OH kg-cat^{-1}. In the multicomponent catalyst, the role of Al$_2$O$_3$ and of ZrO$_2$ was to improve the dispersion of Cu particles in the catalyst, and thus to increase their surface area, while the role of Ga$_2$O$_3$ or Cr$_2$O$_3$ was to increase the specific activity by controlling the ratio of Cu$^+$/Cu$^\circ$ on the Cu surface. The methanol synthesis activity of the above multicomponent catalyst was highly stable, as found in a continuous 3400-h test, during which it lost only 17% of its initial activity. The methanol synthesis activity with the same catalyst and under similar conditions was measured by Ushikoshi et al. in a 50-kg/day test plant at the Research Institute of Innovative Technology for the Earth in Kyoto, Japan. The production rate of methanol was around 600 g L-cat^{-1} h^{-1}, forming methanol of 99.9% purity.[197]

In a comparison of Cu-based catalysts for methanol synthesis from CO$_2$ and H$_2$, Kanaii et al. observed a synergy in methanol synthesis by CO$_2$ hydrogenation over physically mixed Cu/SiO$_2$ and ZnO/SiO$_2$ particles, which had been prereduced at temperatures above 573 K. Such a heat treatment caused migration of ZnO$_x$ moieties from the ZnO/SiO$_2$ particles to the surface of Cu/SiO$_2$, creating active sites for methanol synthesis. These sites were shown by XRD to consist of a Cu–Zn alloy. Enhanced methanol synthesis activity from CO$_2$–H$_2$ (1:3) was discovered with a Pd-doped Cu/ZnO alloy.[198,199]

A highly active catalyst for methanol synthesis from carbon dioxide and hydrogen was prepared by Inui et al. using an "intrinsic uniform gelation method," in which the concentrated mixed nitrate aqueous solution of the catalyst metals was treated with ammonia gas, to form a gel, which was then dried and calcined.[74,157,159] With CuO–ZnO–Cr$_2$O$_3$–Al$_2$O$_3$–Pd (33.9:26.2:1.4:37.8:0.7 wt%) as catalyst, the conversion of CO$_2$–CO–H$_2$ (22:3:68) at 270°C and 80 atm and a space velocity of 18,800 h^{-1} resulted in 19.4% conversion to methanol. The space–time yield of methanol amounted to the high value of 1028 g L^{-1} h^{-1}. The addition of the small amount of Pd enhanced hydrogen spillover. Tests at 250°C indicated that increases in the reaction pressure resulted in improvements in both the yield and the selectivity for methanol.[74,157,159]

Even more efficient conversion of CO$_2$ + H$_2$ to methanol was attained by Inui et al. by partly replacing Al$_2$O$_3$ by Ga$_2$O$_3$ in the catalyst formulation.[73,74,200] The addition of Ga$_2$O$_3$ enhanced the production of methanol and decreased the formation of CO. Over Cu–ZnO–Cr$_2$O$_3$–Al$_2$O$_3$–Ga$_2$O$_3$ (38.1:29.4:1.6:13.1:17.8 wt%) at 290°C and 80 atm, the extremely high space–time yield of methanol of 1483 g L^{-1} h^{-1} was reported, at a space velocity of 18,800, with 25% conversion of CO$_2$ to methanol. The beneficial effect of Ga$_2$O$_3$ was explained by the action of Ga in causing the inverse spillover of hydrogen and thus controlling the oxidation state of the copper component of the composite catalyst.

Pd-impregnated Cu/ZnO/Al$_2$O$_3$ catalysts were applied by Sahibzada et al. to the hydrogenation of CO$_2$ to methanol. The enhancement of methanol production by palladium was explained by hydrogen spillover with the Pd-promoted catalysts. Kinetic studies carried out under typical industrial conditions indicated that the promoting effect of Pd occurred only at high conversions, where inhibition by the product water was significant.[201,202]

Deng et al. developed an improved oxalate gel coprecipitation method for the preparation of ultrafine Cu/ZnO/Al$_2$O$_3$ catalysts for methanol synthesis from CO$_2$ + H$_2$. An optimal catalyst had the composition Cu/Zn/Al (60:30:10 at%). In this catalyst, the crystallite sizes of Cu metal and ZnO were 10.7 and 3.9 nm, respectively. With CO$_2$ + H$_2$ (1:3) at 220°C, 2.0 MPa pressure, and space velocity = 3600 h^{-1}, the CO$_2$ conversion was 14.7%, the selectivities to methanol and CO were 47 and 53%, and the yield of methanol was 6.9%.[203–205]

A mechanical alloying method was found by Fukui et al. to be superior to the conventional wet processing method for the preparation of Cu/ZnO catalysts for methanol synthesis. By milling Cu and ZnO powders together for 120 h, the resulting Cu/ZnO (50/50 wt%) had 50% higher methanol synthesis activity compared with the same catalyst formed by coprecipitation. The milling process induced a mechanical alloying and excellent mixing of the Cu and ZnO particles, reducing the grain size to 20 nm.[206]

10.3.4.2 ZrO$_2$-Supported Catalysts

ZrO$_2$ as a support has advantages for methanol synthesis due to its mechanical and thermal stability and its high specific surface.[207] The mechanism of the catalytic hydrogenation of carbon dioxide over a variety of metal/zirconia catalysts has been studied in detail by Baiker et al.[208,209] On Pd/ZrO_x catalysts, prepared by activation of Pd_1Zr_2 alloys with CO_2–H_2 mixtures, at a reaction temperature of 463 K, the reactant conversion was 8.5%. Methanol, methane, and CO were produced with selectivities of 43, 18, and 39%. In a diffuse reflectance FTIR study of adsorbed species on the activated Pd/ZrO_2 catalyst, the disappearance of surface formate was correlated with the appearance of gas-phase methane.

While Pd/ZrO_2 catalysts favored selectivity to methane, Cu/ZrO_2 catalysts promoted selectivity to methanol.[210–213] With catalysts obtained from an amorphous $Cu_{70}Zr_{30}$ precursor, diffuse reflectance FTIR spectroscopy measurements indicated a mechanism involving (1) rapid adsorption of CO_2 followed by its reduction to surface formate, which was further reduced to methane, and (2) CO_2 on the surface may be hydrogenated to CO by the reverse water–gas-shift reaction. Adsorbed CO in the presence of hydrogen produced π-bonded formaldehyde, which was reduced to methylate and finally to methanol. With amorphous $Au_{25}Zr_{75}$ alloy as precursor, activated *in situ* under CO_2 hydrogenation conditions, the gold was reduced to metallic particles of 8.5-nm mean size, while the zirconia was oxidized to ZrO_2. With this catalyst, the main products of CO_2 hydrogenation were methanol and carbon monoxide. However, the selectivity for methanol is worse on this catalyst than on the Cu/ZrO_2 catalyst. Diffuse reflectance FTIR spectroscopic investigation of adsorbed species during the reaction on the Au/ZrO_2 catalyst indicated the appearance of two types of formate species, with doublets of bands at 1580/1380 cm^{-1} (type I) and at 1600/1360 cm^{-1} (type II).[213]

With Cu/ZrO_2 as catalysts for methanol synthesis from CO_2 + H_2, the activity was found by Nitta et al. to depend markedly on the starting salts used in the formation of the catalyst precursors.[214,215] Such catalysts had usually been prepared by coprecipitation of the Cu and Zr nitrates. By using chloride salts instead, the selectivity for methanol was enhanced, while using sulfate salts enhanced the activity by increasing the dispersion of the Zr species. Highest yields of methanol were obtained with a Cu (50 wt%)/ZrO_2 catalyst prepared from a mixture of copper chloride and zirconium sulfate, coprecipitated with 10% excess of Na_2CO_3. With this catalyst, after prereduction under flowing H_2 at 523 K, the reaction of H_2 + CO_2 (3:1) at 493 K and 0.9 MPa resulted in 3.7% CO_2 conversion, a methanol formation rate of 1.04 mmol g^{-1} h^{-1}, and a methanol selectivity of 53%. The effect of chloride ions on the catalytic activity was explained by the formation of crystalline Cu metal with edge or corner sites decorated with Cl^- ions. An even higher rate of formation of methanol was achieved by Nitta et al. by using a catalyst of Cu/ZrO/ZnO (35:35:30 wt%) produced by coprecipitation from copper chloride, zirconium sulfate, and zinc nitrate.[216] After 3 h on stream, the reaction of H_2 + CO_2 (3:1) at 473 K and 0.9 MPa resulted in 5.9% conversion of CO_2 and methanol and CO formation rates of 1.5 and 1.7 mmol g^{-1} h^{-1}, respectively. In the ternary catalyst, the role of ZrO_2 seemed to enhance the selectivity to methanol, while the addition of ZnO considerably increased the surface area and improved dispersion of Cu.[216]

The addition of 4.8 atom% silver to a Cu/ZrO_2 (Cu/Zr = 1:1 atomic ratio) catalyst for methanol synthesis was tested by Fröhlich et al.[217] The reaction was performed at 493 K in a continuous fixed bed microreactor with $CO_2 + H_2$ (1:3) at a total pressure of 1.7 MPa. The silver promotion of the Cu/ZrO_2 catalyst enhanced the selectivity to methanol (at the expense of CO formation) from 70 to 80%, while leaving the overall activity unchanged.

Copper catalysts supported on zirconia are of particular interest because of their high selectivity and activity for methanol synthesis from CO_2 and H_2. In a careful study by Köppel et al. of several preparation methods for Cu/ZrO_2 catalysts, the procedures tested included impregnation, ion exchange, deposition–precipitation, coprecipitation, and simultaneous precipitation and reduction in the presence of reducing agents.[213] The formation of metallic copper (Cu^o) on the catalysts was found to be essential for good methanol productivity. High activity and selectivity for methanol synthesis from carbon dioxide and hydrogen were observed with catalyst systems that provide large interfacial areas between the CuO and the ZrO_2 components. The reaction mechanism probably involved a common surface intermediate for the two parallel reaction pathways leading to the production of methanol and gaseous CO.[213]

An improved copper catalyst for the hydrogenation of carbon dioxide to methanol was developed by Kieffer by creating a fine dispersion of the metal in a stable pyrochlore matrix.[218–221] The pyrochlore support $La_2Zr_2O_7$, which has good thermal and chemical stability, was prepared by the reactions

$$CuO + La_2O_3 \rightarrow CuLa_2O_4 \tag{10.34}$$

$$CuLa_2O_4 + 2ZrO_2 \rightarrow La_2Zr_2O_7 + CuO \tag{10.35}$$

An optimal catalyst had the composition Cu (50%)/$La_2Zr_{2.5}O_8$. The addition of ZnO enhanced the methanol synthesis activity of this catalyst. At high pressure conditions (6 MPa) and 250 to 300°C, with $CO_2 + H_2$ (1:3), the methanol production rate reached 600 g methanol kg-cat^{-1} h^{-1}. Using this catalyst, high selectivity for methanol formation (vs. the competing water–gas-shift reaction) was obtained by operating at a low reaction temperature, high reaction pressure, high H_2 partial pressure, and high H_2/CO_2 ratio. With cobalt-promoted $Cu/La_2Zr_2O_7$, the formation of hydrocarbons was increased, particularly CH_4.[218–221]

The effect of the additives ZrO_2, MgO, Al_2O_3, or Cr_2O_3 on the activity of CuO–ZnO catalysts for methanol synthesis from CO_2 and H_2 at 13 atm pressure was studied by Xu et al. in a flow reactor.[222] With the catalyst additives Al_2O_3, ZrO_2, and MgO, the activity and selectivity for methanol were higher than with CuO–ZnO alone. With a CuO–ZnO–ZrO_2 catalyst (molar ratio 42:47:11), the highest yield (17.4%) of methanol was at 220°C, and the highest selectivity (94.8%) for methanol was at 160°C.

Raney copper was found by Wainwright and Trimm to catalyze both the water–gas-shift reaction and the methanol synthesis reaction from syngas.[223] CO_2 was proposed to be the major reactant for methanol production, with some formate species as the main intermediate. Various kinds of Raney copper with added metals were tested by Saito et al. as catalysts for methanol synthesis from CO_2 and H_2.[224,225]

The catalysts were prepared by leaching the metal alloys in alkaline sodium zincate solutions. Highest activity was attained with Raney Cu–Zr (1.5 atom%). The initial activity of this catalyst was much higher than that of commercial methanol synthesis catalysts. The effect of sodium zincate leaching was both to increase the surface area and to deposit Zn on the surface of the Cu particles.

Rhenium catalysts supported on ZrO_2 or Nb_2O_5 were found by Iizuka et al. to be effective for methanol synthesis from $CO_2 + H_2$.[226] Re–ZrO_2 gave excellent reactivity and high selectivity (73%) for methanol formation at 160°C and 10 atm pressure. With Re–Nb_2O_5, selectivity for methanol was also good (>50%), but the reactivity was lower. Other CO_2 reduction products were CO, CH_4, and dimethyl ether.

At the very low reactant pressure of 5 atm, CO_2 and H_2 were converted by Xu et al. to methanol on an Re catalyst supported on CeO_2, ZrO_2, and La_2O_3.[227,228] The highest selectivity for methanol formation (76.7%) was observed over Re/CeO_2 at a reaction temperature of only 160°C. In order to distinguish between the direct hydrogenation of CO_2 to methanol and the alternative mechanism of CO formation by the reverse water–gas-shift reaction, experiments were made on mixtures of ^{13}CO, CO_2, and H_2. The extent of ^{13}C incorporation into the methanol indicated the mechanism. Over CuO–ZnO catalysts, methanol was produced almost exclusively by direct hydrogenation of CO_2. Over Re/ZrO_2 and Re/CeO_2, both routes were operative. The reaction involved surface formate and adsorbed formaldehyde as intermediates in methanol synthesis.[227,228]

10.3.4.3   Pd/CeO₂ and Pd/La₂O₃ Catalysts

Ramaroson et al. tested the hydrogenation of CO_2 on several supported Pd catalysts, using $H_2 + CO_2$ (3:1) at 120 bar and 350°C.[229] With the acidic supported catalyst Pd/SiO_2, the reaction favored methanation, with a selectivity for CH_4 of about 93%. On the other hand, with the basic supported Pd/La_2O_3 and Li–Pd/SiO_2, the selectivities for CH_4 were small, and methanol was produced with selectivities of 89 and 84%, respectively. Methanol synthesis from CO_2 was proposed to occur by a direct reaction, and not through prereduction to CO and subsequent hydrogenation.

Pd/CeO_2 and Pd/La_2O_3 prereduced under flowing hydrogen at 500°C were found by Fan and Fujimoto to be promising catalysts for methanol synthesis.[230,231] With $CO_2 + H_2$ (1:3) at 230°C and 30 bar, the CO_2 conversion over these two catalysts was 3.1 and 4.8%. While the selectivity to methanol reached 92 and 83%, the selectivity to CO was 7 and 13%, and the selectivity to methane was only 1 and 4%, respectively. Both catalysts were highly durable for long-term reaction.

10.3.4.4   Cu–Cr–Al and Cu–V–Zn Catalysts

Coprecipitated Cu–Cr–Al catalysts were applied by Kanoun et al. to methanol synthesis.[232] These catalysts were prepared by reacting ammonium bichromate and sodium aluminate with the copper tetramine complex. The precipitate was reduced at 300°C for 3 h under flowing H_2. The optimal composition was Cr–Al–Cu (0.1:0.9:1.0 by weight). Using $CO_2 + H_2$ (1:9) at 32 atm and 300°C, with a GHSV of 3×10^4 h^{-1}, resulted in a methanol formation activity of 880 g kg-Cu^{-1} h^{-1}, a turnover frequency of 2.6×10^3 mol s^{-1} at-Cu^{-1}, and methanol and methane selectivities of 58.5 and 17.5%, respectively. Minor products were dimethylether and CO.[232]

A series of Cu–V–Zn catalysts for methanol synthesis was prepared by Kanoun et al. by reacting ammonium vanadate with Cu and Zn ammonia complexes in concentrated solutions.[233,234] Best results were obtained with V–Cu–Zn (1:0.4:0.6 atomic ratio) which had been reduced under H$_2$ at 300°C. With CO$_2$ + H$_2$ (1:10) at 3.2 MPa and 300°C and GHSV = 3×10^4 h^{-1}, the methanol formation rate reached 2900 g kg-Cu^{-1} h^{-1}, and methanol and methane selectivities reached 41 and 1.5%, respectively.

10.3.4.5   On Cu and Pd Single-Crystal Surfaces

Methanol synthesis from CO$_2$ + H$_2$ on a Cu(100) single-crystal surface was found by Taylor et al. to involve the hydrogenation of adsorbed formaldehyde (or dioxomethylene) as the rate-limiting step. Hydrogenation of formate did not lead directly to methanol.[235]

The synthesis of methanol from CO$_2$ and H$_2$ over Cu(111) and Pd(111) surfaces was analyzed by Shustorovich and Bell using the "bond order conservation–Morse potential" approach.[236] In this semiempirical analysis, the probability of the intermediate reaction mechanisms can be evaluated, based on calculations of the heats of chemisorption of all adsorbed species and of the activation barriers for all elementary reactions assumed to be involved in the synthesis of methanol. In the hydrogenation of CO$_2$ on Cu, the activation barrier for intermediate formate production was shown to be significantly lower than that for the disproportionation into CO$_s$ and OH$_s$. On the other hand, on Pd catalyst, the activation barriers for both reactions were found to be essentially equal. Thus, the formate intermediate route is considered the preferred mechanism for CO$_2$ hydrogenation to methanol on copper.[236]

Considering the importance of Cu/ZnO catalysts for methanol synthesis, Fu and Somorjai studied the interaction of O$_2$, CO, CO$_2$, and D$_2$ with clean Cu(311) and Cu(110) surfaces.[237] Cu(311) was found to be highly reactive for the dissociative adsorption of CO$_2$ and hydrogen, and also for the CO–CO$_2$ exchange reaction, while Cu(110) was unreactive. The equilibrium

$$CO + O_{surface} \leftrightarrow CO_2 \qquad (10.36)$$

as well as the dissociation of H$_2$ occurred readily on Cu(311), but not on Cu(110).

10.3.4.6   Reaction Intermediates and Mechanisms

Isotope tracer experiments by Chinchen et al. suggested the absence of a carbon-containing surface intermediate which was common to methanol synthesis and to the water–gas-shift reaction.[238] This was indicated by the hydrogenation of ^{17}CO/^{14}CO$_2$ mixtures. At low conversions, the ^{14}C label was found by Rozovskii et al.[239] and Kuznetsov et al.[240] to appear in the methanol produced.

On model $Cu-ZnO-Al_2O_3$ catalysts, CO_2 treatment resulted primarily in the generation of adsorbed carbonate and hydrogeno carbonate species. Hydrogenation of these carbonates resulted in the conversion to adsorbed formate species, as indicated by FTIR measurements of Saussey et al.[241] and Deluzarche et al.[242] The proposed mechanism involved this adsorbed formate intermediate:

$$H_2 \rightarrow H^- + H^+ \tag{10.37}$$

$$CO_2 + H^- \rightarrow HCOO^- \tag{10.38}$$

$$HCOO^- + 2H_2 \rightarrow CH_3O^- + H_2O \tag{10.39}$$

The methoxide intermediate was then converted to methanol by hydrolysis or hydrogenolysis.

Ab initio molecular orbital (MO) calculations by Kakumoto and Watanabe on Cu/ZnO catalysts suggested that methanol synthesis initiated mainly on a Cu^+ site, forming a $Cu^+-O=C=O_{ad}$ intermediate, followed by H atom transfer from a metallic Cu site to the C atom of the adsorbed CO_2, producing a formate intermediate. Further H atom transfers led to adsorbed formaldehyde and methoxy intermediates and finally to desorbed methanol.[243]

Froment et al. used DRIFTS to reveal the reaction intermediates formed from CO_2 + H_2 (9:91) at 44 bar and 200°C over an industrial Imperial Chemical Industries methanol synthesis catalyst of $Cu/ZnO/Al_2O_3$ (60:30:10 wt%).[244-246] Strong broad bands at 1600 and 1360 cm^{-1} were assigned to the asymmetric O–C–O and the symmetric stretching frequencies, respectively, of surface-bound formate, while a weak band at 1530 cm^{-1} was assigned to bidentate surface-adsorbed carbonate. The same bands were observed after coadsorption of CO_2 + H_2 on partially oxidized copper and also by adsorption of methanol on the same catalysts. Temperature-programmed desorption (TPD) measurements indicated that this formate observed on a prereduced catalyst and at low adsorption pressures was located at the Cu–Zn interface. At high gas pressures (>8 bars, and as under industrial conditions), copper formate was the predominant species. The key intermediate in methanol synthesis on the $Cu/ZnO/Al_2O_3$ catalysts was therefore proposed to be surface-adsorbed copper formate. The hydrogenation of this formate to a methoxy species acted as the rate-determining step. The main function of ZnO in the catalyst may be to modify the electronic properties of copper.[244-246] Support for this mechanism was provided by Joo et al. with TPD and temperature-programmed hydrogenation studies on copper and zinc formates. The synergistic effect of copper and ZnO was explained by the generation of copper formate as the primary step. The formate then migrated onto the surface of ZnO, while the reduced copper activated hydrogen and transferred to the ZnO for reduction of formate to methanol.[247]

The mechanism of methanol synthesis on the highly effective Cu/ZnO catalyst (Cu/Zn = 3:7) was also studied by Fujita et al. using diffuse reflectance FTIR spectroscopy and TPD.[248] The catalyst was prereduced under H_2, resulting in the formation of metallic copper. When CO_2 + H_2 (1:9) was fed over the catalyst, methanol synthesis occurred, in parallel with the reverse water–gas-shift reaction (producing CO and H$_2$O). The adsorbed species on the catalyst identified after the reaction included zinc methoxide CH$_3$O–Zn, bidentate zinc formate, and copper formate HCOO–Cu. The zinc methoxide species, with IR absorption maxima at 2930 cm^{-1} (asym. CH$_3$ stretching), 2825 cm^{-1} (sym. CH$_3$ stretching), and 1060 cm^{-1} (CO stretching), was shown to be the precursor for methanol synthesis. Methanol was released from the catalyst by hydrolysis with water, which was formed by the reverse water–gas-shift reaction.[248]

The adsorption and decomposition of CO$_2$ on polycrystalline copper were reported by Hadden et al.[249–251] On clean polycrystalline copper, CO$_2$ was only weakly adsorbed. This adsorbed CO$_2$ served as a precursor, dissociating into adsorbed CO and surface O. The oxidized copper surface adsorbed CO$_2$ more strongly. Hydrogenation then resulted in adsorbed formate and methanol. Surprisingly, the rate of decomposition of CO$_2$ increased on decreasing the adsorption temperature. This indicated that the reaction of CO$_2$ on oxide-supported polycrystalline copper was moderated by a precursor state. The activation energy for CO$_2$ decomposition on the surface of the precursor state was smaller than the heat of adsorption of CO$_2$ on the precursor, resulting in a "negative" activation energy.[249–251]

Using both FTIR and TPD to study the interactions of CO$_2$, CO, and H$_2$ on Cu/ZnO/SiO$_2$ methanol synthesis catalysts, Millar et al. observed that the adsorption of CO$_2$ at 295 K produced hydrogen carbonate and carbonate species on copper.[252,253] In the presence of copper, oxygen ion defects were formed on the ZnO surface. Spillover hydrogen from the Cu surface occupied oxygen anion vacancy sites on the ZnO surface. CO$_2$ dissociation was proposed to proceed via a "surface malachite carbonate" intermediate, which was then hydrogenated to adsorbed copper formate. Adsorbed carbonate is also an intermediate in the water–gas-shift reaction.

In an FTIR study by Baiker et al.[254] and Weigel et al.[207,255] on Cu/ZrO$_2$ and Ag/ZrO$_2$ catalysts for methanol synthesis from H$_2$ and either CO or CO$_2$, formate radicals (bands at 1580 and 1380 cm^{-1}) adsorbed on the catalyst surface were found not to be direct precursors of methanol production. The intermediates toward methanol were shown to be surface-bound CO, carbonate (1640 and 1260 cm^{-1}) and hydroxyl groups, followed by π-bound formaldehyde (2820 and 1150 cm^{-1}) and surface-bound methylate, CH$_3$O$^-$ (2930, 2820, and 1050 cm^{-1}). Surface hydroxyl groups (3780 and 3680 cm^{-1}) had been created during the prereduction of the catalyst under hydrogen. During methanol synthesis, the consumption of these hydroxyl groups appears as "negative bands." A variant of the forward and reverse water–gas-shift reactions involving surface-adsorbed hydroxyl groups, operating in parallel with methanol synthesis, enabled the conversion of both CO and CO$_2$ to methanol:

$$CO + 2OH^- = CO_3^{2-} + H_2 \tag{10.40}$$

10.3.4.7 H$_2$ + CO$_2$ at Atmospheric Pressure

Comparative studies by Bardet et al., at atmospheric pressure, of hydrogenation on Cu–ZnO catalysts supported on Al$_2$O$_3$ indicated that the conversion to methanol, according to

$$CO_2 + 3H_2 = CH_3OH(l) + H_2O(l) \qquad \Delta H_{298\,K} = -131 \text{ kJ mol}^{-1} \qquad (10.3)$$

occurred at lower temperatures and with higher yields than the mixture of CO + H_2.[256]

Denise et al. activated Cu/ZnO catalysts by reduction in an H_2 stream at 300°C and performed the hydrogenation of CO_2 and CO to methanol at atmospheric pressure and 225°C. The product selectivity was found to depend on the contact time of the reactants over the catalyst. Prolonged contact times favored the formation of homologous hydrocarbon products.[257]

Efficient catalysis of methanol synthesis from CO_2 + H_2 (0.1:0.9) at atmospheric pressure and a temperature of 463 K was obtained by Fujita et al. over either 25 wt% Cu supported on ZnO, 46 wt% Cu on ZrO_2, or 19 wt% Pd on ZnO and resulted in production of both methanol and carbon monoxide.[248,258] The selectivities with these catalysts for methanol were 30.4, 27.5, and 13.5%, and the yields of conversion of CO_2 to methanol were 0.85, 0.44, and 0.14%, respectively. The catalysts were prepared by coprecipitation from the metal salts, calcination at 623 K, and reduction with H_2. The reactions were carried out in a flow reactor, at a contact time of 0.02 g-cat.min/cm^3. Infrared spectroscopic studies of the surface species produced from CO_2 + H_2 on the support materials ZnO and ZrO_2 indicated the intermediate production of formate groups. Such groups were proposed to play a pivotal role in methanol synthesis on the supported catalysts.[248,258]

10.3.4.8   Process Modeling of Methanol Synthesis

With the assumption that CO_2 is the main carbon source in methanol synthesis from a $CO/CO_2/H_2$ feed over a $Cu/ZnO/Al_2O_3$ catalyst, Vanden Bussche and Froment developed a steady-state kinetic model for both methanol synthesis and the water–gas-shift reaction. This model coupled the rates of both reactions through the common surface oxygen intermediate. The kinetic parameters were determined by experiments on an industrial $Cu/ZnO/Al_2O_3$ catalyst at pressures of up to 51 bar and temperatures in the range of 180 to 280°C.[259]

Since the methanol synthesis is mildly exothermic, it is usually carried out in adiabatic fixed bed reactors under steady-state conditions. Improved operation can be achieved by a reversed flow regime, which has been modeled by Vanden Bussche and Froment.[260] In this regime, the direction of feed gas flow was periodically reversed from the two ends of the tubular reactor. The reaction heat generated was thus trapped inside the catalyst bed, with the temperature profile along the reactor forming a standing wave after several cycles. The reversed flow reactor was predicted to increase slightly the methanol production rate and to improve the economics. A drawback of the simple reversed flow operation was extremely variable exit concentrations upon flow reversal. To overcome this difficulty, the STAR configuration of three reactor beds was proposed by the above authors. In this configuration, each bed was operated cyclically in an inlet, blowout, and exit phase. The predicted exit concentration of methanol was smoothed out and reached a concentration of 3.0 mol%, which was about 10% higher than achieved by the single-bed reversed flow process.[260]

10.3.4.9 Heterogeneous Liquid-Phase Methanol Synthesis

Methanol synthesis from CO, CO$_2$, and H$_2$ has been mainly based on gas–solid-phase reactions on fixed bed reactors, which requires recycling a large fraction of the unconverted gas and in which temperature control is difficult in this very exothermic reaction. Hagihara et al. performed methanol synthesis over Cu/ZnO/Al$_2$O$_3$ (Cu/Zn/Al = 4:3:3 atomic ratio) placed in a catalyst basket-type impeller reactor through which a hydrophobic solvent, n-dodecane (C$_{12}$H$_{26}$), was pumped and into which CO$_2$ + H$_2$ (1:3) was introduced at a total pressure of 15 MPa.[261,262] The solvent, together with the methanol and water produced, was transferred into a liquid–liquid separator in which the lighter solvent separated as the upper phase, while the heavier aqueous methanolic product was the lower phase. The solvent, which contained a small amount of unreacted gas, was continuously recycled to the reactor. The methanol and water were drawn off and separated by distillation. At a reaction temperature of 250°C and a solvent flow rate of 3 L solvent L-cat^{-1} h^{-1}, the yield of methanol (carbon based) was about 95%, and its production rate was 2.25 mol kg-cat^{-1} h^{-1}. For effective separation of the liquid phases, a low temperature (preferably ambient) was required. The reaction heat could thus be used for the distillation of the methanol produced. With the above Cu/ZnO/Al$_2$O$_3$ catalyst, the activity gradually decreased with reaction time, apparently due to sintering of the Cu particles in the catalyst.[261,262] An improved more stable catalyst for the liquid-phase methanol synthesis was developed by Mabuse et al. using Cu/ZnO/ZrO$_2$/Al$_2$O$_3$ modified with hydrophobic materials such as hydrophobic silica. In such a catalyst, after long-term liquid-phase methanol synthesis, a Zn$_2$SiO$_4$ phase was detected.[263]

10.3.5 Coupling Methanol Synthesis with Conversion to Gasoline and Olefins

10.3.5.1 Methanol-to-Gasoline Conversion

An interesting development by Inui et al. was to connect the outlet of the above-described CuO–ZnO–Cr$_2$O$_3$–Al$_2$O$_3$–Pd reactor (Section 10.3.4.1), releasing methanol as well as unreacted excess hydrogen into an H–Fe-silicate reactor at 300°C and 1 atm connected in series, which converted the methanol into a light gasoline.[159,160] Gasoline was obtained with 57% selectivity in a simple one-pass operation. The selectivity to gasoline could be increased by injection of a small amount of propylene into the inlet of the second reactor. Propylene enhanced the autocatalytic conversion of light olefins to gasoline. By replacing the H–Fe-silicate reactor with one packed with a more acidic catalyst of a silico-alumino-phosphate (SAPO-34) with a chabazite structure, and operating at 450°C and 1 atm, Inui et al. could change the selectivity in methanol conversion to almost exclusively produce light olefins: ethene, propene, and butenes.[264]

Instead of connecting the methanol synthesis reactor and the methanol-to-hydrocarbon conversion reactor in series, Inui et al. performed both processes in a single continuous flow reactor, using a composite catalyst of Cu–Cr–Zn/H–ZSM-5.[265] By operating at 320°C, with H$_2$ + CO$_2$ (2.7:1) at 50 atm and a space velocity of 6000 h^{-1}, the CO$_2$ conversion was 30%, and the selectivities to methanol, dimethyl ether, hydrocarbons, and CO were 2.8, 4.8, 10, and 82%, respectively. Of these hydrocarbons, 28% were methane and C_2–C_7 compounds.

An even more efficient process for the direct synthesis of gasoline from CO_2 and H_2 via methanol was achieved by Inui et al. by connecting in series (1) a reactor with a Pd-modified Ga-containing composite catalyst (Cu:ZnO:Cr_2O_3:Al_2O_3:Ga_2O_3 = 38.1:29.4:1.6:13.1:17.8), operated at 270°C, 80 atm, with a space velocity of 18,800 h^{-1}, forming methanol with a space–time yield (STY) of 1410 g L^{-1} h^{-1}, and (2) a second reactor of H–Ga-silicate (Si/Ga = 400), operated at 300°C and 15 atm, which resulted in total conversion of the methanol into hydrocarbons. The STY and the selectivity of gasoline formed were 328 g L^{-1} h^{-1} and 53.6%, respectively.[75]

An alternative process for the hydrogenation of carbon dioxide by Fujiwara et al. used novel hybrid catalysts, which combine the formation of methanol and its conversion to hydrocarbons.[266–269] These catalysts were prepared by physical mixing of the methanol synthesis catalyst, preferably oxides of Cu–Zn–Cr (2:2:1 or 3:3:1) containing hexavalent chromium, with zeolites (such as the synthetic zeolite JRC-Z-HY4.8 with Si/Al = 4.8). Typical conditions of conversion were 400°C, 50 kg cm^{-2}, SV = 3 L $h^{-1}g$-cal^{-1}, H_2/CO_2 = 3. The yields of hydrocarbons (mainly C_2–C_4) reached 8.6%.

With Fe–ZnO (4:1 molar ratio) on the same synthetic zeolite, Fujiwara et al. succeeded in reacting CO_2 + H_2 at 350°C and 50 atm, achieving 13.3% conversion of CO_2, 0.4% conversion to CH_4, and 4.5% conversion to C_2 and higher hydrocarbons, a high proportion of which were the olefins C_2H_4 and C_3H_6.[270–272] The distribution of hydrocarbons, and particularly the low selectivity for methane formation, indicated that the predominant reaction was the methanol-to-gasoline reaction (favored by the zeolite), and not an F-T reaction. If the Fe–ZnO catalyst was used without the zeolite, the hydrocarbon distribution followed the Schulz–Flory rule, with methane the predominant product and only traces of olefins, indicating an F-T mechanism.[270–272]

By physical mixing of a catalyst of Fe/Cu/Na (99:1:1.45) oxides with an HY zeolite (SiO_2/Al_2O_3 = 4.8), Xu et al. carried out the hydrocarbon synthesis from H_2 + CO_2 (3:1) at 250°C and 20 atm. With this composite catalyst, the CO_2 conversion amounted to 8.2%, and the selectivities to C_1 and C_{2+} were 33 and 45%, respectively. Remarkably, about 70% of the C_4, C_5, and C_6 products were branched-chain hydrocarbons. These branched hydrocarbons were proposed to be formed by carbon homologation of the primary olefins on the acidic zeolite.[273]

A selective production of *iso*-butane (2-methylpropane) was achieved by Tan et al. during the hydrogenation of CO_2 over a physically mixed composite catalyst. *Iso*-butane is a precursor in the synthesis of methyl-*tert*-butyl ether. With a catalyst of Fe–Zn–Cr (1:2:1 atomic ratio)/HY, and using H_2 + CO_2 (3:1) at 360°C and 50 atm, the conversion of CO_2 was about 17%, and the distributions of CH_4, *iso*-butane, and other hydrocarbons were 3, 39, and 58%, respectively.[274]

10.3.5.2   Hydrogenation of CO_2 to Olefins and Liquid Hydrocarbons

In the traditional F-T synthesis, converting CO with H_2 resulted mainly in low-molecular-weight hydrocarbons. There have been considerable efforts to find catalysts with improved selectivity for the more useful C$_{5+}$ olefinic hydrocarbons. A process for the hydrogenation of CO$_2$ over iron carbides such as Fe$_5$C$_2$ was patented by Fiato et al. for the Exxon Corporation.[275] A slurry of the carbide catalyst was suspended in a solvent such as octacosane [CH$_3$(CH$_2$)$_{26}$CH$_3$]. Passing through the slurry a mixture of CO$_2$, H$_2$, and N$_2$ at 1000 psig and 270°C resulted in 38% conversion of the CO$_2$ and formation of hydrocarbons, with selectivities for CH$_4$ and C$_{2+}$ products of 11 and 89%, respectively. The primary products of CO$_2$ hydrogenation with these catalysts were shown by Fiato et al. to be α-olefins.[276]

The selective synthesis of lower olefins (C$_2$–C$_4$) was achieved by Kim et al. by the hydrogenation of CO$_2$ over iron catalysts promoted with potassium and supported on ion-exchanged (H, K) zeolite-Y.[277] The reaction was carried out at 573 K with H$_2$ + CO$_2$ (3:1) at 10 atm and 1900 mL g^{-1} h^{-1}. The selectivity to total hydrocarbons increased with increasing iron loading and with the presence of potassium. Highest selectivity to the lower olefins (82%) was attained with 17 wt% Fe on a support containing an Fe:K atomic ratio of 2:1. Another catalyst with high selectivity toward the production of lower olefins and liquid hydrocarbons was reported by Choi et al. to be potassium-promoted Fe/Al$_2$O$_3$ (20 wt% Fe).[278,279]

10.3.6 Methanol and Ethanol Synthesis

10.3.6.1 Heterogeneous Catalysis

Methanol and the higher alcohols ethanol, 1-propanol, and 2-methyl-2-propanol were produced by Calverley and Smith over Cu/ZnO/Cr$_2$O$_3$ catalysts, with an additive of 0.5% K$_2$CO$_3$ as promoter, using CO$_2$, CO, and H$_2$ feed gases at 10 MPa and 285 to 315°C. It was suggested that CO$_2$ and the methanol produced participated directly in higher alcohol synthesis on copper sites.[280]

Silica-supported Ir–Mo catalysts were found by Kishida et al. to promote the hydrogenation of CO$_2$ to methanol, ethanol, some propanol, hydrocarbons (mainly methane), and CO.[281] Highest selectivities for alcohols were obtained with an Ir–Mo (6.5:3.2 wt%)/SiO$_2$ catalyst. By reacting H$_2$–CO$_2$–Ar (6:3:1) at 200°C, 4.9 MPa, and a space velocity of 2000 h^{-1}, the conversion was 7%, and the selectivities for methanol, ethanol, propanol, total hydrocarbons, and CO were 9.7, 3.8, 0.5, 44, and 42%, respectively.

Small yields of ethanol were observed by Ikehara et al. in the reaction of CO$_2$ + H$_2$ (2:1) over Ag–Rh (0.2:1 atomic ratio)/SiO$_2$ catalysts.[282] At 473 K and 2 MPa total pressure, the major product was CO, with more than 90% selectivity. Minor products were methane, acetic acid, ethanol, and methanol.

A systematic study was done by Arakawa et al. to optimize the synthesis of ethanol by hydrogenation of CO$_2$ over various promoted 1 wt% Rh/SiO$_2$ catalysts.[283–285] Without an added promoter, the reaction of H$_2$ + CO$_2$ (3:1) at 260°C and 5 MP pressure resulted in 1% CO$_2$ conversion, with product selectivities for methanol, ethanol, CO, and methane of 11, 6, 68, and 15%, respectively. With 5 wt% Rh–Fe (1:2 atomic ratio)/SiO$_2$ under the same conditions, the CO$_2$ conversion was 26% and the ethanol selectivity was 16%. Dramatic enhancement in the alcohol production was discovered with an Fe- and Li-promoted catalyst, 5 wt% Rh–Fe–Li (1:1:1)/

SiO_2, resulting under the same conditions in 14% CO_2 conversion, but with selectivities for methanol and ethanol of 23 and 34%, respectively. An *in situ* FTIR spectroscopic measurement of a 5 wt% Rh–Li (1:1)/SiO_2 catalyst during the reaction revealed the appearance of one band at 2040 cm^{-1} assigned to linear CO and one at 1860 cm^{-1} assigned to a bridged CO species adsorbed on Rh. Ethanol formation was proposed to occur by insertion of an adsorbed CO species into a CH_3–Rh bond.[283–285] The selectivity of CO_2 reduction products was found by Kusama et al. to depend on the anion of the rhodium salt which had served as precursor of the catalyst.[286] Using 5 wt% Rh/SiO_2, the selectivity for methanol and ethanol production decreased in the order acetate > nitrate > chloride. Such was also the order of increasing Rh particle sizes and of decreasing electron density of Rh (as determined by XPS). This was proposed to explain the preferred selectivity for alcohols with the catalyst prepared from Rh acetate.

An efficient synthesis of ethanol by the hydrogenation of CO_2 was reported by Kurakata et al. using an ($Rh_{10}Se$) catalyst supported on TiO_2.[287] The Rh cluster catalyst was activated by preheating under vacuum at 623 K. By reacting CO_2 + H_2 (1:2) at 47 kPa and 623 K, the initial rates of production of ethanol and methane were 3.7×10^{-3} and 1.4×10^{-3} mol h^{-1} g-cat^{-1}, respectively, and the selectivity to ethanol was 71%. Other Rh complexes tested and other support materials were much less effective for ethanol synthesis.

Very rapid conversion of CO_2 and H_2 into both ethanol and methanol, as well as into CO, was achieved by Yamamoto and Inui using a physical mixture of two catalysts, both supported on γ-Al_2O_3: a Cu-based partial reduction catalyst (Cu:Zn:Al:K = 1:1:1:0.1) which converted CO_2 to CO with a selectivity of more than 70% and a Pd- and Ga-modified Fe-based ethanol synthesis catalyst (Fe:Cu:Al:K = 1:0.03:2:0.7) performing by F-T and alcohol synthesis reactions. With CO_2 + H_2 (1:3) at 80 atm, 350°C, and space velocity = 20,000 h^{-1}, the conversion of CO_2 was 54.5%, the STYs of ethanol and methanol were 476 and 202 g L^{-1} h^{-1}, and the selectivities to ethanol and methanol were 17 and 5 mol%, respectively.[288]

A catalyst composed of K–Cu–Zn–Fe–Cr oxides was reported by Takagawa et al. to provide stable performance in ethanol synthesis. Using CO_2 + H_2 (1:3), GHSV = 20,000 and 7 MPa, resulted in 44% CO_2 conversion, an ethanol STY of 270 g L^{-1} h^{-1}, and 20% ethanol selectivity.[289]

10.3.6.2   Homogeneous Catalysis

Methanol synthesis by homogeneous catalytic reaction of CO_2 + H_2 was achieved by Tominaga et al. using ruthenium complexes in N-methyl-2-pyrrolidone (NMP) solution containing a high concentration of a mineral salt.[290–292] Among various transition metal complexes which were tested, the only active one was $Ru_3(CO)_{12}$, and among various salts, KI was the salt of choice. The presence of the salt was necessary to maintain the Ru complex in solution and to prevent its decomposition at the high reaction temperatures. The process was carried out batchwise by charging an autoclave containing the $Ru_3(CO)_{12}$ and KI in NMP solution with CO_2 + H_2 (1:3) to 80 atm and then heating to 240°C for 3 h. The main carbon product was methanol, followed by CO and methane. In the proposed mechanism, the primary step was a reverse water–gas-shift reaction producing Ru-complexed CO, which was then hydrogenated to methanol. If I$_2$ was used instead of KI, methane was the major product. The time course of the appearance of the products suggested the successive formation of CO, CH$_3$OH, and CH$_4$. Alcohol formation was favored by adding a cobalt complex as cocatalyst. When Co$_2$(CO)$_8$ was used as a cocatalyst, the Ru$_3$(CO)$_{12}$-catalyzed reaction of CO$_2$ (20 kg cm^{-2}) and H$_2$ (100 kg cm^{-2}) in NMP solution containing KI resulted in the production of a mixture of methanol, ethanol, methyl formate, methane, and CO. The Co complex served to stabilize the Ru complex under the conditions of the CO$_2$ hydrogenation.[290–292]

The same bimetallic complex system of Ru$_3$(CO)$_{12}$ and Co$_2$(CO)$_8$ was used by Tominaga et al. for the homologation of alcohols. With methanol as the substrate, dissolved in 1,3-dimethyl-2-imidazolidinone containing LiI as Lewis acid catalyst, under CO$_2$ (20 atm) and H$_2$ (100 atm) at 180°C for 15 h, ethanol was produced in 32% yield.[293]

In a related study on the homogenous hydrogenation of CO$_2$, Isaka and Arakawa used Ru$_3$(CO)$_{12}$ as catalyst, Co$_2$(CO)$_8$ as co-catalyst, tri-n-butylphosphine oxide as solvent, and LiBr as salt additive. With CO$_2$ + H$_2$ (1:5) at an initial pressure (before heating) of 110 kg cm^{-2}, and reacting at 200°C for 18 h, the CO$_2$ conversion was 33%, and the selectivities to methanol, ethanol, CO, and methane were 47, 16, 24, and 14%, respectively.[294]

An advantage of the homogeneous liquid-phase hydrogenation of CO$_2$ to alcohols is the more facile heat disposal and temperature control for such a highly exothermic reaction. However, the requirement of very expensive catalysts may limit industrial application.

10.3.7 Hydrogenation to Formic Acid

A comprehensive survey of the methods of synthesis of formic acid and its industrial uses was provided by Leitner.[295] The annual production of formic acid amounts to about 300 000 tons. Formic acid is an important chemical, useful as a pickling agent, as a reducing agent, in the manufacture of animal silage, and as an intermediate in the production of oxalic acid, formate esters, and amides such as dimethylformamide.[296,297] Formic acid is currently produced industrially by the carbonylation of methanol to methyl formate, followed by hydrolysis. A variety of processes have been developed for the reduction of CO$_2$ to formic acid. Photochemical, electrochemical, photoelectrochemical, and photocatalytic methods are discussed in Chapters 11, 12, 13, and 14, respectively.

10.3.7.1 Heterogeneous Catalysis: Formic Acid

The selective catalytic hydrogenation of bicarbonate ions in aqueous solutions to formate ions was achieved by Stalder et al. using Pd supported on carbon or γ-Al$_2$O$_3$ as catalysts:[298]

$$H_2 + HCO_3^- = HCOO^- + H_2O \qquad (10.41)$$

The reaction did not go to completion; rather, an equilibrium was reached. Thus, with initially 1.0 M NaHCO$_3$/1.7 atm H$_2$, the HCOO$^-$ concentration reached about 0.54 M. The rate of hydrogenation was found by Wiener et al. to increase with the hydrogen gas pressure, until a plateau was reached, in accordance with the Langmuir isotherm law.[299] Because of the common ion effect, the maximal concentrations of formate obtained at 6 atm H$_2$ and 35°C were 2.5 M for sodium formate and 5.8 M for potassium formate. The addition of CO$_2$ gas to the bicarbonate reactant did not result in an appreciable increase in the rate of hydrogenation. Therefore, presumably, the bicarbonate ion was the active species undergoing hydrogenation.[299]

10.3.7.2 Homogeneous Catalysis: Formic Acid and Methyl Formate

The hydrogenation of carbon dioxide to formic acid was accomplished by Tsai and Nicholas in homogeneous solutions, using transition metal catalysts.[296,297] With [Rh(diene)L$_3$]Z (L = R$_3$P) complexes as catalysts, in tetrahydrofuran (THF) solution at room temperature, formic acid was generated with a turnover number of 34 within 3 days. The turnover number could be enhanced by pretreatment of the catalyst with molecular hydrogen and by including some water (0.4% v/v) in the reaction mixture, thus achieving a turnover number of 128 within 2 days. With norbornadiene as the diene ligand, such as with [Rh(NBD)(PMe$_2$Ph)$_3$]BF$_4$ in the above precatalyst, the hydrogenation of CO$_2$ to HCOOH could be performed in THF solution, with CO$_2$ and H$_2$ each pressurized to 700 to 750 psi. The addition of H$_2$ to the precatalysts generated rhodium dihydride complexes, which presumably were the active catalysts.[296,297]

Alternative syntheses of formic acid from carbon dioxide and hydrogen were achieved by Leitner et al. using rhodium phosphane complexes.[300–306] Best results were obtained with the homogeneous {[Rh(cod)Cl]$_2$}–Ph$_2$P(CH$_2$)$_4$PPh$_2$ catalyst system, in which (cod) = cycloocta-1,5-diene. This catalyst was prepared *in situ* from [Rh(cod)Cl]$_2$ and Ph$_2$P(CH$_2$)$_4$PPh$_2$ in dimethyl sulfoxide (DMSO) solution containing triethylamine. With a total initial CO$_2$ + H$_2$ pressure of 40 atm, up to 1150 mol of formic acid per mole of rhodium was formed within 24 h at room temperature. The reaction mechanism involved the insertion of CO$_2$ into the Rh–H bond of intermediate neutral rhodium complexes. With water-soluble rhodium catalysts, Gassner and Leitner succeeded in hydrogenating CO$_2$ to formic acid in aqueous amine mixtures.[303] Most effective was the rhodium complex of the sulfonated phosphane {[(C$_6$H$_4$-m-SO$_3$-Na$^+$)$_3$P]$_3$RhCl} in a homogeneous water–dimethylamine solution. In the reaction of H$_2$ + CO$_2$ (1:1) at 40 atm and room temperature during 12 h, the turnover number reached 3400 mol of HCOOH per mole of rhodium.

A cationic bis(chelate) rhodium complex [Rh(P$^-$O)$_2$][BPh$_4$], where P$^-$O = η^2(O,P)-chelated Cy$_2$PCH$_2$CH$_2$OCH$_3$ ligand, was found by Lindner et al. to serve as catalyst for the hydrogenation of CO$_2$ to formic acid.[307] The reaction was carried out in an autoclave, with a methanolic solution of the complex, in the presence of triethylamine and traces of water, with pressurized CO$_2$ and H$_2$ at 50°C. Formic acid was produced with turnover numbers of up to 1000 within 7 h.

An interesting modification of the synthesis gas conversion to formic acid is the exothermic hydrocondensation of CO$_2$ and H$_2$ in the presence of methanol and anionic ruthenium clusters to produce methyl formate:[308,309]

$$CO_2 + H_2 + CH_3OH \rightarrow HCOOCH_3 + H_2O \quad \Delta H_{298\ K} = -25.37 \text{ kJ mol}^{-1} \quad (10.42)$$

Methyl formate is useful as it can be catalytically isomerized to acetic acid. The production of methyl formate was effectively catalyzed by the complex HRu$_3$(CO)$_{11}^-$, at a temperature of 125°C and pressures of 250 psi CO$_2$ and 250 psi H$_2$. The catalyst turnover reached up to 4 per 24 h. During the reaction, the HRu$_3$(CO)$_{11}^-$ complex was converted into a tetraruthenium cluster, H$_3$Ru$_4$(CO)$_{12}^-$, which was even slightly more active than the original catalyst. In a study of the mechanism of the reaction of HRu$_3$(CO)$_{11}^-$ with carbon dioxide, an insertion product was identified. This carboxylation occurred slowly at 60 psi pressure, but more rapidly at 400 psi pressure of carbon dioxide:[308,309]

$$HRu_3(CO)_{11}^- + CO_2 = HCO_2Ru_3(CO)_{10}^- + CO \quad (10.43)$$

Catalytically active anionic ruthenium complexes for the hydrocondensation of CO$_2$ with methanol to methyl formate were prepared by Süss-Fink et al. from RuCl$_3$ · 3H$_2$O, [N(PPh$_3$)$_2$]Cl, and CO (10 bar).[310] With the complex [N(PPh$_3$)$_2$][Ru(CO)$_3$Cl$_3$] in a methanolic solution of K(OCH$_3$), the reaction of CO$_2$–H$_2$ (1:2) at 60 bar and 160°C during 14 h led to a TON of 170 mol methyl formate per mole Ru. The proposed mechanism involved several intermediate ruthenium complexes.

Solutions of RuCl$_3$ and Ph$_3$P in EtOH–H$_2$O (5:1) were used by Zhang et al. as catalyst precursors for the hydrogenation to formate.[311] The reaction was performed in a stainless-steel autoclave, pressurized at room temperature with CO$_2$–H$_2$ to 6 MPa, and heated at 60°C for 5 h. The turnover number (moles HCOOH per mole of catalyst) reached 200. The active form of the catalyst was determined by IR and NMR spectroscopy to be RuH$_2$(CO)(PPh$_3$)$_3$, which underwent CO$_2$ insertion to a formate complex, HCO$_2$RuH(CO)(PPh$_3$)$_3$. The latter was hydrogenated in a rate-determining step, releasing formate and reforming the active catalyst, RuH$_2$(CO)(PPh$_3$)$_3$.[311]

A more efficient catalyst for the synthesis of methyl formate from methanol, CO$_2$ and H$_2$ was discovered by Kröcher et al.[312] In the presence of [RuCl$_2$(dppe)$_2$], where dppe = Ph$_2$P(CH$_2$)$_2$PPh$_2$, in a solvent-free reaction at 100°C, methyl formate was produced with a turnover frequency of 830 h^{-1}.

Formic acid was produced by Noyori et al. in a very efficient process by using supercritical CO$_2$ (scCO$_2$) as both reactant and solvent, in the presence of triethylamine and either [RuCl$_2$(PMe$_3$)$_4$] or RuCl$_2$(PMe$_3$)$_4$ as catalyst precursor, together with a promoter such as DMSO.[313–316] The reaction was performed at a total pressure of 200 to 220 atm and 50 to 80°C. The main advantage of $scCO_2$ is the high solubility of H_2 in this medium. Formic acid was produced with a turnover frequency of up to 4000 h^{-1}. If methanol was present in the reaction mixture, the products were both formic acid and methyl formate, with a combined yield (TON) of up to 10,000 mol per mole Ru. The methyl formate was produced by thermal esterification from the formic acid.[313–316]

10.3.8 Synthesis of Methylamines and Dimethylformamide

Mixtures of methylamines were obtained by Gredig et al. by passing CO_2 + H_2 + NH_3 (total pressure 0.6 MPa) over Cu/Al_2O_3 catalysts in the temperature range 473 to 573 K.[317,318] Higher reaction temperatures favored monomethylamine (MMA) as the main product, with smaller yields of dimethylamine (DMA) and trimethylamine (TMA), reaching the ratio MMA:DMA:TMA = 1:0.23:0.07. The major by-product was CO. In the absence of ammonia in the feed gas, methanol was formed.

N,N-Dimethylformamide (DMF) is an important industrial solvent. Noyori et al. performed the synthesis of DMF by the reaction of CO_2 and H_2 with DMA in supercritical CO_2 ($scCO_2$) as solvent, using [$RuCl_2(PMe_3)_4$] as catalyst:[313–316]

$$CO_2 + H_2 + Me_2NH \rightarrow Me_2NCHO \tag{10.44}$$

By reacting CO_2 (130 atm) and H_2 (80 atm) with Me_2NH at 100°C for 22 h, DMF was produced with an overall rate of up to 8000 h^{-1} and with TON values of up to 420,000. A drawback of this method is the air sensitivity of the above complex.

In an alternative procedure, Kröcher et al. used for the same synthesis the air- and water-stable complex [$RuCl_2(dppe)_2$], where dppe = $Ph_2P(CH_2)_2PPh_2$.[312] The reaction was carried out in an autoclave at 100°C without added solvent. DMF was produced in up to 100% yield, with the very large turnover frequency of 360,000 h^{-1}.

In order to avoid the inconvenience of homogeneous catalysts in the separation of products after DMF synthesis by the above reaction in $scCO_2$, Kröcher et al. substituted heterogeneous catalysts fixed into a silica matrix.[319] Using a sol-gel process, $Si(OEt)_4$ was co-condensed with silyl ethers of transition complexes. Highest catalytic activity was with an Ru complex containing methylphosphine ligands, {$RuCl_2[PMe_2(CH_2)_2Si(OEt)_3]_3$}. In $scCO_2$ and H_2, at 100°C and a total pressure of 22 MPa, a maximal TON of 110,000 and a selectivity to DMF of 100% were obtained.[319]

10.3.9 Hydrogenation of CO_2 in the Presence of Epoxide

In a unique homogeneous catalytic reaction reported by Sasaki et al, the hydrogenation of CO_2 with H_2 (80 atm) at 140 to 180°C in the presence of ethylene oxide and catalytic amounts of $RuCl_2(PPh_3)_3$ and N-methylpyrrolidine resulted in more than 70% conversion of CO_2, with the formation of CO and ethylene glycol as major products.[320,321] Without the N-methylpyrrolidine, the major product was ethylene carbonate. In a proposed mechanism, CO_2 insertion into the epoxide resulted in the formation of ethylene carbonate, which was then hydrogenated.

10.4 MICROWAVE AND DIELECTRIC BARRIER DISCHARGE-INDUCED REDUCTION

The endothermic carbon dioxide reforming reaction

$$CO_2 + CH_4 \rightarrow 2CO + 2H_2 \tag{10.1}$$

was activated by a plasma discharge, using microwave radiation at 2450 MHz, with an incident power of 80 W. The reaction was performed by Tanaka et al. in a closed circulation system, through a Pyrex glass tube, at a gas pressure of 5 torr.[322] The microwave discharge on methane alone gave acetylene as the main product. On the other hand, with methane + carbon dioxide (1:1), after 4 min, there was 96.5% conversion of CH_4, and the products were CO (46%), H_2 (40%), and $^2C + ^3C$ hydrocarbons (2%).

Microwave-induced catalysis was applied by Wan et al. to promote the reaction of carbon dioxide and water on an Ni/NiO catalyst (which had been activated at 440°C under H_2 + He for 5 to 6 h).[323] This technique enabled the selective absorption of microwave energy and rapid surface heating of the catalyst, with very small energy absorption by the reactants or the support material. The energy source was a 2.45-GHz magnetron providing 168-ms pulses, with 20 s dark times, with an average incident power of 2.2 kW. After a total irradiation time of 29.5 s, using an H_2O:CO_2 ratio of 1:2.5, the product composition was CH_4 (55.1%), C_2H_6 (0.3%), CH_3OH (5.5%), acetone (4.7%), C_3 alcohols (5.8%), and C_4 alcohols (28.4%). No CO_2 reduction occurred in the absence of the catalyst or the microwave radiation. It was proposed that one of the initiating events in the reaction was the microwave-induced decomposition of water, forming H atoms and OH radicals on the catalyst surface.[323]

By passing water vapor and CO_2 through a tubular quartz reactor in which a microwave plasma was created at 2.45 GHz, Ihara et al. obtained oxalic acid and hydrogen peroxide, in yields of up to 0.006 and 0.02%, respectively. In the absence of CO_2, the only product was H_2O_2.[324]

Dielectric barrier discharges (silent discharges) had previously been developed mainly for ozone generation.[325] The silent discharges were produced in an annular gap of 1 mm width between an outer steel cylinder (which served as a grounded electrode) and an inner cylindrical quartz tube. This quartz tube was gold-coated on the inside and was supplied with an AC high voltage of 18 kHz frequency, with power in the range of 50 to 900 W. Bill et al. applied such silent discharges to the reactions of CO_2.[326] The reactant gases were supplied at pressures of up to 50 bar to the annular gap. With only pure CO_2 at 1 bar pressure fed to the silent discharge, at an electric power of 200 W, CO_2 decomposed to CO, O_2, and O_3. The primary reaction presumably was excitation of CO_2 by collisions with electrons, followed by dissociation. With a mixture of $H_2 + CO_2$ (3:1) at a total pressure of 1 bar, the silent discharge led to the formation mainly of CO and H_2 in comparable amounts and of much smaller amounts of CH_4 and methanol. The methanol yield decreased with rising temperature. At 50°C, the maximal yield of methanol was 0.2%. With a mixture of $CH_4 + CO_2$ (2:1), the main products were CO and H_2, and also some methanol.[326] By fitting a commercial methanol synthesis catalyst into the silent discharge gap, Eliasson et al. succeeded in shifting the temperature of maximal catalyst activity from 220°C (without the discharge) to 100°C (with the discharge), but with some loss of the activity. The reaction was performed with a feed of $H_2 + CO_2$ (3:1) at 8 bar, at a power level of 500 W. The yield of methanol was 0.7 to 0.8%. The principal products were CO and CH_4.[327]

A parallel plate dielectric barrier discharge was used by Larkin et al. to convert mixtures of CH_4 and O_2 to organic oxygenates, methanol, formaldehyde, and formic acid. Also in the presence of CO_2, a substantial part of the CO_2 was converted to CO. The presence of O_2 was found necessary to effect a significant activation of methane.[328]

10.5   SONOLYSIS

Ultrasound-initiated redox reactions in aqueous solutions occur only in the presence of monoatomic or diatomic gases. Monoatomic gases such as argon are effective for sonolysis because of their low specific heat, resulting in high temperature rises in the small gas bubbles formed. These gas bubbles are produced by the cavitation due to the ultrasound. During the very rapid adiabatic compression, temperatures of several thousand degrees Kelvin and extreme pressure fluctuations cause water dissociation, forming free radicals, such as ·H and ·OH:

$$H_2O \rightarrow \cdot H + \cdot OH \qquad (10.45)$$

The sonolysis of aqueous CO_2 was studied by Henglein.[329] Pure CO_2 in water did not undergo sonolysis, while with pure water under argon, the only products were H_2 and H_2O_2. With carbon dioxide in water, in the presence of argon, CO was produced, as well as traces of HCOOH; the yield of H_2 decreased and the yield of H_2O_2 increased. Even with only 0.5% CO_2 in the argon gas mixture, the H_2 yield decreased to one-half. A 300-kHz quartz oscillator was used for irradiation at a high-frequency power of 3.5 W cm^{-2}. Carboxylation reactions were observed during the irradiation of 0.05 M ethanol in an Ar–CO_2 atmosphere, resulting in the production of formic, acetic, and lactic acids, while ethanol was also decomposed to acetaldehyde and formaldehyde. The free radical reactions which were proposed to account for the sonolysis products differed from those that occur in photolytic and radiolytic processes in liquid media and were more typical of those that occur in flames. The very high local concentrations of radicals favored radical–radical rather than radical–molecule reactions. Primary reaction steps in the sonolysis of carbon dioxide in water were proposed to be both protonation and direct dissociation:[329]

$$CO_2 + \cdot H \rightarrow \cdot COOH \qquad (10.46)$$

$$CO_2 \rightarrow CO + O \qquad (10.47)$$

10.6 GASIFICATION OF CARBON AND REDUCTION OF CO$_2$ TO CARBON

10.6.1 Gasification of Carbon

The gasification of carbonaceous materials, such as coal by carbon dioxide, could be an attractive method both for the disposal of waste carbon dioxide and for the production of carbon monoxide:

$$C + CO_2 = 2CO \qquad \Delta H_{900\,K} = +171.5 \text{ kJ mol}^{-1} \qquad (10.7)$$

This highly endothermic reaction was found by Yokoyama et al. and Suzuki et al. to be catalyzed by alkali metal ions.[330,331] In the presence of 0.9 wt% potassium carbonate, the above reaction became very rapid at 700°C, with an apparent activation energy E_a of 159 kJ mol^{-1}, while in the absence of potassium carbonate the reaction was just measurable at 700°C, with an apparent activation energy of 360 kJ mol^{-1}.

Molybdenum oxide (MoO$_2$) was used by Carrasco-Marín et al. as a catalyst for the carbon dioxide gasification of activated carbons and chars in the temperature range 773 to 823 K.[332] The E_a and the preexponential factor $A = \ln R_o$ were obtained from the Arrhenius equation, giving values of 41 to 70 kJ mol^{-1} and 7 to 11, respectively. The addition of sulfur to a carbon sample (C/S = 1:5 by wt) resulted in a decrease in the rate of gasification to one-third and an increase in E_a from 41 to 75 kJ mol^{-1}. Thus, sulfur caused a partial poisoning of the MoO$_2$ catalyst.[332]

Sodium lignosulfonate was found by Dhupe et al. to be an effective catalyst for the gasification of active charcoal by carbon dioxide.[333] This catalysis involved decreases in both kinetic parameters, E_a and A. On the other hand, Fe(NO$_3$)$_3$ as catalyst for charcoal gasification caused increases in these kinetic parameters.

Rapid gasification of carbon by CO$_2$ was obtained by Ohme and Suzuki with highly dispersed iron on activated carbon or carbon black. The reaction involved fast oxidation and reduction steps of the iron species.[334]

A low-temperature method for the reduction of CO$_2$ by coal was reported by Akiyama using cesium carbonate as catalyst.[335] Samples of coal in a stainless-steel autoclave were mixed with Cs$_2$CO$_3$, dried under vacuum, and charged at room temperature with CO$_2$ to 1.5 MPa. The autoclave was then heated at 600 to 800°C for 2 h, resulting in the formation of CO. After lowering the temperature to 300°C, the release of CO continued. Regrettably, the less expensive K$_2$CO$_3$ was less effective in promoting the production of CO, and BaCO$_2$ and CaCO$_3$ were completely inactive. It is not clear whether the CO produced actually originated from CO$_2$ or from the coal.[335]

An alternative approach to the reduction of CO_2 by coal was developed by Kodama et al.[336] In the first step of a two-step redox cycle, magnetite (Fe_3O_4) was reduced by pulverized bituminous coal at 800 to 900°C under N_2 to form CO, a reduced iron oxide (mainly FeO), and H_2. In a second step, the reduced iron oxide was reoxidized at 500 to 800°C by CO_2 to magnetite and CO. The overall reaction (where CH_τ = coal) was

$$CH_\tau + CO_2 \rightarrow 2CO + \tau/2H_2 \qquad (10.48)$$

The efficiency of the above first step was considerably enhanced by substituting Ni(II) or In(III) for Fe(II) or Fe(III) in magnetite. Such substituted magnetites were completely reduced to metallic phases by coal at 900°C. This process could be useful with concentrated sunlight as the energy source.[336]

The gasification by CO_2 of single-crystal graphite (a model of the gasification of coal) was compared by Yang and Wong with the gasification by steam.[337] The crystals of natural graphite were cleaved and then etched by exposure to the gases at 600°C and 23 mm pressure. Surface vacancies were thus expanded, creating pits one atomic layer deep. The edges of pits thus formed on the graphite crystal face were decorated with gold nuclei and were examined by TEM. The relative reactivities of H_2O and CO_2 were 11:1. Also, the pits formed from CO_2 were circular, while the pits formed from H_2O were hexagonal.[337]

The carbon–carbon dioxide reaction was measured by Hüttinger and Fritz on a polyvinyl coke, which had been heat pretreated at 1600°C to remove hydrogen.[338] The technique used was the TPD of quenched carbon–oxygen surface complexes. From the Arrhenius plots of the rates of gasification vs. the temperature, an apparent activation energy of 181 kJ mol^{-1} was obtained. This activation energy was equal to the apparent activation energy of the desorption step. This may indicate that the desorption reaction is the rate-limiting step.

The reduction of CO_2 by carbon to form CO was studied by Ono et al. as a method for the chemical storage of concentrated solar energy.[339] The reaction was performed by passing CO_2 at 0.1 MPa pressure over active carbon in a quartz tube and illuminating with sunlight concentrated by a Fresnel lens. Under a flux of 590 W cm^{-2} for about 20 min, the temperature rose to 990 K. The rate of CO evolution initially rose and then decreased during the illumination, reaching a maximal value of 4 mol sec^{-1}.

The gasification of carbon with CO_2 to form CO is one of the steps in the Carnol process for methanol synthesis described by Steinberg et al. The CO thus formed was then reacted with H_2. Both the H_2 and the carbon were obtained from the thermal decomposition of CH_4 at temperatures of 700 to 900°C and pressures of 28 to 56 atm.[340,341]

10.6.2   Reduction of CO_2 to Carbon

Fixation of carbon dioxide to both elementary carbon and carbon monoxide was achieved by Ishihara et al. during catalytic hydrogenation at atmospheric pressure and at temperatures from 573 to 973 K.[342,343] Active catalysts for production of both carbon and CO at 973 K were, in decreasing order of activity, $WO_3 > Y_2O_3 > ZnO > Cr_2O_3 > CeO_2 > Mn_2O_3 > MgO > V_2O_5 > ZrO_2 > MoO_3$. The production of carbon was the predominant reaction below 773 K, while CO production became the favored reaction above 873 K. This is understood on the basis of thermodynamic considerations, as the equilibrium,

$$C + CO_2 = 2CO \qquad (10.7)$$

is shifted to the right with rising temperature.[344,345] Using WO_3 as catalyst, at 973 K, the conversion of CO_2 into C and CO reached 27.6 and 42.3%, respectively. In order to enhance the selectivity for carbon production, mixed oxide catalysts were tested. With some mixed oxide materials, such as $LaFeO_3$, $LaMnO_3$, $LaMn_{0.9}Cu_{0.1}O_3$, and $PbWO_4$, the selectivity for elementary carbon formation was highest, and no CO was produced. The deposited carbon could easily be removed from the catalysts by exfoliation, thus providing an interesting method of carbon fixation.[344,345]

An alternative reagent for the efficient decomposition of carbon dioxide into carbon is oxygen-deficient (H_2-reduced) magnetite. This cation-excess reagent was prepared by Tamaura et al. by flowing H_2 gas through magnetite powder at 290 to 300°C for 2 h.[346–348] The magnetite retained its spinel-type structure and converted CO_2 into carbon with 100% yield at 300°C. The elementary carbon thus formed was readily hydrogenated at 300°C into methane. Similar results of the hydrogenation of carbon dioxide at 300°C to produce mainly carbon black were achieved by Korenaga et al. using strontium ferrite (prepared from 95 parts of FeO and 5 parts of SrO), with a conversion of up to 98%.[349]

Tsuji et al. adapted the reduction of CO_2 to carbon on H_2-reduced magnetite into a method of CO_2 methanation.[350,351] By substituting Ni(II) or Co(II) into ferrites, improved catalysts were obtained both for the decomposition of CO_2 to carbon and for the subsequent hydrogenation of the carbon to methane. Optimal results were obtained with an Ni(II)-bearing ferrite (Ni/Fe = 0.143 molar ratio), hydrogen prereduced at 300°C, which was treated with atmospheric pressure CO_2 at 300°C for 10 min, resulting in 98.8% conversion of CO_2 to carbon. The temperature was then lowered to 200°C, and the CO_2 was replaced by H_2. Up to 87% of the deposited carbon was reduced to methane. Much lower yields in the carbon deposition were obtained with either H_2-reduced magnetite or Co(II)-bearing ferrite. The CO_2 reduction to carbon at 300°C was enhanced by Ni(II)-bearing ferrite activated with Rh or Pt. The noble metal catalysis was explained by the hydrogen and oxygen species spillover from H_2 and CO_2 which had dissociated on the noble metals.[350,351]

Wüstite, which is FeO that has a cubic rock salt structure (similar to NaCl), with Fe^{2+} in the octahedral interstices, was found by Tamaura et al. to be a useful reagent for the decomposition of CO_2 at 300°C.[352,353] The active reagent was deficient in iron and was prepared by heating Fe(II) oxalate under nitrogen for 15 h at 650°C. The active wüstite had the composition $Fe_{0.98}O$ and when reacted with carbon dioxide at 300°C caused its transformation into elementary carbon. Carbon monoxide appeared as an intermediate for several hours, but was then also reduced to carbon. During the

CO_2 decomposition reaction, most of the wüstite was transformed into magnetite (Fe_3O_4).

10.7 CONCLUSIONS ON CO_2 REFORMING AND HYDROGENATION

The formation of synthesis gas by the CO_2 reforming of methane could provide a substantial use for the CO_2 released from fossil-fuel power stations. However, only that part of the synthesis gas converted into chemicals and used, for example, for the production of polymers would be a long-term sink for carbon. As pointed out by Seshan et al., the demand of synthesis gas for chemical production is very much lower than the release of CO_2 from fossil-fuel combustion.[6,7] The conversion of synthesis gas into methanol as a synthetic fuel would meet a much larger demand. However, the carbon contained in the synthetic fuel would be released to the atmosphere after a relatively short delay. The yearly demand for methanol, formaldehyde, and acetic acid amounted to about 24, 8, and 1.9 million tons, respectively, in 1995.

In 1997, the world capacity for methanol production was reported by Peaff to have reached 32 million metric tons, while the demand amounted to about 27 million metric tons.[354] Operating rates of 75 to 85% are necessary to allow for downtime for plant maintenance. Considering the additional 6 million metric tons under construction or planned until the year 2000, the estimated annual demand and capacity growth for methanol amount to about 3.7 and 4.6%, respectively. The current markets for methanol are formaldehyde (35.5%), general chemicals and solvents (30.8%), methyl-*tert*-butyl ether (27.4%), and acetic acid (6.4%). If fuel cells that run on methanol should become commercial, the demand for methanol may increase considerably.

There is a major economic motivation for new uses of CO_2. One significant process may be the aromatization of an LPG fraction (mainly propane) by carbon dioxide over zeolite catalysts (described in Section 10.2.10).

Another interesting reaction is the catalytic hydrogenation of CO_2 to methanol, followed in series by its conversion over a composite catalyst to gasoline-like C_1–C_7 hydrocarbons (described in Section 10.3.5).

The Carnol system, described above[340,341] and in detail in Chapter 7, converts the CO_2 received from power plants by catalytic hydrogenation to produce methanol. The methanol will be used in fuel cells for automotive purposes, significantly reducing CO_2 emissions. The hydrogen could be produced either by the thermal decomposition of methane and storing the carbon or by a nonfossil energy source. The large market for automotive fuel will provide the capacity demand for converting CO_2 to methanol or hydrogen-rich fuels.

These and other processes for the conversion of CO_2 may lead to important industrial products while decreasing the need for wasteful disposal.[355]

REFERENCES

1. Fox, J.M., III, The different catalytic routes for methane valorization: as assessment of processes for liquid fuels, *Catal. Rev. Sci. Eng.*, 35, 169–212, 1993.
2. Krylov, O.V. and Mamedov, A.K., Heterogeneous catalytic reactions of carbon dioxide, *Russ. Chem. Rev.*, 64, 935–959, 1995.
3. Wang, S.B., Lu, G.Q.M., and Millar, G.J., Carbon dioxide reforming of methane to produce synthesis gas over metal-supported catalysts: state of the art, *Energy Fuels*, 10, 896–904, 1996.
4. Schmidt, M., The thermodynamics of CO$_2$ conversion, *Carbon Dioxide Chemistry: Environ. Issues, R. Soc. Chem. Spec. Publ.*, pp. 23–30, 1994.
5. Fischer, F. and Tropsch, H., Conversion of methane into hydrogen and carbon monoxide, *Brennst. Chem.*, 9, 29–46, 1928; *Chem. Abstr.*, 22, 2652.
6. Seshan, K. and Lercher, J.A., Challenges in CH$_4$ + CO$_2$ reforming, *Carbon Dioxide Chemistry: Environ. Issues, R. Soc. Chem. Spec. Publ.*, 153, 16–22, 1994.
7. Ross, J.R.H., van Keulen, A.N.J., Hegarty, M.E.S., and Seshan, K., The catalytic conversion of natural gas to useful products, *Catal. Today*, 30, 193–199, 1996.
8. Gadalla, A.M. and Bower, B., The role of catalyst support on the activity of nickel for reforming methane with carbon dioxide, *Chem. Eng. Sci.*, 43, 3049–3062, 1988.
9. Gadalla, A.M. and Sommer, M.E., Carbon dioxide reforming of methane on nickel catalysts, *Chem. Eng. Sci.*, 44, 2825–2829, 1989.
10. De Groote, A.M. and Froment, G.F., Synthesis gas production from natural gas in a fixed bed reactor with reversed flow, *Can. J. Chem. Eng.*, 74, 735–742, 1996.
11. Turlier, P., Pereira, E.B., and Martin, G.A., Catalytic reforming of methane with carbon dioxide into CO + H$_2$ gas mixture over nickel, in Abstr. Int. Conf. on Carbon Dioxide Utilization, Bari, Italy, September 1993, 119–126.
12. Bradford, M.C.J. and Vannice, M.A., Catalytic reforming of methane with carbon dioxide over nickel catalysts .I. Catalyst characterization and activity, *Appl. Catal. A Gen.*, 142, 73–96, 1996.
13. Teuner, S.C., Make carbon monoxide from carbon dioxide, *Hydrocarbon Process.*, 64, 106–107, 1985.
14. Solymosi, F., Kutsán, Gy., and Erdöhelyi, E., Catalytic reaction of CH$_4$ with CO$_2$ over alumina-supported Pt metals, *Catal. Lett.*, 11, 149–156, 1991.
15. Erdöhelyi, A., Cserényi, J., and Solymosi, F., Activation of CH$_4$ and its reaction with CO$_2$ over supported Rh catalysts, *J. Catal.*, 141, 287–299, 1993.
16. Solymosi, F., Activation and reactions of CO$_2$ on Rh catalysts, *Carbon Dioxide Chemistry: Environ. Issues, R. Soc. Chem. Spec. Publ.*, 153, 44–54, 1994.
17. Bitter, J.H., Hally, W., Seshan, K., Ommen, J.G. van, and Lercher, J.A., The role of the oxidic support on the deactivation of Pt catalysts during the CO$_2$ reforming of methane, *Catal. Today*, 29, 349–353, 1996.
18. Lercher, J.A., Bitter, J.H., Hally, W., Niessen, W., and Seshan, K., Design of stable catalysts for methane–carbon dioxide reforming, *Stud. Surf. Sci. Catal.*, 101, 463–472, 1996.
19. van Keulen, A.N.J., Seshan, K., Hoebink, J.H.B.J., and Ross, J.R.H., TAP investigations of the CO$_2$ reforming of CH$_4$ over Pt/ZrO$_2$, *J. Catal.*, 166, 306–314, 1997.
20. Wang, H.Y. and Au, C.T., Carbon dioxide reforming of methane to syngas over SiO$_2$-supported rhodium catalysts, *Appl. Catal. A Gen.*, 155, 239–252, 1997.

21. Rostrup-Nielsen, J.R. and Hansen, J.H.B., CO_2-reforming of methane over transition metals, *J. Catal.*, 144, 38–49, 1993.

22. Basini, L. and Sanfilippo, D., Molecular aspects in syn-gas production: the CO_2-reforming reaction case, *J. Catal.*, 157, 162–178, 1995.

23. Zhang, Z.L., Tsipouriari, V.A., Efstathiou, A.M., and Verykios, X.E., Reforming of methane with carbon dioxide to synthesis gas over supported rhodium catalysts. 1. Effects of support and metal crystallite size on reaction activity and deactivation characteristics, *J. Catal.*, 158, 51–63, 1996.

24. Efstathiou, A.M., Kladi, A., Tsipouriari, V.A., and Verykios, X.E., Reforming of methane with carbon dioxide to synthesis gas over supported rhodium catalysts. 2. A steady-state tracing analysis: mechanistic aspects of the carbon and oxygen reaction pathways to form CO, *J. Catal.*, 158, 64–75, 1996.

25. Mark, M.F. and Maier, W.F., CO_2 reforming of methane on supported Rh and Ir catalysts, *J. Catal.*, 164, 122–130, 1996.

26. Bhat, R.N. and Sachtler, W.M.H., Potential of zeolite-supported rhodium catalysts for the CO_2 reforming of CH_4, *Appl. Catal. A Gen.* 150, 279–296, 1997.

27. Horiuchi, T., Sakuma, K., Fukui, T., Kubo, Y., Osaki, T., and Mori, T., Suppression of carbon deposition in the CO_2-reforming of CH_4 by adding basic metal oxides to a Ni/Al$_2$O$_3$ catalyst, *Appl. Catal. A Gen.*, 144, 111–120, 1996.

28. Osaki, T., Masuda, H., Horiuchi, T., and Mori, T., Highly hydrogen-deficient hydrocarbon species for the CO_2-reforming of CH_4 on Co/Al$_2$O$_3$ catalyst, *Catal. Lett.*, 34, 59–63, 1995.

29. Osaki, T., Horiuchi, T., Suzuki, K., and Mori, T., Kinetics, intermediates and mechanism for the CO_2-reforming of methane on supported nickel catalysts, *J. Chem. Soc. Faraday Trans.*, 92, 1627–1631, 1996.

30. Osaki, T., Effect of reduction temperature on the CO_2-reforming of methane over TiO$_2$-supported Ni catalyst, *J. Chem. Soc. Faraday Trans.*, 93, 643–647, 1997.

31. Bradford, M.C.J. and Vannice, M.A., Catalytic reforming of methane with carbon dioxide over nickel catalysts. II. Reaction kinetics, *Appl. Catal. A Gen.*, 142, 97–122, 1996.

32. Choudhary, V.R., Uphade, B.S., and Mamman, A.S., Large enhancement in methane-to-syngas conversion activity of supported Ni catalysts due to precoating of catalyst supports with MgO, CaO or rare earth oxide, *Catal. Lett.*, 32, 387–390, 1995.

33. Gronchi, P., Centola, P., and Del Rosso, R., Dry reforming of CH_4 with Ni and Rh metal catalysts supported on SiO$_2$ and La$_2$O$_3$, *Appl. Catal. A Gen.*, 152, 83–92, 1997.

34. Gronchi, P., Centola, P., and Del Rosso, R., CO_2 for petrochemicals feedstock. Conversion to synthesis gas on metal supported catalysts, in Abstr. 4th Int. Conf. on Carbon Dioxide Utilization, Kyoto, Japan, September 1997, O-36.

35. Gronchi, P., Marengo, S., Mazzocchia, C., Tempesti, E., and Del Rosso, R., On the formation of oxygenated products by CO hydrogenation with lanthana promoted rhodium catalysts, *React. Kinet. Catal. Lett.*, 60, 79–87, 1997.

36. Ruckenstein, E. and Hu, Y.H., Carbon dioxide reforming of methane over nickel/alkaline earth metal oxide catalysts, *Appl. Catal. A Gen.*, 133, 149–161, 1995.

37. Hu, Y.H. and Ruckenstein, E., An optimum NiO content in the CO_2 reforming of CH_4 with NiO/MgO solid solution catalysts, *Catal. Lett.*, 36, 145–149, 1996.

38. Ruckenstein, E. and Hu, Y.H., Role of support in CO_2 reforming of CH_4 to syngas over Ni catalysts, *J. Catal.*, 162, 230–238, 1996.

39. Ruckenstein, E. and Hu, Y.H., Interactions between Ni and La$_2$O$_3$ in Ni/La$_2$O$_3$ catalysts prepared using different Ni precursors, *J. Catal.*, 161, 55–61, 1996.

40. Hu, Y.H. and Ruckenstein, E., The characterization of a highly effective NiO/MgO solid solution catalyst in the CO$_2$ reforming of CH$_4$, Catal. Lett., 43, 71–77, 1997.

41. Ruckenstein, E. and Hu, Y.H., The effect of precursor and preparation conditions of MgO on the CO$_2$ reforming of CH$_4$ over NiO/MgO catalysts, Appl. Catal. A Gen., 154, 185–205, 1997.

42. Tomishige, K., Chen, Y., Li, X., Yokoyama, K., Sone, Y., and Fujimoto, K., Development of active and stable nickel–magnesia solid solution catalysts for CO$_2$ reforming of methane, in Abstr. 4th Int. Conf. on Carbon Dioxide Utilization, Kyoto, Japan, September 1997, P-007.

43. Chen, Y.-G., Tomishige, K., and Fujimoto, K., Promotion in activity and stability of nickel–magnesia solid solution catalyst by structural rearrangement via hydration for reforming of CH$_4$ with CO$_2$, Chem. Lett., pp. 999–1000, 1997.

44. Kroll, V.C.H., Swaan, H.M., and Mirodatos, C., Methane reforming reaction with carbon dioxide over Ni/SiO$_2$ catalyst. I. Deactivation studies, J. Catal., 161, 409–422, 1996.

45. Kroll, V.C.H., Delichére, P., and Mirodatos, C., Methane reforming reaction with carbon dioxide over a Ni/SiO$_2$ catalyst — the nature of the active phase, Kinet. Katal., 37, 749–757, 1996; English translation, Kinet. Catal., 37, 698–705, 1996.

46. Bhattacharyya, A. and Chang, V.W., CO$_2$ reforming of methane to syngas: deactivation behavior of nickel aluminate spinel catalysts, Stud. Surf. Sci. Catal., 88, 207–213, 1994.

47. Chang, J.S., Park, S.E., and Chon, H.Z., Catalytic activity and coke resistance in the carbon dioxide reforming of methane to synthesis gas over zeolite-supported Ni catalysts, Appl. Catal. A Gen., 145, 111–124, 1996.

48. Park, S.-E., Chang, J.-S., Park, M.S., Anpo, M., and Yamashita, H., CO$_2$ behavior on supported KNiCa catalyst in the carbon dioxide reforming of methane, in Abstr. 4th Int. Conf. on Carbon Dioxide Utilization, Kyoto, Japan, September 1997, P-016.

49. van Looij, F. and Geus, J.W., The reactivity of carbon-containing species on nickel. Effects of crystallite size and related phenomena, Catal. Lett., 45, 209–213, 1997.

50. Tsuji, M., Kodama, T., Yoshida, T., Kitayama, Y., and Tamaura, Y., Preparation and CO$_2$ methanation activity of an ultrafine Ni(II) ferrite catalyst, J. Catal., 164, 315–321, 1996.

51. Kodama, T., Kitayama, Y., Tsuji, M., and Tamaura, Y., Methanation of CO$_2$ using ultrafine Ni$_x$Fe$_{3-x}$O$_4$, Energy, 22, 183–187, 1997.

52. Wang, H.Y. and Au, C.T., CH$_4$/CD$_4$ isotope effects in the carbon dioxide reforming of methane to syngas over SiO$_2$-supported nickel catalysts, Catal. Lett., 38, 77–79, 1996.

53. Guerrero-Ruiz, A., Rodríguez-Ramos, I., and Sepúlveda-Escibano, A., Effect of the basic function in Co,MgO/C catalysts on the selective oxidation of methane by carbon dioxide, J. Chem. Soc. Chem. Commun., pp. 487–488, 1993.

54. Guerrero-Ruiz, A., Sepúlveda-Escibano, A., and Rodríguez-Ramos, I., Cooperative action of cobalt and MgO for the catalysed reforming of CH$_4$ with CO$_2$, Catal. Today, 21, 545–550, 1994.

55. Lu, Y., Yu, Ch., Xue, J., and Shen, Sh., Studies on the reaction mechanism of CH$_4$/CO$_2$ reforming over Co/γ-Al$_2$O$_3$ catalyst, Chem. Lett., pp. 515–516, 1997.

56. Matsuura, I. and Takayasu, O., Carbon dioxide reforming of methane using transition metal catalysts supported on magnesium oxide, in Proc. Int. Symp. Chemical Fixation of Carbon Dioxide, Nagoya, Japan, December 2 to 4, 1991, 247–252.

57. Takayasu, O., Soman, C., Takegahara, Y., and Matsuura, I., Deactivation of Ni-catalysts and its prevention by mechanically mixing an oxide for the formation reaction of CO + H$_2$ from CO$_2$ + CH$_4$, Stud. Surf. Sci. Catal., 88, 281–288, 1994.

58. Osaki, T., Horiuchi, T., Suzuki, K., and Mori, T., Suppression of carbon deposition in CO_2-reforming of methane on metal sulfide catalysts, *Catal. Lett.*, 35, 39–43, 1995.

59. Osaki, T., Retardation of carbon deposition in CO_2–CH_4 reaction on metal sulfide catalysts, *Am. Chem. Soc. Symp. Ser. Heterogeneous Hydrocarbon Oxidation*, 638, 384–393, 1996.

60. Osaki, T., Horiuchi, T., Suzuki, K., and Mori, T., Catalyst performance of MoS_2 and WS_2 for the CO_2-reforming of CH_4. Suppression of carbon deposition, *Appl. Catal. A Gen.*, 155, 229–238, 1997.

61. Ashcroft, A.T., Cheetham, A.K., Foord, J.S., Green, M.L.H., Grey, C.P., Murrell, A.J., and Vernon, P.D.F., Selective oxidation of methane to synthesis gas using transition metal catalysts, *Nature*, 344, 319–321, 1990.

62. Hickman, D.A. and Schmidt, L.D., Synthesis gas formation by direct oxidation of methane over Pt monoliths, *J. Catal.*, 138, 267–282, 1992.

63. De Groote, A.M. and Froment, G.F., Reactor modeling and simulations in synthesis gas production, *Rev. Chem. Eng.*, 11, 145–183, 1995.

64. De Groote, A.M. and Froment, G.F., Synthesis gas production by catalytic partial oxidation of natural gas. Influence of the operating variables on coke formation, in *Proc. 9th Int. Symp. Large Chemical Plants from 1995 to the Next Decennium*, Antwerp, Belgium, October 1995, 395–434.

65. Wang, D.Z., De Waele, O., De Groote, A.M., and Froment, G.F., Reaction mechanism and role of the support in the partial oxidation of methane on Rh/Al_2O_3, *J. Catal.*, 159, 418–426, 1996.

66. De Groote, A.M. and Froment, G.F., Simulation of the catalytic partial oxidation of methane to synthesis gas, *Appl. Catal. A Gen.*, 138, 245–264, 1996.

67. Ashcroft, A.T., Cheetham, A.K., Green, M.L.H., and Vernon, P.D.F., Partial oxidation of methane to synthesis gas using carbon dioxide, *Nature*, 352, 225–226, 1991.

68. Choudhary, V.R., Rajput, A.M., and Prabhakar, B., Energy-efficient methane-to-syngas conversion with low H_2/CO ratio by simultaneous catalytic reactions of methane with carbon dioxide and oxygen, *Catal. Lett.*, 32, 391–396, 1995.

69. Choudhary, V.R. and Rajput, A.M., Simultaneous carbon dioxide and steam reforming of methane to syngas over NiO–CaO catalyst, *Ind. Eng. Chem. Res.*, 35, 3934–3939, 1996.

70. Choudhary, V.R., Uphade, B.S., and Belhekar, A.A., Oxidative conversion of methane to syngas over $LaNiO_3$ perovskite with or without simultaneous steam and CO_2 reforming reactions: influence of partial substitution of La and Ni, *J. Catal.*, 163, 312–318, 1996.

71. De Groote, A.M. and Froment, G.F., Hydrogen and synthesis gas production by catalytic partial oxidation and steam reforming: fundamental development tools, in *Proc. 5th World Congress Chem. Eng.*, San Diego, July 1996, 893–896.

72. Qin, D.Y., Lapszewicz, J., and Jiang, X.Z., Comparison of partial oxidation and steam–CO_2 mixed reforming of CH_4 to syngas on MgO-supported metals, *J. Catal.*, 159, 140–149, 1996.

73. Inui, T., Saigo, K., Fujii, Y., and Fujioka, K., Catalytic combustion of natural gas as the role of on-site heat supply in rapid catalytic CO_2–H_2O reforming of methane, *Catal. Today*, 26, 295–302, 1995.

74. Inui, T., Rapid CO_2-reforming of methane to syngas and successive conversion to gasoline and/or light olefins, in *Abstr. 3rd Int. Conf. on Carbon Dioxide Utilization*, Norman, OK, May 1995.

75. Inui, T., Highly effective conversion of carbon dioxide to valuable compounds on composite catalysts, *Catal. Today*, 29, 329–337,1996.
76. Hara, H., Takeguchi, T., and Inui, T., Direct synthesis of gasoline from carbon dioxide via methanol as the intermediate, in Abstr. 3rd Int. Conf. on Carbon Dioxide Utilization, Kyoto, Japan, September 1997, P-060.76.
77. Nishiyama, T. and Aika, K., Mechanism of the oxidative coupling of methane using CO_2 as an oxidant over PbO–MgO, *J. Catal.*, 122, 348–351, 1990.
78. Aika, K. and Nishiyama, T., Formation of ethene and carbon monoxide from methane and carbon dioxide, in Proc. Int. Symp. Chemical Fixation of Carbon Dioxide, Nagoya, Japan, December 1991, 413–418.
79. Fraenkel, D., Levitan, R., and Levy, M., A solar thermochemical pipe based on the CO_2–CH_4 (1:1) system, *Int. J. Hydrogen Energy*, 11, 267–277, 1986.
80. Levitan, R., Levy, M., Rosin, H., and Rubin, R., Closed-loop operation of a chemical heat pipe at the Weizmann Institute solar furnace, *Sol. Energy Mater.*, 24, 464–477, 1991.
81. Levy, M., Rubin, R., Rosin, H., and Levitan, R., Methane reforming by direct solar irradiation of the catalyst, *Energy*, 17, 749–756, 1992.
82. Levy, M., Levitan, R., Meirovitch, E., Segal, A., Rosin, H., and Rubin, R., Chemical reactions in a solar furnace. 2. Direct heating of a vertical reactor in an insulated receiver. Experiments and computer simulations, *Sol. Energy*, 48, 395 402, 1992.
83. Levy, M., Levitan, R., Rosin, H., and Rubin, R., Solar energy storage via a closed-loop chemical heat pipe, *Sol. Energy*, 50, 179–189, 1993.
84. Buck, R., Muir, J.F., Hogan, R.E., and Skocypec, R.D., Carbon dioxide reforming of methane in a solar volumetric receiver/reactor: the CAESAR project, *Sol. Energy Mater.*, 24, 449–463, 1991.
85. Muir, J.F., Hogan, R.E., Skocypec, R.D., and Buck, R., Solar reforming of methane in a direct absorption catalytic reactor on a parabolic dish. I. Test and analysis, *Sol. Energy*, 52, 467–477, 1994.
86. Perera, J.S.H.Q., Couves, J.W., Sankar, G., and Thomas, J.M., The catalytic activity of Ru and Ir supported on Eu_2O_3 for the reaction $CO_2 + CH_4$–reversible $2H_2 + 2CO$ — a viable solar–thermal energy system, *Catal. Lett.*, 11, 219–225, 1991.
87. Richardson, J.T. and Paripatyadar, S.A., Carbon dioxide reforming of methane with supported rhodium, *Appl. Catal.*, 61, 293–309, 1990.
88. Meirovitch, E. and Segal, A., Storing concentrated sunlight in the chemical bond: unique stability and internal regulation — a simulative theoretical investigation, *Sol. Energy*, 46, 219–229, 1991.
89. Skocypec, R.D., Hogan, R.E., and Muir, J.F., Solar reforming of methane in a direct absorption catalytic reactor on a parabolic dish. II. Modelling and analysis, *Sol. Energy*, 52, 479–490, 1994.
90. Hattori, T., Komatsuki, M., Satsuma, A., and Murakami, Y., Catalytic reduction of carbon dioxide by lower alkane, *Nippon Kagaku Kaishi*, pp. 648–650, 1991; *Chem. Abstr.*, 115, 34719z.
91. Hattori, T., Yamauchi, S., Komatsuki, M., Satsuma, A., and Murakami, Y., Catalytic reduction of carbon dioxide by using lower hydrocarbons, in Proc. Int. Symp. Chemical Fixation of Carbon Dioxide, Nagoya, Japan, December 2 to 4, 1991, 419–422.
92. Gosling, C.D., Wilcher, F.P., and Pujado, P.R., LPG conversion to aromatics, in Proc. 69th Annu. Conv. Gas Process. Assoc., 1990, 190–194; *Chem. Abstr.*, 114, 46184p.
93. Kitagawa, H., Sendoda, Y., and Ono, Y., Transformation of propane into aromatic hydrocarbons over ZSM-5 zeolites, *J. Catal.*, 101, 12–18, 1986.

94. Hattori, T., Yamauchi, S., Satsuma, A., and Murakami, Y., Catalytic reduction of carbon dioxide on Zn-loaded HZSM-5 accompanying aromatization of propane, *Chem. Lett.*, pp. 629–630, 1992.

95. Hattori, T., Yamauchi, S., Endo, M., Komai, S., Satsuma, A., and Murakami, Y., Catalytic reduction of carbon dioxide by lower alkanes, *Carbon Dioxide Chemistry: Environ. Issues, R. Soc. Chem. Spec. Publ.*, 153, 74–81, 1994.

96. Nishi, K., Endo, M., Satsuma, A., Hattori, T., and Murakami, Y., Effect of carbon dioxide on aromatization of ethane over metal-loaded HZSM-5 catalysts, *Sekiyu Gakkai Shi (J. Jpn. Pet. Inst.)*, 39, 260–266, 1996.

97. Sugino, M., Shimada, H., Turuda, T., Miura, H., Ikenaga, N., and Suzuki, T., Oxidative dehydrogenation of ethylbenzene with carbon dioxide, *Appl. Catal. A Gen.*, 121, 125–137, 1995.

98. Chang, J.-S., Park, S.-E., Park, M.S., Anpo, M., and Yamashita, H., Oxidative dehydrogenation of ethylbenzene with carbon dioxide over ZSM-5-supported iron oxide catalysts, in Abstr. 4th Int. Conf. on Carbon Dioxide Utilization, Kyoto, Japan, September 1997, P-013.

99. Mimura, N., Takahara, I., Saito, M., Hattori, T., Ohkuma, K., and Ando, M., Dehydrogenation of ethylbenzene over iron oxide-based catalysts in the presence of carbon dioxide, in Abstr. 4th Int. Conf. on Carbon Dioxide Utilization, Kyoto, Japan, September 1997, P-024.

100. Mirzabekova, S.R. and Mamedov, A.K., Ethylene conversion to C_4H_6 and C_3H_6 in the presence of carbon dioxide over a tungsten–chromium catalyst, *Kinet. Catal.*, 35, 247–250, 1994; *Chem. Abstr.*, 121, 134844d.

101. Mirzabekova, S.R., Mamedov, A.K., and Rustamov, M.I., Reactivity of oxygen generated by carbon dioxide in heterogeneous catalytic oxidation of organic substances of various classes, *Kinet. Catal.*, 36, 111–116, 1995; *Chem. Abstr.*, 123, 32493n.

102. Aika, K., Study on the oxidative coupling of methane using the oxygen of carbon dioxide, in *Use of CO_2 in the Manufacture of Chemicals*, Chemical Society of Japan, 1993, 20–32.

103. Pillai, S.M., Ohnishi, R., and Ichikawa, M., Catalytic reforming of toluene with carbon dioxide over rare earth promoted Pd/γ-Al_2O_3 catalysts, *React. Kinet. Catal. Lett.*, 48, 247–253, 1992.

104. Olsbye, U., Tangstad, E., and Dahl, I.M., Partial oxidation of methane to synthesis gas in a fluidized bed reactor, *Stud. Surf. Sci. Catal.*, 82 (Natural Gas Conversion II), 303–308, 1994.

105. Slagtern, A., Olsbye, U., Blom, R., Dahl, I.M., and Fjellvåg, H., In situ XRD characterization of La–Ni–Al–O model catalysts for CO_2 reforming of methane, *Appl. Catal. A Gen.*, 145, 375–388, 1996.

106. Slagtern, A., Olsbye, U., Blom, R., and Dahl, I.M., The influence of rare earth oxides on Ni/Al_2O_3 catalysts during CO_2 reforming of CH_4, *Stud. Surf. Sci. Catal.*, 107 (Natural Gas Conversion IV), 497–502, 1997.

107. Ioannides, T. and Verykios, X.E., Application of a dense silica membrane reactor in the reactions of dry reforming and partial oxidation of methane, *Catal. Lett.*, 36, 165–169, 1996.

108. Kikuchi, E., Palladium/ceramic membranes for selective hydrogen permeation and their application to membrane reactor, *Catal. Today*, 25, 333–337, 1995.

109. Kikuchi, E., Steam reforming and related reactions in hydrogen-permselective membrane reactor, *Sekiyu Gakkai Shi (J. Jpn. Pet. Inst.)*, 39, 301–313, 1996.

110. Kikuchi, E. and Chen, Y., Low-temperature syngas formation by CO_2 reforming of methane in a hydrogen-permselective membrane reactor, *Surf. Sci. Catal.*, 107, 547–553, 1997.

111. Taoda, H., Osaki, T., Horiuchi, T., Iseda, K., Tsuge, A., and Yamakita, H., Catalytic reduction of carbon dioxide by supported molybdenum sulphide, in Proc. Int. Symp. Chemical Fixation of Carbon Dioxide, Nagoya, Japan, December 2 to 4, 1991, 379–382.

112. Kaspar, J., Graziani, M., Rahman, A.M., Trovarelli, A., Vichi, E.J.S., and Da Silva, E.C., Carbon dioxide hydrogenation over iron containing catalysts, *Appl. Catal. A Gen.*, 117, 125–137, 1994.

113. Ernst, K.-H., Campbell, C.T., and Moretti, G., Kinetics of the reverse water–gas shift reaction over Cu(110), *J. Catal.*, 134, 66–74, 1992.

114. Román-Martínez, M.C., Cazorla-Amorós, D., Linares-Solano, A., and Salinas-Martínez de Lecea, C., Carbon dioxide hydrogenation catalyzed by alkaline earth- and platinum-based catalysts supported on carbon, *Appl. Catal. A Gen.*, 116, 187–204, 1994.

115. Román-Martínez, M.C., Cazorla-Amorós, D., Linares-Solano, A., and Salinas-Martínez de Lecea, C., CO_2 hydrogenation under pressure on catalysts Pt–Ca/C, *Appl. Catal. A Gen.*, 134, 159–167, 1996.

116. Román-Martínez, M.C., Cazorla-Amorós, D., Salinas-Martínez de Lecea, C., and Linares-Solano, A., Structure sensitivity of CO_2 hydrogenation reaction catalyzed by Pt/carbon catalysts, *Langmuir*, 12, 379–385, 1996.

117. Yoshida, T., Thorn, D.L., Okano, T., Ibers, J.A., and Otsuka, S., Hydration and reduction of carbon dioxide by rhodium hydride compounds. Preparation and reactions of rhodium bicarbonate and formate complexes, and the molecular structure of $RhH_2(O_2COH)(i-Pr)_3)_2$, *J. Am. Chem. Soc.*, 101, 4212–4221, 1979.

118. Choi, P.H., Jun, K.W., Lee, S.J., Choi, M.J., and Lee, K.W., Hydrogenation of carbon dioxide over alumina-supported Fe–K catalysts, *Catal. Lett.*, 40, 115–118, 1996.

119. Schulz, H., Schaub, G., Claeys, M., Riedel, T., and Walter, S., Initial transient rates and selectivities of Fischer–Tropsch synthesis with CO_2 as carbon source, in Abstr. 4th Int. Conf. on Carbon Dioxide Utilization, Kyoto, Japan, September 1997, O-04.

120. Satterfield, C. N., *Heterogeneous Catalysis in Practice*, Mc-Graw Hill, New York, 1980, 308.

121. LeBlanc J.R., Jr., Madhavan, S., Porter, R.E., and Kellogg, P., Ammonia, in *Kirk-Othmer Encyclopedia of Chemical Technology*, Vol. 2, 3rd ed., Wiley-Interscience, New York, 1978, 494.

122. Rostrup-Nielsen, J.R., Schoubye, P.S., Christiansen, L.J., and Nielsen, P.E., The environment and the challenge to reaction engineering, *Chem. Eng. Sci.*, 49, 3995–4003, 1994.

123. Aksoylu, A.E., Akin, A.N., Sunol, S.G., and Önsan, Z.I., The effect of metal loading on the adsorption parameters of carbon dioxide on coprecipitated nickel–alumina catalysts, *Turkish J. Chem.*, 20, 88–94, 1996.

124. Aksoylu, A.E., Akin, A.N., Önsan, Z.I., and Trimm, D.L., Structure/activity relationships in coprecipitated nickel–alumina catalysts using CO_2 adsorption and methanation, *Appl. Catal. A Gen.*, 145, 185–193, 1996.

125. Tahri, A., Amariglio, A., Zyad, M., and Amariglio, H., Elementary steps accompanying the chemisorption of carbon dioxide on a nickel catalyst in relation with the methanation reaction. I. Exposure of the catalyst to carbon dioxide and thermoreaction with hydrogen of the chemisorbed species, *J. Chim. Phys.*, 90, 109–121, 1993.

126. Tahri, A., Amariglio, A., Zyad, M., and Amariglio, H., Elementary steps accompanying the chemisorption of carbon dioxide on a nickel catalyst in relation with the methanation reaction. II. Thermal desorption of adsorbed species after exposure of nickel to carbon dioxide, *J. Chim. Phys.*, 90, 123–137, 1993.

127. Fujita, S., Nakamura, M., Doi, T., and Takezawa, N., Mechanisms of methanation of carbon dioxide and carbon monoxide over nickel/alumina catalysts, *Appl. Catal. A Gen.*, 104, 87–100, 1993.

128. Nakayama, T., Ichikuni, N., Sato, S., and Nozaki, F., Ni/MgO catalysts prepared using citric acid for hydrogenation of carbon dioxide, *Appl. Catal. A Gen.*, 158, 185–199, 1997.

129. Ogura, K., Kawano, M., and Adachi, D., Dark catalytic reduction of CO_2 over photopretreated NiO/kieselguhr catalyst, *J. Mol. Catal.*, 72, 173–179, 1992.

130. Solymosi, F., Erdöhelyi, A., and Lancz, M., Surface interaction between H_2 and CO_2 over palladium on various supports, *J. Catal.*, 95, 567–577, 1985.

131. Erdöhelyi, A., Pásztor, M., and Solymosi, F., Catalytic hydrogenation of CO_2 over supported palladium, *J. Catal.*, 98, 166–177, 1986.

132. Trovarelli, A., De Leitenburg, C., and Dolcetti, G., CO and CO hydrogenation under transient conditions over $Rh–CeO_2$: novel positive effects of metal support interactions on catalytic activity and selectivity, *J. Chem. Soc. Chem. Commun.*, pp. 472–473, 1991.

133. Trovarelli, A., Lutman, A., de Leitenburg, C., and Dolcetti, G., CO_2 methanation over CeO_2 promoted Rh/SiO_2 catalysts: effect of CeO_2 dispersion, in Abstr. Int. Conf. on Carbon Dioxide Utilization, Bari, Italy, September 1993, 371.

134. Trovarelli, A., De Leitenburg, C., Dolcetti, G., and Lorca, J.L., CO_2 methanation under transient and steady-state conditions over Rh/CeO_2 and CeO_2-promoted Rh/SiO_2: the role of surface and bulk ceria, *J. Catal.*, 151, 111–124, 1995.

135. De Leitenburg, C., Trovarelli, A., and Kaspar, J., A temperature-programmed and transient kinetic study of CO_2 activation and methanation over CeO_2 supported noble metals, *J. Catal.*, 166, 98–107, 1997.

136. Fisher, I.A. and Bell, A.T., A comparative study of CO and CO_2 hydrogenation over Rh/SiO_2, *J. Catal.*, 162, 54–65, 1996.

137. Kishida, M., Fujita, T., Umakoshi, K., Ishiyama, J., Nagata, H., and Wakabayashi, K., Novel preparation of metal-supported catalysts by colloidal microparticles in a water-in-oil microemulsion; catalytic hydrogenation of carbon dioxide, *J. Chem. Soc. Chem. Commun.*, pp. 763–764, 1995.

138. Kishida, M., Onoue, K., Tashiro, S., Nagata, H., and Wakabayashi, K., Hydrogenation of carbon dioxide over rhodium catalyst supported on silica, in Abstr. 4th Int.. Conf. on Carbon Dioxide Utilization, Kyoto, Japan, September 1997, P023.

139. Suzuki, T., Saeki, K.-I., Mayama, Y., Hirai, T., and Hayashi, S., Hydrogenation of carbon dioxide over iron oxide catalyst, *React. Kinet. Catal. Lett.*, 44, 489–497, 1991.

140. Baussart, H., Delobel, R., Le Bras, M., and Leroy, J.-M., Hydrogenation of CO_2 over Co/Cu/K catalysts, *J. Chem. Soc. Faraday Trans. I*, 83, 1711–1718, 1987.

141. Fröhlich, G., Kestel, U., Lojewska, J., Lojewski, T., Meyer, G., Voss, M., Borgmann, D., Dziembaj, R., and Wedler, G., Activation and deactivation of cobalt catalysts in the hydrogenation of carbon dioxide, *Appl. Catal. A Gen.*, 134, 1–19, 1996.

142. Lojewska, J. and Dziembaj, R., Model of activation of the cobalt foil as a catalyst for CO_2 methanation, *J. Mol. Catal. A Chem.*, 122, 1–11, 1997.

143. Thampi, K.R., Kiwi, J., and Grätzel, M., Methanation and photo-methanation of CO_2 at room temperature and atmospheric pressure, *Nature*, 327, 506–508, 1987.

144. Grätzel, M., Catalytic and photocatalytic fixation of carbon dioxide, in Proc. Int. Symp. on Chemical Fixation of Carbon Dioxide, Nagoya, Japan, December 2 to 4, 1991, 1–10.

145. Melsheimer, J., Guo, W., Ziegler, D., Wesemann, M., and Schlögl, R., Methanation of carbon dioxide over Ru/Titania at room temperature — exploration for a photoassisted catalytic reaction, *Catal. Lett.*, 11, 157–168, 1991.

146. Prairie, M.R., Renken, A., Highfield, J.G., Thampi, K.R., and Grätzel, M., A Fourier transform infrared spectroscopic study of CO_2 methanation on supported ruthenium, *J. Catal.*, 129, 130–144, 1991.

147. Gupta, N.M., Kamble, V.S., Iyer, R.M., Thampi, K.R., and Grätzel, M., FTIR studies on the CO, CO_2 and H_2 co-adsorption over Ru–Ru$_z$/TiO$_2$ catalyst, *Catal. Lett.*, 21, 245–255, 1993.

148. Gupta, N.M., Kamble, V.S., Kartha, V.B., Iyer, R.M., Thampi, K.R., and Grätzel, M., FTIR spectroscopic study of the interaction of CO_2 and CO_2 + H_2 over partially oxidized Ru/TiO$_2$ catalyst, *J. Catal.*, 146, 173–184, 1994.

149. Gupta, N.M., Kamble, V.S., Thampi, K.R., and Grätzel, M., Direct evidence for simultaneous CO and CO_2 hydrogenation over Ru–Ru$_x$/TiO$_2$ catalyst, *Indian J. Chem. Sect. A*, 33, 374–379, 1994.

150. Gupta, N.M. and Kamble, V.S., Air tolerance of a Ru–RuO$_x$/TiO$_2$ catalyst for CO_2 methanation reaction, *Indian J. Chem.*, 35A, 557–559, 1996.

151. Marwood, M., Doepper, R., Prairie, M., and Renken, A., Transient drift spectroscopy for the determination of the surface reaction kinetics of CO_2 methanation, *Chem. Eng. Sci.*, 49, 4801–4809, 1994.

152. Marwood, M., Doepper, R., and Renken, A., In-situ surface and gas phase analysis for kinetic studies under transient conditions. The catalytic hydrogenation of CO_2, *Appl. Catal. A Gen.*, 151, 223–246, 1997.

153. Weatherbee, G.D. and Bartholomew, C.H., Hydrogenation of carbon dioxide on group VIII metals. II. Kinetics and mechanism of carbon dioxide hydrogenation on nickel, *J. Catal.*, 77, 460–472, 1982.

154. Weatherbee, G.D. and Bartholomew, C.H., Hydrogenation of carbon dioxide on group VIII metals. IV. Specific activities and selectivities of silica-supported cobalt, iron, and ruthenium, *J. Catal.*, 87, 352–362, 1984.

155. Huang, S., Jiang, B.-N., and Yu, W., Kinetics of hydrogenation of CO_2 on supported metallic catalysts, in Proc. Int. Symp. Chemical Fixation of Carbon Dioxide, Nagoya, Japan, December 2 to 4, 1991, 145–150.

156. Dziembaj, R., Makowski, W., and Papp, H., Carbon dioxide hydrogenation of Fe–Mn oxide catalyst doped with Rh and La, *J. Mol. Catal.*, 75, 81–99, 1992.

157. Inui, T., Funabiki, M., and Takegami, Y., Dynamic adsorption studies on nickel-based methanation catalysts using the continuous flow method, *J. Chem. Soc. Faraday Trans. 1*, 76, 2237–2250, 1980.

158. Inui, T., Funabiki, M., and Takegami, Y., Simultaneous methanation of carbon monoxide and carbon dioxide on supported nickel-based composite catalysts, *Ind. Eng. Chem. Prod. Res. Dev.*, 19, 385–388, 1980.

159. Inui, T. and Takeguchi, T., Effective conversion of carbon dioxide and hydrogen to hydrocarbons, *Catal. Today*, 10, 95–106, 1991.

160. Inui, T., Effective conversion of CO_2 and H_2 to hydrocarbons, in Proc. Int. Symp. on Chemical Fixation of Carbon Dioxide, Nagoya, Japan, December 2 to 4, 1991, 159–166.

161. Dubois, J.-L., Sayama, K., and Arakawa, H., CO_2 hydrogenation over carbide catalysts, *Chem. Lett.*, pp. 1115–1118, 1992.

162. Schild, C., Wokaun, A., Köppel, R.E., and Baiker, A., CO_2 hydrogenation over nickel/zirconia catalysts from amorphous precursors: on the mechanism of methane formation, *J. Phys. Chem.*, 95, 6341–6346, 1991.

163. Yoshida, T., Habazaki, H., Yamasaki, M., Akiyama, E., Kawashima, A., and Hashimoto, K., Methanation of carbon dioxide on catalysts derived from amorphous Ni–Zr–rare earth element alloys, in Abstr. 4th Int. Conf. on Carbon Dioxide Utilization, Kyoto, Japan, September 1997, O-24.

164. Hashimoto, K., Recent advances in the catalytic properties of metastable materials, *Mater. Sci. Eng. A*, 226–228, 891–899, 1997.

165. Yamasaki, M., Habazaki, H., Yoshida, T., Akiyama, E., Kawashima, A., Asami, K., Komori, M., and Shimamura, K., Compositional dependence of the CO_2 methanation activity of Ni/ZrO_2 catalysts prepared from amorphous Ni–Zr alloy precursors, *Appl. Catal. A Gen.*, 163, 187–197, 1997.

166. Le Bras, M., Agounaou, M., Baussart, H., Gengembre, L., and Leroy, J., Catalytic performances of the mixed oxides $Bi_xGd_{1-x}VO_4$, catalysts for the hydrogenation of carbon dioxide, *J. Chim. Phys.*, 91, 7–36, 1994.

167. Le Bras, M., Agounaou, M., Gengembre, L., Baussart, H., and Leroy, J.M., Influence of a reduction process on the catalytic performances of $Bi_xGd_{1-x}VO_4$ catalysts for the hydrogenation of carbon dioxide, *J. Chim. Phys.*, 93, 331–354, 1996.

168. Ando, H., Fujiwara, M., Matsumura, Y., Miyamura, H., Tanaka, H., and Souma, Y., Methanation of carbon dioxide over $LaNi_4X$-type intermetallic compounds as catalyst precursor, *J. Alloys Compounds*, 223, 139–141, 1995.

169. Wakuda, K., Habazaki, H., Kawashima, A., Asami, K., and Hashimoto, K., CO_2 methanation catalysts prepared from amorphous Ni-valve metal alloys containing platinum group elements, *Sci. Rep. Res. Inst. Tohoku Univ.*, 38, 76–87, 1993.

170. Tada, T., Habazaki, H., Akiyama, E., Kawashima, A., Asami, K., and Hashimoto, K., Amorphous Fe–valve metal–Pt group metal alloy catalysts for methanation of CO_2, *Mater. Sci. Eng.*, 182, 1133–1136, 1994.

171. Trovarelli, A., Matteazzi, P., Dolcetti, G., Lutman, A., and Miani, F., Nanophase iron carbides as catalysts for carbon dioxide hydrogenation, *Appl. Catal. A Gen.*, 95, L9–L13, 1993.

172. Mori, S., Xu, W.-C., Ishidzuki, T., Ogasawara, N., Imai, J., and Kobayashi, K., Mechanochemical activation of catalysts for CO_2 methanation, *Appl. Catal. A Gen.*, 137, 255–268, 1996.

173. Ohya, H., Fun, J., Kawamura, H., Itoh, K., Ohashi, H., Aihara, M., Tanisho, S., and Negishi, Y., Methanation of carbon dioxide by using membrane reactor integrated with water vapor permselective membrane and its analysis, *J. Membr. Sci.*, 131, 237–247, 1997.

174. Reller, A., Padeste, C., and Hug, P., Formation of organic carbon compounds from metal carbonates, *Nature*, 329, 527–529, 1987.

175. Tsuneto, A., Kudo, A., Saito, N., and Sakata, T., Hydrogenation of solid state carbonates, *Chem. Lett.*, pp. 831–834, 1992.

176. Wade, L.E., Gengelbach, R.B., Trumbley, J. L., and Hallbauer, W.L., in *Kirk-Othmer Encyclopedia of Chemical Technology*, Vol. 15, 3rd ed., Wiley-Interscience, New York, 1981, 398–415.

177. Ziessel, R., Chimie de coordination de la molecule de dioxyde de carbon: activation biologique, chimique, electrochimique et photochimique, *Nouv. J. Chim.*, 7, 613–633, 1983.

178. Inoue, T., Iizuke, T., and Tanabe, K., Support effect of zinc oxide catalysts on synthesis of methanol from CO_2 and H_2, *Bull. Chem. Soc. Jpn.*, 60, 2663–2664, 1987.

179. Klier, K., Chatikavanij, V., Herman, R.G., and Simmons, G.W., Catalytic synthesis of methanol from carbon monoxide/hydrogen. IV. The effect of carbon dioxide, *J. Catal.*, 74, 343–360, 1982.

180. Klier. K., Methanol synthesis, *Adv. Catal.*, 31, 243–313, 1982.

181. Klier, K., Catalytic conversions of CO_2: activation and hydrogenation, in Proc. Int. Symp. Chemical Fixation of Carbon Dioxide, Nagoya, Japan, December 2 to 4, 1991, 139–134.

182. Tagawa, T., Shimakage, M., and Goto, S., Copper-based supported catalysts for methanol synthesis from $CO_2 + H_2$, in Proc. Int. Symp. Chemical Fixation of Carbon Dioxide, Nagoya, Japan, December 2 to 4, 1991, 409–412.

183. Tagawa, T., Nomura, N., and Goto, S., Modification of copper-based catalysts for $CO_2 + H_2$ reaction, in Abstr. Int. Conf. on Carbon Dioxide Utilization, Bari, Italy, September 1993, 369.

184. Tagawa, T., Nomura, N., Shimakage, M., and Goto, S., Effect of supports on copper catalysts for methanol synthesis from $CO_2 + H_2$, *Res. Chem. Intermed.*, 21, 193–202, 1995.

185. Nomura, N., Tagawa, T., and Goto, S., Fe promoted Cu-based catalysts for hydrogenation of CO_2, in Abstr. 4th Int. Conf. on Carbon Dioxide Utilization, Kyoto, Japan, September 1997, P-027.

186. Joo, O.S., Jung, K.D., Han, S.H., and Uhm, S.J., Effects of compositional changes of Cu/ZnO, Cu/Al₂O₃, and Cu/ZnO/Al₂O₃ catalysts on methanol synthesis from CO_2 hydrogenation, *Carbon Dioxide Chemistry: Environ. Issues*, R. Soc. Chem. Spec. Publ., 153, 93–101, 1994.

187. Hirano, M., Akano, T., Imai, T., and Kuroda, K., Methanol synthesis from carbon dioxide on CuO ZnO Al₂O₃ catalysts, in Abstr. 4th Int. Conf. on Carbon Dioxide Utilization, Kyoto, Japan, September 1997, P-062.

188. Sakata, Y., Uddin, M.A., Muto, A., and Imaoka, M., Carbon-supported well-dispersed Cu–ZnO catalysts prepared from sawdust impregnated with [Cu(NO₃)₂ Zn(NO₃)₂] solution. Catalytic activity in CO_2 hydrogenation to methanol, *Microporous Mater.*, 9, 183–187, 1997.

189. Chanchlani, K.G., Hudgins, R.R., and Silveston, P.L., Methanol synthesis from H_2, CO, and CO_2 over Cu/ZnO catalysts, *J. Catal.*, 136, 59–75, 1992.

190. Ghazi, M., Barrault, J., and Ménézo, J.C., CO_2 hydrogenation into methanol on supported nickel–molybdenum catalysts, *Rec. Trav. Chim. Pays-Bas*, 110, 19–22, 1991.

191. Sakurai, H. and Haruta, M., Synergism in methanol synthesis from carbon dioxide over gold catalysts supported on metal oxides, *Catal. Today*, 29, 361–365, 1996.

192. Arakawa, H., Dubois, J.-L., and Sayama, K., Selective conversion of CO_2 to methanol by catalytic hydrogenation over promoted copper catalyst, in Proc. 1st Int. Conf. on Carbon Dioxide Removal, Amsterdam, March 4 to 6, 1992,

193. Arakawa, H., Kusama, H., Sayama, K., and Okabe, K., Effective conversion of carbon dioxide to methanol and dimethylether by catalytic hydrogenation over heteroge-

neous catalysts, in Abstr. Int. Conf. on Carbon Dioxide Utilization, Bari, Italy, September 1993, 95–102.

194. Fujitani, T., Saito, M., Kanai, Y., Kakumoto, T., Watanabe, T., Nakamura, J., and Uchijima, T., The role of metal oxides in promoting a copper catalyst for methanol synthesis, *Catal. Lett.*, 25, 271–276, 1994.

195. Fujitani, T., Saito, M., Kanai, Y., Watanabe, T., Nakamura, J., and Uchijima, T., Comparison between Cu-based and Pd-based catalysts for methanol synthesis from CO_2 and H_2, in Abstr. 3rd Int. Conf. on Carbon Dioxide Utilization, Norman, OK, May 1995.

196. Saito, M., Fujitani, T., Takeuchi, M., and Watanabe, T., Development of copper/zinc oxide-based multicomponent catalysts for methanol synthesis from carbon dioxide and hydrogen, *Appl. Catal. A Gen.*, 138, 311–318, 1996.

197. Ushikoshi, K., Mori, K., Watanabe, T., and Takeuchi, M., A 50 kg/day test plant for methanol synthesis from CO_2 and H_2, in Abstr. 4th Int. Conf. on Carbon Dioxide Utilization, Kyoto, Japan, September 1997, O-40.

198. Kanai, Y., Watanabe, T., and Saito, M., Catalytic conversion of carbon dioxide to methanol over palladium-promoted Cu/ZnO catalysts, *Carbon Dioxide Chemistry: Environ. Issues, R. Soc. Chem. Spec. Publ.*, 153, 102–109, 1994.

199. Kanaii, Y., Watanabe, T., Fujitani, T., Uchijima, T., and Nakamura, J., The synergy between Cu and ZnO in methanol synthesis catalysts, *Catal. Lett.*, 38, 157–163, 1996.

200. Inui, T., Hara, H., Takeguchi, T., and Kim, J., Structure and function of Cu-based composite catalysts for highly effective synthesis of methanol by hydrogenation of CO_2 and CO, *Catal. Today*, 36, 25–32, 1997.

201. Sahibzada, M., Chadwick, D., and Metcalfe, I.S., Hydrogenation of carbon dioxide to methanol over palladium-promoted $Cu/ZnO/Al_2O_3$ catalysts, *Catal. Today*, 29, 367–372, 1996.

202. Sahibzada, M., Metcalfe, I.S., and Chadwick, D., Methanol synthesis from CO_2/H_2 over Pd promoted $Cu/ZnO/Al_2O_3$ catalysts, in Abstr. 4th Int. Conf. on Carbon Dioxide Utilization, Kyoto, Japan, September 1997, O-39.

203. Deng, J.F., Sun, Q., Zhang, Y.L., Chen, S.Y., and Wu, D., A novel process for preparation of a $Cu/ZnO/Al_2O_3$ ultrafine catalyst for methanol synthesis from $CO_2 + H_2$: comparison of various preparation methods, *Appl. Catal. A Gen.*, 139, 75–85, 1996.

204. Zhang, Y.L., Sun, Q., Deng, J.F., Wu, D., and Chen, S.Y., A high activity $Cu/ZnO/Al_2O_3$ catalyst for methanol synthesis: preparation and catalytic properties, *Appl. Catal. A Gen.*, 158, 105–120, 1997.

205. Sun, Q., Zhang, Y.L., Chen, H.Y., Deng, J.F., Wu, D., and Chen, S.Y., Novel process for the preparation of Cu/ZnO and $Cu/ZnO/Al_2O_3$ ultrafine catalysts. Structure, surface properties, and activity for methanol synthesis from $CO_2 + H_2$, *J. Catal.*, 167, 92–105, 1997.

206. Fukui, H., Kobayashi, M., Yamaguchi, T., Kusama, H., Sayama, K., Okabe, K., and Arakawa, H., New preparation method for methanol synthesis from carbon dioxide hydrogenation by mechanical alloying, in Abstr. 4th Int. Conf. on Carbon Dioxide Utilization, Kyoto, Japan, September 1997, P-058.

207. Weigel, J., Köppel, R.A., Baiker, A., and Wokaun, A., Surface species in CO and CO_2 hydrogenation over copper/zirconia: on the methanol synthesis mechanism, *Langmuir*, 12, 5319–5329, 1996.

208. Baiker, A. and Gasser, D., Supported palladium catalyst prepared from amorphous palladium–zirconium, *J. Chem. Soc. Faraday Trans. 1*, 85, 999–1007, 1989.

209. Schild, C., Wokaun, A., and Baiker, A., On the mechanism of CO and CO$_2$ hydrogenation reactions on zirconia-supported catalysts: a diffuse reflectance FTIR study. I. Identification of surface species and methanation reactions on palladium/zirconia catalysts, *J. Mol. Catal.*, 63, 223–242, 1990.

210. Schild, C., Wokaun, A., and Baiker, A., On the mechanism of CO and CO$_2$ hydrogenation reactions on zirconia-supported catalysts: a diffuse reflectance FTIR study. II. Surface species on copper/zirconia catalysts: implications for methanol synthesis selectivity, *J. Mol. Catal.*, 63, 243–254 1990.

211. Schild, C., Wokaun, A., and Baiker, A., On the hydrogenation of CO and CO$_2$ over copper/zirconia and palladium/zirconia catalysts, *Fresenius J. Anal. Chem.*, 341, 395–401, 1991.

212. Köppel, R.A., Baiker, A., Schild, C., and Wokaun, A., Carbon dioxide hydrogenation over Au/ZrO$_2$ catalysts from amorphous precursors — catalytic reaction mechanisms, *J. Chem. Soc. Faraday Trans.*, 87, 2821–2828, 1991.

213. Köppel, R.A., Baiker, A., and Wokaun, A., Copper zirconia catalysts for the synthesis of methanol from carbon dioxide: influence of preparation variables on structural and catalytic properties of catalysts, *Appl. Catal. A Gen.*, 84, 77–102, 1992.

214. Nitta, Y., Fujimatsu, T., Okamoto, Y., and Imanaka, T., Effect of starting salt on catalytic behaviour of Cu–ZrO$_2$ catalysts in methanol synthesis from carbon dioxide, *Catal. Lett.*, 17, 157–165, 1993.

215. Suwata, O., Ikeda, Y., Fujimatsu, T., Okamoto, Y., and Nitta, Y., Preparation effects on Cu–ZrO$_2$ catalysts for the synthesis of methanol from CO$_2$ and H$_2$, *Kagaku Kogaku Ronbunshu*, 21, 1009–1014, 1995; *Chem. Abstr.*, 124, 175374f.

216. Nitta, Y., Suwata, O., Okeda, Y., Okamoto, Y., and Imanaka, T., Copper–zirconia catalysts for methanol synthesis from carbon dioxide. Effect of ZnO addition to Cu–ZrO$_2$ catalysts, *Catal. Lett.*, 26, 345–354, 1994.

217. Fröhlich, C., Köppel, R.A., and Baiker, A., Hydrogenation of carbon dioxide over silver promoted copper/zirconia catalysts, *Appl. Catal. A Gen.*, 106, 275–293, 1993.

218. Kieffer, R., Methanol synthesis from CO$_2$ + H$_2$, in Proc. Int. Symp. Chemical Fixation of Carbon Dioxide, Nagoya, Japan, December 2 to 4, 1991, 151–158.

219. Kieffer, R., Poix, P., Harison, J., and Lechleiter, L., Hydrogenation of CO$_2$ toward methanol on stabilized Cu/La$_2$Zr$_2$O$_7$ and Pd–Cu/La$_2$Zr$_2$O$_7$ catalysts, in Proc. Int. Conf. on Carbon Dioxide Utilization, Bari, Italy, September 1993, 85–94.

220. Kieffer, R. and Udron, L., Methanol and higher alcohol synthesis from CO$_2$–H$_2$. Relation between reaction mechanism and catalyst modelling, in Abstr. 3rd Int. Conf. on Carbon Dioxide Utilization, Norman, OK, May 1995.

221. Kieffer, R., Fujiwara, M., Udron, L., and Souma, Y., Hydrogenation of CO and CO$_2$ toward methanol, alcohols and hydrocarbons on promoted copper rare earth oxide catalysts, *Catal. Today*, 36, 15–24, 1997.

222. Xu, Z., Qian, Z., Mao, L., Tanabe, K., and Hattori, H., Methanol synthesis from CO$_2$ and H$_2$ over CuO–ZnO catalysts combined with metal oxides under 13 atm pressure, *Bull. Chem. Soc. Jpn.*, 64, 1658–1663, 1991.

223. Wainwright, M.S. and Trimm, D.L., Methanol synthesis and water–gas shift reactions on Raney copper catalysts, *Catal. Today*, 23, 29–42, 1995.

224. Saito, M., Fujitani, T., Takeuchi, M., and Watanabe, T., Development of copper/zinc oxide based multicomponent catalysts for methanol synthesis from carbon dioxide and hydrogen, *Appl. Catal. A Gen.*, 138, 311–318, 1996.

225. Toyir, J., Saito, M., Yamauchi, I., Luo, S., Wu, J., Takahara, I., and Takeuchi, M., Development of high performance copper based catalysts for methanol synthesis from

CO_2 and H_2, in Abstr. 4th Int. Conf. on Carbon Dioxide Utilization, Kyoto, Japan, September 1997, O-25.

226. Iizuka, T., Kojima, M., and Tanabe, K., Support effects in the formation of methanol from CO_2 and H_2 over rhenium catalysts, *J. Chem. Soc. Chem. Commun.*, pp. 638–639, 1983.

227. Xu, Z., Qian, Z., Tanabe, K., and Hattori, H., Support effect of Re catalyst on methanol synthesis from CO_2 and H_2 under a pressure of 5 atm, *Bull. Chem. Soc. Jpn.*, 64, 1664–1668, 1991.

228. Xu, Z., Quian, Z.H., and Hattori, H., Mechanistic study of the hydrogenation of carbon dioxide over supported rhenium and copper–zinc catalysts, *Bull. Chem. Soc. Jpn.*, 64, 3432–3437, 1991.

229. Ramaroson, E., Kieffer, R., and Kiennemann, A., Reaction of carbon dioxide and hydrogen on supported palladium catalysts, *J. Chem. Soc. Chem. Commun.*, pp. 645–646, 1982.

230. Fan, L. and Fujimoto, K., Development of an active and stable ceria-supported palladium catalyst for hydrogenation of carbon dioxide to methanol. *Appl. Catal. A Gen.*, 106, L1–L7, 1993.

231. Fan, L. and Fujimoto, K., Hydrogenation of carbon dioxide to methanol by lanthana-supported palladium catalyst, *Chem. Lett.*, pp. 105–108, 1994.

232. Kanoun, N., Astier, M.P., and Pajonk, G.M., Dehydrogenation of ethanol and CO_2–H_2 conversion on new coprecipitated Cu/Cr–Al catalysts, *J. Mol. Catal.*, 79, 217–228, 1993.

233. Kanoun, N., Astier, M.P., and Pajonk, G.M., New vanadium–copper–zinc catalysts, their characterization and use in the catalytic dehydrogenation of ethanol, *Appl. Catal.*, 70, 225–236, 1991.

234. Kanoun, N., Astier, M.P., Lecomte, F., Pommier, B., and Pajonk, G.M., New ternary Cu–V–Zn catalysts for conversion of CO_2 by H_2 into methanol, *Stud. Surf. Sci. Catal.*, 75, 2745–2748, 1993.

235. Taylor, P.A., Rasmussen, P.B., and Chorkendorff, I., Is the observed hydrogenation of formate the rate-limiting step in methanol synthesis? *J. Chem. Soc. Faraday Trans.*, 91, 1267–1269, 1995.

236. Shustorovich, E. and Bell, A.T., An analysis of methanol synthesis from CO and CO_2 on Cu and Pd surfaces by the bond-order-conservation-Morse-potential approach, *Surf. Sci.*, 253, 386–394, 1991.

237. Fu, S.S. and Somorjai, G.A., Interactions of O_2, CO, CO_2, and D_2 with the stepped cupper(311) crystal face: comparison to Cu(110), *Surf. Sci.*, 262, 68–76, 1992.

238. Chinchen, G.C., Denny, P.J., Parker, D.G., Spencer, M.S., and Whan, D.A., Mechanism of methanol synthesis from carbon dioxide/carbon monoxide/hydrogen mixtures over copper/zinc oxide/alumina catalysts: use of carbon-14-labeled reactants, *Appl. Catal.*, 30, 333–338, 1987.

239. Rozovskii, A.Y., Lin, G.I., Liberov, L.G., Slivinskii, E.V., Loktev, S.M., Kagan, Y.B., and Bashkirov, A.N., Mechanism of methanol synthesis from carbon dioxide and hydrogen. III. Determination of the rates of individual steps using carbon-14 monoxide, *Kinet. Katal.*, 18, 691–699, 1977; *Chem. Abstr.*, 87, 117351c.

240. Kuznetsov, V.D., Shub, F.S., and Temkin, M.J., Role of carbon dioxide in the synthesis of methanol in the presence of SNM-1 copper catalyst, *Kinet. Katal.*, 23, 932–935, 1982; *Chem. Abstr.*, 97, 215266e.

241. Saussey, J., Lavalley, J.C., and Bovet, C., Infrared study of carbon dioxide adsorption on zinc oxide. Adsorption sites, *J. Chem. Soc. Faraday Trans. 1*, 78, 1457–1463, 1982.

242. Deluzarche, A., Hindermann, J.P., Kienemann, A., and Kieffer, R., Application of chemical trapping to the determination of surface species and to the study of their evolution under reaction conditions in heterogeneous catalysis, *J. Mol. Catal.*, 31, 225–250, 1985.

243. Kakumoto, T. and Watanabe, T., A theoretical study for methanol synthesis by CO$_2$ hydrogenation, *Catal. Today*, 36, 39–44, 1997.

244. Neophytides, S.G., Marchi, A.J., and Froment, G.F., Methanol synthesis by means of diffuse reflectance infrared Fourier transform and temperature-programmed reaction spectroscopy, *Appl. Catal. A Gen.*, 86, 45–64, 1992.

245. Vanden Bussche, K.M. and Froment, G.F., Nature of formate in methanol synthesis on Cu/ZnO/Al$_2$O$_3$, *Appl. Catal. A Gen.*, 112, 37–55, 1994.

246. Bailey, S., Froment, G.F., Snoeck, J.W., and Waugh, K.C., A DRIFTS study of the morphology and surface adsorbate composition of an operating methanol synthesis catalyst, *Catal. Lett.*, 30, 99–111, 1995.

247. Joo, O.-S., Jung, K.-D., Han, S.-H., Uhm, S.-J., Lee, D.-K., and Ihm, S.-K., Migration and reduction of formate to form methanol on Cu/ZnO catalysts, *Appl. Catal. A Gen.*, 135, 273–286, 1996.

248. Fujita, S., Usui, M., Ohara, E., and Takezawa, N., Methanol synthesis from carbon dioxide at atmospheric pressure over Cu/ZnO catalyst: role of methoxide species formed on ZnO support, *Catal. Lett.*, 13, 349–358, 1992.

249. Hadden, R.A., Vandervell, H.D., Waugh, K.C., and Webb, G., The adsorption and decomposition of carbon dioxide on polycrystalline copper, *Catal. Lett.*, 1, 27–33, 1988.

250. Elliott, A.J., Hadden, R.A., Tabatabaei, J., Waugh, K.C., and Zemicael, F.W., Inverted temperature dependence of the decomposition of carbon dioxide on oxide-supported polycrystalline copper, *J. Catal.*, 157, 153–161, 1995.

251. Hadden, R.A., Sakakini, B., Tabatabaei, J., and Waugh, K.C., Adsorption and reaction induced morphological changes of the copper surface of a methanol synthesis catalyst, *Catal. Lett.*, 44, 145–151, 1997.

252. Millar, G.J., Rochester, C.H., and Waugh, K.C., Infrared study of CO adsorption on reduced and oxidized silica-supported copper catalysts, *J. Chem. Soc. Faraday Trans.*, 87, 1467–1472, 1991.

253. Millar, G.J., Rochester, C.H., Bailey, S., and Waugh, K.C., Combined temperature-programmed desorption and Fourier-transform infrared spectroscopy study of CO$_2$, CO and H$_2$ interactions with model ZnO/SiO$_2$, Cu/SiO$_2$ and Cu/ZnO/SiO$_2$ methanol synthesis catalysts, *J. Chem. Soc. Faraday Trans.*, 89, 1109–1115, 1993.

254. Baiker, A., Kilo, M., Maciejewski, M., Menzi, S., Wokaun, A., Waugh, K.C., Krauss, H.L., Prins, R., Kochloefl, K., Holderich, W., Hojlundnielsen, P.E., Trimm, D.L., Haruta, M., and Rooney, J.J., Hydrogenation of CO$_2$ over copper, silver and gold/zirconia catalysts. Comparative study of catalyst properties and reaction pathways, *Stud. Surf. Sci. Catal.*, 75, 1257–1272, 1993.

255. Weigel, J., Fröhlich, C., Baiker, A., and Wokaun, A., Vibrational spectroscopic study of IB metal/zirconia catalysts for the synthesis of methanol, *Appl. Catal. A Gen.*, 140, 29–45, 1996.

256. Bardet, R., Thivolle-Cazat, J., and Trambouze, Y., Methanolation of carbon oxides on Cu–Zn catalysts at atmospheric pressure, *J. Chim. Phys. (France)*, 78, 135–138, 1981.

257. Denise, B., Sneeden, R.P.A., and Hamon, C., Hydrocondensation of carbon dioxide. IV, *J. Mol. Catal.*, 17, 359–366, 1982.

258. Fujita, S., Usui, M., Hanada, T., and Takezawa, N., Methanol synthesis from CO_2–H_2 and from CO–H_2 under atmospheric pressure over Pd and Cu catalysts, *React. Kinet. Catal. Lett.*, 56, 15–19, 1995.

259. Vanden Bussche, K.M. and Froment, G.F., A steady-state kinetic model for methanol synthesis and the water–gas shift reaction on a commercial $Cu/ZnO/Al_2O_3$ catalyst, *J. Catal.*, 161, 1–10, 1996.

260. Vanden Bussche, K.M. and Froment, G.F., The STAR configuration for methanol synthesis in reversed flow reactors, *Can. J. Chem. Eng.*, 74, 729–734, 1996.

261. Hagihara, K., Mabuse, H., Watanabe, T., Kawai, M., and Saito, M., Effective liquid-phase methanol synthesis utilizing liquid–liquid separation, *Energy Convers. Manage.*, 36, 581–584, 1995.

262. Hagihara, K., Mabuse, H., Watanabe, T., and Saito, M., Liquid-phase methanol synthesis from CO_2 utilizing liquid–liquid separation, *Catal. Today*, 36, 33–37, 1997.

263. Mabuse, H., Watanabe, T., and Saito, M., Development of stable catalysts for liquid-phase methanol synthesis from CO_2 and H_2, in Abstr. 4th Int. Conf. on Carbon Dioxide Utilization, Kyoto, Japan, September 1997, P-056.

264. Inui, T., Matsuda, H., Okaniwa, H., and Miyamoto, A., Preparation of silico-alumino-phosphates by the rapid crystallization method and their catalytic performance in the conversion of methanol to light olefins, *Appl. Catal.*, 58, 155–163, 1990.

265. Inui, T., Kitagawa, K., Takeguchi, T., Hagiwara, T., and Makino, Y., Hydrogenation of carbon dioxide to C_1–C_7 hydrocarbons via methanol on composite catalysts, *Appl. Catal. A Gen.*, 94, 31–44, 1993.

266. Fujiwara, M. and Souma, Y., Hydrocarbon synthesis from carbon dioxide and hydrogen over Cu–Zn–Cr oxide zeolite hybrid catalysts, *J. Chem. Soc. Chem. Commun.*, p. 767, 1992.

267. Fujiwara, M., Kieffer, R., Ando, H., and Souma, Y., Development of composite catalysts made of Cu–Zn–Cr oxide zeolite for the hydrogenation of carbon dioxide, *Appl. Catal. A Gen.*, 121, 113–124, 1995.

268. Fujiwara, M., Ando, H., Tanaka, M., and Souma, Y., Hydrogenation of carbon dioxide over Cu–Zn–chromate/zeolite composite catalysts. The effects of reaction behavior of alkenes on hydrocarbon synthesis, *Appl. Catal. A Gen.*, 130, 105–116, 1995.

269. Souma, Y., Ando, H., Kieffer, R., and Fujiwara, M., Catalytic hydrogenation of CO_2 to methane and hydrocarbons, in Abstr. 3rd Int. Conf. on Carbon Dioxide Utilization, Norman, OK, May 1995.

270. Fujiwara, M., Ando, H., Matsumoto, M., Matsumura, Y., Tanaka, M., and Souma, M., Hydrogenation of carbon dioxide over Fe–ZnO/zeolite composite catalysts, *Chem. Lett.*, pp. 839–840, 1995.

271. Fujiwara, M., Kieffer, R., Ando, H., Xu, Q., and Souma, Y., Change of catalytic properties of Fe–ZnO/zeolite composite catalysts in the hydrogenation of carbon dioxide, *Appl. Catal. A Gen.*, 154, 87–101, 1997.

272. Souma, Y., Fujiwara, M., Kieffer, R., Ando, H., and Xu, Q., Hydrocarbon synthesis from CO_2 over Fe–ZnO/HY catalyst, in Abstr. 4th Int. Conf. on Carbon Dioxide Utilization, Kyoto, Japan, September 1997, O-35.

273. Xu, Q., He, D., Fujiwara, M., Tanaka, M., Matsumura, Y., Souma, Y., Ando, H., and Yamanaka, H., Hydrogenation of carbon dioxide over Fe–Cu–Na/zeolite composite catalysts, in Abstr. 4th Int. Conf. on Carbon Dioxide Utilization, Kyoto, Japan, September 1997, P-026.

274. Tan, Y., Fujiwara, M., Ando, H., Xu, Q., and Souma, Y., Selective formation of iso-butane from carbon dioxide and hydrogen over composite catalysts, in Abstr. 4th Int. Conf. on Carbon Dioxide Utilization, Kyoto, Japan, September 1997, P-029.

275. Fiato, R.A., Soled, S.L., Rice, G.B., and Miseo, S., Preparation of Olefins by Hydrogenation of Carbon Dioxide, U.S. Patent, 5,140,049, 1992; *Chem. Abstr.*, 118, P59276c.

276. Fiato, R.A., Iglesia, E., Rice, G.B., and Soled, S.L., Iron-catalyzed CO$_2$ hydrogenation to liquid hydrocarbons, in Abstr. 4th Int. Conf. on Carbon Dioxide Utilization, Kyoto, Japan, September 1997, O-37.

277. Kim, H., Choi, D.-H., Nam, S.-S., Choi, M.-J., and Lee, K.-W., The selective synthesis of lower olefins (C$_2$–C$_4$) by the CO$_2$ hydrogenation over iron catalysts promoted with potassium and supported on ion exchanged (H,K) zeolite-Y, in Abstr. 4th Int. Conf. on Carbon Dioxide Utilization, Kyoto, Japan, September 1997, P-022.

278. Choi, P.-H., Jun, K.-W., Lee, S.-J., Choi, M.-J., and Lee, K.-W., Hydrogenation of carbon dioxide over alumina-supported Fe–K catalysts, *Catal. Lett.*, 40, 115–118, 1996.

279. Jun, K.-W., Lee, S.-J., Kim, H., Choi, M.-J., and Lee, K.-W., Support effects of the promoted and unpromoted iron catalysts in CO$_2$ hydrogenation, in Abstr. 4th Int. Conf. on Carbon Dioxide Utilization, Kyoto, Japan, September 1997, O-038.

280. Calverley, E.M. and Smith, K.J., The effects of carbon dioxide, methanol, and alkali promoter concentration on the higher alcohol synthesis over a Cu/ZnO/Cr$_2$O$_3$ catalyst, *J. Catal.*, 130, 616–626, 1991.

281. Kishida, M., Yamada, K., Nagata, H., and Wakabayashi, K., CO$_2$ hydrogenation for C$_{2+}$ alcohols over silica-supported Ir–Mo catalysts, *Chem. Lett.*, pp. 555–556, 1994.

282. Ikehara, N., Hara, K., Satsuma, A., Hattori, T., and Murakami, Y., Unique temperature dependence of acetic acid formation in CO$_2$ hydrogenation on Ag-promoted Rh/SiO$_2$ catalyst, *Chem. Lett.*, 2, 263–264, 1994.

283. Arakawa, H., Kusama, H., Sayama, K., and Okabe, K., Effective synthesis of ethanol by catalytic hydrogenation of carbon dioxide over promoted/SiO$_2$ catalysts, in Abstr. 3rd Int. Conf. on Carbon Dioxide Utilization, Norman, OK, May 1995.

284. Kusama, H., Okabe, K., Sayama, K., and Arakawa, H., CO$_2$ hydrogenation to ethanol over promoted Rh/SiO$_2$ catalysts, *Catal. Today*, 28, 261–266, 1996.

285. Kusama, H., Okabe, K., Sayama, K., and Arakawa, H., Ethanol synthesis by catalytic hydrogenation of CO$_2$ over Rh–Fe/SiO$_2$ catalysts, *Energy*, 22, 343–348, 1997.

286. Kusama, H., Okabe, K., Sayama, K., and Arakawa, H., The effect of rhodium precursor on ethanol synthesis by catalytic hydrogenation of carbon dioxide over silica supported rhodium catalysts, in Abstr. 4th Int. Conf. on Carbon Dioxide Utilization, Kyoto, Japan, September 1997, P-028.

287. Kurakata, H., Izumi, Y., and Aika, K.-I., Ethanol synthesis from carbon dioxide on TiO$_2$-supported [Rh$_{10}$Se] catalyst, *Chem. Commun.*, pp. 389–390, 1996.

288. Yamamoto, T. and Inui, T., Highly effective synthesis of ethanol from CO$_2$ on Fe,Cu-based novel catalysts, in Abstr. 4th Int. Conf. on Carbon Dioxide Utilization, Kyoto, Japan, September 1997, P-054.

289. Takagawa, M., Okamoto, A., Fujimura, H., Izawa, Y., and Arakawa, H., Ethanol synthesis from carbon dioxide and hydrogen, in Abstr. 4th Int. Conf. on Carbon Dioxide Utilization, Kyoto, Japan, September 1997, P-057.

290. Sasaki, Y., Tominaga, K.I., and Saito, M., Homogeneous catalytic hydrogenation of carbon dioxide to methanol in the presence of $Ru_3(CO)_{12}$ and iodide salts, in Abstr. Int. Conf. on Carbon Dioxide Utilization, Bari, Italy, September 1993, 103–110.

291. Tominaga, K.I., Sasaki, Y., Saito, M., Hagihara, K., and Watanabe, T., Homogeneous Ru–Co bimetallic catalysis in CO_2 hydrogenation: the formation of ethanol, *J. Mol. Catal.*, 89, 51–56, 1994.

292. Tominaga, K.I., Sasaki, Y., Watanabe, T., and Saito, M., Homogeneous hydrogenation of carbon dioxide to methanol catalyzed by ruthenium cluster anions in the presence of halide anions, *Bull. Chem. Soc. Jpn.*, 68, 2837–2842, 1995.

293. Tominaga, K.I., Sasaki, Y., Watanabe, T., and Saito, M., Homologation of alcohols using carbon dioxide catalyzed by ruthenium–cobalt bimetallic complex system, in Abstr. 4th Int. Conf. on Carbon Dioxide Utilization, Kyoto, Japan, September 1997, P-047.

294. Isaka, M. and Arakawa, H., Methanol and ethanol synthesis by hydrogenation of CO_2 using homogeneous catalyst system, in Abstr. 3rd Int. Conf. on Carbon Dioxide Utilization, Norman, OK, May 1995.

295. Leitner, W., Carbon dioxide as a raw material: the synthesis of formic acid and its derivatives from CO_2, *Angew. Chem. Int. Ed. Engl.*, 34, 2207–2221, 1995.

296. Tsai, J.-C. and Nicholas, K.M., Transition metal-mediated photochemical and catalytic reduction of carbon dioxide: rhodium-catalyzed reduction of carbon dioxide to formic acid, in Proc. Int. Symp. Chemical Fixation of Carbon Dioxide, Nagoya, Japan, December 2 to 4, 1991, 281–286.

297. Tsai, J.-C. and Nicholas, K.M., Rhodium-catalyzed hydrogenation of carbon dioxide to formic acid, *J. Am. Chem. Soc.*, 114, 5117–5124, 1992.

298. Stalder, C.J., Chao, S., Summers, D.P., and Wrighton, M.S., Supported palladium catalysts for the reduction of sodium bicarbonate to sodium formate in aqueous solution at room temperature and one atmosphere of hydrogen, *J. Am. Chem. Soc.*, 105, 6318–6320, 1983.

299. Wiener, H., Blum, J., Feilchenfeld, H., Sasson, Y., and Zalmanov, N., The heterogeneous catalytic hydrogenation of bicarbonate to formate in aqueous solutions, *J. Catal.*, 110, 184–190, 1988.

300. Graf, E. and Leitner, W., Direct formation of formic acid from carbon dioxide and dihydrogen using the $[\{Rh(cod)Cl\}_2]$–$Ph_2P(CH_2)_4PPh_2$ catalyst system, *J. Chem. Soc. Chem. Commun.*, pp. 623–624, 1992.

301. Leitner, W., Rhodium catalysed hydrogenation of carbon dioxide, in Abstr. Int. Conf. on Carbon Dioxide Utilization, Bari, Italy, September 1993, 111–117.

302. Burgemeister, T., Kastner, F., and Leitner, W., CO_2 activation. 2. $[(PP)_2RhH]$ and $[(PP)_2Rh][O_2CH]$ complexes as models for the catalytically active intermediates in the Rh-catalyzed hydrogenation of CO_2 to HCOOH, *Angew. Chem. Int. Ed.*, 32, 739, 1993.

303. Gassner, F. and Leitner, W., CO_2 activation. 3. Hydrogenation of carbon dioxide to formic acid using water-soluble rhodium catalysts, *J. Chem. Soc. Chem. Commun.*, pp. 1465–1466, 1993.

304. Leitner, W., The coordination chemistry of carbon dioxide and its relevance for catalysis: a critical survey, *Coord. Chem. Rev.*, 153, 257–284, 1996.

305. Gassner, F., Dinjus, E., Görls, H., and Leitner, W., CO_2 activation. 7. Formation of the catalytically active intermediate in the hydrogenation of carbon dioxide to formic acid using the $[\{(COD)Rh(\mu\text{-}H)\}_4]$/$Ph_2P(CH_2)_4PPh_2$ catalyst: first direct observation

of hydride migration from rhodium to coordinated 1,5-cyclooctadiene, *Organometallics*, 15, 2078–2082, 1997.

306. Hutschka, F., Dedieu, A., Eichberger, M., Fornika, R., and Leitner, W., Mechanistic aspects of the rhodium-catalyzed hydrogenation of CO_2 to formic acid. A theoretical and kinetic study, *J. Am. Chem. Soc.*, 119, 4432–4443, 1997.

307. Lindner, E., Keppeler, B., and Wegner, P., Catalytic hydrogenation of carbon dioxide with the cationic bis(chelate) rhodium complex [Rh(P⌃O)₂][BPh₄], *Inorg. Chim. Acta*, 258, 97–100, 1997.

308. Darensbourg, D.J., Ovelles, C., and Pala, M., Homogeneous catalysts for carbon dioxide/hydrogen activation. Alkyl formate production using anionic ruthenium carbonyl clusters as catalysts, *J. Am. Chem. Soc.*, 105, 5937–5939, 1983.

309. Darensbourg, D.J., Pala, M., and Waller, J., Potential intermediates in carbon dioxide reduction processes. Synthesis and structure of (μ-formato) decacarbonyl triruthenium and (-acetato) decacarbonyl triruthenium anions, *Organometallics*, 2, 1285–1291, 1983.

310. Süss-Fink, G., Soulié, J.-M., Rheinwald, G., Stoeckli-Evans, H., and Sasaki, Y., Hydrocondensation of carbon dioxide with methanol catalyzed by anionic ruthenium complexes: isolation, structural characterization, and catalytic implications of the dinuclear anion [Ru₂(CO)₄(μ₂-η²-CO₂CH₃)₂-(μ₂-OCH₃)₂]⁻, *Organometallics*, 15, 3416–3422, 1996.

311. Zhang, J.Z., Li, Z., Wang, H., and Wang, C. Y., Homogeneous catalytic synthesis of formic acid (salts) by hydrogenation of CO_2 with H_2 in the presence of ruthenium species, *J. Mol. Catal. A Chem.*, 112, 9–14, 1996.

312. Kröcher, O., Köppel, R.A., and Baiker, A., Highly selective ruthenium complexes with bidentate phosphine ligands for the solvent-free catalytic synthesis of N,N-dimethylformamide and methyl formate, *Chem. Commun.*, pp. 453–454, 1997.

313. Jessop, P.G., Ikariya, T., and Noyori, R., Homogeneous catalytic hydrogenation of supercritical carbon dioxide, *Nature*, 368, 231–233, 1994.

314. Noyori, R., Homogeneous hydrogenation of supercritical carbon dioxide, in Abstr. 3rd Int. Conf. on Carbon Dioxide Utilization, Norman, OK, May 1995.

315. Jessop, P.G., Hsiao, Y., Ikariya, T., and Noyori, R., Methyl formate synthesis by hydrogenation of supercritical carbon dioxide in the presence of methanol, *J. Chem. Soc. Chem. Commun.*, pp. 707–708, 1995.

316. Jessop, P.G., Hsiao, Y., Ikariya, T., and Noyori, R., Homogeneous catalysis in supercritical fluids: hydrogenation of supercritical carbon dioxide to formic acid, alkyl formates, and formamides, *J. Am. Chem. Soc.*, 118, 344–355, 1996.

317. Gredig, S.V., Köppel, R.A., and Baiker, A., Synthesis of methylamines from carbon dioxide and ammonia, *J. Chem. Soc. Chem. Commun.*, pp. 73–74, 1995.

318. Gredig, S.V., Köppel, R.A., and Baiker, A., Comparative study of synthesis of methylamines from carbon dioxide and ammonia over Cu/Al₂O₃, *Catal. Today*, 29, 339–342, 1996.

319. Kröcher, O., Köppel, R.A., and Baiker, A., Sol-gel derived hybrid materials as heterogeneous catalysts for the synthesis of N,N-dimethylformamide from supercritical carbon dioxide, *Chem. Commun.*, pp. 1497–1498, 1996.

320. Sasaki, Y., Tominaga, K.-I., and Saito, M., Ruthenium complex catalyzed hydrogenation of carbon dioxide in the presence of epoxide, in Abstr. 3rd Int. Conf. on Carbon Dioxide Utilization, Norman, OK, May 1995.

321. Tominaga, K., Sasaki, Y., Watanabe, T., and Saito, M., Ethylene oxide-mediated reverse water–gas shift reaction catalyzed by ruthenium complexes, *Energy*, 22, 169–176, 1997.

322. Tanaka, K., Okabe, J., and Aomura, K., A stoichiometric conversion of CO + CH_4 into 2CO + $2H_2$ by microwave discharge, *J. Chem. Soc. Chem. Commun.*, pp. 921–922, 1982.

323. Wan, J.K.S., Bamwenda, G., and Depew, M.C., Microwave induced catalytic reactions of carbon dioxide and water — mimicry of photosynthesis, *Res. Chem. Intermed.*, 16, 241–255, 1991.

324. Ihara, T., Kiboku, M., and Iriyama, Y., Plasma reduction of CO_2 with H_2O for the formation of organic compounds, *Bull. Chem. Soc. Jpn.*, 67, 312–314, 1994.

325. Kogelschatz, U., Eliasson, B., and Egli, W., Dielectric-barrier discharges. Principle and applications, in Proc. Int. Conf. Phenomena in Ionized Gases, Toulouse, France, July 1997, 1–20.

326. Bill, A., Wokaun, A., Eliasson, B., Killer, E., and Kogelschatz, U., Greenhouse gas chemistry, *Energy Convers. Manage.*, 38, S425–S422, 1997.

327. Eliasson, B., Kogelschatz, U., Xue, B., and Zhou, L.M., Application of dielectric-barrier discharges to the decomposition and utilization of greenhouse gases, in Proc. 13th Int. Symp. on Plasma Chemistry, Vol. IV, Beijing, August 1997, 1784–1789.

328. Larkin, D.W., Caldwell, T.A., Lobban, L.L., and Mallinson, R.G., Oxygen pathways and carbon dioxide utilization with methane conversion in ambient temperature electric discharges, in Abstr. 4th Int. Conf. on Carbon Dioxide Utilization, Kyoto, Japan, September 1997, P-005.

329. Henglein, A., Sonolysis of carbon dioxide, nitrous oxide and methane in aqueous solution, *Z. Naturforsch.*, 40B, 100–107, 1985.

330. Yokoyama, S., Miyahara, K., Tanaka, K., Takakuwa, I., and Tashiro, J., Catalytic reduction of carbon dioxide. 1. Reduction of carbon dioxide with carbon carrying potassium carbonate, *Fuel*, 58, 510–513, 1979.

331. Suzuki, T., Ohme, H., and Watanabe, Y., Mechanisms of alkaline–earth metals catalyzed CO_2 gasification of carbon, *Energy Fuels*, 8, 649–658, 1994.

332. Carrasco-Marín, F., Rivera-Utrilla, J., Utrera-Hidalgo, E., and Moreno-Castilla, C., MoO_2 as catalyst in the CO_2 gasification of activated carbons and chars, *Fuel*, 70, 13–16, 1991.

333. Dhupe, A.P., Gokarn, A.N., and Doraiswamy, L.K., Investigations into the compensation effect at catalytic gasification of active charcoal by carbon dioxide, *Fuel*, 70, 839–844, 1991.

334. Ohme, H. and Suzuki, T., Mechanisms of CO_2 gasification of carbon catalyzed with group VIII metals. I. Iron-catalyzed CO_2 gasification, *Energy Fuels*, 10, 980–987, 1996.

335. Akiyama, F., Reduction of carbon dioxide using coal at low temperature, *Chem. Lett.*, pp. 643–644, 1997.

336. Kodama, T., Miura, S., Shimizu, T., Aoki, A., and Kitayama, Y., Efficient thermochemical cycle for CO_2 reduction with coal using a reactive redox system of ferrite, in Abstr. 4th Int. Conf. on Carbon Dioxide Utilization, Kyoto, Japan, September 1997, P-010.

337. Yang, R.T. and Wong, C., Fundamental differences in the mechanisms of carbon gasification by steam and by carbon dioxide, *J. Catal.*, 82, 245–251, 1983.

338. Hüttinger, K.J. and Fritz, O.W., The carbon–carbon dioxide reaction: an extended treatment of the active site concept, *Carbon*, 29, 1113–1118, 1991.

339. Ono, H., Kawabe, M., Amani, H., Tsuji, M., and Tamaura, Y., Solar/chemical energy hybridization via Boudourd reaction for carbon utilization as a solar energy carrier, in Abstr. 4th Int. Conf. on Carbon Dioxide Utilization, Kyoto, Japan, September 1997, P-004.

340. Steinberg, M. and Dong, Y., An analysis of methanol production and utilization with reduced CO_2 emission, in Abstr. Int. Conf. on Carbon Dioxide Utilization, Bari, Italy, September 1993, 127–131.

341. Steinberg, M., The thermal decomposition of methane for hydrogen production and conversion of CO_2 to methanol, in Abstr. 3rd Int. Conf. on Carbon Dioxide Utilization, Norman, OK, May 1995.

342. Ishihara, T., Fujita, T., Mizuhara, Y., and Takita, Y., Fixation of carbon dioxide to carbon by catalytic reduction over metal oxides, *Chem. Lett.*, pp. 2237–2240, 1991.

343. Takita, Y., Ishihara, T., and Fujita, T., Catalytic reduction of carbon dioxide to carbon and carbon monoxide, in Proc. Int. Symp. Chemical Fixation of Carbon Dioxide, Nagoya, Japan, December 2 to 4, 1991, 179–184.

344. Mamantov, G., Molten salt electrolytes in secondary batteries, in *Materials for Advanced Batteries*, Murphy, D.W., Broadhead, J., and Steele, B.C.H., Eds., Plenum Press, New York, 1980, 111–122.

345. Deanhardt, M.L., Stern, K.H., and Kende, A., Thermal decomposition and reduction of carbonate ion in fluoride melts, *J. Electrochem. Soc.*, 133, 1148–1152, 1986.

346. Tamaura, Y. and Tabata, M., Complete reduction of carbon dioxide to carbon using cation-excess magnetite, *Nature*, 346, 255–256, 1990.

347. Tamaura, Y., CO_2 decomposition and conversion into CH_4 at 250–350°C using the H_2-reduced magnetite, in Proc. Int. Symp. Chemical Fixation of Carbon Dioxide, Nagoya, Japan, December 2 to 4, 1991, 167–172.

348. Nishizawa, K., Tabata, M., Kodama, T., Abe, H., Yoshida, T., and Tamaura, Y., Methanation of CO_2 with the oxygen-deficient magnetite at 150–300°C, in Proc. Int. Symp. Chemical Fixation of Carbon Dioxide, Nagoya, Japan, December 2 to 4, 1991, 383–386.

349. Korenaga, T., Kaseno, S., and Takahashi, T., Low-temperature carbon dioxide reduction with ferrite contactor, in Proc. Int. Symp. Chemical Fixation of Carbon Dioxide, Nagoya, Japan, December 2 to 4, 1991, 381–396.

350. Tsuji, M., Kato, H., Kodama, T., Chang, S.G., Hesegawa, N., and Tamaura, Y., Methanation of CO_2 on H_2-reduced Ni(II) or Co(II)-bearing ferrites at 200°C, *J. Mater. Sci.*, 29, 6227–6230, 1994.

351. Tsuji, M., Yamamoto, T., Tamaura, Y., and Kitayama, Y., Catalytic acceleration for CO_2 decomposition into carbon by Rh, Pt or Ce impregnation onto Ni(II)-bearing ferrite, *Appl. Catal. A Gen.*, 142, 31–45, 1996.

352. Akanuma, K., Kodama, T., Tabata, M., Yoshida, T., and Tamaura, Y., Decomposition of CO_2 with wüstite at 300°C, in Proc. Int. Symp. Chemical Fixation of Carbon Dioxide, Nagoya, Japan, December 2 to 4, 1991, 173–178.

353. Kodama, T., Tabata, M., Tominaga, K., Yoshida, T., and Tamaura, Y., Decomposition of CO_2 and CO into carbon with active wüstite prepared from Zn(II)-bearing ferrite, *J. Mater. Sci.*, 28, 547–552, 1993.

354. Peaff, G., Awash in methanol, *Chem. Eng. News*, 75(36), 22–24, 1997.

355. Halmann, M., Comparative evaluation of catalytic, electrochemical, photosynthetic and biomimetic CO_2 fixation — an answer to the greenhouse effect? in Proc. Int. Symp. Chemical Fixation of Carbon Dioxide, Nagoya, Japan, December 2 to 4, 1991, 129–138.

11 Photochemical and Radiation-Induced Activation of CO₂ in Homogeneous Media

11.1 INTRODUCTION

The photochemical fixation of carbon dioxide has stimulated considerable interest in providing models of prebiotic photosynthesis.[1] Intensive efforts have been made during the last decade to mimic the natural process of photosynthesis, using reaction systems which included photosensitizers (such as porphyrin or bipyridine derivatives), electron transfer mediators (such as methyl viologen), and also sacrificial electron donors (such as tertiary amines). The need for sacrificial electron donors is due to the fact that the reduction products of carbon dioxide, such as formic acid, formaldehyde, and methanol, are themselves very effective reducing agents. Earlier work on such artificial photosynthesis was reviewed.[2a] Photochemical fixation of carbon dioxide is also of interest for the potential utilization of this atmospheric constituent on Mars, where it amounts to 95.3% of the lower atmosphere.[2b]

The primary steps in the reactions of carbon dioxide or of the bicarbonate or carbonate ions in solution were identified mainly by pulse radiolysis but also by flash photolysis, electron spin resonance, and electrochemical experiments. The radiolysis of water results in short-lived radicals, ions, and atoms, which may react with certain solutes to produce secondary radicals. If these secondary radicals have intense UV or visible absorption bands, then the rates of their formation and decay can be determined by kinetic spectrophotometry. Their interaction with other solutes may result in increases in the decay rates, from which the second-order rate constants of such interaction can be derived.[3]

11.2　DIRECT PHOTOLYSIS OF CO_2

11.2.1　Gas-Phase Photodissociation

The direct photodissociation of CO_2 in the homogeneous gas phase occurs only by excitation in the vacuum UV region, at $\lambda = 120$ to 167 nm. The reaction steps involved were described by Slanger and Black[4] and Stolow and Lee[5] to occur by the following processes,

$$CO_2(^1\Sigma_g^+) + h\nu \rightarrow CO(^1\Sigma^+) + O(^1D) \tag{11.1}$$

$$CO_2(^1\Sigma_g^+) + h\nu \rightarrow CO(^1\Sigma^+) + O(^3P) \tag{11.2}$$

where Reaction 11.2 is spin-forbidden and Reaction 11.1 occurs with a quantum yield of unity. The oxygen atoms formed were deactivated by collisional relaxation,

$$O(^1D) + CO_2 \rightarrow O(^3D) + CO_2 \tag{11.3}$$

$$2O(^3D) + X \rightarrow O_2 + X \tag{11.4}$$

where X is either the vessel wall or some third body. Thus the quantum yield of the overall reaction,

$$CO_2 \rightarrow CO + 1/2 O_2 \tag{11.5}$$

was also found to be unity.[4,5]

The photolysis of gaseous mixtures of carbon dioxide and water was performed by Kurbanov et al. using vacuum UV irradiation ($\lambda = 147$ nm), leading to carbon monoxide and hydrogen. In the total gas pressure range 4 to 29 kPa, at 0.1% H_2O in CO_2, the total quantum yield $\Phi(CO) + \Phi(H_2)$ was unity.[6]

The vacuum-UV photolysis of CO_2 is a convenient method for generation of the electronically excited $O(^1D)$ atoms, the reaction of which with hydrofluorocarbons (HFCs) has been implicated in the depletion of the ozone layer in the stratosphere. Kojima et al. studied the photolysis of gas mixtures of CO_2 with hydrocarbons and with HFCs by illumination with the F_2 laser line at 158 nm.[7,8] At this wavelength, the HFCs are practically transparent, while CO_2 is strongly absorbing ($\varepsilon = 63$ M^{-1} cm^{-1}). The photodissociation of CO_2 yielded $O(^1D)$ and $O(^3P)$ atoms with quantum yields of 0.94 and 0.06, respectively, as well as CO, which was inactive. The excited $O(^1D)$ atoms reacted with the HFCs by insertion into the C–H bonds, producing hot alcohols, which underwent collisional relaxation. Thus, the photolysis of CO_2 mixed with $CF_3CF_2CH_3$ (1,1,1,2,2-pentafluoropropane) yielded $CF_3CF_2CH_2OH$ with a quantum yield of 0.19.[7,8]

11.2.2　Photolysis in Aqueous Media

Early studies by Getoff et al. on the reduction of aqueous solutions of carbon dioxide by UV light indicated the production of CO, formic acid, and formaldehyde.[9] The yield of the organic products was markedly enhanced in the presence of ferrous ions.[10]

11.2.2.1   The $\cdot CO_3^-$ Radical Anion

For aqueous bicarbonate ions, the UV absorption onset occurs below about 240 nm, resulting in photoionization and formation of the bicarbonate radical, which is in acid–base equilibrium with the carbonate radical anion, $CO_3^-\cdot$,[11]

$$HCO_3^- + h\nu \rightarrow HCO_3\cdot + e_{aq}^- \tag{11.6}$$

$$HCO_3\cdot + H_2O \leftrightarrow CO_3^-\cdot + H_3O^+ \qquad pK_a = 7.9 \tag{11.7}$$

The carbonate radical anion was also formed by photoionization of the carbonate ion:[11]

$$CO_3^{2-} + h\nu \rightarrow \cdot CO_3^- + e^- \tag{11.8}$$

The carbonate radical anion was detected by Behar et al. in flash photolysis by its broad band in the transient absorption spectrum, with $\lambda_{max} = 600$ nm and $\varepsilon = 1.9 \times 10^3 \ M^{-1} \ cm^{-1}$.[3,11,12] This band is convenient for the determination of rate constants of the interactions of $\cdot CO_3^-$ with other species. The rate constant for the recombination of the $\cdot CO_3^-$ radical has a negligible apparent activation energy, which seems to indicate a composite reaction sequence. The carbonate radical anion acts primarily as an oxidant.[11]

The carbonate radical anion, $\cdot CO_3^-$, was also obtained in pulse radiolysis, by oxidation of carbonate and bicarbonate ions by hydroxyl radicals. These reactions were found by Weeks and Rabani to occur with second-order rate constants of 4.2 $\times 10^8$ and $1.5 \times 10^7 \ M^{-1} \ s^{-1}$, respectively:[13]

$$\cdot OH + CO_3^{2-} \rightarrow OH^- + \cdot CO_3^- \tag{11.9}$$

$$\cdot OH + HCO_3^- \rightarrow H_2O + \cdot CO_3^- \tag{11.10}$$

In aqueous ethanol solutions containing duroquinone (DQ), the formation of the $\cdot CO_3^-$ radical anion could be determined by Scheerer and Grätzel using the reaction of carbonate ions with the triplet state of duroquinone,

$$DQ_T + CO_3^{2-} \rightarrow DQ^- + \cdot CO_3^- \tag{11.11}$$

occurring with a rate constant of $7 \times 10^7 \ M^{-1} \ s^{-1}$, to form durosemiquinone (DQ^-).[14] In a micellar solution of dodecyl trimethylammonium chloride in water, the above reaction was followed by subsequent step, in which the $\cdot CO_3^-$ radical reacted with another DQ triplet molecule, producing carbon superoxide, which then decomposed to carbon dioxide and oxygen:[15]

$$\cdot CO_3^- + DQ_T \rightarrow CO_3 + DQ^- \qquad (11.12)$$

$$CO_3 \rightarrow CO_2 + 1/2O_2 \qquad (11.13)$$

The product of the oxidation of CO_2 by the superoxide ion, O_2^-, was shown by Roberts et al. to be the peroxydicarbonate anion, $C_2O_6^{2-}$.[16] The overall chemical reaction was

$$2O_2^- + 2CO_2 \rightarrow C_2O_6^{2-} + O_2 \qquad (11.14)$$

for which the electrochemical equivalent was

$$O_2 + 2e^- + 2CO_2 \rightarrow C_2O_6^{2-} \qquad (11.15)$$

The ion was isolated as its tetramethylammonium salt from acetonitrile solution. The rate of the above charge transfer reaction was measured by rotating ring disk voltammetry in Me_2SO and dimethylformamide (DMF) solutions containing 0.1 M tetraethylammonium perchlorate, in the presence of an excess of CO_2 (hence providing pseudo-first-order kinetics). The second-order rate constant thus derived was $k_2 = 1.4 \times 10^3$ M^{-1} s^{-1}. The proposed primary reaction was the nucleophilic addition of the superoxide ion to carbon dioxide, followed by addition of another CO_2 molecule and further electron transfer from another superoxide ion:[16]

$$O_2^- + CO_2 \leftrightarrow {}^-O-C(O)-OO\cdot \qquad (11.16)$$

$$^-O-C(O)-OO\cdot + CO_2 \leftrightarrow {}^-O-C(O)-O-C(O)-OO\cdot \qquad (11.17)$$

$$^-O-C(O)-O-C(O)-OO\cdot + O_2^- \rightarrow {}^-O-C(O)-O-C(O)-OO^- \qquad (11.18)$$

11.2.2.2   The $\cdot CO_2^-$ Radical Anion

Electron capture by carbon dioxide in aqueous solutions results in formation of the $\cdot CO_2^-$ radical anion,

$$CO_2 + e^-(aq) \rightarrow \cdot CO_2^- \qquad (11.19)$$

which strongly absorbs in the UV region, with λ_{max} = 235 nm and ε = 3000 M^{-1} cm^{-1}.[17] The rate of reaction of the hydrated electron with carbon dioxide is close to the diffusion controlled limit, with $k_2 = 7.7 \times 10^9$ M^{-1} s^{-1}.[18] This radical anion is protonated only in strongly acidic solutions:

$$\cdot COOH = H^+ + \cdot CO_2^- \qquad pK_a = 1.4 \qquad (11.20)$$

With a redox potential of -2.0 V vs. the normal hydrogen electrode (NHE), the $\cdot CO_2^-$ radical anion is strongly reducing.[19] The decay of the $\cdot CO_2^-$ radical followed second-order kinetics. At pH 7, the second-order rate constant for the decay

$$\cdot CO_2^- + \cdot CO_2^- \rightarrow \qquad\qquad (11.21)$$

was $5 \times 10^8 \ M^{-1} \ s^{-1}$.[11]

11.3 PHOTOSENSITIZED REACTIONS

Photochemical reactions of carbon dioxide in the near UV and visible regions are made possible by sensitization. Activation of transition metal complexes of bifunctional supramolecules by UV and visible light sensitizes the insertion of carbon dioxide into such compounds, enabling reaction pathways which are not accessible by thermal activation.

In the presence of uranyl ions, steady-state or flash photolytic illumination ($\lambda >$ 248 nm) of carbonate or bicarbonate solutions resulted in photochromic behavior, which was postulated to be due to the formation of a radical complex intermediate, $UO_2(CO_3)_2(CO_3 \cdot)^{4-}$.[20]

Multielectron transfer in the photoreduction of carbon dioxide to methane was demonstrated by Yamase and Sugeta using dititano-decatungstophosphate as electron transfer agent.[21] With $(PTi_2W_{10}O_{40})^{7-}$ (10 mM) in aqueous CO_2-saturated solutions, containing ethanol (2.5 M) as electron donor, by illumination ($\lambda >$ 270 nm) at its intense UV absorption band ($\lambda_{max} = 255$ nm, $\varepsilon = 3.9 \times 10^4 \ M^{-1} \ cm^{-1}$) resulted in the production of hydrogen, formaldehyde and methane. Quantum yields for illumination at 313 nm were $\Phi_{HCHO} = 4 \times 10^{-3}$ and $\Phi_{CH4} = 6 \times 10^{-4}$.[21]

11.3.1 Reduction to CO and HCOOH with Ru Complexes

The photoinduced reduction of carbon dioxide and water was observed by Lehn et al. to occur under visible light illumination of $[Ru(bpy)^3]^{2+}$ (where bpy $= 2,2'$-bipyridine), together with $CoCl_2$, in CO_2-saturated acetonitrile/water/triethylamine (TEA) solution, yielding simultaneously carbon monoxide and water.[22-27] In this system, the ruthenium complex was the photosensitizer, TEA served as sacrificial electron donor, and $CoCl_2$ acted as electron mediator. The amount of the product synthesis gas, and the CO/H_2 ratio, depended on the composition of the reactants. If TEOA (triethanolamine) was substituted for TEA, the selectivity for CO production was markedly enhanced.[22-27]

However, with the ruthenium complex $Ru(bpy)_3$ as photosensitizer, methyl viologen as electron relay, and TEOA or EDTA as sacrificial electron donor, in carbon-dioxide-saturated aqueous solution, formic acid production occurred with a quantum yield of about 1%.[25,28]

Similarly, with CO_2 in DMF/TEOA solution, using $[Ru(bpy)_3]^{2+}$ as photosensitizer, HCOOH was the major product. The active catalyst in this reaction was proposed by Hawecker et al. to be the intermediate complex $[Ru(bpy)_2(CO)H]^+$.[24]

Carbon dioxide reduction to formate with a maximal quantum yield of 15% was obtained by Lehn and Ziessel using a mixture of $[Ru(bpy)^3]^{2+}$ catalysts with TEOA as a sacrificial electron donor in DMF as solvent.[27]

An analogous photochemical system by Ishida et al. with $[Ru(bpy)_2(CO)_2]^{2+}$ as catalyst in TEOA/DMF (1:4 v/v) under illumination at $\lambda > 400$ nm resulted in a maximal quantum yield of 14% for formate production.[29] A carbonyl carbon of this catalyst was shown to undergo nucleophilic attack by OH⁻ in neutral aqueous solutions:

$$[Ru(bpy)_2(CO)_2]^{2+} + OH = [Ru(bpy)_2(CO)C(O)OH]^+ \qquad (11.22)$$

$$[Ru(bpy)_2(CO)C(O)OH]^+ + OH^- = [Ru(bpy)_2(CO)(COO)] + H_2O \qquad (11.23)$$

The $[Ru(bpy)_2(CO)(COO)]$ species was proposed to be a key intermediate in the reduction of CO_2. A crystal structure determination of the hydrate $[Ru(bpy)_2(CO)(COO)] \cdot 3H_2O$ indicated that the ruthenium atom was octahedrally coordinated by a CO, an h^1-CO_2 in a *cis* position, and four nitrogen atoms of 2,2′-bipyridine ligands.[30]

11.3.2   Reduction to CO with Re Complexes

Kutal et al. observed that $Re(CO)_3(bpy)Br$ in CO_2-saturated DMF/TEOA solution under irradiation at 436 nm produced CO with a quantum yield of up to 15%.[31] A clue to the mechanism came from luminescence quenching studies. Illuminated DMF solutions of $Re(CO)_3(bpy)Br$ had a broad luminescence spectrum, with $\lambda_{max} = 610$ nm and a lifetime of 55 ns. This light emission was assigned to the Re-to-bpy charge-transfer excited state, $ReBr(CO)_3(bpy)^*$. The excited state was reductively quenched by TEOA:

$$ReBr(CO)_3(bpy)^* + TEOA \rightarrow ReBr(CO)_3(bpy)^- + TEOA^{+\cdot} \qquad (11.24)$$

Since the rate of this quenching was the same whether Ar or CO_2 was bubbled through the solution, it was concluded that CO_2 did not react directly with $ReBr(CO)_3(bpy)^*$, but rather with some product of the reaction with TEOA, which could be the anionic complex $ReBr(CO)_3(bpy)^-$.[31]

With $[Re(bpy)(CO)_3Cl]$ photosynthesizer in a similar homogeneous photochemical system, a quantum yield of 14% for the generation of CO was achieved.[26] With the same $[Re(bpy)(CO)_3Cl]$ complex as photosensitizer, in DMF containing tetraethyl ammonium chloride as medium, with TEOA as sacrificial electron donor, the photoreduction of CO_2 to CO occurred with a turnover number of 23 within 4 h. In a similar system, but with Re-*p*-(2,6-di-2-pyridyl-4-pyridyl)phenol as sensitizer, and dimethyl sulfoxide (DMSO) as solvent, the turnover number was 5.32 Another rhenium complex, $ClRe(CO)_3(4\text{-phenylpyridine})_2$, was tested by Ruiz et al.[33] With this complex, in CO_2-saturated acetonitrile and TEOA, under illumination at 350 nm, CO was produced with a quantum yield of 0.73%. However, some of the CO was derived from decomposition of the complex, because CO was also formed in N_2-saturated medium under similar conditions, with a quantum yield of 0.22%.[33]

In short-time photolysis, efficient carbon dioxide reduction to carbon monoxide was achieved by Hukkanen and Pakkanen using $Ru(II)(bipy)_3Cl$ as a photosensitizer and $Re(bipy)(CO)_3Cl$ as a co-catalyst. However, this system ceased to produce CO after 5 to 7 h photolysis and instead released hydrogen.[34]

The intermediates in CO_2 electroreduction catalyzed by rhenium tricarbonyl bipyridyl derivatives were investigated by Christensen et al. in an *in situ* infrared study (see Section 12.3.3.1).[35] Ishitani et al. used fast time-resolved infrared spectroscopy to identify the excited states of rhenium bipyridyl complexes involved in the photoreduction of CO_2 to CO.[36] These experiments were carried out with the complex $\{Re(CO)_2(bpy)[P(OEt)_3]_2Br\}$ in CH_2Cl_2 solution by laser flash illumination at 308 nm (20-ns flashes). Photolysis of this compound caused excitation into a metal-to-ligand charge-transfer state, with a lifetime of ~250 ns, in which the charge was localized on the bpy ligand. An additional step of activation seemed to be involved in the reduction of CO_2.[36] Electron transfer from an electron donor such as TEOA to an excited rhenium complex was proposed by Kutal et al.[31]

A more efficient photocatalyst for the reduction of CO_2 to CO was reported by Hori et al.[37] Using $\{Re(bpy)(CO)_3[P(OEt)_3]\}SbF_6$ in CO_2-saturated DMF–TEOA under illumination at 365 nm with low light intensity (~10^{-7} einstein min^{-1}), the quantum yield of CO formation was 38%, while under approximately similar conditions, the quantum yield using $[Re(bpy)(CO)_3Cl]$ was only 16%. However, with $\{Re(bpy)(CO)_3[P(OEt)_3]\}^+$ as photocatalyst, increasing light intensity caused a decrease in the apparent quantum yield, while the CO production rate rose only slightly. Flash photolysis experiments on $\{Re(bpy)(CO)_3[P(OEt)_3]\}^+$ in acetonitrile solution at 355 nm (10-ns pulses) led to the appearance of a transient UV absorption spectrum, which was assigned to the triplet metal-to-ligand charge-transfer state (^{3}MLCT). In the presence of TEOA, this transient was rapidly quenched, with the appearance of a different transient absorption spectrum, identified as the reduced complex, $\{Re(bpy\cdot^-)(CO)_3[P(OEt)_3]\}$:

$$\{Re^I(bpy)(CO)_3[P(OEt)_3]\}^+ + h\nu \rightarrow {}^*\{Re^I(bpy)(CO)_3[P(OEt)_3]\}^+ \quad (11.25)$$

$$^*\{Re^I(bpy)(CO)_3[P(OEt)_3]\}^+ + TEOA \rightarrow$$

$$\{Re(bpy\cdot^-)(CO)_3[P(OEt)_3]\} + TEOA\cdot^+ \quad (11.26)$$

This one-electron reduced complex, with absorption maxima at about 390 and 506 nm, was quite stable even in the presence of CO_2, with a half-life for decay of 257 s. The same transient absorption spectrum also appeared during the electrochemical reduction of $\{Re(bpy)(CO)_3[P(OEt)_3]\}SbF_6$ in acetonitrile solution containing n-Bu$_4$NClO$_4$ after passing through a flow-through electrochemical cell fitted with a Pt electrode at -2.7 V (vs. Ag/AgCl). The decreasing quantum yield with rising light intensity was explained as possibly due to the inner-filter effect of this stable transient.[37]

Further information on the intermediates of the photochemical CO_2 reduction with rhenium complexes was provided by Hori et al. by using electrospray mass spectrometry (ES).[38,39] This technique has been very useful for molecular weight determination of nonvolatile unstable biomolecules and metal complexes. $[Re(bpy)(CO)_3PPh_3]^+$ solutions in CO_2-saturated DMF/TEOA (5:1 v/v) were illuminated at $\lambda > 330$ nm and injected into the ES probe. Before illumination, the peak with m/e = 689 corresponded to the unchanged $[Re(bpy)(CO)_3PPh_3]^+$ ion. After UV illumination for 10 min, this peak had almost disappeared, and instead the major peak was at m/e = 576, corresponding to the TEOA coordinated complex $[Re(bpy)(CO)_3(TEOA]^+$. Both this complex and the solvent complex $[Re(bpy)(CO)_3(DMF]^+$ have also been observed as intermediates by HPLC after short-term UV irradiation (less than 10 min). Longer illumination led in 52% yield to the formation of the stable formate complex $\{Re(bpy)(CO)_3[OC(O)H]\}$, which was the "real" photocatalyst. The overall quantum yield of CO_2 reduction to CO with this catalytic system was 0.05.[38,39]

11.3.3   Reduction to CO with Macrocyclic Complexes

Tetraaza-macrocyclic complexes are interesting for CO_2 fixation, since it had been observed that one of the active coenzymes in the natural reduction of CO_2 to CH_4 by methanogenic bacteria is a nickel tetrapyrrole.[40] The mechanisms of CO_2 activation by aza-macrocyclic complexes were reviewed by Collin and Sauvage[41] and Costamagna et al.[42]

Tetraaza-macrocyclic cobalt(II) complexes were used by Tinnemans et al. as electron transfer mediators in the photoreduction of carbon dioxide, with $Ru(2,2'-bipyridine)_3^{2+}$ as sensitizer and ascorbic acid as sacrificial electron donor, in aqueous CO_2-saturated solutions (pH 4).[43] Such Co(II) complexes involve coordination of the macrocyclic ligands in a square planar configuration. The CO_2 group may possibly coordinate in the axial sites. Under illumination with daylight lamps, using the complex $[Co(II)(Me_2(14)\text{-}4,11\text{-diene-}N_4]^{2+}$,

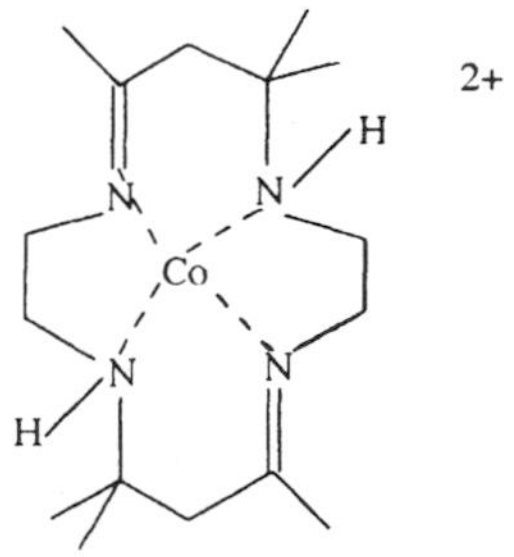

the only observed products were CO and H_2, formed in the ratio 0.27:1. The turnover number (moles of CO + H_2 per mole of the complex) exceeded 500.[43]

With a similar homogeneous photocatalyst system, using $[Ru(bipy)_3]^{2+}$ as photosensitizer and ascorbate buffer as the electron donor, but with $[Ni(cyclam)]^{2+}$ (cyclam = 1,4,8,11-tetraazacyclo tetradecane) as the catalyst, enhanced CO/H_2 ratios were achieved by Grant et al.[40] The CO and H_2 yields depended on the pH of the medium. Optimal CO production rate was at pH 5, providing a CO/H_2 ratio of 0.83. The mechanism was proposed to involve protonation of the reduced Ni species, followed by CO_2 insertion into the Ni–H bond, and finally dissociation to release CO and H_2O.[40] With the nickel cyclam [Ni(14-ane-N_4)]$^{2+}$ as catalyst, Ru(bipy)$_3^{2+}$ as photosensitizer, and ascorbate both as buffer (pH 5) and sacrificial electron donor, at $\lambda = 400$ nm, the quantum yield for CO production was 0.06%.[44] In addition to the problem of low quantum yield, the Ru(bipy)$_3^{2+}$ sensitizer was reported to undergo quite rapid decomposition, 25% within 4 h of photolysis. The primary chemical reaction following the photoexcitation of the sensitizer was proposed to be reductive quenching of the ruthenium excited state by the ascorbate anion.[44]

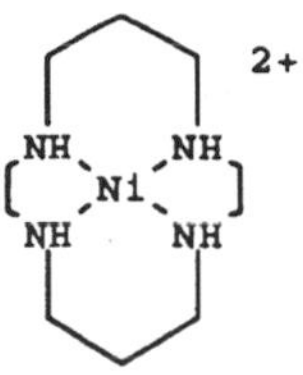

Several pyridine derivatives of Ni(II) cyclam were prepared by Kimura et al.[45] The most effective for CO_2 reduction to CO was the complex

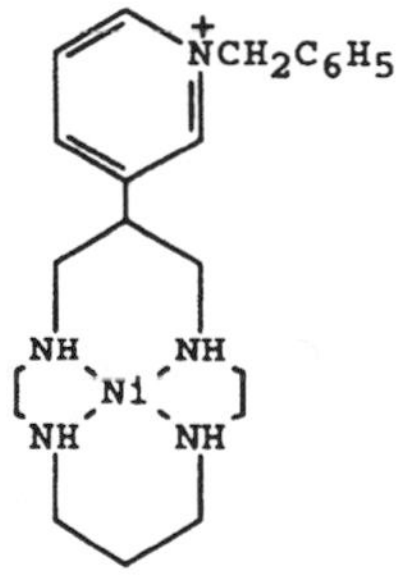

with which, in the presence of Ru(bpy)$_3^{2+}$ as sensitizer, in CO_2-saturated aqueous ascorbate buffer (pH 5.1) under illumination at $\lambda > 350$ nm, the amount of evolved CO was 5.8 times greater than with the underivatized Ni(II)-cyclam.[45]

In order to enhance the catalytic activity for photochemical CO_2 reduction, Mochizuki et al. synthesized a bimacrocyclic Ni(II) complex, [6,6'-bi(5,7-dimethyl-1,4,8,11-tetraazacyclotetradecane)]-dinickel(II) triflate, in which the two dimethylcyclam units were linked by C–C bonds:[46]

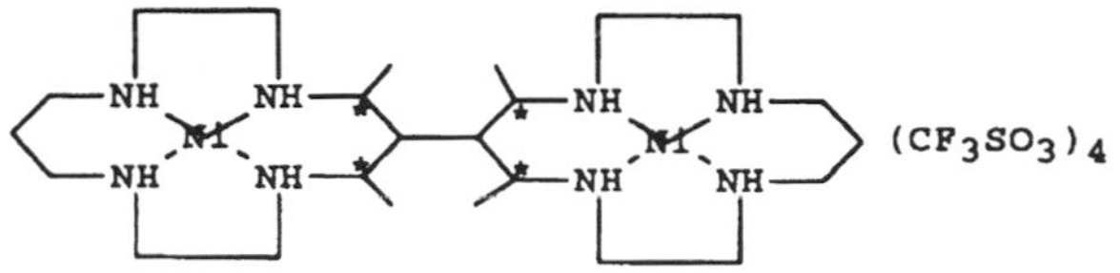

In this complex, the four methyl groups occupied the axial positions of the six-member chelate rings. In such a dimeric structure, by comparison with the monomeric Ni(II)-cyclam, the coordination of CO_2 molecules was enhanced, while the coordination of water molecules was hindered. The photoreduction of CO_2 was carried out in CO_2-saturated aqueous solutions (pH 4), containing $[Ru^{II}(bpy)_3]^{2+}$ as the photosensitizer, ascorbic as a sacrificial electron donor, and the Ni(II) complex as the catalyst, under illumination with a high-pressure Hg lamp. The production rate of CO using the bimacrocyclic complex was about eight times as large as that using the monomacrocyclic $[Ni^I(cyclam)]^{2+}$ salts under similar conditions. Also, the CO/H_2 ratio generated was 15 with the bimacrocyclic complex, while it was only 0.11 with the monomacrocyclic Ni(II) complex. Thus, the selectivity for CO_2 reduction vs. water reduction was dramatically enhanced with the bimacrocyclic complex. Such bimacrocyclic complexes may be considered models of the polynuclear metalloenzymes in biological systems.[46]

11.3.4   Reduction to HCOOH with Aromatic Hydrocarbons

Tazuke and Kitamura proposed a system that mimics the electron transport sensitization in natural photosynthesis.[47] It included an aromatic hydrocarbon such as pyrene or perylene, which transferred electrons from its singlet excited state to an electron acceptor such as 1,4-dicyanobenzene or 9,10-dicyanoanthracene in a polar medium, such as acetonitrile–water (5:1). The acceptor was then presumed to transfer an electron to carbon dioxide, forming the intermediate CO_2^- anion radical, which underwent protonation and further reduction to formate. Hydrogen peroxide was an additional product of the reaction.[47] However, in a reinvestigation of this reaction, Legros and Soumillion observed that if oxygen was carefully excluded from the reaction medium, no formic acid was produced.[48] It was thus necessary to conclude that the origin of the formic acid was not carbon dioxide. Since the photosensitizer was consumed during the illuminations, it was proposed that the origin of the formic acid was the photooxidation of the aromatic sensitizers.

11.3.5   Reduction to CO and HCOOH with p-Terphenyl/Cobalt Macrocycles

The photoreduction of carbon dioxide to formic acid was achieved by Matsuoka et al. even in the absence of an electron mediator, using p-terphenyl as photocatalyst,[49]

p-Terphenyl

in an aprotic polar solvent such as DMF, with TEA as a sacrificial electron donor, leading to a quantum yield of HCOOH production of 7.2% at 313 nm.[49]

Considerable enhancement in the selectivity of p-terphenyl-catalyzed photoreduction of CO_2 to CO and $HCOO^-$ was obtained by Matsuoka et al. with the addition of cobalt(III) cyclam as charge-transfer mediator.[50,51] Using this cyclam, in acetonitrile–methanol solution, in the presence of TEOA or TEA as sacrificial electron donor, the apparent quantum yields at 313 nm for production of CO and HCOOH were 15 and 10%, respectively.

p-Terphenyl (TP) acts as an efficient photosensitizer due to the long lifetime of its anion radical (TP$^{\cdot-}$), 8.3 μs in TEA/tetrahydrofuran (THF) and 2.5 ms in MeCN/MeOH (4:1), monitored by the decay of its absorption maximum at 470 nm.[52–55] This radical anion is a powerful reducing agent (E1/2 = –2.45 V vs. the saturated calomel electrode [SCE] in dimethylamine). The amine-coordinated Co(III)-cyclam was found to suppress the photodegradation of p-terphenyl. The mechanism of photoreduction was elucidated by flash photolysis techniques, by identifying several transient reaction intermediates. Photoexcited p-terphenyl was quenched by the tertiary amine (TEA), forming a quinoid-like radical anion. Direct electron transfer from this photogenerated radical anion of p-terphenyl to a $Co^{II}(cyclam)^+$ complex resulted in a Co^I species. This species reacted with CO_2 with a rate constant of 1.1×10^8 M^{-1} s^{-1} to form a $Co^I(cyclam)(CO_2)^+$ complex, which reacted with TEA or the solvent (S) to form a solvated complex. Further reduction resulted in the release of CO:

$$TP + TEA + h\nu \rightarrow TP^{\cdot-} + TEA^{\cdot+} \tag{11.27}$$

$$TP^{\cdot-} + Co^{II}(cyclam)^{2+} \rightarrow Co^I(cyclam)^+ \tag{11.28}$$

$$Co^I(cyclam)^+ + CO_2 = Co^I(cyclam)(CO_2)^+ \tag{11.29}$$

$$Co^I(cyclam)(CO_2)^+ + S = [S–Co^{III}(cyclam)–(CO_2^{2-})]^+ \tag{11.30}$$

$$[S–Co^{III}(cyclam)–(CO_2^{2-})]^+ \rightarrow Co^{II}(cyclam)^{2+} + CO \tag{11.31}$$

The production of formate was proposed to involve the intermediate formation of a hydride complex, CoIII(cyclam)(H$^-$)(TEA). Cyclam complexes of Ni^{II}, Fe^{II}, and Cu^{II} were ineffective as mediators for CO_2 reduction.[52–55]

Matsuoka et al. compared oligo(p-phenylenes), OPP-n, of different chain lengths as catalysts for the photoreduction of CO_2 to formic acid and carbon monoxide.[52] The most effective catalyst was the tetramer, OPP-4. These reactions were performed in nonaqueous solvents, preferably DMF, in the presence of TEA as a sacrificial electron donor, and without any electron transfer mediator, by illumination at $\lambda > 290$ nm. With OPP-4 as photocatalyst, the turnover number based on the consumed photocatalyst was 45, and the apparent quantum yield at 313 nm was $\Phi_{HCOOH} = 0.084$. The reaction depended on photoexcitation of the oligo(p-phenylenes). For OPP-4, the absorption maximum was at $\lambda_{max} = 313$ nm.[52] Using cyclic voltammetry, the reduction potential of OPP-4 in dimethylamine solution was previously determined by Meerholtz and Heinze to be -2.23 V vs. SCE,[56] so that this oligomer in its reduced form should be able to transfer one electron to CO_2, since $E°$ for the CO_2/CO_2^- couple is -2.21 V in DMF solution.[57]

11.3.6   Reduction to HCOOH and CH$_2$O with Fe(II)-Bipyridyl

UV illumination at $\lambda = 254$ nm of CO_2-saturated aqueous solutions containing $Fe(II)(bpy)_3^{2+}$ resulted in the production of formic acid and formaldehyde. The yield of formaldehyde, based on the amount of Fe(II) that had been consumed, was about 0.06%.[58]

11.4   HOMOGENEOUS PHOTOCARBOXYLATION

Photocarboxylation of cyclohexene was achieved by Morgenstern et al. using a nickel cluster complex, $[Ni_3(\mu_3-I)_2(dppm)_3]$, where dppm = bis(diphenylphosphino)-methane.[59] UV irradiation ($\lambda > 350$ nm) of this complex in a THF/cyclohexene solution (4:1, v/v) under 1 atm of CO_2 caused the oxidation of the complex to $[Ni_3(\mu_3-I)_2(dppm)_3]\cdot^+$, together with the formation of cis-meso and trans-D,L-cyclohexanedicarboxylic acids. The proposed mechanism involved as the primary step the reduction of CO_2 to the radical anion $CO_2\cdot^-$, followed by a Michael-type addition of $CO_2\cdot^-$ to the double bond of cyclohexene. Evidence for this hypothesis came from the results of irradiation of the above Ni cluster complex in CO_2-saturated THF ($\lambda > 290$ nm), producing CO and $CO_3^{2-}\cdot$, presumably by disproportionation of $CO_2\cdot^-$. Photolysis of the Ni complex in dichloromethane solution under CO_2 resulted in the formation of chloroacetic acid. In this case, the primary step may have been the reduction of dichloromethane to the chloromethyl radical, $CH_2Cl\cdot$, followed by addition of CO_2 to this radical, and reduction to chloroacetate.[59]

Carbon dioxide may be incorporated into aromatic compounds such as anthracene, phenanthrene, and pyrene by reductive photocarboxylation in the presence of a sensitizer and an electron donor.[60,61] Solutions of phenanthrene and amines (such as dimethyl- or diethyl-aniline) and CO_2 (pressurized to 4 kg cm^{-2}) in DMSO or DMF solution under UV illumination with a high-pressure Hg lamp yielded 9,10-dihydro-phenanthrene-9-carboxylic acid, with conversions (based on the phenanthrene consumed) of up to 60%, and a quantum yield of 0.17. The yield was enhanced in the presence of proton donors such as cumene or decalin. The primary reaction step was suggested to be electron transfer from the aromatic amine to the photoexcited aromatic hydrocarbon, thus producing a radical anion of the aromatic hydrocarbon. In a subsequent reaction, CO_2 was trapped by the radical anion, producing a stable aromatic carboxylic acid.[60,61]

The mechanism of the reductive photocarboxylation of phenanthrene was further investigated in detail by Nikolaitchik et al.[62] In CO_2-saturated DMSO solution of phenanthrene and dimethyl-aniline, the main products were 9,10-dihydrophenanthrene-9-carboxylic acid and trans-9,10-dihydro-phenanthrene-9,10-dicarboxylic acid, formed in 55 and 11% yield, respectively, as well as traces of some other products. In the presence of 1 M water, the chemical yields of the above two products changed to 68 and 0.5%. The initial photoexcitation of phenanthrene led to a radical anion, PHN$^{\cdot-}$, the formation of which was observed by flash photolysis of solutions of phenanthrene with dimethylaniline in DMSO. The broad transient absorption band with $\lambda = 450$ nm was assigned by Paul et al. to this radical anion.[63] Similar species were previously observed in the adducts of sodium metal with phenanthrene and other aromatic hydrocarbons. The rate of decay of this radical anion transient was enhanced in the presence of CO_2. From the kinetic data, the second-order rate constant for the reaction of the radical anion PHN$^{\cdot-}$ with CO_2 was estimated to be 3.14×10^6 M^{-1} s^{-1}. The initial product of this carboxylation was proposed to be an intermediate 9-carboxy-9,10-dihydrophenanthr-10-yl radical, which then either carried out hydrogen abstraction from a hydrogen donor (e.g., cumene), yielding 9,10-dihydrophenanthene-9-carboxylic, or alternatively was reduced by another PHN$^{\cdot-}$ radical anion, forming a dianion, which trapped a second CO_2 molecule, producing trans-9,10-dihydro-phenanthrene-9,10-dicarboxylic acid. This alternative pathway to the dicarboxylic acid was similar to that occurring in the reductive carboxylation of alkali metal salts of aromatic hydrocarbons.[63]

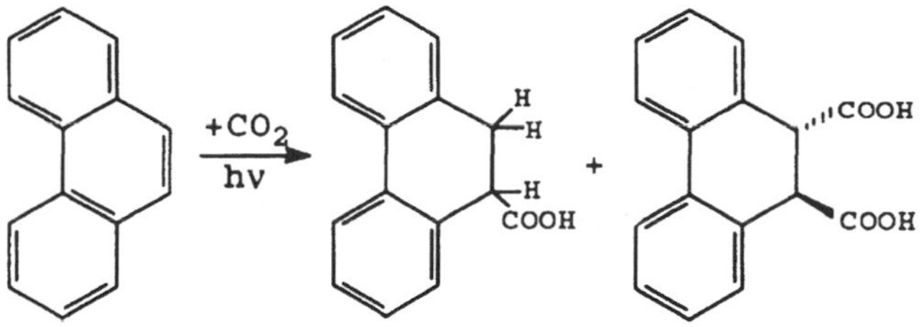

In DMF solution, using N,N-dimethylaniline as donor, naphthalene was carboxylated by carbon dioxide under illumination with visible light, in the presence of phenazine as a sensitizer. The selectivity to naphthoic acids was up to 67%, of which 90% was 1-naphthoic acid.[64,65]

The reductive photocarboxylation of benzophenone was achieved by Ogata et al. using poly(p-phenylene) (PPP) as a photocatalyst.[66] The reaction was performed under illumination at > 400 nm, using CO_2-saturated solutions of benzophenone and PPP in DMF, with TEA as sacrificial electron donor, and Et$_4$NCl as co-catalyst. The yields of the products benzilic acid, benzopinacol, and benzhydrol were 34, 34, and 3%, respectively.

See also the photofixation of CO_2 into benzophenone catalyzed by colloidal CdS microcrystallites (Chapter 14).

A unique reaction discovered by Inoue et al. is the visible-light-induced fixation of carbon dioxide with zinc porphyrins in the presence of secondary amines or alcohols to produce zinc porphyrin carbamates or carbonates.[67] N-Methyltetraphenyl-porphinato-zinc-ethyl in benzene solution in the presence of a secondary amine such as diethylamine reacted with carbon dioxide at room temperature to produce methyltetraphenyl-porphinato-zinc diethylcarbamate. If the amine was replaced by ethanol, the corresponding porphinato-zinc ethyl carbonate was formed. This reaction, which occurred slowly even in the dark, was considerably accelerated by illumination with visible light ($\lambda > 420$ nm).[67]

11.5 HIGH-ENERGY RADIATION-INDUCED CARBOXYLATION

In the γ-radiolysis of gas-phase CO_2 containing 10% H_2, the radiation chemical yield of CO was found by Kurbanov et al. to be independent of temperature up to 400°.[68] Raising the temperature from 500 to 700°C caused a sharp increase in G(CO), as well as in the consumption of H_2 (G-value = number of changed molecules per 100 eV of absorbed energy). At 700°, full conversion of H_2 was observed. A chain reaction was postulated, with a chain length of 10 to 150 in the temperature range 500 to 700°.[68] During the radiolysis of gaseous CO_2 over a ZnO catalyst, the rate of CO formation was faster than in the homogeneous gas-phase decomposition. The initial radiation chemical yield of CO production was G = 4 mol/100 eV. In the irradiated ZnO, paramagnetic Zn^+ and O^- centers were observed by electron spin resonance.[69]

The radiolysis reactions of low levels of hydrocarbons were tested by Norfolk et al. in mixtures of CO_2–CO–CH_4–H_2O–H_2, which are typical of conditions in a gas-cooled nuclear reactor. Reaction intermediates included clustered positive and negative ions, as well as O and H atoms and OH and alkyl radicals.[70]

Ionizing radiation of 40-MeV helium ions from a cyclotron[71,72] or of γ-rays from a ^{60}Co source[9,10] caused reduction of carbon dioxide to carbon monoxide, formic acid, formaldehyde, oxalic acid, and glyoxalic acid. Radiation-induced carboxylation, with $^{14}CO_2$, has been used to convert amines into amino acids.[73] Thus, the carboxylation of methylamine caused the formation of glycine and proline, with G-values of about 0.5 for each product. The carboxylation of ethylamine resulted mainly in alanine and β-alanine, while n-propylamine was converted mainly into α-aminobutyric acid. From iso-amylamine, the main product was leucine.

The radiation-induced carboxylation of methanol with carbon dioxide was tested by Fjodorov and Getoff both in aqueous solutions and in neat methanol.[74] In aqueous solutions containing methanol, the primary reactions were considered to be the reduction of CO_2 by e_{aq}^- and by H atoms to form the CO_2^- radical anion. Organic compounds such as methanol reacted very rapidly with ·OH radicals to form organic radicals, which either dimerized or reacted with the CO_2^- radical anions:

$$RH + \cdot OH \rightarrow R\cdot + H_2O \tag{11.32}$$

$$R\cdot + CO_2^- \rightarrow RCOO^- \tag{11.33}$$

$$2R\cdot \rightarrow RR \tag{11.34}$$

The main radiolytic products of 1 mol dm^{-3} methanol in aqueous solutions under elevated CO_2 pressures from 1 to 13 atm were formic and glycolic acids and ethylene glycol. In pure methanol containing 1.7 mol dm^{-3} CO_2, the final products were formic acid, glycolic acid, ethylene glycol, and formaldehyde, with G-values 3.0, 2.3, 1.6, and 0.7, respectively.[74]

Gamma-irradiation of aqueous suspensions of Mg, Ca, K, Tl, Cu, and Hg carbonates resulted in the formation of oxalic acid as the major product, as well as traces of HCHO, HCOOH, and glycolic acid.[75] In a study of the effects of radiation on aqueous suspensions of semiconductors, the radiolysis of the CO_2–H_2O–TiO_2 system was investigated. The major products were oxalic acid and formaldehyde, as well as traces of formic acid.[76]

The γ-radiolysis of CO_2-saturated aqueous solutions containing sodium phosphomolybdate, $Na_3H_4[P(Mo_2O_7)_6]$, resulted in the production of formaldehyde and oxalic acid.[77]

The carboxylation of linear low-density polyethylene (M_w = 128000, M_w/M_n = 4.4) with gaseous or supercritical CO_2 was achieved by Filardo et al. by γ-irradiation from a ^{60}Co source.[78,79] The polymer was irradiated in the form of films of about 50 μm thickness. Irradiation in the presence of CO_2 resulted in cross-linking, together with grafting of CO_2 into the polymer matrix. Under irradiation at doses of up to 400 KGy and at the subcritical CO_2 pressures of up to 10 MPa, the polymer maintained its ductility and did not become brittle. Increases in CO_2 pressure into the supercritical range, to 15 MPa, at a temperature of 38°C, caused instability in the irradiated polymer at room temperature, a slight decrease in the Young modulus, and an increase of elongation at break. IR absorption spectra of the polymers irradiated in the presence of CO_2 revealed the appearance of additional peaks at about 1720 and 3400 cm^{-1}, indicative of carboxylic and alcoholic and/or hydroperoxide functions, respectively. More detailed identification of these functional groups was attained by derivatization of the irradiated polymers with SF_4 (converting carboxylic groups into acyl fluorides) and with NO (converting alcoholic and hydroperoxide groups into nitrites and nitrates), followed by Fourier transform infrared spectroscopic analysis. The functional groups thus confirmed included carboxylic acids, γ-lactones, esters, ketones, alcohols, and hydroperoxides. The concentrations of these functional groups increased with the radiation dose. Supercritical carbon dioxide by itself (without the polymer) was found to undergo some radiolysis, producing CO and O_2. Therefore, the grafting of carboxylic, alcoholic, and hydroperoxide groups into the polymer irradiated in supercritical CO_2 was proposed to be due to the reactions of CO and oxygen with free radicals formed on the polymer chain.[78,79]

11.6   CONCLUSIONS ON PHOTOCHEMICAL REDUCTION OF CO_2

The main problem with photochemical reactions in homogeneous solutions as a method for CO_2 fixation seems to be the need for sacrificial electron donors, as well as low quantum yields, mainly because of side reactions. Several photocatalytic systems are capable of producing both H_2 and reducing CO_2, but none of them provide sustained production of either synthesis gas or methanol.[80]

REFERENCES

1. Chittenden, G.J.F. and Schwartz, A.W., Prebiotic photosynthetic reactions, *BioSystems*, 14, 15–32, 1981.

2a. Halmann, M., Photochemical fixation of carbon dioxide, in *Energy Resources Through Photochemistry and Catalysis*, Grätzel, M., Ed., Academic Press, New York, 1983, 507–534.

2b. Hepp, A.F., Landis, G.A., and Kubiak, C.P., A chemical approach to carbon dioxide utilization on Mars, in *Resources of Near-Earth Space*, Lewis, J.S., Matthews, M.S., and Guerrieri, M.L., Eds., University of Arizona Press, Tucson, 1993, 799–818.

3. Neta, P., Huie, R.E., and Ross, A.B., Rate constants for reactions of inorganic radicals in aqueous solutions, *J. Phys. Chem. Ref. Data*, 17, 1027–1284, 1988.

4. Slanger, T.G. and Black, G., CO_2 photolysis revisited, *J. Chem. Phys.*, 68, 1844–1849, 1978.

5. Stolow, A. and Lee, Y.T., Photodissociation dynamics of CO_2 at 157.6 nm by photofragment-translational spectroscopy, *J. Chem. Phys.*, 98, 2066–2076, 1993.

6. Kurbanov, M.A., Rustamov, V.R., Mustafaev, I.I., Iskanderova, Z.I., and Gadzhiev, Kh.M., Principles of the photolysis of gaseous mixtures of carbon dioxide and water, *Khim. Vys. Energ.*, 18, 381–382, 1984; *Chem. Abstr.*, 101, 101101t.

7. Kojima, M., Ojima, Y., Nakashima, N., Izawa, Y., Akano, T., and Yamanaka, C., Photolysis of CO_2 with 158 nm (F_2) laser. Reactivity of $O(^1D)$ with CH_4, CF_3H, and CF_3CH_3, *Chem. Lett.*, pp. 1309–1312, 1992.

8. Kojima, M., Ojima, Y., Nakashima, N., Izawa, Y., Yamanaka, C., and Akano, T., F_2 laser (158 nm) photolysis of CO_2 and hydrofluorocarbon systems, *J. Photochem. Photobiol. A Chem.*, 95, 197–202, 1996.

9. Getoff, N., Scholes, G., and Weiss, J., Reduction of carbon dioxide in aqueous solutions under the influence of radiation, *Tetrahedron Lett.*, pp. 17–23, 1960.

10. Getoff, N., Reduction of carbonic acid in aqueous solution under the influence of UV-light, *Z. Naturforsch.*, 17b, 87–90, 1962.

11. Eriksen, T.E., Lind, J., and Merenyi, G., On the acid–base equilibrium of the carbonate radical, *Radiat. Phys. Chem.*, 26, 197–199, 1985.

12. Behar, D., Czapski, G., and Duchovny, I., Carbonate radical in flash photolysis and pulse radiolysis of aqueous carbonate solutions, *J. Phys. Chem.*, 74, 2206–2210, 1970.

13. Weeks, J.L. and Rabani, J., The pulse radiolysis of deaerated aqueous carbonate solutions. I. Transient optical spectrum and mechanism. II. pK for OH radicals, *J. Phys. Chem.*, 70, 2100–2106, 1966.

14. Scheerer, R. and Grätzel. M., Photoinduced oxidation of carbonate ions by duroquinone, a pathway of oxygen evolution from water by visible light, *Ber. Bunsenges. Phys. Chem.*, 80, 979–982, 1976.

15. Scheerer, R. and Grätzel. M., Laser photolysis studies of duroquinone triplet state electron transfer reactions, *J. Am. Chem. Soc.*, 99, 865–871, 1977.

16. Roberts, J.L., Jr., Calderwood, T.S., and Sawyer, D.T., Nucleophilic oxygenation of carbon dioxide by superoxide ion in aprotic media to form the $C_2O_6^{2-}$ species, *J. Am. Chem. Soc.*, 106, 4667–4670, 1984.

17. Neta, P., Simic, M., and Hayon, E., Pulse radiolysis of aliphatic acids in aqueous solutions. I. Simple monocarboxylic acids, *J. Phys. Chem.*, 73, 4207–4213, 1969.

18. Hart, E.J. and Anbar, M., *The Hydrated Electron*, Wiley-Interscience, New York, 1970.

19. Butler, J. and Henglein, A., Elementary reactions of the reduction of thallium(1+) in aqueous solution, *Radiat. Phys. Chem.*, 15, 603–612, 1980.

20. Saini, R.D. and Iyer, R.M., On the reaction of carbonate radical with uranyl ion in aqueous medium: flash photolytic and pulse radiolytic studies, *J. Photochem. Photobiol. A Chem.*, 61, 171–182, 1991.

21. Yamase, T., and Sugeta, M., Photoreduction of CO_2 to CH_4 in water using dititanodecatungstophosphate as multielectron transfer catalyst, *Inorg. Chim. Acta*, 172, 131–134, 1990.

22. Lehn, J.-M. and Ziessel, R., Photochemical generation of carbon monoxide and hydrogen by reduction of carbon dioxide and water under visible light irradiation, *Proc. Natl. Acad. Sci.*, 79, 701–704, 1982.

23. Hawecker, J., Lehn, J.-M., and Ziessel, R., Efficient photochemical reduction of CO_2 to CO by visible light irradiation of systems containing $Re(bpy)(CO)_3X$ or $Ru(bipy)_3^{2+}$-Co^{2+} combination as homogeneous catalysts, *J. Chem. Soc. Chem. Commun.*, pp. 536–538, 1983.

24. Hawecker, J., Lehn, J.M., and Ziessel, R., Photochemical reduction of CO_2 to formate mediated by ruthenium bipyridine complexes as homogeneous catalysts, *J. Chem. Soc. Chem. Commun.*, pp. 56–58, 1985.

25. Ziessel, R., Hawecker, J., and Lehn, J.-M., Photogeneration of carbon monoxide and of hydrogen via simultaneous photochemical reduction of carbon dioxide by visible-light irradiation of organic solutions containing tris (2,2′-bipyridine) ruthenium(II) and cobalt(II) species as homogeneous catalysts, *Helv. Chim. Acta*, 69, 1065–1084, 1986.

26. Hawecker, J., Lehn, J.-M., and Ziessel, R., Photochemical and electrochemical reduction of CO_2 to CO mediated by (2,2′-bipyridine) tricarbonyl chloro rhenium (I) and related complexes as homogeneous catalysts, *Helv. Chim. Acta*, 69, 1990–2012, 1986.

27. Lehn, J.-M. and Ziessel, R., Photochemical reduction of CO_2 to formate catalyzed by 2,2′-bipyridine- or 1,10-phenanthroline-ruthenium complexes, *J. Organomet. Chem.*, 382, 157–173, 1990.

28. Kitamura, N. and Tazuke, S., Photoreduction of carbon dioxide to formic acid mediated by a methylviologen electron relay, *Chem. Lett.*, pp. 1109–1112, 1983.

29. Ishida, H., Terada, T., Tanaka, K., and Tanaka, T., Photochemical CO_2 reduction catalyzed by $[Ru(bpy)_2(CO)_2]^{2+}$ using triethanolamine and 1-benzyl-1,4-dihydronicotinamide as an electron donor, *Inorg. Chem.*, 29, 905–911, 1990.

30. Tanaka, H., Nagao, H., Peng, S.-M., and Tanaka, K., Crystal structure of cis-(carbon monoxide) (h^1-carbon dioxide) bis (2,2′-bipyridyl) ruthenium, an active species in catalytic CO_2 reduction affording CO and $HCOO^-$, *Organometallics*, 11, 1450–1451, 1992.

31. Kutal, C., Weber, M.A., Ferraudi, G., and Geiger, D., A mechanistic investigation of the photoinduced reduction of carbon dioxide mediated by tricarbonylbromo(2,2′-bipyridine)rhenium(I), *Organometallics*, 4, 2161–2166, 1985.

32. Calzaferri, G., Hädener, K., and Li, J., Photoreduction and electro-reduction of carbon dioxide by a novel rhenium(I) p-phenyl-terpyridine complex, *J. Photochem. Photobiol. A Chem.*, 64, 259–262, 1992.

33. Ruiz, G., Wolcan, E., Capparelli, A.L., and Feliz, M.R., Carbon dioxide activation by $ClRe(CO)_3(4\text{-phenylpyridine})_2$: steady state and flash photolysis study, *J. Photochem. Photobiol. A Chem.*, 89, 61–66, 1995.

34. Hukkanen, H. and Pakkanen, T.T., Photochemical catalytic reduction of carbon dioxide by visible light using RuII(bipy)$_3$ and Re(CO)$_3$(bipy)Cl as photocatalysts, *Inorg. Chim. Acta*, 114, L43–L45, 1986.

35. Christensen, P., Hamnett, A., Muir, A.V.G., and Timney, J.A., An in situ infrared study of CO$_2$ reduction catalyzed by rhenium tricarbonyl bipyridyl derivatives, *J. Chem. Soc. Dalton Trans.*, pp. 1455–1463, 1992.

36. Ishitani, O., George, M.W., Ibusuki, T., Johnson, F.P.A., Koike, K., Nozaki, K., Pac, C., Turner, J.J., and Westwell, J.R., Photophysical behavior of a new CO$_2$ reduction catalyst, Re(CO)$_2$(bpy){P(OEt)$_3$}$_2^+$, *Inorg. Chem.*, 33, 4712–4717, 1994.

37. Hori, H., Johnson, F.P.A., Koike, K., Ishitani, O., and Ibusuki, T., Efficient photocatalytic CO$_2$ reduction using [Re(bpy)(CO)$_3$–{P(OEt)$_3$}]$^+$, *J. Photochem. Photobiol. A Chem.*, 96, 171–174, 1996.

38. Hori, H., Ishitani, O., Koike, K., Takeuchi, K., and Ibusuki, T., Electrospray mass spectrometric detection of unstable rhenium complexes as reaction intermediates of photochemical CO$_2$-fixation, *Anal. Sci.*, 12, 587–590, 1996.

39. Hori, H., Johnson, F.P.A., Koike, K.K., Takeuchi, K., Ibusuki, T., and Ishitani, O., Photochemistry of [Re(Bpy)(CO)$_3$(PPh$_3$)]$^+$ (Bpy = 2,2′-bipyridine) in the presence of triethanolamine associated with photoreductive fixation of carbon dioxide: participation of a chain reaction mechanism, *J. Chem. Soc. Dalton Trans.*, pp. 1019–1023, 1997.

40. Grant, J.L., Goswami, K., Spreer, L.O., Otvos, J.W., and Calvin, M., Photochemical reduction of CO$_2$ to CO in water using a Ni(II) tetra-aza macrocycle complex as catalyst, *J. Chem. Soc. Dalton Trans.*, pp. 2105–2109, 1987.

41. Collin, J.-P. and Sauvage, J.-P., Electrochemical reduction of carbon dioxide mediated by molecular catalysis, *Coord. Chem. Rev.*, 93, 245–268, 1989.

42. Costamagna, J., Ferraudi, G., Canales, J., and Vargas, J., Carbon dioxide activation by aza-macrocyclic complexes, *Coord. Chem. Rev.*, 148, 221–248, 1996.

43. Tinnemans, A.H.A., Koster, T.P.M., Thewissen, D.H.M.W., and Mackor, A., Tetraaza–macrocyclic cobalt(II) and nickel(II) complexes as electron-transfer agents in the photo (electro) chemical and electrochemical reduction of carbon dioxide, *Rec. Trav. Chim. Pays-Bas,* 103, 288–295, 1984.

44. Craig, C.A., Spreer, L.O., Otvos, J.W., and Calvin, M., Photochemical reduction of carbon dioxide using nickel tetraazamacrocycles, *J. Phys. Chem.*, 94, 7957–7960, 1990.

45. Kimura, E., Wada, S., Shionoya, M., and Okazaki, Y., New series of multifunctionalized nickel(II)-cyclam (cyclam = 1,4,8,11-tetraaza-cyclotetradecane) complexes. Application to the photoreduction of carbon dioxide, *Inorg. Chem.*, 33, 770–778, 1994.

46. Mochizuki, K., Manaka, S., Takeda, I., and Kondo, T., Synthesis and structure of [6,6′-bi(5,7-dimethyl-1,4,8,11-tetraazacyclo-tetradecane)]dinickel(II) triflate and its catalytic activity for photochemical CO$_2$ reduction, *Inorg. Chem.*, 35, 5132–5136, 1996.

47. Tazuke, S. and Kitamura, N., Photofixation of carbon dioxide to formic acid using water as hydrogen source, *Nature*, 275, 301–302, 1978.

48. Legros, B. and Soumillion, J.Ph., Is carbon dioxide photoreducible by monoelectronic transfers under visible light? *Tetrahedron Lett.*, 26, 4599–4600, 1985.

49. Matsuoka, S., Kohzuki, T., Pac, C.J., and Yanagida, S., Photochemical reduction of carbon dioxide to formate catalyzed by p-terphenyl in aprotic polar solvent, *Chem. Lett.*, pp. 2047–2048, 1990.

50. Matsuoka, S., Yamamoto, K., Pac, C., and Yanagida, S., Enhanced para-terphenyl-catalyzed photoreduction of CO_2 to CO through the mediation of Co(III)-cyclam complex, *Chem. Lett.*, pp. 2099–2100, 1991.

51. Yanagida, S. and Matsuoka, S., Efficient photochemical reduction of CO_2 catalyzed by oligo-(p-phenylenes), in Proc. Int. Symp. Chemical Fixation of Carbon Dioxide, Nagoya, Japan, December 2 to 4, 1991, 19–22.

52. Matsuoka, S., Kohzuki, T., Pac, C.J., Yanagida, S., Takamuku, S., Kusaba, M., Nakashisha, N., and Yanagida, S., Photocatalysis of oligo(para-phenylenes). Photochemical reduction of carbon dioxide with triethylamine, *J. Phys. Chem.*, 96, 4437–4442, 1992.

53. Matsuoka, S., Yamamoto, K., Ogata, T., Kusaba, M., Nakashima, N., Fujita, E., and Yanagida, S., Efficient and selective electron mediation of cobalt complexes with cyclam and related macrocycles in the p-terphenyl-catalyzed photoreduction of CO_2, *J. Am. Chem. Soc.*, 115, 601–609, 1993.

54. Fujita, E., Brunschwig, B.S., Ogata, T., and Yanagida, S., Toward photochemical carbon dioxide activation by transition metal complexes, *Coord. Chem. Rev.*, 132, 195–200, 1994.

55. Ogata, T., Yanagida, S., Brunschwig, B.S., and Fujita, E., Mechanistic and kinetic studies of cobalt macrocycles in a photochemical CO_2 reduction system: evidence of Co–CO_2 adducts as intermediates, *J. Am. Chem. Soc.*, 117, 6708–6716, 1995.

56. Meerholtz, K. and Heinze, J., Multiple reversible electrochemical reduction of aromatic hydrocarbons in liquid alkylamines, *J. Am. Chem. Soc.*, 111, 2325–2326, 1989.

57. Lamy, E., Nadjo, L., and Savéant, J.M., Standard potential and kinetic parameters of the electrochemical reduction of carbon dioxide in dimethylformamide, *J. Electroanal. Chem.*, 78, 403–407, 1977.

58. Åkermark, B., Eklund-Westlin, U., Baeckström, P., and Löf, R., Photochemical, metal-promoted reduction of carbon dioxide and formaldehyde in aqueous solution, *Acta Chem. Scand. B*, 34, 27–30, 1980.

59. Morgenstern, D.A., Wittrig, R.E., Fanwick, P.E., and Kubiak, C.P., Photoreduction of carbon dioxide to its radical anion by [Ni$_3$(μ_3-I)$_2$(dppm)$_3$]: formation of two carbon–carbon bonds via addition of CO_2.$^-$ to cyclohexene, *J. Am. Chem. Soc.*, 115, 6470–6471, 1993.

60. Tazuke, S. and Ozawa, H., Photofixation of carbon dioxide: formation of 9,10-dihydrophenanthrene-9-carboxylic acid from phenanthrene–amine–carbon dioxide systems, *J. Chem. Soc. Chem. Commun.*, pp. 237–238, 1975.

61. Tazuke, S., Kazama, S., and Kitamura, N.J., Reductive photocarboxylation of aromatic hydrocarbons, *J. Org. Chem.*, 51, 4548–5453, 1986.

62. Nikolaitchik, A., Rodgers, M.A.J., and Neckers, D.C., Reductive photodecarboxylation of phenanthrene: a mechanistic investigation, *J. Org. Chem.*, 61, 1065–1072, 1996.

63. Paul, D.E., Lipkin, D., and Weissman, S.I., Reactions of sodium metal with hydrocarbons, *J. Am. Chem. Soc.*, 78, 116–120, 1956.

64. Tagaya, H., Onuki, M., Tomioka, Y., Wada, Y., Karasu, M., and Chiba, K., Photocarboxylation of naphthalene in the presence of carbon dioxide and an electron donor, *Bull. Chem. Soc. Jpn.*, 63, 3233–3237, 1990.

65. Tagaya, H., Onuki, M., Karasu, M., and Chiba, K., Photocarboxylation of an aromatic compound in the presence of carbon dioxide and an electron donor, in Proc. Int. Symp. Chemical Fixation of Carbon Dioxide, Nagoya, Japan, December 2 to 4, 1991, 195–200.

66. Ogata, T., Hiranaga, K., Matsuoka, S., Wada, Y., and Yanagida, S., Visible-light induced photocatalytic fixation of CO_2 into benzophenone using poly(p-phenylene) as a photocatalyst, *Chem. Lett.*, pp. 983–984, 1993.

67. Inoue, S., Nukui, M., and Kojima, F., Light-induced fixation of carbon dioxide with zinc porphyrin, *Chem. Lett.*, pp. 619–622, 1984.

68. Kurbanov, M.A., Rustamov, V.R., Mamedov, Kh.F., Iskenderova, Z.I., and Dzantiev, B.G., Chain transformations during radiolysis of gaseous carbon dioxide–hydrogen mixtures, *Khim Fiz.*, 5, 135–136, 1986; *Chem. Abstr.*, 104, 99290v.

69. Rustamov, V.R., Kerimov, V.K., Kurbanov, M.A., and Ali-Zade, Sh.N., Heterogeneous radiolysis of carbon dioxide on a zinc oxide catalyst, *Khim. Vysok. Energ.*, 19, 350–352, 1985; *Chem. Abstr.*, 103, 96184g.

70. Norfolk, D.J., Skinner, R.F., and Williams, W.J., Hydrocarbon chemistry in irradiated $CO_2/CO/CH_4/H_2O/H_2$ mixtures. I. A survey of the initial reactions, *Radiat. Phys. Chem.*, 21, 307–319, 1983.

71. Garrison, W.M., Morrison, D.C., Hamilton, J.G., Benson, A.A., and Calvin, M., Reduction of carbon dioxide in aqueous solutions by ionizing radiation, *Science*, 114, 416–418, 1951.

72. Garrison, W.M. and Rollefson, G.K., Radiation chemistry of aqueous solutions containing both ferrous ions and carbon dioxide, *Discuss. Faraday Soc.*, pp. 155–161, 1952.

73. Getoff, N., Gütlbauer, F., and De la Paz, L.R., Radiation induced preparation of labelled compounds. III. Incorporation of $^{14}CO_2$ and of $H^{14}COO^-$ into amines, *Kerntechnik*, 14, 75–81, 1972.

74. Fjodorov, V.V. and Getoff, N., Radiation induced carboxylation of methanol under elevated CO_2-pressure, *Radiat. Phys. Chem.*, 22, 841–848, 1983.

75. Lysyak, T.V., Konash, E.A., Kalyazin, E.P., Rudnev, A.V., and Kharitonov, Yu.Ya., Formation of organic products from metal carbonates and water in the presence of ionizing radiation, *Dokl. Akad. Nauk SSSR*, 265, 912–913, 1982; *Chem. Abstr.*, 97, 205662s.

76. Kolomnikov, I.S., Lysyak, T.V., Konash, E.A., Rudnev, A.V., Kalyazin, E.A., and Kharitonov, Yu.Ya., Reduction of CO_2 in aqueous solution in the presence of TiO_2 under gamma radiation, *Zh. Neorg. Khim. SSSR*, 28, 528–529, 1983; *Chem. Abstr.*, 98, 135105g.

77. Lysyak, T.V., Konash, E.A., Rudnev, A.V., Kalyazin, E.P., Kolomnikov, I.S., and Kharitonov, Yu.Ya., Radiolysis of the CO_2–H_2O system in the presence of sodium phosphomolybdate, *Zh. Neorg. Khim. SSSR*, 28, 1603–1604, 1983; *Chem. Abstr.*, 99, 45964c.

78. Filardo, G., Gambino, S., Silvestri, G., Calderaro, E., and Spadaro, G., Carboxylation of a linear low density polyethylene via gamma irradiation in presence of carbon dioxide in subcritical and supercritical conditions, *Radiat. Phys. Chem.*, 44, 597–601, 1994.

79. Dispenza, C., Filardo, G., Silvestri, G., and Spadaro, G., Carboxylation of linear low density polyethylene through gamma-irradiation in presence of supercritical carbon dioxide. Grafted groups analysis via derivatization procedures, *Colloid Polym. Sci.*, 275, 390–395, 1997.

80. Sutin, N., Creutz, C., and Fujita, E., Photoinduced generation of dihydrogen and reduction of carbon dioxide using transition metal complexes, *Comments Inorg. Chem.*, 19, 67–92, 1997.

12 Electrochemical Reduction of CO$_2$

12.1 INTRODUCTION

The equilibrium redox potentials E° (vs. the normal hydrogen electrode [NHE]) for the multielectron transfer reactions of carbon dioxide at pH 7.0 to formic acid, carbon monoxide, formaldehyde, methanol, and methane were calculated from the half-cell reactions:[1]

$$CO_2 + 2H^+ + 2e^- \rightarrow HCOOH \qquad (E° = -0.61 \text{ V}) \qquad (12.1)$$

$$CO_2 + 2H^+ + 2e^- \rightarrow CO + H_2O \qquad (E° = -0.52 \text{ V}) \qquad (12.2)$$

$$CO_2 + 4H^+ + 4e^- \rightarrow HCHO + H_2O \qquad (E° = -0.48 \text{ V}) \qquad (12.3)$$

$$CO_2 + 6H^+ + 6e^- \rightarrow CH_3OH + H_2O \qquad (E° = -0.38 \text{ V}) \qquad (12.4)$$

$$CO_2 + 8H^+ + 8e^- \rightarrow CH_4 + 2H_2O \qquad (E° = -0.24 \text{ V}) \qquad (12.5)$$

Since these multielectron reductions require very much less energy per electron transferred than the direct monoelectronic reduction of carbon dioxide to its radical anion CO_2^- (i.e., -2.1 V vs. the saturated calomel electrode [SCE]),[2] there exists considerable advantage in applying multielectron transfer to the reduction of carbon dioxide.

12.2 REDUCTION IN AQUEOUS SOLUTIONS

A major benefit of the electrochemical reduction of carbon dioxide is the possibility of using water as the proton source. Electrocatalytic reduction of CO_2 in water as medium is much more economic than that in organic solvents and may enable higher current densities. However, the electrochemical reduction reactions of carbon dioxide to organic compounds are difficult to catalyze since they not only require multiple electron transfers but often are also coupled to protonation reactions.[3]

12.2.1   Reduction on Metal Electrodes Leading to CO and HCOOH

One of the first aplications of electrochemical reactions to chemical synthesis was the reduction of aqueous sodium bicarbonate to HCOOH by Beketov[4] and Royer.[5] Since then, the electrochemical reduction of carbon dioxide has been the subject of very many studies in terms of the effect of the electrode materials and electrolyte media on the nature of the products, which include C_1 and sometimes C_2 and higher products. In particular, the nature of the electrochemical interaction of carbon dioxide with metals has been investigated by several physical techniques. The reaction mechanisms leading to CO_2 reduction at solid electrodes, metals, and semiconductors were discussed in detail by Taniguchi[6] and Frese.[7]

The reduction of aqueous CO_2 was developed by Coehn and Jahn to a preparative method, using zinc amalgam electrodes, resulting in 89% faradaic yield of formic acid.[8] Further enhancement in the faradaic yield, to 95%, was achieved by Ehrenfeld with ammonium carbonate as the electrolyte and with the zinc amalgam electrode separated from the platinum anode by a clay diaphragm.[9] On sodium amalgam cathodes with 2% aqueous $NaHCO_3$ saturated with CO_2, at a current density of 16 to 25 mA cm^{-2}, the current yield of formic acid reached by Rabinowitsch and Maschowetz was 86 to 97%.[10] In a useful preparative procedure by Udupa et al., rotating amalgamated copper cathodes in 10% sodium sulfate saturated with CO_2 were separated by a cation-exchange membrane from the lead electrode in the anodic compartment. At a cell potential of 3.5 V, the current density was 20 mA cm^{-2}, the faradaic efficiency was 81%, and the concentration of formic acid reached up to 200 g L^{-1}.[11] The electrochemical reduction of CO_2 on various metal electrodes to formic acid, and further to formaldehyde and methanol, was carefully investigated by Russell et al.[12]

The kinetics and mechanism of CO_2 reduction to formic acid in aqueous solutions were studied by Vassiliev et al. on a wide range of metal electrodes with high and moderate hydrogen overvoltage, such as Sn, In, Bi, Sb, Cd, Zn, Cu, Pb, Ga, Ag, Au, Ni, Fe, W, and Mo, as well as on glassy carbon.[13] Optimal electrodes were those which exhibited CO_2 electroreduction near the zero charge potential, in the range of maximal CO_2 adsorption. The electrode material strongly affected both of the two Tafel regions of the polarization curves. This suggested that both the first and second electrons were transferred to adsorbed species,

$$CO_2(ads) + e^- = CO_2^-(ads) \tag{12.6}$$

$$CO_2^-(ads) + BH + e^- \rightarrow HCOO^- + B^- \tag{12.7}$$

where BH was a proton donor, such as H_2O or H_3O^+. The surface coverage by adsorbed particles was very low on all the electrodes studied, $\phi < 0.1$.

12.2.1.1 Platinum Group Metal Electrodes

Using current potential measurements on platinum metal in 2 N H_2SO_4, carbon dioxide was found by Giner to be adsorbed in a reduced form.[14] In the potential range 0 to 250 mV (vs. NHE), CO_2 reacted with chemisorbed hydrogen on bright platinum to form a chemisorption product of reduced CO_2, which in further work was shown to be CO. The rate of reduction of CO_2 increased markedly with the temperature. The oxidation of this reduction product was irreversible.[14] This reduction of CO_2 by adsorbed hydrogen atoms to adsorbed CO occurred on a platinum electrode without any overpotential and is thus interesting from the standpoint of energy economy.[15-18] The nature of this reduced CO_2 on platinum was further investigated by Beden et al. using an *in situ* technique: electrochemically modulated infrared spectroscopy.[19] The spectral data seemed to show that CO_2 was reduced by reaction with adsorbed hydrogen. Under potential modulation of the Pt electrode between +250 mV and +550 mV, two IR absorption bands were observed. The very strong band centered at about 2060 cm^{-1} was attributed by Schiffrin to a linearly bonded CO species, while the moderately strong band at about 1865 cm^{-1} was assigned to a CO species adsorbed at a surface site of higher coordination number.[20] The above conclusions were confirmed by Aramata et al. in a Fourier transform infrared (FTIR) spectroscopic study.[21]

Anodic stripping voltammetry was used by Taguchi and Aramata to identify the sites of adsorption of the "reduced CO_2" on platinum single-crystal electrodes.[22] On Pt(100) crystal faces in 0.5 M H_2SO_4 in the hydrogen adsorption potential region, most of the "reduced CO_2" formed was found on the terrace sites, and not on edge sites, and was proven by FTIR to be linearly adsorbed CO. The formation of this adsorbed CO on Pt(100) was accompanied by the consumption of strongly bonded adsorbed hydrogen.

The reduction of CO_2 to CO in aqueous 0.1 M $HClO_4$ on platinum single-crystal electrodes was studied by Nicolic et al. using voltammetry and FTIR reflection absorption spectroscopy.[23] On platinum electrodes, the reduction of CO_2 occurred only in the hydrogen adsorption region, by holding the potential for some time at about 0.11 V (vs. the reversible hydrogen electrode [RHE]). The data indicated that on both the Pt(100) and Pt(110) surfaces, CO was the CO_2 reduction product, according to

$$CO_2 + H^+ + 2e^+ \rightarrow CO + H_2O \qquad (12.8)$$

The CO was adsorbed in two bonding states. On Pt(100), both linear CO and a large amount of bridge-bonded CO were observed, while on Pt(110) mainly linear CO and only a small amount of bridge-bonded CO were detected. The Pt(111) face exhibited only low activity for CO_2 reduction.[22] Using cyclic voltammetry and FTIR in 0.5 M H_2SO_4, Rodes et al. identified the linearly bonded CO on Pt(110) by a band at 2048 cm^{-1}, while Pt(100) presented two bands assigned to linear and bridge-bonded CO. At potentials above 0.50 V (vs. RHE), bands in the 1400- to 1600-cm^{-1} region indicated the presence of adsorbed carbonate species.[24]

The rates of reduction of CO_2 on single crystals of Pt group metals depended on the crystal orientation. In 0.5 M H_2SO_4, the order of the catalytic activity for platinum, iridium, and rhodium crystals was found by Hoshi et al. to be[16–18]

$$Pt(110) > Pt(100) > Pt(111)$$

$$Ir(110) >> Ir(100) = Ir(111)$$

$$Rh(100) > Rh(110) > Rh(111)$$

In 0.1 M $HClO_4$, the order of activity for the rhodium crystal surfaces was changed:

$$Rh(110) > Rh(100) > Rh(111)$$

In a comparison of the three Pt group metal electrodes for the reduction of CO_2 to adsorbed CO, the order of activity in 0.1 M $HClO_4$ was

$$Rh(110) > Pt(110) > Ir(110)$$

In 0.5 M H_2SO_4, the rate of formation of CO from CO_2 on crystal surfaces of Pt(110) was ten times higher than that on Pt(111). A clue to the different behavior came with the observation that adsorbed hydrogen atoms were regenerated on Pt(110), but not on Pt(111). The reduction of CO_2 to adsorbed CO was proposed to proceed by reaction with adsorbed hydrogen atoms:

$$CO_2 + 2H_{ad} \rightarrow CO_{ad} + H_2O \qquad (12.9)$$

The difference in reactivities of Pt(110) and Pt(111) was ascribed to different adsorption sites for H_{ad}.[16–18]

The effect of Cl^- impurity on CO_2 reduction on polycrystalline Pt electrodes in 0.1 M $HClO_4$ was determined by Huang. Chloride ions were specifically adsorbed on Pt surfaces and effected a modification in the hydrogen adsorption region, thus interfering in CO_2 reduction.[25]

The adsorption of carbon dioxide on platinum group metals was measured by Zakharyan et al. on smooth Pd, Rh, Ir, Ru, and Os electrodes, in 1 N H_2SO_4, using the method of potentiodynamic pulses, with gold foil as the counterelectrode.[26] On Ir, Pd, Ru, and Os electrodes, no chemisorption of CO_2 was observed. On Rh electrodes, the chemisorption of carbon dioxide was detected in the potential region of hydrogen adsorption. The activation energy for CO_2 adsorption on rhodium was 4.55 kcal. The species adsorbed on the rhodium electrode had the average composition of CH_2 and was so strongly bonded to the electrode surface that it was not desorbed during cathodic polarization.[26] Vassiliev et al. concluded that the electroreduction of CO_2 on the platinum group metals Pt and Rh occurred by chemisorbed CO_2 species reacting with chemisorbed hydrogen on the metal surface.[27] On rhodium electrodes in 0.5 M H_2SO_4, adsorbed CO_2 behaved electrochemically like adsorbed CO.[28]

The "reduced CO_2" species formed on platinum electrodes, mainly the strongly adsorbed formyl radicals (·CHO) and carbonyl groups, cause rapid poisoning of the Pt electrodes and within a few minutes block further CO_2 reduction. Huang and Faguy fabricated reticulated vitreous carbon electrodes modified with Pt black attached onto these electrodes with a Nafion® film. These glassy carbon foams had an effective surface area of 66 cm^2 cm^{-3} and a Pt loading of less than 0.1 mg cm^{-2}. The electrodes were reported to be resistant to poisoning and were able to reduce CO_2 continuously for at least several hours.[29]

On platinum electrodes in dilute aqueous solutions, carbon dioxide was not reduced to organic compounds.[30] However, in saturated carbonate electrolytes at below −5°C in one-compartment cells, there occurred appreciable production of formate ion and formaldehyde. These products were not formed in cells with separated compartments. The explanation given was that the percarbonate ion $C_2O_6^{2-}$ was produced on the anode and was then reduced by hydrogenation on the cathode.[31] See also Section 8.1.2.

12.2.1.2   The CO_2^- Ion Radical

Aylmer-Kelly et al. identified the CO_2^- ion radical by its strong optical absorption peak at 250 nm, using Pb electrodes in 0.1 M Me_4NCl saturated with CO_2, at a potential of −1.0 to −1.8 V (vs. Ag/AgCl), and applying modulated specular reflectance spectroscopy.[32] Photoemission studies by Schiffrin on polarized mercury electrodes in contact with aqueous electrolytes indicated that the subsequent steps leading to formate ions occur on the electrode surface:[20]

$$H_2O + \cdot CO_2^- \rightarrow HCOO\cdot_{ads} + OH^- \qquad k = 7.7 \times 10^2 \ M^{-1} \ s^{-1} \qquad (12.10)$$

$$HCOO\cdot_{ads} + e^- \rightarrow HCOO^- \qquad (12.11)$$

12.2.1.3   Gold Electrodes

On gold cathodes, the reduction of carbon dioxide led to CO as the main product, with faradaic efficiency of about 60%, as well as hydrogen. A detailed kinetic investigation by Noda et al. on the mechanism of CO_2 reduction at the Au electrode in phosphate buffer solutions (pH 2.5 to 6.8) suggested that the primary step, which is rate determining, is an electron transfer from the electrode to a surface-adsorbed CO_2 molecule,[33]

$$CO_2(ads) + e^- \rightarrow CO_2^-(ads) \qquad (12.12)$$

followed by rapid proton transfer from water, releasing carbon monoxide,

$$CO_2^-(ads) + H_2O + e^- \rightarrow CO + 2OH^- \tag{12.13}$$

See also Section 12.2.2.2 on pulsed electroreduction of CO_2 on Au electrodes.

12.2.1.4 Hg, Sn, Zn, In, Pb, Cd, Bi, and Sb Electrodes

On mercury cathodes in aqueous media, the reduction of carbon dioxide led to a polarographic wave with a half-wave potential of 2.1 V vs. SCE.[2] On such electrodes, the primary step of carbon dioxide reduction to formate has been proposed to be electron capture by CO_2 to form the $\cdot CO_2^-$ ion radical in solution:

$$CO_2 + e_{aq}^- \rightarrow \cdot CO_2^- \qquad (E^\circ = -2.21 \text{ V vs. SCE}) \tag{12.14}$$

An evaluation of the reduction potential of the CO_2/CO_2^- couple was made, using a thermodynamic cyle and taking for the electron affinity of CO_2 the value -0.6 eV. Thus the value derived for E° (CO_2/CO_2^-) was -1.93 V vs. NHE.[34]

Using mercury,[18] tin, and indium electrodes,[35] the rate constant for the reaction of the hydrated electron with the CO_2 molecule in aqueous solution was determined by kinetic measurements to be almost diffusion limited, with $k_2 = 7 \times 10^9 \ M^{-1} \ s^{-1}$.

The current–voltage curves on Hg revealed two Tafel regions, with slopes of 0.09 and 0.22 V/decade, at low and high current densities. The changeover occurred at a current density of about 50 μA cm^{-2}. Reactions 12.11 and 12.6 were proposed as the rate-determining steps in the low and high current density regions, respectively.[36–38]

In the reduction of carbon dioxide in neutral aqueous solutions (0.05 M phosphate buffer, pH 6.8) over a mercury pool electrode, the faradaic efficiency for formate production was found by Russell et al. to reach 100% at a cathodic overpotential of 1.15 V and a current density of 1 mA cm^{-2}.[11] However, the energy efficiency (defined as the ratio of the free energy of the organic fuel produced to the electric energy input) was only about 45% (neglecting IR losses). In the low overvoltage region, with a Tafel slope of about 91 mV, the galvanostatic charging curve indicated large coverage by an adsorbed intermediate, presumably the above formate radical $HCOO\cdot_{ads}$. On the other hand, this coverage was small in the high overvoltage region, with the Tafel slope of about 240 mV. The desired further reduction of formic acid to methanol was not possible directly in the above neutral medium. This was achieved in an acid medium, optimally at pH 3.8 (in 0.1 N NaHCO$_3$) in the narrow potential range of -0.68 to -0.72 V (vs. SCE), on an electroetched Sn electrode, resulting in a faradaic efficiency of up to 99.6%, but at a current density of only 3.6 μA cm^{-2}.[11]

Further studies by Babenko et al. on the reduction of carbon dioxide on mercury electrodes, using nanosecond pulsed laser or continuous wave photoelectron emission, showed that the emitted electrons were captured by CO_2, forming CO_2^-, which was adsorbed on the Hg surface. The current density of CO_2 reduction on Hg in neutral buffered solutions was about 1 mA cm^{-2} at -1.6 V vs. SCE, with a reduction rate constant of (3 to 5) $\times$ 10^6 s^{-1}. The transfer coefficient was 0.25 to 0.3 in the potential range -1.5 to -1.8 V.[39]

On tin electrodes, the kinetics of the cathodic reduction of carbon dioxide was studied in more detail by Zakharyan et al. using both steady-state polarization curves recorded with a rotating disk electrode and photoelectron emission measurements at the metal–solution boundary.[40] In contrast to the effect with Hg electrodes, the rate-determining step in reduction of CO_2 to formate on Sn cathodes in the first Tafel region was shown to be the transfer of the second electron. The mechanism was thus represented by:

$$CO_2 + e^- \rightarrow \cdot CO_{2\,ads}^-$$ (12.15)

The high overvoltage encountered with mercury electrodes stimulated the search for other electrode materials. On tin cathodes, optimal conditions for formate production (92% current efficiency) were reported by Kesarev and Fedortsov using a catholyte of 2% $KHCO_3$ continuously saturated with CO_2 (pH below 8.0) at a current density of 3.3 mA cm^{-1}.[41] Zinc, lead, tin, indium, and cadmium were studied by Ito et al. as electrocatalysts for CO_2 reduction. Indium electrodes in 0.1 N lithium carbonate provided the largest current efficiency (92%) for the production of formic acid.[42]

The nature of the electroreduction products of carbon dioxide on zinc electrodes was found by Hattori et al. to depend on the electrolyte medium.[43] In 0.1 M $KHCO_3$, at potentials of -1.5 to -1.7 V (vs. Ag/AgCl), both CO and HCOOH were formed, while in 0.05 M K_2SO_4, the predominant product was CO, in up to 80% faradaic efficiency, with small yields of HCOOH. The explanation for this effect was the higher rate of dissolution of the Zn electrode in the K_2SO_4 solution (pH 4.2) than in the $KHCO_3$ medium (pH 6.8). The dissolved Zn^{2+} ions were found to promote the formation of carbon monoxide.[43]

Highly efficient CO_2 reduction to HCOOH was obtained by Komatsu et al. using Bi-plated glassy carbon electrodes in $KHCO_3$ solutions.[44] At potentials of -1.4 to -1.7 V (vs. SCE) and current densities of about 1 A dm^{-2}, the current efficiency for HCOOH production was close to 100%. On the other hand, with electrodes plated with Sb instead of Bi, yields of HCOOH were very low, and H_2 was the main electrolysis product.

12.2.1.5 Palladium Electrodes

On palladium electrodes, in contrast to mercury electrodes, the cathodic reduction was shown to occur directly on the bicarbonate ion, and not on CO_2.[45–47] The reduction current was dependent on the cation of the electrolyte, decreasing in the order Cs > K > Na > Li. Thus, the current density in 1 M $CsHCO_3$ was nine times higher than in 1 M $NaHCO_3$. The cesium effect in promoting the reduction of HCO_3^- was explained by competitive adsorption, decreasing the amount of adsorbed species from formaldehyde. Formaldehyde, which was presumably the reduction product of formate, was shown to cause the gradual blocking of the reduction of bicarbonate.

An alternative interpretation was that $Cs^+-HCO_3^-$ ion pairs were directly involved in the reaction at the cathode.[37-39,45-47]

In a detailed study by Azuma et al. of the reduction products on a Pd wire electrode in CO_2-saturated aqueous solutions of 0.05 M $KHCO_3$, at -2.0 V vs. SCE, the faradaic yields were 73% H_2, 16.1% HCOOH, 11.6% CO, and 0.083% CH_4. Also, many saturated and unsaturated hydrocarbons, from C_1 to C_6, were detected in very low yields. The current efficiencies for hydrocarbon production increased rapidly with rising temperature in the range 0 to 40°C.[48]

An interesting approach to energy storage is the storage of hydrogen in palladium metal electrodes. On palladium electrodes, preadsorption of hydrogen resulted in increased current efficiencies for reduction of CO_2 in aqueous $KHCO_3$, producing CO, HCOOH, and H_2. Ohkawa et al. suggested that the increased selectivity for the reduction of CO_2 relative to the evolution of H_2 could be due to the interaction of adsorbed hydrogen with adsorbed intermediates of CO_2 reduction. In a nonaqueous electrolyte, 0.1 M Bu_4NClO_4 in acetonitrile, the electroreduction on both Pd and Pd–H electrodes yielded CO and oxalic acid, but no hydrogen.[49,50]

Podlovchenko et al. used Pd electrodeposited on Pt electrodes to electroreduce CO_2 in aqueous solutions of potassium bicarbonate (pH 8 to 10) at potentials more positive than the reversible hydrogen potential. Formic acid was produced in close to 100% faradaic yield.[51]

The selectivity toward production of hydrocarbons was enhanced by using copper-containing Pd–H electrodes. Cu deposition was carried out on hydrogen-adsorbed Pd (atomic ratio 0.36:1) surfaces by electroless plating in dilute aqueous $CuSO_4$. With the Cu-modified Pd–H electrodes in CO_2-saturated 0.1 M aqueous $KHCO_3$ at -1.6 V (vs. SCE), the faradaic efficiencies for the production of HCOOH, H_2, CH_3OH, CO, CH_4, and C_2H_4 were 71, 41, 7, 3, 2, and 1.6 respectively.[52-54]

Online mass spectrometry was used by Kolbe and Vielstich to identify intermediates formed during the electrochemical reduction of CO_2 on electrodes of palladium sputtered on polytetrafluoroethylene foil. In 0.1 M $KHCO_3$, volatile CO was detected at a potential more negative than -1.4 V (vs. Ag/AgCl). However, CO_2 already started to be reduced at the Pd electrode at -1 V, with formation of adsorbed CO (or COOH) and evolution of H_2.[55]

Structural effects on the kinetics of carbon dioxide reduction on single-crystal electrodes of palladium were studied by Hoshi et al.[56]

12.2.1.6 Oxide Semiconductor Electrodes

Nogami et al. studied the mechanism of CO_2 reduction on oxide semiconductors.[57,58] Application of a cathodic step bias resulted in a current transient. Using the current transient technique, the fluorine-doped thin-film SnO_2 electrode was found to undergo degradative reduction in five steps from SnO_2 to Sn. The effects of pH and CO_2 pressure (up to 2.94×10^5 Pa) on the transient were measured. The total charge passed during the transient was linearly related to the square root of the CO_2 pressure and of the pH. X-ray photoelectron spectroscopy (XPS) measurements revealed the formation of HCOOH on the SnO_2 surface. The mechanism of hydrocarbonization of CO_2 on the SnO_2 electrodes in acidic solutions could be represented by:

$$CO_2 + H^+ + 2e^- \rightarrow HCOO_{ads}^- \tag{12.16}$$

Long-term reduction caused the conversion of SnO_2 to Sn. On this metal electrode, a predominant reaction was the reduction of CO_2 to CO:[57,58]

$$Sn + CO_2 \rightarrow SnO + CO \tag{12.17}$$

$$Sn + 2CO_2 \rightarrow SnO + 2CO \tag{12.18}$$

The initial steps in the electroreduction of CO_2 on SnO_2 electrodes were revealed by Aoki and Nogami using a rotating ring disk study with an SnO_2 disk and a Pt ring.[59] The SnO_2 disk was prepared by spray pyrolysis of monobutyltin trichloride on an n-Si substrate. The disk potential was swept with cathodic and anodic pulses of −1.3 and 0.0 V (vs. SCE). Both HCOOH and H_2 were produced and detected on the Pt ring. The HCOOH was released more slowly than the H_2, with a delay of a few seconds. From the dependence of the current on the speed of rotation, it was concluded that the electroreduction of CO_2 on SnO_2 was controlled both by electron transfer, forming some intermediates, and by the mass transport of protons.

At polycrystalline TiO_2 and TiO_2–Ru electrodes in aqueous 0.5 M KCl, cyclic voltammetry measurements by Monnier et al. suggested the existence of a strong interaction between the dissolved CO_2 and the hydrated surface of these semiconductor electrodes.[60] Even stronger irreversible adsorption of CO_2 from aqueous solutions or in the presence of water vapor was observed by Koudelka et al. with platinized TiO_2. In this case, the formation of two distinct products of CO_2 reduction was indicated by the cyclic voltammetry curves.[61] The effect of the pH of the medium on the shape of the voltammetry curves was the subject of some controversy.[62,63]

12.2.1.7 Other Electrode Materials

Titanium diboride (TiB_2) is a ceramic polycrystalline material prepared from the powder by a hot press method. It has metallic character. Experiments using it as a cathode for the reduction of carbon dioxide were performed by Nakabayashi and Kira in aqueous media containing 1 M $NaClO_4$. Cyclic voltammetry measurements showed a cathodic reduction peak, the current intensity of which was proportional to the potential sweep rate. This indicated that the electrode reaction proceeds with surface adsorbed species. The stable reaction products were CO and possibly formic acid.[64,65]

The strong influence of the electrode potential was shown in the presence of tetraalkyl ammonium salts. On glassy carbon electrodes, with 0.1 M aqueous tetramethyl ammonium chloride (pH 8 to 10) as electrolyte, at an electrode potential of −0.9 V vs. SCE, the reduction product of CO_2 by Bennett et al. was oxalate. At a potential of −1.7 V, the product was glyoxalate. No formate was produced.[66] On graphite and mercury electrodes, with aqueous solutions containing tetramethyl ammonium chloride (0.1 M) at pH 9, CO_2 was found by Eggins et al. to be reduced in two steps, yielding oxalic acid at −0.9 V and glyoxylic acid at −1.8 V (vs. SCE). These products were clearly identified by HPLC.[67] In a further study by Eggins et al. of the electrochemical reduction of CO_2 in aqueous solutions (pH 8 to 10) of Et_4NClO_4 on glassy carbon electrodes, the rate-determining step was proposed to be the release of CO_2 from the aqueous bicarbonate.[68] The peak potentials of CO_2 reduction determined by linear sweep voltammetry were very low, −0.74 V (vs. SCE), which was explained by adsorption of the very bulky Et_4N^+ cation on the electrode surface. Earlier work by Hori and Suzuki had observed for CO_2 reduction in 0.1 M solutions of alkali bicarbonates a shift in the electrode potentials toward more positive values with increasing cation radius. Peak potential values for CO_2 reduction with Li^+, Na^+, and K^+ ions were −1.6, −1.52, and −1.45 V (vs. Ag/AgCl), respectively. The stronger adsorption of the larger cations resulted in more positive zeta potentials of the electrode surface.[69]

12.2.1.8 Reduction by Corrosion of Metallic Iron

Fujita et al. suspended iron powder with a large specific surface area (29 m^2/g) in CO_2-saturated water.[71–73] A corrosion reaction resulted in the oxidation of iron and the reduction of water and CO_2 to H_2 (the most abundant product) and to the hydrocarbons methane, ethane, propane, n-butane, ethene (C_2H_4), propene (C_3H_6), and 1-butene (1-C_4H_8). The yields of the hydrocarbons decreased in the order C_1 to C_4. When the above Fe–CO_2–water system was irradiated with γ-rays (from a ^{60}Co source), CO was also formed, in addition to the H_2 and the hydrocarbons. The yield of CO in the radiation-induced reduction of CO_2 was markedly enhanced in the presence of Ni^{2+} or Cu^{2+} ions. The primary action of iron corrosion under γ-irradiation was shown to be a rapid initial decrease both in the pH and the oxidation–reduction potential (ORP) of the solution. These changes in both pH and ORP shifted the state of the solution to a potential (on the Pourbaix diagrams) in which the reduction of CO_2 to CO, CH_4, and higher hydrocarbons became thermodynamically possible.[71–73]

Hardy and Gilham, in a related study, contacted aqueous bicarbonate solutions with metallic iron, producing hydrocarbons from methane to C_5.[74] In batch experiments, simulated groundwater (containing $MgSO_4$, $CaCl_2$, NaCl, and $NaHCO_3$, acidified to pH 6) was stirred in the dark either with commercial iron cuttings (about 90% Fe) or with electrolytic iron (>99% Fe). After about 140 h, the samples contained methane, ethene, ethane, and propene, as well as traces of butenes and pentenes. Similar results were obtained in column experiments, in which simulated groundwater was recirculated for up to 351 days through columns packed with either the commercial or the electrolytic iron. The molecular weight distribution of the hydrocarbon products could be approximately represented by the Anderson–Schultz–Flory (ASF) distribution, which had previously been applied to the composition of products from the Fischer–Tropsch synthesis of hydrocarbons from CO + H_2 over iron or nickel catalysts. In this distribution, the weight fraction m_p of the total hydrocarbons with carbon number p (all isomers included) is related to the probability of chain growth α,

$$m_p = (\ln 2\alpha)p\alpha p \qquad (12.19)$$

or in logarithmic form,

$$\log(m_p/p) = \log(\ln 2\alpha) + p(\log \alpha) \qquad (12.20)$$

Thus, in the ideal ASF distribution, the plot of $\log(m_p/p)$ versus p should be linear. In the above experiments of contacting aqueous bicarbonate with iron, the ASF plots were fairly linear for the C_2 to C_5 compounds, which suggested a mechanism similar to that of the Fischer–Tropsch reaction. Iron by corrosion may have provided electrons for the primary step of CO_2 reduction, and iron also served as a catalyst for the formation and growth of hydrocarbon chains. However, the amounts of methane observed were always much larger than those expected from the ASF distribution.[74] The reason for this discrepancy is not clear. Possibly some other mechanism must be considered for the formation of methane. In such long-term experiments of methanation of CO_2, the action of methanogenic bacteria may not be excluded.

12.2.2   Reduction to Hydrocarbons

The earlier work on the electrochemical reduction of carbon dioxide on metal electrodes searched mainly for the production of formic acid. More recent work by Hori et al.[75–79] and Ito et al.[80–83] showed that the nature of the electrode metal for carbon dioxide reduction in aqueous electrolytes strongly affected the product composition. Electrode metals could be classified into four main groups: On electrodes of Cd, Hg, In, Sn, and Pb, the reduction selectively produced formic acid. On Au, Ag, and Zn, the selectivity was toward carbon monoxide formation. On Cu electrodes, hydrocarbons (mainly methane and ethene), aldehydes, and alcohols were produced. On the other hand, Al, Ga, Pt, Fe, Ni, and Ti have little activity for CO_2 reduction. In a more extensive test by Azuma et al. on 32 metal electrodes, preferential HCOOH production occurred on the heavy metals in the periodic table groups IIB, IIIB, and IV: Cd, In, Sn, Pb, Tl, Hg, Zn, and Pd. Carbon monoxide was preferentially produced on Ti, Ni, Ag, and Au electrodes. Cu was the only metal that favored the efficient production of hydrocarbons.[84]

Carbon dioxide reduction in CO_2-saturated aqueous $KHCO_3$ (0.05 M) on various metal electrodes was measured by Azuma et al. at close to 0°C and compared with room temperature electrolysis. In most cases, the current efficiencies of product formation were dramatically enhanced at low temperatures, probably because of higher solubility of CO_2 and perhaps also because of the longer lifetime of the adsorbed reaction intermediates. For Ni electrodes, at a potential of –2.2 V vs. SCE, the current efficiencies at room temperature and at 2°C were 0.1 and 0.7% for CH_4, 0.6 and 21% for CO, 0.01 and 0.07% for C_2H_4, 0.02 and 0.18% for C_2H_6, and 0.1 and 13.7% for HCOOH, respectively.[85] See also Section 12.2.4 on alloy electrodes.

12.2.2.1 Copper Electrodes

Both copper as electrode material and copper complexes as catalysts have been of particular interest for the electroreduction of CO_2. Also, copper enzymes are involved in the reduction of CO_2 in nature.[86,87] On high-purity copper electrodes in aqueous media, the major products were found by Hori et al. to be methane, ethene, ethanol, n-propanol, carbon monoxide, formic acid, and hydrogen.[75–79,88,89] The yield of methane was highest at 0°C. The sum of the faradaic yields of the carbon dioxide reduction products exceeded 90%. Methanol, formaldehyde, and ethane were not formed. High current yields were achieved only with large overpotentials. The onset potential for methane and ethylene production was –1.5 V vs. SCE. The primary reduction intermediate was proposed to be CO. Carbon monoxide, while weakly adsorbed on the Cu electrode, interfered with cathodic hydrogen production. The adsorbed CO was thus reduced to hydrocarbons and alcohols, with the yields rising with more negative potentials.[21]

Electrochemical reduction of carbon dioxide to methane was also obtained by Ikeda et al. using Cu-coated glassy carbon and platinum electrodes. In 1 M $KHCO_3$ at 0°C, the current efficiency for methane formation was 60%.[90] The selectivity in the products distribution for CO_2 reduction on Cu electrodes was found by Noda et al. to be changed drastically even with small variations in the electrode potentials.[82] In 0.1 M $KHCO_3$ at 25°C, maximal faradaic efficiencies reached 32% at –1.40 V (vs. Ag/AgCl/saturated KCl) for $HCOO^-$, 33% at –1.52 V for CO, 41% at –1.58 V for C_2H_4, and 39% at –1.70 V for CH_4. For the aldehydes and alcohols, the maximal values were 14% for C_2H_5OH at –1.64 V, and at –1.58 V 2% for CH_3CHO, 4.5% for n-C_3H_7OH, and 5% for C_2H_5CHO.[89] On Fe electrodes, which were effective for the reduction of CO to hydrocarbons, CO_2 was *not* reduced.[91]

The surface treatment of the Cu electrodes strongly affected the selectivity of the products formed by CO_2 reduction. Rough surface area electrodes favored high Faradaic yields of CO and $HCOO^-$, while smooth surface electrodes favored CH_4 and C_2H_4 production.[92] Rough surface Cu electrodes were more active than smooth ones in terms of both current efficiency and catalytic electrode life. The electrode deactivation was proposed by Kyriacou and Agnostopoulos to be due to adsorbed organic intermediates.[93] The deposition of cadmium on copper electrodes also modified the product selectivity. With increased Cd deposition, the hydrocarbon formation decreased, and the production of CO and $HCOO^-$ increased.[94]

The effect of the crystal structure of Cu electrodes was determined by Frese using single-crystal Cu electrodes.[7] The order of activity for the rate of formation of methane in CO_2-saturated 0.5 M $KHCO_3$ at pH 7.5 and 22°C was Cu(111) > Cu(110) > Cu(100). On the Cu(111) face at –1.85 V (vs. SCE), the methane formation rate was 5.8×10^{-5} mol cm^{-2} h^{-1}.

On the above very high-purity Cu electrodes, excellent faradaic yields of methane were obtained by Hori et al., but the current densities were rather low, at most 10 mA cm^{-2}.[75,76] Increased current densities were achieved by Cook et al. using less pure copper (nominally 99.9% pure, with various heavy metal impurities) as the cathode, in 0.5 M $KHCO_3$ as the electrolyte, at 0°C, with a Nafion® membrane separating the catholyte from the anolyte. Under continuous CO_2 bubbling through a glass frit, at constant current electrolysis and a current density of 38 mA cm^{-2}, and with the Cu electrode at a potential of -2.29 V vs. SCE, the faradaic yield for methane production was 33%.[95]

In the electrochemical reduction of carbon dioxide on Cu electrodes, the selectivity for methane formation was found by Azuma et al. to be enhanced by the presence of small amounts of methanol. Thus, in an electrolyte of 0.1 M KHCO$_3$, in the absence and presence of 1 mM methanol at an electrode potential of -2.2 V (vs. Hg/Hg$_2$SO$_4$), the current efficiencies for CH$_4$ production were 24 and 47%, respectively. This enhancement of methane formation was at the expense of decreased hydrogen production.[96] The reason for this effect of methanol is not clear.

High current densities, up to 25 mA cm^{-2}, were achieved by Cook et al. on copper deposited *in situ* on glassy carbon electrodes, in aqueous 0.5 M KHCO$_3$ saturated with CO_2, operated at -2.0 V vs. SCE. Onset potentials were -1.6 V vs. SCE. Faradaic yields of methane + ethene reached 79%. The mechanism proposed involved adjacent sites of hydrogen atoms and CO_2 adsorbed on copper,[97,98]

$$H_2O + Cu + e^- \leftrightarrow Cu-H_{ads} + OH^- \qquad (12.21)$$

$$Cu + CO_2 \leftrightarrow Cu(CO_2)_{weakly\ ads} \qquad (12.22)$$

followed by the rate-determining step,

$$Cu-H_{ads} + Cu(CO_2)_{weakly\ ads} \rightarrow Cu-OCHO \qquad (12.23)$$

Also, with Cu deposited on membranes of the solid polymer electrolyte Nafion®, carbon dioxide reduction to ethene and ethane (but not to methane) was obtained from gas-phase carbon dioxide even at ambient temperatures. If, in addition, platinum was deposited on the anodic side of the Nafion® membrane, methane was also produced.[98] This gas-phase heterogeneous electrochemical reduction of CO_2 was further tested by Cook et al. for a variety of other metals, using the configuration CO_2, M/Nafion® 417/Pt, N$_2$ + H$_2$ (90:10), where M was a metal electrocatalyst deposited onto the Nafion® polymer electrode. Carbon dioxide flowed through the cathodic compartment, while the nitrogen–hydrogen mixture flowed through the anodic compartment. With M = Ni, Ru, Pb, Pd, Ag, Re, Os, Ir, Pt, and Au, the faradaic efficiencies for the formation of methane and higher hydrocarbons were very low.[99]

The mechanism of this CO_2 reduction on copper electrodes was proposed by Kim et al. to involve electrochemical splitting of adsorbed CO as the rate-determining step, followed by hydrogenation of surface carbon atoms — in an electrochemical analogue of the Fischer–Tropsch reaction. Methane was produced from carbon dioxide reduction at a 50 times higher rate than from carbon monoxide, which was explained by the approximately 40 times higher solubility of CO_2 than CO in the electrolyte. Since the onset potentials for CH$_4$ and C$_2$H$_4$ formation were the same,

−1.5 V vs. SCE, both reactions were probably limited by the same step of CO dissociation:[100]

$$CO_{ads} + H_2O + e^- \leftrightarrow C_{ads} + OH_{ads} + OH^- \qquad (12.24)$$

The black deposits which were always formed on the copper electrodes were shown by DeWulf et al. using XPS and Auger electron spectroscopy to be graphitic carbon, suggesting the following sequence of reactions:[101]

$$CO_2 \rightarrow CO \rightarrow \text{Surface-bound formyl (Cu–HCO)} \rightarrow$$

$$\text{Surface carbene (Cu=CH}_2) \rightarrow \text{Hydrocarbons} \qquad (12.25)$$

Surface carbenes (methylenes) were previously proposed by Santilli and Castner as the intermediates in the Fischer–Tropsch synthesis of hydrocarbons.[102] The similarity of CO_2 reduction mechanisms on metal electrodes to the Fischer–Tropsch hydrogenation of CO_2 on these metals seems to be indicated by the similarity in the distribution of the reduction products observed by Cook et al., which included CO, alkanes, alkenes, aldehydes, and alcohols.[103] This product distribution was represented by the relationship

$$\log(m_p/p) = \log(\ln^2 \alpha) + p(\log \alpha) \qquad (12.20)$$

The linearity of plots of $\log(m_p/p)$ vs. p for the electroreduction of CO_2 on various metals was proposed as a criterion for a Fischer–Tropsch mechanism, presumably involving intermediate adsorbed CH_x species.[103]

In order to check if formaldehyde, formic acid, or acetaldehyde may possibly be intermediates during the electrochemical reduction of CO_2 to CH_4 and higher hydrocarbons at Cu electrodes, Cook et al. performed the electrolysis of these compounds on Cu in both alkaline and acidic aqueous media. The results showed that HCOOH, HCHO, and CH_3CHO were indeed reduced to methane and may thus be intermediates in the CO_2 reduction. On the other hand, methanol was not reduced to methane under the same conditions and thus probably was not an intermediate.[104]

The mechanism of reduction of carbon dioxide on copper electrodes to methane and ethene was studied by Wasmus et al. using online electrochemical mass spectrometry with rotating electrodes. Electrodeposited copper on glassy carbon was found to be a better catalyst for methane formation than bulk copper. The onset potential for hydrocarbon formation was −1.7 V vs. SCE, detected both by voltammetry and mass spectrometry.[105]

The deactivation of copper electrodes during electroreduction of CO_2-saturated aqueous hydrogen carbonate solution was studied by Friebe et al. using differential electrochemical mass spectrometry.[106] The copper electrode was prepared by vacuum vapor deposition of Cu onto an ethylene–tetrafluoroethylene copolymer membrane. Volatile products diffused through this membrane into a quadrupole mass spectrometer. Deactivation of the electrodes occurred both by deposition of a graphitic blocking layer and by formation of some unidentified water-soluble poison in the electrolyte. In addition to m/e signals assigned to ionization and fragmentation products of methane and ethene, a signal with m/e = 15 was observed in the electrolyte after several minutes of electrolysis, which was proposed to be related to the unknown electrode poison.

The initial product of CO_2 electroreduction on Cu and Ag electrodes was studied by Ogasawara et al. using both *in situ* infrared reflection absorption spectroscopy and surface-enhanced Raman spectroscopy.[107,108] On Cu electrodes, two kinds of linear adsorbed CO species were observed by their vibrational bands at about 2100 and 2000 cm^{-1}. These two bands were assigned to CO adsorbed on adatom defect Cu atoms and terrace Cu atoms, respectively. On the Ag electrodes, a linearly adsorbed CO appeared only by the vibrational band at about 2000 cm^{-1}. This CO was desorbed at potentials above about –0.5 V (vs.the saturated hydrogen electrode [SHE]).

12.2.2.2 Pulsed Electroreduction on Cu, Ag, and Au Electrodes

The three group Ib metals — Cu, Ag, and Au — have in common the advantage of relatively low overpotentials for CO_2 reduction, but also the tendency of their electrodes to become "poisoned" and deactivated after short periods of electrolysis of CO_2 solutions.[109] In the case of continuous electrolysis of CO_2 solutions with copper electrodes, the CO_2 reduction stopped after 20 to 40 min. Surface-enhanced Raman spectroscopic analysis of deactivated copper electrodes by Augustynski indicated the formation of a new band, assigned to some nonstoichiometric Cu_2O species.[110]

As noted above, the electroreduction of CO_2 in aqueous media on copper electrodes resulted in the formation of methane as the major product, as well as formic acid, carbon monoxide, ethene, ethanol, and hydrogen. These electrolysis experiments were normally carried out under either potentiostatic or galvanostatic conditions. The faradaic yields of the products, while initially quite high, decreased rapidly within a few hours. This decrease could be ascribed to the deposition of graphitic carbon on the copper cathodes, presumably obtained by the reduction of the initially produced formate ions.[101] Shiratsuchi et al. discovered that the blocking of the copper electrodes by graphite could be prevented by a regime of pulsed electroreduction.[111,112] The copper electrodes were submitted to periodic pulses of applied bias of 0 and –1.8 V (vs. SCE), at various time intervals. In CO_2-saturated $KHCO_3$ solutions, and using time intervals of 5 s each for the duration of the cathodic and anodic pulses, the faradaic efficiencies for methane and ethene increased gradually and then remained constant. After 15 h, the faradaic efficiencies for CH_4 and C_2H_4 were about 10 and 25%, respectively. For the formation of these products, optimal conditions were at a cathodic bias of –2.6 V (vs. SCE), and the optimal temperature was 10°C, resulting in a total faradaic efficiency for CH_4 + C_2H_4 of about 65%. After the pulsed electrolysis, only traces of CO and HCOOH could be detected. As shown by XPS, the graphitic carbon deposit formed during the cathodic period was reoxidized during the anodic period. A thin film which formed even during the pulsed electrolysis was shown by XPS to be due to Cu_2O.[111,112] By varying the anodic bias, Aoki et al. were able to control the relative yields of CH_4 and C_2H_4. Maximal faradaic yields of CH_4 (30%) and C_2H_4 (10%) were obtained at the anodic bias of –0.1 and 0.0 V, respectively.[113]

In a similar study, Augustynski et al. prevented the deactivation of copper cathodes by applying periodic anodic pulses to the electrode.[114,115] Thus, during the electrolysis for 27 h of 0.5 M $KHCO_3/CO_2$ at 22°C, when imposing on the continuous potential of –1.72 V (vs. NHE) every 5 min three potential scans to +1.3 V and back at a rate of 5 V s^{-1}, the faradaic efficiencies for the production of CH_4, C_2H_4 and C_2H_5OH were 45, 6, and 1.6%, respectively, and the current density (which increased with time) was 75 mA cm^{-2}. This periodic activation enabled prolonged electrolytic reduction of CO_2 with quite high yields of CH_4.

On both silver and gold cathodes, the main product of electroreduction of CO_2 was CO. On polycrystalline silver electrodes immersed in CO_2 saturated 0.1 M $NaClO_4$, the onset of CO formation was observed by Kostecki and Augustynski to occur at –1.1 V (vs. NHE).[109] At a more cathodic potential (–1.32 V), the current density decreased, passed through a minimum after about 30 min, and then rose. Surprisingly, the faradaic efficiency for CO production had a maximum simultaneously with the minimum in current density, followed by decreased CO production with longer electrolysis times, accompanied by formation of trace amounts of CH_4, C_2H_4, and C_3H_6. This inactivation of the silver electrodes was proposed to be due to poisoning by some CO_2 reduction product. The electrodes could be reactivated by carrying out every 5 min a sequence of three anodic/cathodic scans up to +0.6 V (vs. NHE) at a rate of 0.1 V s^{-1}. With such periodic treatment, at a potential of –1.5 V (vs. NHE) and a temperature of 298 K, the faradaic yield of CO production was quasi-stable at close to 100% for at least 5 h.[109] In contrast to this study, Shiratsuchi et al. were able to produce, in addition to CO, substantial amounts of hydrocarbons on Ag electrodes by pulsed electroreduction of aqueous CO_2.[116,117] Maximal faradaic efficiency for the formation of CH_4 (40%) was with cathodic and anodic bias potentials of –2.5 and –0.4 V (vs. Ag/AgCl). With bias potentials of –1.8 and –0.5 V, C_2H_5OH was produced with the remarkably high efficiency of about 20%.

On rotating disk polycrystalline Au electrodes in CO_2-saturated 0.1 M $NaHCO_3$, the onset of formation of CO was observed by Kedzierzawski and Augustynski to occur at the rather low potential of –1.0 V (vs. NHE), with a faradaic yield close to 100%.[118] At more negative potentials, the yields of CO decreased, replaced by the release of hydrogen. The initial high current densities measured at the constant potential reduction of CO_2 decreased rapidly, within a few minutes, presumably due to poisoning of the electrode by some unidentified CO_2 reduction product. The gold electrode could be reactivated by periodic potential scans at 50 mV s^{-1} up to ca. +1.3 V (vs. NHE) carried out in intervals of 15 min. Such anodic treatment brought the electrode to the region of gold oxide formation and stripped off the poisoning species. With this treatment, constant current efficiency of CO production could be maintained for at least several hours.[118]

See also Section 12.2.3.3 on the electrodeduction of CO_2 on pulsed Mo electrodes.

12.2.2.3 Ni vs. Cu Electrodes: Effects of Adsorption

On nickel electrodes, the reduction of aqueous CO_2 yielded mainly H_2, with very low yields of CH_4, C_2H_4, C_2H_6, HCOOH, and CO.[88,119-121] The difference in the behavior of copper and nickel electrodes was correlated with the very strong adsorption of the intermediate CO on nickel, while CO is only moderately adsorbed on copper. A tentative reaction mechanism on both Cu and Ni electrodes is the intermediate formation of adsorbed CO as a precursor for the production of hydrocarbons and alcohols. This hypothesis was supported by the similarity in the electroreduction products of CO to that of CO_2 on Cu and Ni electrodes and was confirmed by observation of the infrared absorption peaks at 2000 and 1900 cm^{-1} due to the linear and bridged structures of adsorbed CO on Ni electrodes. The intermediate CO molecule, which is weakly adsorbed on the copper electrode, is further reduced by adsorbed hydrogen to hydrocarbons and alcohols. On Ni electrodes, on the other hand, CO is very strongly adsorbed, interfering with hydrogen evolution.[119-121] The electrocatalytic activity of Cu, Ni, and Fe electrodes for the reduction of adsorbed CO was found by Hori et al. to be inversely related to the adsorption strength of linearly adsorbed CO on these metals.[122,123]

The dependence of the products distribution on the various metal electrodes was further explained by the different adsorption energies of the CO_2 reduction intermediates.[124,125] These adsorption energies were estimated from the d-band energies of the metals. In group VIII metals, such as Ni, the Fermi levels are situated in the partially filled d-bands. These provide strong covalent s-bonding to adsorbed CO_2^- anion radicals, which have an unpaired electron in a localized sp^2-hybrid orbital. The strong adsorption energy is thus provided by the s-bond energy. Therefore, reduction on such metals produces mainly hydrogen. In the group IB metals Cu, Ag, and Au, the Fermi level is situated in the sp-band. The d-band in these metals is slightly below the Fermi level (e.g., 1.5 eV for Cu) and is completely filled. Upon raising the negative potential on the electrode to that of the Fermi level, an electron will be excited from the d-band to the Fermi level, which thus will obtain an unpaired electron, which can form strong covalent s-bonding with an adsorbed CO_2^- anion radical. The adsorption energy in this case will be the s-bond energy less the excitation energy. Therefore, with the group IB metals, the adsorption energy is much less than with the group VIII metals. Reduction intermediates such as CO_2^- will be further reduced to adsorbed CO, CH_2, and HCO radicals, which will readily desorb to free CO, alcohols, and hydrocarbons. In the case of Ag and Au electrodes, the adsorption energy for the intermediate adsorbed CO is very weak and CO is released as the main reduction product. With Cu electrodes, the adsorption energy for the intermediate CO is slightly larger, and thus the lifetime of the adsorbed CO is longer, and it will undergo further reduction to hydrogenated intermediates, which will eventually be released to free hydrocarbons and alcohols. With the group IIB, IIIB, and IVB metals, such as Hg, In, Sn, and Pb, the d-band is deep below the Fermi level, and electronic excitation from the d-band to the Fermi level is impossible. Intermediate CO_2^- radicals can adsorb only extremely weakly on these metals. The weakly adsorbed CO_2^- radicals undergo further reduction by successive H^+ and electron captures, producing HCOOH as the main product.[124,125]

12.2.3   Reduction to Methanol and Ethanol

There has been considerable effort to produce methanol by direct electrochemical reduction of carbon dioxide. This has been successful with a number of electrode systems, sometimes with high faradaic efficiencies, but the current densities have often been disappointingly low, usually less than 1 mA cm^{-2}.

12.2.3.1   Reduction on Palladium/Pyridine

A remarkable enhancement in the electroreduction of CO_2 to methanol and formaldehyde on palladium electrodes in aqueous electrolytes (0.5 M $NaClO_4$, pH 5.0) was achieved by Bocarsly et al. by using the pyridinium ion (10 mM) as electrocatalyst.[126,127] The Pd electrodes were preloaded with hydrogen by hydrogenation at 1 to 3 mA cm^{-2} in 1 M H_2SO_4. The electrolyses were carried out galvanostatically at current densities of 0.03 to 0.05 mA cm^{-2} in a two-compartment cell separated by a fine glass frit, with a Pt foil as counterelectrode. At a low overpotential (−0.7 V vs. SCE), hydrogen evolution was limited, and the faradaic yield of methanol production reached up to 30%. Experiments with $^{13}CO_2$ proved the formation of $^{13}CH_3OH$. When paraformaldehyde was substituted in the medium instead of CO_2, no methanol was formed, indicating that the CO_2 reduction did not occur via formaldehyde. Also, in the absence of pyridine, or when N-methylpyridine was used instead of pyridine, methanol was not produced, showing that the N–H hydrogen atom of the pyridinium ion was required for the reduction of CO_2. The pyridine was not consumed or degraded during a 19-h run. By cyclic voltammetry, the onset potential for CO_2 reduction at pH 5.4 was found to be at −0.55 V (vs. SCE), which is close to the thermodynamic potential of −0.52 V (SCE) at pH 5.4 for the reduction to methanol. In the proposed mechanism, the pyridinium ions $C_5H_5NH^+$ were electroreduced in a one-electron reduction to the unstable C_5H_5NH radicals, which released ·H atoms into the Pd electrode, while reforming the pyridinium ions. The multielectron reduction of CO_2 then occurred on the Pd–H electrode.[126,127] The use of the inexpensive pyridine as catalyst seems attractive, and this method could be interesting if the reaction could be carried out at much higher current densities and with a less expensive electrode material.

12.2.3.2   Reduction on GaAs

Using n-GaAs single-crystal electrodes (111 As face), in CO_2-saturated Na_2SO_4 solution, Canfield and Frese produced methanol with 100% faradaic yield, at electrode potentials of −1.2 to −1.4 V (SCE), but at current densities of only 0.2 mA cm^{-2}. Much lower yields were observed at (110) and (100) faces and using p-GaAs and p-InP as photocathodes.[128,129]

However, on repeating the reduction of CO_2 on heavily doped p-GaAs, using the (111 As) face, the conversion to methanol was found by Sears and Morrison to occur even at open circuit in the dark, at rates reaching 0.1 μmol h^{-1} cm^{-2}.[130] An even larger rate was found with a moderately doped n-GaAs crystal. On the (Ga) and (100) faces, methanol production was much lower. Closed-circuit electrolysis resulted mainly in formic acid production. The explanation given for the open-circuit effect was the corrosion (dissolution) of the GaAs in carbonic acid.

12.2.3.3   Reduction on Mo, Ru, RuO$_2$, and TiO$_2$ Electrodes

On molybdenum electrodes in CO_2-saturated acidic aqueous solutions (0.2 M Na$_2$SO$_4$, pH 4.2) at –0.7 to –0.8 V vs. SCE, methanol was formed as the major carbon dioxide reduction product, with faradaic efficiencies of 50 to 100% and current densities of up to 0.6 mA cm^{-2}.[131] Minor products included CO and CH$_4$. In 0.05 M H$_2$SO$_4$ at 0.6 V (SCE), efficiencies were 20 to 46%. The rate of formation of methanol on Mo electrodes could be enhanced considerably by cycling the applied voltage between –1.2 and +0.2 V (SCE), thus increasing the faradaic yield of methanol to 370%, which indicated corrosion of the electrode. The proposed mechanism involved the redox pair Mo/MoO$_2$:

$$2CO_2 + 3Mo + 4H_2O = 2CH_3OH + 3MoO_2 \qquad (12.26)$$

$$3MoO_2 + 12e^- + 12H^+ = 3Mo + 6H_2O \qquad (12.27)$$

Alcohols such as methanol, ethanol, and 2-propanol were observed by Nakato and Mori to be formed with high current efficiencies even under low overvoltage on electrodes of Cu, Ni, and RuO$_2$ during the electrochemical reduction of CO_2 in 0.1 M KHCO$_3$.[132] However, the current densities were extremely low, of the order of 20 to 70 μA cm^{-2}. On RuO$_2$ electrodes (coated on Ti metal), at a potential of –0.6 V vs. SCE and a current density of 54 μA cm^{-2}, the faradaic yield for methanol production was 84%. The selective formation of alcohols on these transition metal and metal oxide electrodes was proposed to be due to the large adsorption energies of radical intermediates. The initial reaction steps may be interaction of CO_2 with metal-adsorbed hydrogen (M-H), forming adsorbed carboxylate species (M-COOH), which by further reduction and hydrogenation led to alcohols.

On electroplated ruthenium cathodes in aqueous Na$_2$SO$_4$ saturated with carbon dioxide, Frese et al. observed CO_2 reduction resulting in the formation of both methane and methanol, with current yields of 30 and 35%.[133,134] However, the current densities were less than 1 mA cm^{-2}. The mechanism proposed included hydrogenation of a surface carbon species as the rate-determining step. Ruthenium is known as an active catalyst both for CO_2 methanation and for Fischer–Tropsch-type gas–solid reactions:

$$H_{ads} + CH_x \rightarrow CH_{x+} \qquad \text{where } x = 0 \text{ to } 3 \qquad (12.28)$$

In a comparison of the rates of formation of methane under optimal conditions over Ru and Cu electrodes, Cu electrodes were found by Frese to be much more effective.[7] On electroplated Ru at -1.48 V (vs. SCE), pH 7.6 and 24°C, the methane production rate per geometric area was 7×10^{-9} mol cm^{-2} h^{-1}, while on smooth Cu electrodes under similar conditions, the rate was 1×10^{-7} mol cm^{-2} h^{-1}.

Methanol and acetone formation by CO_2 reduction on ruthenium oxide electrodes in 0.5 M $NaHCO_3$ during 8 h of electrolysis was observed by Popic et al. to be enhanced by Cd and Cu adatoms. This reduction occurred at a low potential of -0.8 V.[135]

On electrodes of ruthenium fixed on poly-hydroquinone/benzoquinone (prepared by electropolymerization of mercapto-hydroquinone), supported on a glassy carbon electrode, the onset potential for CO_2 reduction was found by Arai et al. to be as low as -0.5 V (vs. SCE).[136] At -0.7 V (vs. SCE), in 0.2 M Na_2SO_4, the current efficiency for methanol production attained almost 100%, but at a current density of only 0.2 mA cm^{-2}. Under illumination with visible light, the efficiency for methanol production even exceeded 100%, an effect which may possibly be understood by the corrosion of the electrode.

On Sr-doped TiO_2 electrodes, at -1.7 V (vs. SCE), the faradaic yield for methanol production reached 3.1%, while the yield on the undoped electrodes was only 1.3%.[137]

On RuO_2 + TiO_2 (35:65) coated on titanium foil, at -0.1 V (vs. Hg_2SO_4) in 0.05 M H_2SO_4 (pH 1.2), Bandi achieved the reduction of CO_2 to methanol with 24% current efficiency, and at a current density of 0.52 mA cm^{-2}. The other product was formic acid (2%). Tafel plots of log(current density) vs. electrode potential gave Tafel slopes of 180 to 240 mV.[138] These were similar to those previously observed by Ryu et al. at the higher current densities for CO_2 reduction on Hg electrodes, which had been assigned to the first electron transfer reaction as the rate-determining step.[37] The mechanism proposed involved as a primary fast step the partial reduction of the metal oxide surface,

$$MeO_2 + H^+ + e^- \rightarrow MeOOH \qquad (12.29)$$

followed by a rate-determining step of chemisorption of the CO_2 molecule, with formation of a carbonate intermediate.[138] With mixed TiO_2 + RuO_2 electrodes (ratio 1:3, corresponding to an oxide composition of Ru_3TiO_8, deposited by decomposition of the mixed halides in air at 450°C on Ti plates), the electrolysis of aqueous 0.5 M $KHCO_3$ was performed at a potential of -0.90 to -0.95 V vs. Ag/AgCl and a current density of 5 mA cm^{-2}. In the presence of CO_2, the current efficiency for CO_2 reduction reached a maximal value at potentials just before the steep onset of H_2 evolution. The rate-limiting step was proposed by Bandi and Kühne to be a surface recombination of adsorbed hydrogen with CO_2:[139]

$$CO_2 + H_{ad} + e^- = COOH^-_{ad} \qquad (12.30)$$

Electrodeposition of small amounts of copper on these TiO_2 + RuO_2 electrodes caused very much enhanced yields of methanol, ethanol, and formic acid. Copper deposition shifted the current potential curves for H_2 evolution to more negative potentials, while the CO_2 reduction efficiency increased.

Bandi et al. used pseudocapacity measurements on transition metals and metal oxide electrodes in solutions saturated either with N_2 or CO_2 to determine the strength of adsorption of CO_2 to these electrodes.[140,141] Only electrodes on which CO_2 was weakly adsorbed, such as RuO_2–TiO_2, Au, Cu, Ag, and Ni, and which exhibited high pseudocapacities, were promising electrodes for the electrochemical reduction of CO_2. Since the adsorption of carbon dioxide on the electrode surface is an essential prerequisite for the electron transfer which is the primary reaction step, some effort has been made to correlate the electrochemical activity of different electrode materials with their bond properties. The adsorption properties of several metals and metal oxides for carbon dioxide were determined by pseudocapacity measurements. The surfaces of electrodes with mixed oxides, such as RuO_2–TiO_2 and RuO_2–Co_3O_4–SnO_2–TiO_2 which showed good electrocatalytic activity for CO_2 reduction to methanol, also showed increased pseudocapacity under CO_2 saturation (compared with that under N_2). On the other hand, on electrodes with mixed oxides such as IrO_2–TiO_2 or pure IrO_2, which were inactive for CO_2 reduction, there was no pseudocapacity in CO_2-purged solutions, presumably because the surfaces were totally obstructed by the strong adsorption of CO_2. The electrochemical behavior was correlated with the d-bond character of the different metals. High pseudocapacity for electrodes in CO_2-purged solutions occurred with the metals Cu, Ni, and Au.[140,141]

At electrodes of oxidized copper, prepared by anodizing or thermally air oxidizing copper foil, carbon dioxide reduction to methanol was observed by Frese with onset potentials as low as –0.4 V (SCE).[142] Surprisingly, the faradaic yields for methanol production considerably exceeded 100%, indicating that the reaction was not a simple six-electron reduction. The highest rate of methanol production, 1×10^{-4} mol cm^{-2} h^{-1}, was obtained with anodized copper in 0.5 M $KHCO_3$ at pH = 7.6 and –1.9 V (SCE). Presumably, the reduction occurred on a surface layer of p-Cu_2O. The mechanism proposed to account for the very high faradaic yields (or "open-circuit" CO_2 reduction) involves chemical reduction steps up to HCO_{ads}, followed by three hydrogenation steps to methanol.

12.2.3.4 Selective Ethanol Formation

On Cu electrodes, the product distribution depended on the electrode potential. In 0.1 M $KHCO_3$ at 298 K, ethanol was produced at maximal faradaic efficiency of about 14% at a potential of –1.65V (vs. Ag/AgCl saturated KCl). Other products in smaller yields were CH_3CHO, C_2H_5CHO, and n-C_3H_7OH.[143]

Ikeda et al. reported selective formation of ethanol by electrochemical reduction of CO_2 on CuO/ZnO electrodes in aqueous media.[144,145] Those electrodes were prepared by coprecipitation of the oxides from aqueous solutions of $Cu(NO_3)_2$ and $Zn(NO_3)_2$ with aqueous ammonia. The washed, dried, and calcined precipitate was pressed to tablets at pressures of up to 147 MPa and sintered at 900°C in air for 4 h. The electrodes constructed from these tablets were preelectrolyzed at 0.3 mA cm^{-2} under an N_2 atmosphere, for partial surface reduction of CuO to Cu metal.

Optimal conditions for CO_2 reduction to ethanol were obtained with such electrodes (Cu:Zn atomic ratio 3:7) in 0.1 M KH_2PO_4 at a potential of -1.32 V (vs. Ag/AgCl) and a current density of 0.5 mA cm^{-2}, yielding ethanol, hydrogen, and propane with faradaic efficiencies of 17.3, 20.4, and 0.1%, respectively. Only about 38% of the electricity consumed was devoted to the formation of these products, the balance being "wasted" in reduction of the oxides of the electrode to metallic Cu and Zn.[144,145]

12.2.3.5 Prussian-Blue-Coated Electrodes

Carbon dioxide was reduced by Ogura et al. to methanol at a Prussian blue, $[KFe(III)Fe(II)(CN)_6]$, coated platinum cathode, with an illuminated n-CdS photoanode as energy source and pentacyanoferrate as mediator, in a medium containing 0.02 M methanol.[146–148] The current efficiency for methanol production was reported to be 87%. Other metal complexes were also very effective as mediators, such as 1-nitroso-2-naphthol-3,5-disulfonic acid. The mechanism proposed involves a labile ligand in the metal complex, which is the active site in the catalytic process. A coordination bond is formed between the primary alcohol and the neutral metal atom. Carbon dioxide inserts into this bond, producing a formate-like intermediate, which is reduced by the neighboring surface-bound Fe(II) complex, forming methanol. In the absence of a primary alcohol in the medium, the CO_2 reduction product was HCOOH.[146–148] Instead of a photoelectrode or an external voltage source, a hydrogen fuel cell of Pt gauze was used by Ogura et al. as the anode, at which H_2 was oxidized. CO_2 was reduced by the above metal complex at a Prussian-blue-coated Pt cathode. At short circuit, methanol formation was maximal, reaching 93% current efficiency, but the current density was only about 1 μA cm^{-2}.[149]

Homogeneous catalysis of carbon dioxide electroreduction was achieved by Ogura et al. using a system consisting of cobalt(II)-2-nitroso-1-naphthol 4-sulfuric acid complex with methanol. The electrode was modified by a surface layer of a water-insoluble complex, Everitt's salt, which is the reduced form of Prussian blue, $K_3Fe(III)[Fe(II)(CN)_6]$. With this mediated electrochemical system, CO_2 was reduced to methanol in the potential range -0.15 to -0.35 V (vs. SCE).[150]

12.2.3.6 Reduction on Films of Prussian Blue/Polyaniline

Ogura et al. succeeded in reducing CO_2 in aqueous solutions to C_2 and C_3 compounds by using platinum electrodes coated with a composite film.[151,152] The first layer was the highly conducting Prussian blue (from ferric ferricyanate solution), followed by potential cycling in aniline/HCl aqueous solution, creating the conductive polyaniline polymer, and by surface doping with 2-hydroxy-1-nitrosonaphthalene-3,6-disulfonato-cobalt(II) as mediator. With CO_2-saturated aqueous 0.5 M KCl at pH 2.0, at a constant potential of -0.6 V (vs. SCE) for 24 h, the main products were lactic acid and ethanol, with smaller amounts of acetone and methanol. The total current efficiency for these compounds was about 4%.

In a further study by Ogura et al., bis(1,2-dihydroxybenzene-3,5-disulfonato) ferrate(II) was used as the surface-doping mediator.[153] The onset potential for the reduction of CO_2 was only -0.3 V (SCE), which is close to the thermodynamic potential. The reduction products were determined in potentiostatic runs, at a potential of -0.8 V (SCE). At a surface concentration of 1.1×10^{-8} mol cm^{-2} of the mediator, in CO_2-saturated 0.5 M KCl (pH 3), the total current efficiency for the reduction of CO_2 reached 12.4%, the major product again being lactic acid, with smaller amounts of ethanol, methanol, and acetone and traces of methane and acetaldehyde.[153] When Fe(II)-4,5-dihydroxybenzene-1,3-disulfonate was used as the surface mediator on the Prussian-blue/polyaniline-coated electrodes, the total current efficiency for CO_2 reduction (at -0.6 V vs. SCE in 0.5 M KCl, pH 2) was only 6.9%, but the selectivity for lactic acid production was predominant,[154] and much higher than with the above-described 2-hydroxy-1-nitrosonaphthalene-3,6-disulfonato-cobalt(II) as mediator.[151,152] Using electrodes coated with Prussian blue/polyaniline surface mediated by Fe(II)-bis(1,8-dihydroxynaphthalene-3,6-disulfonate), electroreduction of CO_2 in 0.5 M KCl at pH 3.0 and a potential of -0.8 V (SCE) led to lactic and formic acids as the main products, with smaller amounts of methanol, ethanol, acetone, and 1-propanol, and a total current efficiency of 10%. The observed onset potentials for the production of lactic and formic acids were very close to the thermodynamic values for their formation by the reduction of CO_2, $E^\circ = -0.20$ and -0.41 V (vs. SCE), respectively, indicating very low overpotentials. Ogura et al. observed that formic acid was produced immediately at the start of electrolysis. The other products appeared only after an induction period. This suggested that formic acid (or a surface-adsorbed formyl group) may be a precursor for the other products.[155]

The mechanism proposed by Ogura et al. for these reactions involved as the primary step the electrocatalytic reduction of H$^+$ to H$_{ads}$ during the redox reactions of Prussian blue/Everitt's salt (the reduced form of Prussian blue).[156] The reduction of CO_2 was suggested to occur by hydrogenation by H$_{ads}$. The C_2 and C_3 compounds may then be generated by surface interactions of the adsorbed intermediates. The catalytic site for the activation of CO_2 was proposed to involve bifunctional activation, in which the amino group of the polyaniline binds the electrophilic carbon atom of CO_2, while the central metal atom of the mediator coordinates the basic oxygen atom of CO_2. This activated CO_2 is then hydrogenated by H$_{ads}$, leading to stepwise reduction from formic acid to methanol, and by CO_2 insertion reactions to C_2 and C_3 compounds, such as ethanol, lactic acid, and acetone.[156] In an *in situ* FTIR spectroscopic study on the reduction of CO_2 at a Prussian-blue/polyaniline-modified electrode, Ogura et al. observed that cathodic polarization resulted in capture of CO_2 by the electrode, with the appearance of an absorption band at 1529 cm^{-1} attributed to the amido group, $-CO-NH-$.[149] This indicated binding of the amino nitrogen of polyaniline to the carbon atom of CO_2. Presumably, the basic oxygen atom of CO_2 may coordinate to the central metal atom of the Prussian blue, resulting in a bifunctional activation of CO_2, which facilitated its multielectron reduction.[157]

12.2.3.7 Films of Molybdenum-Blue/Polymethylpyrrole-Coated Electrodes

With composite films of molybdenum blue/polyaniline coatings on Pt electrodes, the electroreduction of CO_2 in 0.5 M KCl at pH 4 and a potential of -0.6 V (vs. SCE) yielded mainly methanol, acetone, and ethanol and some isopropanol. The total current efficiency increased with the number of coatings, reaching about 13% with three coatings.[154]

12.2.4   Alloy Electrodes

On metals with high overpotential for hydrogen production, such as cadmium, indium, lead, mercury, silver, and copper, the current efficiency and product selectivity for carbon dioxide reduction to carbon monoxide, formic acid, or methane and ethane is good, but the overpotentials are high and the energy efficiency is low. On the other hand, on metals with low overpotentials for hydrogen generation, such as from group VIII of the periodic table, both the selectivity and the energy efficiency for carbon dioxide reduction are poor.

Alloy electrodes were introduced by Watanabe et al., thus achieving considerable improvements in the selectivity and energy efficiency of CO_2 reduction.[158–161] These depended on the introduction of low-hydrogen-overpotential metals (such as group VIII metals) on the surface of high-hydrogen-overpotential metals (such as the group IB or IIB metals). Such alloy electrodes enabled catalytic reduction of surface-adsorbed carbon dioxide or reduced carbon intermediates by surface-adsorbed hydrogen atoms.

12.2.4.1   Cu–Ni Alloys

On Cu–Ni alloys (atomic ratio 90.5/9.5), the formation of methanol occurred at almost the reversible potential (onset potential was –0.38 V vs. SHE), with a faradaic efficiency of about 10% at a potential of –0.65 V (vs. SHE). Alloy electrodes were prepared by electroplating from the appropriate salt mixtures onto gold flag electrodes and were tested in CO_2-saturated aqueous 0.05 M $KHCO_3$. For production of CO and formic acid, the alloy electrodes Cu–Sn, Cu–Pb, and Cu–Zn provided low overpotentials and hence high energy efficiency. However, the current densities on these electrodes were low. For methanol and formic acid production, the partial current densities on the Cu–Ni alloy were only 0.008 and 0.28 mA cm^{-2}, respectively.[158–161]

12.2.4.2   Cu–Sn and Cu–Zn Alloys

Watanabe et al. observed that Cu–Sn and Cu–Zn alloys electroplated on gold flag electrodes had different electrocatalytic activities for CO_2 reduction, depending on the electroplating medium.[162,163] Cu–Sn alloy electrodes, which had been prepared by electroplating from a pyrophosphate bath, were more selective for production of HCOOH. Thus, on a Cu–Sn alloy (43% Sn) electrode in CO_2-saturated 0.05 M $KHCO_3$ at 2°C, at a potential of –1.25 V (vs. SCE), the faradaic yield for HCOOH was 60%, and the onset potential was only –0.5 V. In such an alloy, the microcrystalline structure was shown by X-ray diffraction (XRD) to consist of Cu + Cu_6Sn_5 + Sn. On the other hand, alloy electrodes prepared by electroplating from a cyanide bath were more selective for the production of CO. On such a Cu–Sn alloy (14 atom% Sn), at a potential of -1.25 V, the faradaic yield for CO was 67% and about 20% for HCOOH. The alloys originating from the cyanide bath consisted of $Cu_{5.6}Sn$ + Cu_3Sn. The intermetallic compound $Cu_{5.6}Sn$ was found to be a most effective phase for the selective production of CO, requiring however a relatively large overpotential of about 0.7 V. With Cu–Zn alloy (58 atom% Zn) electrodes, prepared from cyanide-plating baths, highly selective CO formation was attained, with a faradaic yield of 80% at -1.50 V (vs. SHE) and a partial current density of about 1 mA cm^{-2}. The overpotential was high, about 1 V, indicating low reversibility. The most active crystalline phase for CO production was found to be Cu_5Zn_8.[162,163]

12.2.4.3 Cu–Ag Alloys

With Cu–Ag (2:3 atomic ratio) alloy electrodes, prepared by electron beam evaporation of a silver layer (about 20 to 30 nm thick) on a clean copper plate (99.999% pure) and annealing under vacuum at 300 to 400°C for about 20 min, much enhanced production of C_2H_4 was obtained by Nakato et al., while the production of CH_4 was slightly lowered.[124] The explanation given for the improved current efficiency for ethylene on the alloy electrodes is that on the Ag grains on the surface of the alloy, CO molecules will be produced (as on pure silver electrodes). In their vicinity, on the Cu grains, chemisorbed H, CO_2^-, CO, HCO, and CH_2 radicals will be formed (as on pure Cu electrodes). Since these grains are presumably randomly distributed, the probability of interaction between the CO molecules on the Ag grains with the H atoms on the Cu grains becomes enhanced, leading to higher production of CH_2 radicals, which dimerize to ethylene.[124]

Using silver electrodes at -1.55 V (vs. NHE) immersed in 0.1 M NaClO$_4$ containing small concentrations of Cu^{2+} ions (10^{-3} to 10^{-6} M), Kostecki and Augustynski observed that methane and ethylene were produced with faradaic efficiencies of 65 and 23%, repectively.[164]

12.2.4.4 Sn–Cd and Sn–Zn Alloys

Cherashev and Khrushch observed that in the electrochemical reduction of CO_2 on Sn–Cd and Sn–Zn alloy electrodes in 1 M NaHCO$_3$, the efficiency of production of formate was higher than on electrodes of the pure metals. This was explained by the formation of an intermetallide species by the alkali metal at dislocations on the surface of the alloy electrodes. This species was active in the reduction of CO_2.[165]

12.2.5 Mediation by Transition Metal Complexes

On most electrode materials, direct electrochemical reduction of carbon dioxide requires a considerable overvoltage, thus decreasing the energy conversion efficiency. Considerable decreases in overvoltage can be attained by using transition metal complexes as electron transfer mediators.[166]

The homogeneous catalysis of electrochemical reaction has been classified into two groups: (1) redox catalysis, in which electron transfer from the electrode occurs to an outer-sphere site of the catalyst, which then shuttles electrons to the substrate, such as CO_2, and (2) chemical catalysis, in which the substrate forms a transient addition product with the reduced form of the catalyst, followed by regeneration of the oxidized form of the catalyst.[167-170] The mechanistic pathways for the electrocatalysis of CO_2 reduction by transition metal complexes complexes were reviewed by Keene and Sullivan.[171]

12.2.5.1   Cobalt Tetraphenylporphin

Metalloporphyrins in aqueous solutions were found by Takahashi et al. to catalyze the electroreduction of carbon dioxide to formic acid.[172] The complexes investigated included tetraphenyl-porphine sulfonates and *meso*-tetracarboxyphenyl-porphyrins. Cobalt, copper, and iron porphyrins were tested in alkaline media (pH 8 to 10), with mercury pool cathodes. Only the cobalt porphyrins catalyzed the reduction of carbon dioxide.

Tezuka and Iwasaki used a cyclic voltammetry study on the electroreduction of CO_2 mediated by cobalt(II) tetraphenylporphin (CoTPP) in homogeneous solution to derive the rate constant for the electron transfer from the CoTPP dianion to CO_2.[173] Cyclic voltammetry of CoTPP on a glassy carbon electrode (in dimethylformamide [DMF] solution containing 0.1 M $Bu_4N \cdot BF_4$) indicated the appearance of two one-electron redox waves. One, at –0.8 V (vs. SCE), was reversible and independent on the absence or presence of CO_2 and was assigned to the reduction of CoTPP to its monoanion, CoTPP⁻. The second redox wave, at –1.9 V, was irreversible and its peak height increased with the concentration of CO_2 in the medium. This wave was proposed to be due to the reduction of CoTPP⁻ to its dianion, which transferred its electron to CO_2, regenerating the CoTPP⁻ monanion. If CO_2 in solution was in large excess, the CoTPP⁻ acted as an electrocatalyst. From the dependence of the peak height of the second redox wave on the potential sweep rate, it was possible to derive the rate constant for the rate-determining step, which was assumed to be the reaction

$$CoTPP^{2-} + CO_2 \rightarrow CoTPP^- + CO_2^- \qquad (12.31)$$

The rate constant at 25°C was 8.2 M^{-1} s^{-1} and the activation energy was 34.5 kJ mol^{-1}.[173]

Carbon dioxide at pressures of 4 to 22 atm was electroreduced by Cao et al. in aqueous phosphate buffer solutions (0.5 M NaH_2PO_4 + NaOH) containing cobalt tetrakis(4-trimethylammonio-phenyl)-porphyrin as mediator, using In, Sn, Pb, and Pb–Hg (lead amalgam) electrodes.[174] Maximal current efficiency for CO production was in the order Pb–Hg > Pb > In > Sn, while the order of electrode activity for HCOOH production was Pb > Pb–Hg > Sn > In. The optimal potential for CO_2 reduction to both CO and HCOOH was –1.2V vs. SCE. Current efficiencies reached >90% with Pb and Pb–Hg electrodes.

Optically transparent thin-layer electrodes were used by Cao et al. to study the electrocatalytic reduction of CO_2, using tetrakis(4-trimethylanilino)porphinato cobalt²⁺ tetraiodide (CoTMAPI) as catalyst.[175] Applying a spectroelectrochemical technique, the electrode potentials were kept at +0.6, –0.3, and –1.0 V. Under these conditions, the UV absorption spectra of the mono-, di-, and trivalent ions were obtained. Controlled potential electrolysis carried out at an electrode potential of −1.0 V in a CO_2-saturated solution of CoTMAPI in the presence of imidazole indicated a catalytic cycle for CO_2 electroreduction.

Zhang et al. developed a method for the determination of CO_2 in aqueous solutions, using a rotating Pt and a coated graphite disk electrode assembly.[176] The graphite disk was coated with the complex tetramethyl-3,4-pyridoporphyrazino-cobalt(II) protected by a Nafion® film. In contact with aqueous solutions, the electroreduction of CO_2 on the coated disk released CO, which was electrooxidized at the Pt ring, thus measuring the CO_2 concentration. At a potential of −1.15 V (vs. SCE), a concentration of 1.9×10^{-4} could be detected.

12.2.5.2 Tetraazamacrocyclic Ni Complexes

Several other metallo macrocycles in aqueous or organic solvents catalyze the reduction of carbon dioxide.[177–180] Nickel-cyclam dichloride (cyclam = 1,4,8,11-tetraazatetradecane),

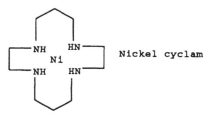

mediates the efficient and selective reduction of CO_2 to CO, according to the overall reaction

$$CO_2 + 2e^- + 2H^+ \rightarrow CO + H_2O \qquad (12.32)$$

At a potential of −1.0 V (vs. NHE), in 0.1 M aqueous KNO_3, the current efficiency for CO production reached 96%. Also, in DMF as solvent, formic acid was produced in 75% current efficiency (see Section 12.3.3.10). The reduction of CO_2 was proposed to involve an Ni(I)carbonyl complex as intermediate. In an initial one-electron reduction step, Ni(cyclam)$^{2+}$ was proposed to be converted to an Ni^{II} species, which underwent coordination by CO_2, followed by protonation on the CO_2 group. In a second one-electron reduction step, CO and OH$^-$ were then released.[181,182] Support for this mechanism was provided by an *ab initio* MO/SD-CI study of model complexes, such as $[Ni^{II}F(NH_3)_4]^+$, serving as simulated intermediates of CO_2 coordination to Ni.[183]

The mechanistic steps for Ni-cyclam-mediated CO_2 reduction on Hg electrodes were suggested by Taniguchi to include (1) adsorption of the reduced form of the mediator on the electrode surface, (2) coordination of CO_2 on the electrode surface, and (3) further electron transfer, from the electrode surface through the catalyst to the CO_2 group.[184]

In the electroreduction of CO_2 to CO on Hg electrodes mediated by Ni(cyclam)$^{2+}$, the catalytically active species was observed by Balazs and Anson to be Ni(cyclam)$^+$ adsorbed on the mercury.[185] A detailed cyclic voltammetry study indicated that Ni(cyclam)$^{2+}$ was only weakly adsorbed at mercury electrodes, and only over a limited potential range, while Ni(cyclam)$^+$ was strongly adsorbed, and over a wide potential range. In its activity as electrocatalyst for CO_2 reduction, an altered configuration of the cyclam ligand may possibly play a role. The decrease in the sustained electroreduction of CO_2 in the Ni(cyclam)$^{2+}$-mediated reduction of CO_2 to CO was shown by Balazs and Anson to be due to the deposition on the electrodes of an insoluble precipitate of a complex of Ni(0), cyclam, and CO (the product of CO_2 reduction).[186]

In a kinetic study by Kelly et al., Ni(cyclam)$^+$ was generated both by pulse radiolysis (with a 12-MeV electron linear accelerator) and by laser flash photolysis (7-ns pulses at 266 nm).[187] The reduction of Ni(cyclam)$^{2+}$ by H· atoms, e$^-$, and CO_2^- radical anions at 22°C occurred with rate constants of 4.1×10^{10}, 5×10^9, and 6.7×10^9 M^{-1} s^{-1}, respectively. Thus, the strongly reducing CO_2^- radical anion interacted with Ni(cyclam)$^{2+}$ at a rate constant close to the diffusion rate limit. The reduction of CO_2 by Ni(cyclam)$^+$ was shown to occur by an inner-sphere mechanism, with addition of CO_2 to an axial coordination site of the nickel complex.

Enhanced electrocatalytic activity for CO_2 to CO in aqueous solutions was obtained by Fujita et al. with some methyl-substituted Ni^{II}(cyclam) complexes, which provided about 30% higher rates of production of CO than the unsubstituted Ni^{II}(cyclam)$^{2+}$ complexes. Large differences in catalytic activities were observed among different geometric isomers of these complexes, which were atributed to differences in adsorption on the mercury electrodes.[188,189]

Monolayers of two amphiphilic nickel(II) complexes with long-alkyl-substituted cyclam derivatives were deposited by Fujihara et al. onto glassy carbon disk electrodes, using the Langmuir–Blodgett technique.[190,191] The complexes were nickel tetrakis(N-hexadecyl)cyclam and nickel N-hexadecyl cyclam. The electrocatalytic activities of these electrodes for the reduction of carbon dioxide were examined by cyclic voltammetry. In aqueous solutions at pH 4.5, the cyclic voltammogram with the monolayer of Ni tetrakis(N-hexadecyl)cyclam or Ni-N-hexadecyl-cyclam-coated electrodes indicated strong rising catalytic currents for the reduction of carbon dioxide at −1.30 V (vs. SCE). This potential was only slightly different from that observed for nickel cyclam in solution, measured on a hanging drop mercury electrode. Since the Ni-N-hexadecylcyclam monolayer was found to be rather unstable, possibly because of appreciable solubility, the Ni(II) complex of a cyclam with a much longer alkyl chain was tested, Ni(II)-N-docosyl-cyclam [docosyl = $CH_3(CH_2)_{20}CH_2^-$]. This cyclam, when deposited on a glassy carbon electrode by the Langmuir–Blodgett technique, provided the electrocatalytic reduction of CO_2 already at an onset potential of only −1.25 V vs. SCE. The first catalyst layer on the electrode was more effective for CO_2 reduction than the subsequent layers. In this first monolayer, the cyclam head group was oriented toward the hydrophilic electrode surface.[192]

Using a nonwetting porous Hg–Au (amalgamated gold mesh) electrode at the polytetrafluoroethylene membrane inlet system of a differential electrochemical mass spectrometer, the electroreduction of CO_2 in an aqueous solution of Ni(II)-cyclam was studied during cyclic voltammetry. The appearance of the cathodic peak at −1.15 V (SCE) coincided with the mass peaks of CO and H_2.[193–195]

Mediation of CO_2 electroreduction by the tetraazamacrocyclic Ni(II) complex $Ni(CR)^{2+}$

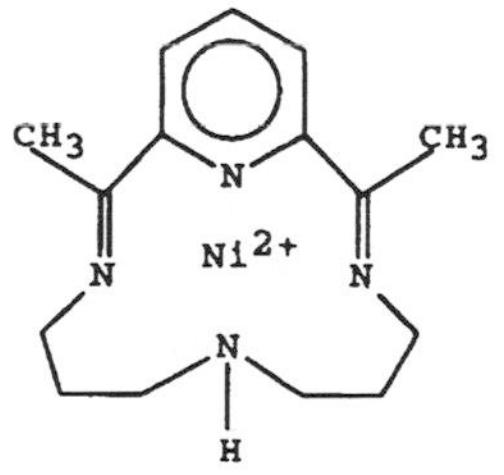

in aqueous solutions over Hg electrodes was found by Bujno et al. to involve isomers of the reduced complex $[Ni(CRH)^{2+}]_{ads}$, according to:

$$[Ni(CRH)^{2+}]_{ads} + CO_2 + e^- + H^+ \rightarrow \{[Ni(CRH)COOH]^{2+}\}_{ads} \qquad (12.33)$$

$$\{[Ni(CRH)COOH]^{2+}\}_{ads} + e^- \rightarrow CO + OH^- \qquad (12.34)$$

In a poisoning reaction, the CO produced reacted irreversibly with the active catalyst, forming an electrode-blocking carbonyl complex, Ni(CRH)CO.[196,197]

12.2.5.3 Co Phthalocyanine

Carbon electrodes (either pyrolytic graphite or carbon cloth) coated with cobalt phthalocyanine were used by Lieber and Lewis in CO_2-saturated aqueous citrate buffer solutions (pH 5).[198] CO and H_2 were produced with about 60 and 30% current efficiency, at −1.0 V (vs. SSCE), with a current density of about 1 mA cm^{-2}. The turnover number (moles of CO produced per mole of catalyst) exceeded 10^5. However, in a related study, with CO_2-saturated acid solutions (pH 3 to 7), on glassy carbon electrodes coated with cobalt or nickel phthalocyanines, Kapusta and Hackerman found that the main product was formic acid, obtained with an overpotential which was smaller by 200 mV than with bare metal electrodes. Current densities reached up to 10 mA cm^{-2}. At lower pH values, methanol was also produced, with a current efficiency of up to 5%.[199] On graphite electrodes impregnated with Co or Ni phthalocyanines, with aqueous tetraalkyl ammonium salts as electrolytes, the products identified by Meshitsuka et al. were oxalic and glycolic acids. Formic acid was not produced. The current potential curve showed a cathodic peak for CO_2 reduction, the height of which was proportional to the square root of the sweep rate. This indicated that the diffusion of carbon dioxide was the rate-limiting step.[200]

The electrocatalytic activity of the metal phthalocyanines was explained by Rollmann and Iwamoto with the observation that these complexes are reduced to their dinegative states at the potentials required for carbon dioxide reduction.[201] It was suggested that the Co and Ni phthalocyanines in their dinegative states have an excess of ligand p-electrons.[202] This may play a role in their electrocatalysis of CO_2 reduction. An interpretation of the differences in CO_2 reduction products obtained with the different metal phthalocyanines was proposed by Furuya and Koide in terms of the theory of Taube, based on the LCAO-MO Hückel method.[203,204]

Tanabe and Ohno developed the plasma-assisted deposition of metal phthalocyanine thin films on glassy carbon electrodes as a promising method for the electrocatalysis of CO_2 reduction.[205] The technique involved vacuum deposition using an intermittent plasma in a gas mixture of argon and hydrogen. The films were uniform and about 400 Å thick. The electrocatalytic activity of the electrodes was tested in aqueous 0.5 M Na_2SO_4 + $NaHCO_3$ (pH 6.65) by measuring current potential curves with rotating disks of these metal-phthalocyanine-coated electrodes. The relative activity depended on the metal of the complexes, in the order Co >> Ni > Fe ~ Mg ~ Mn ~ Zn phthalocyanines.

12.2.5.4 Co Phthalocyanine/Polyvinylpyridine

The rate and selectivity of reduction of CO_2 in aqueous solutions to CO were greatly enhanced by Yoshida et al. using graphite electrodes coated with a film of cobalt phthalocyanine (CoPc) imbedded in poly-4-vinylpyridine (PVP).[206] The advantages of CoPc as mediator are its capacity to lower the required overpotential for CO_2 reduction, as well as the relatively high stability of the phthalocyanine. In the CoPc/PVP film, the CoPc is presumably coordinated axially to the electron-donating nitrogen atoms of the PVP. PVP has a high affinity for CO_2 by acid–base interaction and is effective as a CO_2 gas separation membrane.[207] The CoPc/PVP electrode was prepared by casting a solution of PVP (1%) in DMF containing CoPc (10 μM) on a basal-plane pyrolytic graphite (BPG) electrode. The blue film formed had an absorption maximum at λ = 674 nm, typical of monomeric CoPc. Electroreduction of CO_2-saturated aqueous solutions with the CoPc/PVP-coated electrode caused production of CO and H_2. At an applied potential of -1.20 V (vs. SCE), in solutions of pH 2.3, 4.4, and 6.8, the faradaic yields for CO formation were 34, 72, and 60%, respectively, and the turnover numbers (based on the coverage of the catalytic complex) were 7.2 × 10^5, 2.9 × 10^5, and 4.8 × 10^4, respectively. With CoPc alone (without the PVP) on the graphite electrode, the selectivity for CO production was substantially lower. With the CoPc/PVP-coated membrane, the degradation of the electrode was minimized, enabling sustained electroreduction of CO_2 for at least 10 h. Also, with the CoPc/PVP-coated electrode, less negative potentials were required than with the electrode coated only with pure CoPc. The mechanism of reduction of aqueous CO_2 on CoPc was proposed to involve two fast steps each of electron transfer and proton transfer to the CoPc, followed by coordination of CO_2 with its carbon atom to the highly nucleophilic Co^I atom of the 2e$^-$ reduced CoPc, with liberation of CO as the rate-determining step.[206] Abe et al. used graphite electrodes coated with membranes of poly(4-vinylpyridine) containing cobalt phthalocyanine at a surface concentration of 1.2×10^{-10} mol cm^{-2}.[208] In aqueous 0.1 M NaH_2PO_4 saturated with CO_2 (pH 4.4) at a potential of -1.2 V (vs. Ag/AgCl), the only products were CO and H_2, formed in the ratio 4:1. The turnover number of the catalyst per molecule of the electroactive CoPc for the formation of CO was 5×10^5 h^{-1}.

Even higher activity for the electroreduction of CO_2 in aqueous media was obtained by Abe et al. with a substituted cobalt phthalocyanine.[209] Cobalt octabutoxyphthalocyanine $[CoPc(BuO)_8]$ was coated on a BPG electrode, with the total amount of coating about 1.2×10^{-10} mol cm^{-2}. At -1.30 V (vs. Ag/AgCl) in a CO_2-saturated solution at pH 4.4, the selectivity of the production of CO and H_2 was in the ratio 4.2:1, and the turnover number of the catalyst was 1.1×10^6 h^{-1}, which was 20 times higher than the above-reported value with the unsubstituted cobalt phthalocyanine. The increased activity of the substituted complex was explained by the electron-donating property of the butoxy function, which facilitated both coordination with CO_2 and electron transfer to CO_2.[208,209]

An improved electroreduction of CO_2 to CO with metalloporphyrins as catalysts was achieved by Atoguchi et al. by immobilizing water-insoluble cobalt(II)-tetraphenyl-porphyrin (Co^{II}tpp) on glassy carbon electrodes which had been treated with 4-aminopyridine.[210,211] On these electrodes, at -1.2 V (vs. SCE), in CO_2-saturated aqueous solutions (phosphate buffer, pH 6.86), carbon dioxide reduction yielded as the only products CO and H_2, with current efficiencies of about 50% each. This electroreduction of CO_2 to CO occurred at potentials which were 100 mV more positive than those observed with the water-soluble Co^{II} porphyrins. The turnover number of the immobilized catalyst for CO formation exceeded 10^7.

12.2.5.5 Bipyridine and Terpyridine Complexes

Bipyridine halotricarbonyl complexes of rhenium(I), [fac-Re(I)(bpy)(CO)$_3$Cl] (where bpy = 2.2,-bipyridine), had been shown to be excellent catalysts for the electroreduction of CO_2 in nonaqueous solvents (see Section 12.3.3). The incorporation of such complexes into water-insoluble polymer coatings on metal electrodes enabled their use in aqueous media. Such coatings with poly(pyrrole) films containing the Re(bpy)(CO)$_3$Cl system on platinum electrodes were found by Cosnier et al. to be efficient electrocatalysts for the reduction of CO_2 to CO and CO_3^{2-} as the only products in high current yields.[212–214] Incorporation of electron-withdrawing carboxy ester groups in the bpy ligand of these Re(bpy)(CO)$_3$Cl complexes, both in solution and in polymeric form, strongly stabilized the initial oxidation state Re(I) of the metal center and decreased the electrocatalytic activity toward CO_2 reduction.

In an alternative approach, Re(Bpy)(CO)$_3$Br was incorporated by Yoshida et al. into a Nafion® membrane coated on a BPG electrode.[215] The reduction of CO_2 could thus be performed in an aqueous electrolyte. The Nafion® provided a hydrophobic environment around the complex, thus decreasing the reduction of protons. In a phosphate buffer (pH ~7) saturated with CO_2, at a potential of -1.3 V (vs. SCE), the faradaic efficiencies for the production of HCOOH, CO, and H_2 were 48, 16, and 39%, respectively. The polymer-confined complex was more stable than the homogeneous-phase system, which was deactivated by dimerization of the rhenium complex. Similar coated Nafion® membranes with incorporated [Re(terpy)(CO)$_3$Br] (terpy = 2,2':6',2"-terpyridine) were less active as electrocatalysts for CO_2 reduction.

Metal complexes incorporated into hydrophobic polymer films coated on electrodes were used by Kaneko et al. to achieve electrocatalytic reduction of carbon dioxide.[216] With glassy carbon electrodes coated with a Nafion® membrane, and on these complexes of Ru, Co, or Re, such as Co(terpy)$_2^{2+}$, at –1.55 V (SCE), in aqueous phosphate buffer or KHCO$_3$, formic acid was produced in about 10% faradaic efficiency.

Films of poly[tricarbonyl(vinylbipyridyl)rhenium chloride] were prepared by Christensen et al. by anionic electropolymerization of [(vpy)Re(CO)$_3$Cl] (in which vbpy = 4-vinyl-4'-methyl-2,2'-bipyridyl).[217,218] On such films, grown on platinum or glassy carbon electrodes, the mechanism of electroreduction of CO_2 was studied by ellipsometry and *in situ* FTIR spectroscopy. Intermediate reduced species thus identified on the films were poly[(vbpy)Re(CO)$_3$(COOH)], poly[(vbpy$^{\cdot-}$)Re(CO)$_3$(COOH)], and poly[(vbpy$^{\cdot-}$)Re(CO)$_3$(COO$^-$)]. The advantage of the polymer-bound rhenium complex is its higher stability by comparison with the homogeneous Re(bpy)(CO)$_3$Cl electrocatalyst. On long-term recycling, these films are deactivated mainly by formation of spatially separated regions which are permanently either reduced or oxidized.

In a comprehensive infrared spectroelectrochemical study by Hartl et al. on the detailed pathways of one- and two-electron reduction of [Re(bpy)(CO)$_3$L] and related complexes, the catalytically active species for CO_2 reduction were identified as five-coordinate complexes, the radical [Re(CO)$_3$(BPY)]$^\cdot$ and the anion [Re(CO)$_3$(bpy)]$^-$.[219,220]

In a further study, Sende et al. compared the electrocatalysis of CO_2 reduction mediated by Cr, Ni, Co, Fe, Ru, and Os complexes of electropolymerized 4-vinyl- and 6-vinylterpyridine (4-v-trpy and 6-v-trpy) on glassy carbon electrodes in an aqueous medium, in 0.1 M NaClO$_4$.[221] In this medium, the only identified CO_2 reduction product was formaldehyde, in contrast to the results in a DMF medium, in which the only observed product was formate (see Section 12.3.3).[3] The potentials for the electrocatalytic reduction of CO_2 in the aqueous medium on electropolymerized films of 4-v-trpy complexes of Cr, Co, Fe, Ni, Ru, and Os were –0.86, –0.87, –1.10, –1.12, –1.20, and –1.22 V and for the 6-v-trpy complex of Ru –1.33 V (vs. Ag/AgCl), respectively. These potentials were the peak values for the first redox waves for which the current increased in the presence of CO_2 relative to N_2. Both the Cr(4-v-trpy) and Co(4-v-trpy) complexes mediated CO_2 reduction with very low overpotentials in aqueous media. The current efficiencies for the production of formaldehyde at a potential of –1.10 V in CO_2-saturated aqueous 0.1 M NaClO$_4$ on the Cr(4-v-trpy)- and Co(v-4-trpy)-coated electrodes were 87 and 39%, and the turnover numbers (estimated from the surface coverage of the films) were 6100 and 11,000, respectively. The reduction of CO_2 with these modified electrodes depended strongly on the nature of the supporting electrolyte anions. Thus, when 0.1 M NaH$_2$PO$_4$ was used instead of NaClO$_4$, the reduction of CO_2 was inhibited. The phosphate anion may have been strongly coordinated to the catalytic site, thus blocking the site required for the coordination of CO_2.[221]

Collomb-Dunand-Sauthier et al. observed that the complex $[Ru^{II}(bpy)(CO)_2Cl_2]$ in acetonitrile solution by controlled potential electrolysis at -1.65 V (vs. Ag/Ag^+) on a Pt or carbon electrode underwent electrochemical reduction, forming a deep blue water-insoluble and strongly adhering polymeric film:[222–225]

$$n[Ru^{II}(bpy)(CO)_2Cl_2] + 2ne^- \rightarrow \{[Ru^0(bpy)(CO)_2]_n\} + 2nCl^- \quad (12.35)$$

The composition $\{[Ru^0(bpy)(CO)_2]_n\}$ of the film was deduced by elementary analysis. The structure of this polymer was proposed by Noblat-Chardon et al. to contain a metal–metal backbone:[226,227]

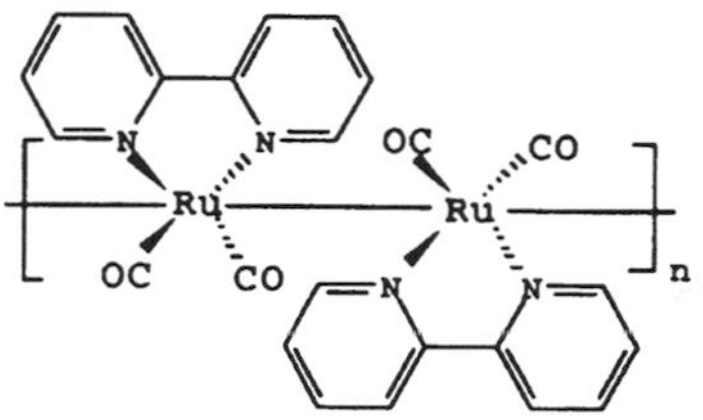

Using an electrode of this polymeric complex deposited on carbon felt in aqueous 0.1 M $LiClO_4$, electroreduction of CO_2 at -1.20 V (vs. SCE) produced CO with >97% current efficiency. The electroactivity of the modified electrodes was presumably due to reduction of the bipyridine moiety to the $bpy^{\cdot-}$ and bpy^{-2} states. Improved stability of these polymeric electrodes was reported by incorporating them into preformed polypyrrolic films. With such electrodes, the electroreduction of CO_2 to CO at -1.35 V (vs. SCE) occurred with 90% chemical yield and 80% current efficiency.[226,227]

Yoshida et al. used BPG electrodes coated with a Nafion® membrane incorporating $Co(terpy)_2^{2+}$ in CO_2-saturated aqueous phosphate buffer (pH 7). At a potential of -1.10 V (vs. SCE), the faradaic efficiencies for production of HCOOH and H_2 were 51 and 13%, respectively.[228]

12.2.5.6 Ti(III)- and Mo(III)-Catechol

The electroreduction of CO_2 to C_1–C_3 hydrocarbons was observed by Petrova and Efimov in the presence of a catalytic system which involved catechol complexes of Ti(III) and Mo(III).[229] The electrolysis was performed in a two-compartment cell, with a ground-glass joint as separator, with a mercury pool cathode at -1.55V vs. SCE, and a catholyte prepared from an aqueous solution of 0.5 M pyrocatechol, 0.05 M $TiCl_3$, and 5 mM Na_2MO_4. Products observed were methane, ethane, ethylene, and C_6 hydrocarbons (in the ratio 4.2:0.45:0.16:0.12), with a total current efficiency of up to 0.2%, as well as copious evolution of hydrogen. The reduction of CO_2 was postulated to occur within the coordination sphere of Mo(III), possibly through carbene complexes such as $Mo=CH_2$. The Ti(III)-catechol complex presumably acted as electron mediator.

12.2.6   Gas Diffusion Electrodes

The challenge of highly efficient electroreduction of carbon dioxide requires the achievement of both high current densities and low overpotentials. The rate of CO_2 electroreduction in liquid-phase electrolysis is often limited by mass transfer. One approach to overcome the low current density in the electrochemical reduction of carbon dioxide, caused by the limited solubility of CO_2 in water, is the application of gas diffusion electrodes. Gas diffusion electrodes enable a considerable enhancement of mass transfer to the triple gas/electrolyte/solid-electrode boundary.[230] Such electrodes had been developed previously for the reduction of oxygen in fuel cells, providing gas–solid charge transfer, usually with the help of metallic catalysts. By comparison with other electrode systems, gas diffusion electrodes seem to offer the most practical approach to the efficient reduction of carbon dioxide, because of the high current densities achieved. An essential further target with the gas diffusion electrodes is to lower the overpotential required and preferably to achieve direct CO_2 reduction to more valuable products, such as alcohols.

12.2.6.1   Reduction to CO, Formic Acid, and Hydrocarbons

Very high current densities of CO_2 reduction to formic acid were obtained by Mahmood et al. using metal-impregnated gas diffusion electrodes. On such electrodes, with lead-impregnated polytetrafluoroethylene- (PTFE) bonded carbon operated at -1.8 V vs. SCE in aqueous solutions (pH 2), HCOOH was produced with nearly 100% current efficiency, at the remarkably high current density of 115 mA/cm^2. The electrocatalytic activity for carbon dioxide reduction to formic acid was in the order Pb > In > Sn.[231]

Since electrodes of pure copper had been shown to favor the production of hydrocarbons in the electroreduction of carbon dioxide in aqueous solutions, an effort was made to optimize the performance of gas diffusion electrodes loaded with copper. Electroreduction of gaseous carbon dioxide on Cu-plated solid polymer electrolyte electrodes was investigated by DeWulf and Bard[232] and Cook et al.[233] Carbon dioxide reduction on a finely divided copper electrode was achieved, using a proton-conducting solid polymer structure. The Cu/Nafion® electrodes were fabricated by an electroless plating method. The gas-phase electrochemical reduction of CO_2 to CH_4 and C_2H_4 was optimal in 1 mM H_2SO_4. At a potential of -2.0 V vs. SCE, the steady-state faradaic efficiency reached about 20%.

Furuya et al. used carbon black gas diffusion electrodes loaded with Pb, Sn, Zn, Cu, or Ag atoms.[234] The CO_2 reduction product was mainly CO. On Cu-loaded gas diffusion electrodes, the reduction of CO_2 to CO at -1.4 to -1.8 V (vs. RHE) occurred with a current efficiency of about 30%. Very high current densities, up to 400 mA cm^{-2}, were obtained by Cook et al. with gas diffusion electrodes containing copper catalysts, leading mainly to the production of methane and ethene.[99]

Cu-loaded gas diffusion electrodes were fabricated by Ito et al. by bonding together a semihydrophilic reaction layer (containing both hydrophilic and hydrophobic carbon black, PTFE, and Cu powder of 5 N purity), and a hydrophobic gas supply layer (hydrophobic carbon black, PTFE, and 20-mesh Cu gauze as current collector).[235-237] With such electrodes in 0.5 M K_2SO_4 supplied with CO_2, the products of reduction were CH_4, C_2H_4, C_2H_5OH, CO, $HCOO^-$, and H_2, formed optimally at a potential of −1.45 V (vs. Ag/AgCl), with a total faradaic efficiency of 70%, and a current density of about 10^3 mA cm^{-2}. The current densities were about 100 times higher than those on pure copper plate electrodes (99.999% Cu). The mechanism of the electroreduction of CO_2 on the Cu electodes was proposed to involve the adsorbed species $CO_{(ads)}$ and $H_{(ads)}$ followed by $CH_{2(ads)}$.

Selective production of ethanol by electroreduction of CO_2 was achieved by Ikeda et al. using gas diffusion electrodes loaded with CuO/ZnO (3:7 molar ratio). In 0.5 M aqueous KH_2HPO_4, the faradaic efficiency of ethanol formation reached a maximal value of 16.7% at a potential of −1.32 V (vs. Ag/AgCl) and a partial current density of 4.2 mA cm^{-2}. CO and HCOOH were minor products.[238]

The electroreduction of carbon dioxide at atmospheric pressure on gas diffusion electrodes containing metal phthalocyanines was pioneered by Mahmood et al.[239] On gas diffusion electrodes impregnated with cobalt(II) phthalocyanine, high rates of reduction of CO_2 to CO were observed, reaching a current density of 137 mA cm^{-2} at a potential of −2.2 V (vs. SCE). On the other hand, on electrodes impregnated with manganese, copper, or zinc phthalocyanine, formic acid was produced at low rates, and no carbon monoxide was formed. The unique effectiveness of cobalt(II) phthalocyanine was explained by a mechanism involving as a primary step the electrochemical reduction of cobalt(II) to cobalt(I). The selective reduction of CO_2 to CO was confirmed by Savinova et al. on gas diffusion electrodes promoted by cobalt phthalocyanine.[230] At a potential of −4.4 V vs. Ag/AgCl, current densities of up to 80 mA cm^{-2} were attained for CO_2 reduction to CO, with faradaic yields of up to 97%, using cobalt phthalocyanine (9%) supported on a gas diffusion electrode, which was prepared from carbon black and Teflon (50%) and supported on carbon paper. Co phthalocyanine was deposited onto the electrode from its solution in DMF.

With gas diffusion electrodes modified by metal phthalocyanines, operated at current densities of up to 100 mA/cm^2, the product distribution was found by Furuya and Matsui to depend on the central atom of the phthalocyanine: For Cu, Ga, and Ti, the main product was CH_4, with current yields of up to 40%, while for Co, Ni, Fe, and Pd, the preferred product was CO, and for Sn, Pb, In, and Al, the major product was HCOOH. Such gas diffusion electrodes were prepared from carbon black, Teflon, and metal phthalocyanines.[240]

Remarkably high production rates of alcohols were achieved by Cook et al. using gas diffusion electrodes based on Teflon-bonded $La_{0.9}Sr_{0.1}CuO_3$, in 0.5 M KOH, with a separator and a Pt counterelectrode.[103] At a current density of 180 mA cm^{-2}, the faradaic yields of ethanol and n-propanol were 30.7 and 10% after 1.25 h of electrolysis. If the copper component in the above perovskite material was replaced by Ni, Co, Fe, Mn, or Pd, the resulting electrodes were inactive for carbon dioxide reduction to alcohols.

Schwartz et al. achieved preferential reduction of CO_2 to ethanol and n-propanol, with current densities of up to 180 mA cm^{-2}, using gas diffusion electrodes catalyzed by the perovskite $La_{1.8}Sr_{0.2}CuO_4$ in 0.5 M KOH as electrolyte at 25°C.[241,242] At a potential maintained at -2.3 to -2.6 V (vs. SCE), and without separation between the gas diffusion electrode and a platinized platinum counterelectrode, the faradaic efficiencies for ethanol and n-propanol were 16.4 and 6.6%, respectively. Methanol was also produced during the first hour, but disappeared after longer electrolysis. In the absence of Cu in the electrocatalyst, no CO_2 reduction was observed.

Very much enhanced current densities were obtained by Hara et al. with gas diffusion electrodes at elevated pressures of carbon dioxide.[243–245] Using as cathode an electrode assembly consisting of a platinum electrocatalyst layer (0.56 mg cm^{-2} Pt, apparent surface area 1 cm^2) on a stainless-steel mesh current collector, 0.5 M $KHCO_3$ as electrolyte, and a Pt wire as anode, pressurized carbon dioxide was applied from the Pt catalyst side. At a CO_2 pressure of 30 atm, the total current density reached 900 mA cm^{-2}. At a bias potential of -1.93 V (vs. Ag/AgCl), the main products identified were CH_4, H_2, $HCOO^-$, C_2H_5OH, CO, and C_2H_4, at faradaic efficiencies of 33, 29, 10, 2, 3, and 0.7%, respectively.[245] In the absence of the Pt electrocatalyst, hydrogen was the main product even at high CO_2 gas pressures.[243] With Ag-catalyzed gas diffusion electrodes under 30 atm of CO_2, with the Ag catalyst layer directed toward the electrolyte phase, an extremely large partial current density of 3.05 A cm^{-2} for CO production was attained.[244] With these gas diffusion electrodes, at atmospheric pressure of carbon dioxide, the predominant reduction product was hydrogen. Increasing CO_2 gas pressure led to enhanced formation of methane, at the expense of decreased hydrogen production.

In order to identify the intermediates in the electrochemical reduction of carbon dioxide, Masheder and Williams applied the technique of Raman spectroelectro-chemistry. *In situ* studies of lead-impregnated PTFE-bonded carbon gas diffusion electrodes revealed that during polarization of the electrode surface, weak Raman bands at 2824 and 2915 cm^{-1} appeared, assigned to free $HCOO^-$ and to HCOOH perturbed at the electrode surface.[246]

Current densities of up to 500 mA cm^{-2} were attained by Takahashi et al. with gas diffusion electrodes containing Cu-impregnated metal oxides, such as ZnO, ZrO_2, TiO_2, Al_2O_3, and Nb_2O_3, which were in a hydrophilic reaction layer in contact with the electrolyte, 0.5 M aqueous KOH.[247] Carbon dioxide entered through a hydrophobic gas diffusion layer. The reaction layer and the gas diffusion layer contained 10 and 20 wt% PTFE, respectively, both mixed with carbon black. Highest current efficiencies for the production of hydrocarbons were obtained with Cu/ZrO_2 (5:95 by weight) as electrocatalyst, yielding 3.3, 2.9, 8.8, and 33% of CH_4, C_2H_4, CO, and HCOOH, at an electrode potential of -1.8 V (vs. SCE). A high selectivity for ethylene production (20%) was achieved on the Cu/ZrO_2 electrocatalyst at a potential of -2.2 V, resulting in 70 mA cm^{-2} partial current density.

Komatsu et al. examined in detail the fabrication of composite electrodes, consisting of solid polymer electrolytes, such as the cation exchanger Nafion® (perfluoroalkanesulfonate, Dupont de Nemours, Inc.) and the anion exchanger

Selemion® (styrene-divinylbenzene copolymer, Asahi Glass Co.).[248] Both membranes were impregnated with Cu by contact with $CuSO_4$ solutions and reduced with either sodium borohydride, hydrazine, or formalin. Best results, in terms of both the adhesion of the copper layer and low surface electric resistance, were obtained by treatment of the solid polymer electrolytes with a combination of $CuSO_4$ $5H_2O$, Rochelle salt (potassium sodium tartrate tetrahydrate), and 10% $NaBH_4$ in aqueous solution. With the Cu–Nafion® working electrode in contact with atmospheric pressure CO_2 on one side and 0.5 M aqueous K_2SO_4 on the other side and with another Nafion® membrane (not Cu treated) separating the cathodic compartment from the anodic compartment fitted with a Pt anode, at a bias of –1.5 V (vs. SCE), the main CO_2 reduction products were C_2H_4, HCOOH, and CO, produced at current efficiencies of 8.8, 5.9, and 2.6%, respectively. With the anion exchanger Selemion®, the product distributions were quite different. HCOOH and CO were formed with current efficiencies of 15 and 10%, respectively. H_2 was produced with both electrodes at current efficiencies of 75 to 90%. The enhanced production of hydrocarbon with the Cu–Nafion® electrode was explained by faster proton transport through this cation exchanger, facilitating the reaction

$$CO_2 + 6H^+ + 6e^- \rightarrow 1/2C_2H_4 + 2H_2O \tag{12.36}$$

while the more sluggish proton transfer through the anion exchanger Selemion® resulted in enhanced production of formic acid and carbon monoxide,

$$CO_2 + 2H^+ + 6e^- \rightarrow HCOOH \tag{12.37}$$

Both the mean current density and the partial current efficiencies with both electrodes only slightly decreased during electrolysis periods of 5 h.[248]

In an investigation by Komatsu et al. on the application of the gas-phase electroreduction of CO_2 to simulated power plant exhaust gas, the effect of CO_2 concentration (in various mixtures with N_2, at total atmospheric pressure) was tested.[248] The exhaust gas from coal-burning electric power plants, after treatment with de-NO_x and de-SO_x equipment, may typically contain 15% CO_2, 90 ppm NO_x, and 30 ppm SO_2. The current efficiencies for the reduction of CO_2 increased rapidly with the CO_2 concentration in the range of 10 to 30% and leveled off at higher CO_2 concentrations. With the Cu–Nafion® electrode, at a bias of –1.7 V (SCE), with 60% CO_2, the mean current density was 22 mA cm^{-2}, and the current efficiencies for C_2H_4 and HCOOH production were 15 and 1.2%, respectively. The presence of 200 ppm NO had no effect on either the current density or the organic products yields. In the presence of SO_2 (170 ppm), the mean current density decreased slightly and the current efficiencies for C_2H_4 and HCOOH changed to 9.3 and 6.1%. The presence of SO_2 also caused marked corrosion of the Cu–Nafion® electrode. For the potential use of gas-phase electroreduction with such solid polymer electrodes, it may be necessary to increase the CO_2 concentration in the power plant effluent to at least 30% and to eliminate the SO_x components.[248]

By using as gas diffusion electrodes high-area Ni electrocatalysts supported on activated carbon fibers containing nanometer-wide pores, Yamamoto et al. achieved current efficiencies of CO_2 reduction at atmospheric pressure to CO in an aqueous alkaline electrolyte of up to 70% at a potential of -1.6 V (vs. SCE). The nanoporous support was proposed to mimic high-pressure conditions.[249]

12.2.6.2 Urea Synthesis on Gas Diffusion Electrode

Industrially, urea is currently produced from carbon dioxide and ammonia under relatively drastic conditions, temperatures of 150 to 210°C and pressures of 120 to 400 atm. Using a Cu-loaded gas diffusion electrode in a medium of aqueous $KHCO_3$ containing either KNO_3 or KNO_2, Shibata et al. succeeded in simultaneously reducing the nitrogen oxyanions to ammonia while reducing CO_2 to a mixture of CO, HCOOH, and H_2NCONH_2.[250] Carbon dioxide was supplied to the cathode gas chamber at a flow rate of 14 mL min^{-1}. The current efficiencies for production of ammonia and urea were much higher with nitrite than with nitrate as the nitrogen source. At an electrode potential of -0.75 V (vs. SHE), the current efficiencies for urea formation from 0.02 M KNO_3 and 0.02 M KNO_2 were 10 and 37%, respectively. Such a method may be useful for converting the NO_x components of fossil-fuel power plant effluents, together with CO_2, to the widely used urea, such as fertilizer and chemical intermediate.

12.2.7 Electroreduction at High CO_2 Pressure

For economical application of electrochemical CO_2 reduction, high current densities are essential. Carbon dioxide reduction in aqueous solutions is enhanced by elevated CO_2 gas pressure, which provides increased solubility. High CO_2 gas pressures over aqueous electrolytes result in elevated CO_2 concentrations in solution. The effect of carbon dioxide gas pressure on the current density was investigated in early studies. Fischer and Prziza used zinc or lead amalgam electrodes, in saturated ammonium or potassium sulfate. With CO_2 pressurized up to 50 atm in a high-pressure electrolysis cell, the conversion of CO_2 to formic acid was practically quantitative, and traces of methanol were also observed.[251]

Ito et al. tested lead, indium, zinc, and tin as electrodes for the high gas pressure reduction of CO_2.[252] In experiments with lithium carbonate as electrolyte, the current density increased with the gas pressure. However, the faradaic efficiency rose rapidly with the pressure only up to 5 kg cm^{-2} and then reached a plateau, while the current densities reached 30 mA cm^{-2}. Tin cathodes required the lowest overpotential, providing maximum current efficiency at a potential of -1.7 V (vs. Ag/AgCl). With the metal electrodes Fe, Co, Ni, Rh, Pt, Mo, and Re, on which there was hardly any reduction of CO_2 at 1 atm pressure, significant reduction was observed by Nakagawa et al. at 50 atm of CO_2.[253,254] Thus, on a nickel electrode in 0.1 M $KHCO_3$ solution, at an electrode potential of -1.8 V (vs. Ag/AgCl), at 1 and 50 atm of CO_2, the faradaic efficiencies were 0 and 28.6% for CO, 0.1 and 12.0% for HCOOH, 0.6 and 1.2% for CH_4, 0.6 and 0.03% for C_2H_4, 0 and 0.7% for C_2H_6, and 0.06 and 0.42% for C_3H_8.

The effects of both temperature and gas presssure on CO_2 reduction with Cu electrodes in aqueous $KHCO_3$ were reported by Mizuno et al.[255] At a potential of -1.6 V (vs. Ag/AgCl), a temperature of 20°C, and pressures of 5 to 15 kg cm^{-2}, the main reduction products were HCOOH (40% current efficiency) and H_2. Raising the temperaturé to 100°C caused a decline in the current efficiency of HCOOH production to only a few percent. Increases in temperature enhanced water electrolysis at the expense of CO_2 reduction.

Kudo et al. investigated the electrochemical reduction of high-pressure CO_2 on nickel electrodes.[256] At a potential of -1.8 V (vs. Ag/AgCl), in 0.1 M $KHCO_3$, the faradaic efficiencies at pressures of 1 and 60 atm CO_2 were 0 and 10.4% for production of CO, 0.1 and 23% for HCOOH, 0.6 and 1.8% for CH_4, 0 and 0.9% for C_2H_6, 0.6 and 0.4% for C_2H_4, and 0.06 and 0.4 for C_3H_8, respectively. Thus, increasing CO_2 pressure led to markedly enhanced formation of organic products. With Ni as well as Fe and Co electrodes, the weight distribution of the hydrocarbons formed agreed with a Schultz–Flory distribution, hinting at a mechanism similar to that of the Fischer–Tropsch synthesis of hydrocarbons from synthesis gas by thermal catalysis, with surface-adsorbed carbene groups (CH_2=) as intermediates.

Hara et al. compared the electrochemical reduction of CO_2 at a pressure of 30 atm and current densities of up to 700 mA cm^{-2} on various electrodes in 0.1 M $KHCO_3$.[257–260] The experiments were performed galvanostatically in an electrochemical autoclave containing a glass cell, with separation of the cathodic and anodic compartments by a Nafion® sheet. At 30 atm CO_2 and 163 mA cm^{-2} current density, on metal electrodes of the group Ti, Nb, Ta, Mo, and Mn, the major reduction product was hydrogen, with about 3 to 8% current efficiency for formic acid. Since hydrogen was also the main product at 1 atm CO_2, the CO_2 pressure did not change the product selectivity much. On electrodes of Rh, Ni, and Pt, which release hydrogen preferentially at 1 atm pressure, the faradaic efficiencies for formic acid at 30 atm CO_2 were of the order of 20 to 50%. With Ag, Au, Zn, and Pb electrodes, which have high overpotentials for hydrogen release at atmospheric pressure, the faradaic efficiencies for CO_2 reduction become substantial at 30 atm CO_2. At such pressure, on Ag and Au electrodes, CO was the main product, with faradaic efficiencies of 76 and 65%, respectively. On Cu electrodes, formic acid, CO, methane, and ethene were the main products, with efficiencies of 54, 20, 10, and 4%, respectively. On electrodes of Zn, Cu, Sn, Pb, In, and Bi, formic acid was produced with efficiencies of 41, 54, 92, 96, 90, and 83%, respectively. Voltammetric measurements with Ag and Zn electrodes revealed a marked shift of the cathodic current in the positive direction with rising CO_2 pressure, indicating with these electrodes the energy-saving reduction of CO_2 at less negative potentials under high CO_2 pressure. Hara et al. found that stirring significantly increased the current density of CO_2 reduction on Cu electrodes at high CO_2 pressures.[257] With stirring, at 30 atm CO_2 and a potential of -1.88 V (vs. Ag/AgCl), at a current density of 900 mA cm^{-2}, the faradaic efficiencies for production of CH_4, C_2H_6, C_2H_4, C_2H_5OH, CO, HCOOH, and H_2 were 39, 0.03, 2.6, 0.5, 1.7, 8.1, and 31.5%, respectively. On glassy carbon electrodes in 0.1 M $KHCO_3$ under 30 atm of CO_2, the product selectivity of CO_2 reduction was found by Hara et al. to depend considerably on the current density.[260]

A maximal total faradaic efficiency of 75% for CO_2 reduction was observed at a current density of 50 mA cm^{-2} and a potential of -1.68 V (vs. Ag/AgCl), with faradaic efficiencies for the formation of CO, HCOOH, H_2, CH_4, C_2H_4, and C_2H_6 of 44, 30, 16, 0.4, 0.04, and 0.03%, respectively. At lower current densities, the selectivities for hydrocarbons were higher, at the expense of the production of CO and HCOOH. The distribution of hydrocarbons followed the Schultz–Flory equation, indicating that the production of the hydrocarbons on the glassy carbon electrode occurred by a mechanism similar to that of the Fischer–Tropsch reaction.

On n-TiO_2 single-crystal electrodes in initially 0.5 M NaOH, pressurized with CO_2 to 8 atm, at a potential of -1.2 V (vs. Ag/AgCl), the electroreduction of CO_2 was found by Halmann and Aurian-Blajeni to yield mainly HCOOH and traces of HCHO, with a total faradaic yield of 22%. Illumination of the electrode with a 150-W xenon lamp had no effect on the faradaic yield, which is understandable since the reduction on n-type TiO_2 involved only majority carriers.[261]

See also Section 12.2.6.1 on high-pressure CO_2 reduction with gas diffusion electrodes.

12.3   REDUCTION IN ORGANIC SOLVENTS

While electron capture by carbon dioxide is generally considered to be the primary step of electroreduction in both aqueous and organic media, subsequent reactions account for the preferred production of formic acid in water and oxalic acid and carbon monoxide in aprotic solvents:[262–264]

$$CO_2 + e^- \rightarrow \cdot CO_2^- \qquad (E^\circ = -2.21 \text{ V vs. SCE}) \qquad (12.38)$$

$$2 \cdot CO_2^- \rightarrow C_2O_4^{2-} \qquad k = 3.2 \times 10^7 \ M^{-1} \ s^{-1} \qquad (12.39)$$

$$2 \cdot CO_2^- \rightarrow O{=}C{-}O(C{=}O){-}O^- \qquad (12.40)$$

$$O{=}C{-}O(C{=}O){-}O^- \rightarrow CO_3^{2-} + CO \qquad k = 3.2 \times 10^3 \ M^{-1} \ s^{-1} \qquad (12.41)$$

In dry DMF solution, the rate constant of electron transfer to CO_2 was shown to be 6×10^3 cm s^{-1}, the transfer coefficient was 0.4, and the second-order rate constant of CO_2^- deactivation was 3.2×10^7 M^{-1} s^{-1}.[265]

12.3.1   Direct Reduction

A comparison of the solvents water, dimethyl sulfoxide (DMSO), acetonitrile (MeCN), propylene carbonate (PC), and DMF for the reduction of carbon dioxide on electrodes of glassy carbon, mercury, platinum, gold, and lead was performed by Eggins and McNeill by measuring the voltammetric reduction waves, with tetramethyl ammonium bromide as the supporting electrolyte.[266] The peak potentials for CO_2 reduction on Hg electrodes in water, MeCN, DMSO, PC, and DMF were -2.16, -2.76, -2.40, -2.59, and -2.21 V (vs. SCE). From an analysis of the voltammetric waves, the apparent number of electrons involved in the reduction of CO_2 was derived, resulting in $n_{app} = 2$ in aqueous solutions on glassy carbon electrodes and $n_{app} = 1$ on Hg and Au electrodes in DMSO solutions, but $n_{app} = 2$ on Pt electrodes in DMSO. In MeCN solutions, the observed n_{app} values with Hg and Pt electrodes were 1 and 2, respectively. The results indicate different reaction mechanism on the various metal electrodes. Improved yields of two- and three-carbon CO_2 reduction products were achieved by Eggins et al. during the electroreduction of CO_2 in methanol solutions.[267]

In aprotic organic solvents such as acetonitrile and DMF, the main carbon dioxide reduction products were usually oxalic acid and carbon monoxide. Metal electrodes which strongly chemisorb carbon dioxide, such as platinum, enhanced the production of CO:

$$2CO_2 + 2e^- \rightarrow CO + CO_3^{2-} \tag{12.42}$$

In DMF solution containing 0.1 M tetrabutyl ammonium perchlorate (TBAP) and 0.02 M CO_2 over an Hg electrode, at 25°C, the faradaic yields of oxalate and CO were found by Isse et al. to be 84 and 1.7%, respectively.[168] Increasing the CO_2 concentration or decreasing the temperature caused decreases in the yields of oxalate. Improved yields of oxalate were obtained by Gennaro et al. with Hg electrodes by adding various compounds active as homogeneous redox catalysts for CO_2 reduction.[169] These were mainly esters of benzoic and phthalic acids and aromatic nitriles. Under optional conditions, with 1.5 mM methyl benzoate in DMF + 0.1 M TBAP at 0°C, the faradaic yield of oxalate was 97%. These redox catalysts were assumed to undergo electrochemical reduction to their anion radicals, which then transferred their electron to CO_2^-, while reforming the neutral electrocatalyst. The electron transfer from the catalyst anion to CO_2 was proposed to involve an initial S_n2-type step of binding of the catalyst anion to the carbon atom of CO_2, followed by homolytic cleavage of the adduct, with release of CO_2^-. Fast dimerization of the CO_2^- radicals produced oxalate. From the cyclic voltammetry measurements of the reduction in the presence and absence of CO_2, Gennaro et al. were able to derive the bimolecular rate constants k of electron transfer from the catalyst to CO_2. In the case of methyl benzoate as catalyst, in DMF with 0.1 M TBAP at 25°C, log k was 4.2 M^{-1} s^{-1}.

On comparing various metals as electrodes for carbon dioxide reduction in nonaqueous solutions, the product composition was found by Ito et al. to depend on the nature of the metal. Oxalic acid was predominantly produced on Hg, Tl, and Pb, while carbon monoxide was the preferred product on Cu, Zn, In, Sn, and Au.[80,81,143]

12.3.1.1 Mechanistic Studies

The mechanism of the electrochemical reduction of carbon dioxide on lead electrodes in acetonitrile and in propylene carbonate solvents was shown by Aylmer-Kelly et al. using modulated specular electroreflectance spectroscopy to involve two intermediate radical species.[32] These were detected by their strong absorption peaks at 285 and at 330 nm in propylene carbonate solution. In acetonitrile, these peaks were blue-shifted to 270 and 315 nm, respectively. The absorption peaks were assigned to the radical anion CO_2^- and the adduct of this radical anion with a neutral CO_2 molecule, $(CO_2)_2^-$:

$$CO_2^- + CO_2 \rightarrow (CO_2)_2^- \qquad (12.43)$$

As noted above (Section 12.2.1.2), in aqueous media only the single peak of CO_2^- at 250 nm was observed.

A mechanistic study of the electroreduction of carbon dioxide in aprotic solvents by Vassiliev et al. used photoemission measurements and steady-state (dark) polarization curves at Hg, Pb, Sn, In, and Pt electrodes.[268,269] In acetonitrile solutions, the overpotentials of CO_2 reduction increased in the order In < Sn < Pb < Hg. This considerable sensitivity of the overpotential to the electrode material indicated that the rate-limiting step of CO_2 electroreduction in aprotic solvents involves species adsorbed on the electrode surface. For all these metals, except platinum, plots of the steady-state currents against potential consisted of two Tafel regions. In the first Tafel region, with a slope of 140 to 180 mV, the rate-limiting step was proposed to be the reaction of the CO_2^- radical anion (formed in a prior fast electron transfer step) with a CO_2 molecule to form the $(CO_2)_2^-$ anion radical:

$$CO_2 + e^- \rightarrow CO_2^- \qquad (12.44)$$

$$CO_2^-(ads) + CO_2(ads) \rightarrow (CO_2)_2^- \qquad (12.45)$$

$$(CO_2)_2^-(ads) + e^- \rightarrow C_2O_4^{2-} \qquad (12.46)$$

In the second Tafel region, with a slope of 600 to 700 mV, the initial electron capture by CO_2 to form the CO_2^- radical anion became the rate-determining step, just as in aqueous solutions. On platinum electrodes, the polarization curve was flat. There appeared only a single Tafel slope, of 600 to 700 mV, corresponding to the second Tafel region of the other metals.[268,269]

Kaiser and Heitz used Hg or Cr–Ni–Mo steel (18:10:2%) cathodes on which CO_2 and its reduction products were only weakly adsorbed.[30] The faradaic yield of oxalic acid was 61%, at a current density of 6 mA cm^{-2}. The reaction was explained by the high concentration near the cathode of CO_2^- radical anions, the dimerization of which resulted in oxalate production:

$$2CO_2 + 2e^- \rightarrow (COO)_2^{2-} \qquad (12.47)$$

The mechanism of electroreduction of carbon dioxide to oxalate on platinum electrodes in acetonitrile solution was studied by Desilvestro and Pons using infrared spectroelectrochemistry. *In situ* FTIR reflectance spectra of the thin electrolyte layer between the cell window and the electrode were measured. Oxalate was formed in at least two different solvation states.[270]

12.3.1.2 High-Pressure CO_2 Reduction in Organic Media

With a high-pressure CO_2 electrolysis cell, using indium, tin, or lead cathodes, with tetraalkyl ammonium salts as electrolytes, Ito et al. attained high efficiency of formic acid production.[271,272] This product reached a concentration of up to 1 M. Also, propionic, n-butyric, and oxalic acids were formed in small yields.

A variety of electrolysis cell types were tested by Goodridge and Presland in an effort to increase the current densities and current yields for oxalic acid production in the electrolysis of carbon dioxide.[273] Lead, lead amalgam, stainless steel, cadmium, aluminium, copper, or zinc cathodes were used, in 0.18 M tetraethylammonium bromide in DMF saturated with CO_2, separated by a cationic membrane from graphite anodes in 1 M NaCl, at atmospheric pressure. The highest current yield for oxalic acid production, 51%, was obtained with the Pb amalgam electrode, at a current density of 10 mA cm^{-2}. Raising the current density to 20 mA cm^{-2} caused a decline in current yield to 29%. Also, at the higher current density, in DMF solution, a black deposit was formed on Pb or Pb amalgam electrodes. With the other electrode materials, the current yields were much lower. Higher production rates of oxalic acid were achieved using a high-pressure cell. With a lead cathode, at a CO_2 pressure of about 7 atm, the current density reached 50 mA cm^{-2} and the current yield was 44%, but the process was plagued by attack on the lead electrode, as well as by the decomposition of DMF to dimethylamine. Stainless-steel electrodes were more stable and at a current density of 20 mA cm^{-2} gave a current yield of 36%.

12.3.1.3 CO_2–Methanol Electrolyte

Methanol is used in the industrial Rectisol process for the absorption and removal of CO_2 from synthesis gas produced by the reforming reaction. The solubility of CO_2 is much higher in methanol than in water. Thus, with CO_2 at 10 atm pressure, methanol at $-30°C$ has about 50 times the loading capacity for CO_2 than water at $35°C$.[274] The electroreduction of CO_2 in a methanolic electrolyte could therefore be an attractive approach to the recovery and conversion of CO_2 released from fossil-fuel power plants.

Hirota et al. devised an interesting assembly to benefit from the specific properties of both solvents.[275] They used a two-compartment cell, with a CO_2–methanol mixture in the catholyte and water in the anolyte and with NH_4Cl as supporting electrolyte. Under galvanostatic conditions at 200 mA cm^{-2}, the main reduction products with Cu cathodes were H_2, CH_4, and CO. With Ag cathodes, CO was produced at close to 100% current efficiency. When a single-compartment cell was used, with methanol as the medium, the current density was much higher, reaching 1 A cm^{-2}, but some of the methanol was oxidized at the anode.

In methanol containing 80 mM benzalkonium chloride, the reduction of CO_2 was carried out by Naitoh et al.[276] and Mizuno et al.[277] using a two-compartment cell with separation by a Nafion® 117 sheet, with a Cu cathode and a Pt anode. At $-15°C$, with a potential at the Cu electrode of -2.3 V (vs. SCE), the faradaic efficiencies for production of CH_4, CO, C_2H_4, and H_2 were 39, 24, 4.4, and 21%, respectively. Under similar conditions, but at $-30°C$, the corresponding efficiencies were 43, 12, 2.1, and

<8%, respectively. Using 0.1 M KOH as the electrolyte in methanol instead of benzalkonium chloride, at −15°C, CO, HCOOH, and CH_4 were produced with efficiencies of 56, 22, and 10%, respectively. At higher temperatures, the yields of organic products were lower.[278]

12.3.2 Electrocarboxylation

12.3.2.1 Sacrificial Metal Anodes

In carboxylation reactions involving carbon dioxide, the use of sacrificial anodes of metals such as Mg, Zn, and Al provides advantages beyond the avoidance of unwanted anodic oxidation of the carbon dioxide reduction products. The Mg^{2+}, Zn^{2+}, or Al^{3+} ions thus introduced into the electrolyte generally have beneficial effects on the yields and selectivities of the reduction products and facilitate the isolation procedures.[279–281] These electrocarboxylation reactions are very useful for transforming organic compounds into organic acids with one additional carbon atom.[282,283]

Relatively high current densities of up to 50 mA/cm^2 for the reduction of CO_2 to CO were achieved by Massebieau et al. using freshly deposited metal cathodes (best results with Cd, Sn, or Ag) and a sacrificial Mg anode, in DMF as solvent, producing CO at current yields of up to 99%, at a bias of −2.2 V vs. SCE.[284] The total reaction was represented by:

$$2CO_2 + Mg \rightarrow CO + MgCO_3 \qquad (12.48)$$

12.3.2.2 Production of Oxalate

A practical process for the production of oxalate was developed by Silvestri et al. by applying elevated pressures of carbon dioxide.[281,285] Using stainless-steel cathodes, sacrificial aluminum anodes, in a diaphragm-less system containing 0.1 M tetrabutyl ammonium bromide in DMF as electrolyte, with CO_2 at pressures of 1.4 to 31 atm, aluminum oxalate was produced with current yields of up to 38%, as well as CO and carbonate. The anodic and cathodic reactions were represented by:

$$Al \rightarrow Al^{3+} + 3e^- \qquad (12.49)$$

$$2CO_2 + 2e^- \rightarrow {}^-OOC–COO^- \qquad (12.50)$$

$$2CO_2 + 2e^- \rightarrow CO + CO_3^{2-} \qquad (12.51)$$

A semi-pilot plant process for carbon dioxide reduction to oxalic acid was developed by Fischer et al. using sacrificial zinc anodes and stainless-steel cathodes in a single-compartment cell, with acetonitrile solutions of tetrabutyl ammonium perchlorate as the electrolyte.[286] At current densities of about 5 to 10 $mA\ cm^{-2}$, faradaic efficiencies of more than 90% were achieved. The product was insoluble zinc oxalate, which was collected by filtration. After hydrolysis of the zinc oxalate, the zinc could be recovered as metal by electrolysis, to be recycled to the carbon dioxide reduction stage. The whole process was performed without any waste products.

12.3.2.3 Carboxylation of Olefins

The electrocarboxylation of olefins was studied in detail by Gambino and Silvestri. Ethylene was carboxylated in DMF–tetrabutyl ammonium bromide solution, with both ethylene and carbon dioxide under pressure, using sacrificial Al anodes and stainless-steel (AISI 316) cathodes.[287] The products were dicarboxylates and monocarboxylates with the general formulae

$$^-OOC(C_2H_4)_nCOO^- \quad \text{and} \quad H(C_2H_4)_nCOO^- \tag{12.52}$$

with n = 0, 1, 2, 3, 4. With an initial ethylene/carbon dioxide molar ratio of 7 in the gas phase, at an overall pressure of about 40 atm, the C_2, C_4, C_6, and C_8 dicarboxylic acids were obtained in appreciable amounts. Under these conditions the reaction appeared to be a radical-induced ethylene telomerization, with $\cdot CO_2^-$ radical anions acting as chain initiators and terminators.[288] In solvents of low proton availability, olefins were reduced at much more positive potentials than CO_2. The main products were the β-carboxylated compounds. In the predominant mechanism, CO_2 seemed to act as an electrophile, adding to the olefin anion radical.[289]

Using sacrificial Al anodes and vitreous graphite cathodes, styrene in anhydrous DMF containing Bu_4NBr underwent dicarboxylation to phenyl succinic acid with yields of up to 85%, as well as producing small amounts of 3-phenyl propionic acid. To prevent the polymerization of styrene, it was necessary to keep the styrene concentration quite low, about 3% of the medium.[265,290,291] The electrocarboxylation of acenaphthylene in anhydrous media using sacrificial Al anodes with a "gas-lift" electrochemical cell resulted in *trans*-acenaphthene-1,2-dicarboxylic acid

Acenaphthylene →(CO_2)→ Acenaphthene-1,2-dicarboxylic acid

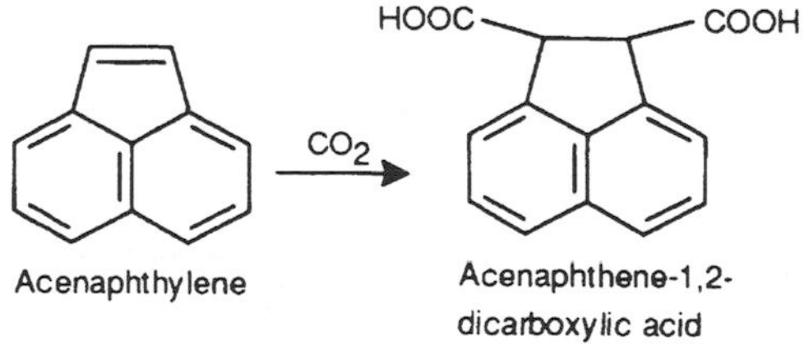

in 81% yield, at a current density of 40 mA cm^{-2}, with more than 80% current efficiency.[291]

The electrocarboxylation of ethylene with CO_2 was found by Aresta et al. to be promoted by the rhodium complex $\{[(C_2H_4)_2Rh(\eta^6\text{-}Ph)]_2BPh_2\}O_3\text{-}SCF_3$ in tetrahydrofuran (THF) solution at 293 K. With a 1:1 mixture of CO_2 and ethylene (0.1 MPa overall pressure), at a bias of −1.8 V (vs. SCE), and using a consumable Mg anode, propionic acid was produced in 16% yield, as well as smaller amounts of styrene, benzene, ethylbenzene, and diphenyl.[292]

The electrocatalytic carboxylation of perfluoroalkyl olefins by Chiozza et al. led to perfluoroalkyl carboxylic acids, involving incorporation of CO_2 on the C=C bond and double-bond migration. These amphiphilic compounds are expected to be useful intermediates to pharmaceuticals.[293]

12.3.2.4   Carboxylation of Carbonyl Derivatives

The electrocarboxylation of aldehydes and ketones is of considerable practical interest, as it leads to some α-arylpropionic acids, which are used as nonsteroidal anti-inflammatory agents (NSAIs).[294] These 2-aryl propionic acids act by cyclooxygenase inhibition, stopping the arachidonic acid cascade to prostaglandins and thromboxane A_2, which are responsible for the inflammation mechanism.[295] Using sacrificial anodes, 6-methoxyacetonaphthone was carboxylated to 2-hydroxy-2-(6-methoxynaphthyl)-propionic acid, which was then hydrogenated to 2-(6-methoxynaphthyl)-propionic acid (Naproxen®), which is a most active NSAI compound.[281,296,297] Chan et al. described the electrosynthesis of Naproxen® by the carboxylation of 6-methoxy-2-acetylnaphthalen (6-methoxyacetonaphthone) to 2-(6-methoxy-2-naphthyl)lactic acid, an intermediate in the synthesis of Naproxen®:[298]

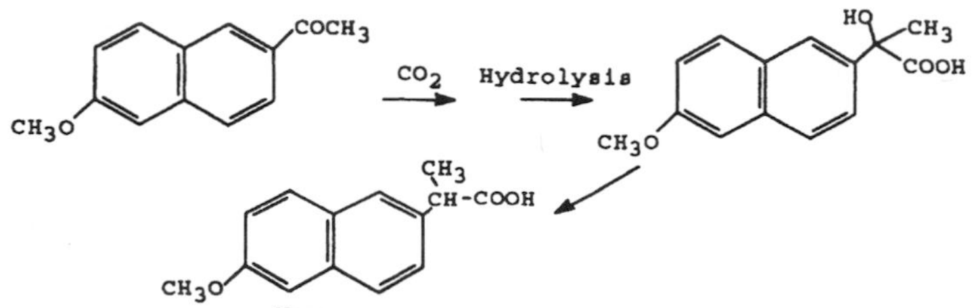

Naproxen

The process was performed in a single-cell reactor, using a sacrificial Al foil anode and a lead foil cathode, with DMF + 0.1 M Bu$_4$NBr as electrolyte. For continuous production, the reaction was carried out in a flow cell reactor, operated at CO_2 of up to 60 psig. Yields of the electrocarboxylation were up to 91%.[298]

In a study of the mechanism of electrocarboxylation of alkyl aryl ketones, Pletcher and Slevin carried out the carboxylation of 2-acetylnaphthalene in an undivided cell with a sacrificial Mg anode and a steel gauze cathode in CO_2-saturated DMF + 0.1 M Bu$_4$NBF$_4$.[299] The initial step was shown to be formation of the radical anion, followed either by dimerization to the pinacol or carboxylation:

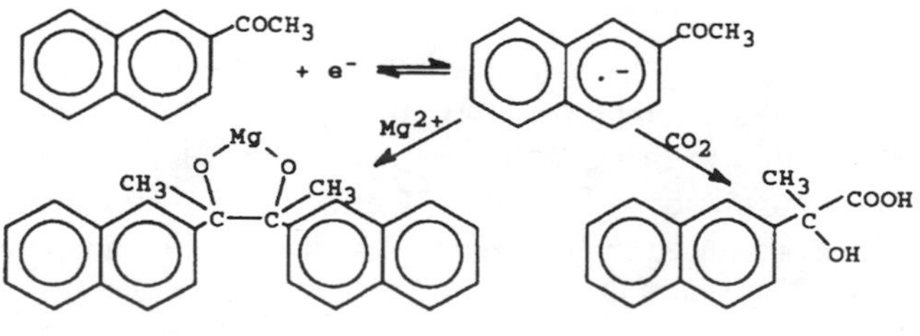

Electrocarboxylation was also applied to aliphatic and aromatic acyl chlorides, forming α-oxoacids in one-step reactions. Koshechko et al. performed the reduction of acetyl chloride at a current density of 3 mA cm^{-2} in an undivided cell with a Zn anode and a Pt cathode in CO_2-saturated DMF + 0.1 M Bu$_4$NBF$_4$. The yield of pyruvic acid was 60 to 62%:[300,301]

$$RCOCl + CO_2 + 2e^- \rightarrow RCOCOO^- + Cl^- \qquad (12.53)$$

Similarly, benzoyl chloride was converted to phenylglyoxylic acid (benzoylformic acid, PhCOCOOH) in 39% yield.

12.3.2.5 Carboxylation of Benzyl Chlorides

The use of sacrificial anodes for the conversion of organic halides into carboxylic acids was reviewed by Chaussard et al., who also describe the operation of pilot plant experiments for the development of industrial applications.[282]

In a route to the very important NSAI agent ibuprofen (p-isobutylhydratropic acid), Filardo et al. used a homogeneous charge transfer catalyst (HCTC) to mediate the carboxylation of the corresponding benzyl chloride:[302]

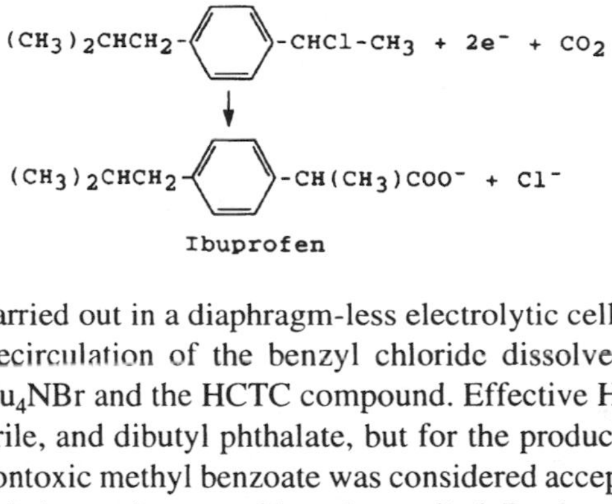

Ibuprofen

The reaction was carried out in a diaphragm-less electrolytic cell, with sacrificial Al anodes and with recirculation of the benzyl chloride dissolved in CO_2-saturated DMF with 0.1 M Bu$_4$NBr and the HCTC compound. Effective HCTCs were methyl benzoate, benzonitrile, and dibutyl phthalate, but for the production of the pharmaceutical, only the nontoxic methyl benzoate was considered acceptable. Also, for this reason, the toxic nickel complexes could not be applied. Product yields of 70 to 85% were achieved, with oxalic acid formed as a by-product. In the absence of the HCTC, passivation of the cathode interfered with the electrolysis. The mechanism proposed proceeded by reduction of the HCTC to a radical anion, which transferred its charge both to the benzyl chloride and to CO_2:[302]

$$HCTC + e^- = HCTC^{-\cdot} \qquad (12.54)$$

$$Benzyl\text{-}Cl + 2HCTC^{-\cdot} + CO_2 \rightarrow Benzyl\text{-}COO^- + Cl^- + 2HCTC \qquad (12.55)$$

12.3.2.6 Nickel-Complex-Mediated Carboxylation of Alkenes

The incorporation of carbon dioxide into the carbon–carbon double bond under mild conditions was achieved by Dérien et al. using electrogenerated Ni(0) complexes as electrocatalysts.[303–306] These complexes were generated *in situ* by two-electron reduction of the stable Ni(II) salts, in an electrolyte of tetrabutylammonium tetrafluoroborate in DMF solution, using carbon fiber cathodes and sacrificial magnesium anodes. In the nickel-complex-catalyzed electrocarboxylation, the anodic reaction is the oxidation of magnesium to Mg^{2+} ions, while the cathodic reaction is the reduction of Ni(II) to a Ni(0) species, and is carried out at a potential of -1.2 to -1.4 V (vs. SCE):

$$Mg \rightarrow Mg^{2+} + 2e^- \tag{12.56}$$

$$LNi^{2+} + 2e^- \rightarrow LNi(0) \tag{12.57}$$

The intermediate LNi(0) complex coordinates both with CO_2 and with an unsaturated hydrocarbon, forming a new carbon–carbon bond. Thus, with the bpy ligand, the transition is $Ni(bpy)_3^{2+}/Ni(bpy)_2$. The Ni(0) species coordinates both with the carbon–carbon double bond and with CO_2, forming a nickelacyle. An example of this carboxylation is the reaction with norbornene (norbornylene), carried out with $Ni(bpy)_3(BF_4)_2$ as catalyst precursor, with 5 atm CO_2, resulting (after hydrolytic workup) in a monocarboxylic acid as the major product, with small amounts of a dicarboxylic acid:

The electrocarboxylation was also successful with arylalkenes yielding monocarboxylic acids, with 1,2-alkenes (allenes) yielding monocarboxylated unsaturated acids, and with 1,3-dienes such as 1,4-diphenylbutadiene yielding dicarboxylated unsaturated acids.[303,304]

12.3.2.7 Carboxylation of Acetylenic Compounds

A general procedure for the electrocarboxylation of both terminal and internal alkynes with carbon dioxide was developed by Dérien et al. with the octahedral complex of nickel(II), $Ni(bpy)_3(BF_4)_2$ (bpy = 2,2′-bipyridine) in 1 mM Bu$_4$N · BF$_4$ in DMF solution as catalyst.[307,308] In this reaction, alkynes were converted to α,β-unsaturated acids. When carried out in one-compartment cells, with sacrificial magnesium anodes, the process was catalytic with respect to the nickel complex. In a preparative procedure, the carboxylation of 4-octyne carried out at a constant current of 50 mA and a cathode potential of about -1.2 V (SCE), with Mg/carbon fiber electrodes, at atmospheric pressure of carbon dioxide, yielded (E)-2-propyl-2-hexenoic acid in 80% yield,

$$n\text{-}C_3H_7\text{-}C{\equiv}C\text{-}n\text{-}C_3H_7 + CO_2 \rightarrow n\text{-}C_3H_7\text{-}CH{=}C(n\text{-}C_3H_7)COOH \tag{12.58}$$

The magnesium ions released by the oxidation of the magnesium anode catalyzed the cleavage and recycling of an intermediate nickelacyle:

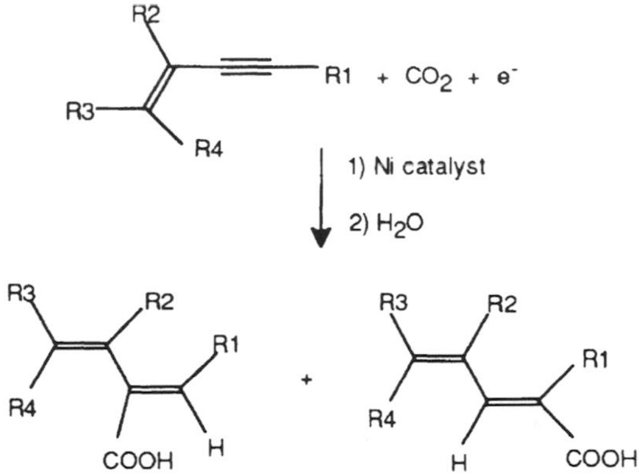

$$RC \equiv CR + CO_2 + Ni\text{-bpy-complex} \longrightarrow bpyNi$$

where the R substituents represent H atoms or various alkyl groups.

With diyne that contain either two conjugated or two nonconjugated triple bonds, the Ni-complex-catalyzed carboxylation was shown by Dérien et al. to lead to the formation of unsaturated carboxylic acids.[305] With 1,7-octadiyne as an example of α,ω-diynes, at a CO_2 pressure of 5 atm and using (pentamethyl-diethylenetriamine (PMDTA) as the Ni ligand, a monocarboxylic acid was produced with 98% selectivity:

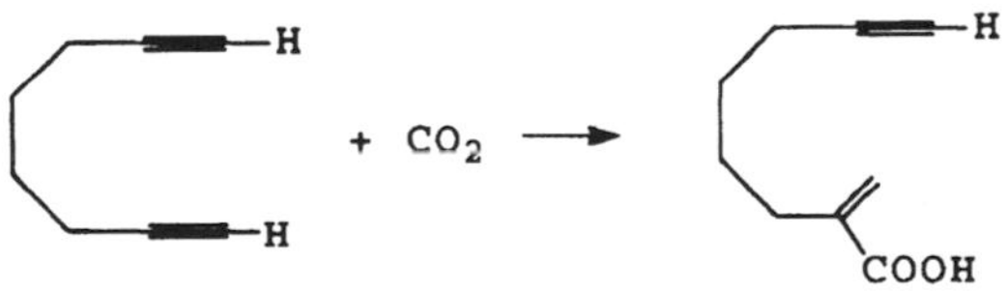

With diynes that have both an internal and a terminal triple bond, the electrocarboxylation occurred only at the terminal triple bond, and with regioselectivity at the 2-position. The carboxylation of conjugated diynes was best catalyzed by Ni–PMDTA, resulting in a selective monocarboxylation reaction. Thus, with the following conjugated diyne, in which $R = C_5H_{11}$, at 5 atm CO_2, two products were obtained after hydrolysis, with cis-addition of H and COOH only across a single triple bond:

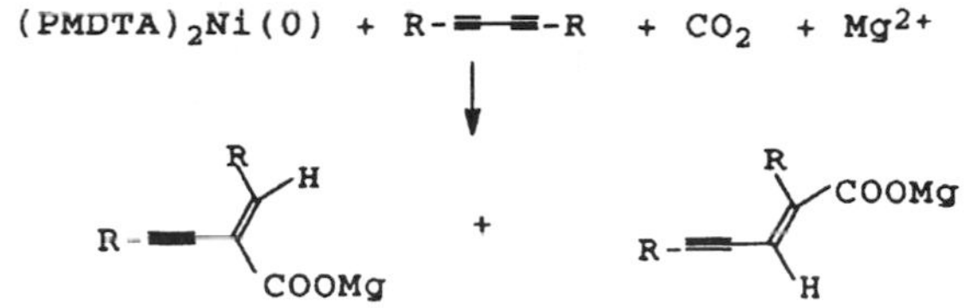

12.3.2.8   Carboxylation of Epoxides

The incorporation of atmospheric pressure CO_2 into terminal epoxides to form cyclic carbonates was achieved by Tascedda and Duñach in an electrochemical reaction, using Ni(cyclam)Br$_2$ in DMF solution as electrocatalyst.[309] Cyclic carbonates are useful polar solvents and valuable intermediates in the production of polycarbonates. Their formation by a route that does not require the use of toxic phosgene is a major advantage. The reactions were carried out with sacrificial Mg anodes and with stainless-steel cathodes. In the cathodic reaction, the $Ni^{II}(cyclam)^{2+}$ was reduced to $Ni^{I}(cyclam)^{+}$:

$$Ni^{II}(cyclam)^{2+} + e^- \rightarrow Ni^{I}(cyclam)^{+} \tag{12.59}$$

which was the catalytically active species. The overall reaction was

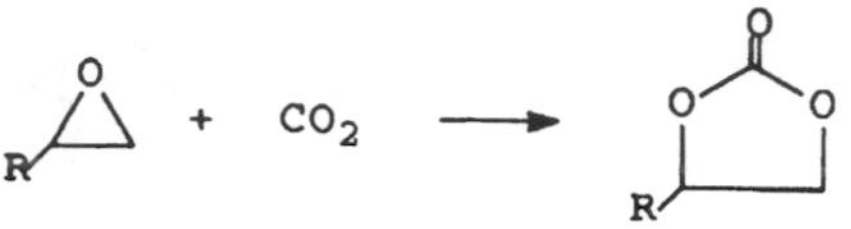

Aromatic, benzylic, and aliphatic terminal epoxides were converted to the corresponding carbonates in high yields. The presence of the epoxides in the electrolysis medium inhibited the reduction of CO_2 to CO, which was the predominant Ni(cyclam)$^{2+}$-catalyzed reaction in the absence of the epoxides.[166,178]

12.3.2.9   Carboxylation of Vinylic, Allylic, Aromatic, and Polymeric Halides

The electroreductive carboxylation of aryl-substituted vinyl bromides on sacrificial Mg anodes was found by Kamekawa et al. to lead to aryl-substituted 2-alkenoic acids.[310] Thus, the electrolysis of β-bromostyrene in CO_2-saturated DMF containing 0.1 M Bu$_4$NBF at 5°C resulted in a 30:70 mixture of E- and Z-cinnamic acids in up to 76% yield:

$$PhCH=CHBr + CO_2 \rightarrow PhCH=CHCOOH$$

This reaction also provided an alternative route to the important NSAI agent ibuprofen, by electrocarboxylation of 1-bromo-1-(p-isobutylphenyl)ethene to 2-(p-isobutylphenyl)-propenoic acid, followed by hydrogenation to ibuprofen. The electrocarboxylation of alkyl-substituted vinyl bromides was performed by Kamekawa et al. with atmospheric pressure CO_2 in 0.1 M Bu$_4$NBF$_4$–DMF containing 20 mol% of the Ni(II)Br$_2$-bpy (bpy = 2,2′-bipyridine) at –10°C, using Mg anodes and Pt cathodes.[311] Various cyclic and noncyclic vinylic bromides were converted in good yields to the corresponding α,β-unsaturated carboxylic acids. Without the nickel(II) complex, the yields were much lower.

Regioselective electrocarboxylation was carried out by Tokuda et al. by electrolysis of γ-mono- and disubstituted allylic bromides with sacrificial Mg or Al anodes and Pt cathodes.[312] Such reactions lead to β,γ-unsaturated carboxylic acids, which are useful as synthetic intermediates. Thus, prenyl bromide, in CO_2-saturated DMF solution with 0.1 M Et_4NClO_4 at –20°C, when electrolyzed with an Mg anode at 26 mA cm^{-2} was carboxylated, yielding 4-methyl-3-pentenoic acid (a) and 2,2-dimethyl-3-butenoic acid (b), in an isomer ratio of 91:9:

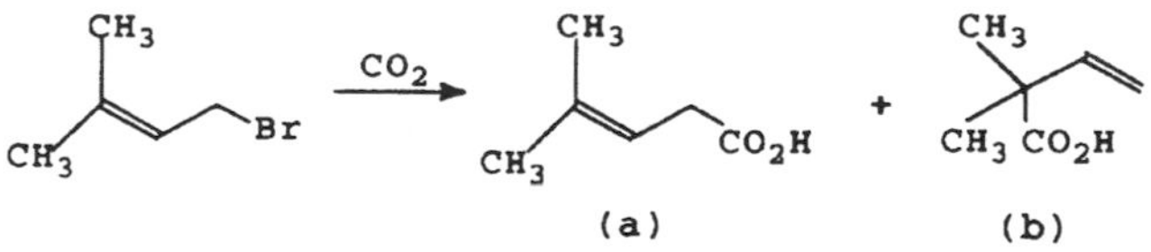

CH$_3$ $\xrightarrow{CO_2}$ CH$_3$ CO$_2$H   +   CH$_3$ CO$_2$H

(a)   (b)

The electrochemical reductive cyclization of allyl *o*-halophenyl ethers with atmospheric presssure CO_2 at room temperature in the presence of Ni(cyclam)$^{2+}$ in DMF solution was performed by Duñach and Olivero in one-compartment cells with sacrificial Mg anodes and carbon fiber cathodes:[313]

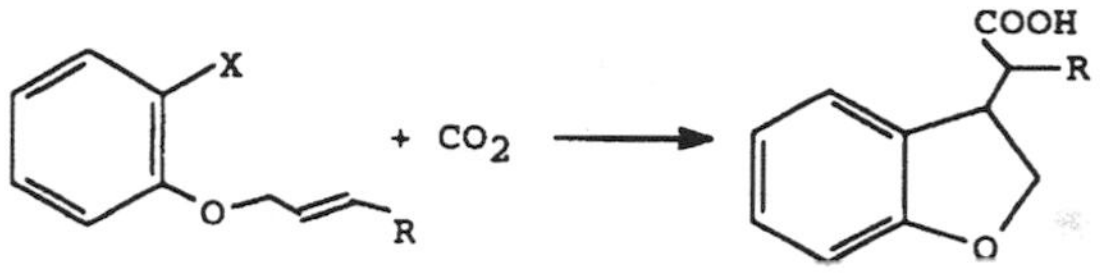

X + CO$_2$ → COOH, R

X = Cl, Br, I

In this cyclization–carboxylation reaction, the bicyclic monocarboxylic acids produced were substituted dihydrobenzofuran and benzopyran derivatives. The initial step in the process was proposed to be the cathodic reduction of Ni(cyclam)$^{2+}$ to Ni(cyclam)$^+$, followed by insertion of Ni(cyclam)$^+$ into the C–X bond of the halophenyl ether to produce an aryl-Ni(III) intermediate complex, which underwent reductive carboxylation with CO_2. In the presence of Mg^{2+} ions in the medium, the Mg-carboxylate of the bicyclic product was formed, releasing the Ni(cyclam)$^{2+}$ for further catalytic action.

An electrochemical carboxylation of aromatic halides somewhat reminescent of the Grignard reaction was reported by Amatore and Jutand.[314,315] The electrocarboxylation of bromobenzene by CO_2 was carried out in CO_2-saturated THF/hexamethylphosphoramide + 0.1 M n-Bu$_4$NBF$_4$ containing a catalytic amount of the complex Ni^{II}(dppe)Cl$_2$, where dppe = 1,2-bis(diphenyl-phosphino)ethane, using a mercury pool cathode at a constant potential of –2 V (vs. SCE). Benzoate was produced in quantitative yield, according to the stoichiometry

$$PhBr + CO_2 + 2e^- \rightarrow PhCOO^- + Br^- \qquad (12.60)$$

as well as traces of biphenyl. Cyclic voltammetry measurements indicating a mechanism passing through a squence of intermediates of Ni(0), Ni(I), Ni(II), and Ni(III). The proposed catalytic cycle involved as the active catalyst the intermediate Ni^0(dppe), which reacted with the aryl halide PhX to form the intermediate Ph–Ni^{II}X(dppe). Insertion of CO_2 led to the intermediate species Ph–CO_2–Ni^{II}(dppe), which underwent further one-electron reduction, splitting off the carboxylate anion $PhCOO^-$ and reforming the active catalyst $Ni^0(dppe)$.[314,315]

The electrocarboxylation of aromatic halides by CO_2 was found by Torii et al. to be catalyzed also by palladium–tertiary phosphine complexes.[316] Bromo- and iodobenzene and various p-substituted derivatives in CO_2-saturated DMF containing catalytic amounts (10%) of $Pd^{II}Cl_2(PPh_3)_2$ underwent carboxylation to the corresponding carboxylic acids in high yields, while aryl chlorides were unreactive. Vinyl bromide under similar conditions underwent dicarboxylation:

$$Ph-CH=CH-Br + CO_2 + 2e^- \rightarrow Ph-CH_2-CH(COOH)_2 \qquad (12.61)$$

A detailed mechanism of the catalytic cycle with the Pd complex was proposed by Amatore et al.[317] In the initial step, two-electron reduction of $Pd^{II}Cl_2(PPh_3)_2$ led to a $Pd^0(PPh_3)_2$ intermediate, which underwent oxidative addition of the aryl halide to form an aryl-palladium(II) intermediate. Further two-electron reduction of this species led to halide ion release and formation of the $ArPd^0(PPh_3)_2^-$ intermediate, which then split to a "free" aryl anion Ar^- while reforming the catalytic $Pd^0(PPh_3)_2$ intermediate. In the product-forming step, the aryl anion reacted as a nucleophile with CO_2 to produce the aromatic carboxylate:

$$Pd^{II}Cl_2(PPh_3)_2 + 2e^- \rightarrow Pd^0(PPh_3)_2 + 2Cl^- \qquad (12.62)$$

$$Pd^0(PPh_3)_2 + ArX \rightarrow ArPd^{II}X(PPh_3)_2 \qquad (12.63)$$

$$ArPd^{II}X(PPh_3)_2 + 2e^- \rightarrow ArPd^0(PPh_3)_2^- + X^- \qquad (12.64)$$

$$ArPd^0(PPh_3)_2^- \rightarrow Ar^- + Pd^0(PPh_3)_2 \qquad (12.65)$$

$$Ar^- + CO_2 \rightarrow ArCOO^- \qquad (12.66)$$

A Pd-catalyzed synthesis of α,β-unsaturated carboxylic acids was achieved by Jutand and Négri by electrocarboxylation of vinyl triflates (vinyl trifluoromethane sulfonates). The reaction was carried out in CO_2-saturated DMF containing 0.1 M Bu_4NBF_4 and $PdCl_2(PPh_3)_2$, with a carbon rod cathode at -2.0 V (vs. SCE) and an Mg anode. Thus, the following vinyl triflate was converted into the carboxylate in 85% yield:[318]

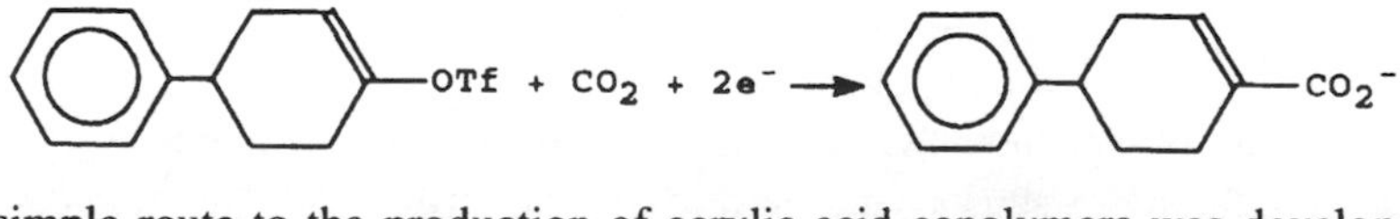

A simple route to the production of acrylic acid copolymers was developed by Koshechko et al., based on the electrocarboxylation of poly(vinyl halides)s and polybutadiene.[319] The reactions were performed in CO_2-saturated DMF solutions of the polymers containing 0.1 M Et_4NBr, using Pt cathodes and sacrificial Zn anodes in undivided cells. At a constant potential of -2.4 V (vs. Ag/AgCl), poly(vinyl chloride) was carboxylated with a current efficiency of about 10% and with a yield (relative to the number of monomer units in the original polymer) of about 8%. The same reaction was also successful with poly(vinyl bromide) and with polybutadiene. The proposed mechanism involved as a primary step the cathodic reduction of sections of the polymer chain to carbanions, which then interacted with CO_2-forming carboxylate groups.

p-Anisic acid (4-methoxybenzoic acid) was obtained by Murcia and Peters by electroreductive carboxylation of *p*-iodoanisole in CO_2-saturated DMF containing Et_4NClO_4 over an Hg pool cathode. By controlled potential electrolysis at -1.05 V (vs. SCE), *p*-anisic acid and anisole were produced in 50 and 40% yield, respectively.[320]

12.3.2.10 Symmetrical Ketones from Organohalides and CO_2

Organic halides, such as alkyl bromides and benzyl and allyl halides, were converted by Garnier et al. into symmetrical ketones.[321] The electrosynthesis was performed in CO_2-saturated N-methylpyrrolidone or DMF containing Bu_4NBF_4, in the presence of Ni(bpy) and bpy, using sacrificial Mg anodes. Thus, benzyl bromide was transformed in 85% yield into $PhCH_2COCH_2Ph$. The reactive intermediate reductant was proposed to be Ni^0(bpy).

12.3.2.11 Metalloporphyrin-Promoted Carboxylation

Zheng et al. observed the catalytic effect of metalloporphyrins on the electrocarboxylation of benzyl chloride with CO_2 in organic solvents, yielding phenylacetic acid and benzyl phenylacetate:[322]

$$C_6H_5CH_2Cl + CO_2 \rightarrow C_6H_5CH_2COOH \rightarrow C_6H_5CH_2COOCH_2C_6H_5 \quad (12.67)$$

At a potential of -1.6 V (vs. SCE), benzyl chloride was converted with about 50% current efficiency into benzyl phenylacetate. Most active catalysts were those metalloporphyrins in which the central metal easily formed M(I) complexes, such as CoTPP and FeTPP (H_2TPP = 5,10,15,20-tetraphenylporphyrin). In the presence of CoTPP, the positive potential required for the electrocarboxylation by CO_2 of alkyl halides, alkenes, and ketones was markedly decreased. Styrene was thus electrocarboxylated to phenylpropionic acid,[323-325]

$$C_6H_5CH=CH_2 + CO_2 \rightarrow C_6H_5CH_2CH_2COOH \quad (12.68)$$

and naphthalene to 1,4-dicarboxy-1,4-dihydronaphthalene,

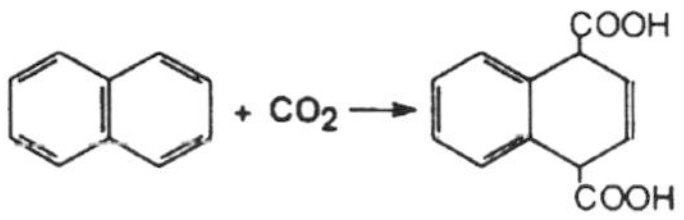

Such electrocarboxylations were performed in dry CO_2-saturated DMSO as solvent, with tetrabutylammonium perchlorate (0.5 M) as supporting electrolyte, in the presence of CoTPP (1 mM), with mercury or platinum cathodes and aluminum counterelectrodes. The primary electrochemical step was proposed by Zheng et al. to be the electroreduction of the central atom of CoTPP from Co(II) to Co(I), forming a strongly nucleophilic complex anion,[326]

$$Co(II)TPP + e^- \rightarrow Co(I)TPP^- \quad E_{1/2} - -1.17 \text{ V} \qquad (12.69)$$

followed by reaction of this anion complex with the alkyl halide, yielding a Co(III) complex, which was electroreduced to a Co(II) complex:

$$Co(I)TPP^- + RX \rightarrow RCo(III)TPP + X^- \qquad (12.70)$$

$$RCo(III)TPP + e^- \rightarrow RCo(II)TPP^- \qquad (12.71)$$

A shift of the alkyl group from the central atom to one of the nitrogen atoms of the porphyrin yields a rearranged complex, which dissociates to form an alkyl anion, which then reacts with CO_2 to yield the carboxylated product:[326]

$$R^- + CO_2 \rightarrow RCO_2^- \qquad (12.72)$$

$$RCO_2^- + RX \rightarrow RCOOR \qquad (12.73)$$

12.3.2.12   Co(salen)-Promoted Carboxylation

The carboxylation of some substituted benzyl chlorides was also performed without sacrificial anodes. Several arylmethyl chlorides were converted to the corresponding carboxylic acids and hydrocarbons by Isse et al. using as electrocatalyst Co(salen) [H_2salen = N,N'-bis(salicylidene)-ethane-1,2-diamine] in a medium of CO_2-saturated acetonitrile containing 0.1 M Bu$_4$NClO$_4$.[327] The reduction reactions were carried out on mercury cathodes. With benzyl chloride, 4-methoxybenzyl chloride, diphenylmethyl chloride, and 4-(trifluoromethyl)benzyl chloride, the yields of the carboxylic acids reached up to 44, 38, 93, and 84%, respectively. The reactions presumably involved one-electron reduction of the Co^{II}L complexes (L = salen) to Co^I complexes, which reacted with the benzyl chlorides to give organometallic complexes of Co^{III}, further reduction of which leads to [Co^{II}L(R)]$^-$:

$$[Co^{II}L] + e^- \rightarrow [Co^IL]^- \qquad (12.74)$$

$$[Co^IL]^- + RCl \rightarrow [Co^{III}L(R)] + Cl^- \qquad (12.75)$$

$$[Co^{III}L(R)] + e^- \rightarrow [Co^{II}L(R)]^- \qquad (12.76)$$

The decomposition of the unstable $[Co^{II}L(R)]^-$ species in the presence of CO_2 or a proton source (e.g., water) resulted in the formation of carboxylic acids or hydrocarbons. The details of the mechanism are yet unclear. The reaction of $[Co^{II}L(R)]^-$ with CO_2 may be either a direct displacement yielding $RCOO^-$ or may involve prior dissociation of $[Co^{II}L(R)]^-$ to release either the carbanion R^- or the radical $R\cdot$, both of which would react rapidly with CO_2.[327]

12.3.2.13 Carboxylation of Quinones

Semiquinones and quinone dianions in acetonitrile solutions, obtained by electro-reduction from the quinones, react with CO_2, forming addition products. Thus, Mizen and Wrighton found that 9,10-phenanthrene-quinone in $CH_3CN/(n$-$Bu_4N)BF_4$, when electroreduced, added two molecules of CO_2 to produce the bis(carbonate) dianion:[328]

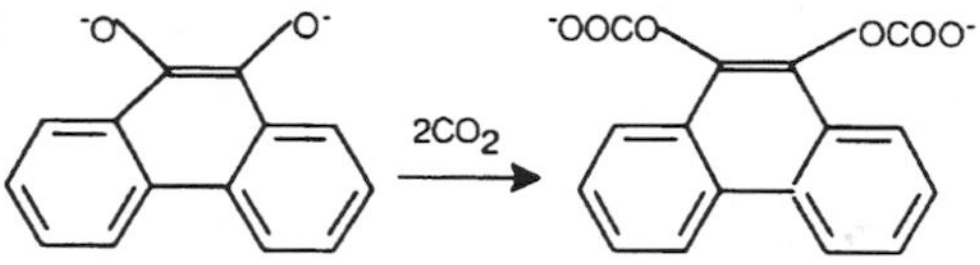

The reaction was found to be chemically reversible and may therefore in principle be a basis for the reversible abstraction of carbon dioxide from the atmosphere. Cyclic voltammetry measurements indicated reductive coupling of CO_2 to the semiquinone anions:[328]

$$Q + e^- = \cdot Q^- \tag{12.77}$$

$$\cdot Q^- + CO_2 = \cdot QCO_2^- \tag{12.78}$$

$$\cdot QCO_2^- + e^- = Q(CO_2)^{2-} \tag{12.79}$$

$$Q(CO_2)^{2-} + CO_2 = Q(CO_2)_2^{2-} \tag{12.80}$$

A similar reversible reductive carboxylation was observed by Tanaka et al. with 2,3,5,6-tetramethylquinone in acetonitrile or methanol solutions, which led to formation of doubly carboxylated species.[329] The scope of the reaction was extended by Nagaoka et al. to several chloro-, dichloro-, tetrafluoro-, hydroxy-, and dihydroxy-1,4-benzoquinones, 9,10-anthraquinones, and 1,4-naphthoquinones.[330]

12.3.2.14 Electrosynthesis of Oxazolidine-2,4-Diones

Casadei et al. used electrocarboxylation to produce various derivatives of oxazolidine-2,4-dione, which have important therapeutic uses, particularly as anticonvulsants and antiepileptics.[331,332] Using Hg or vitreous carbon cathodes and sacrificial anodes, NH-protic carboxamides bearing a leaving group at the 2-position were carboxylated by CO_2-forming oxazolidine-2,4-diones. Thus the important anticonvulsant Trimethadione® (3,5,5-trimethyloxazolidine-2,4-dione) was obtained in about 60% yield by the reaction of a bromoamide with CO_2 in the presence of methyl iodide in DMF/0.1 M Et_4NClO_4 solution:

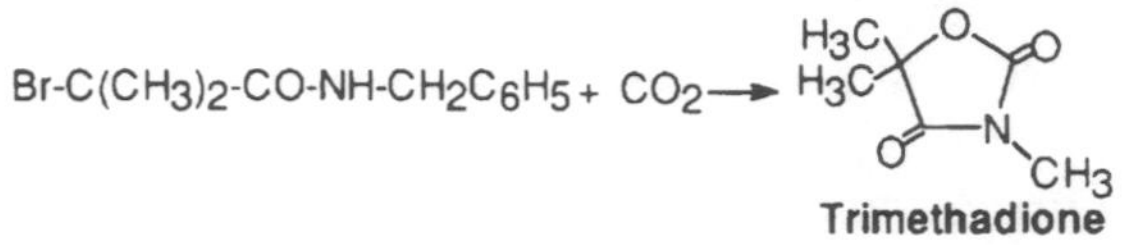

$$Br-C(CH_3)_2-CO-NH-CH_2C_6H_5 + CO_2 \longrightarrow$$

Trimethadione

These reactions carried out under mild conditions enable the synthesis of such heterocycles without requiring the application of the toxic and hazardous isocyanate intermediates used in the conventional production methods.

12.3.3   Reduction with Macrocycles and Metal Complexes

12.3.3.1   Mediation by Rhenium and Rhodium Complexes

Very high selectivity for CO formation was attained by Hawecker et al. with the rhenium(I) complex $Re(bpy)(CO)_3Cl$ (bpy = 2,2′-bipyridine) as mediator. In DMF–H_2O (9:1) solution, on glassy carbon electrodes at a potential of −1.2 V (vs. NHE), carbon monoxide was produced with 98% current efficiency.[333,334]

Using the rhenium complex fac-$Re(bpy)(CO)_3Cl$ in acetonitrile solution, Sullivan et al. performed the electrocatalytic reduction of CO_2 on a platinum electrode at −1.5 V (vs. SCE), resulting in the production of CO (current efficiency about 98%) and CO_3^{2-}, in agreement with the equation[335]

$$2CO_2 + 2e^- \rightarrow CO + CO_3^{2-} \tag{12.81}$$

The cyclic voltammetry study suggested the occurrence of two electrocatalytic pathways for the reduction of CO_2 to CO, involving either initial one-electron or two-electron reduction of the above rhenium complex.

The mechanism of the electrochemical reduction of CO_2 to CO at glassy carbon electrodes in acetonitrile solutions containing NEt_4BF_4 and fac-$[Re(dmbipy)(CO)_3Cl]$, where dmbipy = 4,4′-dimethyl-2,2′-bipyridyl, was studied by Christensen et al. using an *in situ* infrared analysis.[336] In the nominal absence of water, the one-electron reduction of the complex followed by CO_2 addition led to the formation of $[Re(dmbipy)(CO)_3(CO_2H)]$. In the presence of added water (10%), this complex was protonated to $[Re(dmbipy)(CO)_3(CO_2H_2)]^+$, which reacted with acetonitrile to form $[Re(dmbipy)(CO)_3(MeCN)]^+$ and CO. In these acetonitrile–water solutions (with NBu_4Cl as electrolyte), the rate of CO production was much higher than in the above anhydrous acetonitrile.

Films of $[Re(CO)_3(v-bpy)Cl]$, where v-bpy is 4-vinyl-4′-methyl-2,2′-bipyridine, were deposited by electropolymerization on Pt and glassy carbon electrodes. In acetonitrile containing tetra-n-butyl ammonium perchlorate, CO_2 was electroreduced to CO with a current efficiency of over 95% and a turnover number of about 600.[337,338]

The rhodium complex $Rh(diphos)_2Cl_3$ (diphos = 1,2-bis(diphenyl phosphino)ethane) was found by Slater and Wagenknecht to catalyze the electroreduction of CO_2.[339] Using a mercury pool cathode, at about −1.5 V (vs. a silver wire reference), in 0.1 M tetraethylammoniun perchlorate in acetonitrile solution, separated by a glass frit from a Pt anode, at a current density of about 8 mA cm^{-2}, the main product was the formate anion, obtained in up to 42% current efficiency, as well as small amounts of cyanoacetate. The mechanism proposed involved the reduction of the starting rhodium complex to a neutral species $Rh(diphos)_2^0$, which then interacted with CO_2.

12.3.3.2 Ni-Bipyridine and -Phenanthroline Complexes

With $[Ni(bpy)_3]^{2+}$ in acetonitrile solutions containing tetrabutyl ammonium carbonate and saturated with CO_2, electroreduction on a glassy carbon electrode did produce carbon monoxide and carbonate ions.[340] However, due to a side reaction, a large part of these products was not released. The primary product of the two-electron reduction of $[Ni(bpy)_3]^{2+}$ is a reversible reaction leading to $[Ni^0(bpy)_2]$. This intermediate was proposed not only to react with CO_2 to form CO and carbonate,

$$[Ni^0(bpy)_2] + 2CO_2 \rightarrow [Ni(bpy)_3]^{2+} + CO + CO_3^{2-} \qquad (12.82)$$

but also with CO to form a carbonyl complex,

$$[Ni^0(bpy)_2] + 2CO \rightarrow [Ni(CO)_2(bpy)] + bpy \qquad (12.83)$$

Highest current efficiency, up to 55% for CO production, was achieved in constant current experiments in a single-compartment electrolysis cell.

The mechanism of electroreduction of CO_2 mediated by nickel(0)-4,4'-dimethyl 2,2'-bipyridine and nickel(0)-1,10-phenanthroline complexes on Pt or glassy carbon electrodes in CO_2-saturated acetonitrile containing 0.1 M Et_4NBF_4 was studied by Christensen et al. using *in situ* FTIR spectroscopy.[341] The $[Ni(Me_2bpy)_3]^+$ complex mediated the reduction of CO_2 to CO and CO_3^{2-} in aprotic media. Electrogenerated $[Ni(0)(Me_2bpy)_2]$ reacted slowly with CO_2, forming $[Ni(Me_2bpy)-(CO)_2]$ and CO_3^{2-}. Another reduction step formed the electroactive catalyst, the complex $Ni(Me_2Bpy^{\cdot-})(CO)_2$.[321] Analogous CO_2 reduction reactions occurred with $[Ni(Phen)_3](ClO_4)_2$ (Phen = phenanthroline) as mediator. Major products were CO_3^{2-} and CO. The active catalytic intermediate species was identified by FTIR to be the $[Ni(Phen^{\cdot-})(CO)_2]$ radical anion complex.[341]

12.3.3.3 Fe, Ni, and Co Poly(vinyl)terpyridine Complexes

Film-coated bis(vinylterpyridine)Co^{2+} (v-trpy) electrodes were prepared by Potts et al. by electropolymcrization of v-trpy in acetonitrile solutions.[342] Hurrell et al.

observed that with electropolymerized films of v-trpy in CO_2-saturated DMF containing 0.1 M Bu_4ClO_4, CO_2 was already reduced to HCOOH at a potential of -0.90 V (vs. sodium saturated calomel electrode [SSCE]).[343]

Arana et al. compared the electrocatalytic activity for CO_2 reduction in acetonitrile or DMF solutions on either bare Pt or glassy carbon electrodes with Fe, Ni, and Co complexes of monomeric terpyridine (trpy) in solution or on the same electrodes coated with Fe, Ni, and Co complexes of polymerized 4'-vinyl-terpyridine (v-trpy).[3] Both in homogeneous solutions and with film-coated electrodes, the presence of the metal–trpy complexes lowered the overpotential and strongly catalyzed the reduction of CO_2. Much higher activity was observed only with the Fe and Co film-coated electrodes. The potentials for the electrocatalytic reduction of CO_2 in DMF/0.1 M Bu_4NClO_4 with the Fe, Co, and Ni complexes of trpy in solution were -1.13, -1.56, and -1.20 V (vs. SSCE), while with the corresponding metal–v-trpy-coated electrodes the potentials were only -0.95, -0.90, and -1.22 V, respectively. The reduction potential with the electropolymerized Co(v-trpy) electrode was thus substantially less negative than with the $Co(trpy)_2^+$ complex in solution, resulting in a considerable energy saving. The major product from the $Co(trpy)_2^+$-mediated reduction was formate. The mechanism of reduction of CO_2 involved both metal-localized and ligand-localized processes. The reduction of CO_2 at -1.56 and -1.91 V mediated by $Co(trpy)_2^+$ was assigned to ligand-based processes. The electropolymerized films were resistant to washing with water, acetone, acetonitrile, and DMF, but slowly lost their activity (almost 10%) during 6 h of continuous potential cycling. Among the above three metal complexes, the highest electrocatalytic activity was with the cobalt complexes, followed by the iron and nickel complexes. The exceptional activity of the Co complexes was attributed to their acting as two-electron donors through the Co(III)/Co(I) couple.[3]

12.3.3.4 Multielectron Reduction with Ru Bipyridine and Terpyridine Complexes

Both the electrochemical[344–346] and photochemical[347] reduction of CO_2 are catalyzed by the $[Ru(bpy)_2(CO)]_2^{2+}$ complex, which in aqueous/organic media exists in rapid acid–base equilibria:

$$[Ru(bpy)_2(CO)]_2^{2+} + OH^- = [Ru(bpy)_2(CO)(C(O)OH)]^+ \qquad (12.84)$$

$$[Ru(bpy)_2(CO)(C(O)OH)]^+ + OH^- = [Ru(bpy)_2(CO)(CO_2)] + H_2O \qquad (12.85)$$

In dry CO_2-saturated acetonitrile, the $[Ru(bpy)_2(CO)]_2^{2+}$ complex transforms by two-electron reduction into $[Ru(bpy)_2(CO)(CO_2)]$, with evolution of CO:[348,349]

$$[Ru(bpy)_2(CO)]_2^{2+} + CO_2 + 2e^- \rightarrow [Ru(bpy)_2(CO)(CO_2)] + CO \qquad (12.86)$$

Controlled potential electrolysis over Hg pool electrodes of solutions of $[Ru(bpy)_2(CO)_2]^{2+}$ or $[Ru(bpy)_2(CO)Cl]^+$ in CO_2-saturated H_2O (pH 6.0)/DMF (9:1 v/v) at -1.50 V vs. SCE produced both CO and hydrogen. On the other hand, in alkaline media, H_2O (pH 9.5)/DMF (9:1 v/v), the products were $HCOO^-$ (current efficiency 34%), CO, and H_2. The mechanism proposed involved irreversible two-electron reduction to an unstable penta-coordinated Ru(0) complex $[Ru(bpy)_2(CO)]^0$ from both of the above Ru complexes. This intermediate coordinated CO_2 and then underwent protonation and dehydroxylation, finally releasing either CO or $HCOO^-$. Much improved selectivity for formate production was achieved by Ishida et al. using the $[Ru(bpy)_2(CO)_2]^{2+}$ complex in acetonitrile solution containing methyl-amines or phenol as proton sources. In a two-compartment electrolysis cell, with an Hg pool cathode and a Pt plate anode separated by a Nafion® membrane, containing the Ru complex (0.5 mM) and Bu_4NClO_4 as electrolyte, at a potential of -1.5 V vs. SCE, the current yields for the reduction of CO_2 to formate in the presence of 0.2 M $Me_2NH \cdot HCl$ or PhOH were 84 and 81%, respectively, at a current density of 2 to 3 mA cm^{-2}.[344-346]

Nagao et al. observed that in the electrochemical reduction of CO_2 mediated by $[Ru(bpy)(trpy)(CO)](PF_6)_2$ (bpy = 2,2'-bipyridine, trpy = 2,2',6,2'-terpyridine) in aqueous ethanol at low temperature, the reduction products included not only CO and HCOOH as the major products, but also formaldehyde, methanol, glyoxylic acid, and glycolic acid, the formation of which must have involved four- and six-electron reduction processes.[350] Under controlled potential electrolysis at -1.70 V (vs. Ag/Ag$^+$) in CO_2-saturated ethanol/water (8:2 v/v) at $-20°C$, the products CO, HCOOH, H_2, and CH_3OH were formed with current efficiencies of 35, 20, 20, and 0.3%, respectively. When the electrolysis was performed at -1.75 V, the C_2 compounds $HOCH_2COOH$ and $HC(O)COOH$ were also formed — in small yields. Tracer experiments with $^{13}CO_2$ proved that the organic products were derived from CO_2. A key intermediate in the multielectron reduction of CO_2 was proposed to be the formyl complex $[Ru(bpy)(trpy)(CHO)]^+$, formed by two-electron reduction from the above initial complex. A confirmation for the possible role of this intermediate complex in the CO_2 reduction reactions was obtained by Toyohara et al. with a nonelectrochemical synthesis of this intermediate from the above starting complex $[Ru(bpy)(trpy)(CO)](PF_6)_2$ in CD_3CN solution at $-40°C$ with $LiBEt_3H$. The resulting formyl complex was identical (by ^1H-NMR) to the electrochemically formed intermediate. Its reaction with CO_2 produced formic acid in 60% yield and regenerated the starting complex:[351,352]

$$[Ru(bpy)(trpy)(CHO)]^+ + CO_2 \rightarrow [Ru(bpy)(trpy)(CO)]^+ + HCOO^- \quad (12.87)$$

The reduction of CO_2 mediated by mononuclear homogeneous catalysts usually enabled only two-electron reduction reactions, leading to CO, HCOOH, and oxalate. Haines proposed the advantage of metal cluster compounds for reducing CO_2 beyond the two-electron stage and in particular to achieve the six-electron reduction to methanol.[353] By reacting triruthenium dodecacarbonyl [$Ru_3(CO)_{12}$] with 2,2′-bipyridine in cyclohexane under reflux, a bipyridine-chelated triruthenium complex was obtained. With this complex, in acetonitrile solution containing a trace of water, the electroreduction of CO_2 on a vitreous carbon electrode at −1.6 V (vs. SCE) yielded CO, $HCOO^-$, and CH_3OH with the remarkably high current efficiencies of 40, 15, and 30%, respectively. The catalytic effect was observed to be due to the deposition of an insoluble intensely blue material on the electrode, the elementary composition of which was [$Ru_3(CO)_8(bpy)_2$]. This was thus actually a case of heterogeneous electrocatalysis.

Three- and four-carbon compounds, in addition to CO and HCOOH, were obtained by Nakajima et al. during the electrochemical reduction of CO_2 mediated by the [$Ru(bpy)_2(qu)(CO)$]$^{2+}$ (qu = quinoline) complex, when carried out in CH_3CN/DMSO (1:1 v/v) in the presence of $(CH_3)_4NBF_4$.[354,355] The reduction was performed on a glassy carbon working electrode at a potential of −1.5 V (vs. Ag/AgCl) in CO_2-saturated solvent, producing carbon monoxide, acetone, acetoacetic acid (CH_3COCH_2COOH), and formic acid with current efficiencies of 42, 16, 6, and 7%, respectively, as well as carbonate. The methyl groups of the methylated products were shown to be derived from the $(CH_3)_4NBF_4$, while in a further step of electrocarboxylation, acetone was converted to acetoacetate:

$$2CO_2 + 4e^- + 2(CH_3)_4N^+ \rightarrow CH_3COCH_3 + CO_3^{2-} + 2(CH_3)_3N \qquad (12.88)$$

$$CH_3COCH_3 + 2e^- + 2CO_2 \rightarrow CH_3COCH_2COO^- + HCOO^- \qquad (12.89)$$

Similar methylation products were obtained when using CH_3I instead of $(CH_3)_4NBF_4$ as the methylating agent. When $LiBF_4$ was used as the supporting electrolyte, the only products were CO and carbonate, produced by a reductive disproportionation of CO_2:

$$2CO_2 + 2e^- \rightarrow CO + CO_3^{2-} \qquad (12.90)$$

The proposed tentative mechanism of methylation of the carbonyl ligand of the ruthenium complex in the presence of $(CH_3)_4NBF_4$ or CH_3I involved as the primary step the two-electron reduction of the initial [$Ru(bpy)_2(qu)(CO)$]$^{2+}$ complex, leading to intermediate complexes such as [$Ru(bpy)_2(qu)(CO)$]^0 and [$Ru(bpy)_2(qu)(h^1(C)-CO_2)$]. Double methylation of the carbonyl ligand may possibly occur via [$Ru(bpy)_2(qu)(C(O)CH_3)$]$^+$.[354,355]

In the electrochemical reduction of CO_2 in acetonitrile containing added water, the complex cis-[$Ru(bpy)_2(CO)H$]$^+$ caused the production of formic acid and carbon monoxide. This same complex was previously proposed by Hawecker et al. to be the active catalyst in the photochemical reduction of CO_2 using [$Ru(bpy)_3$]$^{2+}$ as sensitizer.[356] The electrochemical HCOOH production was shown by Pugh et al. to be due to a catalytic cycle involving (1) one-electron reduction of the above catalyst; (2) insertion of CO$_2$ into the Ru–H bond to form a once-reduced formato complex, cis-[Ru(bpy)$_2$(CO)(OC(O)H)]^0; and (3) one-electron reduction of this complex, releasing HCOO$^-$ and recycling the complex to cis-[Ru(bpy)$_2$(CO)H]$^+$ by the reduction of H$_2$O.[357]

By linking together covalently two bipyridine complexes of transition metals, an analogy to the molecular assemblies active in natural photosynthesis may be created. The covalently linked polypyridine Re(I)/Ru(II) complex[358,359]

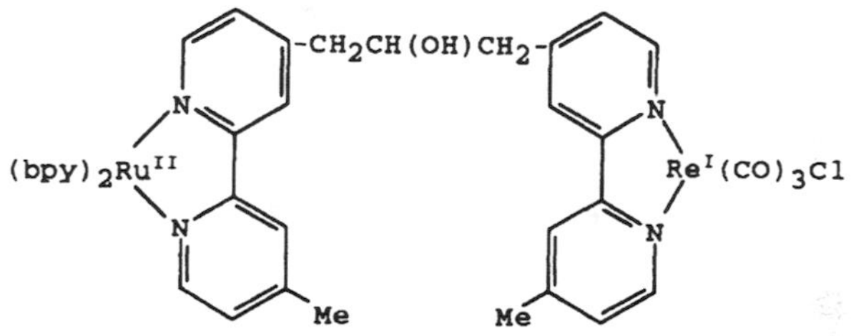

in CO$_2$/acetonitrile solution showed a catalytic wave at −1.3 V vs. SCE, while Re(bpy)(CO)$_3$Cl under similar conditions revealed a catalytic wave only at −1.65 V vs. SCE. The only detected reduction product was CO. The pathway for CO$_2$ electrochemical reduction involved an intramolecular electron transfer step from Ru(I) to Re(I).

12.3.3.5 Cu, Os, Rh, and Ir Bipyridine Complexes

A dicopper complex, [Cu$_2$(μ-PPh$_2$bipy)$_2$(CH$_3$CN)$_2$](PF$_6$)$_2$, where PPh$_2$bpy is 6-(diphenylphosphino)-2,2'-bipyridyl, was developed by Field et al.[360] and Haines et al.[353] as a two-electron catalyst for the electroreduction of CO$_2$. The μ-phosphinobipyridyl group has the unique property of combining bipyridyl and bridging phosphine ligands. In reduction reactions involving these ligands, the electrons become localized at the π*-orbitals of the bipyridyl groups, which serve as electron reservoirs for the catalytic action. With this dicopper complex, in CO$_2$-saturated acetonitrile solution, and using a platinum gauze electrode at −1.7 V (vs. Ag/AgCl), the products formed were CO and carbonate, in agreement with a reductive disproportionation of CO$_2$:

$$2CO_2 + 2e^- = CO + CO_3^{2-} \qquad (12.91)$$

The binding of CO$_2$ to a reduced form of the dicopper complex was proposed to be a key step in the process.

Electrocatalysis of carbon dioxide reduction was also obtained by Bruce et al. with the complex cis-[Os(bpy)$_2$(CO)H][PF$_6$] in acetonitrile containing 0.1 M tetra-n-butyl ammonium hexafluoro phosphate, using a Pt mesh electrode.[361,362] With anhydrous

CH_3CN, at potentials of -1.4 to -1.6 V (vs. SSCE), CO was produced with up to 90% current efficiency, as well as small amounts of formic acid. Adding water to the electrolysis medium caused an increase in the formic acid yield, which reached a maximal current efficiency of 25% with about 0.3 M H_2O in the medium. The net reactions may be represented by:

$$3CO_2 + 2e^- + H_2O \rightarrow CO + 2HCO_3^- \qquad (12.92)$$

$$2CO_2 + 2e^- + H_2O \rightarrow HCO_2^- + HCO_3^- \qquad (12.93)$$

Presumably even the *dry* acetonitrile contained enough water for the above equations. The proposed mechanistic pathway involved the association of CO_2 with the di-reduced dipyridyl complex $[Os(bpy)_2(CO)H]^-$ by coordination sphere expansion, forming a reactive intermediate, which either dissociated to CO or, in the presence of sufficient water, produced formic acid.[361,362]

Rhodium and iridium bipyridyl complexes, *cis*-$[Rh^{III}(bpy)_2Cl_2]^+$ and *cis*-$[Ir^{III}(bpy)_2Cl_2]^+$, were found by Bolinger et al. to be active catalyst precursors for CO_2 reduction.[363] With *cis*-$[Rh^{III}(bpy)_2Cl_2]^{2+}$ in acetonitrile solution containing $[(n-Bu)_4N](PF_6)$ as supporting electrolyte, using a carbon cloth electrode at -1.55 V (vs. SSCE), CO_2 reduction to HCOOH occurred with 80% current efficiency. The turnover number reached up to 12. The proton of the formic acid was presumably derived from the decomposition of the supporting electrolyte. The active electrocatalysts were proposed to be the hydride complexes (e.g., *cis*-$[Ir^{III}(bpy)_2H_2]^+$). Carbon monoxide was not observed as a product. The electron reservoir character of 2,2'-bipyridine was suggested to promote a metal-centered reduction of carbon dioxide.

Mixed-metal trimetallic complexes containing two ruthenium atoms and one iridium atom were used by Nallas and Brewer to electroreduce CO_2.[364] In this trimetallic system, $\{[(bpy)_2Ru(BL)]_2IrCl_2\}^{5+}$, BL was the bridging ligand 2,3-bis(2-pyridyl)quinoxaline (dpq) or 2,3-bis(2-pyridyl)benzoquinoxaline (dpb). In contrast with the above-noted results with the monometallic $[Ir^{III}(bpy)_2Cl_2]^+$, in which HCOOH was the only product,[363] the reduction of CO_2 in the trimetallic system led with up to 100% current efficiency to CO.[364]

12.3.3.6   Ag, Pd, Fe, and Co Porphyrins

In methylene chloride solutions, with Ag(II) or Pd(II) metaloporphyrins as homogeneous catalysts, using a glassy carbon electrode at about -1.5 V (vs. Ag reference electrode) and Pt gauze as anode, at a current density of about 3 mA cm^{-2}, oxalic acid was obtained by Becker et al. as the only product of CO_2 reduction.[365]

The electrochemical reduction of carbon dioxide was found by Savéant et al. to be catalyzed by iron(0) porphyrins, resulting in carbon monoxide as the main product.[366–368] With tetraalkyl ammonium salts as supporting electrolyte, in dimethylformaldehyde solution, the porphyrins were rapidly degraded, possibly by carboxylation. Much improved stability of the iron(0) porphyrin electrocatalysts was achieved by addition of anhydrous magnesium perchlorate to the medium and by operation of the electrolysis at a low temperature, –40°C. The low temperature also resulted in enhanced solubility of carbon dioxide, thus increasing the catalytic efficiency. The proposed mechanism involved the insertion of one carbon dioxide molecule into the Fe coordination sphere, which was followed at room temperature by carbon–oxygen bond breakage, releasing carbon monoxide — a process which was assisted by the electrophilic Mg^{2+} ions. At low temperatures, two carbon dioxide molecules were found to be involved in the reactive intermediate complex. The catalytic process may depend on the intermediate production of a carbenoid-type complex, with electron transfer from the iron atom to the carbon dioxide group.[366-368] Other Lewis acid cations, such as Ca^{2+}, Ba^{2+}, and Na$^+$, also had a synergistic effect on the iron(0)-porphyrin-catalyzed reduction of CO$_2$ in DMF solution.[170]

Bhugun et al. observed that with iron(0) tetraphenylporphyrin as electrocatalyst, the reduction of CO$_2$ to CO was dramatically enhanced by the presence of weak Brönsted acids, such as trifluoroethanol, 1-propanol, and 2-pyrrolidone.[170,369] With 1 mM iron tetraphenylporphyrin and 0.55 M CF$_3$CH$_2$OH in DMF solution, using a mercury pool as working electrode, the faradaic yield of CO$_2$ reduction to CO was 94%, while H$_2$ was not detectable. By cyclic voltammetry, the catalytic efficiency of the reduction was shown to have increased by more than 130 times in the presence of CF$_3$CH$_2$OH. The turnover number of CO per hour initially reached the relatively high value of about 40, but unfortunately decreased substantially within a few hours, due to degradation of the porphyrin. With 1-propanol, the reaction was less selective, yielding 60% CO and 35% formate. The function of the acid synergist may be to stabilize the initial Fe^{II}CO$_2^{2-}$ complex by hydrogen bonding.

Cobalt(II) tetraphenylporphyrin (Co^{II}TPP) in tetra-n-butyl ammonium tetrafluoroborate dissolved in DMF was found by Atoguchi et al. to mediate the electrochemical reduction of CO$_2$, using glassy carbon or platinum as working electrodes.[370] At a controlled potential of –1.5 V vs. SCE, the current density was 0.1 mA cm^{-1}, and the main reduction product was formic acid, obtained with a current efficiency of about 10%. The electrochemistry of Co^{II}TPP was studied by Tezuka et al. using cyclic voltammetry on glassy carbon electrodes in hexamethylphosphoramide solutions containing tetra-butylammonium tetrafluoroborate as electrolyte.[371] Under an argon atmosphere, the cobalt complex underwent two reversible one-electron reduction waves, corresponding respectively to the transitions Co(II) → Co(I), followed by Co(I) → Co(0). Under a CO$_2$ atmosphere, the second transition was markedly changed: the reduction current was considerably enlarged, while the reoxidation peak disappeared. The explanation proposed was that the increase in the reduction current was due to the reduction of carbon dioxide according to

$$Co(0)TPP + CO_2 \rightarrow Co(I)TPP + CO_2^- \qquad (12.94)$$

12.3.3.7 Pd(triphosphine) Complexes

Electrochemical reduction of carbon dioxide to carbon monoxide was the main reaction pathway when catalyzed by some palladium triphosphine complexes in acidic DMF or acetonitrile solutions. A variety of square-planar Pd complexes of the type [Pd(triphos)L](BF_4)$_2$ (where triphos is a triphosphine ligand, such as $PhP(CH_2CH_2PPh_2)_2$, and L = CH_3CN, $P(OMe)_3$, PEt_3, $P(CH_2OH)_3$ and PPh_3),

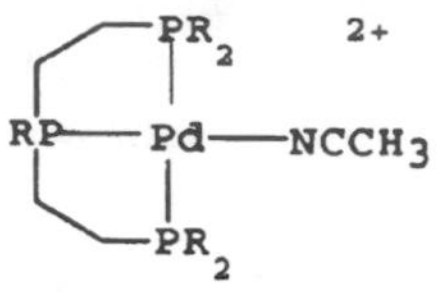

were found by DuBois and Miedaner to be active catalysts in the electrochemical reduction of CO_2, while the analogous Ni and Pt complexes were inactive under similar conditions.[372] Since the addition of the free phosphine ligand caused an inhibitory effect, decreasing the rate of CO_2 reduction, it was concluded that dissociation of a phosphine ligand from the complex was necessary for CO_2 insertion. Using complexes of [Pd(etpC)(Ch_3CN)](BF_4)$_2$, where etpC represents bis(dicyclohexyl phosphinoethyl)phenylphosphine, effective reduction of carbon dioxide to carbon monoxide was achieved, with current efficiencies of up to 85% and turnover number (moles of CO produced per mole of catalyst) of up to 130. Upon prolonged electroreduction of CO_2, the catalyst was degraded, due to formation of an inactive Pd–Pd dimerization product. Controlled potential electrolysis measurements showed that only one electron per palladium was consumed in the presence of CO_2, indicating that a Pd(I) species was the active intermediate in the reactions with CO_2. The dependence of the catalytic current on acid concentration was biphasic: At low acid concentration, the current followed first-order kinetics, with cleavage of the C–O bond as the rate-determining step, followed by protonation of a hydroxycarbonyl intermediate. At higher acid concentration (the normal reaction conditions), the current was independent of acid concentration, and the rate-determining step was the reaction of a monomeric Pd(I) intermediate with CO_2. Mechanistic studies led DuBois et al.[373–376] to the proposition of a five-coordinate palladium(I) complex as the transition state for the rate-determining step at high acid concentrations. The approach of CO_2 to this Pd(I) species may occur along the axis perpendicular to the plane formed by the catalyst:

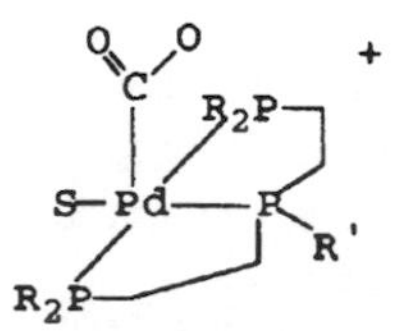

When the central phosphorus atom of the tridentate ligand in the above Pd complexes was substituted by C, N, O, As, and S, the CO_2 reduction was inhibited.[377]

In order to prevent the deactivation of the [Pd(triphosphine)(CH₃CN)](BF₄)₂ complexes, which had been shown to be due to formation of Pd(I) dimers with bridging triphosphine ligands, Wander et al. prepared a series of triphosphine ligands, in which the number of methylene groups bridging the phosphorus atoms was varied.[378] With the complex [Pd(ttpE)(CH₃CN)](BF₄), where ttpE = bis(3-(diethylphosphino)propyl)phenyl-phosphine, in a CO_2-saturated acidic DMF solution at a potential of −1.4 V (vs. ferrocene/ferrocenium), the current efficiency of CO production reached 95%, with a turnover number of 120. Using this complex, containing two trimethylene bridges, the formation of the inactive Pd(I) dimers was inhibited. However, the catalyst lifetime was not improved, because another degradation pathway appeared, with formation of a catalytically inactive [Pd(triphosphine)(H)](BF₄) complex.

Analogous palladium complexes were prepared by Miedaner et al. with organophosphine dendrimers.[379] These dendrimers were obtained by the free-radical-initiated addition of either primary phosphines to diethyl vinylphosphonate or of secondary phosphines to tetravinylsilane. The palladium complexes formed with such dendrimers, containing up to 15 phosphorus atoms, were active as electrocatalysts for the reduction of CO_2 to CO in DMF solution. However, the current efficiencies and turnover numbers achieved were inferior to those attained with the simpler monomeric [Pd(triphosphine)(CH₃CN)] · (BF₄)₂ complexes.

In an effort to enhance the activity of the Pd(triphosphine) electrocatalysts for the reduction of CO_2 to CO, Steffey et al. synthesized a dinuclear palladium complex which contained a bridging hexaphosphine ligand:[380]

$$(Et_2PCH_2CH_2)_2PCH_2P(CH_2CH_2PEt_2)_2 = (eHTP) \qquad (12.95)$$

With the active electrocatalyst [Pd₂(CH₃CN)₂(eHTP)](BF₄)₄ in acidic solutions of DMF (0.1 M HBF₄), using a glassy carbon electrode at a potential of −1.3 V (vs. the ferrocene/ferrocenium couple), CO was produced with a current efficiency of 85%. However, the turnover number was only 8, due to very rapid degradation of the catalyst. The kinetic measurements indicated that both palladium atoms of the complex were involved in CO_2 reduction.

Water-soluble derivatives of the above [Pd(triphosphine)(CH₃CN)](BF₄)₂ complexes were prepared by Herring et al. by introducing polar functional groups such as hydroxypropyl and dialkylamino substituents into the triphosphine moiety.[381] The use of an aqueous medium instead of the organic solvents for the electroreduction would be advantageous both environmentally and for higher electric conductivity. Unfortunately, the current efficiencies for the reduction of CO_2 to CO were substantially lower in water than in the organic solvents. Thus, with the water-soluble catalyst

$$HOCH_2CH_2CH_2P \overbrace{\underbrace{\begin{array}{c} PEt_2 \\ Pd-NCCH_3 \\ PEt_2 \end{array}}}^{+}$$

the current efficiencies for CO production in water and DMF were 40 and 49%, respectively. The turnover number was only 10. The reason for the different efficiencies may be the higher solubility of CO_2 in DMF than in water.

12.3.3.8   Reduction to HCOOH with Pd-Pyridine and -Pyrazole Complexes

The electrochemical reduction of bicarbonate to formate was achieved by Stalder et al. with high current efficiency using Pd-impregnated polymers. The polymer was obtained from the bipyridine-containing monomer,[382]

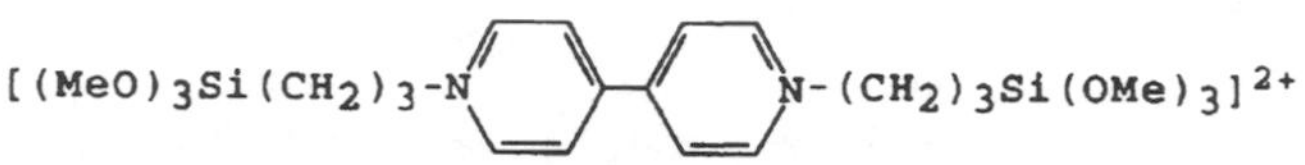

$$[(MeO)_3Si(CH_2)_3-N \quad \quad N-(CH_2)_3Si(OMe)_3]^{2+}$$

on metals (W or Pt). Highest current efficiency (85%) for the production of formate was obtained using 7 M CsHCO$_3$ as the catholyte, at a current density of about 0.1 mA cm^{-2} and a potential which was close to the thermodynamic potential $E°$ (CO$_3$H$^-$/HCO$_5^-$) = -0.76 V vs. SCE. In order to optimize the electroreduction of bicarbonate to formate on the redox polymers, André and Wrighton determined the relative electrostatic binding of bicarbonate and formate ions on such polymers.[383] The binding constants were measured by FTIR spectroscopy of the polymer derived from an N,N'-dialkyl-4,4'-bipyridinium monomer immobilized on a single-crystal Si electrode surface. When both ions were at high concentrations, the CO$_3$H$^-$ ion was more strongly bound. Thus, at 3.0 M concentration, the bicarbonate ion was about seven times more firmly bound than the formate ion.

With several (PdCl$_2$L$_2$) complexes (where L = substituted pyridine or pyrazole) as mediators, in acetonitrile containing water (4% by volume) and 0.1 M Et$_4$NClO$_4$, on glassy carbon or Pt electrodes, Hossain et al. succeeded in reducing CO$_2$ to formic acid as the only observed product.[384,385] Carbon monoxide was not formed. At a potential of -1.10 V (vs. Ag/Ag$^+$), with the active electrocatalysts (PdCl$_2$L$_2$) in which L = pyrazole, 4-methylpyridine, and 3-methylpyrazolé, the current efficiencies for production of HCOOH were 10, 20, and 10%, respectively, and for hydrogen production were 31 to 54%. In dry acetonitrile, there was no production of HCOOH. Thus, water served as the proton source for HCOOH formation. In the proposed tentative mechanism, the primary steps were the two-electron reduction of (PdCl$_2$L$_2$) to (PdL$_2$)°, in which the pyridine and pyrazole ligands acted as electron acceptor sites, followed by CO$_2$ binding to carbon atoms of the electron-rich ligands.

12.3.3.9   Reduction to CO Mediated by Co(salen) Complexes

A landmark discovery was the observation by Gambarotta et al. that CO$_2$ can be reversibly activated by the bifunctional complexes [Co(R-salen)M], in which R-salen = substituted salen ligand, H₂salen = N,N′-bis(salicylidene)ethylene-1,2-di-amine, and M = Li, Na, K, Cs:[386]

Co(salen) R = -CH₂CH₂- , Co(salophen) R = o-C₆H₄-

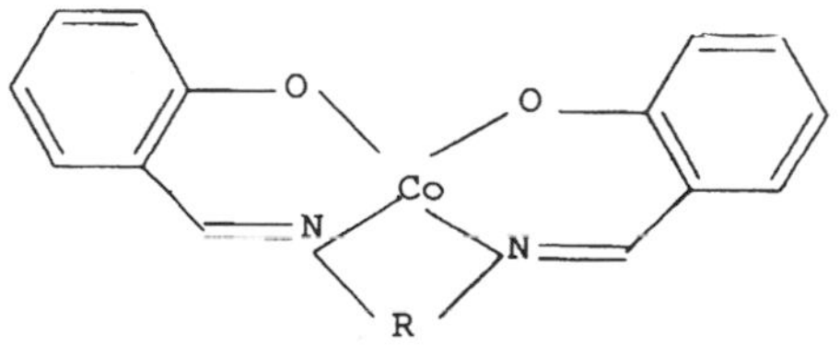

In the presence of alkali cations, the electrocatalytic reduction of carbon dioxide to carbon monoxide and carbonate was mediated by the cobalt complexes Co(salen) and Co(salophen), carried out either in anhydrous or aqueous acetonitrile, with an alkali metal perchlorate as the base electrolyte.[168,387,388]

In the case of Co(salen) catalysis, this reaction was enhanced by the presence of water. The reaction was performed on an Hg pool, at −1.4 V vs. SCE, at a current density of about 2 mA cm⁻², yielding both CO and HCO₃⁻. Replacing NaClO₄ by LiClO₄ as the base electrolyte caused a positive shift in the CO₂ reduction potential of 110 mV.[387]

With Co^I(salophen)Li, where H₂salophen = N,N′-bis(salicylidene)-o-phenylenediamine, the addition of water decreased the selectivity to CO formation and caused increased production of hydrogen.[388] In anhydrous acetonitrile solution, with lithium perchlorate as base electrolyte, on a mercury electrode at −1.5 V vs. SCE, the overall reduction of CO₂ resulted in formation of CO and carbonate, which precipitated as Li₂CO₃,

$$2CO_2 + 2e^- \rightarrow CO + CO_3^{2-} \tag{12.96}$$

The current efficiency for CO production was 29%, with a turnover number based on the catalyst of 22. The mechanism was represented by the reduction of Co(II) to Co(I) as the primary process (with a color change in the complex from brownish-red to green), followed by reversible stepwise binding of two CO₂ molecules to the Co(I) complex. The lithium ion played an essential role in the reaction, and the intermediate complex was assumed to contain a head-to-tail CO₂ dimer, C(O)OCO₂, which was C-bonded to cobalt and stabilized by Li⁺.

12.3.3.10 Reduction to CO with Tetraazamacrocyclic Compounds

Fisher and Eisenberg discovered that tetraazamacrocyclic compounds of Co and Ni dissolved in water/acetonitrile mixtures, with Hg cathodes at about −1.5 V (vs. SCE), mediated the reduction of CO₂. The products were CO and H₂ (in the ratio 1:1) and formic acid at a total current yield of up to 98%. Carbon dioxide reduction occurred at a potential which was closer to the thermodynamic value than in the absence of the mediator.[166]

Several tetraazamacrocyclic Co(II) and Ni(II) complexes were tested by Tinnemans et al. as electron transfer agents in the electroreduction of CO_2 on mercury electrodes, in both aqueous and organic solvents.[389] The products determined were CO and H_2. Highest current efficiency for CO, up to 66%, was achieved with the complex $\{Co(II)[Me_2\text{-}Pyo(14)trieneN_4]\}^{2+}$,

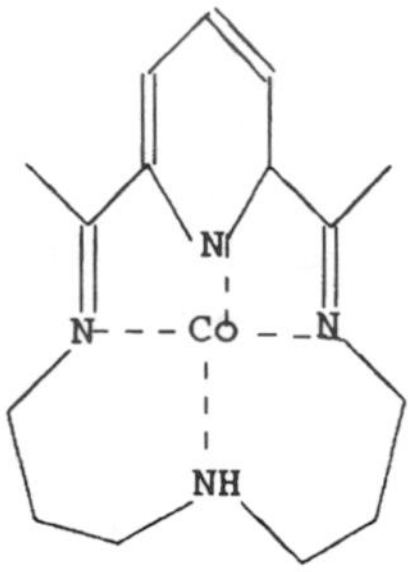

using a DMF/H_2O medium (95:5 v/v) containing 0.1 M Et_4NCl under a CO_2 atmosphere and with the Hg electrode at a potential of -1.30 V (vs. SCE).

Cyclic voltammetric studies by Gangi and Durand of the electroreduction of the cobalt macrocycle,[390]

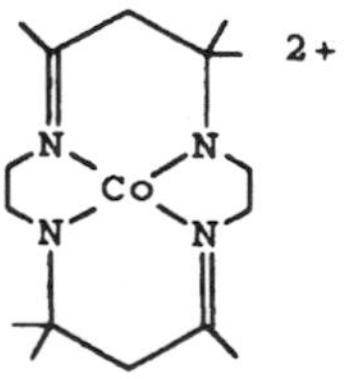

2+

in CO_2-saturated DMSO with 0.1 M Et_4NClO_4 indicated an equilibrium constant of 7×10^4 for CO_2 binding to the complex, with cobalt in the Co^I oxidation state. The reaction was proposed to occur by formation of a carbon–metal bond. The affinity of the corresponding Ni complex to CO_2 was much lower.

The electrocatalytic reduction of CO_2 to CO with cobalt macrocycles was carried out by Fujita et al. in CO_2-saturated water over an Hg electrode, yielding CO, H_2, and some formate.[391] Best yields and selectivities for CO production were obtained with the complexes

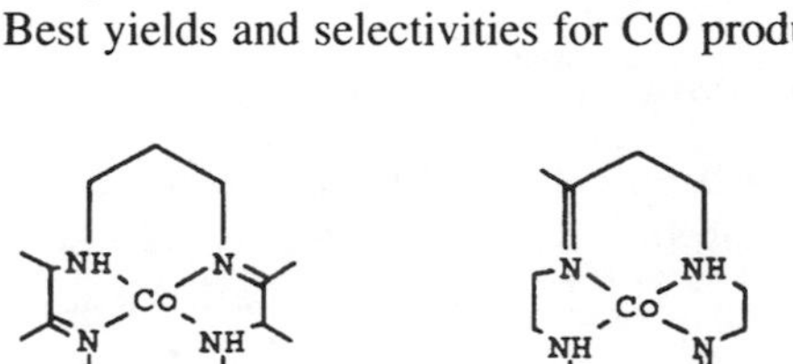

In the proposed mechanism, the one-electron reduction of the Co^{II} complex to a Co^I complex was followed by CO_2 association, forming a carbon–metal bond:

$$CoL^{2+} + e^- \rightarrow CoL^+ \tag{12.97}$$

$$CoL^+ + CO_2 \rightarrow CoL\text{–}CO_2^+ \tag{12.98}$$

$$CoL\text{–}CO_2^+ + H^+ = CoL\text{–}CO_2H^{2+} \tag{12.99}$$

$$CoL\text{–}CO_2H^{2+} + e^- \rightarrow CoL^{2+} + CO + OH^- \tag{12.100}$$

In a search for mediators for carbon dioxide reduction to formic acid, terdentate transition metal coordination complexes were found by Arana et al. to be more stable than those with bidentate and tetradentate ligands.[392] With DMF or acetonitrile as solvent, Fe, Co, and Ni complexes were tested. With the cobalt complex of dapa,

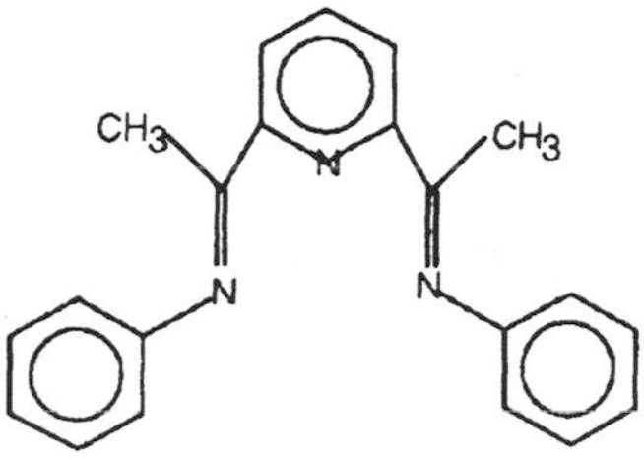

dapa

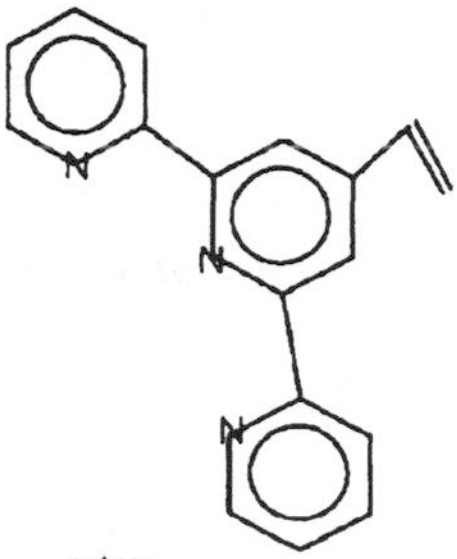

v-tpy

$Co(dapa)_2(PF_6)_2$ in DMF solution, the main CO_2 reduction product was formic acid, produced with a current efficiency of more than 60%. In case the ligand contained an olefinic group, such as in Co(v-tpy) and Ni(v-tpy) (v-tpy = vinyl-terpyridine), the complex was deposited on the electrode surface by electropolymerization, forming an electroactive polymeric film of the complex. Such films were effective for the electrocatalytic reduction of carbon dioxide. Thus, the reduction potential of $Co(v\text{-}tpy)_2(PF_6)_2$ was only –0.9 V (vs. SSCE). On carrying out rotating disk electrode experiments, and applying a Koutecky–Levich plot, the second-order rate constant for the electrocatalytic reduction of carbon dioxide was determined to be $k_2 = 40$ M^{-1} s^{-1}.

12.3.3.11   Reduction with Metal Cluster Complexes

A marked decrease in the overpotential required for CO_2 reduction in organic solvents was achieved by mediation with metal sulfur clusters. Rhodium and iridium complexes with triply bridging sulfido ligands were prepared by Nishioka and Isobe.[393] Thus $[RhC_p^*(\mu_3\text{-}S)_2]^{2+}$ (where $C_p^* = \eta^5$-pentamethylcyclopentadienyl) was obtained by reacting $[RhC_p^*(NCMe)_3]^{2+}$ with Na_2S in MeCN solution. These rhodium complexes have an equilateral triangle Rh core and two triply bridged sulfido ligands on both sides of the plane formed by the three rhodium atoms:

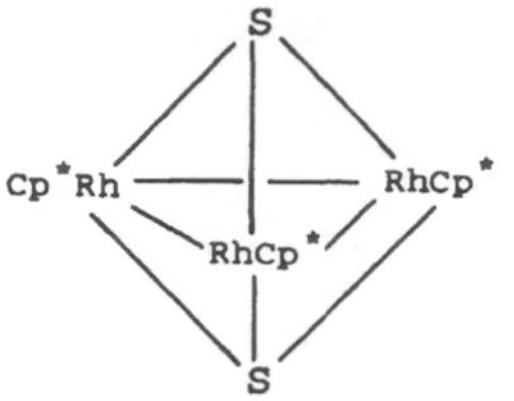

$$
\begin{array}{c}
S \\
Cp^*Rh \longrightarrow RhCp^* \\
RhCp^* \\
S
\end{array}
\quad \text{Rh/S Cluster}
$$

Use of $[RhC_p^*(\mu_3-S)_2]^{2+}$ by Kushi et al. as a homogeneous catalyst in the presence of $LiBF_4$ enabled the electroreduction of CO_2 to oxalate at a potential of only -1.50 V (vs. SCE).[394,395] This potential is much lower than that required for the direct reduction of CO_2, which for the CO_2/CO_2^- couple has the value $E^\circ = -2.21$ V. The reaction was performed on a glassy carbon electrode in CO_2-saturated acetonitrile, producing a precipitate of lithium oxalate, with a current efficiency of 60%. When Bu_4NBF_4 was used instead of $LiBF_4$, formate was produced, also with 60% current efficiency.[353]

A trinuclear nickel complex, $[Ni_3(\mu_3-CNMe)(\mu_3-I)(dppm)_3][PF_6]$ (where dppm = $Ph_2PCH_2PPh_2$), was found by Ratliff et al. to be an electrocatalyst for CO_2 reduction in nonaqueous solutions.[396] This complex has a potential for one-electron reduction at $E_{1/2} = -1.09$ V vs. Ag/AgCl. Using 0.1 M $NaPF_6$/MeCN as medium, on a Pt gauze electrode at the same potential of -1.09 V vs. Ag/AgCl, the reduction of CO_2 with this electrocatalyst produced CO and CO_3^{2-}. Seven turnovers were reached within 3 h, indicating the catalytic nature of the reaction. The low reduction potential may indicate the formation of an adduct with the reduced form of the trinickel complex cluster. A second CO_2 molecule may then insert into this adduct in a "head-to-tail" configuration. After a second electron transfer, the CO_2 dimer may then disproportionate to CO and CO_3^{2-}.

Electrochemical reduction of CO_2 to CO, HCOOH, and traces of CH_4 in nonaqueous solvents was found by Simpson and Durand to be catalyzed by complexes of Co(II), Ni(II), Fe(II), and Cu(II) with 1,10-o-phenanthroline:

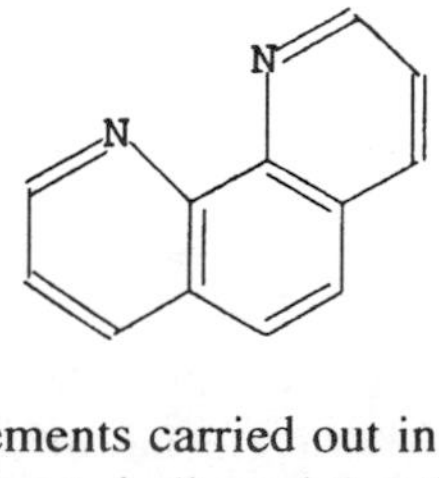

Cyclic voltammetry measurements carried out in DMSO solution containing 0.1 M tetrabutyl ammonium perchlorate indicated that the reduction of CO_2 was dependent upon both ligand and metal reduction on the complexes.[397]

12.3.3.12 Reduction with CuCl₂ Tertiary Phosphine Complexes

Using complexes formed by $CuCl_2$ and the tertiary phosphines Ph_3P and Bu_3P in acetonitrile solution, Fujiwara and Nonaka observed electroreduction of CO_2 to $HCOOH$, $(COOH)_2$, and CO. A total current efficiency of 73% of CO_2 reduction was obtained with $CuCl_2 + Ph_3P$ (ratio 1:5). The electrochemically active catalyst was proposed to be a Cu(0) complex.[398]

12.3.3.13 Reduction with Co(II) Polypyrrole Schiff Bases

Electropolymerized films of a polypyrrole Co(II) Schiff base on glassy carbon or gold electrodes were used by Losada et al.[399] The electrolysis of CO_2 in solutions of DMF containing Bu_4NClO_4 resulted in a two-electron reduction.

12.4 REDUCTION IN MOLTEN SALT MEDIA

In molten salt media, electrochemical reduction may be carried out at elevated temperatures, with lowered overpotentials. Advantages of reactions in molten salts include high solubilities for many inorganic compounds and fast reaction rates because of high temperatures.[400] Most of the interest until now in the electrochemistry of carbon dioxide or carbonates has been due to their potential application in alkaline fuel cells or for oxygen recovery from CO_2 in life support systems.[401]

In the molten alkali carbonate eutectic, (Li_2CO_3–Na_2CO_3–K_2CO_3 at 43.5:31.5:25.0 mol%, liquidus temperature, 397°C) at 600 to 700°C, the predominant reactions are both metal deposition,

$$M^+ + e^- \rightarrow M \qquad (12.101)$$

and the direct cathodic reduction of carbon dioxide to CO,

$$2CO_2 + 2e^- \rightarrow CO + CO_3^{2-} \qquad (12.102)$$

Below 600°C, carbon dioxide is mainly reduced to elementary carbon:

$$CO_3^{2-} + 4e^- \rightarrow C + 3O^{2-} \qquad (12.103)$$

In K_2SO_4 melts, metal deposition is the predominant reaction at all temperature ranges. Of technical importance is the cathodic deposition of refractory carbides from molten carbonates, leading to silicon, tantalum, and tungsten carbides:[402,403]

$$2CO_3^{2-} + 10e^- \rightarrow C_2^{2-} + 6O^{2-} \qquad (12.104)$$

Above 700°C, the change in the equilibrium,

$$C + CO_2 = 2CO \qquad (12.105)$$

leads to carbon monoxide as the major product.[404,405]

Deposition of carbon at Au or Pd cathodes in (Li–Na–K)CO_3 (molar ratio 4:3:3) melts at 550°C at potentials of –2.5 to –2.7 V (vs. Ag electrode in Ag_2SO_4 melt) was observed by Dubois and Buvet.[406]

Most previous studies on reactions of carbonate in molten salts were concerned with the reduction of carbonate ions in fuel cells. The main cathodic process in fuel cells in the presence of both oxygen and carbon dioxide is the reaction of carbonate with oxygen leading to peroxide,

$$O_2 + 2CO_3^{2-} \rightarrow 2O_2^{2-} + 2CO_2 \qquad (12.106)$$

followed by conversion of the carbon dioxide with part of the peroxide to carbonate,

$$2O_2^{2-} + 2CO_2 + 2e^- \rightarrow 2CO_3^{2-} \qquad (12.107)$$

Thus, the overall reaction is

$$O_2 + 2e^- \rightarrow O_2^{2-} \qquad (12.108)$$

Smooth gold and NiO electrodes were used in order to determine the exchange current density of the fuel cell cathodic oxygen reduction in the $Li_2CO_3 + K_2CO_3$ eutectic at 650°C.[407,408]

The anodic reaction at noble metal electrodes is oxidation of the carbonate ion:[409]

$$CO_3^{2-} \rightarrow CO_2 + 1/2O_2 + 2e^- \qquad (12.109)$$

Nickel electroplated on the refractory oxides Al_2O_3, $SrTiO_3$, and $LiAlO_2$ was used for the preparation of porous nickel plates, to serve as stable anodes in molten carbonate fuel cells.[410]

An interesting application of molten alkali carbonates is in life support systems, in the recovery of oxygen from carbon dioxide (e.g., in spacecraft or submarines).[401,411] For this purpose, the anodic and cathodic reactions are

$$CO_3^{2-} \rightarrow CO_2 + 1/2O_2 + 2e^- \qquad (12.110)$$

$$2CO_2 + 2e^- \rightarrow CO + CO_3^{2-} \qquad (12.111)$$

Hence, the overall reaction is oxygen recovery:

$$CO_2 \rightarrow CO + 1/2O_2 \qquad (12.112)$$

In the $Li_2CO_3 + Na_2CO_3 + K_2CO_3$ eutectic at 680°C, at smooth gold electrodes, the cathodic process of oxygen reduction was shown to be diffusion controlled, forming peroxide (O_2^{2-}), while oxide ions (O^{2-}) were formed in the anodic reaction.[412,413]

Higher current densities in carbonate melts were achieved by using alkali halide–carbonate mixtures.[414] In a KCl + NaCl eutectic at 700°C saturated with carbon dioxide under pressure, the current–voltage relationship suggested the following two-stage reduction of carbon dioxide, leading to elementary carbon:[415,416]

$$CO_2 + 2e^- \rightarrow CO_2^{2-} \tag{12.113}$$

$$CO_2^{2-} + 2e^- \rightarrow C + 2O^{2-} \tag{12.114}$$

The above mechanism was further confirmed, mainly with Ni electrodes in LiCl–KCl eutectic melts at 450°C, using the potential sweep method. The two cathodic peaks observed at 0.8 and 0.5 V (vs. Li/Li$^+$) were ascribed respectively to the reduction of the carbonate ion and the intercalation of reduced lithium into the electrodeposited carbon layer. The electrodeposited films were identified by ESCA and XRD to be amorphous carbon.[417] The two-step mechanism of carbonate reduction was proven by the appearance of two anodic peaks in the voltammogram, indicating the reoxidation of reduced intermediates. The rate-determining two-electron transfer,

$$CO_3^{2-} + 2e^- \rightarrow CO_2^{2-} + O^{2-} \tag{12.115}$$

was followed by the carbon-forming reaction,

$$CO_2^{2-} + 2e^- \rightarrow C + 2O^{2-} \tag{12.116}$$

The overall reaction was thus proposed to be represented by:

$$CO_3^{2-} + 4e^- \rightarrow C + 3O^{2-} \tag{12.117}$$

The diffusion coefficient of the carbonate ion at 450°C was estimated to be 1.66×10^{-5} cm^2 s^{-1}.[418]

In a study by Halmann and Zuckerman of the temperature dependence of the rate of reduction of CO_2 to CO in the alkali carbonate eutectic mixture, in a cell of Ni/(Li–Na–K)CO$_3$/Pt, with the Ni cathode at –0.01 V (vs. Ag), at a constant current density of 5.3 mA cm^{-2}, the onset of CO evolution was at about 700°C. The CO production increased with rising temperature, reaching a rate of 15 µmol min^{-1} at 885°C.[419] Silver and nickel electrodes were found to be relatively resistant to alkali carbonates and alkali halides at high temperatures. A comparison of the rates of CO_2 reduction to CO with various combinations of electrodes in an LiCl + KCl eutectic (58.5 to 41.5 mol%) in the temperature range 840 to 890°C indicated the following order of decreasing activity: Pt/Ti–LaRhO$_3$ > Pt/stainless steel > Pt/Ni > Ni/Pt. With stainless-steel electrodes in LiCl + KCl melts, extremely high faradaic yields of conversion of CO_2 to CO were obtained, much larger than could be accounted for by the simple two-electron reduction process. This was accompanied by strong corrosion of the electrodes, proving the participation of additional redox reactions.

While it is feasible to reduce carbon dioxide to carbon monoxide in molten salts as electrolytes, the practical application is limited by the considerable technical problems of working with molten salts. Alkali carbonates, unless very thoroughly dried (such as by passing dry HCl through the melt), are corrosive even to quartz, porcelain, and alumina vessels. Alkali halides, such as LiCl + KCl, undergo sublimation at high temperatures and are corrosive not only to the above containers but also to stainless steel, copper, and platinum.

12.5   SOLID-PHASE ELECTROLYTES

Using ceramic oxide electrolytes, such as calcium- or yttrium-stabilized zirconia, the electrolyte serves as an oxide ion conductor, effecting the cathodic release of hydrogen and the anodic release of oxygen. High-temperature proton-conductive solid electrolytes based on Y- or Yb-doped $SrCeO_3$ have been used for fuel cells at 800 to 1000°C[420] and for steam electrolysis at 800°C.[421]

The mechanism of ionic transport in calcia- or yttria-stabilized zirconia at high temperatures depends on oxygen vacancies, which enable the migration of oxide ions O^{2-} through the lattice toward the anode, at which the oxide ion may be oxidized to molecular oxygen.[422] The lowest resistivities of cubic solid solutions of yttria or calcia in zirconia at about 1000°C were obtained when they contained the minimum amount of the lower valent oxide needed to stabilize the solution.[423] Electrolysis of water at high temperatures with solid electrolytes was investigated in numerous studies.[424,425] Using ceramic oxide electrolytes, such as calcium- or yttrium-stabilized zirconia, the electrolyte served as an oxide ion conductor, effecting the cathodic release of hydrogen,

$$H_2O + 2e^- \rightarrow O^{2-} + H_2 \tag{12.118}$$

and the anodic release of oxygen,

$$O^{2-} \rightarrow 1/2O_2 + 2e^- \tag{12.119}$$

so that the overall process is

$$H_2O \rightarrow H_2 + 1/2O_2 \tag{12.120}$$

In a somewhat analogous process, using nickel electrodes and yttria-stabilized zirconia electrolyte, Gür and Huggins achieved the methanation of carbon monoxide or carbon dioxide with high efficiency at 600 to 700°C.[426,427]

12.5.1   Closed-Cycle Life Support Systems

The regeneration of oxygen from carbon dioxide was developed for closed-cycle life support systems. Using Th, Y, or La oxide solid electrolytes at 400 to 900°C, the cathodic reaction was[411]

$$2CO_2 + 2e^- \rightarrow CO + CO_3^{2-} \qquad (12.121)$$

while the anodic reaction was

$$CO_3^{2-} \rightarrow CO_2 + 1/2O_2 + 2e^- \qquad (12.122)$$

so that the net reaction was

$$CO_2 \rightarrow CO + 1/2O_2 \qquad (12.123)$$

12.5.2 Carbon Dioxide Sensors

In the reverse of the above processes, oxygen gas sensors based on the measurement of the electromotive force of cells using stabilized zirconia were applied in the steel industry and in combustion control. Similar gas sensors were developed for the detection and assay of CO_2, using as a solid electrolyte either K_2CO_3[428] or Na_2CO_3 coated on sodium β/β''-alumina or $Na_3Zr_2Si_2PO_{12}$, which involved cationic conductance.[429,430]

In an improved carbon dioxide sensor, a sodium ion conductor was applied by Miura et al. using a binary carbonate auxiliary electrode.[431] The sensor consisted of a disk of a sodium ion conductor (NASICON, $Na_3Zr_2Si_2PO_{12}$). The CO_2-sensing surface was coated with platinum black and platinum mesh followed by a layer of a mixture of $BaCO_3$ + Na_2CO_3 (46 atom% Ba), which was fixed to the disk by melting and quenching. This sensor had a quick CO_2 response time, less than 8 s at 550°C, and was not affected by water vapor. Also, the sensor followed the Nernst equation, that is, had excellent linear relationship between the logarithm of the CO_2 concentration, in the range of 250 to 2000 ppm CO_2, and the electromotive force. The high selectivity of this sensor for CO_2 was indicated by its being unaffected by the presence of CO (up to 1000 ppm) and NO (up to 50 ppm). Similarly improved CO_2 sensors were constructed using $SrCO_3$ or $CaCO_3$ instead of $BaCO_3$ in the binary carbonate. XRD analysis of the $BaCO_3$–Na_2CO_3 melt of the sensor indicated the absence of free Na_2CO_3 and instead the appearance of microneedle deposits. These microneedles, containing both Na and Ba atoms, were proposed to be responsible for the excellent resistance to water vapor.

β-Alumina solid electrolytes operated at 450°C were employed by Liu and Weppner for the potentiometric determination of carbon dioxide partial pressures in the presence of oxygen.[432] These sensors depend on fast sodium ion conductance. A surface layer of Na_2CO_3 on the solid electrolyte served to relate the activity of sodium to the CO_2 activity of the gas. Another Na^+ conductor (NASICON) solid electrolyte CO_2 sensor used an Li-based binary carbonate auxiliary electrode containing the Li_2CO_3–$CaCO_3$ (1.8:1) eutectic mixture, operated at 500°C.[433] Lithium conductance in a solid electrolyte enabled lowering of the operating temperature of CO_2 gas sensors.[434]

The anodic reaction in CO_2–O_2 atmospheres is

$$Na_2CO_3 \rightarrow 2Na^+ + CO_2 + 1/2O_2 + 2e^- \qquad (12.124)$$

In $CO–CO_2$ atmospheres, the anodic reaction was found to be

$$Na_2CO_3 + CO \rightarrow 2Na^+ + 2CO_2 + 2e^- \qquad (12.125)$$

The Na^+ ions migrate through the solid electrolyte to the cathode, at which the reaction is

$$2Na^+ + 1/2O_2 + 2e^- \rightarrow Na_2O \qquad (12.126)$$

A different type of CO_2 sensor depends on the high sensitivity of certain mixed oxide capacitors on the CO_2 concentration. Optimal performance was reported for $CuO–BaSnO_3$ operated at 830 K.[435]

12.6   ELECTROCHEMICAL CONCENTRATION

Alkali treatment is one of the possible methods for recovery of carbon dioxide in low concentration from exhaust and waste gases. However, for efficient reduction on most metal electrodes, the carbonate or bicarbonate must be acidified to release the electroreducible carbon dioxide. An ingenious electrode assembly to achieve this was demonstrated by Fujiwara et al. by anodically generating acid at a small platinum plate.[436] Carbon dioxide was cathodically reduced at the surface of a mercury pool electrode. The anolyte (1 mol dm^{-3} H_2SO_4) was separated from the catholyte (0.8 mol dm^{-3} $NaHCO_3$ through which CO_2 was bubbled) by a glass frit. At a cathodic current density of 10 mA cm^{-2}, with magnetic stirring of the catholyte, at 0°C, the current efficiency for reduction of sodium hydrogen carbonate to formic acid was 75%.

In order to recover carbon dioxide at its low concentration in the atmosphere, a process was proposed by Dubois et al. which depended on the cathodic reduction of a carrier molecule (RO) to a negatively charged state (RO^-) which then binds CO_2:[374,437,438]

$$RO + e^- = RO^- \qquad (12.127)$$

$$RO^- + CO_2 = RCO_2^- \qquad (12.128)$$

This was then circulated to the anodic compartment, where it was oxidized to the neutral species (RCOO), which may have a lower binding constant for carbon dioxide and thus may release the gas at a higher pressure than the initial pressure:

$$RCO_2^- = RCO_2 + e^- \qquad (12.129)$$

$$RCO_2 = RO + CO_2 \qquad (12.130)$$

Among various chemical systems tested as potential carriers of carbon dioxide, only substituted benzoquinones, such as tetrachlorobenzoquinone, had the desired properties of reversible reduction and oxidation reactions and CO_2 binding constants. However, although these quinones were able to pump CO_2 efficiently, they were too unstable for practical applications.

12.7 LIQUID CO_2 AS MEDIUM FOR ELECTROCHEMICAL REDUCTION

Liquid carbon dioxide could be an ideal medium for the electroreduction of CO_2, since in this medium the reduction current would not be limited by the mass transfer of CO_2. However, liquid carbon dioxide by itself is nonpolar and not ionically conducting. Saeki et al. succeeded in using liquid CO_2 by introducing CO_2 at elevated pressures, up to 60 atm, into methanol containing 0.1 M tetrabutylammonium tetrafluoroborate, thus obtaining an electrochemically conducting solution.[439,440] Electrolyses were carried out in a one-compartment electrochemical autoclave, fitted with a Cu working electrode, a Pt anode, and an Ag quasi-reference electrode. At a pressure of 40 atm and a bias potential of -2.3 V (vs. Ag), the total current density was 500 mA/cm^2, and the faradaic yields of the main products CO, HCOOCH$_3$, H$_2$, CH$_4$, and C$_2$H$_4$ were 46.6, 34.6, 4.0, 3.4, and 2.6, respectively. In the methyl formate (HCOOCH$_3$) produced, $^{13}CO_2$ experiments proved that the formyl group HCO was derived from the carbon dioxide, while the methyl group CH$_3$ originated from the methanol of the medium. Formic acid was a possible primary CO_2 reduction product, which was then rapidly esterified to methyl formate. Presumably, some of the methanol was also oxidized at the anode. For practical application of liquid carbon dioxide, it should be necessary to find a less valuable sacrificial reductant. Various other metal were also tested as electrodes for the reduction of CO_2 in the methanol–CO_2 medium. The highest current efficiencies for the production of methyl formate were obtained with Pb and Sn electrodes.

12.8 CO_2 OXIDATION TO CARBONATE BY AROMATIC NITRO COMPOUNDS

The carbon-dioxide-promoted electrochemical reduction of nitro aromatic compounds to azoxy and azo compounds was discovered by Ohba et al. to result in the conversion of CO_2 to carbonate.[441] Thus, the reduction of nitrobenzene in CO_2-saturated acetonitrile solution (containing 0.1 M Bu$_4$NClO$_4$) over an Hg pool cathode at -1.4 V (vs. SCE) caused the production of Bu$_4$N(CO$_3$)$_2$, azoxybenzene, and azobenzene. The stoichiometry was accounted for by the equations

$$2C_6H_5-NO_2 + 3CO_2 + 6e^- \rightarrow C_6H_5-N(O)=N-C_6H_5 + 3CO_3^{2-} \qquad (12.131)$$

$$2C_6H_5-NO_2 + 4CO_2 + 8e^- \rightarrow C_6H_5-N=N-C_6H_5 + 4CO_3^{2-} \qquad (12.132)$$

The same products were obtained during the reduction of nitrosobenzene. Analogous reactions occurred with several p-substituted nitrobenzenes. Such processes lead to valuable aromatic intermediates (e.g., in the dye industry), while simultaneously fixing CO_2 to carbonate.

12.9   CONCLUSIONS ON ELECTROCHEMICAL REDUCTION

An interesting development is the electroreduction of CO_2 in liquid CO_2–methanol mixtures, at high CO_2 pressures, reaching current densities of 0.5 amp cm^{-2} and yielding CO and methyl formate as the main products (see Section 12.7).

Remarkably high current densities of up to 3 amp cm^{-2} for CO production have been achieved using Ag-catalyzed gas diffusion electrodes with elevated pressures of CO_2, in aqueous $KHCO_3$ as electrolyte (see Section 12.2.6.1).

A comparison of the effectiveness of electrochemical vs. thermal catalytic reduction of CO_2 to methanol was suggested by Frese.[7] With Mo electrodes at pH 4.2, cycled between -1.2 and $+0.2$ V (vs. SCE), the rate of formation of methanol reached 3.4×10^{-6} mol cm^{-2} h^{-1}. The Cu/ZnO-catalyzed formation rate of methanol from CO/ H_2 (50/50) at 20°C and 50 atm was only 2×10^{-8} mol cm^{-2} h^{-1}. The rate at conditions more relevant to industrial production on Cu/ZnO from $H_2/CO/CO_2$ (70/24/6) at 250°C and 75 atm was 1.4×10^{-7} mol cm^{-2} h^{-1}.

An estimate of the energy efficiency (defined as the ratio of the free energy of the organic fuel produced to the electric energy input) for the reduction of carbon dioxide to formate in neutral aqueous solutions over an Hg electrode was provided by Russell et al.[12] At a cathodic overpotential of 1.15 V and a current density of 1 mA cm^{-2}, the faradaic efficiency for formate production reached 100%. However, the energy efficiency was at best only about 45%.[11] Higher energy efficiency may be achieved on metal electrodes with lower overpotentials (see Section 12.2.1.4).

The electrocarboxylation of organic compounds has considerable potential for the synthesis of fine chemicals. Several processes have reached the stage of pilot plant testing and may become competitive with gas–solid-phase catalytic reactions (see Section 12.3.2).

REFERENCES

1. Ziessel, R., Chimie de coordination de la molecule de dioxyde de carbon: activation biologique, chimique, electrochimique et photochimique, *Nouv. J. Chim.*, 7, 613–633, 1983.
2. Teeter, T.E. and Van Rysselberghe, P., Reduction of carbon dioxide on mercury cathodes, in *Proc. 6th Meet. Int. Committee of Electrochemical Thermodynamics and Kinetics,* Poitiers, Butterworths, London, 1954, 538–542.
3. Arana, C., Keshavarz, M., Potts, K.T., and Abruña, H.D., Electrocatalytic reduction of CO_2 and O_2 with electropolymerized films of vinyl-terpyridine complexes of Fe, Ni and Co, *Inorg. Chim. Acta*, 225, 285–295, 1994.
4. Beketov, N.N., *Zh. Russ. Fiz. Khim. Obshch.*, 1, 33, 1869.

5. Royer, E., Reduction of carbonic acid into formic acid, *C.R. Acad. Sci.*, 70, 731–735, 1870.
6. Taniguchi, I., Electrochemical and photoelectrochemical reduction of carbon dioxide, *Mod. Aspects Electrochem.*, 20, 327–400, 1989.
7. Frese, K.W., Jr., Electrochemical reduction of CO_2 at solid electrodes, in *Electrochemical and Electrocatalytic Reactions of Carbon Dioxide*, Sullivan, B.P., Krist, K., and Guard, H.E., Eds., Elsevier, Amsterdam, 1993, chap. 6.
8. Coehn, A. and Jahn, S., On the electrolytic reduction of carbonic acid, *Chem. Ber.*, 37, 2836–2842, 1904.
9. Ehrenfeld, R., On the electrolytic reduction of carbonic acid, *Chem. Ber.*, 38, 4138–4143, 1905.
10. Rabinowitsch, M. and Maschowetz, A., Electrochemical production of formate from carbonic acid, *Z. Elektrochem.*, 36, 846–850, 1930.
11. Udupa, K.S., Subramanian, G.S., and Udupa, H.V.K., The electrolytic reduction of carbon dioxide to formic acid, *Electrochim. Acta*, 16, 1593–1598, 1971.
12. Russell, P.G., Kovac, N., Srinivasan, S., and Steinberg, M., The electrochemical reduction of carbon dioxide, formic acid, and formaldehyde, *J. Electrochem. Soc.*, 124, 1329–1338, 1977.
13. Vassiliev, Yu.B., Bagotzky, V.S., Osetrova, N.V., Khazova, O.A., and Mayorova, N.A., Electroreduction of carbon dioxide. I. The mechanism and kinetics of electroreduction of CO_2 in aqueous solutions on metals with high and moderate hydrogen overvoltages, *J. Electroanal. Chem.*, 189, 271–294, 1985.
14. Giner, J., Electrochemical reduction of CO_2 on platinum electrodes in acid solutions, *Electrochim. Acta*, 8, 857–865, 1963.
15. Hoshi, N., Mizumura, T., and Hori, Y., Significant difference of the reduction rates of carbon dioxide between Pt(111) and Pt(110) single crystal electrodes, *Electrochim. Acta*, 40, 883–887, 1995.
16. Hoshi, N., Ito, H., Suzuki, T., and Hori, Y., CO_2 reduction on Rh single crystal electrodes and the structural effect, *J. Electroanal. Chem.*, 395, 309–312, 1995.
17. Hoshi, N., Suzuki, T., and Hori, Y., Step density dependence of CO_2 reduction rate on Pt(S) [n(111)x(111)] single crystal electrodes, *Electrochim. Acta*, 41, 1647–1653, 1996.
18. Hoshi, N., Suzuki, T., and Hori, Y., Catalytic activity of CO_2 reduction on Pt single crystal electrodes: Pt(S)-[n(111)x(111)], Pt(S)-[n(111)x(100), and Pt(S)-[n(100)x(111)], *J. Phys. Chem. B*, 101, 8520–8524, 1997.
19. Beden, B., Bewick, A., Razak, M., and Weber, J., On the nature of reduced CO_2. An IR spectroscopic investigation, *J. Electroanal. Chem.*, 139, 203–206, 1982.
20. Schiffrin, D.J., Application of photo-electrochemical effects to the study of the electrochemical properties of radicals: CO_2^- and CH_3, *Faraday Discuss. Chem. Soc.*, 56, 75–95, 1974.
21. Aramata, A., Enyo, M., Koga, O., and Hori, Y., FT-IR spectrometry of the reduced CO_2 at Pt electrode and anomalous effect of Ca^{2+} ions, *Chem. Lett.*, pp. 749–752, 1991.
22. Taguchi, S. and Aramata, A., Surface-structure sensitive reduced CO_2 formation on Pt single-crystal electrodes in sulfuric acid solution, *Electrochim. Acta*, 39, 2533–2537, 1994.
23. Nikolic, B.C., Huang, H., Gervasio, D., Lin, A., Fierro, C., Adzic, R.R., and Yeager, E.B., Electroreduction of carbon dioxide on platinum single crystal electrodes: electrochemical and in situ FTIR studies, *J. Electroanal. Chem.*, 295, 415–423, 1990.

24. Rodes, A., Pastor, E., and Iwasita, T., Structural effects on CO_2 reduction at Pt single-crystal electrodes. 1. The Pt(110) surface, *J. Electroanal. Chem.*, 369, 183–191, 1994.

25. Huang, H., Effect of impurities on the electro-reduction of carbon dioxide on platinum electrodes in acid solutions, *J. Electrochem. Soc.*, 139, 55C–58C, 1992.

26. Zakharyan, A.V., Osetrova, N.V., and Vasil'ev, Yu.B., Adsorption of CO_2 on platinum metals, *Elektrokhimiya*, 12, 1854, 1976; *Chem. Abstr.*, 86, 129782m.

27. Vassiliev, Yu.B., Bagotzky, V.S., Osetrova, N.V., and Mikhailova, A.A., Electroreduction of carbon dioxide. III. Adsorption and reduction in aprotic solvents, *J. Electroanal. Chem.*, 189, 311–324, 1985.

28. Marcos, M.L., González-Velasco, J., Bolzán, A.E., and Arvia, A.J., Comparative electrochemical behaviour of CO_2 on Pt and Rh electrodes in acid solution, *J. Electroanal. Chem.*, 395, 91–98, 1995.

29. Huang, M. and Faguy, P.W., Carbon dioxide reduction on platinum/ Nafion® /carbon electrodes, *J. Electroanal. Chem.*, pp. 219–222, 1996.

30. Kaiser, U. and Heitz, E., On the mechanism of the electrochemical dimerization of CO_2 to oxalic acid, *Ber. Bunsenges. Phys. Chem.*, 77, 818–823, 1973.

31. Osetrova, N.V., Vasil'ev, Yu.B., Bagotskii, V.S., Sadkova, R.G., Cherashev, A.F., and Krushch, A.P., Role of percarbonate in the electroreduction of carbon dioxide on platinum, *Elektrokhimiya*, 20, 286, 1984; English translation, p. 272.

32. Aylmer-Kelly, A.W.B., Bewick, A., Cantrill, P.R., and Tuxford, A.M., Studies of electrochemically generated reaction intermediates using modulated specular reflectance spectroscopy, *Faraday Discuss. Chem. Soc.*, 56, 96–105, 1974.

33. Noda, H., Ikeda, S., Yamamoto, A., Einaga, H., Yoshida, H., and Ito, K., Kinetics of the electroreduction of carbon dioxide on a Au cathode in phosphate buffer solution, in Proc. Int. Symp. Chemical Fixation of Carbon Dioxide, Nagoya, Japan, December 2 to 4, 1991, 327–332.

34. Koppenol, W.H. and Rush, J.D., Reduction potential of the CO_2/CO_2^- couple. A comparison with other C_1 radicals, *J. Phys. Chem.*, 91, 4429–4430, 1987.

35. Kapusta, S. and Hackerman, N., The electroreduction of carbon dioxide and formic acid on tin and indium electrodes, 130, 607–613, 1983.

36. Jordan, J. and Smith, P.T., Free-radical intermediate in the electroreduction of carbon dioxide, *Proc. Chem. Soc.*, pp. 246–247, 1960.

37. Ryu, J., Andersen, T.N., and Eyring, H., The electrode reduction kinetics of carbon dioxide in aqueous solution, *J. Phys. Chem.*, 76, 3278–3286, 1972.

38. Augustynski, J., Electrochemical reduction of carbon dioxide in aqueous solution, *Chimia*, 42, 172–175, 1988.

39. Babenko, S.D., Benderskii, V.A., Krivenko, A.G., and Kurmaz, V.A., Photocurrent kinetics of the electron emission from a metal into electrolyte solution. VII. Absolute rate constants of CO_2 electrochemical reduction on mercury, *J. Electroanal. Chem.*, 159, 163–181, 1983.

40. Zakharyan, A.V., Rotenberg, Z.A., Osetrova, N.V., and Vasil'ev, Yu.B., Electroreduction of carbon dioxide on a tin electrode, *Elektrokhimiya*, 14, 1520–1527, 1978; English translation, pp. 1317–1323.

41. Kesarev, V.V. and Fedortsov, V.F., Electrochemical reduction of carbon dioxide on zinc and cadmium electrodes, *Zh. Prikl. Khim.*, 42, 707–709, 1969; English translation, 42, 673–675, 1969.

42. Ito, K., Murata, T., and Ikeda, S., Electrochemical reduction of carbon dioxide to organic compounds, *Bull. Nagoya Inst. Technol.*, 27, 209–214, 1975.

43. Hattori, A., Ikeda, S., Maeda, M., Einaga, H., and Ito, K., Electroreduction behavior of carbon dioxide on zinc and zinc oxide electrodes, in Proc. Int Symp. Chemical Fixation of Carbon Dioxide, Nagoya, Japan, December 2 to 4, 1991, 323–326.

44. Komatsu, S., Yanagihara, T., Hiraga, Y., Tanaka, M., and Kunugi, A., Electrochemical reduction of CO$_2$ at Sb and Bi electrodes in KHCO$_3$ solution, *Denki Kagaku*, 63, 217–224, 1995; *Chem. Abstr.*, 122, 225084t.

45. Spichiger-Ulmann, M. and Augustynski, J., Electrochemical reduction of bicarbonate ions at a bright palladium electrode, *J. Chem. Soc. Faraday Trans. I*, 81, 713–716, 1985.

46. Spichiger-Ulmann, M. and Augustynski, J., Specific cation effect upon the cathodic reduction of bicarbonate anion at palladium, *Nouv. J. Chim.*, 10, 487–491, 1986.

47. Spichiger-Ulmann, M. and Augustynski, J., Remarkable enhancement of the rate of cathodic reduction of hydrocarbonate anions at palladium in the presence of caesium cations, *Helv. Chim. Acta*, 69, 632–634, 1986.

48. Azuma, M., Hashimoto, K., Watanabe, M., and Sakata, T., Electrochemical reduction of carbon dioxide to higher hydrocarbons in a KHCO$_3$ aqueous solution, *J. Electroanal. Chem.*, 294, 299–303, 1990.

49. Ohkawa, K., Hashimoto, K., Fujishima, A., Noguchi, Y., and Nakayama, S., Electrochemical reduction of carbon dioxide on hydrogen-storing materials. I. The effect of hydrogen absorption on the electrochemical-behavior on palladium electrodes, *J. Electroanal. Chem.*, 345, 445–456, 1993.

50. Ohkawa, K., Noguchi, Y., Nakayama, S., Hashimoto, K., and Fujishima, A., Electrochemical reduction of carbon dioxide on hydrogen-storing materials. 4. Electrochemical behavior of the Pd electrode in aqueous and nonaqueous electrolyte, *J. Electroanal. Chem.*, 369, 247–250, 1994.

51. Podlovchenko, B.I., Kolyadko, E.A., and Lu, S., Electroreduction of carbon dioxide on palladium electrodes at potentials higher than the reversible hydrogen potential, *J. Electroanal. Chem.*, 373, 185–187, 1994.

52. Fujishima, A., Electrochemical carbon dioxide reduction using solar energy, in Proc. Int. Symp. Chemical Fixation Carbon Dioxide, Nagoya, Japan, December 2 to 4, 1991, 11–18.

53. Fujishima, A., Electrochemical and photoelectrochemical reduction of CO$_2$, in Proc. Int. Conf. on Carbon Dioxide Utilization, Bari, Italy, September 1993, 303–310.

54. Ohkawa, K., Noguchi, Y., Nakayama, S., Hashimoto, K., and Fujishima, A., Electrochemical reduction of carbon dioxide on hydrogen-storing materials. II. Copper-modified palladium electrode, *J. Electroanal. Chem.*, 348, 459–464, 1993.

55. Kolbe, D. and Vielstich, W., Adsorbate formation during the electrochemical reduction of carbon dioxide at palladium — a DEMS study, *Electrochim. Acta*, 41, 2457–2460, 1996.

56. Hoshi, N., Noma, M., Suzuki, T., and Hori, Y., Structural effect on the rate of CO$_2$ reduction on single crystal electrodes of palladium, *J. Electroanal. Chem.*, 412, 15–18, 1997.

57. Shiratsuchi, R., Hongo, K., Nogami, G., and Ishimura, S., Reduction of CO$_2$ on fluorine-doped SnO$_2$ thin-film electrodes, *J. Electrochem. Soc.*, 139, 2544–2549, 1992.

58. Nogami, G., Aikoh, Y., and Shiratsuchi, R., Investigation of fixation mechanism of carbon dioxide on oxide semiconductors by current transients, *J. Electrochem. Soc.*, 140, 1037–1041, 1993.

59. Aoki, A. and Nogami, G., Rotating-ring-disk electrode study on the fixation mechanism of carbon dioxide, *J. Electrochem. Soc.*, 142, 423–427, 1995.

60. Monnier, A., Augustynski, J., and Stalder, C., On the electrolytic reduction of carbon dioxide at TiO_2 and TiO_2–Ru cathodes, *J. Electroanal. Chem.*, 112, 383–385, 1980.

61. Koudelka, M., Monnier, A., and Augustynski, J., Electrocatalysis of the cathodic reduction of carbon dioxide on platinized titanium dioxide film electrodes, *J. Electrochem. Soc.*, 131, 745–750, 1984.

62. Tinnemans, A.H.A., Koster, T.P.M., Thewissen, D.H.W.M., De Kreuk, C.W., and Mackor, A., On the electrolytic reduction of carbon dioxide at TiO_2 and other titanates, *J. Electroanal. Chem.*, 145, 449–456, 1983.

63. Augustynski, J., Comments on the paper "On the electrolytic reduction of carbon dioxide at TiO_2 and other titanates" by Tinnemans, A.H.A., Koster, T.P.M., Thewissen, D.H.W.M., De Kreuk, C.W., and Mackor, A., *J. Electroanal. Chem.*, 145, 457–460, 1983.

64. Nakabayashi, S. and Kira, A., An electrochemical reduction of CO_2 on conductive ceramics, in Proc. Int. Symp. Chemical Fixation of Carbon Dioxide, Nagoya, Japan, December 2 to 4, 1991, 291–294.

65. Nakabayashi, S. and Kira, A., Electrochemical reduction of carbon dioxide on titanium diboride, *J. Electroanal. Chem.*, 319, 381–385, 1991.

66. Bennett, E.M., Eggins, B.R., McNeill, J., and McMullan, E.A., Recycling carbon dioxide from fossil fuel combustion, *Anal. Proc.*, pp. 356–359, 1980.

67. Eggins, B.R., Brown, E.M., McNeill, E.A., and Grimshaw, J., Carbon dioxide fixation by electrochemical reduction in water to oxalate and glyoxylate, *Tetrahedron Lett.*, 29, 945–948, 1988.

68. Eggins, B.R., Bennett, E.M., and McMullan, E.A., Voltammetry of carbon dioxide. 2. Voltammetry in aqueous solutions on glassy carbon, *J. Electroanal. Chem.*, 408, 165–171, 1996.

69. Hori, Y. and Suzuki, S., Electrolytic reduction of carbon dioxide at mercury electrode in aqueous solution, *Bull. Chem. Soc. Jpn.*, 55, 660–665, 1982.

70. Fujita, N., Morita, H., Matsuura, C., and Hiroishi, D., Radiation-induced CO_2 reduction in an aqueous-medium suspended with iron-powder, *Radiat. Phys. Chem.*, 44, 349–357, 1994.

71. Fujita, N., Fukuda, Y., Matsuura, C., and Hiroishi, D., CO_2 reducing reaction induced by the corrosion of iron, *Corros. Eng.*, 43, 322–330, 1994; *Chem. Abstr.*, 121, 185876g.

72. Fujita, N., Fukuda, Y., Matsuura, C., and Saigo, K., Changes in pH and redox potential during radiation-induced CO_2 reduction in an aqueous solution containing iron powder, *Radiat. Phys. Chem.*, 47, 543–549, 1996.

73. Fujita, N., Fukuda, Y., Matsuura, C., and Saigo, K., Radiation enhanced H^+ generation in iron-containing solution saturated with CO_2, *Radiat. Chem. Phys.*, 48, 297–304, 1996.

74. Hardy, L.O. and Gillham, R.W., Formation of hydrocarbons from the reduction of aqueous CO_2 by zero-valent iron, *Environ. Sci. Technol.*, 30, 57, 1996.

75. Hori, Y., Kikuchi, K., and Suzuki, S., Production of CO and CH_4 in electrochemical reduction of CO_2 at metal electrodes in aqueous hydrogen carbonate solution, *Chem. Lett.*, pp. 1695–1698, 1985.

76. Hori, Y., Kikuchi, K., Murata, A., and Suzuki, S., Production of methane and ethylene in electrochemical reduction of carbon dioxide at copper electrode in aqueous hydrogen carbonate solution, *Chem. Lett.*, pp. 897–898, 1986.

77. Hori, Y., Murata, A., Kikuchi, K., and Suzuki, S., Electrochemical reduction of CO_2 to CO at a gold electrode in aqueous $KHCO_3$, *J. Chem. Soc. Chem. Comm.*, pp. 728–729, 1987.

78. Hori, Y., Murata, A., and Takahashi, R., Formation of hydrocarbons in the electrochemical reduction of carbon dioxide at a copper electrode in aqueous solution, *J. Chem. Soc. Faraday Trans. I*, 85, 2309–2326, 1989.

79. Hori, Y., Murata, A., and Yoshinami, Y., Adsorption of CO, intermediately formed in electrochemical reduction of CO_2 at a copper electrode, *J. Chem. Soc. Faraday Trans.*, 87, 125–128, 1991.

80. Ito, K., Ikeda, S., Yamauchi, N., Iida, T., and Takagi, T., Electrochemical reduction products of carbon dioxide at some metallic electrodes in nonaqueous electrolytes, *Bull. Chem. Soc. Jpn*, 58, 3027–3028, 1985.

81. Ikeda, S., Takagi, T., and Ito, K., Selective formation of formic acid, oxalic acid, and carbon monoxide by electrochemical reduction of carbon dioxide, *Bull. Chem. Soc. Jpn.*, 60, 2517–2522, 1987.

82. Noda, H., Ikeda, S., Oda, Y., and Ito, K., Potential dependencies of the products on electrochemical reduction of carbon dioxide at a copper electrode, *Chem. Lett.*, pp. 289–292, 1989.

83. Noda, H., Ikeda, S., Oda, Y., Imai, K., Maeda, M., and Ito, K., Electrochemical reduction of carbon dioxide at various metal electrodes in aqueous potassium hydrogen carbonate solution, *Bull. Chem. Soc. Jpn*, 63, 2459–2462, 1990.

84. Azuma, M., Hashimoto, K., Hiramoto, M., Watanabe, M., and Sakata, T., Electrochemical reduction of carbon dioxide on various metal electrodes in low-temperature aqueous $KHCO_3$ media, *J. Electrochem. Soc.*, 137, 1772–1778, 1990.

85. Azuma, M., Hashimoto, K., Hiromoto, M., Watanabe, M., and Sakata, T., Carbon dioxide reduction at low temperatures on various metal electrodes, *J. Electroanal. Chem.*, 260, 441–415, 1989.

86. Haines, R.J., Carbon dioxide — pollutant or potential chemical feedstock? *S. Afr. Tydskr. Chem.*, 47, 112–124, 1994.

87. Linder, M.C. and Hazeghazam, M., Copper biochemistry and molecular biology, *Am. J. Clin. Nutr.*, 63, S797–S811, 1996.

88. Hori, Y. and Murata, A., Electrochemical evidence of intermediate formation of adsorbed CO in cathodic reduction of CO_2 at a Ni electrode, *Electrochim. Acta*, 35, 1777–1780, 1990.

89. Murata, A. and Hori, Y., Product selectivity affected by cationic species in electrochemical reduction of CO_2 and CO at a Cu electrode, *Bull. Chem. Soc. Jpn.*, 64, 123–127, 1991.

90. Ikeda, S., Amakusa, S., Noda, H., Saito, Y., and Ito, K., Photo-electrochemical and electrochemical formation of methane from carbon dioxide at copper coated electrodes, in Proc. Electrochem. Soc., Photoelectrochemistry and Electrosynthesis on Semiconducting Materials, Vol. 88–14, 1988, 130–136.

91. Murata, A. and Hori, Y., Formation of hydrocarbons in electrochemical reduction of carbon monoxide at an Fe electrode in connection with electrochemical reduction of carbon dioxide, *Denki Kagaku*, 59, 499–503, 1991; *Chem. Abstr.*, 115, 265251g.

92. Koga, O., Nakama, K., Murata, A., and Hori, Y., Effects of surface state of copper electrode on the selectivity of electrochemical reduction of CO_2, *Denki Kagaku*, 57, 1137–1140, 1989; *Chem. Abstr.*, 112, 65318k.

93. Kyriacou, G. and Anagnostopoulos, A., Electro-reduction of CO_2 on differently prepared copper electrodes — the influence of electrode treatment on the current efficiencies, *J. Electroanal. Chem.*, 322, 233–246, 1992.

94. Koga, O., Murata, A., and Hori, Y., Effect of cadmium deposition on electroreduction of carbon dioxide by a copper electrode, *Nippon Kagaku Kaishi*, pp. 873–878, 1991; *Chem. Abstr.*, 115, 101388f.

95. Cook, R.L., Mac Duff, R.C., and Sammells, A.F., Electrochemical reduction of carbon dioxide to methane at high current densities, *J. Electrochem. Soc.*, 134, 1873–1874, 1987.

96. Azuma, M., Kawasaki, Y., and Tamura, H., Electrochemical reduction of carbon dioxide in an aqueous $KHCO_3$ solution with small amounts of methanol, in Proc. Int. Symp. Chemical Fixation of Carbon Dioxide, Nagoya, Japan, December 2 to 4, 1991, 287–290.

97. Cook, R.L., Macduff, R.C., and Sammells, A.F., Efficient high rate CO_2 reduction to methane and ethylene at in situ electrodeposited copper electrode, *J. Electrochem. Soc.,* 134, 2375, 1987.

98. Cook, R.L., Mac Duff, R.C., and Sammells, A.F., On the electrochemical reduction of carbon dioxide at in situ electrodeposited copper, *J. Electrochem. Soc.*, 135, 1320–1326, 1988.

99. Cook, R.L., MacDuff, R.C., and Sammells, A.F., Gas-phase CO_2 reduction to hydrocarbons at metal/solid polymer electrolyte interface, *J. Electrochem. Soc.*, 137, 187–189, 1990.

100. Kim, J.J., Summers, D.P., and Frese, K.W., Jr., Reduction of CO_2 and CO to methane on Cu foil electrodes, *J. Electroanal. Chem.*, 245, 223–244, 1988.

101. DeWulf, D.W., Jin, T., and Bard, A.J., Electrochemical and surface studies of carbon dioxide reduction to methane and ethylene at copper electrodes in aqueous solutions, *J. Electrochem. Soc.*, 136, 1686–1691, 1989.

102. Santilli, D.S. and Castner, D.G., Mechanism of chain growth and product formation for the Fischer–Tropsch reaction over iron catalysts,, *Energy Fuels*, 3, 8–15, 1989.

103. Cook, R.L., MacDuff, R.C., and Sammells, A.F., Electrochemical Fischer–Tropsch reduction of carbon dioxide to hydrocarbons and alcohols, in Proc. Int. Symp. Chemical Fixation of Carbon Dioxide, Nagoya, Japan, December 2 to 4, 1991, 39–48.

104. Cook, R.L., MacDuff, R.C., and Sammells, A.F., Evidence for formaldehyde, formic acid and acetaldehyde as possible intermediates during electrochemical CO_2 reduction at copper electrodes, *J. Electrochem. Soc.*, 136, 1982–1984, 1989.

105. Wasmus, S., Cattaneo, E., and Vielstich, W., Reduction of carbon dioxide to methane and ethene — an on-line MS study with rotating electrodes, *Electrochim. Acta*, 35, 771–775, 1990.

106. Friebe, P., Bogdanoff, P., Alonso-Vante, N., and Tributsch, H., A real-time mass spectrometric study of the (electro)chemical factors affecting CO_2 reduction at copper, *J. Catal.*, 168, 374–385, 1997.

107. Ogasawara, H., Inukai, J., and Ito, M., Potential-induced migration of top-layer atoms and molecules on Pt(110) electrode surface studied by infrared absorption spectroscopy, *Chem. Phys. Lett.*, 198, 389–394, 1992.

108. Oda, I., Ogasawara, H., and Ito, M., Carbon monoxide adsorption on copper and silver electrodes during carbon dioxide electroreduction studied by infrared reflection absorption spectroscopy and surface-enhanced Raman spectroscopy, *Langmuir*, 12, 1094–1097, 1996.

109. Kostecki, R. and Augustynski, J., Electrochemical reduction of CO_2 at an activated silver electrode, *Ber. Bunsenges. Phys. Chem.*, 98, 1510–1515, 1994.
110. Augustynski, J., Electrochemical reduction of CO_2 at metallic electrodes, in Abstr. 4th Int. Conf. on Carbon Dioxide Utilization, Kyoto, Japan, September 1997, KL-4.
111. Shiratsuchi, R., Aikoh, Y., and Nogami, G., Pulsed electroreduction of CO_2 on copper electrodes, *J. Electrochem. Soc.*, 140, 3479–3482, 1993.
112. Nogami, G., Itagaki, H., and Shiratsuchi, R., Pulsed electro-reduction of CO_2 on copper electrodes. II, *J. Electrochem. Soc.*, 141, 1138–1142, 1994.
113. Aoki, A., Shiratsuchi, R., and Nogami, G., Pulsed electroreduction of CO_2 on Cu and SnO_2 electrodes, in Abstr. 3rd Int. Conf. on Carbon Dioxide Utilization, Norman, OK, May 1995.
114. Augustynski, J., Carroy, A., Feiner, A.S., Jermann, B., Link, J., Kedzierzawski, P., and Kostecki, R., Enhanced electrochemical CO_2 reduction at copper, silver and gold electrodes, in Proc. Int. Conf. on Carbon Dioxide Utilization, Bari, Italy, September 1993, 331–338.
115. Jermann, B. and Augustynski, J., Long-term activation of the copper cathode in the course of CO_2 reduction, *Electrochim. Acta*, 39, 1891–1896, 1994.
116. Shiratsuchi, R., Liu, A., and Nogami, G., Electrochemical reduction of CO_2 on noble metal electrodes, in Abstr. 3rd Int. Conf. on Carbon Dioxide Utilization, Norman, OK, May 1995.
117. Shiratsuchi, R. and Nogami, G., Pulsed electroreduction of CO_2 on silver electrodes, *J. Electrochem. Soc.*, 143, 582–586, 1996.
118. Kedzierzawski, P. and Augustynski, J., Poisoning and activation of the gold cathode during electroreduction of CO_2, *J. Electrochem. Soc.*, 141, L58–L60, 1994.
119. Hori, Y., Electrochemical reduction of CO_2 at metallic electrodes, in Proc. Int. Symp. Chemical Fixation of Carbon Dioxide, Nagoya, Japan, December 2 to 4, 1991, 107–116.
120. Koga, O. and Hori, Y., Reduction of adsorbed CO on a Ni electrode in connection with the electrochemical reduction of CO_2, *Electrochim. Acta*, 38, 1391–1394, 1993.
121. Koga, O. and Hori, Y., Infrared spectroscopic and electrochemical approach of adsorbed CO intermediately formed from CO_2 at a nickel electrode, *Denki Kagaku*, 61, 812–813, 1993.
122. Koga, O., Matsuo, T., Yamazaki, H., and Hori, Y., Infrared spectroscopic study of CO_2 and CO reduction at metal electrodes, in Abstr. 4th Int. Conf. on Carbon Dioxide Utilization, Kyoto, Japan, September 1997, P-072.
123. Hori, Y., Takahashi, R., Yoshinami, Y., and Murata, A., Electrochemical reduction of CO at a copper electrode, *J. Phys. Chem. B*, 101, 7075–7081, 1997.
124. Nakato, Y., Yano, S., Yamaguchi, T., and Tsubomura, H., Reactions and mechanism of the electrochemical reduction of carbon dioxide on alloyed copper–silver electrodes, *Denki Kagaku*, 59, 491–498, 1991.
125. Nakato, Y., The mechanism of electrochemical reduction of carbon dioxide at metal electrodes, in Proc. Int. Symp. Chemical Fixation of Carbon Dioxide, Nagoya, Japan, December 2 to 4, 1991, 117–122.
126. Bocarsly, A.B., Seshadri, G., and Chao, L., The electroreduction of CO_2 to methanol using an aqueous pyridinium catalyst, in Proc. Int. Conf. on Carbon Dioxide Utilization, Bari, Italy, September 1993, 381.
127. Seshadri, G., Lin, C., and Bocarsly, A.B., A new homogeneous electrocatalyst for the reduction of carbon dioxide to methanol at low overpotential, *J. Electroanal. Chem.*, 372, 145–150, 1994.

128. Canfield, D. and Frese, K.W., Jr., Reduction of carbon dioxide to methanol on n-GaAs and p-GaAs and p-InP. Effect of crystal face, electrolyte and current density, *J. Electrochem. Soc.*, 130, 1772–1773, 1983.

129. Frese, K.W., Jr. and Canfield, D., Reduction of CO_2 on n-GaAs electrodes and selective methanol synthesis, *J. Electrochem. Soc.*, 131, 2518–2522, 1984.

130. Sears, W.M. and Morrison, S.R., Carbon dioxide reduction on gallium arsenide electrodes, *J. Phys. Chem.*, 89, 3295–3298, 1985.

131. Summers, D.P., Leach, S., and Frese, K.W., Jr., The electrochemical reduction of aqueous carbon dioxide to methanol at molybdenum electrodes with low overpotentials, *J. Electroanal. Chem.*, 205, 219–232, 1986.

132. Nakato, Y. and Mori, T., Electrochemical reduction of carbon dioxide into alcohols on transition metal electrodes, in Environmental Aspects of Electrochemistry and Photoelectrochemistry, *Proc. Electrochem. Soc.*, 93–18, 95–103, 1993.

133. Frese, K.W., Jr. and Leach, S., Electrochemical reduction of carbon dioxide to methane, methanol and CO on Ru electrodes, *J. Electrochem. Soc.*, 132, 259–260, 1985.

134. Summers, D.P. and Frese, K.W., Jr., Electrochemical reduction of CO_2. Characterization of the formation of CH_4 at Ru electrodes in CO_2 saturated aqueous solutions, *Langmuir*, 4, 51–57, 1988.

135. Popic, J.P., Avramovivic, M.L., and Vukovic, N.B., Reduction of carbon dioxide on ruthenium oxide and modified ruthenium oxide electrodes in 0.5 M $NaHCO_3$, *J. Electroanal. Chem.*, 421, 105–110, 1997.

136. Arai, G., Harashina, T., and Yasumori, I., Selective electrocatalytic reduction of carbon dioxide to methanol on Ru-modified electrode, *Chem. Lett.*, pp. 1215–1218, 1989.

137. Okada, G., Kobayashi, K., and Kumanotani, J., Electrochemical reduction of carbon dioxide at metal doped TiO_2 electrodes, *Denki Kagaku*, 56, 651–652, 1988.

138. Bandi, A., Electrochemical reduction of carbon dioxide on conductive metallic oxides, *J. Electrochem. Soc.*, 137, 2157–2160, 1990.

139. Bandi, A. and Kühne, H.-M., Electrochemical reduction of carbon dioxide in water. Analysis of reaction mechanism on ruthenium–titanium oxide, *J. Electrochem. Soc.*, 139, 1605–1610, 1992.

140. Schwarz, J., Maier, C.U., and Bandi, A., Adsorption of CO_2 on different transition metals and oxides, in Proc. Int. Symp. Chemical Fixation of Carbon Dioxide, Nagoya, Japan, December 2 to 4, 1991, 439–442.

141. Bandi, A., Schwarz, J., and Maier, C.U., Adsorption of CO_2 on transition metals and metal oxides, *J. Electrochem. Soc.*, 140, 1006, 1993.

142. Frese, K.W., Electrochemical reduction of CO_2 at intentionally oxidized copper electrodes, *J. Electrochem. Soc.*, 138, 3338–3344, 1991.

143. Ikeda, S. and Ito, K., Artificial photosynthetic systems for carbon dioxide fixation, in Proc. Int. Symp. Chemical Fixation of Carbon Dioxide, Nagoya, Japan, December 2 to 4, 1991, 23–30.

144. Ikeda, S., Tomita, Y., Hattori, A., Ito, K., Noda, H., and Sakai, M., Selective ethanol formation by electrochemical reduction of carbon dioxide on electrodes comprised of the mixtures of copper and zinc oxides, *Denki Kagaku*, 61, 807–809, 1993.

145. Ikeda, S., Tomita, Y., Hattori, A., and Ito, K., Electrochemical fixation of CO_2 using oxide electrodes, in Abstr. Int. Conf. on Utilization of Carbon Dioxide, Bari, Italy, September 1993, 383.

146. Ogura, K. and Yoshida, I., Electrocatalytic reduction of carbon dioxide to methanol in the presence of 1,2-dihydroxybenzene–3,5–disulphonate ferrate (III) and ethanol, *J. Mol. Catal.*, 34, 67–72, 1986.

147. Ogura, K. and Yoshida, I., Catalytic conversion of CO and CO_2 into methanol with a solar cell, *J. Mol. Catal.*, 34, 309–311, 1986.

148. Ogura, K. and Takagi, M., Electrocatalytic reduction of carbon dioxide to methanol. IV. Assessment of the current-potential curves leading to reduction, *J. Electroanal. Chem.*, 206, 209–216, 1986.

149. Ogura, K., Migita, C.T., and Imura, H., Catalytic reduction of CO_2 with a hydrogen fuel cell, *J. Electrochem. Soc.*, 137, 1730–1732, 1990.

150. Ogura, K., Migita, C.T., and Wadaka, K., Homogeneous catalysis in the mediated electrochemical reduction of carbon dioxide, *J. Mol. Catal.*, 67, 161–173, 1991.

151. Ogura, K., Mine, K.-I., Yano, J., and Sugihara, H., Electrocatalytic reduction of CO_2 at a composite film electrode with a surface-confined metal complex, *Denki Kagaku*, 61, 810–811, 1993.

152. Ogura, K., Mine, K.-I., Yano, J., and Sugihara, H., Electrocatalytic generation of C_2 and C_3 compounds from carbon dioxide on a cobalt complex-immobilized dual-film electrode, *J. Chem. Soc. Chem. Commun.*, pp. 20–21, 1993.

153. Ogura, K., Hisaga, M., Yano, J., and Endo, N., Electroreduction of CO_2 to C_2 and C_3 compounds on bis(4,5-dihydroxybenzene-1,3-disulphonato)ferrate(II)-fixed polyaniline Prussian blue modified electrode in aqueous solutions, *J. Electroanal. Chem.*, 379, 373–377, 1994.

154. Ogura, K., Sugihara, H., Yano, J., and Higasa, M., Electrochemical reduction of carbon dioxide on dual-film electrodes modified with and without cobalt(II) and iron(II) complexes, *J. Electrochem. Soc.*, 141, 419–424, 1994.

155. Ogura, K., Yamada, M., Nakayama, M., and Endo, N., Electrocatalytic reduction of CO_2 to more worthier compounds on a functional dual-film electrode with a solar cell as the energy source, in Abstr. 4th Int. Conf. on Carbon Dioxide Utilization, Kyoto, Japan, September 1997, O-14.

156. Ogura, K., Endo, N., Nakayama, M., and Ootsuka, H., Mediated activation and electroreduction of CO_2 on modified electrodes with conducting polymer and inorganic conductor films, *J. Electrochem. Soc.*, 142, 4026–4032, 1995.

157. Ogura, K., Nakayama, M., and Kusumoto, C., In situ FTIR spectroscopic study on the catalytic reduction of CO_2 on the modified electrode with conducting polymer and inorganic conductor films, in Abstr. 3rd Int. Conf. on Carbon Dioxide Removal, Cambridge, MA, September 1996, 96–97.

158. Watanabe, M., Shibata, M., Kato, A., Azuma, M., and Sakata, T., Design of alloy catalysts for CO_2 reduction. III. The selective and reversible reduction of CO_2 on Cu alloy electrodes, *J. Electrochem. Soc.*, 138, 3382–3389, 1991.

159. Watanabe, M., Shibata, M., and Katoh, A., Design of alloy catalysts for energy-efficient and selective reduction of CO_2, in Proc. Int. Symp. Chemical Fixation of Carbon Dioxide, Nagoya, Japan, December 2 to 4, 1991, 123–128.

160. Watanabe, M., Shibata, M., Katoh, A., Sakata, T., and Azuma, M., Design of alloy electrocatalysts for CO_2 reduction. Improved energy efficiency, selectivity, and reaction rate for the CO_2 electroreduction on Cu alloy electrodes, *J. Electroanal. Chem.*, 305, 319–328, 1991.

161. Watanabe, M., Shibata, M., Katoh, A., Azuma, M., and Sakata, T., Design of alloy electrocatalysts for CO_2 reduction. 1. The selective and reversible reduction of CO_2

at Cu–Ni alloy electrodes, *Denki Kagaku*, 59, 508–516, 1991; *Chem. Abstr.*, 115, 265252h.

162. Watanabe, M., Shibata, M., and Katoh, A., Design of electrocatalyst for CO_2 reduction, *Interface*, Vol. 2, Abstr. 948, The Electrochem. Society Meeting, Honolulu, May 1993.

163. Katoh, A., Uchida, H., Shibata, M., and Watanabe, M., Design of electrocatalyst for CO_2 reduction. V. Effect of the microcrystalline structures of Cu–Sn and Cu–Zn alloys on the electrocatalysis of CO_2 reduction, *J. Electrochem. Soc.*, 141, 2054–2058, 1994.

164. Kostecki, R. and Augustynski, J., Electrochemical reduction of CO_2 on copper modified silver electrode, in Abstracts Int. Conf. on Carbon Dioxide Utilization, Bari, Italy, September 1993, 382.

165. Cherashev, A.F. and Khrushch, A.P., The electrochemical reduction of carbon dioxide at the tin–cadmium and tin–zinc alloys, *Russ. J. Electrochem.*, 33, 181–185, 1997.

166. Fisher, B.J. and Eisenberg, R., Electrocatalytic reduction of carbon dioxide by using macrocycles of nickel and cobalt, *J. Am. Chem. Soc.*, 102, 7361–7363, 1980.

167. Andrieux, C.P., Hapiot, P., and Savéant, J.-M., Fast kinetics by means of direct and indirect electrochemical techniques, *Chem. Rev.*, 90, 723–738, 1990.

168. Isse, A.A., Gennaro, A., Severin, M.G., and Vianello, E., Carbon dioxide reduction by heterogeneous and homogeneous electrocatalysis, in Proc. Int. Conf. on Carbon Dioxide Utilization, Bari, Italy, September 1993, 287–294.

169. Gennaro, A., Isse, A.A., Savéant, J.-M., Severin, M.-G., and Vianello, E., Homogeneous electron transfer catalysis of the electrochemical reduction of carbon dioxide. Do aromatic anion radicals react in an outer-sphere manner? *J. Am. Chem. Soc.*, 118, 7190–7196, 1996.

170. Bhugun, I., Lexa, D., and Savéant, J.-M., Catalysis of the electro-chemical reduction of carbon dioxide by iron(O) porphyrins: synergystic effect of weak Brönsted acids, *J. Am. Chem. Soc.*, 118, 1769–1776, 1996.

171. Keene, F.R. and Sullivan, B.P., Mechanisms of the electrochemical reduction of carbon dioxide catalyzed by transition metal complexes, in *Electrochemical and Electrocatalytic Reactions of Carbon Dioxide,* Sullivan, B.P., Krist, K., and Guard, H.E., Eds., Elsevier, Amsterdam, 1993.

172. Takahashi, K., Hiratzuka, K., Sasaki, H., and Toshima, S., Electrocatalytic behavior of metal porphyrins in the reduction of carbon dioxide, *Chem. Lett.*, pp. 305–308, 1979.

173. Tezuka, M. and Iwasaki, M., Voltammetric study on CO_2 reduction electrocatalyzed by cobalt tetraphenylporphine in DMF solution, *Chem. Lett.*, pp. 427–430, 1993.

174. Cao, X., Mu, Y., Wang, M., and Luan, L., The electrocatalytic reduction of carbon dioxide using cobalt tetrakis (4-trimethyl ammonio phenyl)-porphyrin under high pressure, *Huaxue Xuebao (Acta Chim. Sinica)*, 44, 220–224, 1986; *Chem. Abstr.*, 104, 195348r.

175. Cao, X., Zheng, G., and Teng, Y., Electrocatalytic reduction of carbon dioxide. IV. Studies on the mechanism of the electrocatalytic reaction by optically transparent thin layer electrode (OTTLE), *Huaxue Xuebao (Acta Chim. Sinica)*, 47, 575–582, 1989; *Chem. Abstr.*, 111, 122682e.

176. Zhang, J., Pietro, W.J., and Lever, A.B.P., Rotating ring-disk electrode analysis of CO_2 reduction electrocatalyzed by a cobalt tetramethyl pyridoporphyrazine on the disk and detected as CO on a platinum ring, *J. Electroanal. Chem.*, 403, 93–100, 1996.

177. Beley, M., Collin, J.-P., Ruppert, R., and Sauvage, J.-P., Nickel (II)-cyclam: an extremely selective electrocatalyst for reduction of CO_2 in water, *J. Chem. Soc. Chem. Commun.*, pp. 1315–1316, 1984.

178. Beley, M., Collin, J.-P., Ruppert, R., and Sauvage, J.-P., Electrocatalytic reduction of CO_2 by Ni-cyclam^{2+} in water. Study of the factors affecting the efficiency and the selectivity of the process, *J. Am. Chem. Soc.*, 108, 7461–7467, 1986.

179. Collin, J.-P., Jouaiti, A., and Sauvage, J.-P., Electrocatalytic properties of Ni(cyclam)$^{2+}$ and Ni$_2$(biscyclam)$^{4+}$ with respect to CO_2 and H_2O reduction, *Inorg. Chem.*, 27, 1990–1993, 1988.

180. Hay, R.W., Crayston, J.A., Cromie, T.J., Lightfoot, P., and Dealwis, D.C.L., The preparation, chemistry and crystal structure of the nickel(II) complex of N-hydroxyethylazacyclam [3-(2′-hydroxyethyl)]-1,3,5,8,12-penta-azacyclotetradecane nickel(II) perchlorate. A new electrocatalyst for CO_2 reduction, *Polyhedron*, 16, 3557–3563, 1997.

181. Collin, J.-P. and Sauvage, J.-P., Electrochemical reduction of carbon dioxide mediated by molecular catalysis, *Coord. Chem. Rev.*, 93, 245–268, 1989.

182. Collin, J.P., Sauvage, J.P., and Sakaki, S., Can CO_2 coordinate to a Ni(I) complex? An ab initio MO/SD-CI study, *J. Am. Chem. Soc.*, 112, 7813–7814, 1990.

183. Sakaki, S., An ab initio MO/SD-CI study of model complexes of intermediates in electrochemical reduction of CO_2 catalyzed by NiCl$_2$(cyclam), *J. Am. Chem. Soc.*, 114, 2055–2062, 1992.

184. Taniguchi, I., Electrocatalytic reduction of greenhouse gases using biofunctional metal complexes, in Proc. Int. Symp. Chemical Fixation of Carbon Dioxide, Nagoya, Japan, December 2 to 4, 1991, 81–88.

185. Balazs, G.B. and Anson, F.C., The adsorption of Ni(cyclam)$^+$ at mercury electrodes and its relation to the electrocatalytic reduction of CO_2, *J. Electroanal. Chem.*, 322, 325–345, 1992.

186. Balazs, G.B. and Anson, F.C., Effects of CO on the electrocatalytic activity of Ni(cyclam)$^{2+}$ towards the reduction of CO_2, *J. Electroanal. Chem.*, 361, 149–157, 1993.

187. Kelly, C.A., Mulazzani, Q.G., Venturi, M., Blinn, E.L., and Rodgers, A.J., The thermodynamics and kinetics of CO_2 and H^+ binding to Ni(cyclam)$^+$ in aqueous solution, *J. Am. Chem. Soc.*, 117, 4911–4919, 1995.

188. Fujita, E., Haff, J., Sanzenberger, R., and Elias, H., High electrocatalytic activity of RRSS-[Ni^{II}HTIM](ClO$_4$)$_2$ and [Ni^{II}DMC](ClO$_4$)$_2$ for carbon dioxide reduction (HTIM = 2,3,9,10-tetramethyl-1,4,8,11-tetraazacyclotetradecane, DMC = C-meso-5,12-di-methyl-1,4,8,11-tetraazacyclotetradecane), *Inorg. Chem.*, 33, 4627–4628, 1994.

189. Fujita, E., Haff, J., Sanzenberger, R., Elias, H., and Kobiro, K., Higher electrocatalytic activity of some nickel(II) macrocycle complexes than that of nickel cyclam(ClO$_4$)$_2$ for carbon dioxide reduction: structural considerations, in Abstr. 3rd Int. Conf. on Carbon Dioxide Utilization, Norman, OK, May 1995.

190. Hirata, Y., Suga, K., and Fujihara, M., Electrocatalytic reduction of CO_2 on modified electrodes with alkylcyclam–metal complex Langmuir–Blodgett films, *Thin Solid Films*, 179, 95–101, 1989.

191. Akiba, U., Nakamura, Y., Suga, K., and Fujihira, M., Electrocatalytic reduction of CO_2 on Langmuir–Blodgett film modified electrode with Ni(II) complexes of amphiphilic cyclam, in Proc. Int. Symp. Chemical Fixation of Carbon Dioxide, Nagoya, Japan, December 2 to 4, 1991, 339–342.

192. Akiba, U., Nakamura, Y., Suga, K., and Fujihira, M., Electrocatalytic reduction of CO_2 on a modified electrode with Langmuir–Blodgett films of nickel(II) complex with long chain alkyl substituted cyclam, *Thin Solid Films*, 210, 381–383, 1992.

193. Hirata, Y., Suga, K., and Fujihira, M., In-situ analysis of products in electrocatalytic reduction of CO_2 with Ni-cyclam by differential electrochemical mass spectrometry during cyclic voltammetry on an amalgamated-gold mesh electrode, *Chem. Lett.*, pp. 1155–1158, 1990.

194. Fujihira, M., Nakamura, Y., Hirata, Y., Akiba, U., and Suga, K., Electrocatalytic reduction of carbon dioxide by nickel(II) complexes of cyclam and C-alkylated and N-alkylated cyclams, *Denki Kagaku*, 59, 532–539, 1991.

195. Fujihira, M. and Noguchi, T., In situ analysis of electrochemical reduction products of CO_2 by DEMS, in Proc. Int. Symp. Chemical Fixation of Carbon Dioxide, Nagoya, Japan, December 2 to 4, 1991, 97–102.

196. Bujno, K., Bilewicz, R., Siegfried, L., and Kaden, T., Electrochemical behaviour of isomers of tetraazamacrocyclic Ni(II) complex in solutions saturated with argon, CO, and CO_2, *J. Electroanal. Chem.*, 407, 131–140, 1996.

197. Bujno, K., Bilewicz, R., Siegfried, L., and Kaden, T., Electroreduction of CO_2 catalyzed by Ni(II)tetraazamacrocyclic complexes. Reasons of poisoning of the catalytic surfaces, *Electrochim. Acta*, 42, 1201–1206, 1997.

198. Lieber, C.M. and Lewis, N.S., Catalytic reduction of CO_2 at carbon electrodes modified with cobalt phthalocyanine, *J. Am. Chem. Soc.* 106, 5033–5034, 1984.

199. Kapusta, S. and Hackerman, N., Carbon dioxide reduction at a metal phthalocyanine catalyzed carbon electrode, *J. Electrochem. Soc.*, 131, 1511–1514, 1984.

200. Meshitsuka, S., Ichikawa, M., and Tamaru, K., Electrocatalysis by metal phthalocyanines in the reduction of carbon dioxide, *J. Chem. Soc. Chem. Commun.*, pp. 158–159, 1974.

201. Rollmann, L.D. and Iwamoto, R.T., Electrochemistry, electron paramagnetic resonance, and visible spectra of cobalt, nickel, copper, and metal-free phthalocyanines in dimethyl sulfoxide, *J. Am. Chem. Soc.*, 90, 1455–1463, 1968.

202. Taube, R., The electronic structure of anionic phthalocyanine complexes of some 3d elements, *Z. Chem.*, 6, 8–21, 1966.

203. Furuya, N. and Koide, S., Electroreduction of carbon dioxide by metal phthalocyanines, *Electrochim. Acta*, 36, 1309–1313, 1991.

204. Taube, R., New aspects of the chemistry of transition metal phthalocyanines, *Pure Appl. Chem.*, 38, 427–438, 1974.

205. Tanabe, H. and Ohno, K., Electrocatalysis of metal phthalocyanine thin film prepared by the plasma–assisted deposition on a glassy carbon in the reduction of carbon dioxide, *Electrochim. Acta*, 32, 1121–1124, 1987.

206. Yoshida, T., Kamato, K., Tsukamoto, M., Iida, T., Schlettwein, D., Wöhrle, D., and Kaneko, M., Selective electrocatalysis for CO_2 reduction in the aqueous phase using cobalt phthalocyanine/poly-4-vinylpyridine modified electrodes, *J. Electroanal. Chem.*, 385, 209–225, 1995.

207. Yoshikawa, M., Ezaki, T., Sanui, K., and Ogata, N., Selective permeation of carbon dioxide through synthetic polymer membranes having pyridine moiety as a fixed carrier, *J. Appl. Polym. Sci.*, 35, 145–154, 1988.

208. Abe, T., Yoshida, T., Tokita, S., Taguchi, F., Imaya, H., and Kaneko, M., Factors affecting electrocatalytic CO_2 reduction with cobalt phthalocyanine incorporated in a polyvinylpyridine membrane coated on a graphite electrode, *J. Electroanal. Chem.*, 412, 125–132, 1996.

209. Abe, T., Taguchi, F., Yoshida, T., Tokita, S., Schnurpfeil, G., Wöhrle, D., and Kaneko, M., Electrocatalytic CO_2 reduction by cobalt octabutoxyphthalocyanine coated on graphite electrode, *J. Mol. Catal. A Chem.*, 112, 55–61, 1996.
210. Atoguchi, T., Aramata, A., Kazusaka, A., and Enyo, M., Cobalt(II)-tetraphenyl porphyrin–pyridine complex fixed on a glassy carbon electrode and its prominent catalytic activity for reduction of carbon dioxide, *J. Chem. Soc. Chem. Commun.*, pp. 156–157, 1991.
211. Atoguchi, T., Aramata, A., Kazusaka, A., and Enyo, M., Electrocatalytic activity of $Co^{II}TPP$-pyridine complex modified carbon electrode for CO_2 reduction, *J. Electroanal. Chem.*, 318, 309–320, 1991.
212. Cosnier, S., Deronzier, A., and Moulet, J.-C., Electrochemical coating of a platinum electrode by a poly(pyrrole) film containing the fac-(2,2'-bipyridine) tricarbonyl chlororhenium system. Application to electrocatalytic reduction of carbon dioxide, *J. Electroanal. Chem.*, 207, 315–321, 1986.
213. Cosnier, S., Deronzier, A., and Moulet, J.-C., Electrocatalytic reduction of CO_2 on electrodes modified by (2,2'-bipyridine)-$(CO)_3Cl$ complexes bonded to polypyrrole films, *J. Mol. Catal.*, 45, 381–391, 1988.
214. Cosnier, S., Deronzier, A., and Moulet, J.-C., Substitution effects on the electrochemical behaviour of the (2,2'-bipyridine) tricarbonyl chlororhenium (I) complex in solution or in polymeric form and their relation to the catalytic reduction of carbon dioxide, *New J. Chem.*, 14, 831–839, 1990.
215. Yoshida, T., Tsutsumida, K., Teratani, S., Yasufuku, K., and Kaneko, M., Electrocatalytic reduction of CO_2 in water by [Re(bpy)$(CO)_3$Br] and [Re-(terpy)$(CO)_3$Br] complexes incorporated into coated Nafion membrane (bpy = 2,2'-bipyridine, terpy = 2,2':6',2''-terpyridine), *J. Chem. Soc. Chem. Commun.*, pp. 631–633, 1993.
216. Kaneko, M., Lin, R.-J., and Yoshida, T., Electrocatalytic reduction of CO_2 by metal complexes incorporated into coated polymer membranes, in Proc. Int. Symp. Chemical Fixation of Carbon Dioxide, Nagoya, Japan, December 2 to 4, 1991, 103–106.
217. Christensen, P., Hamnett, A., Muir, A.V.G., Timney, J.A., and Higgins. S., Growth and electrochemical behavior of a poly [tricarbonyl(vinylbipyridyl)rhenium chloride] film. Heterogeneous reduction of CO, *J. Chem. Soc. Faraday·Trans.*, 90, 459–469, 1994.
218. Hamnett, A., Christensen, P.A., and Higgins, S.J., Analysis of electrogenerated films by ellipsometry and infrared spectrometry, *Analyst*, 119, 735–747, 1994.
219. Stor, G.J., Hartl, F., Van Outersterp, J.W.M., and Stufkens, D.J., Spectroelectrochemical (IR, UV/Vis) determination of the reduction pathways for series of [Re(CO)3(a-diimine)L']$^{0/+}$ (L = Halide, Otf⁻, THF, MeCN, n-PrCN, PPh3, P(OMe)3) complexes, *Organometallics*, 14, 1115–1131, 1995.
220. Johnson, F.P.A., George, M.W., Hartl, F., and Turner, J.J., Electrocatalytic reduction of CO_2 using the complexes [Re(bpy)$(CO)_3$L]^n (n = +1, L = P(OEt)$_3$, CH_3CN. n = 0, L = Cl⁻, OtF⁻, bpy = 2,2'-bipyridine, Otf = (F_3SO_3)) as catalyst precursors: infrared spectroelectrochemical investigation, *Organometallics*, 15, 3374–3387, 1996.
221. Sende, J.A.R., Arana, C.R., Hernández, L., Potts, K.T., Keshevarz-, K.M., and Abruña, H.D., Electrocatalysis of CO_2 reduction in aqueous media at electrodes modified with electropolymerized films of vinylterpyridine complexes of transition metals, *Inorg. Chem.*, 34, 3339–3348, 1995.

222. Collomb-Dunand-Sauthier, M.-N., Deronzier, A., and Ziessel, R., Electrochemical behaviour of $[Ru^{II}(L)(CO)_2Cl_2]$, $[Ru^{II}(L)(CO)Cl_3][Me_4N]$ and $[Ru^{II}(L)(CO)_2$ $(MeCN)_2][CF_3SO_3]_2$ complexes (L = 2,2'-bipyridine or 4,4'-isopropoxy-carbonyl-2,2'-bipyridine), *J. Electroanal. Chem.*, 350, 43–55, 1993.

223. Collomb-Dunand-Sauthier, M.-N., Deronzier, A., and Ziessel, R., Electrochemical elaboration of thin films of poly$[Ru^{II}(L)(CO)_2Cl_2]$ (L = 4-(2-pyrrol-1-ylethyl)-4'-methyl-2,2'-bipyridine or 4,4'-bis((3-pyrrol-1-ylpropyloxy) carbonyl)-2,2'-bipyridine). Photochemical properties and photoimaging, *J. Phys. Chem.*, 97, 5973–5979, 1993.

224. Collomb-Dunand-Sauthier, M.-N., Deronzier, A., and Ziessel, R., Electrocatalytic reduction of CO_2 in water on a polymeric $[\{Ru^0(bpy)(CO)_2\}_n]$ (bpy = 2,2'-bipyridine) complex immobilized on carbon electrodes, *J. Chem. Soc. Chem. Commun.*, pp. 189–191, 1994.

225. Collomb-Dunand-Sauthier, M.-N., Deronzier, A., and Ziessel, R., Electrocatalytic reduction of carbon dioxide with mono-(bipyridine)carbonylruthenium complexes in solution or as polymeric thin films, *Inorg. Chem.*, 33, 2961–2967, 1994.

226. Chardon-Noblat, S., Collomb-Dunand-Sauthier, M.-N., Deronzier, A., Ziessel, R., and Zsoldos, D., Formation of polymeric $[\{Ru^0(bpy)(CO)_2\}_n]$ films by electrochemical reduction of $[Ru(bpy)_2(CO)_2](PF_6)_2$: its implication in CO_2 electrocatalytic reduction, *Inorg. Chem.*, 33, 4410–4412, 1994.

227. Noblat-Chardon, S., Deronzier, A., and Ziessel, R., Photo- and electrocatalytic reduction of CO_2 with monobipyridine carbonyl ruthenium complexes in solution and as polymeric films, in Abstr. 3rd Int. Conf. on Carbon Dioxide Utilization, Norman, OK, May 1995.

228. Yoshida, T., Iida, T., Shirasagi, T., Lin, R.J., and Kaneko, M., Electrocatalytic reduction of carbon dioxide in aqueous medium by bis(2,2'-6',2''-terpyridine)cobalt(II) complex incorporated into a coated polymer membrane, *J. Electroanal. Chem.*, 344, 355–362, 1993.

229. Petrova, G.N. and Efimov, O.N., Electrocatalytic reduction of CO_2 to C_1–C_3 hydrocarbons, *Elektrokhimya*, 19, 978, 1983; English translation, p. 875.

230. Savinova, E.R., Yashnik, S.A., Savinov, E.N., and Parmon, V.N., Gas-phase electrocatalytic reduction of CO_2 to CO on carbon gas-diffusion electrode promoted by cobalt phthalocyanine, *React. Kinet. Catal. Lett.*, 46, 249–254, 1992.

231. Mahmood M.N., Masheder, D., and Harty, C.J., Use of gas-diffusion electrodes for high-rate electrochemical reduction of carbon dioxide. I. Reduction at lead, indium- and tin-impregnated electrodes, *J. Appl. Electrochem.*, 17, 1159–1170, 1987.

232. Dewulf, D.W. and Bard, A.J., The electrochemical reduction of carbon dioxide to methane and ethene at copper/Nafion electrodes (solid polymer electrolyte structures), *Catal. Lett.*, 1, 73–79, 1988.

233. Cook, R.L. MacDuff, R.C., and Sammells, A.F., Ambient temperature gas phase CO_2 reduction to hydrocarbons at solid polymer electrolyte cells, *J. Electrochem. Soc.*, 135, 1470–1471, 1988.

234. Furuya, N., Matsui, K., and Motoo, S., Utilization of gas-diffusion electrodes as cathode for carbon dioxide reduction. III, *Denki Kagaku*, 56, 980–984, 1988.

235. Ito, T., Ikeda, S., Maeda. M., Yosida, H., and Ito, K., Electrochemical reduction of carbon dioxide on Cu-loaded gas diffusion electrodes, in Proc. Int. Symp. Chemical Fixation of Carbon Dioxide, Nagoya, Japan, December 2 to 4, 1991, 313–318.

236. Ikeda, S., Ito, T., Azuma, K., Ito, K., and Noda, H., Electrochemical mass reduction of carbon dioxide using Cu-loaded gas diffusion electrodes. I. Preparation of electrode and reduction products, *Denki Kagaku*, 63, 303–309, 1995.

237. Ikeda, S., Ito, T., Azuma, K., Nishi, N., Ito, K., and Noda, H., Electrochemical mass reduction of carbon dioxide using Cu-loaded gas diffusion electrodes. II. Proposal of reaction mechanism, *Denki Kagaku*, 64, 69–75, 1996.

238. Ikeda, S., Siozaki, S., Suzuki, J., Ito, K., and Noda, H., Electroreduction of CO_2 using Cu/Zn oxides loaded gas diffusion electrodes, in Abstr. 4th Int. Conf. on Carbon Dioxide Utilization, Kyoto, Japan, September 1997, O-17.

239. Mahmood M.N., Masheder, D., and Harty, C.J., Use of gas-diffusion electrodes for high-rate electrochemical reduction of carbon dioxide. II. Reduction at metal phthalocyanine-impregnated electrodes, *J. Appl. Electrochem.*, 17, 1223–1227, 1987.

240. Furuya, N. and Matsui, K., Electroreduction of carbon dioxide on gas-diffusion electrodes modified by metal phthalocyanines, *J. Electroanal. Chem.*, 271, 181–191, 1989.

241. Schwartz, M., Cook, R.L., Kehoe, V.M., MacDuff, R.C., Patel, J., and Sammells, A.F., Carbon dioxide reduction to alcohols using perovskite-type electrocatalysts, *J. Electrochem. Soc.*, 140, 614–618, 1993.

242. Schwartz, M., Vercauteren, M.E., and Sammells, A.F., Fischer–Tropsch electrochemical CO_2 reduction to fuels and chemicals, *J. Electrochem. Soc.*, 141, 3119–3127, 1994.

243. Hara, K., Kudo, A., Sakata, T., and Watanabe, M., High efficiency electrochemical reduction of carbon dioxide under high pressure on a gas diffusion electrode containing Pt catalysts, *J. Electrochem. Soc.*, 142, L57–L59, 1995.

244. Hara, K. and Sakata, T., Large current density CO_2 reduction under high pressure using gas diffusion electrodes, *Bull. Chem. Soc. Jpn.*, 70, 571–576, 1997.

245. Hara, K. and Sakata, T., Electrocatalytic formation of CH_4 from CO_2 on a Pt gas diffusion electrode, *J. Electrochem. Soc.*, 144, 539–545, 1997.

246. Masheder, D. and Williams, K.P.J., Raman spectro-electrochemistry. I. In situ Raman studies of the electrochemical reduction of CO_2 at lead-impregnated PTFE-bonded carbon gas diffusion electrode, *J. Raman Spectrosc.*, 18, 387–390, 1987.

247. Takahashi, K., Hashimoto, K., Fujishima, A., Omata, K., and Kimura, N., Electrochemical CO_2 reduction using gas diffusion electrode containing Cu catalyst supported on various metal oxides, *Chem. J. Chin. Univ.*, 16, 189–192, 1995.

248. Komatsu, S., Tanaka, M., Okumura, A., and Kungi, A., Preparation of Cu–solid polymer electrolyte composite electrodes and application to gas-phase electrochemical reduction of CO_2, *Electrochim. Acta*, 40, 745–753, 1995.

249. Yamamoto, T., Tryk, D.A., Hashimoto, K., Fujishima, A., and Okawa, M., Electrochemical reduction of CO_2 in micropores, in Abstr. 4th Int. Conf. on Carbon Dioxide Utilization, Kyoto, Japan, September 1997, P-077.

250. Shibata, M., Yoshida, K., and Furuya, N., Electrochemical synthesis of urea on reduction of carbon dioxide with nitrate and nitrite ions using Cu-loaded gas-diffusion electrode, *J. Electrochem. Chem.*, 387, 143–145, 1995.

251. Fischer, F. and Prziza, O., On the electrolytic reduction of carbon dioxide and carbon monoxide dissolved under pressure, *Chem. Ber.*, 47, 256–260, 1914.

252. Ito, K., Ikeda, S., and Okabe, M., Electrochemical reduction of carbon dioxide under high pressure. I. In an aqueous solution of inorganic salt, *Denki Kagaku*, 48, 247–252, 1980.

253. Nakagawa, S., Kudo, A., Azuma, M., and Sakata, T., Effect of pressure on the electrochemical reduction of CO_2 on group-VIII metal electrodes, *J. Electroanal. Chem.*, 308, 339–343, 1991.

254. Nakagawa, S., Kudo, A., Azuma, M., and Sakata, T., Effect of pressure on electrochemical reduction of CO_2, in Proc. Int. Symp. Chemical Fixation of Carbon Dioxide, Nagoya, Japan, December 2 to 4, 1991, 319–322.

255. Mizuno, T., Sasaki, A., Ohta, K., Akano, T., and Hirano, M., Effect of temperature and pressure on electrochemical reduction of CO_2, in Abstr. Int. Conf. on Carbon Dioxide Utilization, Bari, Italy, September 1993, 384.

256. Kudo, A., Nakagawa, S., Tsuneto A., and Sakata, T., Electrochemical reduction of high-pressure CO_2 on Ni electrodes, *J. Electrochem. Soc.*, 140, 1541–1545, 1993.

257. Hara, K., Tsuneto, A., Kudo, A., and Sakata, T., Electrochemical reduction of CO_2 on a Cu electrode under high-pressure — factors that determine the product selectivity, *J. Electrochem. Soc.*, 141, 2097–2103, 1994.

258. Hara, K., Kudo, A., and Sakata, T., Electrochemical reduction of carbon dioxide under high pressure on various electrodes in an aqueous electrolyte, *J. Electroanal. Chem.*, 391, 141–147, 1995.

259. Todoroki, M., Hara, K., Kudo, A., and Sakata, T., Electrochemical reduction of high pressure CO_2 at Pb, Hg, and In electrodes in an aqueous $KHCO_3$ solution, *J. Electroanal. Chem.*, 394, 199–203, 1995.

260. Hara, K., Kudo, A., and Sakata, T., Electrochemical CO_2 reduction on a glassy carbon electrode under high pressure, *J. Electroanal. Chem.*, 421, 1–4, 1997.

261. Halmann, M. and Aurian-Blajeni, B., Electrochemical reduction of carbon dioxide at elevated pressure on semiconductor electrodes in aqueous solution, *J. Electroanal. Chem.*, 375, 379–382, 1994.

262. Gressin, J.C., Michelet, D., Nadjo, L., and Savéant, J.M., Electrochemical reduction of carbon dioxide in weakly protic medium, *Nouv. J. Chim.*, 3, 545–554, 1979.

263. Amatore, C. and Savéant, J.-M., Mechanism and kinetic characteristics of the electrochemical reduction of carbon dioxide in media of low proton availability, *J. Am. Chem. Soc.*, 103, 5021–5023, 1981.

264. Amatore, C., Nadjo, L., and Savéant, J. M., A propos de la reduction electrochimique du dioxide de carbone, *Nouv. J. Chim.*, 8, 565–566, 1984.

265. Lamy, E., Nadjo, L., and Savéant, J.M., Standard potential and kinetic parameters of the electrochemical reduction of carbon dioxide in dimethylformamide, *J. Electroanal. Chem.*, 78, 403–407, 1977.

266. Eggins, B.R. and McNeill, J., Voltammetry of carbon dioxide. I. A general survey of voltammetry at different electrode materials in different solvents, *J. Electroanal. Chem.*, 148, 17–24, 1983.

267. Eggins, B.R., Ennis, C., McConnell, R., and Spence, M., Improved yields of oxalate, glyoxylate and glycolate from the electrochemical reduction of carbon dioxide in methanol, *J. Appl. Electrochem.*, 27, 706–712, 1997.

268. Vassiliev, Yu.B., Bagotzky, V.S., Khazova, O.A., and Mayorova, N.A., Electroreduction of carbon dioxide. II. The mechanism of reduction in aprotic solvents, *J. Electroanal. Chem.*, 189, 295–309, 1985.

269. Mayorova, N.A., Khazova, O.A., and Vassiliev, Yu.B., Electroreduction of carbon dioxide in aprotic solvents, *Elektrokhimiya*, 22, 1196–1204, 1986; English translation, pp. 1122–1129.

270. Desilvestro, J. and Pons, S., The cathodic reduction of carbon dioxide in acetonitrile. An electrochemical and infrared spectroelectrochemical study, *J. Electroanal.Chem.*, 267, 207–220, 1989.

271. Ito, K., Ikeda, S., Iida, T., and Niwa, H., Electrochemical reduction of carbon dioxide dissolved under pressure. II. In aqueous solutions of tetraalkylammonium salts, *Denki Kagaku*, 49, 106–112, 1981.

272. Ito, K., Ikeda, S., Iida, T., and Nomura, A., Electrochemical reduction of carbon dioxide dissolved under high pressure III. In nonaqueous electrolytes, *Denki Kagaku*, 50, 463–469, 1982.

273. Goodridge, F. and Presland, G., The electrolytic reduction of carbon dioxide and monoxide for the production of carboxylic acids, *J. Appl. Electrochem.*, 14, 791–796, 1984.

274. Hochgesand, G., Rectisol and purisol. Efficient acid gas removal for high pressure hydrogen and syngas production, *Ind. Eng. Chem.*, 62, 37–43, 1970.

275. Hirota, K., Saeki, T., Hashimoto, K., Fujishima, A., Omata, K., and Kimura, N., Electrochemical reduction of highly concentrated CO_2 using water as an electron donor, in Abstr. 3rd Int. Conf. on Carbon Dioxide Utilization, Norman, OK, May 1995.

276. Naitoh, A., Ohta, K., Mizuno, T., Yoshida, H., Sakai, M., and Noda, H., Electrochemical reduction of carbon dioxide in methanol at low temperature, *Electrochim. Acta*, 38, 2177–2178, 1993.

277. Mizuno, T., Naitoh, A., and Ohta, K., Electrochemical reduction of CO_2 in methanol at −30°, *J. Electroanal. Chem.*, 391, 199–201, 1995.

278. Mizuno, T., Ohta, K., Kawamoto, M., and Saji, A., Electrochemical reduction of CO_2 in 0.1 *M* KOH–methanol, in Abstr. 3rd Int. Conf. on Carbon Dioxide Utilization, Norman, OK, May 1995.

279. Silvestri, G., Gambino, S., and Filardo, G., Electrochemical synthesis involving carbon dioxide, in *Enzymatic and Model Carboxylation and Reduction Reactions for CO_2 Utilization*, NATO ASI Series, Vol. 314, Aresta, M. and Schloss, J.V., Eds., Reidel, Dordrecht, 1990, 101–127.

280. Silvestri, G., Gambino, S., and Filardo, G., Electrochemical synthesis involving carbon dioxide, in *Carbon Dioxide Fixation and Reduction in Biological and Model Systems*, Bränden, C.-I. and Schneider, G., Eds., Oxford University Press, 1994, 185–209.

281. Silvestri, G., Gambino, S., and Filardo, G., Use of sacrificial anodes in synthetic electrochemistry. Processes involving carbon dioxide, *Acta Chem. Scand.*, 45, 987–992, 1991.

282. Chaussard, J., Folest, J.C., Nedelec, J.Y., Périchon, J., Sibille, S., and Troupel, M., Use of sacrificial anodes in electrochemical functionalization of organic halides, *Synthesis*, pp. 369–381, 1990.

283. Tomilov, A.P., Electrochemical synthesis employing sacrificial anodes, *Russ. J. Electrochem.*, 32, 25–36, 1996.

284. Massebieau, M.-C., Duñach, E., Troupel, M., and Perichon, J., Efficient electrochemical reduction of carbon dioxide on freshly coated metal electrodes, *New J. Chem.*, 14, 259–260, 1990.

285. Silvestri, G., Electrochemical synthesis of carboxylic acids from carbon dioxide, *NATO ASI Ser., Ser. C.*, 206, 339–369, 1987.

286. Fischer, J., Lehmann, Th., and Heitz, E., The production of oxalic acid from CO_2 and H_2O, *J. Appl. Electrochem.*, 11, 743–750, 1981.

287. Gambino, S. and Silvestri, G., On the electrochemical reduction of carbon dioxide and ethylene, *Tetrahedron Lett.*, 32, 3025–3028, 1973.

288. Silvestri, G., Gambino, S., and Filardo, G., Electrocarboxylation of ethylene. Synthesis of industrially significant dicarboxylic acids, The Electrochemical Society Meeting, Washington, D.C., May 2 to 7, 1976, pp. 698–699.

289. Lamy, E., Nadjo, L., and Savéant, J.M., On the electrochemical carboxylation of activated olefins, *Nouv. J. Chim.*, 3, 21–29, 1979.

290. Gambino, S., Gennaro, A., Filardo, G., Silvestri, G., and Vianello, E., Electrochemical carboxylation of styrene, *J. Electrochem. Soc.*, 134, 2172–2175, 1987.

291. Gambino, S., Filardo, G., and Silvestri, G., Electrochemical carboxylation of organic substrates. Synthesis of carboxylic derivatives of acenaphthalene, *J. Appl. Electrochem.*, 12, 549–555, 1982.

292. Aresta, M., Quaranta, E., Tommasi, I., Dérien, S., and Duñach, E., Tetraphenylborate anion as a phenylating agent: chemical and electrochemical reactivity of BPh_4–Rh complexes toward monoene and diene and carbon dioxide, *Organometallics*, 14(7), 3349–3356, 1995.

293. Chiozza, E., Desigaud, M., Greiner, J., and Duñach, E., Incorporation of CO_2 into perfluoroalkyl derivatives by electrochemical methods, in Abstr. 4th Int. Conf. on Carbon Dioxide Utilization, Kyoto, Japan, September 1997, O-15.

294. Ikeda, Y. and Manda, E., Synthesis of benzilic acids through electro-chemical reductive carboxylation of benzophenones in the presence of carbon dioxide, *Bull. Chem. Soc. Jpn.*, 58, 1723–1726, 1985.

295. Rieu, J.P., Boucherle, A., Cousse, H., and Mouzin, G., Methods for the synthesis of antiinflammatory 2-aryl propionic acids, *Tetrahedron*, 42, 4095–4131, 1986.

296. Silvestri, G., Gambino, S., and Filardo, G., U.S. Patent No. 4,708,780, 1987.

297. Maspero, F., Piccolo, O., Romano, U., and Gambino, S., Eur. Patent Appl. No. 286, 944, 1988.

298. Chan, A.S.C., Huang, T.T., Wagenknecht, J.H., and Miller, R.E., A novel synthesis of 2-aryllactic acids via electrocarboxylation of methyl aryl ketones, *J. Org. Chem.*, 60, 742–744, 1995.

299. Pletcher, D. and Slevin, L., Influence of magnesium(II) ions on cathodic reactions in aprotic solvents — carboxylation of methyl aryl ketones, *J. Chem. Soc. Perkin Trans. 2*, pp. 217–220, 1996.

300. Koshechko, V.G., Titov, V.E., Lopushanskaja, V.A., and Pokhodenko, V.D., Electrochemical synthesis of α-oxoacids using carbon dioxide, in Abstr. 3rd Int. Conf. on Carbon Dioxide Utilization, Norman, OK, May 1995.

301. Pokhodenko, V.D., Koshechko, V.G., Titov, V.E., and Lopushanskaja, V.A., A convenient electrochemical synthesis of α-oxoacids, *Tetrahedron Lett.*, 36, 3277–3278, 1995.

302. Filardo, G., Galia, A., Gambino, S., and Silvestri, G., Homogeneous charge transfer catalysis in preparative electrocarboxylation of benzyl halides, in Proc. Int. Conf. on Carbon Dioxide Utilization, Bari, Italy, September 1993, 279–286.

303. Dérien, S., Clinet, J.C., Duñach, E., and Périchon, J., Electrochemical incorporation of carbon dioxide into alkenes by nickel complexes, *Tetrahedron*, 48, 5235–5248, 1992.

304. Dérien, S., Clinet, J.-C., Duñach, E., and Périchon, J., New C–C bond formation through the nickel-catalyzed electrochemical coupling of 1,3-enynes and carbon dioxide, *J. Organometal. Chem.*, 424, 213–224, 1992.

305. Dérien, S., Clinet, J.-C., Duñach, E., and Périchon, J., Activation of carbon dioxide: nickel-catalyzed electrochemical carboxylation of diynes, *J. Org. Chem.*, 58, 2578–2588, 1993.

306. Duñach, E., Dérien, S., Clinet, J.C., and Périchon, J., Coupling of unsaturated hydro-carbons and carbon dioxide by electrogenerated nickel complexes, in Proc. Int. Conf. on Carbon Dioxide Utilization, Bari, Italy, September 1993, 271–278.

307. Dérien, S., Duñach, E., and Périchon, J., From stoichiometry to catalysis: electroreductive coupling of alkynes and carbon dioxide with nickel-bipyridine complexes. Magnesium ions as the key for catalysis, *J. Am. Chem. Soc.*, 113, 8447–8454, 1991.

308. Dérien, S., Clinet, J.-C., Duñach, E., and Périchon, J., First example of direct carbon dioxide incorporation into 1,3-diynes: a highly regio selective and stereo-selective nickel catalysed electrochemical reaction, *J. Chem. Soc. Chem. Commun.*, pp. 549–550, 1991.

309. Tascedda, P. and Duñach, E., Novel electrochemical reactivity of $Ni(cyclam)Br_2$: catalytic carbon dioxide incorporation into epoxides, *J. Chem. Soc. Chem. Commun.*, pp. 43–44, 1995.

310. Kamekawa, H., Senboku, H., and Tokuda, M., Facile synthesis of aryl-substituted 2-alkenoic acids by electroreductive carboxylations of vinylic bromides using a mag-nesium anode, *Electrochim. Acta*, 42, 2117–2123, 1997.

311. Kamekawa, H., Kudoh, H., Senboku, H., and Tokuda, M., Synthesis of α,β-unsatur-ated carboxylic acids by nickel(II)-catalyzed electrochemical carboxylation of vinyl bromides, *Chem. Lett.*, pp. 917–918, 1997.

312. Tokuda, M., Kabuki, T., Katoh, Y., and Suginome, H., Regioselective synthesis of β,γ-unsaturated acids by the electrochemical carboxylation of allylic bromides using a reactive-metal anode, *Tetrahedron Lett.*, 36, 3345–3348, 1995.

313. Duñach, E. and Olivera, S., Nickel catalysis in electrochemical carbon–carbon bond forming reactions involving carbon dioxide, in Abstr. 3rd Int. Conf. on Carbon Dioxide Utilization, Norman, OK, May 1995.

314. Amatore, C. and Jutand, A., Activation of carbon dioxide by electron transfer and transition metals. Mechanism of nickel-catalyzed electro-carboxylation of aromatic halides, *J. Am. Chem. Soc.*, 113, 2819–2825, 1991.

315. Amatore, C. and Jutand, A., Mechanism of nickel-catalyzed electron transfer of aromatic halides. 2. Electrocarboxylation of bromobenzene, *J. Electroanal. Chem.*, 306, 141–156, 1991.

316. Torii, S., Tanaka, H., Hamatani, T., Morisaki, K., Jutand, A., Pflüger, F., and Fauvarque, J. F., Pd(0)-catalyzed electroactive carboxylation of aryl halides, β-bromostyrene, and allyl acetates with CO_2, *Chem. Lett.*, pp. 169–172, 1986.

317. Amatore, C., Jutand, A., Khalil, F., and Nielsen, M.F., Carbon dioxide as a C_1 building block. Mechanism of palladium-catalyzed carboxylation of aromatic halides, *J. Am. Chem. Soc.*, 114, 7076–7085, 1992.

318. Jutand, A. and Négri, S., Palladium-catalyzed carboxylation of vinyl triflates. Electrosynthesis of α,β-unsaturated carboxylic acids, *Synlett.*, pp. 719–721, 1997.

319. Koshechko, V.G., Titov, V.E., and Sednev, D.V., New route for producing acrylic acid copolymers, based on electrochemical carboxylation of poly(vinyl halide)s and polybutadiene, *Polymer*, 35, 1787–1788, 1994.

320. Murcia, N.S. and Peters, D.G., Electroreductive carboxylation of halobenzenes. Pro-duction of p-anisic acid by reduction of p-iodoanisole at mercury in dimethylformamide saturated with carbon dioxide, *J. Electroanal. Chem.*, 326, 69–79, 1992.

321. Garnier, L., Rollin, Y., and Périchon, J., Electrosynthesis of symmetrical ketones from organic halides and carbon dioxide catalyzed by 2,2′-bipyridine-nickel, *J. Organomet. Chem.*, 367, 347–358, 1989.

322. Zheng, G.D., An, Q., Quing, D., Han, X., and Cao, X., Electrocarboxylation of organic compounds with carbon dioxide catalyzed by metalloporphyrins. I. Catalytic activities of metalloporphyrins for electrocarboxylation of benzyl chloride, *Chin. Chem. Lett.*, 3, 97–98, 1992; *Chem. Abstr.*, 117, 120333s.

323. Zheng, G.D., An, Q., Liu, Y.,Wang, Y. and Cao, X., Electrocarboxylation of organic compounds with carbon dioxide catalyzed by metalloporphyrins. II. Electrosynthesis of some organic acids with carbon dioxide catalyzed by CoTPP, FeTPP, *Chin. Chem. Lett.*, 3, 265–266, 1992; *Chem. Abstr.*, 117, 130845z.

324. Zheng, G.D., Liu, Y.W., An, Q.D., and Cao, X.Z., Electrocarboxylation of organic compounds with carbon dioxide catalyzed by metalloporphyrins. III. Effect of axial ligands on catalytic activity of CoTPP, *Chin. Chem. Lett.*, 3, 355–356, 1992; *Chem. Abstr.*, 118, 59374h.

325. Zheng, G.D., Ding, Q., An, Q.D., and Cao, X.Z., Electrocarboxylation of organic compounds with carbon dioxide catalyzed by metalloporphyrins. IV. Effects of various factors on electrocarboxylation, *Chin. Chem. Lett.*, 3, 357–358, 1992; *Chem. Abstr.*, 118, 59375j (see also *Chem. Abstr.*, 121, 310315t and 122, 91381b).

326. Zheng, G.D., Yan, Y., Gao, S., Tong, S.L., Gao, D., and Zhen, K.J., The reaction mechanism of alkyl halides with carbon dioxide catalyzed by 5,10,15,20-tetraphenyl porphyrin cobalt (CoTPP), *Electrochim. Acta*, 41, 177–182, 1996.

327. Isse, A.A., Gennaro, A. and Vianello, E., Electrochemical carboxylation of arylmethyl chlorides catalysed by [Co(salen)] [H$_2$salen=N,N′-bis(salicylidene)ethane-1,2-diamine], *J. Chem. Soc. Dalton Trans.*, pp. 1613–1618, 1996.

328. Mizen, M.B. and Wrighton, M.S., Reductive addition of CO$_2$ to 9,10-phenanthrenequinone, *J. Electrochem. Soc.*, 136, 941–946, 1989.

329. Tanaka, H., Nagao, H., and Tanaka, K., Evaluation of acidity of CO$_2$ in protic media. Carboxylation of reduced quinone, *Chem. Lett.*, pp. 541–544, 1993.

330. Nagaoka, T., Nishii, N., Fujii, K., and Ogura, K., Mechanisms of reductive addition of CO$_2$ to quinones, *J. Electroanal. Chem.*, 322, 383–389, 1992.

331. Casadei, M.A., Cesa, S., Inesi, A., and Moracci, F.M., Electrochemical studies on haloamides. 11. Electrocarboxylation of carboxamides, *J. Chem. Res. S*, p. 166, 1995.

332. Casadei, M.A., Cesa, S., and Inesi, A., Electrochemical studies on haloamides. 12. Electrosynthesis of oxazolidine-2,4-diones, *Tetrahedron*, 51, 5891–5900, 1995.

333. Hawecker, J., Lehn, J.-M., and Ziessel, R., Electrocatalytic reduction of carbon dioxide mediated by Re(bipy)(CO)$_3$Cl (bipy = 2,2′-bipyridine), *J. Chem. Soc. Chem. Commun.*, pp. 328–330, 1984.

334. Hawecker, J., Lehn, J.-M., and Ziessel, R., Photochemical and electro-chemical reduction of carbon dioxide to carbon monoxide mediated by (2,2′-bipyridine) tricarbonyl chlororhenium (I) and related complexes as homogeneous catalysts, *Helv. Chim. Acta*, 69, 1990–2012, 1986.

335. Sullivan, B.P., Bolinger, C.M., Conrad, D., Vining, W.J., and Meyer, T.J., One- and two-electron pathways in the electrocatalytic reduction of CO$_2$ by fac-Re (bipy) (CO)$_3$Cl (bipy = 2,2′-bipyridine), *J. Chem. Soc. Chem. Commun.*, pp. 1414–1416, 1985.

336. Christensen, P., Hamnett, A., Muir, A.V.G., and Timney, J.A., An in situ infrared study of CO$_2$ reduction catalyzed by rhenium tricarbonyl bipyridyl derivatives, *J. Chem. Soc. Dalton Trans.*, pp. 1455–1463, 1992.

337. Cabrera, C.R. and Abruña, H.D., Electrocatalysis of CO$_2$ reduction at surface modified metallic and semiconducting electrodes, *J. Electroanal. Chem.*, 209, 101–107, 1986.

338. O'Toole, T.R., Sullivan, B.P., Bruce, M.R.-M., Margerum, L.D., Murray, R.W., and Meyer, T.J., Electrocatalytic reduction of CO_2 by a complex of rhenium in thin polymeric films, *J. Electroanal. Chem.*, 259, 217–239, 1989.

339. Slater, S. and Wagenknecht, J.H., Electrochemical reduction of CO_2 catalyzed by Rh(diphos)$_2$Cl, *J. Am. Chem. Soc.*, 106, 5367–5368, 1984.

340. Daniele, S., Ugo, P., Bontempelli, G., and Fiorani, M., An electro-analytical investigation on the nickel-promoted electrochemical conversion of CO_2 to CO, *J. Electroanal. Chem.*, 219, 259–271, 1987.

341. Christensen, P.A., Hamnett, A., Higgins, S.J., and Timney, J.A., An in-situ Fourier transform infrared study of CO_2 electroreduction catalysed by Ni(0)-4,4'-dimethyl-2,2'-bipyridine and Ni(0)-1,10-phenanthroline complexes, *J. Electroanal. Chem.*, 395, 195–209, 1995.

342. Potts, K.T., Usifer, D.A., Guadalupe, A., and Abruña, H.D., 4-Vinyl, 6-vinyl, and 4'-vinyl-2,2':6',2''-terpyridinyl ligands: their synthesis and the electrochemistry of their transition-metal coordination complexes, *J. Am. Chem. Soc.*, 109, 3961–3967, 1987.

343. Hurrell, H.D., Mogstad, A.-L., Usifer, D.A., Potts, K.T., and Abruña, H.D., Electrocatalytic activity of electropolymerized films of bis (vinylterpyridine)cobalt^{2+} for the reduction of carbon dioxide and oxygen, *Inorg. Chem.*, 28, 1080–1084, 1989.

344. Ishida, H., Tanaka, K., and Tanaka, T., The electrochemical reduction of CO_2 catalyzed by ruthenium carbonyl complexes, *Chem. Lett.*, pp. 405–406, 1985.

345. Ishida, H., Tanaka, K., and Tanaka, T., Electrochemical CO_2 reduction catalyzed by [Ru(bpy)$_2$(CO)$_2$(CO)$_2$]$^{2+}$ and [Ru(bpy)$_2$(CO)Cl]$^+$. The effect of pH on the formation of CO and HCOO$^-$, *Organometallics*, 6, 181–186, 1987.

346. Ishida, I., Tanaka, H., and Tanaka, K., Tanaka, T., Selective formation of HCOO$^-$ in the electrochemical CO_2 reduction catalysed by [Ru(bpy)$_2$(CO)$_2$]$^{2+}$, *J. Chem. Soc. Chem. Commun.*, pp. 131–132, 1987.

347. Ishida, H., Terada, T., Tanaka, K., and Tanaka, T., Photochemical CO_2 reduction catalyzed by [Ru(bpy)$_2$(CO)$_2$]$^{2+}$ using triethanolamine and 1-benzyl-1,4-dihydronicotinamide as an electron donor, *Inorg. Chem.*, 29, 905–911, 1990.

348. Tanaka, H., Nagao, H., Peng, S.M., and Tanaka, K., Crystal-structure of cis-(carbon monoxide) (η^1-carbon dioxide)-bis(2,2'-bipyridyl)ruthenium, an active species in catalytic CO_2 reduction affording CO and HCOO$^-$, *Organometallics*, 11, 1450–1451, 1992.

349. Tanaka, H., Tzeng, B.-C., Nagao, H., Peng, S.-M., and Tanaka, K., Comparative study on crystal structures of [Ru(bpy)$_2$(CO)$_2$(PF$_6$)$_2$, [Ru(bpy)$_2$ (CO)(C(O)OCH$_3$)]-B(C$_6$H$_5$)$_4$.CH$_3$N, and [Ru(bpy)$_2$(CO)$_2$(η^1-CO$_2$)].3H$_2$O (bpy = 2,2'-bipyridyl), *Inorg. Chem.*, 32, 1508–1512, 1993.

350. Nagao, H., Mizukawa, T., and Tanaka, K., Carbon–carbon bond formation in the electrochemical reduction of carbon dioxide catalyzed by a ruthenium complex, *Inorg. Chem.*, 33, 3415–3420, 1994.

351. Toyohara, K., Nagao, H., Mizukawa, T., and Tanaka, K., Ruthenium formyl complexes as the branch point in 2-electron and multielectron reductions of CO_2, *Inorg. Chem.*, 34, 5399–9400, 1995.

352. Toyohara, K., Tsuge, K., and Tanaka, K., Comparison of Ru–C bond characters involved in successive reduction of Ru–CO_2 to Ru–CH$_2$OH, *Organometallics*, 14, 5099–5103, 1995.

353. Haines, R.J., Wittrig, R.E., and Kubiak, C.P., Electrocatalytic reduction of carbon dioxide by the binuclear copper complex [Cu$_2$(6-(diphenylphosphino)-2,2'-bipyridyl)$_2$(MeCN)$_2$][PF$_6$]$_2$, *Inorg. Chem.*, 33, 4723–4728, 1994.

354. Nakajima, H., Mizukawa, T., Nagao, H., and Tanaka, K., Catalytic formation of ketones via double alkylation of carbon monoxide resulting from reductive disproportionation of carbon dioxide by [Ru(bpy)$_2$(qu)(CO)]$^{2+}$ (bpy = 2,2'-bipyridine, qu = quinoline), *Chem. Lett.*, pp. 251–252, 1995.

355. Nakajima, H., Kushi, Y., Nagao H., and Tanaka, K., Multistep CO$_2$ reduction catalyzed by [Ru(bpy)2(qu)(CO)]$^{2+}$ (bpy = 2,2'-bipyridine, qu = quinoline). Double methylation of the carbonyl moiety resulting from reductive disproportionation of CO$_2$, *Organometallics*, 14, 5093–5098, 1995.

356. Hawecker, J., Lehn, J.M., and Ziessel, R., Photochemical reduction of CO$_2$ to formate mediated by ruthenium bipyridine complexes as homogeneous catalysts, *J. Chem. Soc. Chem. Commun.*, pp. 56–58, 1985.

357. Pugh, J.R., Bruce, M.R.M., Sullivan, B.P., and Meyer, T.J., Formation of a metal hydride bond and the insertion of CO$_2$. Key steps in the electrocatalytic reduction of carbon dioxide to formate anion, *Inorg. Chem.*, 30, 86–91, 1991.

358. Furue, M., Yoshidzumi, T., Kinoshita, S., Kushida, T., Nozakura, S., and Kamachi, M., Intramolecular energy transfer in covalently linked polypyridine ruthenium(II)/osmium(II) binuclear complexes, *Bull. Chem. Soc. Jpn.*, 64, 1632–1640, 1991.

359. Furue, M., Maruyama, K., Naiki, M., and Kamachi, M., CO$_2$ reduction by covalently-linked binuclear complexes comprising polypyridine Re(I)/Ru(II) complexes, in Proc. Int. Symp. Chemical Fixation of Carbon Dioxide, Nagoya, Japan, December 2 to 4, 1991, 435–436.

360. Field, J.S., Haines, R.J., Kubiak, C.P., Landsberg, C., Campbell, P., Sookraj, S., and Wittrig, R., Dinuclear complexes bridged by phosphorus polypyridine ligands as electrocatalysts for the reduction of carbon dioxide, in Proc. Int. Conf. on Carbon Dioxide Utilization, Bari, Italy, September 1993, 311–316.

361. Bruce, M.R.M., Megehee, E., Sullivan, B.P., Thorp, H., O'Toole, T.R., Downard, A., and Meyer, T.J., Electrocatalytic reduction of CO$_2$ by associative activation, *Organometallics*, 7, 238–240, 1988.

362. Bruce, M.R.M., Megehee, E., Sullivan, B.P., Thorp, H., O'Toole, T.R., Downard, A., Pugh, J.R., and Meyer, T.J., Electrocatalytic reduction of carbon dioxide based on 2,2'-bipyridyl complexes of osmium, *Inorg. Chem.*, 31, 4864–4873, 1992.

363. Bolinger, C.M., Story, N., Sullivan, B.P., and Meyer, T.J., Electrocatalytic reduction of carbon dioxide by 2,2'-bipyridine complexes of rhodium and iridium, *Inorg. Chem.*, 27, 4582–4587, 1988.

364. Nallas, G.N.A. and Brewer, K.J., Electrocatalytic reduction of carbon dioxide by mixed-metal trimetallic complexes of the form {[(bpy)$_2$Ru(BL)]$_2$IrCl$_2$}$^{5+}$ where bpy = 2,2'-bipyridine and BL = 2,3-bis(2-pyridyl)quinoxaline (dpq) or 2,3-bis(2-pyridyl)-benzo-quinoxaline (dpb), *Inorg. Chim. Acta*, 253, 7–13, 1996.

365. Becker, J.Y., Vaina, B., Eger, R., and Kaufman, L., Electrocatalytic reduction of CO$_2$ to oxalate by Ag^{II} and Pd^{II} porphyrins, *J. Chem. Soc. Chem. Commun.*, pp. 1471–1473, 1985.

366. Hammouche, M., Lexa, D., and Savéant, J.-M., Catalysis of the electro-chemical reduction of carbon dioxide by iron("0")porphyrins, *J. Electroanal. Chem.*, 249, 347–351, 1988.

367. Hammouche, M., Lexa, D., Momenteau, M., and Savéant, J.-M., Chemical catalysis of electrochemical reactions — homogeneous catalysis of the electrochemical reduction of carbon dioxide by iron ("0") porphyrins — role of the addition of magnesium cations, *J. Am. Chem. Soc.*, 113, 8455–8466, 1991.

368. Savéant, J.-M., Molecular catalysis of the electrochemical reduction of carbon dioxide. Iron "0" porphyrins, in Proc. Int. Symp. Chemical Fixation of Carbon Dioxide, December 2 to 4, 1991, Nagoya, Japan, 49–54.

369. Bhugun, I., Lexa, D., and Savéant, J.-M., Ultraefficient selective homogeneous catalysis of the electrochemical reduction of carbon dioxide by an iron (0) porphyrin associated with a weak Brönsted acid cocatalyst, *J. Am. Chem. Soc.*, 116, 5015–5016, 1994.

370. Atoguchi, T., Aramata, A., Kazusaka, A., and Enyo, M., Electrochemical reduction of CO$_2$ mediated by CoIITPP/CoITPP redox complex on GC and Pt in DMF, *Denki Kagaku*, 59, 526–527, 1991.

371. Tezuka, M., Iwasaki, M., and Yajima, T., Voltammetric study on the electro-catalytic reduction of CO$_2$ in aprotic media, in Proc. Int. Symp. Chemical Fixation of Carbon Dioxide, Nagoya, Japan, December 2 to 4, 1991, 303–306.

372. DuBois, D.L. and Miedaner, A., Mediated electrochemical reduction of CO$_2$. Preparation and comparison of an isoelectronic series of complexes, *J. Am. Chem. Soc.*, 109, 113–117, 1987.

373. DuBois, D.L., Miedaner, A., and Haltiwanger, R.C., Electrochemical reduction of CO$_2$ catalyzed by [Pd(triphosphine) (solvent)] (BF$_4$)$_2$ complexes: synthetic and mechanistic studies, *J. Am. Chem. Soc.*, 113, 8753–8764, 1991.

374. Bernatis, P., Curtis, C.J., Herring, A., Miedaner, A., and DuBois, D.L., Development of homogeneous catalysts for electrochemical CO$_2$ concentration and reduction, in Proc. Int. Symp. Chemical Fixation of Carbon Dioxide, Nagoya, Japan, December 2 to 4, 1991, 89–96.

375. Bernatis, P.R., Miedaner, A., Haltiwanger, R.C., and DuBois, D.L., Exclusion of six-coordinate intermediates in the electrochemical reduction of CO$_2$ catalyzed by [Pd(triphosphine)(CH$_3$CN)](BF$_4$)$_2$ complexes, *Organometallics*, 13, 4835–4843, 1994.

376. DuBois, D., Development of electrocatalysts for carbon dioxide reduction using polydentate ligands to probe structure–activity relationships, in Abstr. 4th Int. Conf. on Carbon Dioxide Utilization, Kyoto, Japan, September 1997, PL-4.

377. Steffey, B.D., Miedaner, A., Maciejewski-Farmer, M.L., Bernatis, P.R., Herring, A.M., Allured, V.S., Carperos, V., and DuBois, D.L., Synthesis and characterization of palladium complexes containing tridentate ligands with PXP (X = C, N, O, S, As) donor sets and their evaluation as electrochemical CO$_2$ reduction catalysts, *Organometallics*, 13, 4844–4855, 1994.

378. Wander, S.A., Miedaner, A., Noll, B.C., Barkley, R.M., and DuBois, D.L., Chelate bite effects for [Pd(triphosphine)(solvent)](BF$_4$)$_2$ in electrochemical CO$_2$ reduction and the heterolytic cleavage of molecular hydrogen, *Organometallics*, 15, 3360–3373, 1996.

379. Miedaner, A., Curtis, C.J., Barkley, R.M., and DuBois, D.L., Electro-chemical reduction of CO$_2$ catalyzed by small organophosphine dendrimers containing palladium, *Inorg. Chem.*, 33, 5482–5490, 1994.

380. Steffey, B.D., Curtis, C.J., and DuBois, D.L., Electrochemical reduction of CO$_2$ catalyzed by a dinuclear palladium complex containing a bridging hexaphosphine ligand: evidence for cooperativity, *Organometallics*, 14, 4937–4943, 1995.

381. Herring, A.M., Steffey, B.D., Miedaner, A., Wander, S.A., and DuBois, D.L., Synthesis and characterization of water-soluble (Pd(triphosphine) (CH$_3$CN))(BF$_4$)$_2$ complexes for CO$_2$ reduction, *Inorg. Chem.*, 34, 1100–1109, 1995.

382. Stalder, C.J., Chao, S., and Wrighton, M.S., Electrochemical reduction of aqueous bicarbonate to formate with high current efficiency near the thermodynamic potential at a chemically derivatized electrode, *J. Am. Chem. Soc.*, 106, 3673–3675, 1984.

383. André, J.-F. and Wrighton, M.S., Electrostatic binding of bicarbonate and formate in viologen-based redox polymers: importance in catalytic reduction of bicarbonate to formate, *Inorg. Chem.*, 24, 4288–4292, 1985.

384. Hossain, A.G.M.M., Nagaoka, T., and Ogura, K., Electrocatalytic reduction of carbon dioxide by substituted pyridine and pyrazole complexes, *Electrochim. Acta*, 41, 2773–2780, 1996.

385. Hossain, A.G.M.M., Nagaoka, T., and Ogura, K., Palladium and cobalt complexes of substituted quinoline, bipyridine and phenanthroline as catalysts for electrochemical reduction of carbon dioxide, *Electrochim. Acta*, 42, 2577–2585, 1997.

386. Gambarotta, S., Arena, F., Floriani, C., and Zanazzi, P.F., Carbon dioxide fixation: bifunctional complexes containing acidic and basic sites working as reversible carriers, *J. Am. Chem. Soc.*, 104, 5082–5092, 1982.

387. Pearce, D.J. and Pletcher, D., A study of the mechanism for the electrocatalysis of carbon dioxide reduction by nickel and cobalt square planar complexes in solution, *J. Electroanal. Chem.*, 197, 317–330, 1986.

388. Isse, A.A., Gennaro, A., Vianello, E., and Floriani, C., Electrochemical reduction of carbon dioxide catalyzed by [Co^l(salophen)Li], *J. Mol. Catal.*, 70, 197–208, 1991.

389. Tinnemans, A.H.A., Koster, T.P.M., Thewissen, D.H.M.W., and Mackor, A., Tetraazamacrocyclic cobalt(II) and nickel(II) complexes as electron-transfer agents in the photo (electro) chemical and electrochemical reduction of carbon dioxide, *Rec. Trav. Chim. Pays-Bas*, 103, 288–295, 1984.

390. Gangi, D.A. and Durand, R.R., Binding of carbon dioxide to cobalt and nickel tetraazamacrocycles, *J. Chem. Soc. Chem. Commun.*, pp. 697–699, 1986.

391. Fujita, E., Binding of carbon dioxide to metal macrocycles: toward a mechanistic understanding of electrochemical and photochemical carbon dioxide reduction, in Proc. Int. Conf. on Carbon Dioxide Utilization, Bari, Italy, September 1993, 231–240.

392. Arana, C., Yan, S., Potts, K.T., and Abruña, H.D., Electro-catalysis of CO$_2$ reduction with transition metal complexes incorporating terdentate coordination, in Proc. Int. Symp. Chemical Fixation of Carbon Dioxide, Nagoya, Japan, December 2 to 4, 1991, 73–80.

393. Nishioka, T. and Isobe, K., Synthesis and crystal structures of triangular rhodium and iridium complexes with triply bridging sulfido ligands, *Chem. Lett.*, pp. 1661–1664, 1994.

394. Kushi, Y., Nagao, H., Nishioka, T., Isobe, K., and Tanaka, K., Oxalate formation in electrochemical CO$_2$ reduction catalyzed by rhodium–sulfur clusters, *Chem. Lett.*, pp. 2175–2178, 1994.

395. Kushi, Y., Nagao, H., Nishioka, T., Isobe, K., and Tanaka, K., Remarkable decrease in overpotential of oxalate formation in electrochemical CO$_2$ reduction by a metal–sulfide cluster, *J. Chem. Soc. Chem. Commun.*, pp. 1223–1224, 1995.

396. Ratliff, K.S., Lentz, R.E., and Kubiak, C.P., Carbon dioxide chemistry of the trinuclear complex [Ni$_3$(μ_3-CNMe)(μ_3-I)(dppm)$_3$][PF$_6$]. Electro-catalytic reduction of carbon dioxide, *Organometallics*, 11, 1986–1988, 1992.

397. Simpson, T.C. and Durand, R.R., Jr., Ligand participation in the reduction of CO$_2$ catalyzed by complexes of 1,10-o-phenanthroline, *Electrochim. Acta*, 33, 581–583, 1988.

398. Fujiwara, H. and Nonaka, T., Cyclic voltammetric study on carbon dioxide reduction using Cu complexes as electrocatalysts, *J. Electroanal. Chem.*, 332, 303–307, 1992.

399. Losada, J., del Peso, I., Beyer, L., Hartung, J., Fernández, V., and Möbius, M., Electrocatalytic reduction of O$_2$ and CO$_2$ with electropolymerized films of polypyrrole cobalt(II) Schiff-base complexes, *J. Electroanal. Chem.*, 398, 89–93, 1995.

400. Volkov, S.V., Chemical reactions in molten salts and their classification, *Chem. Soc. Rev.*, 19, 21–28, 1990.

401. Weaver, J.L. and Winnick, J., The molten carbonate carbon dioxide concentrator: cathode performance at high CO$_2$ utilization, *J. Electrochem. Soc.*, 130, 20–28, 1983.

402. Bartlett, H.E. and Johnson, K.E., Electrolytic reduction and Ellingham diagrams for oxyanion systems, *Can. J. Chem.*, 44, 2119–2129, 1966.

403. Stern, K.H. and Gadomski, S.T., Electrodeposition of tantalum carbide coatings from molten salts, *J. Electrochem. Soc.*, 130, 300–305, 1983.

404. Mamantov, G., Molten salt electrolytes in secondary batteries, in *Materials for Advanced Batteries,* Murphy, D.W., Broadhead, J., and Steele, B.C.H., Eds., Plenum Press, New York, 1980, 111–122.

405. Deanhardt, M.L., Stern, K.H., and Kende, A., Thermal decomposition and reduction of carbonate ion in fluoride melts, *J. Electrochem. Soc.*, 133, 1148–1152, 1986.

406. Dubois, J. and Buvet, R., Electroactivity range and pO^{2-} scale determination in a medium of alkali carbonate melts, *Bull. Soc. Chim. France*, pp. 2522–2526, 1963.

407. Uchida, I., Nishina, T., Mugikura, Y., and Itaya, K., Gas electrode reactions in molten carbonate media. I. Exchange current density of oxygen reduction in (Li + K)CO$_3$ eutectic at 650°C, *J. Electroanal. Chem.*, 206, 229–239, 1986.

408. Uchida, I., Mugikura, Y., Nishina, T., and Itaya, K., Gas electrode reactions in molten carbonate media. II. Oxygen reduction kinetics on conductive oxide electrodes in (Li + K)CO$_3$ eutectic at 650°C, *J. Electroanal. Chem.*, 206, 241–252, 1986.

409. Janz, G.J. and Conte, A., Potentiostatic polarization studies in fused carbonates. I. The noble metals, silver and nickel, *Electrochim. Acta,*, 9, 1269–1278, 1964.

410. Iacovangelo, C.D., Stability of molten carbonate fuel cell nickel anodes, *J. Electrochem. Soc.*, 133, 2410–2416, 1986.

411. Chandler, H.W., Design of a test model for a solid electrolyte carbon dioxide reduction system, *Sci. Tech. Aerospace Rep.*, 4, 1008, 1966; *Chem. Abstr.*, 66, 25343q.

412. White, S.H. and Twardoch, U.M., The influence of gas composition on the oxygen electrode reaction in the molten Li$_2$CO$_3$–Na$_2$CO$_3$–K$_2$CO$_3$ eutectic mixture, *Electrochim. Acta*, 27, 1599–1607 1982.

413. White, S.H. and Twardoch, U.M., The behavior of water in molten salts, *J. Electrochem. Soc.*, 134, 1080–1088 1987.

414. Randin, J.-P., Carbon, in *Encyclopedia of Electrochemistry of the Elements,* Vol. VII, Bard, A.J., Ed., Marcel Dekker, New York, 1976, 1–291.

415. Delimarskii, Yu.K., Shapoval, V.I., and Vasilenko, V.A., Significance of a kinetic process during the electroreduction of carbonate ions in fused KCl–NaCl, *Elektrokhimiya*, 7, 1301–1304, 1971; English translation, pp. 1255–1258.

416. Shapoval, V.I., Kushkhov, K.B., and Soloviev, V.V., Cation catalysis of carbonate-ion electroreduction against a background of melted chlorides, *Ukr. Khim. Zh.*, 51, 1263, 1985; *Chem. Abstr.*, 104, 118490d.

417. Shimada, T., Yoshida, N., and Ito, Y., Cathodic reduction of carbonate ion in LiCl–KCl melt. I. Electrodeposition of carbon, *Denki Kagaku*, 59, 701–706, 1991.

418. Shimada, T., Yoshida, N., and Ito, Y., Cathodic reduction of carbonate ion in LiCl–KCl melt. II. Mechanism of cathodic reduction, *Denki Kagaku*, 60, 194–199, 1992.

419. Halmann, M. and Zuckerman, K., Electroreduction of carbon dioxide to carbon monoxide in molten LiCl + KCl, LiF + KF + NaF, Li_2CO_3 + Na_2CO_3 + K_2CO_3 and $AlCl_3$ + NaCl, *J. Electroanal. Chem.*, 235, 369–380, 1987.

420. Iwahara, H., Uchida, H., and Tanaka, S., High temperature-type proton conductive solid oxide fuel cells using various fuels, *J. Appl. Electrochem.*, 16, 663–668, 1986.

421. Iwahara, H., Uchida, H., and Yamasaki, I., High-temperature steam electrolysis using $SrCeO_3$-based proton conductive solid electrolyte, *Int. J. Hydrogen Energy*, 12, 73–77, 1987.

422. Wagner, C., The mechanism of electric conduction in the Nernst glower, *Naturwissenschaften*, 31, 265–268, 1943.

423. Dixon, J.M., LaGrange L.D., Merten U., Miller, C.F., and Porter, J.T., II, Electrical resistivity of stabilized ZrO_2 at elevated temperatures, *J. Electrochem. Soc.*, 110, 276–280, 1963.

424. Obayashi, H. and Kudo, T., High temperature electrolysis/fuel cells: materials problems, in *Solid State Chemistry of Energy Conversion and Storage*, Advances in Chemistry Series No. 163, Goodenough, J.B. and Whittingham, M.S., Eds., American Chemical Society, Washington, D.C., 1977, 316–363.

425. Barbi, G.B. and Mari, C.M., High temperature electrochemical reduction of the water molecule at cerium dioxide electrodes, *Solid State Ionics*, 15, 335–343, 1985.

426. Gür, T.M. and Huggins, R.A., Methane synthesis by a solid-state ionic method, *Science*, 219, 967–969, 1983.

427. Gür, T.M. and Huggins, R.A., Methane synthesis over transition metal electrodes in a solid state ionic cell, *J. Catal.*, 102, 443–446, 1986.

428. Gauthier, M. and Chamberland A., Solid-state detectors for the potentiometric determination of gaseous oxides. I. Measurement in air, *J. Electrochem. Soc.*, 124, 1579–1583, 1977.

429. Maruyama T., Sasaki, S., and Saito, Y., Potentiometric gas sensor for carbon dioxide using solid electrolytes, *Solid State Ionics*, 23, 107–112, 1987.

430. Maruyama, T., Ye, X.-Y., and Saito, Y., Electromotive force of the $CO–CO_2–O_2$ concentration cell using Na_2CO_3 as a solid electrolyte at low oxygen partial pressures, *Solid State Ionics*, 23, 113–117, 1987.

431. Miura, N., Yao, S., Shimizu, Y., and Yamazoe, N., Carbon dioxide sensor using sodium ion conductor and binary carbonate auxiliary electrode, *J. Electrochem. Soc.*, 139, 1384–1388, 1992.

432. Liu, J. and Weppner, W., Potentiometric CO_2 gas sensor based on Na-β/β″-alumina solid electrolyte at 400°C, *Eur. J. Solid State Inorg. Chem.*, 28, 1151–1160, 1991.

433. Yao, S., Hosohara, S., Shimizu, Y., Miura, N., Futata, H., and Yamazoe, N., Solid electrolyte CO_2 sensor using NASICON and Li-based binary carbonate electrode, *Chem. Lett.*, pp. 2069–2072, 1991.

434. Imanaka, N., Murata, T., Kawasato, T., and Adachi, G., The operating temperature lowering for CO_2 gas sensor with a lithium conducting solid electrolyte, *Chem. Lett.*, pp. 103–106, 1992.

435. Ishihara, T., Kometani, K., Mizuhara, Y., and Takita, Y., Mixed oxide capacitor of $CuO–BaSnO_3$ as a sensor for CO_2 detection over a wide range of concentration, *Chem. Lett.*, pp. 1711–1714, 1991.

436. Fujiwara, H., Konno, A., and Nonaka, T., Carbon dioxide fixation by electrolysis of aqueous hydrogen carbonate solution. Reduction to formic acid at a mercury cathode, *Chem. Lett.*, pp. 1843–1846, 1991.

437. Bell, W.L., Miedaner, A., Smart, J.C., DuBois, D.L., and Verostko, C.E., Synthesis and evaluation of electroactive CO$_2$ carriers, in SAE Tech. Paper Ser. No. 881078, 18th Intersoc. Conf. on Environ. Systems, San Francisco, July 11 to 13, 1988.

438. DuBois, D.L., Miedaner, A., Bell, W., and Smart, J.C., Electrochemical concentration of carbon dioxide, in *Electrochemical and Electrocatalytic Reactions of Carbon Dioxide*, Sullivan, B.P., Krist, K., and Guard, H.E., Eds., Elsevier, Amsterdam, 1993, 94–117.

439. Saeki, T., Hashimoto, K., Noguchi, Y., Omata, K., and Fujishima, A., Electrochemical reduction of liquid CO$_2$, *J. Electrochem. Soc.*, 141, L130–L132, 1994.

440. Saeki, T., Hashimoto, K., Kimura, N., Omata, K., and Fujishima, A., Electrochemical reduction of CO$_2$ with high current density in a CO$_2$ plus methanol medium at various metal electrodes, *J. Electroanal. Chem.*, 404, 299–302, 1996.

441. Ohba, T., Ishida, H., Yamaguchi, T., Horiuchi, T., and Ohkubo, K., Carbon dioxide-promoted electrochemical reduction of aromatic nitro compounds to azoxy compounds in acetonitrile, *J. Chem. Soc. Chem. Commun.*, pp. 263–264, 1994.

13 Photoelectrochemical Reduction of CO₂

13.1 INTRODUCTION

A drawback of the direct electrochemical reduction of carbon dioxide is the high electrical bias potential required for the first step in its reduction:

$$CO_2 + e^- \rightarrow CO_2^- \tag{13.1}$$

With semiconductor materials as electrodes, and illuminating with light more energetic than the bandgap of these materials, the electrochemical reduction of carbon dioxide may be achieved at much lower bias potentials than in the "dark" electrochemical reduction, resulting in considerable saving in the input of electrical energy. Using semiconductor electrodes as photoelectrodes, some of the overpotential required for the reduction of carbon dioxide may be gained by the photopotential produced. Unfortunately, most of the semiconductors which have a bandgap overlapping the visible part of the solar spectrum (about 300 to 800 nm), and which could thus be useful for solar energy conversion, are unstable in aqueous electrolytes. The general principles of photoelectrochemistry and photocatalysis at semiconductors were reviewed by Bard,[1] and their applications to the reduction of carbon dioxide were discussed by Halmann[2] and Taniguchi.[3]

In photocatalytic reactions, the illuminated semiconductor is dispersed in fine particles in the solution. Each particle, which may be as small as a few nanometers in diameter (colloidal particles), acts as a minute photoelectrolysis cell (see Chapter 14).

13.2 PHOTOELECTROLYSIS IN AQUEOUS MEDIA

Fujishima and Honda pioneered the use of illuminated single-crystal n-TiO₂ semiconductor electrodes for the photolysis of water.[4] Irradiation of semiconductors with light of wavelengths shorter than the bandgap causes excitation of electrons from the valence band to the conductance band. With p-type semiconductors, electrons from the conductance band may interact with electron-deficient molecules such as CO_2 adsorbed on the semiconductor surface, causing reduction of these molecules. In most photoelectrochemical systems, the major reduction product was HCOOH. With some electrode and mediator combinations, selectivity to CO or methanol was obtained.

13.2.1   Photoreduction on Bare Semiconductors

The photoelectrolysis by Halmann et al. of carbon dioxide bubbled through an aqueous phosphate buffer solution (pH 6.8) in a cell containing an illuminated p-GaP photocathode and a carbon counterelectrode resulted after 18 h in the production of formic acid as the main product, in 1.2×10^{-2} M concentration, as well as traces of formaldehyde and methanol.[6,7] The cathode was maintained at a bias of -1.0 V (vs. the saturated calomel electrode [SCE]), resulting in a current density of 1 mA cm^{-2}. In similar experiments by Inoue et al. with either an illuminated p-GaP photocathode or an n-TiO$_2$ electrode in the dark, at a potential of -1.5 V vs. SCE, in a medium of aqueous 0.5 M H$_2$SO$_4$, the carbon dioxide reduction products were formic acid, formaldehyde, and methanol.[8] In subsequent work by Ito et al. on p-GaP photocathodes in aqueous electrolytes, the only CO_2 reduction product observed was formic acid.[9]

The light intensity effect on the reduction of CO_2 to HCOOH on p-GaP was studied by Ito et al. using a xenon lamp.[10,11] Maximal faradaic efficiency of 70% was achieved at a light flux of 320 mW cm^{-2}, at an electrode potential of -0.9 V (vs. Ag/AgCl saturated KCl) in 0.1 M KHCO$_3$ saturated with CO_2. At constant electrode potential, the faradaic efficiency for HCOOH production as a function of light intensity reached a maximum at about 200 to 300 mW cm^{-2} and above that declined. The reason for this decline at high light intensity was explained by diffusion limitation for the supply of CO_2 to the electrode and by competition with H$_2$ production.[10,11]

Ikeda et al. tested the effects of surface treatment of p-GaP by different etchants on the photoelectrochemical reduction of CO_2 in 0.1 M KHCO$_3$.[12] On electrodes of single-crystal p-GaP (111 face), after etching treatment, the order of photocurrent activity at a potential of -1.0 V (vs. Ag/AgCl) was aqua regia > alkaline potassium hexacyanoferrate(III) > conc. HCl > conc. HNO$_3$ > as polished. After etching with aqua regia, the photocurrent onset potentials were zero, both with the (111) face and with the (100) face p-GaP electrodes. With aqua-regia-etched electrodes of p-GaP (111) face, on constant-current photoelectrolysis at a current density of 5 mA cm^{-2} and a potential of -0.8 to -1.0 V (vs. Ag/AgCl), the faradaic efficiencies for the production of HCOO$^-$, CO, and H$_2$ were 66, 6, and 29%, respectively. With p-GaP (100) face electrodes under similar conditions, the faradaic efficiencies were 56, 12, and 20%, respectively. With the other etchants, the efficiencies of formate production were much lower, while the production of hydrogen increased. Etching electrodes revealed that etching mainly removed oxygen atoms bound to the Ga and P atoms of p-GaP. The strongly oxidizing aqua regia also removed the electrode surface lattice defects.[12]

In a comparison by Yoneyama et al. of single-crystal p-CdTe (111 face) and p-InP (100 face) electrodes for the photoelectrolysis of CO_2-saturated aqueous solutions, the p-InP was found to be more sensitive to corrosive deterioration than the p-CdTe.[13] The nature of the supporting electrolyte strongly affected the product selectivity. Carbonate salts favored the production of formic acid. Other salts, such as sulfates, phosphates, and perchlorates, favored the formation of CO. Highest current efficiency for CO_2 reduction was obtained at the p-CdTe electrode in aqueous solutions in the presence of tetraalkyl ammonium salts.[13]

13.2.2 Photoreduction on Metal-Modified Semiconductors

The current efficiency for CO_2 reduction on p-GaP photocathodes was enhanced by Ikeda et al. by electroplating or sputtering of Pb or Zn on the electrodes.[14] Using 0.1 M aqueous Et_4NClO_4 or Et_4NBr as electrolytes, the only products observed were HCOOH and CO. While the current efficiency for HCOOH production was only 4.8% on the bare p-GaP electrode, it reached 48.2% on the Pb-coated p-GaP electrode, at a potential of only -1.2 V (vs. Ag/AgCl/saturated KCl).[14]

Metal-coated p-GaP and p-InP were tested by Ito et al. for the photoelectrochemical reduction of CO_2 in aqueous and nonaqueous electrolytes. Methane was produced by illumination of Cu-coated p-GaP electrodes. At a potential of -1.6 V (vs. Ag/AgCl), the faradaic efficiency was 7% at 1°C.[15-17]

Electrodes of p^+/p^- Si were compared by Chu et al. in CO_2 and in N_2-saturated 0.5 M Na_2SO_4 solutions.[18] The current-potential curves indicated that the rate of hydrogen generation in the CO_2-saturated solutions was larger than the rate of carbon dioxide reduction. However, when these electrodes were coated with thin films of TiO_2 rutile, grown by metal organic chemical vapor deposition, the photocurrent in CO_2-saturated solutions was positively shifted by about 200 mV relative to N_2-saturated solutions. The reduction products were formic acid and formaldehyde.[18] Liu and Baozhu observed that the photoreduction of CO_2 in aqueous solutions to HCOOH on p^+/p^- Si photocathodes was enhanced by a coating of Pb. Also, the yield and efficiency of HCOOH production depended on the pH of electrolyte. It was low in the acidic range and increased markedly above pH 6.[19]

Nakato and Tsubomura modified silicon photoelectrodes by deposition of ultrafine metal islands, thus markedly improving their photovoltage and photocatalytic activity.[20] Hinogami et al. applied this technique to prepare p-Si electrodes, on which Cu particles 10 to 1000 nm in diameter were deposited by photoelectrochemical reduction from aqueous $CuSO_4$.[21] Using such electrodes, in CO_2-saturated aqueous 0.1 M $KHCO_3$, under illumination with a tungsten–halogen lamp, the photocurrent onset occurred at a 0.4-V more positive potential than that with a Cu metal electrode. Photoelectrolysis of such a solution at a constant potential of -1.2 V (vs. SCE), at a current density of 2.3 mA cm^{-2}, resulted in the production of HCOOH, CO, CH_4,

C_2H_4, and H_2 with current efficiencies of 32, 19, 1.6, 3.6, and 1.6%, respectively. With either naked p-Si electrodes or p-Si electrodes covered with a continuous layer of Cu, the yields of CO_2 reduction products were much lower, while increasing the production of H_2.[21] These Cu-modified p-Si electrodes were also effective in nonaqueous media. Nakamura et al. performed the photoelectrolysis of CO_2 on such electrodes in acetonitrile solutions containing 3 M water and 0.1 M Bu_4NClO_4, producing methane, formic acid, carbon monoxide, and hydrogen.[22]

13.2.3   High-Pressure CO_2 Photoelectrolysis

The yields of CO_2 photoreduction products were enhanced by Aurian-Blajeni et al. using elevated CO_2 gas pressures.[23] With an electrochemical autoclave, illuminated with a 150-W Xe lamp through a side window of quartz, using a p-GaP photocathode at –1.0 V (vs. SSE), in 0.1 M aqueous $HClO_4$, separated by a cation exchange membrane from a platinum counterelectrode, at 3 atm pressure of CO_2, the products identified were formic acid, formaldehyde, and methanol, produced in a molar ratio of 75:6:19, and with a total faradaic yield of 18%. In a similar experiment, but in a medium of 0.5 M $NaHCO_3$, at 7.5 atm CO_2, at a cathodic bias of –1.4 V vs. SSE, the HCOOH, HCHO, and CH_3OH product ratio was 89:3:8, with a total faradaic yield of 38%.[23]

13.2.4   Methanol and Methane Formation

Selective production of methane and methanol in the photoelectrochemical reduction of aqueous carbon dioxide was achieved by Fujishima using p-type copper oxide and silicon carbide.[24,25] The copper oxide electrode was prepared by anodic oxidation of copper metal. Under illumination (500-W xenon lamp), with a cathodic bias of –0.3 V (vs. SCE), in CO_2-saturated 0.1 M $KHCO_3$, the current efficiencies for production of methane and methanol were 1.0 and 22%, respectively. During this process, the CuO underwent reduction to Cu_2O and finally to metallic copper, at which stage the photoreduction of carbon dioxide stopped. With single-crystal p-SiC, no photoreduction products of carbon dioxide could be detected. However, using a microcrystalline SiC film electrode (which had been deposited on indium tin oxide by a mercury-assisted chemical vapor deposition method), in 0.05 M Na_2SO_4, at –1.0 V (vs. SCE), the current efficiencies for the production of H_2, CO, CH_4, and CH_3OH were 80.0, 1.6, 1.3, and 16.7%, respectively.[24,25]

On a Cu_2S compound semiconductor electrode (prepared by electroplating on a Cu plate from an aqueous solution of 0.1 M CuCl, 0.6 M $Na_2S_2O_3$, and 1 M $NaHSO_3$ at 10 mA cm^{-2}), CH_4 was produced selectively by Saeki et al. under illumination at a potential of –0.5 V (vs. SCE), with a current efficiency of 5.3%.[26]

13.3   PHOTOELECTROLYSIS IN ORGANIC MEDIA

The chief drawback of aqueous media is the competition of CO_2 reduction with the electrolysis of water, yielding H_2 as the major reduction product. This may be overcome by using nonaqueous solvents such as dimethylformamide (DMF), dimethyl sulfoxide (DMSO), or acetonitrile, with supporting electrolytes such as tetraalkylammonium salts. The disadvantages of these organic media are lower electrical conductivity and higher cost relative to aqueous media.

13.3.1 Photoreduction on Bare Semiconductors

Taniguchi et al. illuminated a p-CdTe electrode in DMF containing 5% water and 0.1 M tetrabutyl ammonium perchlorate in a CO_2 atmosphere with visible light.[27] The photoreduction of CO_2 to CO occurred at potentials which were at least 0.7 V less negative than at metal electrodes. At a potential of -1.6 V vs. SCE, at a current density of about 1.8 mA cm^{-2}, the current efficiency for CO production was about 70%, while the photocurrent quantum efficiency (calculated for 600 nm) was close to unity.[27] In a comparison by Taniguchi et al. of p-type Si, In, GaP, GaAs, and CdTe as electrodes for the reduction of carbon dioxide in DMF solutions containing 5% water and 0.1 M tetrabutyl ammonium perchlorate, best results were obtained with the (100) plane of CdTe.[28,29] At a potential of -1.6 V (vs. SCE), the only products were CO and H_2. By controlled potential photoassisted electrolysis, the faradaic efficiency for CO production was 70 to 80%, while that for hydrogen formation was 0.5%. Similar results were obtained using as solvents DMSO (+5% H_2O) or propylene carbonate (+5% H_2O), while much lower current yields for CO production were found with acetonitrile (+5% H_2O) as solvent.[28,29]

Characterization of carbon dioxide adsorbed on the electrode surface improved considerably with application of Fourier transform infrared spectrometry by Bockris et al. to the solid/solution interface. On a p-CdTe electrode in acetonitrile containing 0.1 M tetrabutyl ammonium perchlorate, CO_2 was reduced, forming an adsorbed CO_2^- ion radical, which gave rise to IR absorption peaks at 2018, 2041, 2065, and 2090 cm^{-1} [30,31]

13.3.2 Photoelectrocarboxylation

An interesting application of photoelectrochemistry is the electrocarboxylation of organic substrates. This was carried out by Ueda and Uosaki using p-GaP as a photocathode, illuminated with a 500-W xenon lamp, acetonitrile as solvent, Bu$_4$NBF$_4$ as supporting electrolyte, and phenanthrene as the reactant:[32]

During photoelectrolysis at a potential of -1.65 V (vs. SCE) under a CO_2 atmosphere, phenanthrene was carboxylated. The proposed mechanism involved prior formation of the CO_2^- radical anion, which then interacted with the aromatic compound.[32]

Ueda et al. performed the photoelectrochemical carboxylation of benzyl chloride to phenylacetic acid using a sacrificial Mg anode and illuminated p-GaP or p-GaAs as photocathode:[33]

$$C_6H_5Cl + CO_2 + 2e^- \rightarrow C_6H_5CH_2COO^- + Cl^- \qquad (13.2)$$

The reaction was performed in CO_2-saturated acetonitrile in a single-compartment cell at 25°C without applied potential (i.e., at quasi short-circuit conditions, with an average photocurrent of 3 to 4 μA). After 20 h of photoelectrolysis, the current efficiencies for the production of phenylacetic acid using p-GaP and p-GaAs were 44 and 42%, respectively. In this reaction, the proposed initial step was the reduction of benzyl chloride on the p-type semiconductor electrode, presumably forming a benzyl radical anion, which then captured a CO_2 molecule to produce the phenylacetate anion.[33]

13.3.3 CO_2–Methanol Electrolyte

In the electrochemical reduction of CO_2, high current densities were achieved using solutions of pressurized CO_2 in methanol. At 40 atm pressure, the mole fraction of CO_2 is about 0.3, and its mass transport is no longer a limiting factor (see also Section 12.3.1.3).[34] Similarly, the benefits of highly concentrated methanolic solutions of CO_2 were applied by Hirota et al. to the photoelectrochemical reduction.[35] Using p-InP as photocathode, in a stainless-steel pressure chamber fitted with a quartz window, CO_2 was brought to 40 atm pressure in an electrolyte of Bu_4NClO_4 in methanol. Under visible light illumination ($\lambda > 370$ nm), the galvanostatic electrolysis at 100 mA cm^{-2} resulted in formation of CO with 90% current efficiency, with methyl formate and H_2 as minor products. Using p-GaP as photocathode, the current efficiency of CO under similar conditions was only 48%.

13.4   MEDIATION BY MACROCYCLES AND METAL COMPLEXES

In analogy with the electrochemical systems, the photoelectrochemical reduction of carbon dioxide may be enhanced by electron transfer mediators such as complexes of transition elements. Bradley et al. used tetraazamacrocyclic metal complexes such as $[(Me6[14]aneN_4)Ni^{11}]^{2+}$ with a p-Si electrode in a CO_2-saturated solution of acetonitrile–H_2O–$LiClO_4$ (1:1:0.1 M). The main products were CO and H_2 formed in a 2:1 ratio, with close to 100% current efficiency. The potentials required were about 1.3 V less negative than those without the mediator.[36]

Low-bandgap semiconductors, such as p-Si, p-GaAs, p-InP, and p-CdTe, are optimally matched to the solar spectrum. However, they are often unstable in aqueous media due to corrosion. This corrosion problem was overcome by Zafrir et al. with some of such low-bandgap electrodes by using the vanadium redox couple V(II)–V(III), providing efficient electron transfer from cathodically polarized semiconductors.[37] With p-GaAs photocathodes and carbon counterelectrodes, in acidic media containing vanadous chloride, carbon dioxide was photoreduced to formic acid, formaldehyde, and methanol with faradaic efficiencies of up to 4.1, 1.5, and 1.0%, respectively. The efficiencies increased with both rising temperature and light intensity.

In aqueous media, using p-GaP as photocathode, Li_2CO_3 as electrolyte, in the presence of 15-crown-5, at –0.95 V vs. SCE, carbon dioxide was reported by Taniguchi et al. to be reduced to methanol, formic acid, and formaldehyde with current efficiencies of 44, 15, and 4%, respectively. The initial step was proposed to be electron transfer from deposited Li to CO_2, forming the CO_2^- radical anion.[38]

Selective reduction of CO_2 to CO was achieved by Beley et al. in aqueous media by using Ni-cyclam with p-type semiconductors. Using p-GaP or p-GaAs photoelectrodes, in 0.1 M $NaClO_4$, onset potentials were only –0.2 or –0.8 V vs. the normal hydrogen electrode (NHE), with faradaic yields close to 100%. However, the turnover numbers were very low, possibly because of carbon deposits on the photoelectrodes.[39–41]

Various organic mediators were tested by Bockris et al. for CO_2 reduction on p-CdTe photocathodes.[42,43] The best combination for methanol production used 15-crown-5 or 18-crown-6 in DMF–Bu4NClO4–5% H_2O, where the current efficiencies for CO and methanol at ambient temperature were 85 and 13%, at a current density of about 9 mA/cm². The onset potential was extremely low, only –0.1 V vs. NHE, which was due to an anodic shift of 410 mV relative to that of the bare CdTe electrode. The mechanism of action of the crown ethers was proposed to be due to their adsorption on the electrode, displacing the adsorbed solvent molecules and coordinating the tetraalkylammonium cations. These then were adsorbed on the inner Helmholtz layer on the semiconductor. Electron transfer from the electrode to the tetraalkylammonium cation resulted in a tetraalkylammonium radical, which mediated electron transfer to CO_2:

$$NR_4^+ + e^- = \cdot NR_4(ads) \tag{13.3}$$

$$\cdot NR_4(ads) + CO_2(ads) = NR_4^+ + CO_2^-(ads) \tag{13.4}$$

$$CO_2^-(ads) + H^+ + e^- + hn \rightarrow CO(ads) + OH^- \tag{13.5}$$

$$OH^- + H^+ \rightarrow H_2O \tag{13.6}$$

13.5 PHOTOREDUCTION ON POLYMER-COATED SEMICONDUCTORS

In an alternative approach to overcome the corrosion sensitivity of low-bandgap semiconductors, polymer coatings on the semiconductor surfaces have been applied. Conductive polyaniline coatings by electropolymerization of aniline were originally developed for Pt electrodes and later for semiconductor electrodes.[44,45] p-Type silicon was coated by Aurian-Blajeni et al. with polyaniline by electropolymerization at a fixed potential of +0.8 V vs. SCE from a solution of 0.1 M aniline in 0.1 M

H_2SO_4, resulting in a coating of about 50 nm thickness.[46] With this electrode, in aqueous 0.1 M LiClO$_4$ saturated with CO_2, the onset potential of the photocurrent was displaced by about +0.3 V relative to that of the bare p-Si electrode. The reduction products identified were formaldehyde and formic acid, in a total faradaic yield of up to 28% and an energy conversion efficiency (energy output/energy input) of about 4%.

Cabrera and Abruña deposited on p-Si and thin-film p-WSe$_2$ electrodes by electropolymerization films of [Re(CO)$_3$(v-bpy)Cl], where v-bpy is 4-vinyl-4'-methyl-2,2'-bipyridine.[47] Carbon dioxide reduction to carbon monoxide at essentially unit current efficiency was observed upon illumination of these electrodes, in acetonitrile containing tetra-n-butyl ammonium perchlorate, with onset potentials of −0.65 V (vs. SSCE) and turnover numbers of about 450.

13.6   CONCLUSIONS ON PHOTOELECTROCHEMICAL REDUCTION OF CO$_2$

Until now, most photoelectrochemical CO_2 reduction systems suffered from low current densities, probably due to the high resistances of the semiconductor electrodes. Also, the long-term stability of these systems for CO_2 reduction has not yet been demonstrated.

REFERENCES

1. Bard, A.J., Photoelectrochemistry and heterogeneous photocatalysis at semiconductors, *J. Photochem.*, 10, 59–75, 1979.
2. Halmann, M., Photochemical fixation of carbon dioxide, in *Energy Resources Through Photochemistry and Catalysis,* Grätzel, M., Ed., Academic Press, New York, 1983, 507–534.
3. Taniguchi, I., Electrochemical and photoelectrochemical reduction of carbon dioxide, in *Modern Aspects of Electrochemistry*, Vol. 20, Bockris, J. O'M. and Conway, R.E., Eds., Plenum Press, New York, 1989, 323–400.
4. Fujishima, A. and Honda, K., Electrochemical photolysis of water at a semiconductor electrode, *Nature*, 238, 37–38, 1972.
5. Ulman, M., M.Sc. thesis, Weizmann Institute of Science, Rehovot, Israel, 1982.
6. Halmann, M., Photoelectrochemical reduction of aqueous carbon dioxide on p-type gallium phosphide in liquid junction solar cells, *Nature*, 275, 115–116, 1978.
7. Halmann, M. and Aurian-Blajeni, B., Semiconductor electrolyte solar cells for the photoelectrochemical reduction of carbon dioxide to solar fuel, in Proc. 1979 Photovoltaic Solar Energy Conf., Berlin, April 23 to 26, 1979, 682.
8. Inoue, T., Fujishima, A., Konishi, S., and Honda, K., Photo-electrocatalytic reduction of carbon dioxide in aqueous suspensions of semiconductor powders, *Nature*, 277, 637–638, 1979.
9. Ito, K., Ikeda, S., Yoshida, M., Ohta, S., and Iida, T., On the reduction products of carbon dioxide at a p-type gallium phosphide photocathode in aqueous electrolytes, *Bull. Chem. Soc. Jpn.*, 57, 583–584, 1984.

10. Noda, H., Yamamoto, A., Ikeda, S., Maeda, M., and Ito, K., Influence of light intensity on photoelectroreduction of CO_2 at a p-GaP photocathode, *Chem. Lett.*, pp. 1757–1760, 1990.

11. Ikeda, S. and Ito, K., Artificial photosynthetic systems for carbon dioxide fixation, in Proc. Int. Symp. Chemical Fixation of Carbon Dioxide, Nagoya, Japan, December 2 to 4, 1991, 23–30.

12. Ikeda, S., Yamamoto, A., Noda, H., Maeda, M., and Ito, K., Influence of surface treatment of the p-GaP photocathode on the photoelectrochemical reduction of carbon dioxide, *Bull. Chem. Soc. Jpn.*, 66, 2473–2477, 1993.

13. Yoneyama, H., Sigumura, K., and Kuwabata, S., Effects of electrolytes on the photoelectrochemical reduction of CO_2 at illuminated p-type CdTe and p-type InP electrodes in aqueous solutions, *J. Electroanal. Chem.*, 249, 143–153, 1988.

14. Ikeda, S., Yoshida, M., and Ito, K., Photoelectrochemical reduction products of carbon dioxide at metal coated p-GaP photocathodes in aqueous electrolytes, *Bull. Chem. Soc. Jpn.*, 58, 1353–1357, 1985.

15. Ikeda, S., Amakusa, S., Noda, H., Saito, Y., and Ito, K., Photoelectrochemical and electrochemical formation of methane from carbon dioxide at copper coated electrodes, in Proc. Electrochem. Soc., Photoelectrochemistry and Electrosynthesis on Semiconducting Materials, Vol. 88-14, 1988, 130–136.

16. Noda, H., Ikeda, S., Saito, Y., Nakamura, T., Maeda, M., and Ito, K., Photoelectrochemical reduction of carbon dioxide at metal-coated p-InP photocathodes, *Denki Kagaku*, 57, 1117–1120, 1989; *Chem. Abstr.*, 112, 65316h.

17. Ikeda, S., Saito, Y., Yoshida, M., Noda, H., Maeda, M., and Ito, K., Photoelectrochemical reduction products of CO_2 at metal coated p-GaP photocathodes in non-aqueous electrolytes, *J. Electroanal. Chem.*, 260, 335–345, 1989.

18. Chu, C.Y., Peihai, W., Junfu, L., and Shiyong, Z., Catalysis of TiO_2 thin film for photo-electrochemical reduction of carbon dioxide, in Proc. Int. Symp. Chemical Fixation of Carbon Dioxide, Nagoya, Japan, December 2 to 4, 1991, 31–38.

19. Liu, J.-F. and Baozhu, C.Y., Photoelectrochemical reduction of carbon dioxide on a p^+/p^- Si photocathode in aqueous electrolyte, *J. Electroanal. Chem.*, 324, 191–200, 1992.

20. Nakato, Y. and Tsubomura, H., Silicon photoelectrodes modified with ultrafine metal islands, *Electrochim. Acta*, 37, 897–907, 1992.

21. Hinogami, R., Mori, T., Yae, S.J., and Nakato, Y., Efficient photoelectrochemical reduction of carbon dioxide on a p-type silicon (p-Si) electrode modified with very small copper particles, *Chem. Lett.*, pp. 1725–1728, 1994.

22. Nakamura, Y., Hinogami, R., Yae, S., and Nakato, Y., Photoelectrochemical reduction of CO_2 at a metal-particle modified p-Si electrode in non-aqueous solution, in Abstr. 4th Int. Conf. on Carbon Dioxide Utilization, Kyoto, Japan, September 1997, P-071.

23. Aurian-Blajeni, B., Halmann, M., and Manassen, J., Electrochemical measurements on the photoelectrochemical reduction of aqueous carbon dioxide on p-gallium phosphide and p-gallium arsenide semiconductor electrodes, *Sol. Energy Mater.*, 8, 425–440, 1983.

24. Fujishima, A., Electrochemical carbon dioxide reduction using solar energy, in Proc. Int. Symp. Chemical Fixation of Carbon Dioxide, Nagoya, Japan, December 2 to 4, 1991, 11–18.

25. Fujishima, A., Electrochemical and photoelectrochemical reduction of CO_2, in Proc. Int. Conf. on Carbon Dioxide Utilization, Bari, Italy, September 1993, 303–310.

26. Saeki, T., Hashimoto, K., and Fujishima, A., Electrochemical reduction of CO_2 on bimetal electrodes — bimetal, compound semiconductor and highly dispersed metal electrodes, in Proc. Int. Symp. Chemical Fixation of Carbon Dioxide, Nagoya, Japan, December 2 to 4, 1991, 297–302.

27. Taniguchi, I., Aurian-Blajeni, B., and Bockris, J. O'M., Photo-aided reduction of carbon dioxide to carbon monoxide, *J. Electroanal. Chem.*, 157, 179–182, 1983.

28. Taniguchi, I., Aurian-Blajeni, B., and Bockris, J. O'M., The mediation of the photoelectrochemical reduction of carbon dioxide by ammonium ions, *J. Electroanal. Chem.*, 161, 385, 1984.

29. Taniguchi, I., Aurian-Blajeni, B., and Bockris, J. O'M., The reduction of carbon dioxide at illuminated p-type semiconductor electrodes in nonaqueous media, *Electrochim. Acta*, 29, 923–932, 1984.

30. Aurian-Blajeni, B., Habib, M.A., Taniguchi, I., and Bockris, J. O'M., The study of adsorbed species during the photoassisted reduction of carbon dioxide at a p-CdTe electrode, *J. Electroanal. Chem.*, 157, 399–404, 1983.

31. Habib, M.A. and Bockris, J. O'M., FT-IR spectrometry for the solid–solution interface, *J. Electroanal. Chem.*, 180, 287–306, 1984.

32. Ueda, J. and Uosaki, K., Photoelectrochemical fixation of CO_2 at semiconductor electrodes, in Proc. Int. Symp. Chemical Fixation of Carbon Dioxide, Nagoya, Japan, December 2 to 4, 1991, 343–346.

33. Ueda, J.-J., Nakabayashi, S., Ushizaki, J.-I., and Uosaki, K., A photoelectrochemical fixation of carbon dioxide. Spontaneous up quality conversion of organic compound, *Chem. Lett.*, pp. 1747–1750, 1993.

34. Fujishima, A., New approaches in CO_2 reduction, in Abstr. 4th Int. Conf. on Carbon Dioxide Utilization, Kyoto, Japan, September 1997, PL-3.

35. Hirota, K., Tryk, D.A., Hashimoto, K., Fujishima, A., and Ohkawa, M., Photoelectrochemical reduction of highly concentrated CO_2, in Abstr. 4th Int. Conf. on Carbon Dioxide Utilization, Kyoto, Japan, September 1997, P-078.

36. Bradley, M.G., Tysak, T., Graves, D.J., and Vlachopoulos, N.A., Electrocatalytic reduction of carbon dioxide at illuminated p-type silicon semiconducting electrode, *J. Chem. Soc. Chem. Commun.*, pp. 349–350, 1983.

37. Zafrir, M., Ulman, M., Zuckerman, Y., and Halmann, M., Photo-electrochemical reduction of carbon dioxide to formic acid, formaldehyde and methanol on p-gallium arsenide in an aqueous V(II)–V(III) chloride redox system, *J. Electroanal. Chem.*, 159, 373–389, 1983.

38. Taniguchi, Y., Yoneyama, H., and Tamura, H., Photo-electrochemical reduction of carbon dioxide at p-type gallium phosphide electrodes in the presence of a crown ether, *Bull. Chem. Soc. Jpn.*, 55, 2034–2039, 1982.

39. Beley, M., Collin, J.-P., Sauvage, J.-P., Petit, J.-P., and Chartier, P., Photoassisted electro-reduction of CO_2 on p-GaAs in the presence of Ni cyclam^{2+}, *J. Electroanal. Chem.*, 206, 333–339, 1986.

40. Petit, J.-P., Chartier, P., Beley, M., and Sauvage, J.-P., Selective photoelectrochemical reduction of CO_2 to CO in an aqueous medium on p-GaP, mediated by Ni cyclam^{2+}, *New J. Chem.*, 11, 751–752, 1987.

41. Petit, J.-P., Chartier, P., Beley, M., and Deville, J.-P., Molecular catalysts in photoelectrochemical cells. Study of an efficient system for the selective

photoelectroreduction of CO_2: p-GaP or p-GaAs/Ni(cyclam)$^{2+}$, aqueous medium, *J. Electroanal. Chem.*, 269, 267–281, 1989.

42. Bockris, J. O'M. and Wass J.C., The photoelectrocatalytic reduction of carbon dioxide, *J. Electrochem. Soc.*, 136, 2521–2528, 1989.

43. Bockris, J. O'M. and Gonzáles-Martin, A., The photoelectrochemical reduction of carbon dioxide, in Proc. Int. Symp. Chemical Fixation of Carbon Dioxide, Nagoya, Japan, December 2 to 4, 1991, 63–72.

44. Diaz, A.F. and Logan, J.A., Electroactive polyaniline films, *J. Electroanal. Chem.*, 111, 111–114, 1980.

45. Noufi, R., Nozik, A.J., White, J., and Warren, L.F., Enhanced stability of photoelectrodes with electrogenerated polyaniline films, *J. Electrochem. Soc.*, 129, 2261–2265, 1982.

46. Aurian-Blajeni, B., Taniguchi, I., and Bockris, J. O'M., Photo-electrochemical reduction of carbon dioxide using polyaniline-coated silicon, *J. Electroanal. Chem.*, 149, 291–293, 1983.

47. Cabrera, C.R. and Abruña, H.D., Electrocatalysis of CO_2 reduction at surface modified metallic and semiconducting electrodes, *J. Electroanal. Chem.*, 209, 101–107, 1986.

14   Heterogeneous Photoassisted Reduction of CO$_2$

14.1   INTRODUCTION

The ease of experiments on the photoassisted reduction of CO_2 over suspended or colloidal semiconductors has encouraged much interest in such work. In principle, illuminated semiconductor particles may be considered to act as tiny photoelectrolysis cells.

14.2   PRIMARY REACTION STEPS ON METALS AND SEMICONDUCTORS

Carbon dioxide adsorbs only weakly on clean surfaces of platinum group metals. However, as described in Section 8.1.2, alkali metals deposited on such metals cause markedly enhanced adsorption, with electron donation from the alkali metal to the p-orbital of CO_2, causing a change in the linear structure of CO_2 to the bent form of CO_2^-.

Solymosi et al. observed that UV illumination of clean Rh(111) surfaces covered with a multilayer of K adatomes at 90 K in the presence of CO_2 at 10^{-9} Pa caused a very considerable increase in the formation of the adsorbed CO_2^- species.[1-3] This was detected by high-resolution electron energy loss spectroscopy (HREELS), which revealed the appearance on the illuminated surface of loss features at 1610 to 1620, 1320, and 830 cm^{-1}. These were assigned to the asymmetric stretch, the symmetric stretch, and the bending modes of the bent CO_2^- anion, possibly as a $K^{\delta+}-CO_2^{\delta-}$ species. Also, coupling between neutral CO_2 and CO_2^- may yield the cluster compound $CO_2 \cdot CO_2^-$. Postirradiation temperature-programmed desorption released both CO_2 and CO. Annealing of the irradiated layer above 355 K led to the appearance of loss bands at 1430 to 1470 cm^{-1}, which were attributed to the carbonate species. CO was produced at low K coverage on Rh(111) by the reaction

$$CO_2^- = CO(ads) + O(ads) \qquad (14.1)$$

while at high K coverage, the predominant reaction was also formation of carbonate:

$$2CO_2^- = CO(ads) + CO_3^{2-}(ads) \qquad (14.2)$$

From the wavelength dependence of the photoeffects, it was concluded that photoelectrons released from the K-covered Rh caused the photoactivation of CO_2.[1-3]

Rh, Pt, Ir, and Ru supported on TiO_2 are important catalysts for a variety of photochemical reactions. Raskó and Solymosi performed the photoinduced activation of CO_2 at 190 K in gas–solid reactions on both TiO_2 and Rh/TiO_2 under UV illumination with an Hg lamp.[4] Fourier transform infrared spectroscopy was used to identify the surface-adsorbed species produced. On pure TiO_2, the bands appearing at 1630 to 1640 and 1218 cm^{-1} were due to bent CO_2 adsorbed on the catalyst. With prereduced Rh/TiO_2, strong additional bands appeared at 2020 and 2040 cm^{-1}, assigned to CO bonded to Rh. With oxidized samples of Rh/TiO_2, no photoeffects were observed.[4] Similar processes occurred on titania-supported Rh, Pt, Ir, and Ru and also on clean Cu and Ni surfaces after the illumination of adsorbed CO_2.[3]

Semiconductor photocatalysts have been used as powdered substances, dispersed as slurries or colloids in aqueous or nonaqueous solvents, or as catalyst beds in gas–solid reactions. The particles of such materials were considered by Bard as minute electrolysis cells.[5] Under illumination at wavelengths shorter than the bandgap, these particles may cause photocatalytic reactions, including the reduction of carbon dioxide to carbon monoxide and to several organic compounds, such as formic acid and methanol and even two- and three-carbon compounds. The simplicity of operating such photocatalytic reactions has stimulated considerable recent efforts to improve the efficiency and selectivity of such processes to desired products. In some cases, high quantum yields of formic acid, and in a few cases also methanol and ethanol, were reported.

Photocatalytic reactions on semiconductor particles depend on a primary step of chemisorption. One tool to clarify the structure of carbon dioxide chemisorbed on TiO_2 is infrared absorption spectroscopy. In the case of reduced Pt/TiO_2, the adsorbed species was proposed by Tanaka and White to form a bidentate carbonate structure, which was identified by its absorption bands at 1245 and 1673 cm^{-1}.[6]

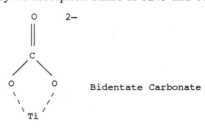

Bidentate Carbonate

Following the primary step of adsorption of carbon dioxide on the semiconductor particles, electrons and holes may be transferred from the photoexcited semiconductor to surface-adsorbed molecules. Interaction of these excited surface molecules with the medium may result in the release of highly reactive intermediate species in solution. Some of these shortlived transient species, such as $\cdot OH$ and $\cdot O_2H$ radicals, can be stabilized by transforming them into "spin adducts" with much longer lifetimes.[7] Spin adducts based on nitrones were detected by Harbour and Hair[8] and Jaeger and Bard[9] using electron spin resonance (ESR). In experiments with illuminated suspensions of tungsten oxide in aqueous sodium hydrogen carbonate containing the spin trap α-phenyl-N-*tert*-butylnitrone, a spin adduct was detected by Aurian-Blajeni et al., which was assigned as the $\cdot CO_3^-$ radical anion.[10]

14.3 PHOTOCATALYSIS IN AQUEOUS SLURRIES

In a comparison by Inoue et al. of the effectiveness of various semiconductors, highest yield of photoreduction of CO_2 was obtained with SiC.[11] After 7 h of illumination, the concentrations of formaldehyde and methanol were 1.1 and 0.23 mM with TiO_2, 1.2 and 0.35 mM with ZnO, 2.0 and 1.2 mM with CdS, 1.0 and 1.2 mM with GaP, 1.0 and 5.4 mM with SiC, and 0.0 and 0.0 mM with WO_3. The yields correlated with the level of the conduction band of the semiconductor used, the most negative conduction band level being that of SiC, about -1.6 V vs. the normal hydrogen electrode (NHE). Using SiC as photocatalyst, the quantum yields for production of formaldehyde and methanol were 0.05 and 0.45%, respectively.[11] In the photoassisted reduction of carbonated aqueous suspensions of several powdered semiconductors by illumination with high-pressure Hg lamps, the production of formaldehyde and methanol was measured by Aurian-Blajeni et al. The order of activity of the semiconductors used was $SrTiO_3 > WO_3 > TiO_2$, resulting in absorbed energy conversion efficiencies of 6, 5.9, and 1.2%, respectively.[12]

In efforts to enhance the yields of CO_2 photoreduction products, the effect of surface treatment with transition metal oxide additives on powdered strontium titanate was tested by Ulman et al.[13] The organic products observed included formic acid, formaldehyde, methanol, acetaldehyde, and ethanol. Surface doping of $SrTiO_3$ with lanthanum chromite ($LaCrO_3$ [0.5 mol%]) caused a decrease in the production of formic acid, but enhanced the formation of methanol. With the same $LaCrO_3/SrTiO_3$ catalyst, the photoreduction of CO_2 was compared in water and in 0.1 M $LiHCO_3$. The production rates of formaldehyde and methanol were enhanced in the $LiHCO_3$ medium, while HCOOH production was faster in water as the medium.[13]

Formaldehyde production was also reported by Tennakone in a study of the effect of different metals coated on TiO_2. Highest yields of HCHO were obtained by illumination of CO_2-saturated aqueous suspensions of Hg-coated TiO_2.[14]

In the photoassisted reduction of carbon dioxide passed through aqueous suspensions of TiO_2 containing 0.1% RuO_2 and 570 ppm Cr in 0.1 M Li_2CO_3, during 47 h of illumination with a 75-W high-pressure Hg lamp, the production rates of formic acid and methanol were considerably increased by the presence of 1 mM Na_2S, as

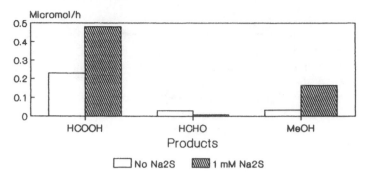

FIGURE 14.1 CO_2 photoreduction on aqueous TiO_2 containing 0.1% RuO_2 and 570 ppm Cr. Sulfide ion effect.

shown in Figure 14.1. Presumably the sulfide ions serve as sacrificial hole traps or as electron donors.[15]

The presence of hydrogen sulfide was found by Aliwi and Al-Jubori to also enhance the photoreduction of carbon dioxide to formic acid and formaldehyde in aqueous suspensions of amorphous n-Bi_2S_3 or n-CdS semiconductors.[16]

With a CdS–ZnS mixture (1.4:1.0 ratio) suspended in 0.5 M K_2CO_3 containing 0.1 M Na_2S as hole scavenger, thermostatted at 61°C and under argon as carrier gas, illumination during 34 h with a 150-W xenon lamp caused the production of methanol and formaldehyde at rates of 1.5×10^{-9} and 0.2×10^{-9} mol h^{-1} cm^{-2} (per illuminated area). Formic acid was not detected.[17]

The high-bandgap semiconductors barium titanate and lithium niobate, doped with the rare earths europium, neodynium, and samarium, in aqueous carbonated suspensions, were tested by Ulman et al. as photocatalysts for the reduction of carbon dioxide. The yields of formic acid and formaldehyde were markedly enhanced by the presence of the rare earth dopants. The order of effectiveness of the rare earths was $Nd_2O_3 > Sm_2O_3 >> Eu_2O_3$.[18]

Another high-bandgap semiconductor is ZrO_2 ($E_g = 5.0$, where E_g is the bandgap). Sayama and Arakawa observed that UV-irradiated suspensions of ZrO_2 in aqueous solutions of $KHCO_3$ (1 M, pH 8.4) were very effective in both splitting water to H_2 and O_2 and releasing CO from the bicarbonate.[19] Under illumination with a 400-W high-pressure Hg lamp, the rates of evolution of H_2, O_2, and CO were 294, 157, and 2.0 µmol h^{-1} g-cat^{-1}, respectively. There was no production of HCOOH, CH_3OH, and CH_4. The effectiveness of ZrO_2 may be understood to be due to its very negative flat-band potential ($E_{fb} = -1.0$ eV vs. NHE at pH 0). There was no reaction with ZrO_2 dispersed in aqueous Na_2CO_3.[19]

By illumination of a CO_2-saturated aqueous slurry of $CaFe_2O_4$ powder in aqueous dispersion (pH 5.8) with a high-pressure Hg lamp, Matsumoto et al. observed the production of methanol and formaldehyde.[20] $CaFe_2O_4$ is a p-type semiconductor with a relatively small bandgap of 1.9 eV, which was prepared by heating a mixture of Fe_2O_3 and $CaCO_3$ powders at 1100°C for 10 h. The quantum yield for methanol production at 300 nm was estimated to be about 0.01%. The yield of methanol was somewhat enhanced in the presence of NaH_2PO_2 (0.2 M), presumably acting as a hole trap. Further yield improvement was attained by adding $BaCO_3$ into the reaction slurry.[20]

An interesting approach to enhance the efficiency of reduction of carbon dioxide was to use a semiconductor which is also a sacrificial hole trap. With an aqueous suspension of hydrous cuprous oxide ($Cu_2O \cdot xH_2O$), selective and enhanced production of methanol and formaldehyde was achieved by Tennakone et al., reaching maximal concentrations of 24.0 and 3.5 μM after about 45 min.[21]

A marked dependence on the size of particles was found by Yamamura et al. in a study of powdered SiC.[22] Thus, with 100-mesh SiC, the photoreduction of carbon dioxide in aqueous media yielded only traces of methanol and no detectable amounts of ethanol. However, using 1000-mesh SiC, in aqueous sulfuric acid (pH 2.9), during 8 h of illumination, ethanol and methanol were produced in amounts of 18 and 12 μmol, respectively.[22]

14.4 METAL-TREATED SEMICONDUCTORS

Metal loading of TiO_2 in aqueous suspensions in some cases enhanced its efficiency as photocatalyst for the reduction of carbon dioxide to organic compounds. Ishitani and Ibusuki observed that palladium loading increased the activity of methane production by more than 10 times relative to that of bare TiO_2. The order of activity of several metal deposits in accelerating methane production was Pd > Rh > Pt > Ru.[23]

Dzhabiev and Uskov tested strontium titanate doped with both Pt and Ru in aqueous suspensions for the photoreduction of CO_2.[24] With $SrTiO_3$/Pt (0.5 wt%)/Ru (0.5 wt%), under illumination with an ultra-high-pressure Hg lamp, the gaseous products were CO, H_2, and CH_4, formed at the relative initial rates of 6:1.2:1. Thus, CO was produced preferentially. From the temperature dependence of the rate of formation of CO, the activation energy was derived to be about 8 kcal mol^{-1}.[24]

While formic acid was the predominant product of the photoreduction of CO_2 in aqueous solutions in the presence of TiO_2/Pt, other organic products detected by Ulman were formaldehyde, methanol, acetaldehyde, and ethanol.[25]

With platinized TiO_2 doped with Cr or Mn (0.5 atom%) suspended in 1 M Na_2CO_3 in a nitrogen atmosphere, illuminated with a medium-pressure mercury lamp, a photoreduction was observed by Rophael and Malati yielding carbon as well as smaller amounts of formaldehyde and methanol.[26,27] In similar experiments, but with TiO_2 (rutile) coated with Fe^{2+} or Co^{2+} phthalocyanines (2% surface coverage) suspended in aqueous sodium carbonate under nitrogen, enhanced yields of methanol and formaldehyde were reported.[28,29] Malati et al. compared the efficiency of Pt/ TiO_2, MoS_2, and ZnTe for the photocatalytic reduction of aqueous carbonate (pH 6 to 7).[30] Under UV illumination, methanol was produced with all these catalysts, and with MoS_2 formaldehyde was also produced. Interestingly, when filtering off these catalysts after the reaction, washing, and drying at 850°C, a weight increase was observed, which was attributed to the formation of carbon. With MoS_2 as photocatalyst, the yield of this carbon was much higher than that of the volatile products. Thus, after 5 h of illumination of 0.1 M carbonate with a high-pressure Hg lamp, the rates of methanol, formaldehyde, and carbon formation were 12, 15, and 93 μmol h^{-1}, respectively.[30]

Yamamura et al. tested the effect of metal loading on 1000-mesh SiC particles.[31] Highest concentrations of organic products were obtained with SiC/Pd, over which CO_2 was photoreduced to HCOOH, HCHO, methanol, acetaldehyde, and ethanol, in an overall energy conversion efficiency of 0.013%. This efficiency was defined as:

Effic. (%) = 100(Heat of combustion of products)/(Incident light energy)

Cook et al. used a mixture of p-SiC (325 mesh) and copper powder (100 mesh) in 0.5 M aqueous $KHCO_3$ (pH 5). CO_2 photoreduction resulted in production rates of methane, ethylene, and ethane of 1.49, 0.39, and 0.23 μL h^{-1}, respectively.[32]

While the illumination of carbon dioxide in the presence of aqueous suspensions of TiO_2 alone is quite ineffective for the reduction of CO_2, appreciable reduction was attained by Hirano et al. when using mixtures of TiO_2 and metallic copper.[33] With TiO_2 (0.5 g) and Cu powder (0.3 g, mean particle diameter 200 μm) dispersed in CO_2-saturated water, illumination with a xenon lamp resulted in the production of small amounts of CO, formic acid, formaldehyde, and methanol, in concentrations gradually increasing in time. The concentrations of formaldehyde and methanol reached a maximum at about 9 to 10 min and then declined. Even larger amounts of methanol were produced when the TiO_2–Cu mixture was illuminated in aqueous sodium bicarbonate. Free metallic copper was required for this photocatalyzed reduction of CO_2 in TiO_2 powder suspensions. No reduction was observed on copper supported on TiO_2. Presumably, the metallic copper acts both as an effective catalyst for the reduction (as in the electrochemical reduction on copper electrodes) and as a sacrificial reducing species, as it is oxidized by the positive holes generated on the illuminated semiconductor.[33]

Instead of the above results with mixtures of metallic copper and TiO_2, loading of Cu ions on TiO_2 resulted in the preferential photocatalytic reduction of CO_2 into hydrocarbons. The Cu-loaded catalyst was prepared by Adachi et al. by soaking TiO_2 powder in copper chloride solution, followed by reduction under H_2 at 400°C for 2 h.[34] The photocatalyst was dispersed in an aqueous solution of CO_2 pressurized to about 28 atm, resulting in a pH of 5.45, and was illuminated with a 450-W Xe lamp. With an optimal Cu loading of 5 wt% on the TiO_2, the major products were H_2, C_2H_4, CH_4, and C_2H_6. The production rates for C_2H_4 and CH_4 were 0.92 and 0.65 μL g^{-1} h^{-1}, respectively. Methanol and formaldehyde were not formed.[34]

In a careful study of the photocatalytic reduction of CO_2 in aqueous suspensions of TiO_2, Solymosi and Tombácz found that by doping the TiO_2 with both Rh (1%) and WO_3 (2%), the main organic product was methanol. Under illumination with a 500-W high-pressure Xe lamp, the production rate of methanol was about 3 μmol h^{-1} g-cat^{-1}. Only traces of HCOOH and HCHO were formed.[35,36]

TiO$_2$, with its relatively high bandgap, requires UV irradiation and is thus quite inefficient for use with sunlight as the energy source. In order to switch its photocatalytic activity to the visible range of the solar spectrum, Heleg and Willner prepared a Pd–TiO$_2$ photocatalyst to which the dye eosin was bound covalently.[37] This was achieved by silanating the surface hydroxy groups of TiO$_2$ with aminopropyl triethoxysilane. Under visible light illumination ($\lambda > 435$ nm), in CO$_2$-saturated aqueous bicarbonate, containing EDTA as sacrificial electron donor, the only product observed was formate. The quantum yield for HCOO$^-$ production was 0.8%. If the eosin was instead homogeneously dissolved in the medium, the rate of formate production was only one-fifth that with the covalently bound dye. Evidently, the immobilized dye under photoexcitation sensitized electron transfer to the semiconductor, followed by electron tunneling to Pd atoms and then to adsorbed CO$_2$ species for reduction to CO$_2^-$.[37]

Premkumar and Ramaraj adsorbed cobalt or zinc phthalocyanines onto acidtreated Nafion® membranes and dipped these membranes into CO$_2$-saturated aqueous solutions of triethanolamine (0.1 M) and HClO$_4$ (0.1 M). Under illumination with visible light, formic acid was obtained as the only CO$_2$ reduction product. These metal phthalocyanines acted as p-type semiconductors. The cobalt phthalocyanine was a more active photocatalyst than the zinc phthalocyanine.[38]

14.5 SOLAR PHOTOREDUCTION

A photochemical solar collector was built for testing the photoassisted reduction of carbon dioxide under illumination with natural sunlight.[39] The device consisted of a rectangular aluminum plate, which was painted with an aqueous slurry of the powdered semiconductor to be studied, immersed in water through which CO$_2$ was bubbled and which was surrounded by an aluminum frame, and exposed to sunlight through a cover plate of glass or methyl methacrylate (1 × 0.5 m^2 area). The outflowing gas was passed through a series of two to three traps cooled to 0°C, to collect the organic reduction products (see Figure 14.2).

The products were mainly formic acid, as well as smaller amounts of formaldehyde, methanol, ethanol, and acetaldehyde. Results for experiments conducted in carbonated aqueous 0.1 M Li$_2$CO$_3$ are presented in Figure 14.3, in which the production rates of the organic products are expressed on a logscale as 10^3 × µmol/KJ (normalized by the light dose). The semiconductor materials tested were TiO$_2$, plain SrTiO$_3$, and SrTiO$_3$ doped with 0.5 mol% LaCrO$_3$. The TiO$_2$ used was the least effective of the three materials. The predominant product with all photoactive agents was formic acid. Methanol production was highest with the plain SrTiO$_3$.[39]

14.6 COLLOIDAL PARTICLES AS PHOTOCATALYSTS

Quantum effects in light absorption and chemical reactivity were discovered by Rossetti et al. when the sizes of these particles were decreased to the dimensions of colloidal particles.[40] The quantization effects were explained to be due to the con-

FIGURE 14.2 Photochemical solar collector.

finement of the charge carriers, electrons and holes, in the very small particles. This confinement results in perturbations of the semiconductor band structure, causing increases in the effective bandgaps, and leading to a series of discrete states in the conduction and valence bands. The increased effective bandgap results in enhanced redox potentials for the photoexcited electrons and holes. Thus, these colloidal particles can photocatalyze reactions which are not possible with large-particle semiconductors.

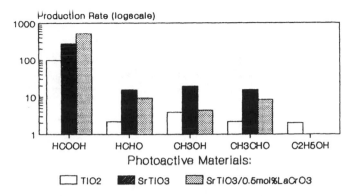

FIGURE 14.3 Photoreduction of carbon dioxide in photochemical solar collector.

14.6.1 Colloidal TiO_2

With colloidal dispersions of platinized TiO_2 (prepared by acid hydrolysis of titanium tetraisopropoxide) in 1 M aqueous Na_2CO_3 under illumination at $\lambda > 300$ nm, the only organic product detected by Chandrasekaran and Thomas was formaldehyde.[41] Flash photolysis with a 337-nm N_2 laser of a similar dispersion of colloidal TiO_2 in 0.1 M Na_2CO_3 produced a short-lived transient absorption peak at 600 nm, decaying exponentially with a half-life of 0.9 μs. This species was assigned as the carbonate radical anion, formed from surface-adsorbed carbonate anions which injected electrons into the holes:

$$TiO_2 + h\nu \rightarrow e^- + h^+ \qquad (14.3)$$

$$CO_3^{2-} + h^+ \rightarrow \cdot CO_3^- \qquad (14.4)$$

In the presence of even traces of oxygen, there may be charge transfer to form the neutral carbonate radical, which may decompose into surface-adsorbed CO and oxygen:

$$\cdot CO_3^- + O_2 \rightarrow \cdot CO_3 + O_2^- \qquad (14.5)$$

$$\cdot CO_3 \rightarrow CO_{ad} + O_2 \qquad (14.6)$$

The adsorbed CO presumably was further reduced on the semiconductor, releasing formaldehyde. Methanol formation was not detected in this reaction.[41]

A remarkable improvement in the photocatalytic reduction of CO_2 was achieved by Inoue et al. using quantum-sized TiO_2 particles embedded in SiO_2 matrices as photocatalysts.[42] These were prepared by a sol-gel technique, by hydrolyzing aqueous ethanol solutions of $Si(OEt)_4$ and $Ti(OEt)_4$, casting the sol on glass plates, drying, stripping off, and heating the gel formed at 120°C for 2 h under vacuum. The resulting transparent flake films were dispersed in CO_2-saturated water containing 2-propanol (1 M) as hole scavenger and were illuminated with a high-pressure Hg arc ($\lambda > 270$ nm). The size of the TiO_2 particles in the SiO_2 matrices decreased with decreasing ratio of Ti/Si in the original sol. Decreasing TiO_2 particle sizes caused increased bandgaps and a blue shift in the UV absorption spectra of aqueous dispersions of these quantum-sized particles. The activity for the reduction of CO_2 very markedly increased with the smaller particle sizes. With the smallest TiO_2 particles, with a Ti/Si ratio of 0.027 and an average particle size of about 3.5 nm, the quantum efficiencies at 280 nm for the production of $HCOO^-$, CH_4, and C_2H_4 were 14.2, 3.4, and 2.3%, respectively, while with a mixture of bulk TiO_2/SiO_2, the corresponding efficiencies were only 2.4, 0.03, and 0%, respectively. Presumably, the SiO_2 matrices prevented aggregation of the TiO_2 particles.[42]

In the above CO_2 photoreduction on quantized TiO_2 in SiO_2 matrices, no methanol was formed. Selective production of methanol was however attained by Kuwabata et al. on quantized TiO_2 particles (2 to 4 nm in diameter) immobilized in a poly(vinylpyrrolidine) film.[43] By laying such a film on the surface of CO_2-saturated propylene carbonate, containing 2-propanol (1 M) as a hole scavenger, with CO_2 also in the gas space above the film, and illuminating with a high-pressure Hg arc ($\lambda > 300$ nm), the products were acetone (from the oxidation of 2-propanol), methanol, and H_2. $^{13}CO_2$-labeling experiments proved that the methanol was derived from CO_2. Surprisingly, methanol was the only CO_2 reduction product which could be detected. The amount of methanol produced under these conditions was 84 times that formed with a dispersion of commercial TiO_2 (Degussa P25).[43]

The solvent effects on the product distribution of the photocatalytic reduction of carbon dioxide were tested by Torimoto et al. using both TiO_2 nanocrystals (Q-TiO_2) embedded in SiO_2 matrices and bulk CdS powder as catalysts.[44] CO_2-saturated dispersions of the catalysts in various solvents and containing 1 M 2-propanol as hole scavenger were illuminated ($\lambda > 300$ nm). The products were HCOOH and CO. With both catalysts, the fraction of formate production decreased with decreasing dielectric constant of the solvent, in the order H_2O > propylene carbonate > sulfolane > MeCN > CCl_4. The preferential production of formate in the more polar medium was explained by stabilization of the CO_2^- intermediate by the solvent molecules, resulting in reaction of protons with the carbon of the CO_2^- to produce formate. In the low polar solvents, the CO_2^- is strongly adsorbed on the photocatalyst surface, and the reaction with protons occurred with an oxygen atom of the CO_2^- radical anion, releasing CO.[44]

14.6.2   Colloidal ZnS

Very high quantum yields, up to 80%, were reported by Henglein for CO_2 reduction to HCOOH in aqueous media, using microcrystalline ZnS, which was stabilized by colloidal SiO_2, with sulfite as a hole acceptor.[45,46] The mechanism proposed involved a two-electron reduction. Colloidal ZnS was also found to catalyze the photoreduction of carbon dioxide in the presence of methanol or 2-propanol. 2-Propanol was much more efficient than methanol in promoting the production of formic acid, resulting in a quantum yield of up to 80%.[45,46] In more recent studies by Inoue et al. with ZnS, much lower quantum yields were observed, maximally reaching 30%. The quantum yield was found to increase with the $(Zn^{2+})/(S^{2-})$ ratio used in the preparation of the ZnS colloids.[47] With suspensions of powdered ZnS in water/2,5-dihydrofuran, the photoreduction of carbon dioxide to formic acid was observed by Kisch and Twardzik to occur with a quantum yield of only 0.1% (at $\lambda = 300$ nm).[48] The 2,5-dihydrofuran functioned as reducing agent and was itself oxidized to dehydro dimers. The mechanism proposed involved two-electron transfer from zinc sulfide to adsorbed hydrogen carbonate. Low quantum yields (0.3%) for CO_2 reduction were also found by Albers and Kiwi using Pd/TiO_2 dispersions.[49]

The above widely different results with ZnS may possibly be understood by different methods of preparing the photocatalyst. Defect-free colloidal ZnS crystallites were prepared by Yanagida et al. by mixing equivalent amounts of aqueous $ZnSO_4$ and Na_2S at 0°C under argon.[50,51] With these quantum crystallites, in an aqueous mixture of Na_2S and NaH_2PO_2 saturated with CO_2 at pH 7, UV illumination at $\lambda > 290$ nm resulted in production of formic acid, carbon monoxide, and hydrogen at initial rates of 75, 1.7, and 86 μmol h^{-1}, respectively. The apparent quantum yield for formic acid was $\Phi_{1/2HCOOH} = 0.24$ at 313 nm. In this reaction, hypophosphite ions served as sacrificial electron donors, while sulfide ions served as suppressors of surface defects. Since these quantized microcrystallites of ZnS had in their reflectance spectra a steep onset, at shorter wavelength than with ordinary preparations of ZnS, these microcrystallites presumably have a lower density of surface defects, which could be the reason for their improved effectiveness for CO_2 reduction.[50,51]

In a more detailed study by Kanemoto et al., colloidal solutions of monodispersed ZnS nanocrystallites were prepared by reacting zinc perchlorate with H_2S in either dimethylformamide (DMF), acetonitrile (AN), or methanol (MeOH) solutions.[52] The preparation of the ZnS nanocrystallites in aprotic organic solvents had the advantage of not requiring stabilizers for the colloidal solutions. The hexagonal ZnS–DMF microcrystallites, which had a narrow size distribution of 1.6 to 2.7 nm in diameter, had a steep UV absorption onset at 315 nm. Using the ZnS–DMF as photocatalyst in CO_2-saturated DMF solution containing triethylamine (TEA) as a sacrificial electron donor, under illumination with a high-pressure Hg arc ($\lambda > 290$ nm), the main products were formate, H_2, and CO. Oxalate, formaldehyde, and methanol were not detected. At $\lambda = 302$ nm, the apparent quantum yield for formate production was $\Phi_{1/2HCOOH} = 0.14$. Experiments with $^{13}CO_2$ proved that the formate and CO were derived from the CO_2. The H_2 was derived from traces of water in the medium. ZnS–MeOH was also active as photocatalyst for CO_2 reduction to formate, while ZnS–AN was inactive. These results were markedly different from the electrochemical reduction of CO_2 in DMF solution, in which the main products were oxalate and CO (see Chapter 12). The mechanism postulated for the photocatalyzed reaction assumed electron transfer from the conduction band of the photoexcited ZnS–DMF nanocrystallites to adsorbed CO_2, forming the CO_2^- radical anion, which reacted with weakly adsorbed H atoms, producing formate ions.[52]

High quantum yields of CO_2 photoreduction to formate were achieved by Inoue et al. using aqueous colloidal dispersions of cadmium-loaded ZnS microcrystals in the presence of 2-propanol (1 M) and 0.1 mM Zn^{2+}.[53] These Cd-loaded microcrystals were photodeposited by UV illumination of argon-flushed solutions of the ZnS colloidal solutions containing cadmium perchlorate. The highest quantum yield for formate production, 32.5%, was obtained with a Cd loading of 0.025 mol% on the ZnS microcrystals. This was about twice the quantum yield obtained with bare ZnS microcrystals. Loading of the ZnS with Pb, Ni, Ag, or Cu metals did not result in increased photocatalytic activity. Formate was the only CO_2 reduction product observed.[53]

With colloidal ZnS in CO_2-saturated aqueous solutions containing tetramethyl-ammonium chloride (0.1 M), under illumination with a medium-pressure arc lamp (λ > 320 nm), Eggins et al. observed the production not only of the one-carbon HCOOH and HCHO, but also of the two-carbon oxalic, glycolic, and glyoxylic acids and the four-carbon tartaric acid. The latter may have been formed by a pinacol-type dimerization from glyoxylic acid. Oxalic acid was the most abundant product, presumably formed by dimerization of the CO_2^- radical anion.[54,45]

14.6.3   Colloidal CdS and CdSe

With colloidal CdS microcrystallites, the selective photoreduction of CO_2 to CO was achieved by Kanemoto et al. even by visible light, in DMF solution containing 1 v/v% water, with TEA as sacrificial electron donor.[56] These hexagonal microcrystallites were prepared from $Cd(ClO_4)_2$ and H_2S in DMF solution and were shown by high-resolution electron microscopy to range in size from 3 to 5 nm in diameter. The apparent quantum yield was $\Phi_{1/2CO}$ = 0.098 at λ > 400 nm. Ordinary powdered CdS was inactive.[56] Size quantization effects were proposed to account for the photocatalytic activity, in analogy to the previous results with microcrystallites of ZnS.[50,51] Fujiwara et al. discovered that the yield of the photocatalytic reduction of CO_2 to CO in DMF/TEA solutions containing quantized CdS nanocrystallites was markedly enhanced in the presence of an excess of Cd^{2+} ions.[57] The photoproduction of CO was maximal when 0.2 equivalents of excess Cd^{2+} (relative to the CdS) was added, thus approximately doubling the yield of CO, which was the only observed product. This effect was explained on the basis of emission and EXAFS measurements by the formation of sulfur vacancies on the CdS–DMF surface due to the excess of Cd^{2+}. Strong adsorptive interaction of CO_2 with these sulfur vacancies may have led to an assumed intermediate $Cd^{2+}OCOCO_2$ complex, which was further reduced, with elimination of CO and carbonate ions.[57]

When In^{3+} ions were also present in the above colloidal CdS/DMF/TEA system, formic acid was formed in addition to CO. Highest yields of HCOOH were observed by Kanemoto et al. at 0.3 mM concentration of In^{3+}, resulting in quantum yields (at λ > 400 nm) of $\Phi_{1/2CO}$ = 0.061 and $\Phi_{1/2HCOOH}$ = 0.039, respectively.[58]

To prevent agglomeration of the microcrystals of CdS, Inoue et al. added ionic long-chain molecules as stabilizers.[59] The CdS microcrystals (less than 5 nm in diameter) were prepared under an N_2 atmosphere by mixing aqueous $Cd(ClO_4)_2$ containing the stabilizer (0.8 mM monomeric concentration) with Na_2S. Such colloidal solutions were stable for at least 3 weeks. By illumination at $\lambda > 310$ nm of CO_2-saturated solutions (pH 5.5) of the stabilized CdS, in the presence of 2-propanol as a hole scavenger, CO_2 reduction occurred. When using the negatively charged stabilizers polyacrylic acid, poly(sodium vinyl-sulfonate), and poly(sodium 4-styrenesulfonate), the only reduction products were formate and H_2. On the other hand, when using the positively charged stabilizers poly(allylamine hydrochloride), polyethyleneimine, and polybrene, the only observed reduction products were CO and H_2. Acetone was formed in all cases, from the oxidation of the 2-propanol. The selective photoproduction of formate with the negatively charged stabilizer surrounding the CdS particles was explained by adsorption of CO_2^- ion radicals with their carbon atoms on the CdS surface, resulting in an end-on adsorption. This configuration favored reaction with hydrogen atoms to produce formate. With the positively charged stabilizers, the adsorption of CO_2^- occurred by both the carbon atom and the O^- of CO_2^-, in a side-on adsorption, which favored CO production.[59]

While CO_2 reduction in water as solvent yielded mainly formic acid, the addition of tetraalkylammonium salts modified the semiconductor surface.[60] In water alone as medium, with colloidal CdS as photocatalyst, CO_2 was photoreduced to formic acid, methanol, and formaldehyde in the ratio 87:75:11, with a total quantum yield of 0.035%. In the presence of tetramethylammonium chloride, with the addition of sulfite to the medium, the product composition changed to formate and glyoxylate, in the ratio 22:12, with a total quantum yield of 0.1%.

Nedeljkovic et al. observed that with CdSe colloids (diameter Dp < 50 Å, prepared from $Cd(ClO_4)_2$ and H_2Se gas in water–alcohol mixture at –20°C and stabilized by SiO_2 colloids), CO_2 in aqueous solution was photoreduced to formic acid. No photoreduction of CO_2 was detected with CdSe of larger particle sizes.[61]

14.6.4 Colloidal Pd, Ru, and Os

Significantly improved yields of formic acid were obtained by Willner et al. using Pd colloids stabilized by β-cyclodextrin as photocatalysts.[62,63] In aqueous $NaHCO_3$ as medium, with deazariboflavin as photosensitizer, methyl viologen (MV^{2+}) as electron relay, and oxalate as sacrificial electron donor, the quantum yield for formate production with visible light reached 110%. This photocatalytic system may be considered as an abiological model for the enzymatic formate dehydrogenase system.

Selective photoreduction of carbon dioxide to formate was also obtained by Goren et al. using aqueous suspensions or colloids of Pd/TiO_2, stabilized by polyvinyl alcohol and β-cyclodextrin, in the presence of sacrificial oxalate. Quantum yields for formate production were about 1.4%.[64]

Visible light photoreduction of CO_2 to CH_4, C_2H_4, and C_2H_6 in aqueous solutions was obtained by Willner et al. using Ru or Os colloids as catalysts.[65–67] In one system, $Ru(II)(bpy)_3^{2+}$ was the photosensitizer, triethanolamine (TEOA) the electron donor, and one of several bipyridinium compounds served as mediators (charge relays). With N,N'-bis-(3-sulfonatopropyl)-3,3'-dimethyl-4,4'-bipyridinium as mediator, in 0.1 M aqueous bicarbonate solution (pH 7.8), with Ru colloid as catalyst, the quantum yields Φ were 2.6×10^{-3} for H_2, 5.7×10^{-4} for CH_4, and 1.9×10^{-5} for C_2H_4. Even lower yields were obtained with the Os colloid. In a second system, Ru(II)-tris(bipyrazine), Ru(bpz)$_3^{2+}$, was the photosensitizer, TEOA the electron donor, and colloidal Ru the catalyst, in 0.05 M NaHCO$_3$–ethanol (2:1, pH 7.8) as medium. With this sensitizer, the production of H_2 was inhibited. Values of Φ were 4.0×10^{-4} CH$_4$, 7.5×10^{-5} C$_2$H$_4$, and 4×10^{-5} C$_2$H$_6$. The turnover number based on the Ru(bpz)$_3^{2+}$ sensitizer was 15. This sensitizer absorbs strongly in the visible region, with $\lambda_{max} = 443$ nm, $\varepsilon = 15000$ M^{-1} cm^{-1}. It has a long-lived excited state ($\tau = 1.04$ μs). Its reduced photoproduct, Ru(bpz)$_3^+$, is a strong reducing agent, with E° [Ru(bpz)$_3^+$/ Ru(bpz)$_3^{2+}$] = –0.86 V vs. SCE, thus adequate to reduce CO$_2$ to CH$_4$.[65–67]

14.6.5   Colloidal Hydroxy-Oxobis(8-Quinolyloxo)Vanadium(V)

Hydroxy-oxobis(8-quinolyloxo)vanadium(V) was found by Aliwi to act as photocatalyst in the reduction of CO$_2$ to formic acid and formaldehyde.[68] The compound was prepared by reacting 8-hydroxyquinoline with ammonium metavanadate. The bandgap of the black water-insoluble compound was determined by its reflectance spectrum to be 1.5 eV. Optimal conditions for CO$_2$ reduction were with aqueous colloidal dispersions of the compound at pH 5 (citrate buffer) in the presence of methyl viologen as electron mediator and EDTA as electron donor.

14.7   GAS–SOLID REACTIONS

In most experiments on the photocatalytic reduction of carbon dioxide on semiconductor powders in aqueous slurries, the main products have been formic acid or carbon monoxide. A much higher degree of reduction was observed by Hemminger et al. upon illuminating a single crystal of strontium titanate (111 crystal face), which was in contact with a platinum foil, in an atmosphere of carbon dioxide and water vapor. Methane was detected by gas chromatography, indicating an eight-electron reduction of carbon dioxide.[69]

By illuminating TiO$_2$ (Degussa P25, mainly anatase) dispersed over quartz wool with a 1000-W Xe arc lamp, while recirculating over it a mixture of CO$_2$ and water vapor at an approximate temperature of 383 K, Saladin et al. observed the release of CO, H$_2$, and CH$_4$. No CO$_2$ reduction was found in the dark.[70]

Using a highly dispersed CeO$_2$–TiO$_2$ catalyst (0.5 wt% CeO$_2$, prepared by coprecipitation from titanium(IV) sulfate and cerium(III) nitrate), under illumination with visible light ($\lambda > 370$ nm), Ogura et al. performed the gas–solid-phase photolysis of an H$_2$O–CO$_2$ mixture.[71] With initial gas pressures of 100 torr CO$_2$ and 25 torr H$_2$O, at 25°C, the products were H$_2$, O$_2$, and CH$_4$, formed at initial rates (during the first hour) of 36, 14, and 3.4 μmol h^{-1} g-cat^{-1}. When the catalyst had been pretreated in H$_2$, the yield of methane was linearly related to the CO$_2$ pressure. It was thus proposed that the primary reaction was the photoassisted dissociation of water to hydrogen and that some of the hydrogen carried out a methanation reaction on the carbon dioxide. Considering each methane molecule to contain four hydrogen atoms, the maximal quantum yield for the formation of hydrogen ($H_2 + 2CH_4$) was estimated to be 0.05%.[71]

Anpo et al. prepared highly dispersed anchored TiO_2 particles by reacting $TiCl_4$ with the surface hydroxyl groups on transparent porous Vicor glass.[72,73] The bonding of the catalyst was performed in the gas–solid phase at 453 to 473 K, followed by hydrolysis with water vapor and then calcination at up to 873 K. By illumination of this glass-anchored TiO_2 with UV light ($\lambda > 280$ nm) in the presence of CO_2 and H_2O vapor at 275 K, both CH_4 and CH_3OH were produced, as well as CO, O_2, and traces of C_2H_4 and C_2H_6. Considerable amounts of additional methane and methanol were released by heating the catalyst after the illumination to 673 K, presumably desorbing strongly adsorbed products. An ESR study of the reaction at 77 K led to the appearance of signals assigned to Ti^{3+} and O_2^- species and also graphitic carbon.[72,73]

Improved photocatalysts for the reduction of CO_2 with water to CH_4 and CH_3OH were obtained by Anpo et al. using a hydrothermal synthesis leading from $(EtO)_4Si$ and $(BuO)_4Ti$ to mesoporous zeolites.[74,75] Most effective was a cubic titanium–zeolite (Ti–MCM-48), over which the selectivity for methanol production reached 30%, while over bulk TiO_2 (Degussa P25) it was only 1.4%. The remarkable reactivity and selectivity for methane and methanol formation were explained by the high dispersion of the Ti–O moieties in the zeolite cavities and by the large pore size (>20 Å) in the three-dimensional channel structure.

A comparison of the photocatalytic activity of several highly dispersed samples of TiO_2 for the gas–solid-phase reduction of CO_2 by H_2O was carried out by Yamashita et al.[77] Illumination with a high-pressure Hg lamp ($\lambda > 290$ nm) at 275 K led to the formation of CH_4, as well as to traces of C_2H_4 and C_2H_6. Highest photoactivity was achieved with a standard anatase TiO_2 catalyst (JRC–TiO-4) which had a high bandgap of 3.50 eV and a surface area of 49 m^2 g^{-1}. The increased bandgap shifted the conduction band edge to higher energy, and thus the reductive potential to a more negative value. The yield of CH_4 increased with an increasing H_2O/CO_2 ratio. With an initial ratio of 3:1, the production rate of methane on the above catalyst was 0.17 μmol h^{-1} g-cat^{-1}. The ESR spectrum formed under UV illumination of this TiO_2 sample in the presence of CO_2 and H_2O at 77 K revealed signals assigned to TiO^{3+} ions, H radicals, and CH_3 radicals, which were proposed to be key intermediates in the photoreduction of CO_2. Loading of the above catalyst with Cu^{2+} ions (by impregnation of the TiO_2 in an aqueous solution of $CuCl_2 \cdot 2H_2O$) resulted in a decrease in the photocatalytic activity of CO_2 reduction to CH_4 and instead to the formation of CH_3OH. Using a Cu^{2+} loading of 0.3 to 1.0 wt% on the TiO_2 enabled a methanol production rate of about 1 nmol h^{-1} g-cat^{-1}. X-ray photoelectron spectroscopy of the Cu^{2+}/TiO_2 indicated that the active copper species was Cu^+.[77]

Insight into the detailed mechanism of the gas–solid-phase photoreduction of CO_2 by water was obtained by Yamashita et al. by performing the reaction over wafers of single-crystal rutile TiO_2.[78] Under illumination with a high-pressure Hg lamp ($\lambda > 290$ nm) of CO_2 and H_2O vapor (in a molar ratio of 1:3) at 275 K over the crystal surface of $TiO_2(100)$, both CH_4 and CH_3OH were formed, at the remarkably high production rates of 3.5 and 2.4 nmol h^{-1} g-cat^{-1}, respectively. Using instead the surface of $TiO_2(110)$, no methane was formed, while methanol was produced only at the much smaller rate of 0.8 nmol h^{-1} g-cat^{-1}. The photocatalytic activity of the single-crystal $TiO_2(100)$, per unit surface area, was also substantially higher than that of small-particle TiO_2 powders, which are widely used as catalysts. The activities for $CO_2 + H_2O$ photoreduction on $TiO_2(100)$ and on the TiO_2 powder were 12 μmol h^{-1} m^{-2} and 3.5 nmol h^{-1} m^{-2}, respectively. The reaction intermediates formed on the UV-illuminated $TiO_2(100)$ surface after the reaction with CO_2 and H_2O were detected by HREELS. Energy loss peaks at 2920 and 3630 cm^{-1} were assigned to the C–H stretching vibration of some CHx species and to the O–H stretching vibration of surface hydroxyl groups, respectively. It was proposed that UV irradiation formed a reductive surface on TiO_2, facilitating the reduction of CO_2, possibly by electrons localized on surface Ti atoms of photoexcited $TiO_2(100)$ acting as reductive sites.[78]

14.8   PHOTOASSISTED HYDROGENATION

A variety of metal oxide catalysts were tested by Lichtin et al. for the gas–solid-phase photoassisted reduction of CO_2 by H_2 in the absence and presence of water vapor under visible light.[79,80] In experiments at 30°C and 1 atm pressure, highest formate production was observed with $Pt/LaNiO_3$ as catalyst, followed by Co–Mo/Al_2O_3 and α-Fe_2O_3. In the absence of water vapor in the dark, $CO_2 + H_2$ (1:2) over all catalysts produced CO as the major product and small amounts of CH_4. In the presence of water vapor and under illumination, yields of HCOOH became appreciable, using Zn–Fe-oxide or Co–Mo/Al_2O_3 as catalysts.[79,80]

Photoassisted methanation of CO_2 to CH_4 occurred by using as catalyst a modified zeolite, in which some of the silicon atoms had been replaced by titanium atoms. Yamagata et al. prepared titanium silicalite (TS-1) by a hydrothermal procedure.[81] In this material, with Si/Ti = 33.2, the Ti was incorporated in the zeolite framework and was dispersed in isolation. Under UV illumination of the TS-1 catalyst with an Hg lamp through a quartz window (λ < 380 nm), an H_2 and CO_2 mixture (3:1) was converted into CH_4. No reaction occurred in the dark. Bands appeared in the infrared absorption spectrum of the photocatalyst after the reaction at around 2900 cm^{-1}, assigned to –CH_3 and –CH_2– species, which may be intermediates in the formation of methane. The total amount of the carbon species adsorbed on the photocatalyst was determined by oxidation under pure oxygen. This amount was approximately equivalent to the amount of Ti in the photocatalyst, indicating that practically all Ti atoms in the zeolite served as catalytic sites. There was no photoactivity with the unmodified silicalite (without Ti).[81]

In a gas–solid reaction, Kohno et al. irradiated with an ultra high-pressure Hg lamp a mixture of $CO_2 + H_2$ (3:1; total pressure 25 kPa) in contact with powdered ZrO_2. The only CO_2 reduction product identified was CO, formed in low yield.[82] Under similar conditions, the UV irradiation of $CO_2 + CH_4$ (3:1) over ZrO_2 resulted in the formation mainly of CO, with much smaller amounts of H_2.[83] Experiments with $^{13}CH_4$ showed that the CO produced was derived only from the CO_2. Presumably, the CH_4 was converted into a carbonaceous deposit on the catalyst. Since ZrO_2 is a wide-bandgap semiconductor (5.0 eV), only short-wave UV light is effective for these reactions ($\lambda < 290$ nm).

14.9 PHOTOREDUCTION OF CO_2 BY NH_3

The photoassisted reduction of CO_2 by NH_3 was carried out by Ogura et al. in a gas–solid reaction at 5°C, using preadsorbed ammonia on silica-supported iron (prepared by impregnation of silica gel with ferric nitrate).[84,85] The catalyst, which was pretreated with H_2, contained 5 wt% Fe/SiO_2. The reaction was carried out by illumination at 185 nm with a 15-W low-pressure Hg lamp. The iron was shown by X-ray diffraction to be in the form of Fe_3O_4. The major products were CO, CH_4, H_2, and N_2. The quantum yield of CO formation increased with the amount of adsorbed NH_3, reaching a plateau of $\Phi = 0.5$ at higher values of adsorbed NH_3. It was proposed that coadsorbed CO_2 and NH_3 were involved in the initiation of the photolysis. This was followed by the reaction of adsorbed CO with adsorbed H atoms, leading to the production of methane.[84,85]

14.10 DARK CO_2 REDUCTION ON PREILLUMINATED CATALYST

Catalytic reduction in the dark of CO_2 was observed by Uchida et al. by recirculating atmospheric pressure CO_2 and water vapor (at its vapor pressure at 25°C) over solid Prussian-blue-coated TiO_2 particles, which had been preilluminated with a 500-W xenon lamp.[86] The products were methanol, ethanol, acetaldehyde, and acetone. In the proposed mechanism, during the illuminated stage, Prussian blue was photoreduced by electrons released from the photoexcited TiO_2 to Everitt's salt (the reduced form of Prussian blue), which interacted with adsorbed water molecules to produce adsorbed hydrogen atoms, while reforming Prussian blue. During the dark reaction, CO_2 underwent stepwise hydrogenation to C_1, C_2, and C_3 species.

14.11 PHOTOREDUCTION WITH TiO_2 IN SUPERCRITICAL AND HIGH-PRESSURE CO_2

The photocatalytic carbon dioxide reduction in supercritical CO_2 fluid was examined by Mizuno et al. using either TiO_2 or Cu powders or a mixture of both as photocatalyst.[87] A small amount of water dissolved in the supercritical fluid was necessary for the reaction, with water presumably acting as a hole scavenger. The process was performed by pressurizing CO_2 in a stainless-steel autoclave to 8.0 MPa and illuminating with a xenon arc through a glass window ($\lambda > 340$ nm). When only TiO_2 was dispersed as photocatalyst, the products were mainly H_2 and much smaller amounts of HCOOH and CH_3OH. Addition of Cu powder resulted in enhanced yields of HCOOH, but had little effect on the yield of methanol. Surprisingly, Cu powder alone dispersed in the supercritical CO_2 was also effective for the formation of the same products. This was explained by the photocatalytic effect of an assumed surface layer of some copper oxide on the Cu metal particles.[87]

The effect of pressure on the photocatalytic reduction of CO_2 in suspensions either in water or in aqueous alkali was measured by Mizuno et al. at pressures up to 2.5 MPa.[88] While no gaseous CO_2 reduction products were detected with water as the medium at atmospheric pressure of CO_2, increasing pressure caused the formation of substantial amounts of methane and smaller amounts of ethylene and ethane. In the aqueous phase, methanol and formic acid were produced at the high pressures, in yields which were an order of magnitude larger than the gaseous products. The yield of methanol as a function of CO_2 pressure had a maximum at a pressure of 1 MPa. When 0.2 N NaOH was the initial medium, the yields of methanol produced under CO_2 pressure were smaller than with water as the initial medium, but the two-carbon compounds acetaldehyde and ethanol were formed in considerable amounts.[88]

14.12 PHOTOCARBOXYLATION

CdS colloidal microcrystallites in CO_2-saturated DMF solution, containing TEA as sacrificial electron donor, were found by Yanagida et al. to be effective for the photofixation of CO_2 into aromatic compounds. Thus, benzophenone was converted into benzilic acid in about 10% yield,[89–91]

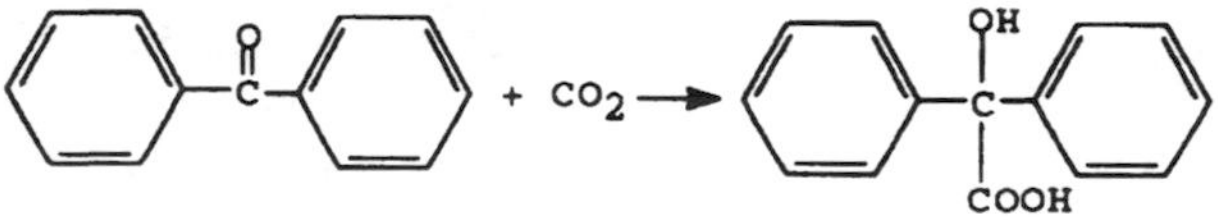

as well as into benzhydrol and benzopinacol. Illumination was at $\lambda > 400$ nm. Surprisingly, highest yields of benzilic acid were obtained after preirradiation of the reaction mixture without the benzophenone (1 h), followed by irradiation with the benzophenone (0.5 h). Experiments with $^{13}CO_2$ proved that the origin of the carboxylic function in the benzilic acid formed was CO_2. In the proposed mechanism, benzilic acid was produced by reaction of the CO_2 anion radical with the ketyl radical of benzophenone. Under similar conditions, benzyl bromide was carboxylated with CO_2 to phenylacetic acid,

$$C_6H_5CH_2Br + CO_2 \rightarrow C_6H_5CH_2COOH \qquad (14.7)$$

and acetophenone was carboxylated to atrolactic acid ([methyl(phenyl)glycolic acid]), without requiring enzymatic electron relays. The intermediate formation of the CO_2 anion radical, CO_2^-, was proven by ESR, using the spin trap 5,5'-dimethyl-1-pyrroline-N-oxide. In the proposed mechanism of the photofixation, benzyl radicals from benzyl halides couple with the CO_2^- radial anion on the surface of the CdS nanocrystallites.[89–91]

The photocatalytic carboxylation of phenol with CO_2 to salicylic acid (2-hydroxybenzoic acid) was achieved by Sclafani et al. using polycrystalline semiconductors in aqueous slurries as catalysts. The reaction was carried out in deaerated aqueous solutions of phenol at pH 5.85 saturated with CO_2 under illumination with a 1500-W Xe lamp. Among several semiconductors tested, the most effective for the carboxylation was anatase TiO_2, which caused only minor formation of the hydroxylation product catechol.[92]

14.13 CONCLUSIONS ON HETEROGENEOUS PHOTOASSISTED REDUCTION

Heterogeneous photoreduction of CO_2 or carbonate has attracted much interest because of the ease of setting up laboratory experiments. Regrettably, most of the photocatalytic CO_2 reduction reactions hitherto required the presence of sacrificial hole traps (or electron donors), such as n-propanol, tertiary amines, or EDTA, in order to achieve substantial yields. These compounds are usually more valuable than the CO_2 reduction products. In the absence of such hole traps, the yields of CO_2 reduction products were abysmally low. In a few cases, sulfide or sulfite ions were active as hole traps. Sulfide and sulfite ions are often available as waste products. Thus, in the petrochemical and metallurgical industries and from fossil-burning fuel stations, huge amounts of either H_2S or SO_2 and SO_3 are released. Sulfides are also produced in domestic effluents. The oxidation of these obnoxious compounds is environmentally beneficial. Therefore, a major benefit could be the development of high-yield photocatalytic processes for CO_2 reduction using sulfides or sulfites as hole scavengers.

REFERENCES

1. Solymosi, F. and Klivényi, G., HREELS study of photoinduced formation of CO_2 anion radical on Rh(111) surface, *Catal. Lett.*, 22, 337–342, 1993.
2. Solymosi, F. and Klivényi, G., Photoinduced generation of CO_2^- anion radical on K-promoted Rh(111) surface, *J. Phys. Chem.*, 98, 8061–8066, 1994.
3. Solymosi, F., Thermal and photo-induced activation and dissociation of CO_2 on clean and promoted metal surfaces, in Abstr. 3rd Int. Conf. on Carbon Dioxide Utilization, Norman, OK, May 1995.
4. Raskó, J. and Solymosi, F., Infrared spectroscopic study of the photoinduced activation of CO_2 on TiO_2 and Rh/TiO_2 catalysts, *J. Phys. Chem.*, 98, 7147–7152, 1994.
5. Bard, A.J., Photoelectrochemistry and heterogeneous photocatalysis at semiconductors, *J. Photochem.*, 10, 59–75, 1979.
6. Tanaka, K. and White, J.M., Dissociative adsorption of CO_2 on oxidized and reduced Pt/TiO_2, *J. Phys. Chem.*, 86, 3977–3980, 1982.
7. Evans, C.A., Spin trapping, *Aldrichimica Acta*, 12, 23–29, 1979.
8. Harbour, J.R. and Hair, M.L., Superoxide generation in the photolysis of aqueous cadmium sulfide dispersions. Detection by spin trapping, *J. Phys. Chem.*, 81, 1791–1793, 1977.
9. Jaeger, C.D. and Bard, A.J., Spin trapping and electron spin resonance detection of radical intermediates in the photodecomposition of water at TiO_2 particulate systems, *J. Phys. Chem.*, 83, 3146–3152, 1979.

10. Aurian-Blajeni, B., Halmann, M., and Manassen, J., Radical generation during the illumination of aqueous suspensions of tungsten oxide in the presence of methanol, sodium formate and bicarbonate. Detection by spin trapping, *Photochem. Photobiol.*, 35, 157–162, 1982.

11. Inoue, T., Fujishima, A., Konishi, S., and Honda, K., Photoelectro-catalytic reduction of carbon dioxide in aqueous suspensions of semiconductor powders, *Nature*, 277, 637–638, 1979.

12. Aurian-Blajeni, B., Halmann, M., and Manassen, J., Photoreduction of carbon dioxide and water into formaldehyde and methanol on semiconductor materials, *Sol. Energy*, 25, 165–170, 1980.

13. Ulman, M., Tinnemans, A.H.A., Mackor, A., Aurian-Blajeni, B., and Halmann, M., Photoreduction of carbon dioxide to formic acid, formaldehyde, methanol, acetaldehyde and ethanol using aqueous suspensions of strontium titanate with transition metal additives, *Int. J. Sol. Energy*, 1, 213–222, 1982.

14. Tennakone, K., Photoreduction of carbonic acid by mercury coated n-titanium oxide, *Sol. Energy Mater.*, 10, 235–238, 1984.

15. Halmann, M., Katzir, V., Borgarello, E., and Kiwi, J., Photoassisted carbon dioxide reduction on aqueous suspensions of titanium dioxide, *Sol. Energy Mater.*, 10, 85–93, 1984.

16. Aliwi, S.M. and Al-Jubori, K.F., Photoreduction of carbon dioxide by metal sulfide semiconductors in presence of hydrogen sulfide, *Sol. Energy Mater.*, 18, 223–229, 1989.

17. Halmann, M. and Zuckerman, K., Photoassisted reduction of carbon and nitrogen compounds with semiconductors, in *Homogeneous and Heterogeneous Photocatalysis*, Pelizzetti, E. and Serpone, N., Eds., NATO ASI Ser., Sr. C., Vol. 174, Reidel, Dordrecht, 1986, 521–532.

18. Ulman, M., Aurian-Blajeni, B., and Halmann, M., Photoassisted carbon dioxide reduction to organic compounds using rare earth doped barium titanate and lithium niobate as photoactive agents, *Isr. J. Chem.*, 22, 177–179, 1982.

19. Sayama, K. and Arakawa, H., Photocatalytic decomposition of water and photocatalytic reduction of carbon dioxide over ZrO_2 catalyst, *J. Phys. Chem.*, 97, 531–533, 1993.

20. Matsumoto, Y., Obata, M., and Hombo, J., Photocatalytic reduction of carbon dioxide on p-type $CaFe_2O_4$ powder, *J. Phys. Chem.*, 98, 2950–2951, 1994.

21. Tennakone, K., Jayatissa, A.H., and Punchihewa, S., Selective photoreduction of carbon dioxide to methanol with hydrous cuprous oxide, *J. Photochem. Photobiol. A Chem.*, 49, 369–375, 1989.

22. Yamamura, S., Kojima, H., Iyoda, J., and Kawai, W., Formation of ethyl alcohol in the photocatalytic reduction of carbon dioxide by SiC and ZnSe/metal powders, *J. Electroanal. Chem.*, 225, 287–290, 1987.

23. Ishitani, O. and Ibusuki, T., Metal-loaded TiO_2 photocatalyzed reduction of CO_2 to hydrocarbons, in Abstr. XVIth Int. Conf. on Photochemistry, Paris, July 28 to August 2, 1991, VII-1.

24. Dzhabiev, T.S. and Uskov, A.M., Photocatalytic reduction of carbon dioxide in semiconducting material suspensions, *Zh. Fiz. Khim. SSSR*, 65, 1039–1046, 1991; *Chem. Abstr.*, 114, 256787p.

25. Ulman, M., M.Sc. thesis, Weizmann Institute of Science, Rehovot, Israel, 1982.

26. Rophael, M.W. and Malati, M.A., The photocatalysed reduction of aqueous sodium carbonate to carbon using platinised titania, *J. Chem. Soc. Chem. Commun.*, pp. 1418–1420, 1987.

27. Rophael, M.W. and Malati, M.A., The photocatalyzed reduction of aqueous sodium carbonate using platinized titania, *Photochem. Photobiol.*, 46, 367–377, 1989.
28. Khalil, L.B., Youssef, N.S., Rophael, M.W., and Moawad, M.M., Reduction of aqueous carbonate photocatalysed by treated semiconductors as an application of solar energy conversion, in Abstr. 8th Int. Conf. Photochemical Conversion and Storage of Solar Energy, Palermo, Italy, July 15 to 20, 1990, 214.
29. Khalil, L.B., Youssef, N.S., Rophael, M.W., and Moawad, M.M., Reduction of aqueous carbonate photocatalysed by treated semiconductors, *J. Chem. Tech. Biotech.*, 55, 391–396, 1992.
30. Malati, M.A., Attubato, L., and Beaney, K., Efficient photo-catalysts for the reduction of aqueous carbonate and Cr(VI), *Sol. Energy Mater. Sol. Cells*, 40, 1–4, 1996.
31. Yamamura, S., Kojima, H., Iyoda, J., and Kawai, W., Photocatalytic reduction of carbon dioxide with metal-loaded SiC powders, *J. Electroanal. Chem.*, 247, 333–337, 1988.
32. Cook, R.L., MacDuff, R.C., and Sammells, A.F., Photoelectro-chemical carbon dioxide reduction to hydrocarbons at ambient temperature and pressure, *J. Electrochemical Soc.*, 135, 3069–3070, 1988.
33. Hirano, K., Inoue, K., and Yatsu, T., Photocatalyzed reduction of CO_2 in aqueous TiO_2 suspension mixed with copper powder, *Photochem. Photobiol. A Chem.*, 64, 255–258, 1992.
34. Adachi, K., Ohta, K., and Mizuno, T., Photocatalytic reduction of carbon dioxide to hydrocarbon using copper-loaded titanium dioxide, *Sol. Energy*, 53, 187–190, 1994.
35. Solymosi, F. and Tombácz, I., Photocatalytic reaction of H_2O + CO_2 over pure and doped Rh/TiO_2, *Catal. Lett.*, 27, 61–65, 1994.
36. Solymosi, F., Activation and reactions of CO_2 on Rh catalysts, *Carbon Dioxide Chemistry: Environ. Issues, R. Soc. Chem. Spec. Publ.*, 153, 44–54, 1994.
37. Heleg, V. and Willner, I., Photocatalysed CO_2-fixation to formate and H_2 — evolution by eosin-modified $Pd–TiO_2$ powders, *J. Chem. Soc. Chem. Commun.*, pp. 2113–2114, 1994.
38. Premkumar, J. and Ramaraj, R., Photoreduction of carbon dioxide by metal phthalocyanine adsorbed Nafion membrane, *Chem. Commun.*, pp. 343–344, 1997.
39. Halmann, M., Ulman, M., and Aurian-Blajeni, B., Photochemical solar collector for the photoassisted reduction of aqueous carbon dioxide, *Sol. Energy*, 31, 429–431, 1983.
40. Rossetti, R., Nakahara, S., and Brus, L.E., Quantum size effects in the redox potentials, resonance Raman spectra, and electronic spectra of CdS crystallites, *J. Chem. Phys.*, 79, 1086–1088, 1983.
41. Chandrasekaran, K. and Thomas, J.K., Photochemical reduction of carbonate to formaldehyde on TiO_2 powder, *Chem. Phys. Lett.*, 99, 7–10, 1983.
42. Inoue, H., Matsuyama, T., Liu, B.-J., Sakata, T., Mori, H., and Yoneyama, H., Photocatalytic activities for carbon dioxide reduction of TiO_2 microcrystals prepared in SiO_2 matrices using a sol-gel method, *Chem. Lett.*, pp. 653–656, 1994.
43. Kuwabata, S., Uchida, H., Ogawa, A., Hirao, S., and Yoneyama, H., Selective photoreduction of carbon dioxide to methanol on titanium dioxide photocatalysts in propylene carbonate solution, *J. Chem. Soc. Chem. Commun.*, pp. 829–830, 1995.
44. Torimoto, T., Liu, B.-J., and Yoneyama, H., Effect of solvents on photocatalytic reduction of carbon dioxide using semiconductor photocatalysts, in Abstr. 4th Int. Conf. on Carbon Dioxide Utilization, Kyoto, Japan, September 1997, P-065.

45. Henglein, A., Catalysis of photochemical reactions by colloidal semiconductors, *Pure Appl. Chem.*, 56, 1215–1224, 1984.

46. Henglein, A., Gutiérrez, M., and Fischer, Ch.-H., Photochemistry of colloidal metal sulfides. 6. Kinetics of interfacial reactions at zinc sulfide particles, *Ber. Bunsenges. Phys. Chem.*, 88, 170–175, 1984.

47. Inoue, H., Torimoto, T., Sakata, T., Mori, H., and Yoneyama, H., Effects of size quantization of zinc sulfide microcrystallites on photocatalytic reduction of carbon dioxide, *Chem. Lett.*, pp. 1483–1486, 1990.

48. Kisch, H. and Twardzik, G., Zinc sulfide catalyzed photoreduction of carbon dioxide, *Chem. Ber.*, 124, 1161–1162, 1991.

49. Albers, P. and Kiwi, J., Photochemical generation of formate via HCO_3^-/CO_2 reduction on Pd dispersions, *New J. Chem.*, 14, 135–139, 1990.

50. Yanagida, S. and Kanemoto, M., Effective photoreduction of carbon dioxide to formate catalyzed by defect-free ZnS quantum crystallites in water, in Proc. Int. Symp. Chemical Fixation of Carbon Dioxide, Nagoya, Japan, December 2 to 4, 1991, 359–364.

51. Kanemoto, M., Shiragami, T., Pac, C.J., and Yanagida, S., Semiconductor photocatalysis. Effective photoreduction of carbon dioxide catalyzed by ZnS quantum crystallites with low density of surface effects, *J. Phys. Chem.*, 96, 3521–3526, 1992.

52. Kanemoto, M., Hosokawa, H., Wada, Y., Murakoshi, K., Yanagida, S., Sakata, T., Mori, H., Ishikawa, M., and Kobayashi, H., Semiconductor photocatalysis. 20. Role of surface in the photoreduction of carbon dioxide catalysed by colloidal ZnS nanocrystallites in organic solvent, *J. Chem. Soc. Faraday Trans.*, 92, 2401–2411, 1996.

53. Inoue, H., Moriwaki, H., Maeda, K., and Yoneyama, H., Photoreduction of carbon dioxide using chalcogenide semiconductor microcrystals, *J. Photochem. Photobiol. A Chem.*, 86, 191–196, 1995.

54. Eggins, B.R., Robertson, P.K.J., Stewart, J.H., and Woods, E., Photoreduction of carbon dioxide on zinc sulfide to give four-carbon and two-carbon acids, *J. Chem. Soc. Chem. Commun.*, pp. 349–350, 1993.

55. Eggins, B.R., Irvine, J.T.S., Murphy, E., Woods, E., and Robertson, P.K.J., The photocatalytic fixation of carbon dioxide in aqueous tetraalkylammonium chloride using semiconductor colloids, in Abstr. 4th Int. Conf. on Carbon Dioxide Utilization, Kyoto, Japan, September 1997, P-O64.

56. Kanemoto, M., Ishihara, K.-I., Wada, Y., Sakata, T., Mori, H., and Yanagida, S., Visible light induced effective photoreduction of CO_2 to CO catalyzed by colloidal CdS microcrystallites, *Chem. Lett.*, pp. 835–836, 1992.

57. Fujiwara, H., Hosokawa, H., Murakoshi, K., Wada, Y., Yanagida, S., Okada, T., and Kobayashi, H., Effect of surface structures on photocatalytic CO_2 reduction using quantized CdS nanocrystallites, *J. Phys. Chem. B*, 101, 8270–8278, 1997.

58. Kanemoto, M., Nomura, M., Wada, Y. Akano, T., and Yanagida, S., Effect of In^{3+} in nano-scale CdS-catalyzed photoreduction of CO_2, *Chem. Lett.*, pp. 1687–1688, 1993.

59. Inoue, H., Nakamura, R., and Yoneyama, H., Effect of charged conditions of stabilizers for cadmium sulfide microcrystalline photocatalysts on photoreduction of carbon dioxide, *Chem. Lett.*, pp. 1227–1230, 1994.

60. Eggins, B.R., Irvine, J.T.S., Murphy, E.P., and Grimshaw, J., Formation of two-carbon acids from carbon dioxide by photoreduction on CdS, *J. Chem. Soc. Chem. Commun.*, pp. 1123–1124, 1988.

61. Nedeljkovic, J.M., Nenadovic, M.T., Micic, O.I., and Nozik, A.J., Enhanced photoredox chemistry in quantized semiconductor colloids, *J. Phys. Chem.*, 90, 12–13, 1986.

62. Mandler, D. and Willner, I., Effective photo-reduction of CO_2/HCO_3^- to formate using visible light, *J. Am. Chem. Soc.*, 109, 7884, 1987.

63. Willner, I. and Mandler, D., Characterization of Pd-β-cyclodextrin colloids in the photosensitized reduction of bicarbonate to formate, *J. Am. Chem. Soc.*, 111, 1330–1336, 1989.

64. Goren, Z., Willner, I., Nelson, A.J., and Frank, A.J., Selective photoreduction of CO_2/HCO_3^- to formate by aqueous suspensions and colloids of Pd–TiO_2, *J. Phys. Chem.*, 94, 3784–3790, 1990.

65. Maidan, R. and Willner, I., Photoreduction of CO_2 to CH_4 in aqueous solutions using visible light, *J. Am. Chem. Soc.*, 108, 8100–8101, 1986.

66. Willner, I., Maidan, R., Mandler, D., Dürr, H., Dörr, G., and Zengerle, K., Photosensitized reduction of CO_2 to CH_4 and H_2 in the presence of ruthenium and osmium colloids: strategies to design selectivity of products distribution, *J. Am. Chem. Soc.*, 109, 6080–6086, 1987.

67. Willner, I. and Willner, B., Photochemical CO_2-fixation by heterogeneous catalysis: design of photosystems for the selective formation of methane or formate, in Abstr. Int. Conf. on Carbon Dioxide Utilization, Bari, Italy, September 1993, 256–269.

68. Aliwi, S.M., Photofixation of carbon dioxide by a hydroxo-oxobis(8-quinolyloxo)-vanadium(V)-MV^{2+} ethylenediaminetetraacetic acid dispersion system, *J. Photochem. Photobiol. A Chem.*, 67, 329–336, 1992.

69. Hemminger, J.C., Carr, R., and Somorjai, G.A., The photoassisted reaction of gaseous water and carbon dioxide adsorbed on the $SrTiO_3(111)$ crystal face to form methane, *Chem. Phys. Lett.*, 57, 100–104, 1978.

70. Saladin, F., Forss, L., and Kamber, I., Photosynthesis of CH_4 at a TiO_2 surface from gaseous H_2O and CO_2, *J. Chem. Soc. Chem. Commun.*, pp. 533–534, 1995.

71. Ogura, K., Kawano, M., Yano, J., and Sakata, Y., Visible-light-assisted decomposition of H_2O and photomethanation of CO_2 over CeO_2–TiO_2 catalyst, *J. Photochem. Photobiol. A*, 66, 91–97, 1992.

72. Anpo, M. and Chiba, K., Photocatalytic reduction of CO_2 on anchored titanium oxide, *J. Mol. Catal.*, 74, 207–212, 1992.

73. Anpo, M., Yamashita, H., Ichihashi, Y., and Ehara, S., Photo-catalytic reduction of CO_2 with H_2O on various titanium oxide catalysts, *J. Electroanal. Chem.*, 396, 21–26, 1995.

74. Zhang, S.G., Fujii, Y., Yamashita, H., Koyano, K., Tatsumi, T., and Anpo, M., Photocatalytic reduction of CO_2 with H_2O on Ti–MCM-41 and Ti–MCM-48 mesoporous zeolites at 328 K, *Chem. Lett.*, pp. 659–660, 1997.

75. Anpo, M., Yamashita, H., Fujii, Y., Ikeue, K., Ichihashi, Y., Zhang, S.G., Park, D.R., Ehara, S., Park, S.-E., Chang, J.-S., and Yoo, J.W., Photocatalytic reduction of CO_2 with H_2O on titanium oxides anchored within zeolites, in Abstr. 4th Int. Conf. on Carbon Dioxide Utilization, Kyoto, Japan, September 1997, O-08.

76. Anpo, M., Yamashita, H., Ichihashi, Y., Fujii, Y., and Honda, M., Photocatalytic reduction of CO_2 with H_2O on titanium oxides anchored within micropores of zeolites. Effects of the structure of the active sites and the addition of Pt, *J. Phys. Chem. B*, 101, 2632–2636, 1997.

77. Yamashita, H., Nishiguchi, H., Kamada, N., and Anpo, M., Photo-catalytic reduction of CO_2 with H_2O on TiO_2 and Cu/TiO_2 catalysts, *Res. Chem. Intermed.*, 20, 815–823, 1994.

78. Yamashita, H., Kamada, N., He, H., Tanaka, K.-I., Ehara, S., and Anpo, M., Reduction of CO_2 with H_2O on $TiO_2(110)$ and $TiO_2(110)$ single crystals under UV-irradiation, *Chem. Lett.*, pp. 855–858, 1994.

79. Vijayakumar, K.M. and Lichtin, N.N., Reduction of CO_2 by H_2 and water vapor over metal oxides assisted by visible light, *J. Catal.*, 90, 173–177, 1984.

80. Lichtin, N.N., Vijayakumar, K.M., and Rubio, B.I., Photoassisted reduction of CO_2 by H_2 over metal oxides in the absence and presence of water vapor, *J. Catal.*, 104, 246–251, 1987.

81. Yamagata, S., Nishijo, M., Murao, N., Ohta, S., and Mizoguchi, I., CO_2 reduction to CH_4 with H_2 on photoirradiated TS-1, *Zeolites*, 15, 490–493, 1995.

82. Kohno, Y., Tanaka, T., Funabiki, T., and Yoshida, S., Photoreduction of carbon dioxide with hydrogen over ZrO_2, *Chem. Commun.*, pp. 841–842, 1997.

83. Kohno, Y., Tanaka, T., Funabiki, T., and Yoshida, S., Photoreduction of carbon dioxide with methane, *Chem. Lett.*, pp. 993–994, 1997.

84. Ogura, K., Seno, A., and Kawano, M., Photo-assisted catalytic reduction of CO_2 with pre-adsorbed ammonia on silica supported iron, *J. Mol. Catal.*, 73, 225–235, 1992.

85. Ogura, K., Koreishi, T., Yano, J., and Mine, K., Effect of temperature on photoassisted reduction of CO_2 with preadsorbed ammonia on silica-supported iron, *J. Mol. Catal.*, 79, 47–54, 1993.

86. Uchida, H., Sasaki, T., and Ogura, K., Dark catalytic reduction of CO_2 over Prussian-blue-deposited TiO_2 and the photo-reactivation of the catalyst, *J. Mol. Catal.*, 93, 269–277, 1994.

87. Mizuno, T., Tsutsumi, H., Ohta, K., Saji, A., and Noda, H., Photo-catalytic reduction of CO_2 with dispersed TiO_2/Cu powder mixtures in supercritical CO_2, *Chem. Lett.*, pp. 1533–1536, 1994.

88. Mizuno, T., Adachi, K., Ohta, K., and Saji, A., Effect of CO_2 pressure on photocatalytic reduction of CO_2 using TiO_2 in aqueous solutions, *J. Photochem. Photobiol. A Chem.*, 98, 87–90, 1996.

89. Kanemoto, M., Ankyu, H., Wada, Y., and Yanagida, S., Visible-light induced photofixation of CO_2 into benzophenone catalyzed by colloidal CdS microcrystallites, *Chem. Lett.*, pp. 2113–2114, 1992.

90. Kanemoto, M., Wada, Y., and Yanagida, S., Photofixation of CO_2 catalyzed by nano-scale CdS crystallites under visible-light irradiation, in Abstr. Int. Conf. on Carbon Dioxide Utilization, Bari, Italy, September 1993, 388.

91. Fujiwara, H., Kanemoto, M., Ankyu, H., Murakoshi, K., Wada, Y., and Yanagida, S., Visible-light induced photofixation of carbon dioxide into aromatic ketones and benzyl halides catalyzed by CdS nanocrystallites, *J. Chem. Soc. Perkin Trans. 2*, pp. 317–321, 1997.

92. Sclafani, A., Palmisano, L., and Farneti, G., Synthesis of 2-hydroxybenzoic acid from CO_2 and phenol in aqueous heterogeneous photocatalytic systems, *Chem. Commun.*, pp. 529–530, 1997.

15 Conclusion on Mitigation Strategies

CHAPTER 1: THE SCIENCE AND THE SOURCE OF THE GREENHOUSE EFFECT AND GLOBAL CLIMATE CHANGE

The two hard physical facts about global climate change are that (1) the radiative gases CO_2, CH_4, and N_2O have been increasing in the atmosphere since the Industrial Revolution and (2) more than two-thirds of the radiative forcing is due to the combustion of fossil fuels. The consensus is that a discerning increase in mean global temperature increase has been detected ranging from 0.3 to 0.6°C over the past 130 years, and it is predicted that the temperature will rise by an average of 2°C due to a doubling of CO_2 concentration in the atmosphere over the next century. However, because of positive and negative forcing factors, there are significant uncertainties in predicting specific effects of global temperature rise on sea level rise and climate change. The application of CO_2 mitigation technologies in developed countries could benefit the developing countries in constructing their evolving energy industries with reduced CO_2 emissions.

CHAPTER 2: ADAPTIVE VS. MITIGATIVE RESPONSE STRATEGIES FOR GLOBAL WARMING

The adaptive approach assumes that life on earth will adapt to global warming effects. The adaptive response includes relocation of population, managed land areas, construction of water walls around coastal populated areas, changing agricultural patterns, special health care, special economic assistance, and conservation. The mitigative response deals with reducing CO_2 emissions from the use of fossil fuels.

CHAPTER 3: CONCEPTS FOR CONTROLLING ATMOSPHERIC CARBON DIOXIDE

A suggested logic for guiding CO_2 mitigation and control technologies includes (1) selection of control points, (2) removal processes, (3) identification of disposal alternatives for captured CO_2, (4) CO_2 recovery process selection, (5) identification of disposal alternatives for recovered CO_2, and (6) identification of CO_2 reuse alternatives. The criteria for evaluating systems for mitigating the atmospheric CO_2 concentration depend on (1) sufficient scientific and technical information on the proposed mitigating process, (2) the magnitude of the CO_2 removal capacity of the proposed control process, (3) the mass and energy balances of the process, (4) the capital and operating cost of the process, and (5) environmental effects of the CO_2 mitigation scheme. The mitigating process systems are categorized as physical, chemical, physical/chemical, and biological. CO_2 disposal schemes include (1) ocean, (2) terrestrial, and (3) extraterrestrial. A number of process systems are given and evaluated in this chapter. Utilization of CO_2 in current material commodities markets is a very small fraction of the total CO_2 emissions. Future growth of CO_2 utilization in the materials market would have to grow by factors of tens to hundreds to have an impact on CO_2 emission reduction. CO_2 utilization for enhanced oil recovery and conversion to liquid transportation fuels are the main potentially significant markets for impacting CO_2 emissions reduction.

CHAPTER 4: TECHNOLOGIES FOR IMPROVING EFFICIENCY

A significant factor for reducing CO_2 emissions from the use of fossil fuels is to improve the conversion efficiency of fossil fuels to useful end-use energy and to improve the efficiency of utilization of the converted energy. There is a considerable potential for improving energy conversion and utilization efficiency from present conditions to what can reasonably be achieved approaching the theoretical maximum limits. Based on estimates of the major energy savings technologies in the major energy-consuming sectors of the economy (residential and commercial, industrial, transportation, and electricity generation from fossil fuels) in the nine major regional areas of the world, together with estimates of global energy demand projections to the year 2050 (during which time a doubling in CO_2 concentration is projected), the fossil-fuel energy demand and the reduction in CO_2 emission have been estimated. The energy savings show an approximate 60% reduction in fossil-fuel energy demand by the year 2050, projected from the base year 1975. A concomitant 60% reduction in CO_2 emission would also result, thus essentially stabilizing the CO_2 concentration in the atmosphere. Further reductions are projected by considering increasing use of nonfossil-fuel forms. Estimates of the capital cost for turning over the capital stock from current to the improved technologies, although huge (estimated in the tens of trillions of U.S. dollars), could result in a 10% return on the investment, purely based on savings in fuel costs alone. Improving efficiency thus pays for itself and emphasizes its importance in mitigating the global greenhouse problem.

CHAPTER 5: REMOVAL, RECOVERY, AND DISPOSAL OF CARBON DIOXIDE

Because of the large central sources of concentrated CO_2, fossil-fuel-fired power plant stack gases lend themselves as control points for removal, recovery, and disposal of CO_2. The technology for removal and recovery from power plant stack gases is fairly well developed, especially absorption/stripping with liquid solvent absorbents, especially alkanolamines. Other promising methods of separation include adsorption and membrane technologies. The problem lies with disposal and sequestering of CO_2. The greatest capacity for sequestering of CO_2 is in the ocean below the thermocline, allowing hundreds of years of disposal of the world's emissions. Depleted gas wells and deep aquifers also have significant sequestering capacity but are an order of magnitude less than the ocean. Estimates of the energy penalty for capture and disposal of CO_2 vary from over 30% of the power plant capacity using current technology to less than 20% for future improved technology. Most of the energy is expended in removal, recovery, and liquefying the CO_2, and for near-shore plants less than 10% of the energy is used for sequestering in the ocean. The latter increases for inland plants. The cost of producing electricity when removing and disposing of CO_2 from power plants can increase by as much as 100% above current production costs. Before CO_2 disposal in the ocean can be seriously considered, adverse environmental effects, among which are effects on marine life, must be investigated, as must public opposition to putting anything in the ocean. On a worldwide basis, only 32% of the total CO_2 emitted to the atmosphere from central power plant stacks can be sequestered either in the ocean or on land, thus limiting the extent of mitigation that this method can offer.

CHAPTER 6: BIOCONVERSION OF CO_2 AND BIOMASS PROCESSES

The natural photosynthetic process provides renewable energy through the production and utilization of biomass with essentially zero net emission of CO_2. A variety of biomass feedstocks can supply the energy and fuel conversion process. The feedstocks include tree wood, agricultural crops, waste, and algae. The conversion processes include combustion, gasification, hydrogasification, and fermentation. Coprocessing with fossil fuels has some economic advantages and assists in the utilization of fossil fuels with reduced CO_2 emissions.

A major advantage of the bioconversion of carbon dioxide by comparison with the gas–solid-phase catalytic hydrogenation (see Chapter 10) is that, in addition to methane, the methanogenic bacteria also produce valuable materials, such as enzymes and amino acids. Hydrogen sulfide, which is poisonous to many catalysts, is actually beneficial and is consumed during the bioconversion. The application of thermophilic methanogens should be particularly attractive for the treatment of blast furnace and converter gases released during the production of iron and steel. These gases contain hydrogen, carbon oxides, and the toxic and obnoxious hydrogen sulfide.

The culture of marine microalgae in seawater mixed with nutrient-rich wastewater directly supplied with power plant stack gases could be an attractive option, with some of the costs offset by the production of useful products from the biomass formed (e.g., proteins, vitamins, β-carotene).

Enzyme-catalyzed reactions are interesting since they lead directly to larger molecules of considerable complexity, which may be more difficult to access by purely synthetic methods.

The search for models that mimic the action of the natural enzyme carbonic anhydrase could be attractive in the context of the disposal of unused CO_2. The uncatalyzed hydration of CO_2 is slow (except in highly acidic water), while the gas–liquid exchange of CO_2 between the atmosphere and water is fast. Therefore, an efficient and inexpensive catalyst for the hydration of CO_2 to bicarbonate could be a method for the fixation of CO_2 from the atmosphere.

CHAPTER 7: DECARBONIZATION OF FOSSIL FUELS AND CONVERSION TO ALTERNATIVE FUELS

Decarbonization of fossil fuels applied to postcombustion requires sequestration of CO_2. Decarbonization applied to fossil fuels prior to combustion is directed toward producing hydrogen-rich fuels. The carbon can be removed and sequestered as elemental carbon or as CO_2. Production of hydrogen from fossil fuels and from biomass is key to the production of hydrogen-rich fuel for reducing CO_2 emission. Natural gas can be reformed to produce hydrogen for stationary power or for transportation fuel, but the CO_2 produced must be sequestered. Coal can be gasified with steam and oxygen to produce hydrogen and the CO_2 sequestered. Biomass can be gasified with steam and oxygen to produce hydrogen, which does not require CO_2 sequestration because biomass is CO_2 neutral.

An alternative to CO_2 sequestration for hydrogen production is the thermal decomposition of natural gas to carbon and hydrogen and either storing or using the carbon as a materials commodity. The process energy for producing hydrogen by this method is the least, while substantially eliminating formation and emission of CO_2.

When hydrogen is used in conjunction with recycled CO_2 from biomass or from power plant stack gases to produce methanol as a liquid transportation fuel, significant CO_2 emission reduction can be realized. The Carnol system converts the CO_2 recovered from coal-fired power plant stack gases with hydrogen produced from the thermal decomposition of methane to methanol by catalytic hydrogenation. The methanol can be used in conventional internal combustion engines or, more efficiently, in fuel cells for transportation purposes. The system uses the carbon from coal twice, once for producing electricity and a second time for producing a liquid transportation fuel. A combined system reduction of CO_2 from the power generation sector and the transportation sector approaching 80% could be realized. With transportation fuels contributing over 30% to worldwide CO_2 emissions, this system could obtain a very significant worldwide reduction in greenhouse gas. The promising large resource of methane hydrates could provide the long-term supply of fuel required by this system. Realization of these results awaits further development work.

CHAPTER 8: THERMOCHEMICAL REACTIONS

The combined thermochemical reforming of methane to synthesis gas by CO_2 produced *in situ* during the production of metals from their ores and of lime or cement from limestone is another approach to decreasing of CO_2 emissions. The coproduction of the commodity metals iron, aluminum, zinc, and cement and synthesis gas, particularly with the application of solar energy as the source of process heat, provides a promising approach to significantly reducing CO_2 emission. It will be necessary to confirm whether such processes may be achieved economically on an industrial scale, especially using concentrated sunlight as the energy source.

CHAPTER 9: CARBOXYLATION BY CO_2 INSERTION

Among the promising products of CO_2 mitigation by chemical conversion, polycarbonate plastics have an outstanding potential, because of their excellent properties. The world consumption of polycarbonates in 1979 was only 150,000 metric tons. However, in 1996, the production capacity of polycarbonates worldwide was already about 1 million tons per year, and the demand for this commodity has been growing by more than 10% per year. About one-third of the volume of sales of polycarbonates has been for building products. For polyurethanes, annual sales in the U.S. amounted to more than 2 million tons. Increased production of polycarbonates and polyurethanes by processes that replace phosgene with carbon dioxide will (1) contribute directly to enhanced use of CO_2 and (2) indirectly provide considerable mitigation of CO_2 release to the atmosphere by the substitution of the highly energy-intensive metal and cement production by plastic production. In order to have a significant effect on CO_2 mitigation, the demand for these plastics should increase by at least two orders of magnitude.

Other materials, the enhanced use of which may contribute to CO_2 mitigation, are fuels and fuel additives, such as methanol and methyl-*tert*-butyl ether, and solvents, such as dimethyl carbonate and ethylene carbonate.

CHAPTER 10: CO_2 REFORMING AND HYDROGENATION

The formation of synthesis gas by the CO_2 reforming of methane could provide a substantial use for the CO_2 released from fossil-fuel power stations. New methods of combined partial oxidation and CO_2 reforming of methane have considerably improved the energy economy and yield of this reaction. However, only that part of the synthesis gas converted into chemicals and used, for example, for the production of polymers would be a long-term sink for carbon. The demand of synthesis gas for chemicals production is very much lower than the release of CO_2 from fossil-fuel combustion. The conversion of synthesis gas into methanol as a synthetic fuel would meet a much larger demand. The carbon contained in the synthetic fuel would be released to the atmosphere after having been used twice. Current methanol markets are very small compared to total worldwide carbon emission. However, if methanol or its derivatives should become useful as a transportation fuel either in internal combustion engines or in fuel cells, the potential for recycling and fixation of carbon can increase significantly.

There exists a major economic motivation for new uses of CO_2. One significant process may be the aromatization of a liquid petroleum gas fraction (mainly propane) by carbon dioxide over zeolite catalysts. Another interesting reaction is the catalytic hydrogenation of CO_2 to methanol followed in series by its conversion over a composite catalyst to gasoline and diesel-like C_1–C_2 hydrocarbons.

CHAPTER 11: PHOTOCHEMICAL AND RADIATION-INDUCED ACTIVATION OF CO_2 IN HOMOGENEOUS MEDIA

The main problem with photochemical reactions in homogeneous solutions as a method for CO_2 fixation seems to be the need for sacrificial electron donors as well as low quantum yields, mainly because of side reactions. Several photocatalytic systems are capable of both producing H_2 and reducing CO_2, but none of them provides substantial conversion to either synthesis gas or methanol.

CHAPTER 12: ELECTROCHEMICAL REDUCTION OF CO_2

In the electrochemical reduction of CO_2, the activation of this inert molecule is achieved by an input of electric energy. In the last few years, considerable progress has been made in enhancing the efficiency of electrochemical processes by applying gas diffusion electrodes operating at high pressures of CO_2 or by carrying out the electrolysis in supercritical CO_2 (s-CO_2). With these methods, current densities of the order of Amp/cm^{-2} have been attained. By introducing alloy metal electrodes or transition metal complex electrodes, the selectivity of the CO_2 reduction products has been directed to favor alcohols and olefins.

The electrocarbonoxylation of organic compounds has considerable advantage for the synthesis of fine chemicals. Several processes have reached the stage of pilot plant testing and may become competitive with gas–solid-phase catalytic reactions.

CHAPTER 13: PHOTOELECTROCHEMICAL REDUCTION OF CO_2

Until now, most photoelectrochemical CO_2 reduction systems suffered from low current densities, probably due to the high resistances of the semiconductor electrodes. Also, the long-term stability of these systems for CO_2 reduction has not yet been demonstrated.

CHAPTER 14: HETEROGENEOUS PHOTOASSISTED REDUCTION OF CO_2

Heterogeneous photoreduction of CO_2 or of carbonate has attracted much interest because of the ease of setting up laboratory experiments. Regrettably, most of the photocatalytic CO_2 reduction reactions hitherto required the presence of sacrificial hole traps (or electron donors), such as n-propanol, tertiary amines, or EDTA, in order to achieve substantial yields. These compounds are usually more valuable than the CO_2 reduction products. In the absence of such hole traps, the yields of CO_2 reduction products were abysmally low. In a few cases, sulfide or sulfite ions were active as hole traps. Sulfide and sulfite ions are often available as waste products. Thus, in the petrochemical and metallurgical industries and from fossil-fuel-burning power plants, huge amounts of either H_2S or SO_2 and of SO_3 are increased. Sulfides are also produced in domestic waste effluents. The oxidation of these obnoxious compounds is environmentally beneficial. Therefore, a major benefit could be the development of high-yield photocatalytic processes for CO_2 reduction using sulfides or sulfites as hole scavengers.

CONCLUDING REMARKS

The ever-increasing anthropogenic consumption of fossil fuel together with defor-estation of the earth's tropical rain forests have caused a very marked increase in the global concentration of greenhouse gases, especially carbon dioxide, in the atmo-sphere since the advent of the Industrial Revolution. The international scientific community has recognized an increase in global warming due to increasing concen-trations of carbon dioxide in the atmosphere and has alerted the world as to the potential consequences. However, due to the complexity of factors contributing to climate change, uncertainties exist concerning the effects of global warming in causing changes in the earth's ecology. Because of these uncertainties and the different positions and states of economic development among the nations of the world, it is very difficult to reach a uniform global agreement concerning reduction of emission of carbon dioxide into the atmosphere. In the U.S., without a regulation or tax incentive (an energy or carbon tax) to limit carbon dioxide emission, mitiga-tion technologies including decarbonization, sequestration, and utilization probably will not be readily implemented. The developing countries are reluctant to limit carbon dioxide emissions because of the fear that their economic growth will be stunted. Ongoing national and international discussions will continue searching for an acceptable global agreement. If global warming is to be taken seriously, the major developed countries will have to take the lead in introducing CO_2 mitigation technologies and offering them to the developing countries.

A positive effect resulting from the concern about global warming is that it has stimulated research on not only terrestrial, atmospheric, and oceanographic science, but also on developing novel physical, chemical, and biological processes and technologies that deal with energy and carbon dioxide utilization. The spinoff ben-efits that will undoubtedly result from application based on the science and technol-ogy of CO_2 mitigation will certainly justify the world's continued investment in the necessary research and development in this field.

Index